International Conference on Applied Sciences, Information and Technology 2019

AA007354

IOP Conference Series: Materials Science and Engineering
Volume 846

Padang, Indonesia
1 - 3 November 2019

Editors:

Rikki Vitria
Ihsan Lumasa Rimra

ISBN: 978-1-7138-4729-8
ISSN: 1757-8981

Printed from e-media with permission by:

Curran Associates, Inc.
57 Morehouse Lane
Red Hook, NY 12571

Some format issues inherent in the e-media version may also appear in this print version.

Institute of Physics
Dirac House, Temple Back
Bristol BS1 6BE UK

Phone: 44 1 17 929 7481
Fax: 44 1 17 920 0979

techtracking@iop.org

Additional copies of this publication are available from:

Curran Associates, Inc.
57 Morehouse Lane
Red Hook, NY 12571 USA
Phone: 845-758-0400
Fax: 845-758-2633
Email: curran@proceedings.com
Web: www.proceedings.com

TABLE OF CONTENTS

IOP Conf. Series: Materials Science and Engineering **846** (2020) 011001 doi:10.1088/1757-899X/846/1/011001

Preface

Politeknik Negeri Padang is pleased to conduct ICo-ASCNITech where for the the third for ICo-ASCNITech. As a leading higher vocational educational institution in Indonesia, Politeknik Negeri Padang must develop its international reputation to achieve long term vision as one of the reputable vocational education in South East Asia by 2025. This international event is part of our strategic plans toward that vision which is also to strengthen our cooperation with Politeknik Ibrahim Sultan, Malaysia.

This proceeding contains the written versions of the International Conference on Applied Sciences, Information and Technology (ICo-ASCNITech) 2019. The Conference is considered as a good opportunity for participants from overseas such as Malaysia, Taiwan, Japan, Thailand, Philippines, Vietnam, Nigeria and Indonesia itself to present and discuss topics in their respective research areas.

This annual event is jointly organized by Politeknik Negeri Padang (Indonesia) and Politeknik Ibrahim Sultan (Malaysia). There are 156 titles written by more than 200 authors had been participate in this prestigious conferences representing more than 20 local and international higher education institutions as well as research bodies. These are a great experience to share information of applied research, technology and innovation among researchers and academicians, students, government, industries, and other stakeholders. Hopefully this wonderful agenda would be revolutionary contribution to the world which has become the theme of this conference, "Revolutionary Contribution of Applied Science and Technology to Society in Industrial Era".

We would like to thank advisory board, scientific committee, organizing committee, presenters and volunteers for their contributions to the conference's program and to the proceeding and had a great and fruitful experiences during the conferences. We sincerely express our gratitude to IOP for publishing the proceeding of ICo-ASCNITech 2019.

The Editors:

Rikki Vitria, S.S.T., M.Sc. Eng

Chairman of the ICo-ASCNITech 2019

Ihsan Lumasa Rimra, SST.,M.Sc DECN

Peer review statement

All papers published in this volume of *IOP Conference Series: Materials Science and Engineering* have been peer reviewed through processes administered by the proceedings Editors. Reviews were conducted by expert referees to the professional and scientific standards expected of a proceedings journal published by IOP Publishing.

Fabrication by Electrospinning Technique and Characterization of Curcuma Mangga Val Reinforced PVA Fibrous Membranes

Harini Sosiati*, Muhammad Dirga Rianto, Aris Widyo Nugroho, and Sudarisman

Department of Mechanical Engineering, Faculty of Engineering, Universitas Muhammadiyah Yogyakarta, Jl. Brawijaya, Kasihan, Bantul, Yogyakarta 55183, Indonesia.

*Email: hsosiati@ft.umy.ac.id

Abstract. Antimicrobial and anti-diabetic of Curcuma mangga Val properties have attracted research interest. Curcuma mangga Val extract (CME) and poly (vinyl alcohol) (PVA) were blended and fabricated to the fibrous membranes by the electrospinning method. The effects of CME concentration on the fiber morphology and tensile properties of CME/PVA membranes are the goal of the current study. Scanning electron microscopy (SEM) of the membranes revealed bead-free fibers with straightly/continuously orientation formed in all CME/PVA membranes. The average fiber diameter increased from 186 nm to 297 nm, with the CME concentration from 0 to 3 wt.%, respectively. The addition of 1% CME resulted in a relatively high tensile strength of 24.96 ± 0.20 MPa, which is the highest among the CME/PVA membrane specimens. However, very high tensile modulus and low elongation showed in these results lead to reducing the functionality of the CME/PVA fibrous membrane due to the brittleness. The CME properties may contribute to those shortages.

1. Introduction

Curcuma mangga Val is well known in Asia, especially Indonesia and, also in Africa and Australia. Curcuma mangga Val is a kind of plant that possesses beneficial properties for medicine materials such as antimicrobial, anti-diabetic, antioxidant, and non-toxic [1, 2]. Therefore, numerous studies on Curcuma mangga Val have been carried out for medicine mostly in Indonesia. Those properties have appealed to biomedical research interest. It can be potential as an alternative wound dressing-based material. In this case, Curcuma mangga Val extract (CME) would be useful as reinforcing material for the polymer as chitosan and aloe Vera to fabricate fibrous membranes by electrospinning technique.

Chitosan and aloe Vera are the natural materials having the properties of antimicrobial activity, anti-inflammation, and non-toxic [3, 4]. Those properties had also dramatically increased the interest to use them as based materials for wound healing, wound dressing, scaffold, and tissue engineering applications. The fibrous membranes made of chitosan and aloe Vera combined with poly (vinyl alcohol) (PVA) [5–8], polyethylene oxide (PEO) [9] and polycaprolactone [10] had been extensively investigated. However, no study of CME-based fibrous membranes fabricated by electrospinning method was found. Thus, it is a great challenge to initially create the work to make the CME-based fibrous membrane by electrospinning. It would be a benefit in biomaterial science.

PVA is a synthetic polymer having a conductive property and easily blends with chitosan as a natural polymer. The chitosan combined the PVA solution in various ratios of chitosan to PVA produced the chitosan/PVA nanofiber mats with differences in fiber structures corresponding to the ratio. A rise of chitosan content tended to form beaded fibers. A PVA/chitosan ratio of 50:50 showed severe bead formation [11]. The applied voltage in this study was also varied: i.e., 15 kV and 20 kV, and the fiber diameter changed irregularly in both conditions. The smallest fiber diameter (~ 148 nm) was achieved by the PVA/chitosan ratio of 70:30 at 20 kV. By a similar proportion and applied voltage of 20 kV, the chitosan/PVA exhibited the tensile strength of 5.26 ± 0.53 MPa, which was lower than that of neat PVA 7.4 ± 0.37 MPa [12].

On the other hand, the hybrid fibrous membranes of chitosan nano-emulsion blended with aloe Vera/PVA were prepared by the electrospinning method at an applied voltage of 15 kV. The membrane at 15% chitosan nano-emulsion concentration yielded the highest tensile strength of 6.18 ± 0.51 MPa with an average fiber diameter of 180 nm [6]. Except for the viscosity as a function of concentration, the electrical conductivity of the polymer solution has played a crucial role in the formation of fiber morphology as well [13–17]. Insufficient volume charge density in the polymer solution, the fiber would not form any fiber. As mentioned earlier, PVA is one of the conductive polymers. Based on our experience, the high molecular weight (Mw) of PVA was more challenging to blend with the reinforcing materials in comparison with the low Mw of PVA.

Due to no information of study on the Curcuma mangga Val based electrospun fibrous membranes, in this preliminary study, CME concentrations were varied and incorporated into the PVA solution. To avoid the difficulty of polymer solution preparation, PVA used in this work was low Mw of PVA, which rarely used in other studies. Changes in the characteristics of CME/PVA fibrous membranes due to different CME concentrations were discussed. Besides, the results are compared to other fibrous membranes made of antibacterial natural materials reinforced PVA.

2. Materials and Methods

Curcuma mangga Val (GMV) (Fig.1a) used in this research was a commercial Curcuma mangga Val extract purchased at "Obat Herbal Alami Jogja" shop, Yogyakarta, Indonesia. PVA Gohsenol (PVOH/PVA, Mw: ~ 22,000 g/mol) purchased at CV. Multi Kimia, Yogyakarta, Indonesia, was used.

The manufacturing of the fibrous membranes was begun by preparing a PVA solution with 10% concentration: i.e., by dissolving 10 g PVA powder into 100 g H_2O by continuously magnetic stirred at 80 ± 2°C for one hour. The CME powder with a particle size of 400 mesh (Fig.1b) was blended into the PVA solution with varying CME concentrations of 0%, 1%, 2%, and 3%, and continuously magnetic stirred at 200 rpm and 50°C for one hour. After that, the CME_PVA solutions were left for four days (Fig.2) and used for the spinning solution in the electrospinning process. The viscosity and electrical conductivity of the spinning solutions at each CME concentration were measured. The CME/PVA fibrous membranes were fabricated at the optimized condition at an applied voltage of 18 kV, a distance from the tip to the collector (TCD) of 15 cm, the diameter of spinneret 0.6 mm and flow rate of 0.003 ml/min. The fibrous membrane specimens were designated as CME-0/PVA, CME-1/PVA, CME-2/PVA, and CME-3/PVA. In this case, the optimization of applied voltage and TCD was carried out by fabricating thin fibrous membranes at various applied voltage and TCD. The fiber structure formed in the membranes was then observed with an optical microscope and optimized. The optimum condition was selected based on the uniformity of fiber orientation with free fiber-defects or free from beads formation. Figure 3 confirmed that an applied voltage of 18 kV and TCD of 15 cm are the optimum condition.

The characterization of fiber morphology built-in CME/PVA fibrous membranes was performed by scanning electron microscopy (SEM, Hitachi SU-3500). The measurement of fiber diameter was conducted using the ImageJ digital image analysis on at least one hundred points.

The tensile test was conducted on all fibrous membrane specimens according to ASTM 882 using a universal testing machine (UTM, Zwick Z0.5 Germany) at a crosshead speed of 10 mm/min and a specimen gauge length of 20 mm. In each case, eight tensile specimens were prepared.

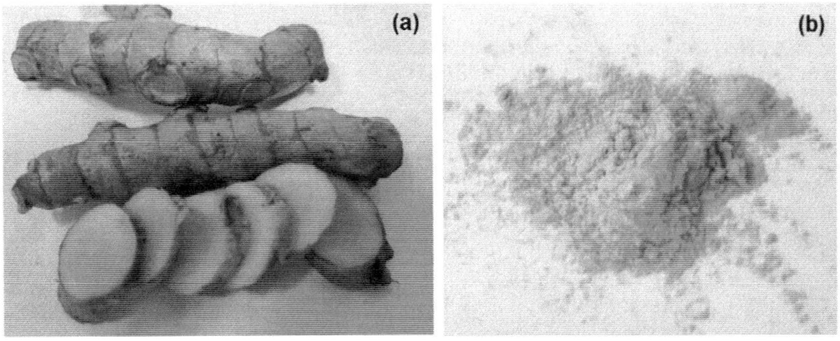

Figure 1. The photographs of Curcuma manga Val (a), and Curcuma manga Val extract (CME). (b).

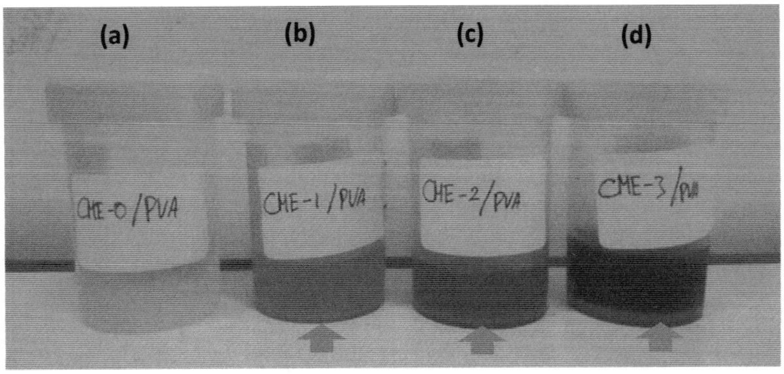

Figure 2. The spinning solutions of PVA (a), and CME_PVA with different CME concentrations, 1% (b), 2% (c) and 3% (d) showing the presence of precipitation (see, red arrows).

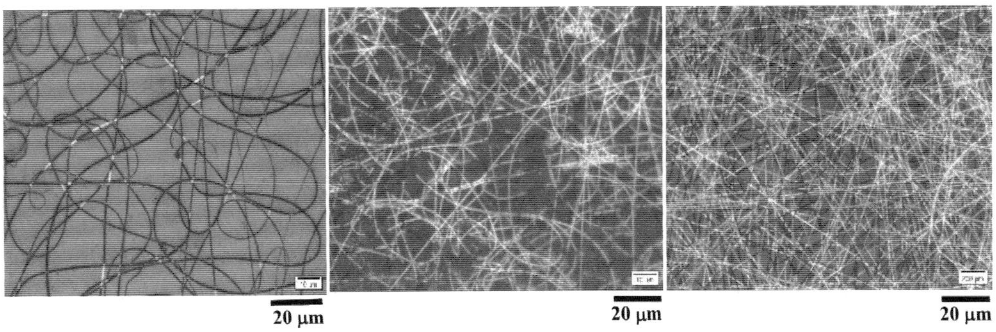

Figure 3. Optical micrographs of the fibrous membranes resulting from the electrospinning process at different applied voltage and TCD. (a) 15 kV and 13 cm, (b) 15 kV and 15 cm, and (c) 18 kV and 15 cm.

3. Results and Discussion

SEM images of all fibrous membrane specimens (Fig.4) exhibit a relatively similar fiber structure: i.e., bead-free and straightly oriented. However, the average fiber size appeared differently. A rise in CME concentration increases the fiber size, as exhibited in the distribution of fiber size (Fig.5). The fiber size in a neat PVA fibrous membrane (CME-0/PVA) is mostly present in the range of 100 – 200 nm.

IOP Conf. Series: Materials Science and Engineering **846** (2020) 012001 doi:10.1088/1757-899X/846/1/012001

The fibrous membranes of CME-1/PVA and CME-2/PVA show nearly similar fiber size distribution in the range between 200 – 300 nm, but a portion of fiber size at the position between 100 – 200 nm in the CME-1/PVA membrane is higher than that in the CME-2/PVA membrane. The largest fiber size is mostly in the range between 200 – 300 nm and 300 – 400 nm demonstrated in a CME-3/PVA membrane. Enhancement of a mean fiber size by increasing the CME concentration is associated with an increase of the viscosity of the spinning solution (Table 1). It also shows that an increase in CME concentration decreases the electrical conductivity. It might be due to the un-ionic nature of the CME. The trend of those results is consistent with that reported by Shahreen et al. [15]. In this study, however, the increasing viscosity and reducing the electrical conductivity affected by an increase of CME concentration did not lead to the formation of beads. The existence of beads in the fiber would inhibit the interaction between the fibers, which usually lead to reducing the tensile strength of the membrane [18].

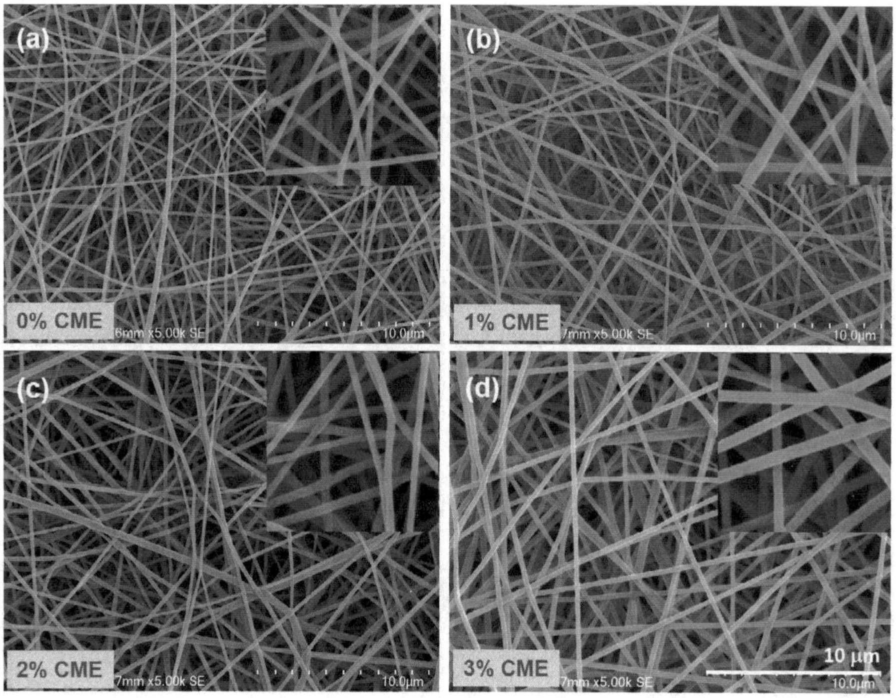

Figure 4. SEM images of CME/PVA nanofibrous membranes with differences in CME concentrations. (a) 0% CME, (b) 1% CME, (c) 2% CME and (d) 3% CME.

In the electrospinning process, the viscosity and electrical conductivity are the vital process parameters affecting the fiber formation. The polymer solution will be suitable for electrospinning when the polymer solution viscosity is present between 100 cP – 2000 cP [13]. Based on our experience, the solution having low viscosity leads to droplet occurrence and then usually produces the beaded fibers. Too high viscosity, the polymer solution will be difficult to pump through the syringe needle, and the solution may dry before starting the electrospinning process. Usually, the fiber size increased.

Thus, the excellent fiber structure resulted in the current work might be caused by the viscosity of all CME_PVA solutions that are in the range between 100 cP and 2000 cP. However, the formation of bead-free and continuous fibers in polyethylene oxide (PEO) fibrous membrane occurred at the viscosity of 1250 cP [19]. The bead-free and continuous fibers built in the hybrid membrane of CSNe/AV_PVA with the average fiber size of 242 nm formed by the polymer solution viscosity of

379.9 cP [5]. Those in the $PdCl_2$/polyvinylpirolydone (PVP)/titanium (IV) isopropoxide (TTIP) fibrous membranes revealed at the viscosity of 65.8 cP with 256 nm fiber size [15]. Besides, a study on the fibrous membranes made of mixed (PVA and chitosan) solutions has verified the changes in both the average fiber diameter and fiber structure. Those are affected by the volume ratio of PVA to chitosan, and, also the applied voltage. In this case, the viscosity of mixed PVA and chitosan solutions are a function of the volume ratio, although their effect did not change linearly [11]. A correlation between the polymer solution viscosity and concentration is strictly dependent on the nature of the polymer and the intermolecular interactions within the polymer solution [20]. In principle, to produce the fibers by electrospinning technique is that the polymer solution should have sufficient molecular weight, viscosity, and also electrical conductivity [13, 14].

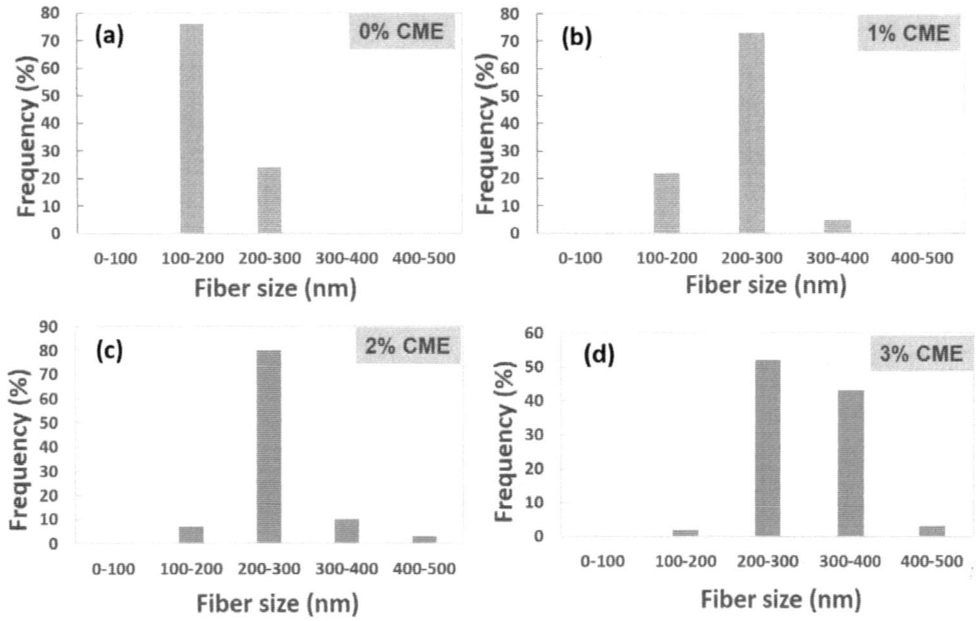

Figure 5. Fiber size distribution in (a), (b), (c) and (d) are related to the SEM images shown in Fig.4a, 4b, 4c and 4d, respectively.

Table 1. Viscosity, electrical conductivity and average fiber size

No	Specimen	Viscosity (cP)	Electrical conductivity (µs/cm)	Average fiber size (nm)
1	CME-0/PVA	436	451	186.21
2	CME-1/PVA	452	423	227.07
3	CME-2/PVA	466	393	251.04
4	CME-3/PVA	482	340	296.67

On the other hand, the fiber morphology resulted in all fibrous membranes may correspond to their tensile properties (Fig.6 and Fig.7). Insertion of 1% CME in the PVA solution improves the tensile strength of CME-1/PVA fibrous membrane, but further increases to 2% and 3% CME reduced the tensile strength of the fibrous membranes (Fig.6). The tensile strength obtained from this result is quite high; it is higher compared to other findings [5, 12, 17, 19]. However, the tensile modulus of all fibrous membranes is also remarkably high. They are too high; thus, the membranes tend to be brittle, although the tensile strengths are considerably high. According to the result of the tensile- stress versus strain at break (Fig.7), it indicates that maximum elongation reached by the neat PVA fibrous

IOP Conf. Series: Materials Science and Engineering **846** (2020) 012001 doi:10.1088/1757-899X/846/1/012001

membrane. The addition of CME decreased elongation. The higher the CME concentration, the lower the elongation.

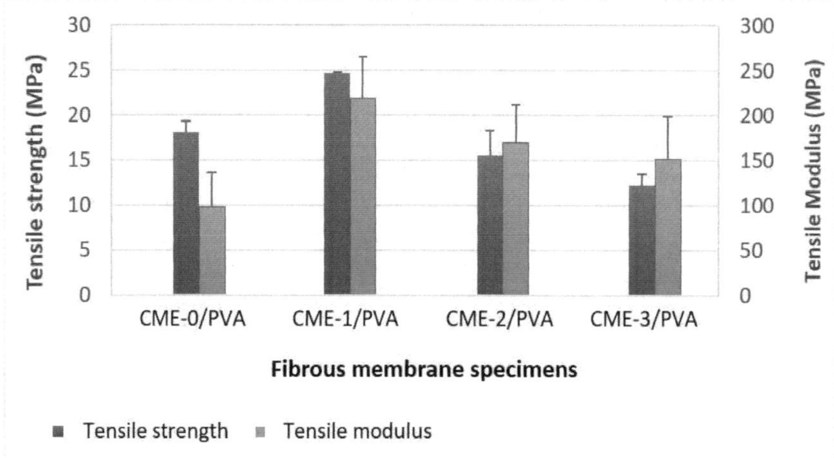

Figure 6. Tensile strength and modulus of CME reinforced PVA fibrous membranes with various CME concentrations.

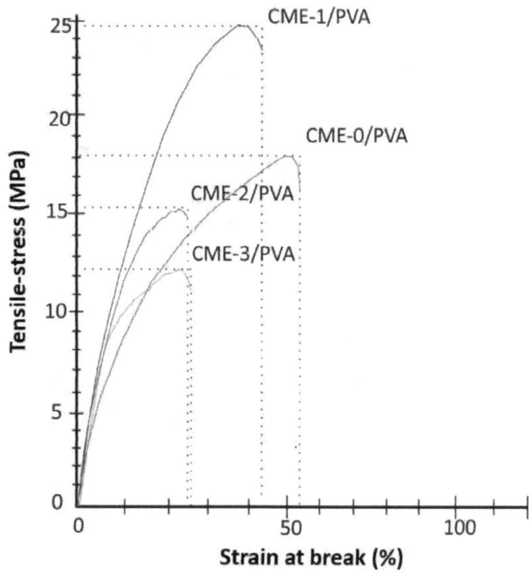

Figure 7. Tensile-stress versus strain at break

In comparison with a study on the fibrous membranes made of aloe Vera extract (AVE) reinforced PVA [5], at the similar concentration of 2% AVE, the tensile properties showed 5.74 MPa for tensile strength, 33.99 MPa for tensile modulus, and 135% for elongation at break with the average fiber size of around 342 nm. The fiber structures resulted from the current work are much better, and the average fiber size is also smaller compared to those resulting from the AVE/PVA fibrous membrane.

Based on those results, the predominant factor influencing the tensile properties of the CME/PVA membrane, especially the improvement of the tensile modulus and reducing the elongation, is of the properties of CME particles rather than the fiber size and structure. The commercial CME particle size might be too large and impure, leading to form the precipitation (Fig.2). In this case, CME could not be dissolved in the PVA solution completely. Thus, the particle size should be down to a nanometer to disperse the CME particles in the PVA solution uniformly. Another propose for an alternative way is to prepare CME by making the Curcuma mangga Val filtrate and then blend with the PVA solution. Therefore, further research on the electrospun fibrous of CME/PVA is necessary because of the valuable properties of Curcuma mangga Val. All CME/PVA fibrous membranes produced from this preliminary study using CME, however, could not be recommended as an alternative wound dressing material. The stiffness and brittleness are quite high, although the current work achieved very high tensile strength and excellent fiber morphology built in the fibrous membranes.

4. Conclusions

These experimental results have provided new information on the electrospun fibrous membrane made of Curcuma mangga Val reinforced PVA. The fibrous membranes have successfully fabricated and yielded comparatively small fiber size with bead-free continuous fibers. Those contributed to the improvement of the tensile strength (24.96 ± 0.20 MPa) of CME/PVA fibrous membranes, especially on the membrane, by adding 1% CME. However, all the produced CME/PVA membranes are being less functional material due to their very high tensile modulus and low strain. These conditions made the membrane tend to be brittle. The properties of CME might be the cause. The shortcoming of these results would be corrected by developing the preparation technique of CME in the next research.

Acknowledgment

This experimental research was supported by a research grant of "Penelitian Dasar Unggulan Perguruan Tinggi"2018, contract no: 227/SP2H/LT/DRPM/2019 by the Directorate General of Higher Education (DIKTI), Ministry of Research, Technology and Higher Education, the Republic of Indonesia. The authors would like to grate appreciate Mr Kunto Wandono for his assistance related to the modification of the electrospinning machine.

References

[1] D. Pujimulyani, W. A. Yulianto, A. Setyowati, S. Arumwardana, and R. Rizal, "Antidiabetic and antioxidant potential of Curcuma mangga Val extract and fractions," *Asian J. Agric. Biol.*, vol. 6, no. 2, pp. 162–168, 2018.

[2] A. Rohman, H. Widodo, E. Lukitaningsih, M. Rafi, N. A.F., and A. Windarsih, "Review on in vitro antioxidant activities of Curcuma species commonly used as herbal components in Indonesia," *Food Res.*, vol. 4, no. 2, pp. 286–293, 2019.

[3] L. Ammayappan and J. Jeyakodi Moses, "Study of antimicrobial activity of aloe Vera, chitosan, and curcumin on cotton, wool, and rabbit hair," *Fibers Polym.*, vol. 10, no. 2, pp. 161–166, 2009.

[4] Y. Pang, A. Qin, X. Lin, L. Yang, Q. Wang, Z. Wang, Z. Shan, S. Li, J. Wang, S. Fan, and Q. Hu, "Biodegradable and biocompatible high elastic chitosan scaffold is cell-friendly both in vitro and in vivo," *Oncotarget*, vol. 8, no. 22, pp. 35583–35591, 2017.

[5] H. Sosiati, A. N. Widodo, and A. W. Nugroho, "The influence of aloe Vera concentration on morphology and tensile properties of electrospun aloe Vera-PVA nanofiber," *Jurnal Sains Materi Indonesia*, vol. 19, no. 4, pp. 157–162, 2018.

[6] H. Sosiati, W. N. Fatihah, Yusmaniar, and M. B. Nur Rahman, "Characterization of the properties of electrospun blended hybrid poly (vinyl alcohol)_aloe Vera/chitosan nano-emulsion nanofibrous membranes," *Key Eng. Mater.*, vol. 792, pp. 74–79, 2018.

[7] İ. Uslu, S. Keskin, A. Gül, T. C. Karabulut, and M. L. Aksu, "Preparation and properties of electrospun poly (vinyl alcohol) blended hybrid polymer with aloe Vera and HPMC as wound dressing," *Hacettepe J. Biol. Chem.*, vol. 38, no. 1, pp. 19–25, 2010.

[8] D. M. Escobar-Sierra and Y. P. Perea-Mesa, "Manufacturing and evaluation of chitosan, PVA and aloe vera hydrogels for skin applications," DYNA, vol. 84, no. 203, pp. 134–142, 2017.

[9] M. Pakravan, M. C. Heuzey, and A. Ajji, "A fundamental study of chitosan/PEO electrospinning," *Polymers*, vol. 52, no. 21, pp. 4813–4824, 2011.

[10] P. Carter, S. M. Rahman, and N. Bhattarai, "Facile fabrication of aloe vera containing PCL nanofibers for barrier membrane application," *J. Biomater. Sci. Polym.* Ed., vol. 27, no. 7, pp. 692–708, 2016.

[11] R. P. Gonçalves, W. H. Ferreira, R. F. Gouvêa, and C. T. Andrade, "Effect of chitosan on the properties of electrospun fibers from mixed poly (vinyl alcohol)/Chitosan solutions," *Mater. Res.*, vol. 20, no. 4, pp. 984–993, 2017.

[12] M. Koosha and H. Mirzadeh, "Electrospinning, mechanical properties, and cell behavior study of chitosan/PVA nanofibers," *J. Biomed. Mater. Res. - Part A*, vol. 103, no. 9, pp. 3081–3093, 2015.

[13] N. Amariei, L. R. Manea, A. P. Bertea, A. Bertea, and A. Popa, "The influence of polymer solution on the properties of electrospun 3D nanostructures," *IOP Conf. Ser. Mater. Sci. Eng.*, vol. 209, no. 1, 2017.

[14] C. J. Angammana and S. H. Jayaram, "Analysis of the effects of solution conductivity on electrospinning process and fiber morphology," *IEEE Trans. Ind. Appl.*, vol. 47, no. 3, pp. 1109–1117, 2011.

[15] L. Shahreen and G. G. Chase, "Effects of electrospinning solution properties on formation of beads in TiO_2 fibers with PdO particles," *J. Eng. Fiber. Fabr.*, vol. 10, no. 3, pp. 136–145, 2015.

[16] Y. Sun, S. Cheng, W. Lu, Y. Wang, P. Zhang, and Q. Yao, "Electrospun fibers and their application in drug controlled release, biological dressings, tissue repair, and enzyme immobilization," *RSC Adv.*, vol. 9, no. 44, pp. 25712–25729, 2019.

[17] P. Korycka, A. Mirek, K. Kramek-Romanowska, M. Grzeczkowicz, and D. Lewinska, "Effect of electrospinning process variables on the size of polymer fibers and bead-on-string structures established with a 23 factorial design," *Beilstein J. Nanotechnol.*, vol. 9, no. 1, pp. 2466–2478, 2018.

[18] B. Tarus, N. Fadel, A. Al-Oufy, and M. El-Messiry, "Effect of polymer concentration on the morphology and mechanical characteristics of electrospun cellulose acetate and poly (vinyl chloride) nanofiber mats," *Alexandria Eng. J.*, vol. 55, no. 3, pp. 2975–2984, 2016.

[19] V. Pillay, C. Dott, Y. E. Choonara, C. Tyagi, L. Tomar, P. Kumar, L. C. du Toit, and V. M. K. Ndesendo, "A review of the effect of processing variables on the fabrication of electrospun nanofibers for drug delivery applications," *J. Nanomater.*, vol. 2013, 2013.

[20] S. Huan, G. Liu, G. Han, W. Cheng, Z. Fu, Q. Wu, and Q. Wang, "Effect of experimental parameters on morphological, mechanical and hydrophobic properties of electrospun polystyrene fibers," *Materials*, vol. 8, no. 5, pp. 2718–2734, 2015.

IOP Conf. Series: Materials Science and Engineering **846** (2020) 012002 doi:10.1088/1757-899X/846/1/012002

Sub chronic diagnosis of administration with *Scurrula atropurpurea* to blood biochemichal analysis

N Athiroh[*], **Y Purnomo, N J Mubarakati**

University of Islam Malang East Java, Kota Malang, Indonesia

*E-mail: *nur_athiroh_mlg@yahoo.co.id*

Abstract. This research is to evaluate whether *Scurrula atropurpurea* has a protective effect on the liver caused by its sub-chronic treatment. The rats consist of four groups: normotensive group and three groups administered methanolic extract of *S. atropurpurea* (MESA) at a dosage of 250; 500; and 1000 mg/kg BW. Serum liver and kidney function were assayed. Serum cholesterol was checked. Histopathology of the brain, liver, cardiac, lung, and kidney was identified using hematoxylin-eosin staining. The administration of MESA250, MESA500 and MESA1000 caused no differences among the groups compared with the control group. Structural evaluations on the brain and lungs showed that the MESA administration carried no changes. In liver and heart, MESA1000 led to repair tissue damage. But for the kidney, MESA250 was safe for the tissue. The results of this study revealed that the methanolic extract of *S. atropurpurea* at *antihypertension* dose is safe even when taken for a more extended period. At a higher dose, the extract may have the potential to increase some hematological indices but may induce tissue repair in the liver and heart. The lowest dose is safe for kidney tissue.

1. Introduction

The mistletoe of *Scurrula atropurpurea* (BL.) Dans is a hemiparasitic tea plant, especially on the island of Java. The leaves and stems of *Scurrula atropurpurea* (BL.) Dans has been empirically and potential used for therapeutic treatment. Previous studies showed that this plant has a function in blood pressure lowering [1][2]. Furthermore, the pathway in which this plant can reduce blood pressure is in oxidative stress inhibition and endothelial cell protection.

Scurrula atropurpurea (BL.) Dans is a growing determination that comprehension of conventional medicine is potential not only for its essential as curative herbs but also for its socioeconomic status and cultural ingredients [3]. Although many health issues have been investigated by using medicinal herbs, acute, sub chronic, and chronic tests of some plants are also reported [4].

Nevertheless, there is no scientific report on the safety of using *Scurrula atropurpurea* (BL.) Dans as a herbal cure. The present study is, consequently, pointed to evaluate the security of *Scurrula atropurpurea* (BL.) Dans by exploring the chronic toxicity of this herb extract.

2. Methods

2.1 Preparation tea parasite crude extract

Determination and identification of the crude extract characteristic of botanical determination were performed at the Indonesian Scientific Institute (LIPI) at Purwodadi, Pasuruan, East Java. "One hundred milligrams of dry leaf powder was steeped in methanol in a 1000 ml-erlenmeyer flask. The mixture was shaken for 30 min to distribute the powder homogenously in methanol. Subsequently, the mixture was left to stand overnight to precipitate. The supernatant, a combination of methanol and the active constituents, was subjected to evaporation. The extract was labeled and stored in a freezer. The methanolic extract of *S. atropurpurea* (MESA) was administered daily by the oral gavage (2 ml) using a catheter for 90 days" [1] [2] [5].

The procedure of the experimental animal model was done by the guidelines of Indonesia legislations on the ethical clearance of laboratory animals. The research protocol was approved by the Animal Review Board of the Brawijaya University, Malang, Indonesia. The number of rats, which is 40 tail, aged 8-10 weeks and weighing 180-200 g. The treatment groups were divided into the normotensive group, the MESA treatment at dosages of 250, 500, and 1000 mg/kgBW. The rats were assigned randomly into the groups; each group contained ten rats. After 90 days of treatment, the blood and tissue were determined [5].

2.2 Analysis of liver function and lipid profiles
After sacrifice, blood samples were collected by left ventricular blood. The liver function of the animal model was recognized based on changes in plasma levels of glutamic-oxaloacetic transaminase (GOT) and glutamic-pyruvic transaminase (GPT). Plasma levels of albumin, total protein, alkaline phosphatase (ALP), total cholesterol, triglyceride, low-density lipoprotein cholesterol (LDL- C), high-density lipoprotein cholesterol (HDL-C) were measured using Spotchem EZ SP 4430 (ARKRAY Inc, Kyoto, Japan).

2.3 Histological examination
The brain, liver, heart, and kidney were placed in 25 mM KCl & Phosphate-buffered saline (PBS) to relaxing buffer, then will keep in neutral-buffered formaldehyde 40% at room temperature. Hematoxylin and eosin (H&E) stained sections (~5 μm) were prepared to measure histopathology [6]. The sections were photographed at 400x magnification using an Olympus (Tokyo, Japan). Microscope lighting, focus, and field selection were optimized for the distinction of cell boundaries. Images were opened in Image J and after setting the threshold, analyzed. Data from all the fields were combined and then analyzed.

2.4 Data Analysis
Data are performed as mean ± SEM. The research data analyzed by one-way analysis of variance (ANOVA), and continued with a post hoc analysis was evaluated using Fisher's least significant difference (LSD) test. A *P*-value of less than 0.05 was considered to be statistically significant.

Tabel 1. Effect of subchronic administration of MESA in blood serum levels of rats.

Group	Protein	Albumin	Globulin	SGOT	SGPT	Cholesterol	Triglycerida	BUN	Creatinine	Bilirubin	Urea
C	7.34	3.84	3.51	239.67	147.67	75.00	66.00	22.50	0.45	0.02	48.10
G1	7.81	3.44	4.37	198.56	90.33	51.67	44.78	24.40	0.54	0.07	51.69
G2	8.37	3.38	4.93	216.50	88.00	59.83	63.67	26.52	0.62	0.03	56.88
G3	8.59	3.64	4.93	210.22	104.00	54.00	43.78	24.20	0.58	0.03	51.78

3. Results

3.1 Effect of subchronic administration of MESA on blood serum measurement
Oral administrations of MESA at 250, 500, and 1000 mg/kg BW did not cause any mortality in rats during the observation period. Furthermore, there were no unconcealed signs and symptoms of

indication of toxicity in study groups. Base on table 1, the subchronic administration of MESA in all groups did not significantly influence of the examined blood serum indicator to the control. Previous studies revealed that the administration of MESA caused no toxic effect in the biochemical indices [5][7][8][9][10]. Figure 1 (a-k) reveals that MESA did not cause dysfunction in liver and kidney function. Furthermore, lipid profile was standard in comparison with the control.,Cholesterol profiles are found no differences in female rats with subchronic administration of MESA [11]. Furthermore, treatment with MESA subchronically did not affect cell necrosis in the liver because MESA contains many active substances such as quercetin that inhibit tissue damage-induced by free radical [12].

3.2 Effect of subchronic administration of MESA to wet organ weights
Necropsy macroscopic observation of the research studied visceral organs, such as the brain, pulmo, cor, hepar, and kidneys for any feasible changes in position, size, shape, and color, did not disclose any obvious abnormality. Wet weight (in g) and respective organ weight (in g per 100 g body weight) of brain, hepar, lung, heart, and kidney of both extract-treated and normotensive groups. No significant difference was investigated in absolute and relative organ weights of MESA and normotensive groups [6]. No significant differences between absolute and relative organ weights of the kidneys were investigated in normotensive and MESA.

3.3 Effect of subchronic administration of MESA on histopathology of brain and lung
Histological examinations of liver sections of rats treated with 250, 500, and 1000 mg/kg b.w MESA showed a standard architecture of the brain and lung in comparison with the control group.

3.4 Effect of subchronic administration of MESA on histopathology of the hepar
Histological observation of hepar frontal sections of rats used with 250 and 500 mg/kg b.w MESA appeared a normal features with normal appearance of the central vein and hepatic sinusoids lined with endothelial and *Kupffer* cells similar to the normotensive groups. The hepatocytes showed normal in shape and size, and no available of vacuoles were prominent in the cytoplasm. In the rats' group used with 1000 mg/kg b.w MESA, however, there was cellular damage of hepatocytes. Furthermore, central veins and portal triads were infiltrated and enlarge by mononuclear leukocytic cells, indicating mild hepatitis [6].

3.5 Effect of subchronic administration of MESA on histopathology of the heart
Histological observations of hepar frontal sections of rats treated with 500 and 1000 mg/kg b.w MESA appeared a normal features with normal appearance of the cardiomyocytes compared to the control group. There were a few cardiomyocytes damage showed in rats administered with 250 mg/kg b.w MESA.

3.6 Effect of subchronic administration of MESA on histopathology of the kidney
Histological study of kidney frontal sections of rats administered with 250 mg/kg b.w MESA revealed normal features with normal appearance of the nephrons compared to the control group. Moreover, this normal architecture was found both in the right and left kidney. There was damage in glomerular structure showed in rats treated with 500 and 1000 mg/kg b.w MESA. These mean that the subchronic administration of MESA should be less than 500 mg/kg b.w MESA to protect structural damage of the kidney.

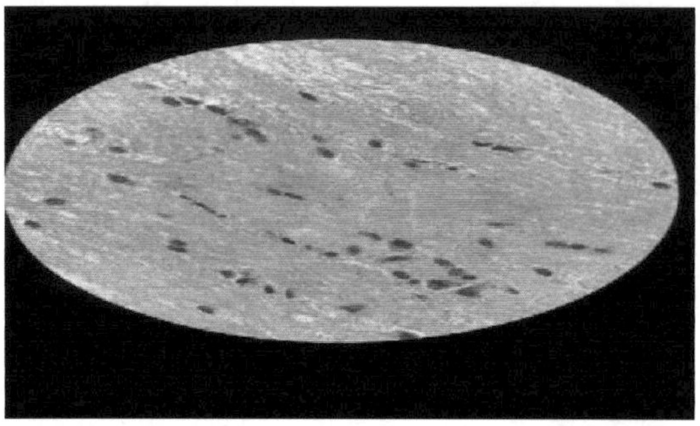

Figure 1. Histopathology of the brain after administration of MESA (Olympus CX21, 40x10).

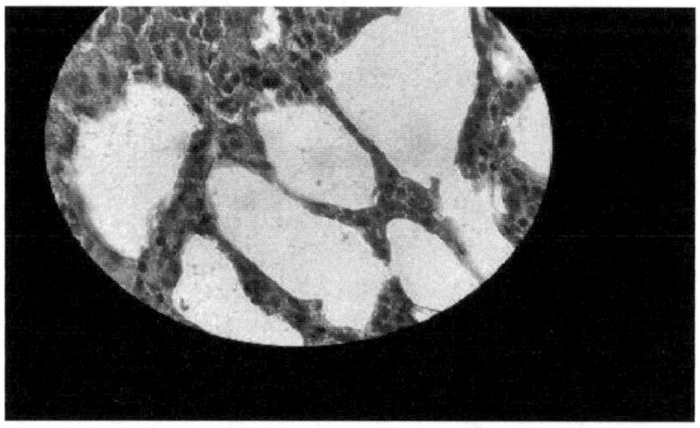

Figure 2. Histopathology of the lung after administration of MESA (Olympus CX21, 40x10).

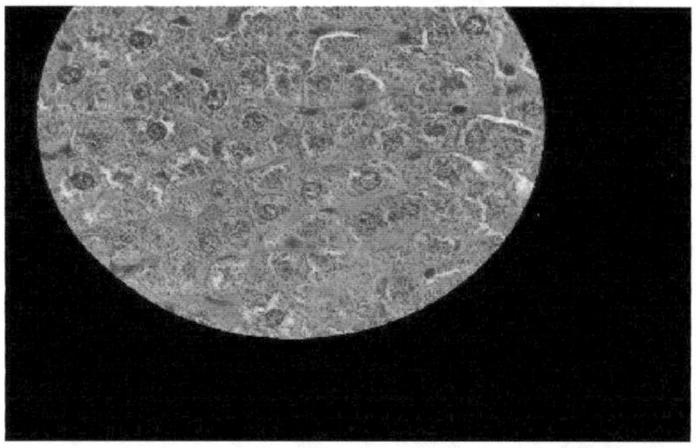

Figure 3. Histopathology of heart after administration of MESA (Olympus CX21, 40x10).

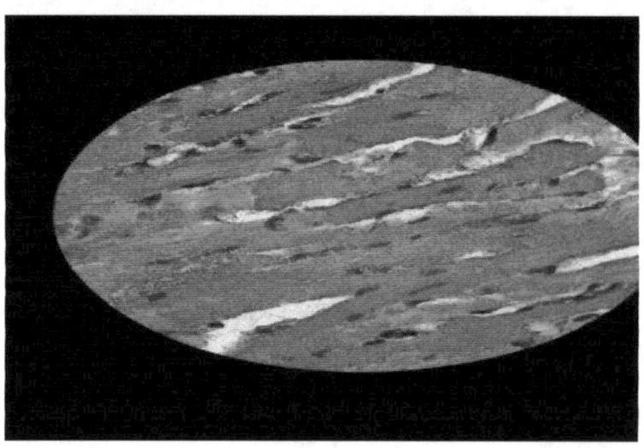

Figure 4. Histopathology of the kidney after administration of MESA (Olympus CX21, 40x10).

4. Discussion

Study in female rat with subchronic administration of MESA cause no tissue damage in the heart [13], lung [14]. Therefore, a subchronic experiment in this research also showed that the MESA was well tolerated by all animal models, as there were mortality and MESA-correlated marker and symptoms of toxicity test were examined all round the study phase. The MESA had a non-toxic or damaging effect on the body growth marking of all treatment groups. The function of whole blood as an indicator of physiological and pathological category in animals and humans is well recorded [6] [15]. In acute study tests and sub chronic tests, changes in biochemistry analysis, in addition, biochemical indicators, are usually treated as the lead of harmful or toxicities test. In the current study, MESA treatment of 250, 500 or 1000 mg/kg b.w of MESA for 90 days did not appear the standar of blood serum indicator. The non-clinical security study guidance for herb products that have better cause for a treat or calculated for a long time of use. In accordance with these principles, levels of serum total protein, ALP, ALT, and AST were measured in the present study and used as liver function tests, while creatinine and urea standar were used as kidney function study [16].

The biological characteristics; the correlation between body weight and organ weight is that in the major part of the study, except for the brain. Besides, all of the organs are commonly observed to decide whether the measurements of the organ has changed, especially concerning the weight of the all animal as an indicator. It is for this reason that this study investigated the effect of the subchronic experiment of MESA on absolute and relative organ weights. Histopathological indicator of brain, lung, heart, liver, and kidneys. The current study, no scientific confirmation, has ever specified the effects of MESA on organ weights of rats [6].

The major target organ is hepar. Toxicity of the hepar organ, damage to the hepar cause-effect of hepatocytes causing to release the membrane-bound enzymes (e.g., ALT and AST), injury to the hepato-biliary system thereby leading to release the essential enzymes (e.g., ALP), and/or harmful the catabolic capacity and biosynthetic of the liver function. Thus appear histopathological damage [17].

The concentration level of the kidney function measured such as serum of urea creatinine, and BUN levels were significantly different in the normotensive and MESA group. Other than, histological investigation contrast of the kidney frontal sections of the investigated rat and normotensive rat. There are no significant differences in the MESA experiment. In all treatment, groups showed that the tubular renal and glomerular were the intact and normal features. Consequently, not found glomerulonephritis [18] after exposure with the MESA. Thus MESA was not causing toxic in the kidney function.

Therefore, this study suggests that the subchronic administration of *S. atropurpuera* is safe for adjuvant with the treatment of herbal medicine. On further research, there needs to be a clinical trial in humans — however, nonappearance of any mortality in the animal model. In the long term, MESA as a remedy antihypertensive herbs.

5. Conclusion

The results of this study indicated that the methanolic extract of *S. atropurpurea* at antihypertension dose is safe for a long term period. In high doses, MESA will trigger damage to organ tissue and increase levels of blood biochemistry. Another hand, the lowest dose is safe for organ tissue.

6. Acknowledgment

Thanks to *the Directorate for Research and devotion Society,* Ministry of Research, Technology and the Higher Education Republic of Indonesia, Nomor: 022/SP2H/K2/KM/2017; 12/SP2H/LT/MULTI/L7/2019; 112/G164/U.LPPM/K/B.07/IV/2019, which has supported grant funding for research Penelitian Skim Unggulan Perguruan Tinggi (PTUPT) with the title "Kombinasi Herbal Benalu sebagai Produk Fitofarmaka suatu Kandidat Alternatif Obat Antihipertensi Alami Tradisional Indonesia." Thanks also to the team of research.

7. References

[1] Athiroh N, Permatasari N, Sargowo D, and Widodo MA 2014 Antioxidative and blood pressure-

lowering effects of Scurrula atropurpurea on deoxycorticosterone acetate-salt hypertensive rats *Biomarkers Genomic Med.* **6** 32–36

[2] Athiroh N, Permatasari N, Sargowo D, and Widodo A 2014 Effect of Scurrula atropurpurea on nitric oxide, endothelial damage, and endothelial progenitor cells of DOCA-salt hypertensive rats *Iran J Basic Med Sci.* **17** 622–625

[3] Alves RRN, Rosa IML 2007 Biodiversity, traditional medicine and public health: where do they meet? *Journal of Ethnobiology and Ethnomedicine* **3** 1746-4269

[4] Steenkamp V. and Gouws M. C. 2006 Cytotoxicity of six South African medicinal plant extracts used in the treatment of cancer. *South African Journal of Botany* **72** 630-633

[5] Athiroh, N and Sulistyowati E 2013 Scurrula atropurpurea Increases Nitric Oxide and Decreases Malondialdehyde in Hipertensive Rats *J. Nas. Terakreditasi J. Universa Med* **32** 44–50

[6] Nigatu, TA, 2017 Toxicological investigation of acute and chronic treatment with Gnidia stenophylla Gilg root extract on some blood parameters and histopathology of spleen, liver and kidney in mice. *BMC research notes* **10** 625

[7] Mahyan A, Athiroh N, Santoso H 2016 Paparan 28 Hari Ekstrak Metanolik Benalu Tehm (Scurrula atropurpurea Bl. Dans) terhadap Kadar SGPT Tikus Betina. *Jurnal Ilmiah Biosaintropis* **2** 53-58

[8] Hikmah U, Athiroh N, and Santoso H 2017 Kajian Subkronik Ekstrak Metanolik Scurrula atropurpurea (Bl.) Dans terhadap Serum Glutamic Oxaloacetic Transaminase Tikus Wistar Betina Sub-Chronic Study of Scurrula atropurpurea (Bl.) Dans Methanolic Extract toward SGOT Level in Female Wistar Rats, *J. Ilm. BIOSAINTROPIS (BIOSCIENCE-TROPIC* **2** 30-35

[9] Prastika NI, Athiroh N, and Santoso H 2017 Pengaruh Pemberian Subkronik Ekstrak Metanolik Scurrula atropurpurea (Bl) Dans Terhadap Kadar Kreatinin Tikus Wistar Sub-Chronic Present Effect of Scurrula atropurpurea (Bl) Dans Methanolic Extract toward Creatinine Levels in Wistar rats *J. Ilm. BIOSAINTROPIS (BIOSCIENCE-TROPIC* **2** 42-48

[10] Sammad FHA, Athiroh N, and Santoso H 2017 Pemberian Ekstrak Metanolik Scurulla atropurpurea (Bl) Dans Secara Subkronik Terhadap Protein Total Dan Albumin Tikus Betina Injection of Scurrula atropurpurea (Bl) Dans Metanolic Extract in Sub-chronic to Total Protein and Albumin of Female Mice *J. Ilm. BIOSAINTROPIS (BIOSCIENCE-TROPIC* **2** 49-54

[11] Zakia JF, Athiroh N 2017 Studi Sub-Kronik 90 Hari: Pengaruh Ekstrak Scurrula atropurpurea Terhadap Kadar Kolesterol Tikus (Rattus norvegicus) Betina Galur Wistar Study of Sub Chronic during 90 Days: Effect of Scurrula atropurpurea Extract toward Cholesterol Level in Rat (Rattus norvegicus)," *BIOSCIENCE-TROPIC*, vol. **3** 37-43

[12] Mihmidati L, Athiroh N 2016 Pengaruh Ekstrak Metanolik (Scurrula atropurpurea (Bl.) Dans) Yang Diberikan Secara Subkronik 90 Hari Pada Tikus Betina (Rattus norvegicus) Terhadap Necrosis Otak The Effect of Extraction of Metanolic (Scurrula atropurpurea (Bl.) Dans) Which Is Given Sub-chronic 90 Days on Female Rats (Rattus norvegicus) toward Necrosis of Brain *J. Ilm. BIOSAINTROPIS (BIOSCIENCE-TROPIC* **1** 16-23

[13] Mareta D, Hidayati N, Athiroh N 2017 Profil Histopatologi pada Jaringan Jantung Tikus Subkronik 90 Hari Menggunakan Ekstrak Metanolik Scurrula atropurpurea (Bl.) Dans Histopathologic Profile in the Heart Tissue of 90-Days Sub-Chronic Rats Using Metanolic Extract of Scurrula atropurpurea (Bl.) Dans *J. Ilm. BIOSAINTROPIS (BIOSCIENCE-TROPIC* **3** 30-36

[14] Fahmiy R, Athiroh N 2017 Studi Sub-Kronik 90 Hari: Profil Histopatologi Pada Jaringan Pulmo Tikus (Rattus norvegicus) Akibat Paparan Ekstrak Scurrula atropurpurea (Bl.) Dans. Study of 90 Days Sub Chronic: The Histopathology Profile in The Pulmo Tissue Rats Exposured Scurrula atropurpurea (Bl.) Dans. Extraction *J. Ilm. BIOSAINTROPIS (BIOSCIENCE-TROPIC* **3** 39-45

[15] Adedapo AA 2007 Effects of some plants of the spurge family on haematological and biochemical parameters in rats. *Veterinarski arhiV* **77** 29-38

[16] Gaskill C, Miller LM, Mattoon JS 2005 Liver histopathology and liver and serum alanine aminotransferase and alkaline phosphatase activities in epileptic dogs receiving phenobarbital *Veterinary Pathology* **42** 147-160
[17] Kumar RS, Gupta M, Mazumdar UK 2005 Effects of methanol extracts of Caesalpinia bonducella and Bauhinia racemosa on hematology and hepatorenal function in mice *The Journal of toxicological sciences* **30** 265-274
[18] Ebaid H, Dkhil MA, Danfour MA, Tohamy A 2007 Piroxicam-induced hepatic and renal histopathological changes in mice *Libyan Journal of Medicine* **2** 82-89

IOP Conf. Series: Materials Science and Engineering **846** (2020) 012003 doi:10.1088/1757-899X/846/1/012003

Contribution of Activated Carbon Based on Cacao Peels (*Theobroma cacao L.*) to Improve the Well Water Quality (COLOR, BOD, and COD)

Yuli Yetri1[1]*, Dwi Astuti Marantika[2], Takdir Alamsyah[1], Matlal Fajri Alif[2], Rahmiana Zein[2]*

[1]Politeknik Negeri Padang, Padang, Indonesia
[2]Faculty of Mathematics and Natural Sciences, Andalas University

*yuliyetri@pnp.ac.id
*mimiedison@yahoo.co.id

Abstract. In this research, we have made activated carbon from the cacao peels (*Theobroma cacao L.*) for the purpose of water purification. The cacao peels activated using 50% H_3PO_4at the temperature of 400 °C for 1 hour. Scanning Electron Microscopy was applied for a surface analysis of activated carbon, and the functional groups identified by Fourier Transform Infra-Red (FTIR). The adsorption process was carried out by a column methodwith a flow rate variation of 5 mL/min and 10 mL/min. The results showed that by using 2 grams of actived carbon with a flow rate of 5 mL/min is the optimum condition, with the obtained absorption efficiency of 91.8% BOD, 90.2% COD, and 96.5% color. Morphological analysis has determinedthe changes in pore on the surface of activated carbon. Analysis of functional groups showed that the produced activated carbon hada pattern of absorption with OH, CH, and CO type of bond. Based on the results, it can be concluded that the activated carbon of cacao peels can improve the quality of dirty well water, and make it consumable for the daily use.

1. Introduction

In line with the increasing population growth, industrial activity is also increased. However, industrial growth has adverse effects on society, because some of these industries dispose of their waste directly into sewers/water bodies without prior management. As a result, the water source can be polluted because of the exceeding use on its capacity to make it renewable [1].

To this day, many people still take well as their source for clean water supply. Nevertheless, at recent time, the water from the well has been contaminated, either organic or inorganic which causes the color of the water becoming unclear. The ideal clean water must have the characteristics of clear, colorless, tasteless, and odorless, does not contain chemical substances that can alter the function of the body, and harmless to human health, and others. It is noteworthy to prevent the occurrence and spread of water-borne disease [2].

To reduce the impact on the water, the well water must be purified in order to maintain the quality of the clean water so it can be consumed in the daily basis. To achieve the above mentioned objectives, the alternatives for natural and environmentally friendly water purification should be determined. Some studies on water purification have been carried out using the LMM (Layer Multi Media) method on local ingredients-basis. Taken as the example on the wastewater noodle industry, and peat water have been able to be used in constructing the pH value, Color, Chemical Oxygen Demand (COD), Biologycal Oxygen Demand (BOD), Total Suspended Solids (TSS), Total Disolved Solids (TDS), tomeet the Minister of Environment Regulationby utilizing the activated carbon [3]. Additionally it has done manufacture of activated carbon from bark peels to a solution of methylene blue dyesand gumitir plant stems on water content, volatile matter, total ash content, caron content, absorption of methylene blue, and absorption capacity of iodine [4], [5].

The price of commercialized activated carbon is very expensive, so it will cost a lot using it for water purification. Therefore,the the alternative for activated carbon like the one derived from abundant natural-ingredients should be made. One of natural resources which has not been maximally used is cacao peels. Some studies have utilized cacao peels for corrosion inhibitors [6], [7], [8], [9], mechanical properties repair [10], fodder [11], hydrofobic properties [12], thin layer [13],[14], and particle boards [15]. Using cacao peels directly is not good as an adsorbent for water purification due to many of cacao peels recently are attacked by pests so that they become rotten, and damage the environment [16]. While in Indonesia, cacao plantsare widely grown by farmers,and generate solid waste which ruin the the aesthetics of environment.

Cacao peels produce 21.06% hemicellulose content, 20.15% cellulose, and 55.11% lignin. Such compounds are polymers of carbon elements so that the cacao peels can be used as activated carbon with a large porous, and deep surface to make its absorption capacity becoming high [17]. Therefore, to make the waste from cacao peels does not pollute the environment, it can be processed into activated carbon which can be used as an adsorbent to purify the water.

2. Materials And Methods

Chemicals and reagents

Well water, cacao peels, filter paper, glass woll, distilled water, alkali iodide azide, H_3PO_4 concentrated, $KMnO_4$, 97%H_2SO_4, Potassium Iodide, $Na_2S_2O_3$, HCl, NaOH, $MnSO_4$, $HgSO_4$, 1% Amilum/ tapioca, $K_2Cr_2O_7$, distilled water, K_2PtCl_6, $CoCl_2.6H_2O$, pH 4, 7, and 9 of buffer solution are analytical grade and obtained from E-Merck (Germany) unless otherwise note.

The tools used in this research is a glass column with a length of 25 cm, and 1.5 cm in diameter, standards, clamps, well water container, Front Laboratory Analytical Balance peristaltic pump (Kern & Sohn GmbH), Miyako Blender, Furnace Carbolite CWF 1200, pH meter (HANNA HI 98 127), mortar and pestle, FTIR (Mattson Unican 7000 Mod FTIR), SEM (HITACHI S-3400 N), UV-Vis Spectrophotometer (Thermo Scientific GENESYS 10S Series), a sieve with $\leq$ 160 μm particle size.

Preparation of Activated Carbon

Cacao peels washed with water, then wind dried, carbonized for 1 hour at 400°C. Cacao peels charcoal was then mixed with 50% H_3PO_4 with a ratio of 1: 4, and allowed to stand for 24 hours. The mixture was washed with water until reaching neutral pH, and put in the oven at the temperature of 105 °C for 3 hours. Activated carbon was sifted until resulting $\leq$ 160 μm particle size, and it was ready to be used. At the end, obtained activated carbon was characterized by SEM and FTIR [18].

Treatments of Sample

The adsorbent was inserted into the column glass with mass variation of 1 gram, and 5 mL/minute and 10 mL/minute flow rates using a peristaltic pump. After obtaining the best flow rate, the treatment of 2 grams of perlite mass was performed. Sample of groundwater was channeled to the column, and the flowed sample volumes were calculated to see to what extent this material can still be used. The water that had been flowed through the adsorbent was then analyzed for its color, BOD, and COD values.

Determination of Color

The determination of the water color was carried out using a spectrophotometer with a spectrophotometric method by wavelength of 456 nm. The used standard solution is platinum-cobalt.

Determination of Biological Oxygen Demand (BOD)

The determination of the BOD value can be done by reducing the values of $DO_{(0)}$ with $DO_{(5)}$. Whereas, the determination of $DO_{(0)}$ was done by taking the sample that has been prepared. 1 mL of $MnSO_4$ and 1 mL of alkaline azide iodide were added by the tip of pipette right above the surface of the solution. Closed the solution immediately, and homogenized it until a perfect lump is formed. Next, left the lumps for 5 to 10 minutes. Afterward, added 1 mL of concentrated H_2SO_4, then covered and homogenized it until the deposits were completely dissolved. Then, took 50 mL by using pipette, and put into 150 mL of erlenmeyer. Titrated it with $Na_2S_2O_3$ by using the starch indicator until the blue color disappeared. The determination on DO($_5$) was stored for 5 days, after 5 days it was titrated in the same way as the determination of DO($_0$).

Determination of Chemical Oxygen Demand (COD)

Organic and inorganic compounds, especially organic, in the sample test were oxidized by MnO_4^- in closed reflux for 1 hour producing Mn^{2+}. The excess of unreduced potassium dichromate was titrated with sodium thiosulfate solution using starch indicator. The amount of needed oxidant was stated in oxygen equivalent (O_2 mg/L).

The procedure in determining the COD was by pipetting 50 mL of the sample volume, adding ± 0.1 gram of $HgSO_4$, and 5 mL of $KMnO_4$ into erlenmeyer. Then closed and heated the erlenmeyer for ± 1 hour in a heater at the temperature of 70 °C. Let it slowly cooled down until it reached room temperature. After it was cold, 5 mL of potassium iodide and 10 mL of 4N H_2SO_4 were added. Then it was titrated with 0.025 N $Na_2S_2O_3$ until the color turned to light yellow, then 1 mL of starch indicator was added. Continued the titration until an obvious changed of color occured from blue to clear or until the blue color disappeared. Then, recorded the volume of 0.025 N $Na_2S_2O_3$ solution which was used. These steps were also done towards organic free water as blank. Likewise, the volume of $Na_2S_2O_3$ solution used was recorded as well.

3. Results And Discussion

Characterization of Activated Carbon

The activated carbon absorption on the iodine was set in order to determine the ability of activated carbon in absorbing colored solutions. The obtained value of Iodine absorption in the study can be seen in Figure 1.

Notably, Iodine absorption value obtained in this research was 1012.662 mg/g, where the Iodine absorption capacity for commercial activated carbon was 1247.85 mg/g. To some extent, this obtained absorption capacity has met the SNI 06-3730-1995 standards (at least 750 mg/g). Hence, this Iodine absorption capacity was influenced by the activation and carbonization process. The immersed samples in H_3PO_4 produced samples with high iodine absorption, where the phosphoric acid cleansed the pores, expanded the carbon surface, provided active group in order to enlarge the absorption of activated carbon. The increase of activation temperature also raised the absorption capacity of activated carbon against iodine. It was due to the higher the temperature, the more the carbon plates were shifted that encouraged hydrocarbons, tar, other organic compounds to release during the activation time.

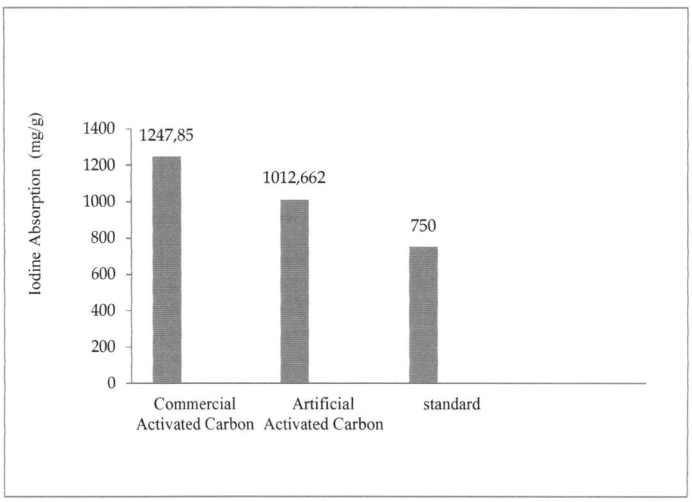

Figure 1. Effect of Activated Carbon onto Iodine Absorption

Characterization and FTIR analysis was conducted to determine the functional groups contained in the cacao peels so that it can be used as activated carbon in order to improve the water quality. Shown in Figure 2 is carbon absorption band in the cacao peels before and after the adsorbed change. The main component of activated carbon from peels are cellulose, hemicellulose, and lignin [19].

FTIR analysis (Figure 2) shows the content of cellulose in the presence of hydroxyl OH bond at wave number 3347.13 cm^{-1} and shifted to a wave number 3232.33 cm^{-1} after the adsorption process - OH functional group on adsorbents experienced deprotonation, so the functional groups became negatively charged, which was very reactive in absorbing Ca^{2+} ions, and other metal cations [20].

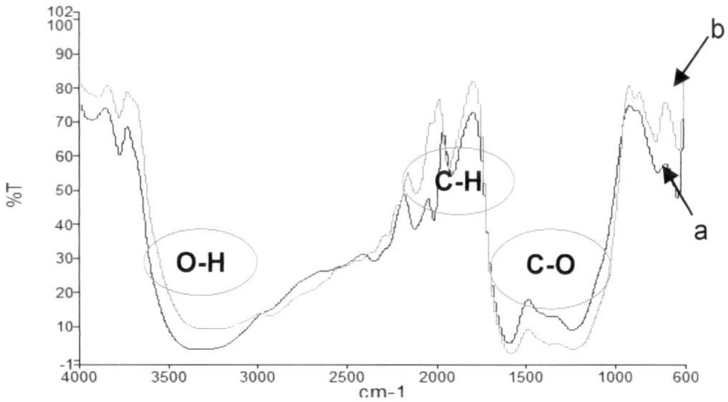

Figure 2. The FTIR spectrum Activated Carbon before absorption (a), and after absorption (b)

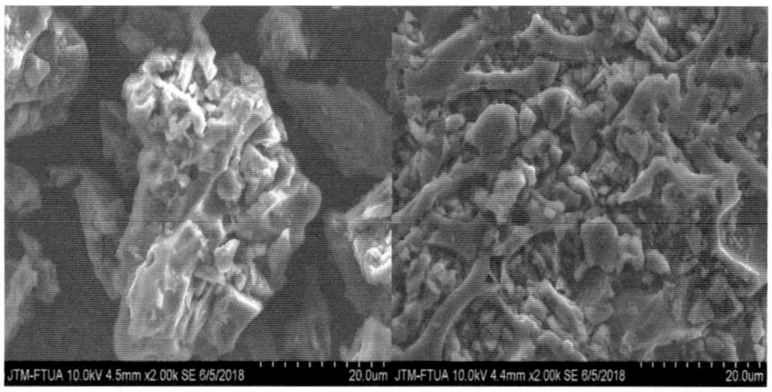

Figure 3. Surface morphology of activated carbon (a) prior to adsorption, (b) after adsorption (magnification 2000x)

At the early stage of activated carbon was seen a peak around a wavelength of 2346.79 cm-1 indicating the presence of C≡N group. The surface containing nitrogen group on the activated carbon surface increased the ability to absorb acid gases [21]. The carboxyl (Aliphatic C-O stretching) function at wave number 1235.96 cm^{-1} indicated the presence of hemicellulose content, and after adsorption there was a shift to the wave number 1236.01 cm^{-1}. At wave number 1916.18 cm^{-1} showed lignin content, and undergone a shifting to wave number 1935.37 cm^{-1} marked by the presence of aromatic C-H group [22]. The hydroxyl and carboxyl groups took part in adsorbing metal cations. The produced activated carbon had an absorption pattern with O-H, C-H, and C-O bond types [19].

The SEM analysis results at 400°C of combustion temperature is shown in Figure 3. Its shows that before experiencing the adsorption, activated carbon has a rough surface, and porous. Meanwhile, after the absorption, the pores are filled and becomes denser. It is due to the occurence of metal ions adsorption, and other organic material containing in the well water into the activated carbon pores. The activation usingacid phosphate at 50% concentration has made the pores becoming larger. The surface area and pore volume of activated carbon is increasing in line with the increase on the concentration of H_3PO_4 solution as an activator. The increase in porosity is also caused by the amount of impregnating agent used, the more activator used the more pores formed on the activated carbon [23], [24], [4].

IOP Conf. Series: Materials Science and Engineering **846** (2020) 012003 doi:10.1088/1757-899X/846/1/012003

Effect of BOD, COD, and Color Before and After Adsorption

The color of water is caused by the presence of organic and inorganic substances, plankton, humus and metal ions (iron and manganese), and other materials [25], [26], [27]. Colored water indicates that the quality of the water is low, so it is not consumable. The color concentration of well water before being flowed is 1.1718 TCU. The changes of the color concentration of this well water after being flowed can be seen in Figure 4.

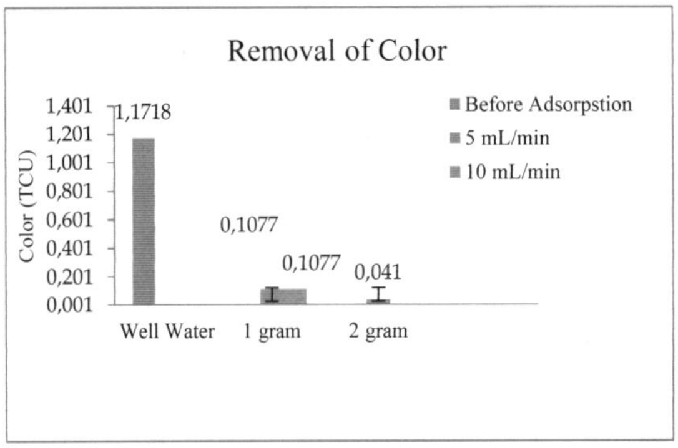

Figure 4. The Effect of Well Water Flow Rate and Adsorbent Mass on Changes in Color Values After Adsorption

Figure 4 shows the effect of activated carbon mass on the color of well water. The well water that has been drained using the activated carbon went through a significant change in color concentration. Whereas the well water taken in this research was the water with yellow color as a consequence of the high level of iron in the water. After passing through the activated carbon, the cacao peels became clear. The decrease in color concentration occured because the well water that had passed through the activated carbon received a filtration process so that the deposits in the water were lost, and the iron content was disappeared. Thus,the obtained well water after flowing it with activated carbon became colorless and in accordance with the water standards for daily consumption. Color reduction efficiency using 1 gram activated carbon with a flow rate of 5 mL/minute and 10 mL/minute was 90.8%, whereas with 2 grams of activated carbon in the 5 mL/minute, the color reduction efficiency was 96.5%.

The analysis of Biochemical Oxygen Demand (BOD) in this research used the winkler method, and conducted at the Chemical Laboratory for Environmental Analysis. The whole measurement process was carried out in dark conditions, in order to avoid the occurence of photosynthesis which produced oxygen.$DO_{(0)}$ was measured immediately right after sampling collected, then continued to measure the dissolved oxygen content in the sample that had been incubated for 5 days. Measurements for $DO_{(5)}$ was expected only set the occurence of decomposition process by microorganisms, so only the use of oxygen ensued.

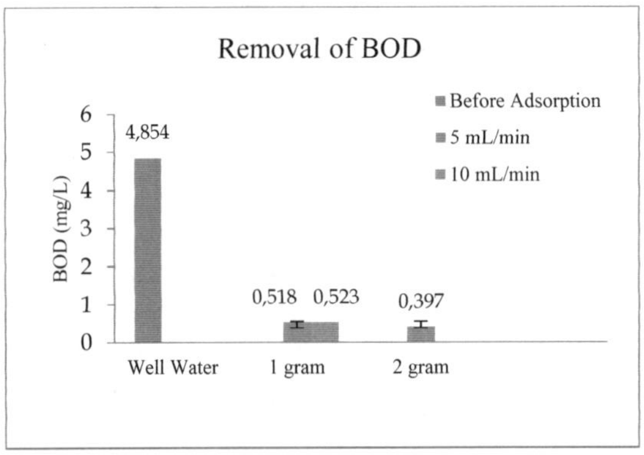

Figure 5. Effect of Well Water Flow Rate and Adsorbent Mass on Changes in BOD Values After Adsorption

The BOD concentration of the well water before flowing was 4.854 mg/L. The changes in BOD concentration from the water after being flowed can be seen in Figure 5. The data in Figure 5 points out that there is a significant decrease in BOD value. The decrease efficiency of BOD in activated carbon mass of 1 gram with a flow rate of 5 mL/min and 10 mL/min was 89.2%, while in 2 grams activated carbon mass, the decrease efficiency in BOD was 91.8%. The decrease in BOD value indicated that the number of microorganisms contained in the water was decreased as well. It was due to the reduced supply of food (organic matter) for microorganisms, so that it increased the dissolved oxygen, and reduced the levels of organic matter in the water [20].

BOD reduction efficiency is also affected by differences in flow rate. Where the higher the flow rate, the lesser the reduction efficiency. As a result of high flow rates, the contact time became shorter so it reduced the decomposition rate of organic substances in the water. Conversely, the efficiency of reduction increased when the flow rate decreased, by the reason of the contact time in the column lasted longer, and the decomposition rate by microorganisms occured slowly, and ran more impeccable [28].

Chemical Oxygen Demand (COD) describes the total amount of the presented organic substances. The COD concentration of well water before being flowed was 9,114 mg/L. The changes in COD concentration of the well water after being flowed can be seen in Figure 6. Taken as a note, there is a significant decrease in COD value where the efficiency of COD reduction in activated carbon mass of 1 gram with a flow rate of 5 mL/min and 10 mL/min respectively was 88.0% and 77.1%, while on 2 grams activated carbon mass the efficiency of COD reduction was 90.2%.

COD reduction efficiency also represented that the lower the flow rate, the longer the contact time, therefore, the efficiency of COD reduction increased. The decrease in COD value was due to the solid material began to deposit, thus the organic substances within the water also decreased. In addition, the decrease made some organic materials being oxidized, whereas some others being absorbed by activated carbon reducing the value of COD [20]. The COD value is usually higher than the BOD value because many of waste material can be oxidized through a chemical process rather than through biological process. High COD values occured due to the presence of influencing environmental factors such as dissolved oxygen content in the well water which was sufficient to help the bacteria breaking down the pollutant compounds [29].

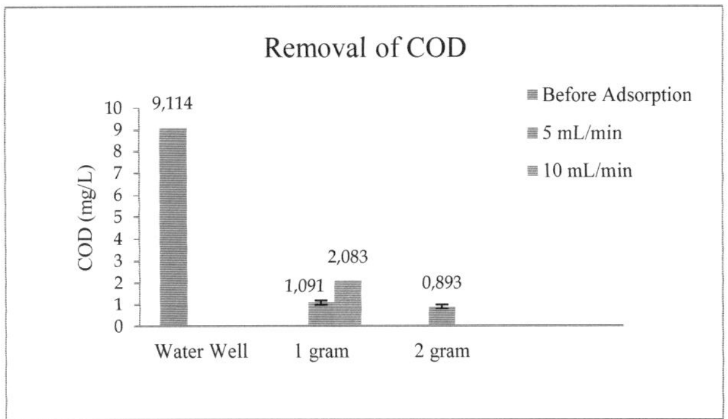

Figure 6. The Effect of Well Water Flow Rate and Adsorbent Mass on Changes in COD Values After Adsorption

4. Conclusion

Dirty water well can be purified by using activated carbon from cacao peels (*Theobroma cacao L.*) which is activated with 50% H_3PO_4 at 400 ^oC for 1 hour at 5 mL/ minute of flow rate optimum conditions, and 2 grams mass, with 96.5% decrease efficiency on the color value, 91.8% BOD, and

90.2% COD. The obtained analysis results determines that cacao peels are applicable to be used as activated carbon in the purification process on the dirty well water becoming clean water, in accordance with RI Minister of Health Regulation No. 492/Menkes/Per/IV/ 2010. Morphological analysis using scanning electron microscope (SEM) has indicated pore changes on the surface of activated carbon. Analysis of functional groups using FTIR spectroscopy shows that the produced activated carbon has an absorption pattern with O-H, C-H, and C-O bonds.

5. Acknowledgment

The author would like to thank Politeknik Negeri Padang for funding this research through The Unggulan Perguruan Tinggi Research Grant with contract Number: 230 / PL9.1.4 / LT / 2018.

References

[1] K. A. Adegoke and O. Solomon, "Dye Sequestration Using Agricultural Wastes As Adsorbents," *Water Resour. Ind.*, vol. 12, pp. 8–24, 2015.

[2] M. Sulyman, J. Namiesnik, and A. Gierak, "Low-cost Adsorbents Derived from Agricultural By-products / Wastes for Enhancing Contaminant Uptakes from Wastewater : A Review," *Pol. J. Environ. Stud.,* vol. 26, no. 2, pp. 479–510, 2017.

[3] R. Zein, R. Suhaili, L. Novita, S. Ningsih, Mukhlis, N. Swesty and H. Novrian, "Novel Two Stage Vertical Flow Biofilter System for Efficiency Treatment of September – October," *Research Journal of Pharmaceutical, Biological and Chemical Sciences* vol. 7, no. 71, pp. 71–79, 2016.

[4] P. Taylor, H. A. Al-aoh, R. Yahya, M. J. Maah, and M. R. Bin Abas, "Adsorption Of Methylene Blue On Activated Carbon Fiber Prepared From Coconut Husk : Isotherm , Kinetics And Thermodynamics Studies," *Desalination And Water Treatment,* vol. 1, no. 13, pp. 37–41, 2013.

[5] H. M. B. Eustáquio, C. W. Lopes, R. S. Rocha, and B. D. Cardoso, "Modification of Activated Carbon for the Adsorption of Humic Acid," *Adsorption and Science Technology,* vol. 2, no. 1, pp. 117–126, 2014.

[6] Y. Yetri, Emriadi, N. Jamarun and Gunawarman, "Corrosion Behavior of Environmental Friendly Inhibitor of Theobroma Cacao Peels Extract for Mild Steel in NaCl 1.5 M," *EnvironmentAsia,* vol. 9, no. 1, pp. 45–59, 2016.

[7] Y. Yetri and Sukatik, "Green Inhibitor for Mild Steel in Acidic Solution by Using Crude Extract and Polar Extract of Theobroma cacao Peels," *Oriental Journal of Chemistry,* vol. 33, no. 4, pp. 2071-2079, 2017.

[8] Y. Yetri, Gunawarman and N. Jamarun, "Inhibitory Action Of Theobroma Cacao Peels Extract On Corrosion Of Mild Steel In Different Media," *Rasayan Journal of Chemistry,* vol. 9, no. 4, pp. 716–727, 2016.

[9] Y. Yetri, Gunawarman, Emriadi, and N. Jamarun "Theobroma Cacao Extract Peels (TCPE) Green Inhibitor To Recovery The Mechanical Properties Of Mild Steel After Corrosion," *ARPN Journal of Enginering and Applied Science,* vol. 12, no. 18, pp. 5325–5332, 2017.

[10] Gunawarman, Y. Yetri, Emriadi, N. Jamarun, Ken-Cho, M. Nakai, and M. Niinomi, "Effect Of Polar Extract Of Cocoa Peels Inhibitor On Mechanical Properties And Microstructure Of Mild Steel Exposed in Hydrocloric Acid," *Applied Mechanic and Materials,* vol. 776, pp. 193–200, 2015.

[11] E. O. K. Oddoye, S. W. A. Rhule, V. Anchirinah, and F. O. Ansah, "Fresh Cocoa Pod Husk As An Ingredient In The Diets Of Growing Pigs," *Scienific Research adn Essays,* vol. 5, no. 10, pp. 1141–1144, 2010.

[12] M. Ritaliah, D. Dahlan, and Y. Yetri, "Potentiality Of Theobroma Cacao Peels Extract As Additive On Water-Based Paint To Hydrophobic Properties," *Rasayan Journal of Chemistry* vol. 11, no. 3, pp. 1277–1285, 2018.

[13] N. P. Tissos, D. Dahlan, and Y. Yetri, "Synthesis of Cuprum (Cu) Layer by Electrodeposition Method with Theobroma cacao Peels as Corrosion Protector of Steel," *International Journal on Advanced Science Engineering Information Technology,* vol. 8, no. 4, pp. 1290–1295,

2018.

[14] S. A. Mahaputri, D. Dahlan, and Y. Yetri, "The Use of Cacao Peels Extract (Theobrema Cacao) as the Corrosion Inhibitor on Steel Layers Electrodeposition," *Recent Advance in Petrochemical Science,* vol. 5, no. 3, pp. 28–32, 2018.

[15] Y. Yetri, Sukatik, R. Sumiati and R. Hidayati, "The Effect of Size and Composition Variation of Cacao (Theobroma Cacao) Peels Powder with Bagasse Fillers on Physical And Mechanical Properties of Particle Board," *International Journal of Applied and Physical Sciences,* vol. 4, no. 2, pp. 50–58, 2018.

[16] SNI.06-3730-1995: Technical Activated Carbon. National Standardization Agency , Jakarta., 1975.

[17] R. M. Ali, H. A. Hamad, M. M. Hussein, and G. F. Malash, "Potential Of Using Green Adsorbent Of Heavy Metal Removal From Aqueous Solutions : Adsorption Kinetics , Isotherm , Thermodynamic , Mechanism And Economic Analysis," *Ecol. Eng.*, vol. 91, pp. 317–332, 2016.

[18] L. Ding, B. Zou, W. Gao, Q. Liu, Z. Wang, Y. Guo, X. Wang and Y. Liu., "Colloids and Surfaces A : Physicochemical and Engineering Aspects Adsorption of Rhodamine-B from Aqueous Solution Using Treated Rice Husk-Based Activated Carbon," *Colloids Surfaces A Physicochem. Eng. Asp.*, vol. 446, pp. 1–7, 2014.

[19] A. K. Kushwaha, N. Gupta, and M. C. Chattopadhyaya, "Removal of Cationic Methylene Blue And Malachite Green Dyes From Aqueous Solution By Waste Materials Of Daucus Carota," *J. Saudi Chem. Soc.*, vol. 18, no. 3, pp. 200–207, 2014.

[20] A. Rotimi and G. Okeoghene, "Sorption and Desorption Studies on Toxic Metals From Brewery Effluent Using Eggshell as Adsorbent," *Adv. Nat. Sci,* vol. 7, no. 2, pp. 15–24, 2014.

[21] D. Pathania, S. Sharma, and P. Singh, "Removal of Methylene Blue By Adsorption Onto Activated Carbon Developed From Ficus Carica Bast," *Arab. J. Chem.*, vol. 10, pp. S1445–S1451, 2017.

[22] A. M. Aljeboree, A. N. Alshirifi, and A. F. Alkaim, "Kinetics and equilibrium Study For The Adsorption Of Textile Dyes On Coconut Shell Activated Carbon," *Arab. J. Chem.*, vol. 10, pp. S3381–S3393, 2017.

[23] L. Hsu and H. Teng, "Influence of Different Chemical Reagents On The Preparation Of Activated Carbons From Bituminous Coal," *Fuel Proccess. Technol.,* vol. 64, pp. 155-166, 2000.

[24] S. Velten, D. R. U. Knappe, J. Traber, and H. Kaiser, "Characterization of Natural Organic Matter Adsorption In Granular Activated Carbon Adsorbers," *Water Research,* vol. 45, pp. 3951-3959, 2011.

[25] N. Jacob, "Adsorption and Treatment of Organic Contaminants using Activated Carbon from Waste Nigerian Bamboo," *J. Appl. Sci. Environ. Manage.,* vol. 13, no. 3, pp. 39-47, 2009.

[26] O. H. Tukki, J. T. Barminas, S. A. Osemeahon, J. C. Onwuka, and R. A. Donatus, "Adsorption of Colloidal Particles Of Moringa Oleifera Seeds On Clay For Water Treatment Applications," *J. Water Supply Res. Technol.,* vol. 65 pp. 75–86, 2016.

[27] B. N. Thomas, S. C. George, and C. O. Shell, "Production of Activated Carbon from Natural Sources," *Review Article, iMedPub Journals,* vol. 1, no. 1:7, pp. 1–5, 2015.

[28] Sy. Salmaraiza, et al., "The Effect of MSL Reactor Influent Flow Rate on Reduction of BOD, COD, TSS, and Oils/Fats of Edible Oil Industry Wastewater," *Jurnal Litbang Industri*, vol. 7, no. 1, 2017.

[29] R. Han, L. Zhang, C. Song, M. Zhang, H. Zhu, and L. Zhang, "Characterization of Modified Wheat Straw , Kinetic And Equilibrium Study About Copper Ion And Methylene Blue Adsorption In Batch Mode," *Carbohydr. Polym.*, vol. 79, no. 4, pp. 1140–1149, 2010.

Utilization of ZSM-5/TiO_2 Powder and Membrane to Reduce Concentration Cu(II) and Cr(VI) Ions in Water

Ana Hidayati Mukaromah*, **Umrotun Chasanah, Ilma Rizqi Assyifa, Mifbakuddin, and Sri Sinto Dewi**

Universitas Muhammadiyah Semarang, Semarang, Indonesia

*Corresponding Author: ana_hidayati@unimus.ac.id

Abstract: Cancer is an effect caused by excessive concentrations of Cu(II) and Cr(VI) metal ions in the body. Including health effects of heavy metal ion poisoning industries arise when humans consume fish from various wastes, dyes, electroplating, paint, and battery. Techniques and processes have been developed to separate heavy metal ions which were very dangerous from water by ion exchange, chemical precipitation, and adsorption. Adsorption has no harmful side effects to health, simple, economical and easy to do. Natural materials can be used as adsorbents including ZSM-5, ZSM-5/TiO_2 powder, and membrane. The purpose of this study is to determine the initial, final concentration and the percentage decrease concentration of Cu(II) and Cr(VI) ions in water by ZSM-5/TiO_2 powder and membrane with variations gauze type and irradiation time. The results showed that the highest percentage reduction Cu(II) ion and Cr(VI) ion with 0.25% w/v ZSM-5/TiO_2 powder during UV irradiation 75 minutes were 74.99% and 16.99%. The highest percentage reduction in Cu (II) and Cr (VI) concentration after passing through the ZSM-5/TiO_2 membrane with gauze support 304-400 during 90 minute UV irradiation were 57.72% and 54.65%. As the conclusion ability of ZSM-5/TiO_2 powder is greater than the ZSM-5/TiO_2 membrane in reducing the concentration of Cu(II) and Cr(VI).

Keywords: Cu(II), Cr(VI), ZSM-5/TiO_2 powder, ZSM-5/TiO_2 membrane, irradiation time

1, Introduction

Cancer is a disease caused by excessive concentrations of Cu (II) and Cr (VI) metal ions in the body. Poisoning, nausea, vomiting, and causing damage to the liver and kidneys also caused by excess Cu (II) and Cr (VI) metal ions because it is carcinogenic, very toxic and very active in water [1, 2]. According to the Minister of Health Regulation No. 492 Menkes /Per /IV /2010 concerning the quality of drinking water, the maximum permissible Cuprum level is 2 mg/L and the maximum chromium of drinking water is 0.05 mg/L [3]. Techniques and processes have been developed to separate heavy metal ions which were very dangerous from water by ion exchange, chemical precipitation, and adsorption. Adsorption was the safest method, has no harmful side effects to health, simple, cheap and easy to do. The application of adsorption technology has been widely developed using biomaterials as well as synthetic materials to reduce the heavy metal content of water bodies (biosorption) such as using rice husks. Other biomaterials can be used as adsorbents, there were silica gel, chitin, chitosan, humic acid, and natural zeolite. Synthetic materials that can be used as adsorbents were synthetic TiO_2, ZSM-5, ZSM-5/TiO_2 powder and ZSM-5/TiO_2 membrane [1, 4, 5]. Many naturally-derived polymers have been assessed for application as biomaterials specifically collagen, chitosan, keratin, and silk fibroin [6]. The purpose of this study is to determine the initial concentration of Cu (II) and Cr (VI) ions in artificial water, measure the levels of Cu (II) and Cr (VI) ions and the percentage decrease in the levels of Cu (II) and Cr (VI) ions in water and analyze the effectiveness of decreasing the concentration of Cu(II) and Cr(VI) ions by ZSM-5/TiO_2 powder and ZSM-5/TiO_2 membrane with variations gauze type and irradiation time.

1.1 Cu (II) and Cr(VI) Ion

Cu (II) ion is a component of the enzyme needed to produce energy, anti-oxidation, and synthesis of the hormone adrenaline, as well as for the formation of connective tissue. Cu (II) is an essential metal ion, and if excess will be detrimental to health, resulting in poisoning [1, 7]. Cr (III) is an essential nutrient very important for sugar metabolism and some enzyme reactions, Heavy metals can have health effects on humans depending on which part of the heavy metal is bound in the body. Chromium (VI) in chromate and dichromic form is very toxic which can cause skin and respiratory tract cancer [2, 5, 8].

1.2 Titanium Dioxide (TiO_2)

TiO_2 has a white crystalline shape, has a molecular weight of 79.8886 g/mol, the density of 4.23 g/cc, the melting point of 1843°C in the absence of oxygen and 1892 °C in the presence of oxygen, and has a boiling point of 2972 °C. TiO_2 crystals are insoluble in water, hydrochloric acid, dilute sulfuric acid and alcohol, TiO_2 has several advantages, namely the economical price, non-toxic and the most important is the stability and activity when subjected to light. Titanium dioxide is widely used as a photocatalyst because it is stable, corrosion-resistant, safe, has amphiphilic properties, and is cheap and is stable at pH 4.5-8. TiO_2 is the most effective semiconductor catalyst because it has a relatively large band energy (3.2 eV) suitable for photocatalysts, non-toxic, and has a good absorbency to ultraviolet light [4, 9, 10, 11].

1.4 Zeolite Socony Mobile (ZSM-5)

ZSM-5 is a porous solid crystal with a three-dimensional structure containing aluminosilicate, built from $(SiO_4)_4$- and $(AlO_4)_5$- and has a tetrahedral bond with an oxygen atom forming a pentasil chain. ZSM-5 is a zeolite that is synthesized with high silica content, has pores of 2-50 nm and has two types of channels, the namely straight channel with size 5,3 Å × 5,6 Å and winding channel (zig-zag) with size 5.1 Å × 5.5 Å that connects one channel straight to another channel [12, 13, 14, 15].

1.5 ZSM-5 impregnated TiO_2

TiO_2 is the most effective semiconductor catalyst because it has a relatively large band energy (3.2 eV) suitable for photocatalysts, non-toxic, and has a good absorbency to ultraviolet light [9, 10, 11]. Photocatalyst TiO_2 is less than optimal if used in the pure state because it has a relatively low surface area. Thus, TiO_2 needs to be embedded in an adsorbent. The commonly used adsorbent is Zeolite ZSM-5 because it has an active group of silica-alumina ($SiO_2.Al_2O_3$), a large surface area, has a channel that can filter the ion or molecule to absorb heavy metals especially Cr (VI) metal [2, 5]. ZSM-5/TiO_2 membrane work can be improved by impregnating TiO_2 into supporting media ZSM-5 [1, 2, 4, 7, 8].

2. **Research Methods**

2.1 Tools and Materials

Materials of this study were Gauze of AISI 316-180, 304-200, 304-400, TiO_2, distilled water, $K_2Cr_2O_7$, $CuSO_45H_2O$, TPABr 0.1 M, $NaAlO_2$, ludox HS-40%, diphenylcarbazide, Na diethyl dithiocarbamate, HCl, NaOH, H_2SO_4, NH_4OH, and absolute ethanol. The tools used are analytic Balances, and Technical, UV-Vis spectrophotometer, flask, bowl, filter paper, pH meter, magnetic stirrer, sieve, funnel, stirring rod, rotator, stative, burette, mortar, furnace and Reactor with UV lamp.

Object of this study was to use a solution 50 mg/L of Cr (VI) and Cu (II), then a decrease in Cr(VI) and Cu(II) levels in water using ZSM-5/TiO_2 powder and membrane with supports were type AISI 316-180, 304-200 and 304-400 and UV irradiation times are 30, 60 and 90 minutes.

2.2 Procedure

2.2.1 Gauze Treatment

Nine gauzes with 3 cm x 3 cm size on AISI 316-180 type is soaked in 15% w/v NaOH for 20 minutes (to remove oil/organic substances), then immersed in 15% v/v HCl solution for 20 minutes (to remove organic substances), then washed with aqua dest in ultrasonic for 20 minutes. Furthermore, it was oxidized electrochemically in H_2SO_4 20% v/v with a constant voltage of 3-5 V and a current of 0,01Ampere, then dried at 110ºC for 1 hour. The procedure was repeated for 304-400 and 304-200 mesh gauze support [16, 17].

2.2.2 Synthesis of ZSM-5/TiO_2 Powder and Membrane

2.2.2.1 Preparation of ZSM-5 Zeolite at low temperature (90 °C)

Preparation of ZSM-5 zeolite by mixing 0.136 g $NaAlO_2$ and 1.390 g NaOH 50% w/v in polypropylene bottle 1. TPABr 1.549 g dissolved with 7.3802 g water into propylene 2 bottle and stirred with a magnetic stirrer for 5 minutes, then put into a propylene 1 bottle and added 24.940 g of Ludox HS-40% and stirring for 6 hours. Furthermore, the propylene 1 bottle is inserted in an oven with a temperature of 90 °C for 4 days. The white precipitate was washed and inserted into the oven at 60 °C for 24 h to obtain maximum drying and was heated in a furnace at 550 °C for 6 hours is crushed and sieved with a mesh size of 100 [11, 15].

2.2.2.2 Synthesis of ZSM-5/TiO_2 Powder

ZSM-5/TiO_2 precursors synthesis with 20 g of ZSM-5 powder is mixed with 1 g of TiO_2 and added with 20 mL of absolute ethanol and the mixture was stirred by using a magnetic stirrer for 5 hours, ZSM-5/TiO_2 precursors was dried in oven at 120 °C for 5 hours [1, 4, 5, 7, 8].

2.2.2.3 Synthesis of ZSM-5/TiO_2 Membrane

After treatment, AISI 316-180, 304-200, and 304-400 mesh gauze are embedded in ZSM-5/TiO_2 precursors in polypropylene plastic containers with a surface area to reactor volume ratio of 1,44 [16] and then heated in an oven at 120ºC for 5 hours [4]. ZSM-5/TiO_2 membrane was washed with distilled water and heated at 60 °C for 3 hours [13, 15, 18, 19].

2.2.3 Decrease of Cr (VI) and Cu(II) Ions Concentration with ZSM-5/TiO_2 Powder and Membrane

Cr (VI) and Cu(II) sample solution 50 mg/L were channeled through the ZSM-5/TiO_2 powder with UV irradiation for 30 minutes and separated for filtrate. The resulting filtrate was calculated at Cr (VI) and Cu(II) final concentration. The treatment was repeated for time variations of 60 and 90 minutes [1, 2, 5, 6]. This procedure is repeated too or membrane with AISI 316-180 gauze, 304-400 and 304-200 gauze support [14, 15, 18, 19].

2.2.4 Determination of Cr (VI) and Cu(II) Ion concentration with Spectrophotometer

The filtrate solution resulting from a decrease Cr (VI) and Cu(II) concentration which has been flowed into the ZSM-5/TiO_2 powder and membrane with irradiation time 30, 60, and 90 minutes was taken 5.0 ml and put into 50.0 ml volumetric flask and then added ± 35 ml of distilled water, Diphenylcarbazide solution 2.5 ml is added Cr(VI) solution sample [2, 5, 8] and 5.0 mL NH_4OH 5% and 5.0 mL Na- diethyl dithiocarbamate (Cu (II) solution sample) which is adjusted with distilled water to mark the markers and homogenized, then the absorbance is read using a spectrophotometer at a wavelength of 540 nm (Cr(VI), 480 nm (Cu (II) sample and a stability time of 5 minutes [1, 7, 18, 19].

2.2.5 Data Analysis

In this study, data were obtained from experiments and analyzed statistically using the Anova One Way test to determine whether there was an effect of variations in gauze type and UV irradiation time with ZSM-5/TiO_2 membranes on decreasing Cr (VI) and Cu (II) ion concentration in water.

3. Result and Discussion

3.1 Decrease of Cu(II) and Cr(VI) Concentration (%) with ZSM-5/TiO_2 Powder

Decrease percentage of Cu(II) concentration (%) with 0.25 % w/v TiO_2, ZSM-5, ZSM-5/TiO_2 powder by UV irradiation time in Table 1 and decrease percentage of Cr(VI) concentration By ZSM-5/TiO_2 Powder in Table 2.

Table 1. Decrease Percentage of Cu(II) Concentration (%) with 0.25 % w/v TiO_2, ZSM-5, ZSM-5/TiO_2 Powder

UV Irradiation Time	Persentage Decrease Cu^{2+} Concentration (%)		
(minute)	TiO_2	ZSM-5	TiO_2 – ZSM-5
15	46.69	67.13	68.30
30	47.81	67.82	69.81
45	49.45	68.39	71.46
60	51.52	69.01	73.43
75	54.29	69.49	74.99

Table 2. Decrease Percentage of Cr(VI) Concentration (%) with 0,75 % w/v TiO_2, ZSM-5, ZSM-5/TiO_2 Powder

Irradiation Time	The initial Cr (VI) Ion Concentration	Decrease Cr (VI) Ion Concentration (%)		
(Minute)	(mg/L)	TiO_2	ZSM-5	ZSM-5/TiO_2
15	42.64	15.21	13.33	17.01
30	42.15	16.31	14.73	17.99
45	41.83	17.40	15.72	18.99
60	36.68	18.59	16.81	28.73
75	32.92	22.76	22.07	36.20

Table 1 and Table 2 show the highest percentage decrease Cu (II) ion concentration with TiO_2, ZSM-5, and ZSM-5/TiO_2 powder (0.25 %w/v) and Cr(VI) concentration (0.75 %w/v) at UV irradiation time 75 minutes in a row is 74.99 and 36.20 %, The longer contact UV irradiation time of the ZSM-5/TiO_2, so the more photon energy is absorbed by TiO_2 photocatalysts and the •OH more formed in photocatalysts and more contact between the •OH and Cu (II) and Cr(VI) ion substrate also increases, As a comparison, the addition of ZSM-5/TiO_2 0,25% w/v during the irradiation time of 75 minutes to decrease the concentration of Cu (II) and Cr (VI) ions was 74.99 and 1699%, respectively (Table 3).

Table 3. Decrease Percentage of Cr(VI) Concentration (%) with variation Concentration ZSM-5/TiO_2 Powder on Irradiation Time 75 minutes

Concentration Powder	Decrease Cr (VI) Ion Concentration (%)		
(%w/v)	TiO_2	ZSM-5	ZSM-5/TiO_2
0.25	6.16	10.60	16.99
0.50	9.68	13.01	18.84
0.75	12.09	14.59	24.83
1.00	14.39	16.88	30.82
1.25	17.28	21.44	36.14

Table 3 shows that the adsorption capability of ZSM-5 can be increased by impregnation with the supporting medium such as TiO_2 to decrease Cr (VI) in water more optimally than using TiO_2 and ZSM-5 only. These results are consistent with Agusty's research (2012) that the greater the concentration of ZSM-5/TiO_2 and the UV irradiation time, the greater the decrease in congo red color concentration [4]. The decrease Cu (II) ion concentration with the same ZSM-5/TiO_2 concentration (0.25% w/v) is greater than the reduction in Cr (VI). This is due to the radius of Cu (II) ions greater than Cr (VI) ions so that the ability of ZSM-5/TiO_2 to adsorption Cu (II) metal ions is higher than Cr (VI) ions concentration.

3.2 Decrease of Cu(II) and Cr(VI) Concentration (%) with ZSM-5/TiO_2 Membrane

3.2.1 Decrease of Cu(II) Concentration Ion (%) with ZSM-5/TiO_2 Membrane at time

Decrease percentage of Cu(II) concentration (%) with TiO_2, ZSM-5, ZSM-5/TiO_2 Membrane in Table 4 and Figure 1.

Table 4. Decrease Percentage Cu (II) Ion Concentration with Initial Concentration 47.13±0.53 mg/L

Type of Gauze	Time (minute)	Gauze (%)	ZSM-5 / TiO_2 membrane (%)	Membrane weight(g)	Membran zeolite ZSM-5/TiO_2 UV (%)
AISI 316-180	30	8.86	13.13	0.1252	26.47
	60	9.46	13.54	0.1143	33.20
	90	12,50	15.38	0.1294	37.26
304-200	30	9.80	15.98	0.0909	31.26
	60	11.50	16.80	0.1024	39.49
	90	13.13	18.23	0.0977	41.72
304-400	30	11.09	18.01	0.1349	40.51
	60	13.54	20.05	0.1284	48.56
	90	13.94	21.08	0.1330	57.72

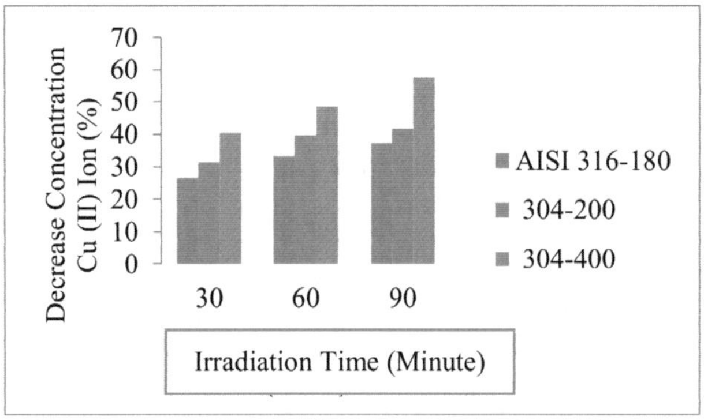

Figure 1 Decrease of Cu(II) Ion Concentration (%) with ZSM-5/TiO_2 Membrane from initial Cu(II) Ion Concentration 47.13±0.53 mg/L [19].

Table 4 and Figure 1 show that the highest percentage of reduction in Cu (II) ion concentration was ZSM-5/TiO_2 membrane with 304-400 gauze support with 90 minutes irradiation of 57.72%, This is due to the pore size in the mesh (1 inch = 2.54 cm x2.54 cm^2) there are 400 holes, then in 1 inch the distance between holes in the gauze 304-400 is getting tighter so that the ZSM-5/TiO_2 precursors attached stronger to the gauze. Thus, the decrease in Cu (II) ion concentration after geting through ZSM-5/TiO_2 membrane with 304-400 gauze is higher than AISI 316-180 and 304-200 gauze supports.

3.3.2 Decrease of Cr(VI) Concentration (%) with ZSM-5/TiO_2 Membrane

Table 5 and Figure 2 show that the highest percentage reduction in Cr(VI) ion concentration after getting through ZSM-5/TiO_2 membrane with 304-400 gauze supports for 90 minutes irradiation time of 54.65%. The highest decrease in Cr(VI) ion levels was obtained with a UV irradiation time of 90 minutes. The longer the irradiation time the more photon energy is absorbed by the TiO_2 photocatalyst so that the •OH greater formed in the photocatalyst. Contact between •OH and Cr (VI) ion substrate can increase the effectiveness of decreasing Cr (VI) ion levels.

Decrease of Cr(VI) Concentration (%) with ZSM-5/TiO_2 Membrane in Table 5 and Figure 2,

Table 5. Decrease of Cr(VI) Concentration (%) with ZSM-5/TiO_2 Membrane

Type of Gauze	Time (%)	Gauze (%)	ZSM-5 / TiO_2 membrane (%)	Membrane weight (g)	ZSM-5/TiO_2 Membrane with UV irradiation (%)
AISI 316-180	30	11.96	15.53	0.2459	16.51
	60	12.30	16.42	0.2390	22.92
	90	12.49	16.78	0.2182	29.65

304-200	30	12.49	34.97	0.1944	34.25
	60	13.38	35.52	0.1915	40.33
	90	13.56	36.22	0.1898	41.75
304-400	30	13.74	37.83	0.2694	49.53
	60	14.09	38.18	0.2688	51.39
	90	14.27	38.91	0.2680	54.65

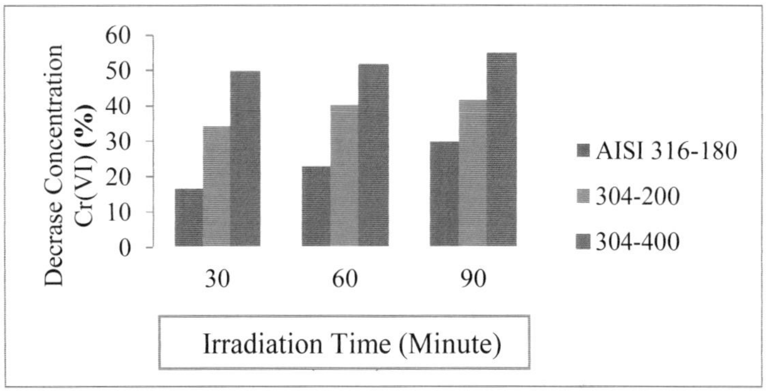

Figure 2. Decrease of Cr(VI) Concentration (%) with ZSM-5/TiO_2 Membrane from initial Cr(VI) Concentration 50.73±0.18 mg/L [18].

The results of this study are consistent with the results of Azizah *et al.* (2016) research study). The effectiveness of ZSM-5 membrane size and contact time to decrease carbon monoxide gas levels from cigarette smoke with a time of 5, 10, 20 and 30 minutes also increased. In this study using ZSM-5/TiO_2 membrane with variations, the gauze supports and UV irradiation time also increased [20]. Decrease Cu(II) ion concentration with ZSM-5/TiO_2 powder and membrane is higher for Cr(VI) ion concentration because the Cu (II) ion radius is greater than the Cr (VI) ion radius so that the ZSM-5 TiO_2 membrane's ability to adsorb Cu (II) ions is also increased. The ability of ZSM-5/TiO powder is greater than ZSM-5/TiO_2 membrane in reducing the concentration of Cu (II) and Cr (VI) ions because ZSM-5/ TiO_2 powder has a larger surface area so that when exposed to UV light then the amount of radicals OH is also more produced, so the ability to adsorb is also greater.

4. Conclusion

The highest percentage reduction in Cu (II) ion and Cr(VI) ion levels with 0.25% w/v ZSM-5/TiO_2 powder during UV irradiation for 75 minutes was 74.99% and 16.99%. The highest percentage of reduction in Cu (II) and Cr (VI) levels after passing through the ZSM-5/TiO_2 membrane with gauze buffer 304-400 during 90-minute UV irradiation is 57.72% and 54.65%. The ability of ZSM-5/TiO_2 powder is greater than ZSM-5/TiO_2 membrane in reducing the concentration of Cu (II) and Cr (VI).

References

[1] Ismania EZ, Mukaromah AH, Ethica SN 2018 Pemanfaatan Zeolit ZSM-5 Terimpregnasi TiO2 Untuk Menurunkan Kadar Ion Cu (II) dengan Variasi Waktu Penyinaran UV dalam Air Prosiding Nasional Seminar Unimus. 1(1). 204-211

[2] Faridah F, Mukaromah AH, Sulistyaningtyas AR 2018 Photocatalytic Degradation Cr (VI) By ZSM-5 Impregnated TiO_2 In Various UV-Irradiation Time Proceeding of the National Seminar Edusainstech Unimus (6 October 2018), 47-53

[3] Departement of Health RI 2010 Minister of Health Regulation RI No. 492/ MenKes/Per/IV/ Drinking Water quality requirements

[4] Agusty IP 2012 Use of TiO_2 Impregnated Zeolites to Degrade Congo Red Dyes Doctoral Dissertation Universitas Airlangga

[5] Alfiani Y, Mukaromah AH, Sulistyaningtyas AR 2018 Photodegradarion Of Cr (VI) In Various Concentration Of ZSM-5 Impregnated TiO_2 Proceeding Seminar Nasional Edusainstek FMIPA UNIMUS 2018, 84-92

[6] Purnomo, Setyarini PH and Mukaromah AH 2018 The potential of silk fibroin as a polymer composite reinforcement for bone implant materials *J MATEC Web of Conferences* 204, 05017

[7] Rochmatun S, Mukaromah AH, Wardoyo FA 2018 Penurunan Kadar ion Cu (II) Dalam Air Menggunakan Zeolit ZSM-5 Terimpregnasi TiO_2 Dengan Variasi Waktu perendaman. Karya Tulis Ilmiah Unimus

[8] Nurprihandayani S, Ethica SN, Mukaromah AH 2018 Utilization Of TiO_2 Impregnated Zeolit ZSM-5 To Decrease Concentration Of Cr (VI) In Solution At pH Variance Proceeding International Seminar on Education and Development of Asia, 1st INseIDEA (Saturday, July 14th 2018) 216-227

[9] Mukaromah A H, Wahyuni E T, Siswanta D 2016 Decreased Concentration of p-Chlorophenol with TiO_2 Photocatalyst and Ion Fe(III). Proceeding in The Field of Mathematics and Natural Sciences The 3rd University of Research Colloqium, 107-112.

[10] Poernomo JB. Mukaromah AH. Widiyandari H. Marwoto P 2016 Characterization of Nephelium Lappaceum Peel Extract as a Dye Sensitized Solar Cell *Journal of Physics: Conference Series* 739. 012070

[11] Joshi, K.M., Shirivastva, V. S. 2010. Removal of Hazardous Textile dyes From Aqueous Solution by Using Commercial Activated Carbon With TiO_2 and ZnO as Photocatalyst *International Journal of Chem Tech Research*, (2), 427-435.

[12] Ali I, Hassan A, Shabaan S, El-Nasser K 2017 Synthesis and characterization of composite catalysts Cr/ZSM-5 and their effects toward photocatalytic degradation of p-nitrophenol *Arabian Journal of Chemistry* **10** S2106-S2114

[13] Utami RA, Mukaromah AH, and Yusrin Y 2017 Sintesis Membran Zeolit ZSM-5 Secara Coating Pada Suhu 90°C Berdasarkan Variasi Perlakuan Kasa Stainless Steel AISI 316-180 Mesh Dalam Menurunkan Kadar Gas CO Proceeding International and National Seminar Unimus **1** (1) 52-58

[14] Mukaromah AH, and Ariyadi T 2017 A Novel Method for the Growth of ZSM-5 Membranes on stainless steel Supports by coating on Reactor Vessel with Surface-to-volume Ratio of Synthesis Governing Low Temperature Crystallization of ZSM-5. The 3 rd International Seminar On Education and Technology ISET Collaborate Graduate Schools Conference, 88-94

[15] Mukaromah AH, Kadja GTM, Mukti RR, Pratama IR, Zulfikar M A, & Buchari B 2016 The Surface-to-volume Ratio of the Synthesis Reactor Vessel Governing the Low Temperature Crystallization of ZSM-5. *Journal of Mathematical and Fundamental Sciences ITB*, 48(3), 241-251

[16] Gao Y, Chen M, Zhang T, and Zheng X 2011 A novel method for the growth of ZSM-5 zeolite membrane on the surface of stainless steel *J Materials letters. 65*(17-18). 2789-2792

[17] Mukaromah AH, Azizah IH, Ariyadi T, dan Mifbakhuddin 2018 Karakterisasi Membran Zeolit ZSM-5 Berdasarkan Variasi Jenis Dan Ukuran Kasa Dengan Pre-Treatment Direndam Dalam NaOH, HCl dan elektro-oksidasi Dengan H_2SO_4, Prosiding Seminar Nasional Unimus (1), 194-205

[18] Chasanah U, Mukaromah AH 2019 Penurunan Kadar Ion Cr (VI) Dalam Air Menggunakan Membran ZSM-5/TiO_2 yang Disentesis Dengan Variasi Kasa dan Waktu Penyinaran UV Prosiding Seminar Mahasiswa Unimus 2, 52-58

[19] Assyifa IR, Mukaromah AH 2019 Penurunan Kadar Ion Cu (II) Dalam Air Menggunakan Membran ZSM5/TIO2 yang Disintesis dengan Variasi Kasa dan Waktu Penyinaran UV Decrease Prosiding Seminar Mahasiswa Unimus 2, 88-94

[20] Azizah IH, Mindhumalid T, Mukaromah AH 2016 The Strategy of Adsorption of Carbon Monoxide (CO) on ZSM-5 Membrane In Smoking Area *J El-Hayah,* 6(2), 41-44

IOP Conf. Series: Materials Science and Engineering **846** (2020) 012005 doi:10.1088/1757-899X/846/1/012005

Chemical Separation and Antibacterial Activity of Nutmeg seed Essential Oil against Shigella sp. and Escherichia coli ATCC 25922

Hery Muhamad Ansory[1*], Ika Nur Fitriani[2], Anita Nilawati1[1]

[1]Universitas Setia Budi Surakarta, Indonesia
[2]Universitas Islam Negeri Walisongo Semarang

*hery.ansory89@gmail.com

Abstract. The essential oil of the nutmeg seeds has antibacterial activity and contains some major compounds: sabinene, α-pinene, 2-β-pinene, myristicin, safrole, and 1-4-terpineol. This essential oil was extracted from the nutmeg seeds by separation process using the distillation method. The purpose of this study is to separate compounds in essential oils of nutmeg seed and to evaluate the antibacterial activity against two bacterial species. Nutmeg essential oils compound separation was conducted by distillation at 95 mmHg with different temperature 408, 428, 438, and 503 K. The separation results analyzed physical properties, chemical compounds with GC-MS, and the antibacterial activity test of Shigella sp. and Escherichia coli ATCC 25922 using dilution and diffusion method. The samples antibacterial activity were evaluated by the determination of minimum inhibitory concentration. The GC-MS analysis of nutmeg essential oils shows the presence of 25 compounds. Four fractions of fractional distillation under reduced pressure have 57.12%; 17.00%; 1.40%; and 19.68% of yield with different physical properties as well as the number of different compounds in each fraction. Fraction 1 has 14 compounds with α-pinene (27, 44%), sabinene (23,78%), dan 2-β-pinene (18,29%) as major compound. Fraction 2 has 15 compounds with sabinene (16,18%), 2-β-pinene (14,27%), β-phellandrene (12,99%), dan γ-terpinene (14,48%) as major compound. Fraction 3 has 12 compounds with α-terpinene (11,24%), para-cymene (23,74%), limonene (13,23%), γ-terpinene (23,46%), dan α-terpinolene (15,01%) as major compound. Fraction 4 has 12 compounds with 1-4-terpineol (40,65%), safrol (20,34%) danmyristicin (16,06%) as major compound. Results showed that MIC and MBC on Shigella sp. nutmeg essential oils had 12.11mm inhibitory effects at 1.25%; fraction 2 has 5.88mm inhibitory effects at 2.5%, and fraction 4 has 9.67mm inhibitory effects at 1.25%. MIC and MBC in Escherichia coli ATCC 25922 nutmeg essential oils have 10.89mm inhibitory effects at 2.5%; fraction one has 10.22mm inhibitory effects at 10%, and fraction 4 has 10.11mm inhibitory effects at 2.5%.

Keywords—Nutmeg essential oil, Separation compounds of nutmeg essential oil, Antibacterial

1. Introduction

Nutmeg essential oil can be used as an antibacterial [1]. The main contents of nutmeg essential oil are myristicin (22.22%), 4-terpineol (14.45%), safrole (6.94%), sabinene (5.87%), α-pinene (5.45%), γ-limonene (3.88) [2]. Separation of compounds in nutmeg essential oil can be done by fractionation at low pressure [3]. Another method that can be applied is the process of steam distillation [4] [5].

Diarrhea is one of the health problems in the community, and the cause is often found in the field of clinical diarrhea caused by infection and poisoning [6]. Shigellasp and Escherichia coli ATCC 25 922 is a bacterial infection-causing bacteria. Society general treating diseases frequent use of antibiotics synthesis. Overuse of antibiotics and less directed can cause resistance to microorganisms.

One alternative to reduce the consumption of synthetic antibiotics is by consuming natural antibiotic derived from plants. One of the plants that potential as natural antibiotics are nutmeg because it has some terpenoids and phenolic compounds that can be applied as antibacterial [7][8]. Separation of compounds Separation in the essential oil of nutmeg using fractional distillation method under reduced pressure. Analysis of nutmeg essential oil content and results of separation were performed using GC-MS and test the antibacterial activity of essential oils of nutmeg fraction of the bacteria *Shigella* sp. and *Escherichia coli* ATCC 25 922 the diffusion and dilution methods.

2. Method

2. 1. Chemical separation

Nutmeg essential oil separation performed at 0.125atm pressure with temperature differences 408, 428, 438, and 503K.

2. 2. Identification of essential oil fractions and fractions nutmeg essential oil nutmeg

Essential oils of nutmeg and compound fractions were analyzed by GC-MS with the condition of the columns used is the type of RTX-5ms with programmed column temperature of 70-250°C with an increase of 2 °C / min and using an FID detector [9] [10], all sample analyze with the same method.

2. 3. Antibacterial activity test

Essential oils of nutmeg and nutmeg essential oil fractions tested the antibacterial activity using diffusion and dilution method at a concentration of 20; 10; 5; 2.5; and 1,25% with aqua dest solvent and addition of tween 80 Starting antibacterial testing as 100μL.

3. Results and Discussion

3.1. Chemical separation

Nutmeg essential oil testing physical properties, the results are as **Table 1**. Comparison test the physical properties of essential oils of nutmeg with nutmeg oil quality requirements SNI. It tells us that the sample is good enough to use for this research.The main focus in this study remains to dig up information on how to separate chemical compounds in nutmeg essential oils in the best way, so of course, the number of compound components found in nutmeg essential oil becomes the determinant, compared to the physical properties which are focused on SNI.

Table1. Nutmeg essential oils of physical properties comparation with SNI.

type of test	Nutmeg oil quality requirements	Essential oils of nutmeg
Color	Bening-yellow	Bening yellowish
Smell	Typical nutmeg oil	Typical nutmeg oil
Solubility in ethanol 96%	1: 1-1: 3	1: 1
Specific gravity	0.880 to 0.910	0.903
The refractive index	1.475 to 1.485	1,476
The content myristicin	At least 10%	16.34%

The comparison showed that the quality of essential oil of nutmeg to meet the standard requirements of ISO nutmeg oil.

Nutmeg essential oil fractionation is done using fractional distillation under reduced pressure. The principle of this separation is the difference in the vapor pressure of each component that causes the boiling point. Pressure reduction used to be the compounds that want to be separated to evaporate at a lower boiling point and to avoid the use of heating is too high.

At pressures 0.125atm, four fractions obtained as on **Table 2**. It showed that fraction 1 has a big yield, as the author predicted, it could cause a major compound composition with a lower boiling point like pinene and sabinene.

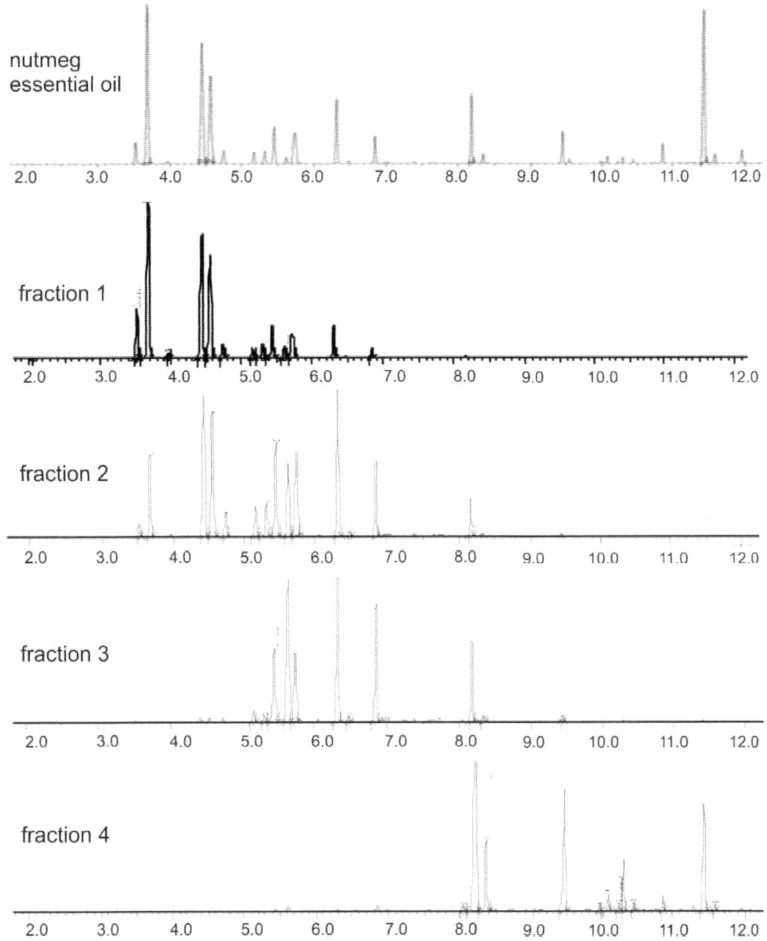

Figure 1. Results GC essential oils of nutmeg, fraction 1, fraction 2, fraction 3, and fraction 4.

The separation process which is carried out starting from low to high temperatures also allows this compound to still be present in the second fraction, and this can also be caused by unstable heating temperatures that use cooking oil as a conductor of heat.The author suggests that refraction get better results than this.

Table 2. The results of reduced pressure distillation fractionation

No.	Fraction	Heating temperature (K)	The yield (%)
1.	fraction 1	408	57.12
2.	fraction 2	428	17:00
3.	fraction 3	438	1:40
4.	fraction 4	503	1968

3.2. Identification of essential oil fractions and fractions nutmeg essential oil nutmeg.

The results of the analysis of physical properties testing compared with the test results of Nutmeg essential oils is as follows:

Table 3. Comparative testing physical properties of the fraction 1 to fraction 4 with essential oils of nutmeg

	Essential oils of nutmeg	fraction 1	fraction 2	fraction 3	fraction 4
Color	Bening yellowish	Clear	Clear	Clear	Bening yellowish
Smell	Typical nutmeg oil	Typical nutmeg oil	Typical nutmeg oil	Typical nutmeg oil	Typical nutmeg oil
Solubility in ethanol	1: 1	1: 1	1: 1	1: 1	1: 1
Specific gravity	.903	0.841	.842	-	1,220
The refractive index	1,476	1, 471	1,470	1.475	1,492

Table 4. The result of the separation of compounds

No.	Compound name	Retention Time (Minutes)	Composition (%)				
			Nutmeg Seed Essential Oil	fraction 1	fraction 2	fraction 3	fraction 4
1.	2-pentene	1.52		0:23	0:22	0:23	
2.	α-thujene	3:52		6:45	0.91		
3.	α-pinene	3.69	17.60	27.44	7:15		
4.	Camphene	3.96		0:47			
5.	Sabinene	4:44	16:39	23.78	16:18		
6.	2- β-pinene	4:56	11.89	18.69	14:27		
7.	β-myrcene	4.74		2:08	2:14		
8.	1-phellandrene	5:16		1:49	2.87	1:46	
9.	Delta-3-Caren	5:31		2:13	3:42	0.77	
10.	α-terpinene	5:44		4.71	9:31	11:24	
11.	*para*-cymene	5.61		1.71	7:03	23.74	
12.	Limonene	5.72		5:24		13:23	
13.	β-phellandrene	5.74			12.99		
14.	γ-terpinene	6:32			14:48	23:46	
15.	Trans-sabinene	6:47			0:45	0.76	
16.	α-terpinolene	6.84			5.94	15:01	
17.	Bisiklo [2.2.1] heptane-2-ol	8:05					0.96
18.	1-4-terpineol	8:17	6:16		2.63	8.91	40.65
19.	3-cyclohexane-1-methanol	8:34				0.63	9.95
20.	Safrole	9:44				0:57	20.34
21.	Citronellyl Acetate	9.96					0.62
22.	Eugenol	10:01					1.92
23.	Geraniol acetate	10:23					0.95
24.	α-copaene	10:29					5:49
25.	Methyl eugenol	10:44					0.81
26.	Isoeugenol	10.86					1.60
27.	Myristicin	11:43	16:34				16:06
28.	Cis-asaron	11:58					0.65

INFORMATION: FIGURES RETENTION TIME TO TWO BEHIND THE DIFFERENCES IN EACH SAMPLE.

The results of the analysis of the nutmeg essential oil content and fractions, as in **figure 1**, the same retention time shows the same compound, because the analysis is done by the same method for all samples.

3.3. Antibacterial Activity Testing

Antibacterial activity test conducted by diffusion and dilution methods. The bacteria used are and *Shigella*sp with SSA media and *Escherichia coli* ATCC 25 922 with EA media.

Table 5. The test results diffusion and dilution*Shigella*sp and *Escherichia coli* ATCC 25 922

Compound	Trial	diffusion test		dilution test	
		*Shigella*sp	*Ecshericia coli* ATCC 25 992	*Shigella*sp	*Ecshericia coli* ATCC 25 992
	control +	16.44	15,22	+	+
	control -	0	0	-	-
Essential oils of nutmeg	20%	11.67	9.00	-	-
	10%	11,00	8.56	-	-
	5%	12.22	10.56	-	-
	2.5%	11,00	10.89	-	-
	1.25%	12.11	0	-	+
	control +	15.55	14,11	+	+
	control -	0	0	-	-
	20%	0	9.67	+	-
Fraction 1	10%	0	10,22	+	-
	5%	0	0	+	+
	2.5%	0	0	+	+
	1.25%	0	0	+	+
	control +	15,22	16.45	+	+
	control -	0	0	-	-
	20%	9.33	0	-	+
Fraction 2	10%	8.55	0	-	+
	5%	7.22	0	-	+
	2.5%	7.89	0	-	+
	1.25%	5.88	0	+	+
	control +	15,22	15.89	+	+
	control -	0	0	-	-
	20%	9.89	10.44	-	-
Fraction 4	10%	9.89	10.01	-	-
	5%	7.89	9.78	-	-
	2.5%	8.56	10,11	-	-
	1.25%	9.67	0	-	+

INFORMATION:
DIFFUSION: CONTROL + : DIKS DISCS GENTAMICIN
CONTROL - : DISC DISK + SOLVENT
DIFFUSION: POSITIVE CONTROL : BACTERIAL CULTURE
NEGATIVE CONTROL: SOLVENTS
+: NO BACTERIAL GROWTH
- : NO BACTERIAL GROWTH

The result of dilution and diffusion addressed that there is an excellent antibacterial activity in fraction 4 containing compound propanoicterpenoids and phenyl which is the phenol group, as compared with the fraction one and fraction two containing always compounds terpenoids. The results of the best antibacterial activities contained in the essential oils of nutmeg compared with the fraction four due to their compound and phenylpropanoidterpenoids which are groups of phenols and terpenoids more complex compounds. Terpenoids as an antibacterial mechanism are reacted with Porin (trans-membrane proteins) on the outer membrane of the bacterial cell wall, forming an active bond polymer destroying Porin. Damage Porin which is the entry and exit of the compounds will reduce the permeability of the bacterial cell wall which will result in a bacterial cell would be a lack of nutrients, thereby inhibited bacterial growth or death [11]. Phenolic compounds may function as an antimicrobial for their OH groups that are toxic to microbes. The mechanism of antibacterial phenolic compounds at low concentrations in the cytoplasmic membrane damage and can lead to leakage of the cell nucleus, whereas at high concentrations of phenolic compounds with cellular proteins coagulate. Activity is beneficial when the bacteria in the cleavage stage wherein the phospholipid layer around the cell is in a state that is so thin that it can easily damage the phenol content of the cell [12]. Essential oils of nutmeg and Fraction have better antibacterial activity shown in the bacteria *Shigella* sp. compared with *Escherichia coli* ATCC 25 922.

4. Conclusion

Based on the research conclusions can be drawn with the dominant compound in the fraction 1 contained three compounds, namely α-pinene (27.44%), sabinene (23.78%), and 2-β-pinene (18.69%); 2 there are 4 compound fraction that is sabinene(16.18%), γ-terpinene (14.48%), 2-β-pinene (14.27), and β-phellandrene (12.99%); 3 fractions which contained the compound 5 cymene (23.74%), γ-terpinene (23.46%), α-terpinolene (15.01%), limonene (13.23%), and α-terpinene (11, 24%); 4 fractions contained 3 compound is 1-4-terpineol (40.65%), safrole (20.34%), and myristicin (16.06%). Essential oils of nutmeg possess antibacterial activity in bacteria *Shigella*sp. and *Escherichia coli* ATCC 25 922IM ore fraction better than the target because the essential oil compounds are much more complicated so much better against bacterial.

Acknowledgments

Thank DPRM Kemenristek DIKTI that has funded this research.

Reference

[1] A. Tanendri, S. Sumiati, and M. Meliansyah, "Aktivitas Antibakteri Daging Buah dan Daun Pala (Myristica Fragrans) Terhadap Escherichia Coli," *J. Vokasi Kesehat.*, vol. 4, no. 2, pp. 81–84, 2018.

[2] D. P. Wibowo, Y. Febriana, H. Riasari, and D. L. Auilifa, "Essential Oil Composition, Antioxidant and Antibacterial Activities of Nutmeg (Myristica fragrans Houtt) from Garut West Java," *Indones. J. Pharm. Sci. Technol.*, vol. 5, no. 3, p. 82, 2018.

[3] H. M. Ansory, H. Sastrohamidjoyo, and B. Purwono, "Perbandingan Kualitas Minyak Pala Hasil Isolasi Dari Bagian-Bagian Buah Pala Berdasarkan Kadar Miristisin," *J. Farm. Indones.*, vol. 12, no. 2, pp. 127–136, 2015.

[4] R. Astuti, "Pengaruh Waktu Distilasi Minyak Biji Pala (Myristica fragrans) dengan Metode Distilasi Uap dan Identifikasi Komponen Kimiawi," *Indones. J. Lab.*, vol. 1, no. 2, pp. 36–40, 2019.

[5] S. Ketaren, S. Ngudiwaluyo, and A. Friyadi, "Isolasi Miristisin dari Minyak Pala (Myristica fragrans) dengan Metode Penyulingan Uap," *J. Teknol. Ind. Pertan.*, vol. 17, no. 1, pp. 23–28, 2007.

[6] Menteri Kesehatan, *Keputusan Menteri Kesehatan Republik Indonesia No.1216/MenKes/SK/XI/2001 tentang Pedoman Pemberantasan Penyakit Diare*. 2001.

[7] T. I. Purwantiningsih, Y. Y. Suranindyah, and W. (Widodo), "Aktivitas Senyawa Fenol dalam Buah Mengkudu (Morinda citrifolia) sebagai Antibakteri Alami Untuk Penghambatan Bakteri Penyebab Mastitis," *Bul. Peternak.*, vol. 38, no. 1, p. 59, 2014.

[8] U. Rastuti, S. Widyaningsih, D. Kartika, and D. R. Ningsih, "Aktivitas Antibakteri Minyak Atsiri Daun Pala dari Banyumas terhadap Staphylococcus aureus dan Escherichia colli serta Identifikasi Senyawa Penyusunnya," *J. Ilm. Kim.*, vol. 8, no. 2, pp. 197–203, 2013.

[9] N. 'Aini Hidayah, H. M. Ansory, P. K. K. Putri, and A. Nilawati, "Analisis Senyawa Minyak Atsiri Fuli Pala Secara GC-MS dan Uji Aktivitas Antibakteri Terhadap Ecschericia coli dan Staphylococcus aureus," *Maj. Farm.*, vol. 13, no. 2, p. 56, 2018.

[10] Sinambela, SrivitaEfi. 2012. Isolasidananalisikimiaminyakatsiridaritemulwak (Curcuma

xanthorizaRoxb) dengan Gas Kromatografi-Spektrometer Massa (GC-MS) danujiaktifitasantibakteri. Universitas Sumatra Utara.
[11] Cowan, M.M., 1999, Plant Products as Antimicrobial Agents, Clinical Microbiology Reviews s12, No.4: 564-82.
[12] Volk dan Wheeler 1984 MikrobiologiDasarEdisiKelimaJilid I (Erlangga. Jakarta.)

Improvement Development of Bill Acceptor for Laundry Application with Internet-of-Thing as Notification.

Edzwan Ismail[1], Shamsul Fakhar Abd Gani[1], Amar Faiz Zainal Abidin[1], Tengku Nadzion Tengku Ibrahim[2*], Saifuddin Semail[2*],

[1]Faculty of Electrical & Electronic Engineering Technology, Universiti Teknikal Malaysia Melaka, Malaysia
[2]Department of Electrical Engineering, Politeknik Ibrahim Sultan, Malaysia

*tengku.pjb@gmail.com, saifuddinsemail@gmail.com

Abstract. Nowadays, self-laundry kiosk service has been widely available in Malaysia. Although the technology has been progressing rapidly especially the technology of Internet-of-Thing, most of the system used by the self-laundry machines are outdated. One difficulty faced by the owners of the self-laundry kiosk, they need to check frequently by themselves at their cashbox machines in order to check whether the cashbox filled or emptied. This leads to inefficiency and can be addressed if self-laundry machine is integrated with Internet-of-Thing technology. Thus, this project attempts to improvise the current system by developing the bill acceptor that gives the owner to monitoring the flow of current amount inserted by customer in real time by displaying it in apps at owner smartphone. The project uses Arduino Mega as a controller, bill acceptor machine as token receiver, Wi-fi module to send the data to apps, LCD and Blynk apps to display the current amount and Bluetooth module to communicate between the Arduino and the washing machine. A checklist analysis and scenario analysis are implemented to verify the functionality of the proposed prototype. The result shows that the proposed prototype achieve all the objectives required.

1. Introduction

The self-laundry kiosk service has been growing rapidly in Malaysia. According to [1] [1], this rapid growth is due to Malaysians are keen to accept and adopt the self-service technology. Most of these self-laundry machines implemented are still similar to their predecessor in early 2000s where each self-laundry machine is independent and are not connected to internet in anyway. Most of the system used by the self-laundry machines does not connected with internet. This leads to difficulty for the owners for them to access real-time information of their machines. Especially when it is dealing with the inefficiency of travelling to their self-laundry kiosks frequently in order to check manually their cashbox machines whether the cashbox filled or emptied. This inefficiency created many losses: 1) monetary loss due to downtime because the cashbox is filled therefore it cannot receive cash note anymore or emptied token because the machine can no longer convert cash note to token for the machine; 2) loss of time and increase in cost due to unnecessary travelling to check the self-laundry kiosks.

As the technology has been progressing rapidly with the term Industry 4.0 being introduced, there are increasing trend in integrating internet with all machines and devices. This trend is called Internet-of-Thing (IoT) and has shown many benefits especially in the area of notifying, reporting, and predicting based on the data its collected. Few examples of the implementations of IoT are: 1) in the water quality monitoring system [2] where the data of temperature, PH, turbidity and the flow values can be access by the user in real-time via internet, home automation and reporting [3], 3) real-time health care monitoring system [4]; 4) monitoring the irrigation and tank in India [5]; 5) monitoring weather parameters [6]; monitoring air pollution [7];

Noticing the benefit of integrating real-time of notification and reporting to the current self-laundry service machine in addressing the problems mentioned in earlier chapter, this project attempts to do exactly that by employing Arduino Mega as a controller, bill acceptor machine as token receiver, Wi-fi module to send the data to apps, LCD and Blynk apps to display the current amount and Bluetooth module to communicate between the Arduino and the washing machine. The use of Arduino as the main microcontroller due to its cost and programmer-friendly where the success of its implementation can be seen in several areas: educational kit [8-10], electronic game board [11-12], can crusher [13], etc.

The remaining of this project will explain the methodology of the project, the result and discussion of the project, and conclusion of this project.

2. Methodology

This project used a 36cm X 29cm X 36cm cube black box as the main container for the prototype. Figure 1 shows the actual physical construction in the project. Noticeable at the front panel of the proposed prototype, the bill acceptor machine is on the left side which accept cash bill note from the user. A push button is located at the middle as an emergency button where the user can press if it need to alarm the owner if there is problem faced when using the proposed prototype. The top right is the liquid crystal display (LCD) which is used to display the amount inserted by the user and other menu for the user to respond by using the keypad on the bottom right of the front panel. The relationship between the components is illustrated in Figure 2.

Figure 1: Actual physical construction of the proposed prototype

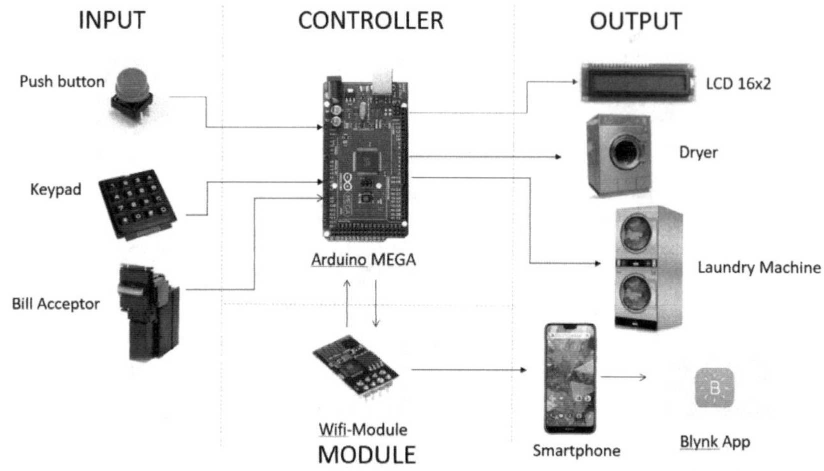

Figure 2: Block diagram of the proposed prototype

Figure 3 shows about the flowchart of the proposed prototype. Bill acceptor will accept the notes either in RM1, RM5 or RM10. After that, sensor that located in the bill acceptor will detect the notes inserted before. When the sensors detected the present of notes, the LCD will display the amount of money. Then, the total amount of money will also be displayed in Blynk apps. If the money is equal or more than RM5, then LCD will appear new menu which is if customers choose number 1 at keypad, motor that has be set as washing machine will be activated. If customer choose number 2 at keypad, motor that has be set as dryer machine will be activated.

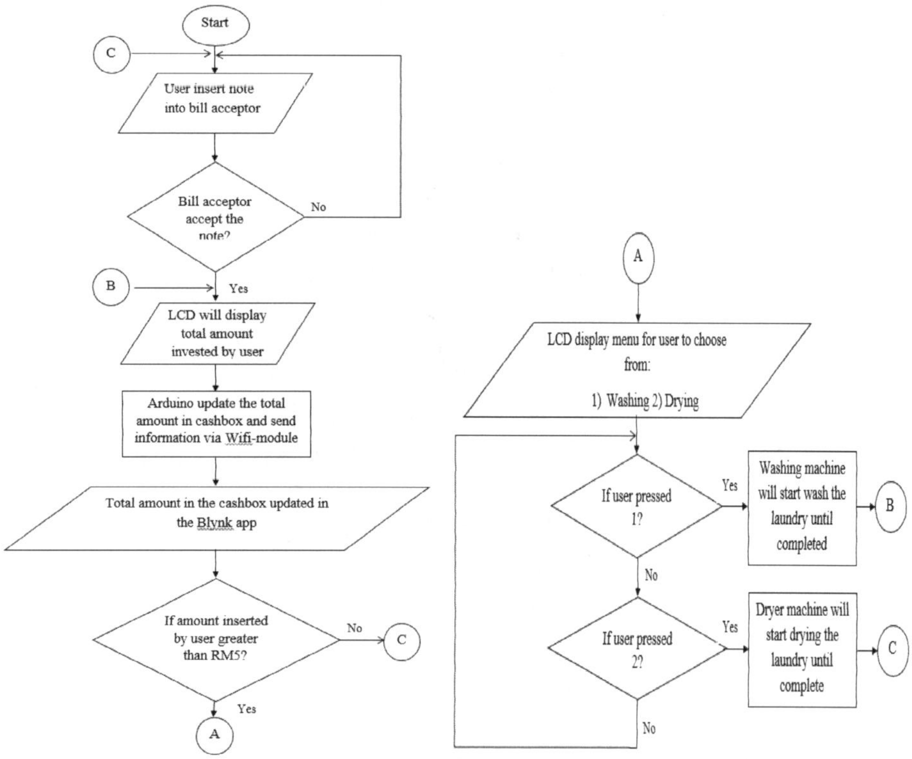

Figure 3: Flowchart of the proposed prototype

Figure 4 shows the schematic diagram of this project. Two motors are replacing the role of dryer and washing machine. And serial module is used to replace the Wi-fi module. This project use Arduino Mega as the microcontroller that will act as a brain for this system. Bill acceptor will be used as an input which is money will be inserted through it. After that Arduino mega will do the process to display the amount inserted at LCD and via Blynk apps. Keypad 4x4 was used to select the machine that want to be use by customer which "1" for washing machine and "2" for dryer machine.

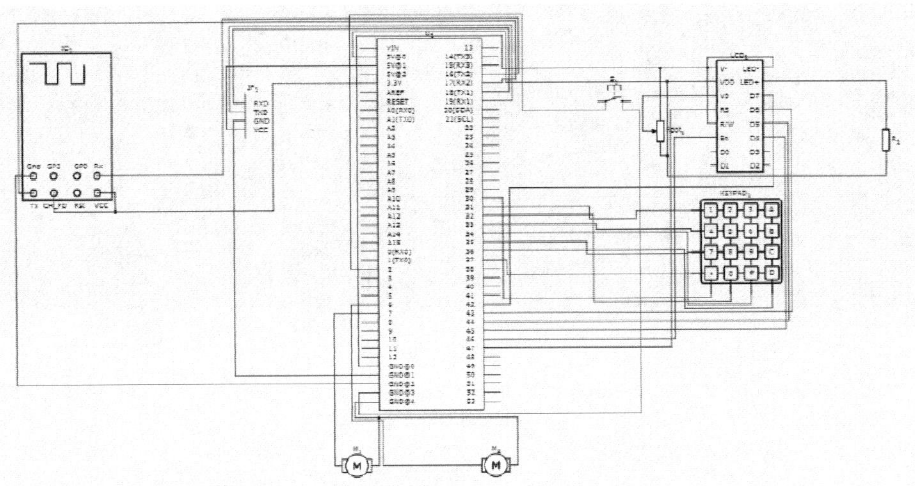

Figure 4: Schematic diagram of the proposed prototype

3. Result and Discussion

A checklist analysis has been done in order to test the functionality of this bill acceptor system project. By referring to Table 1, all the expected functionality of the proposed prototype has been able to be executed on the actual proposed prototype. This shows that the proposed prototype able to function according to the expectation when employed.

Table 1. Checklist Analysis

Number	Statement	Expected	Actual
1	Accept bill RM 1	√	√
2	Accept bill RM 2	X	X
3	Accept bill RM 5	√	√
4	Accept bill RM 10	√	√
5	Accept bill RM 20	X	X
6	Accept bill RM 50	X	X
7	Accept bill RM 100	X	X
8	Display amount of money inserted into the bill acceptor at LCD	√	√
9	Display and monitoring real time the incoming cash flow into the machine via website/apps.	√	√
10	Controlling washing/dryer machine from the coin changer machine.	√	√

For project analysis, this project has been rigorously tested for one month in order to test the performance reliability of the project systems. Figure 5(a) below show before bill acceptor activate and after bill acceptor already activates and figure 5(b) ready to accept the bill notes. To activate the bill acceptor, user needs to press the push button at the box. This is to avoid buffer from bill acceptor if the power is on. After that LCD will display the current total which is RM0.00 due to there still no money insert in it. Figure 5(c) below show the current money already inside the box.

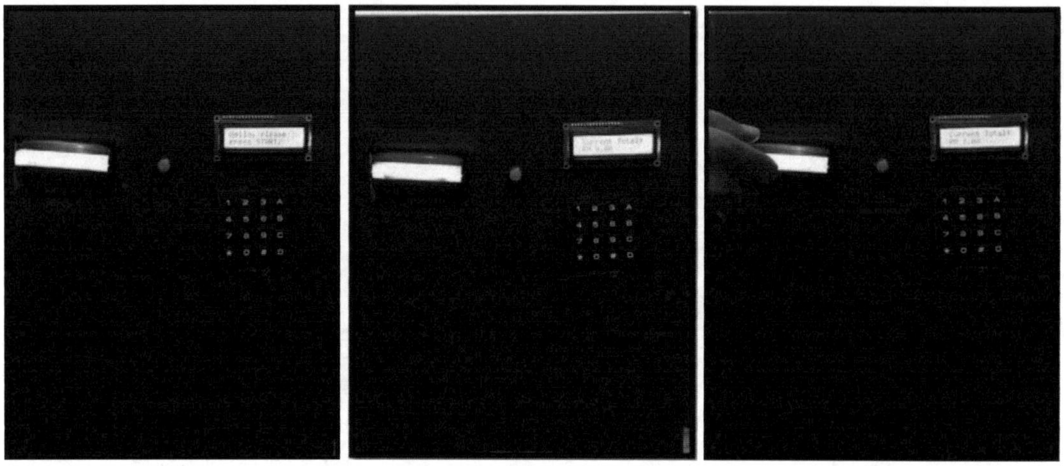

(a) (b) (c)

Figure 5. (a) Before Bill Acceptor Activate, (b) After Bill Acceptor Activate, and (c) Current Money Inside the Box

After that user need to insert money until the RM5.00 to proceed to other menu which is menu to choose between using washing machine or dryer machine. Since this bill acceptor system can be monitored and display the amount of money inside of it in real time, the display on LCD and Blynk application need to be tally. Figure 6(a) shows the current amount of money inside the box is equivalent with the Blynk apps in Figure 6(b). Figure 6(c) shows the LCD when user insert the money to bill acceptor after reach to RM5.00. Then, new menu will display in LCD which is 1 for washing machine and 2 for dryer machine.

(a) (b) (c)

Figure 6. (a) LCD Display the Current Amount, (b) Current Amount Display in Blynk Apps, and (c) New Menu to Select Laundry Application

After done the selected process, LCD will display back the main menu which is displaying the default amount RM0.00 regarding the first user already done using the machine. But in the Blynk apps, the amount of money inserted before will be remain. Figure 7(a) shows Blynk apps record the amount inserted before even the first process already done. This program will keep looping until the power is off. When the power is off, all the data send to Blynk apps will also reset. Figure 7(b) below shows when the power is off, the current amount display in Blynk Apps is reset to RM0.00.

(a) (b)

Figure 7. (a) Money Inserted in First Process Remain in Blynk Apps, (b) Value in Blynk Apps Already Reset When Power is Off

Since ThingSpeak is a cloud based system for this project, so it will record all of the flow money inserted even the Blynk apps already reset when the power is off. Figure 8 below shows all of the flow that has been recorded in ThingSpeak since this system implemented.

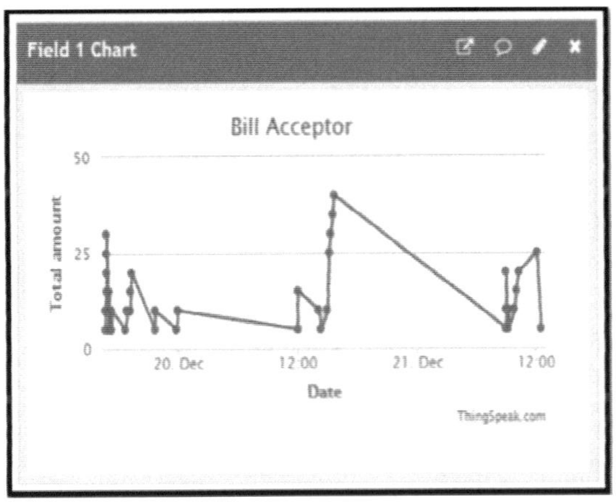

Figure 8. Data Recorded in ThingSpeak Since System Has Been Implemented

4. Conclusion

This project is focused on the development of a Bill Acceptor for Laundry Application with Internet-of-Thing as Notification. The block diagram, schematic diagram, flowchart of the proposed system has been explained concisely. The result shows the proposed prototype fulfil the expected result.

REFERENCES

[1] M. C, "Perceived Factors Influencing the Acceptance and Adoption of Self-Service Technology," UniTAR Malaysia, Kuala Lumpur, 2015.

[2] G. M. A and D. V. V, "Water Quality Monitoring System Based on IOT," *Advance in Wireless and Mobile Communication,* vol. 10, no. 5, pp. 1107-1116, 2017.

[3] K. S. S and R. P, "Real Time Wireless Patient Monitoring System Based On IOT," *Imperial Journal of Interdisciplinary Research (IJIR),* vol. 2, no. 12, 2016.

[4] S. S and G. K, "Real-TimeHealth Care Monitoring System using IoT," *International Journal of Engineering & Technology,* vol. 7, no. 2, pp. 484-488, 2018.

[5] K. P. H and K. P. D, "Internet Of Things Based System For Remote Monitoring Of Weather Parameters And Applications," *International Journal of Advances in Electronics and Computer Science,* vol. 3, no. 2, 2016.

[6] S. I, P. B and R. S, "Wi-Fi Based Portable Weather Station for Monitoring Temperature, Relative Humidity, Pressure, Precipitation, Wind Speed and Direction," in *Conference in Information and Communication Technology for Sustainable Development*, Goa, 2018.

[7] P. S. M, V. M and T. R, "IoT Based Solar Energy Monitoring System," in *International Conference on Energy, Communication, Data Analytics and Soft Computing (ICECDS)*, Tamilnadu, 2017.

[8] A. P. U. Siahaan and R. Rahim, "Dynamic Key Matrix of Hill Cipher Using Genetic Algorithm," *International Journal of Security and its Applications,* vol. 10, no. 8, pp. 173-180, 2016.

[9] R. Rahim, "128 Bit Hash of Variable Length in Short Message Service Security," *International Journal of Security and Its Applications,* vol. 11, no. 1, pp. 45-58, 2017.

[10] R. Rahim and A. Ikhwan, "Study of Three Pass Protocol on Data Security," *International Journal of Science and Research (IJSR),* vol. 5, no. 11, pp. 102-104, 2016.

[11] D. Y. Pawade, A. Lahigude and D. Reja, "Review Report On Security Breaches Using Keylogger And Clickjacking," *International Journal of Advance Foundation and Research in Computer (IJAFRC),* vol. 2, no. Special Issue (NCRTIT 2015), pp. 55-59, 2015.

[12] M. Dadkhah, M. D. Jazi, C. Ana-Maria and E. Barati, "An Introduction to Undetectable Keyloggers with Experimental Testing," *International Journal of Computer Communications and Networks,* vol. 4, no. 3, pp. 1-5, 2014.

[13] R. Venkatesh and R. K. Sekhar, "User Activity Monitoring Using Keylogger," *Asia Journal of Information Technology,* vol. 15, no. 23, pp. 4758-4762, 2015.

The Model of Artificial Neural Network and Nonparametric MARS Regression for Indonesian Composite Index

Dodi Devianto[1*], Putri Permathasari[2], Mutia Yollanda[3], Afridian Wirahadi Ahmad[4]

[1,2,3]Department of Mathematics, Andalas University, Padang 25163, Indonesia
[4]Padang State Polytechnic, Padang 25163, Indonesia

*Corresponden Email: ddevianto@sci.unand.ac.id

Abstract. The Indonesian composite stock price index is an indicator of changes in stock prices as a guide for investors to invest in reducing risk. The regression model for Indonesian Composite Index (ICI) has the response variable as stock prices with fluctuation behavior and several financial predictor variables, the model tends to violate the assumptions of normality, homoscedasticity, autocorrelation and multicollinearity. This problem can be overcome by modeling the composite stock price index by using the Artificial Neural Network (ANN) and nonparametric regression of Multivariate Adaptive Regression Spline (MARS). In this study, the time series data from the composite stock price index starting from April 2003 to March 2018 with its predictor variables are crude-oil prices, interest rates, inflation, exchange rates, gold prices, Dow Jones price, and Nikkei 225 Index. The both methods give better goodness of fit, where the coefficient of determination ANN is 0.98925 and the MARS determination coefficient is 0.99427. While based on the Mean Absolute Percentage Error (MAPE) of ANN was obtained 6.16383 and the MAPE value of MARS is 4.51372. This means that the ANN method and nonparametric MARS regression method have good performance to forecast the value of the Indonesian composite stock price index in the future, but in this case of data the nonparametric MARS regression method shows the accuracy of the model is slightly better than ANN.

1. Introduction

The measures of the relationship between the predictors and response variables are necessary to obtain the regression model. The constant values as the coefficient in the regression model, which has not known are called parameters, and these values also have the role in prediction a model [1]. In each regression model, there is a residual obtained from the difference between the observed value and the model. A numerical approach can overcome the regression models that violate the residual assumptions. The numerical method is developed to modeling the data without regard to obey the classical statistical assumptions, where the Artificial Neural Network (ANN) and nonparametric regression of Multivariate Adaptive Regression Spline (MARS) are the alternatives model used to approach the model of Indonesian Composite Index (ICI).

The Indonesian composite index is an investment instrument that can be modeled in a regression equation with a numerical approach. An investment that is needed in running a business and some risks have to be accepted by investors as explained in [2]. To minimize risks in investing, the forecasting of stock prices is to be one of the alternatives to solve the risks. There are many things that influence stock prices; they are Dow Jones price, Nikkei 225 index, inflation, interest rate, exchange rate, gold price, earning per-share, crude-oil price and many other observable variables. The historical data of stock prices and their fluctuation has discovered as described in [3][4]. Analyzing historical data of stock prices could be predicting the price fluctuation on the stock market.

There are many machine learning techniques that have been expanded to forecast stock prices as explained in [5] and also artificial neural networks [6]. While the nonparametric model of MARS was first popularized by [7] and it is one of the nonparametric regression methods which could be used to make a linear model. Regarding the forecast of stock prices index as in [8], they are comparing intelligent systems using a multilayer perceptron presents more reliable results than the other intelligent systems to forecast the stock exchange index. In this study, the Indonesian composite index would be forecasted by using ANN and nonparametric MARS regression as a response variable by using seven predictor variables. The methods will be compared based on their goodness of fit and Mean Absolute Percentage Error (MAPE) value to get the best model between them.

2. Methodology

A. Data Source

This research has used the historical data from April 2003 until March 2018, that is 180 data. The data were accessed formally via https://yahoofinance.com and http://www.bi.go.id. In this study, there is one response variable (Y) namely Indonesian Composite Index and there are seven predictor variables, namely crude-oil price (X_1), interest rate (X_2), inflation (X_3), exchange rate (X_4), gold price (X_5), Dow Jones price (X_6) and Nikkei 225 index (X_7).

B. Stage of Research for Artificial Neural Network

Artificial neural networks were made in 1943 by McCulloch and Pitts as a mathematical model of the neuron for a basic switching element of the brain as in [7]. The neural networks are composed of nodes or units connected by directed links for providing to propagate the activation element of x_i from the unit i to j. Each link has weight w_{ij} associated to the element of x_i, which determines the strength and sign of the connection. Each unit j first computes a weighted sum of its inputs:

$$in_j = \sum_{i=0}^{n} x_i w_{ij} \tag{1}$$

Then the activation function g is applied to this sum to obtain the output $z_j = g(in_j)$ and continued to other neurons as a new set of input data.

According to [8], there are feed-forward and recurrent networks to make the network after deciding on the mathematical model of a neuron. The output of their layers is obtained through activation functions which unique after this transformation. These activation functions have to scale the output of the neural network into proper ranges. There are some activation functions provided which are explained in [9], wherein this study, it is used a sigmoid activation function defined as follows:

$$f(x) = \frac{1}{1 + \exp(-x)} \tag{2}$$

The result of the sigmoid function is a value on the interval (0,1) where the sigmoid function only returns positive values. Based on [9], there are two steps to construct a model of the artificial neural network, they are training and validation. In the training neural network, the methods generally fall into the categories of supervised, unsupervised, and various hybrid approaches. The validating a neural network is a final step in ANN to determine if the additional training is required where the validation data is completely separate from the training data.

According to [10], during training neural network, the networks attempt to minimize the error. Error for k-th training data can be expressed in the following equation.

$$Err_k(W) = \frac{1}{2}\sum_{j=1}^{p}(t_{kj} - y_{kj})^2 \tag{3}$$

where t_{kj} is j-th output value which wants to be achieved or target value for k-th data and y_{kj} is j-th actual output for k-th data. The updating output layer for ANN can apply the chain rule to factorize the derivative $Err_k(W)$ with respect to w^h_{jp}, which is explained in [10] and [11]. The weights which have updated can be used in the next steps if the stopping criterion is not available on the process.

The backpropagation algorithm was the learning algorithm, and it uses gradient descent as the core learning mechanism. Based on [12], the learning network consists of the following steps. First, initialization of the network; the weights are generally initialized to be determined with random weights. Second, feed-forward; information process is passed forward through the network from input to hidden and output layer via node activation functions and weights. Third, assess the error; the output of the network is assessed relative to known output, if the error is below a pre-specified threshold the network is trained, and the algorithm terminated. Forth, propagation; the error at the output layers is used to re-modify the weights where the algorithm propagates the error backwards through the network and computes the gradient of the change in error with respect to changes in the weight values. Fifth, adjustment; adjustment is necessary to determine the weights by using the gradients of change to reduce the error.

The model of ANN for ICI is obtained in the following steps:
1. Define the input and output data that want to be obtained.
2. Normalization of data
3. Construct the architecture of the artificial neural network.
4. Training the networks of 134 data which consists of initialization, feed-forward, error assessment, propagation, and adjustment.
5. Validation the networks of 46 data to determine whether the process will be continued or be repeated the training process to get the error less than a threshold value.
6. Build the ICI model by using an artificial neural network.

C. Stage of Research for Multivariate Adaptive Regression Spline

The first step formation of the MARS model begins by determining the knots and basis functions of each predictor variable by plotting each predictor variable with the response variable. A good MARS model is determined by the location and the optimum number of knots. The MARS model is focused on overcoming high dimensional problems and discontinuities in the data. Besides, the MARS model is a development of the Recursive Partition Regression (RPR) approach that still has weaknesses, where the model that is later produced by this method is not continuous on knots. The MARS model can be written [13][14] as follows

$$\hat{f}(x) = a_0 + \sum_{m=1}^{M} a_m \prod_{k=1}^{K_m} (S_{km}(x_{v(k,m)} - t_{km}))_+ \tag{4}$$

where a_0 is regression constant and a_m is the coefficient of the m-th basis function. The coefficient a_m is determined by the least squares method, where M is the maximum basis function. It is noted Km as degree of interaction for m-th basis function, where $S_{km} = +1$ if the knot is located on the right subregion and $S_{km} = -1$ if the knot is left, the notation $x_{v(k,m)}$ is the predictor variable of v for determining the k-th variable and m-th subregion. The t_{km} is the knot value of the predictor variable $x_{v(k,m)}$. The MARS equation model in Equation (4) can be rewritten for simplicity understanding into the following equation

$$\hat{f}(x) = a_0 + a_1 BF_1 + a_2 BF_2 + \ldots + a_M BF_M \tag{5}$$

where BF_M is the m-th basis function.

The best model is determined based on the minimum Generalized Cross Validation (GCV) criteria, where the GCV function is defined as follows:

$$GCV(M) = \frac{\frac{1}{N}\sum_{i=1}^{N}(y_i - \hat{f}_M(x_i))^2}{(1 - C(M^*)/N)^2} \tag{6}$$

where M is the number of basis functions, x_i is predictor variable and y_i is response variable, N is the number of observations, and complex model function of $C(M^*)$ is defined as $C(M^*) = C(M) + dM$ for d as the smoothing parameter and $C(M)$ as the number of parameter on the model.

There are a number of measures of model accuracy, the most frequently used accuracy measures are Mean Square Error (MSE) and Mean Absolute Percentage Error (MAPE) that can be formulated respectively as follows:

$$MSE = \frac{1}{N}\sum_{i=1}^{N}(t_i - y_i)^2 \tag{7}$$

$$MAPE = \frac{1}{N}\sum_{i=1}^{N}\frac{|t_i - y_i|}{t_i} \times 100\% \tag{8}$$

where N is the number of observational data, y_i is actual output data, and t_i is the output data using a particular method as in [15]. Beside these two measures, the coefficient of determination is also used to measure the accuracy of the model using certain forecasting methods in a given data set on the regression model [16]. The coefficient of determination (R^2) can be formulated as follows:

$$R^2 = \frac{1}{N}\frac{\sum_{i=1}^{N}(t_i - \bar{y})^2}{\sum_{i=1}^{N}(y_i - \bar{y})^2} \times 100\% \tag{9}$$

The model of MARS for ICI is obtained in the following steps:

1. Identification the behavior of the response variable and predictor variables.
2. Testing parameters for nonparametric regression models.
3. Standardization each variable which is involved in having the same value scale, and then determining the maximum number of basis functions (BF), maximum interaction (MI), minimum observation (MO). In this study it is selected BF = 14, 21, 28, MI = 1, 2, 3, and MO = 0,1,2,3. If more than three interactions, it will lead to a very complex interpretation of the model.
4. Determining the MARS model based on the smallest value of *GCV* and test of significance of the MARS.
5. Interpretation of the best MARS model and influential variables in the model.

3. Results and Discussion

A. The Model of Artificial Neural Network

In determining a nonlinear model of the artificial neural network with backpropagation method, after defining the input and output data and normalization of data. The next step is building the architecture of the artificial neural network. Based on the data, the architecture of the artificial neural network, which used is 7-4-3-1, it means that the network consists of seven input units, four hidden units in the first hidden layer, three hidden units in the second hidden layer, and one output unit.

The next step is training neural networks. Learning algorithms of the artificial neural network consist of initialization, feed-forward, error assessment, propagation, and adjustment.

1. ***Initialization.*** In the training process, the network has 134 data, learning rate (η) is 0.01, activation function which used is sigmoid function, and error assessment of 0.001 as a threshold value so far algorithm can be stopped.
2. ***Feed-forward.*** Each input unit (x_{ki}, =1, 2, 3, ..., 7) accepts input data and distributes the data into each hidden unit. Based on the first data which is taken on April 2003 input values that are

normalized as follows x_{11} = 0.11600, x_{12} = 0.99294, x_{13} = 0.29681, x_{14} = 0.09430, x_{15} = 0.00000, x_{16} = 0.09364, x_{17} = 0.01694 and output value which want to be achieved or target value is t_{11} = 0.00000.

3. ***Error assessment.*** Because $| t_{k1} - y_{k1} | > 0$ for $k = 1$, then stopping criterion is not available, thus the training of the network has not finished being continued.
4. ***Propagation.*** The back propagation method used to find out the error information on the output unit and find out errors on each hidden unit. In the hidden layer, the propagation is used to find out the correction value of weights and bias.
5. ***Adjustment.*** The last step in learning algorithms of the artificial neural network, these are two steps to adjust the weights to get new weights, which will be used in the next process.

Before doing the next step, repeat the process of feed-forward, error assessment, propagation, until the error between the actual output or target and predicted output is less than a value that has determined. The following step is the validation neural network. In the validation process, data which used to test a network are 46 data. The following table is a model summary of the training and the testing process.

Table 1. Model Summary

Model Summary		
Training	Sum of Squares Error	0.8259
	Relative Error	0.0126
Testing	Sum of Squares Error	0.1826
	Relative Error	0.0065
Dependent Variable: ICI.		

Table 2. Parameter Estimates

Predictor		Predicted							
		Hidden Laye 1				Hodden Layer 2			Output Layer
		H(1:1)	H(1:2)	H(1:3)	H(1:4)	H(2:1)	H(2:2)	H(2:3)	ICI
Input Layer	(Bias)	1.086	-.030	-.318	.436				
	Crude-Oil Price	.288	.391	-.420	-.227				
	Interest Rate	-.280	-.153	.325	-.247				
	Inflation	-.246	.160	-.205	.009				
	Exchange Rate	.584	.032	-.064	.977				
	Gold Price	1.179	-.217	-.231	-.320				
	Dow Jones Price	-.200	.237	-.534	-.965				
	Nikkei 225 index	-.087	.099	.131	-.474				
Hidden Layer 1	(Bias)					.273	.150	-.283	
	H(1:1)					1.311	-.932	-.363	
	H(1:2)					.338	-1.033	-.520	
	H(1:3)					.427	1.239	.729	
	H(1:4)					-.185	.939	1.029	
Hidden Layer 2	(Bias)								.966
	H(2:1)								2.092
	H(2:2)								-2.729
	H(2:3)								-1.402

Based on Table 1, the MSE value is obtained by using Equation (7) for the testing process, 0.0038 is less than the training process, 0.0063. It means that the artificial neural network has the ability to recognize new data. Finally, the obtained model can be used to forecast composite stock price index value. The last step is the modeling of the artificial neural network. After learning algorithms have finished, obtained new weights which will be used to make a nonlinear model of Indonesian composite index. Parameters estimation for the nonlinear model of ANN are shown in Table 2. The model of ANN is built by using two kinds of predicted hidden layer, the first hidden layer consists H(1:1), H(1:2), H(1:3) and H(1:4), while the second hidden layer consists H(2:1), H(2:2) and H(2:3), these structure are used to make prediction of Indonesian composite index as the output layer.

Based on the data, after forecasting the Indonesian composite index, it is obtained MAPE value of 6.16383, this is less than 10 and the coefficient of determination of 0.98925 approaches 1. It means that the model that has been obtained having a good performance to forecast composite stock price index in the future because it is almost approaching the real value.

B. The Model of MARS

After getting information about descriptive statistics and data behavior, furthermore, the regression model assumption is done, namely normality, homoscedasticity, autocorrelation, and multicollinearity test. This assumption is used as the strategy for testing parameters for nonparametric regression suitability. If there is an assumption that is not fulfilling the MARS method, which is a nonparametric regression model can be an alternative to solve the problem. A regression model assumption can be seen in the following Table 3.

Table 3. Regression Model Assumptions

	p-value	ABS RES	Durbin Watson	*VIF*
Indonesian Composite Index	0.051			
Crude-Oil Price	0.001	0.000		14.068
Interest Rate	0.001	0.617		3.219
Inflation	0.000	0.609	0.556	1.698
Exchange Rate	0.000	0.000		4.477
Gold Price	0.001	0.546		5.277
Dow Jones Price	0.000	0.000		17.442
Nikkei 225 Index	0.003	0.000		8.774

Based on Table 3, for normality test, it is obtained p-value (2-tailed) for ICI (Y) is greater than $\alpha = 0.05$, which means that residuals are normally distributed. While, p-value (2-tailed) for the crude-oil price (X_1), interest rate (X_2), inflation (X_3), exchange rate (X_4), gold price (X_5), Dow Jones Index (X_6) and Nikkei 225 Index (X_7) are lower than $\alpha = 0.05$ which means that residuals are not normally distributed. For the homoscedasticity test, the significant value of interest rate (X_2), inflation ($X3$) and gold price (X_5) are greater than significant level 5% in order that there is homoscedasticity. For autocorrelation test, if $d<dL$ or $d > 4\text{-}dL$ then there is exist an autocorrelation. If $dU< 4\text{-}d$, then there is no autocorrelation. If $dL<d<dU<d<4$ then the conclusion could not be taken. Based on Table 4, it is obtained the value of d is 0.865 where from Durbin Watson's table ($N = 180$, $k = 7$), it obtained dU= 1.8374 and dL= 1.6761 in order that $d < dL$. Based on a significant level 5% can be taken a conclusion that there is autocorrelation between residuals for each observation. For multicollinearity test, *VIF* value of X_1 and X_6 are greater than 10, which means that they are any correlation between the predictor variables, there is exist multicollinearity.

The classical regression model on the response variable of the Indonesian composite index does not meet the assumptions of autocorrelation and multicollinearity, so that an alternative MARS model can be used as a nonparametric statistical approach.

Table 4. Results of Selection Model Using GCV

Model	BF	MI	MO	GCV	Model	BF	MI	MO	GCV
1	14	1	0	0.01555	19	21	2	2	0.01333
2	14	1	1	0.01563	20	21	2	3	0.01325
3	14	1	2	0.01560	21	21	3	0	0.01252
4	14	1	3	0.01562	22	21	3	1	0.01210
5	14	2	0	0.01613	23	21	3	2	0.01253
6	14	2	1	0.01555	24	21	3	3	0.01248
7	14	2	2	0.01572	25	28	1	0	0.01445
8	14	2	3	0.01574	26	28	1	1	0.01390
9	14	3	0	0.01415	27	28	1	2	0.01440
10	14	3	1	0.01548	28	28	1	3	0.01389
11	14	3	2	0.01545	29	28	2	0	0.01221
12	14	3	3	0.01547	30	28	2	1	0.01244
13	21	1	0	0.01477	31	28	2	2	0.01241
14	21	1	1	0.01418	32	28	2	3	0.01241
15	21	1	2	0.01491	33	28	3	0	0.01201
16	21	1	3	0.01506	34	28	3	1	0.01017
17	21	2	0	0.01290	35	28	3	2	0.01173
18	21	2	1	0.01279	36	28	3	3	0.01059

In determining the MARS model using minimum *GCV* value, it is obtained by trial and error by combining the number of basis functions (*BF*), maximum interaction (*MI*), and minimum observation (*MO*) as the results in Table 4.

The best model of MARS based on the smallest value of *GCV* is 0.01017 with model equations as follows:

$$Y = -0.763939 + 0.548287 * BF_3 - 0.180626 * BF_5 - 0.18591 * BF_6 + 0.0887371 * BF_9 - 0.515422 * BF_{10} + 0.395404 * BF_{11} + 1.6566 * BF_{14} - 0.502737 * BF_{17} - 0.264626 * BF_{18} + 0.810037 * BF_{19} - 0.129654 * BF_{20} + 0.110508 * BF_{21} - 0.317845 * BF_{23} + 0.343702 * BF_{24} + 0.122742 * BF_{25} - 0.0379905 * BF_{26} + 0.223046 * BF_{27} + 0.123024 * BF_{28}.$$

where we have the basis function as follows

$BF_1 = \max(0, ZX_1 + 0.0720768)$, $BF_3 = \max(0, ZX_6 + 1.63949)$,
$BF_4 = \max(0, ZX_5 - 0.641147)$, $BF_5 = \max(0, 0.641147 - ZX_5)$,
$BF_6 = \max(0, ZX_3 + 0.314867) * BF_3$, $BF_9 = \max(0, 1.4938 - ZX_4) * BF_4$,
$BF_{10} = \max(0, ZX_6 - 1.79433) * BF_1$, $BF_{11} = \max(0, 1.79433 - ZX_6) * BF_1$,
$BF_{14} = \max(0, ZX_1 - 1.09198) * BF_5$, $BF_{17} = \max(0, -0.293419 - ZX_4) * BF_5$,
$BF_{18} = \max(0, ZX_4 + 0.539429)$, $BF_{19} = \max(0, -0.539429 - ZX_4)$,
$BF_{20} = \max(0, ZX_7 + 0.03686) * BF_{18}$, $BF_{21} = \max(0, -0.03686 - ZX_7) * BF_{18}$,
$BF_{23} = \max(0, 0.00555283 - ZX_5) * BF_3$, $BF_{24} = \max(0, ZX_4 - 0.712028) * BF_6$,
$BF_{25} = \max(0, 0.712028 - ZX_4) * BF_6$, $BF_{26} = \max(0, ZX_2 + 1.42696) * BF_3$,
$BF_{27} = \max(0, -1.42696 - ZX_2) * BF_3$, $BF_{28} = \max(0, ZX_7 + 1.54936) * BF_3$.

From the best model which just have obtained, predictor variables that have effect Indonesia Composite Index (ICI) using the minimum GCV are X_1, X_2, X_3, X_4, X_5, X_6 and X_7. Based on the data, after forecasting ICI value, MAPE value of 4.53172 is less than 10, and also the coefficient of determination of 0.99427 approaches 1. It means that the model that has been obtained having good performance to forecast composite stock price index in the future. Furthermore, the forecasting value of the Indonesia composite stock price index in the future is almost approaching the real value.

4. Conclusion

The both methods of ANN and nonparametric MARS regression can be used to forecast Indonesian composite index with better accuracy and goodness of fit respectively based on the MAPE value less than 10, and also the coefficient of determination approach 1, even though the nonparametric MARS regression method shows the accuracy of the model is slightly better than ANN. In order that the time series modeling by using ANN and nonparametric MARS regression methods can be an alternative to forecasting Indonesian composite index in the future.

References

[1] J. O. Rawlings, S. G. Pantula, and D. A. Dickey, *Applied Regression Analysis*. USA: Springer, 2001.

[2] F. K. Reilly and K. C. Brown, *Investment Analysis and Portofolio Management*. USA: Cangenga Learning Product, 2019.

[3] F. Liu and J. Wang, "Fluctuation prediction of stock market index by Legendre neural network with random time strength function," *Neurocomputing*, pp. 12–21, 2012.

[4] D. Devianto, Maiyastri, and D. R. Fadhilla, "Time series modeling for risk of stock price with value at risk computation," *Appl. Math. Sci.*, vol. 9(56), pp. 2779–2787, 2015.

[5] J. Patel, S. Shah, P. Thakkar, and K. Kotecha, "Predicting stock and stock price index movement using trend deterministic data preparation and machine learning techniques," *Expert Syst. Appl.*, pp. 259–268, 2015.

[6] M. Yollanda, D. Devianto, and H. Yozza, "Nonlinear Modeling of IHSG with Artificial Intelligence," *IEEE Explor. ICAITI 2018*, pp. 85–90, 2019.

[7] J. H. Friedman, "Multivariate Adaptive Regression Spline (With Discussion)," *Ann. Stat.*, vol. 19, pp. 1–141, 1991.

[8] E. Guresen, G. Kayakutlu, and T. U. Daim, "Using artificial neural network models in stock market index prediction," *Expert Syst. Appl.*, pp. 10389–10397, 2011.

[9] K. Wong, T. A. Bodnovich, and Y. Selvi, "A bibliography of neural network business applications research: 1988–September 1994," *Expert Syst. Appl.*, pp. 253–262, 1995.

[10] S. Russel and P. Norvig, *Artificial Inteligence: A Modern Approach*. New Jersey: Pearson Education, 2019.

[11] J. Heaton, *Introduction to Neural Networks for C*. America: Heaton Research, Inc, 2018.

[12] J. L. McClelland and D. E. Rumelhart, *Parallel Distributed Processing: Explorations in the Microstructure of Cognition*. Cambridge: MIT Press, 2016.

[13] J. A. Freeman and D. M. Skapura, *Neural Network: Algoritms, Applications, and Programming Techniques*. America: Company, Addison-Wisley Publishing, 2014.

[14] N. D. Lewis, *Build Your Own Neural Network Today!: With Step by Step Instructions Showing You How to Build Them Faster than You Imagined Possible Using R*. Create Space Independent Publishing Platform, 2015.

[15] J. D. Andres, F. Sanchez-Lasheras, P. Lorca, and F. J. De Cos Juez, "A hybrid device of self organizing maps (SOM) and multivariate adaptive regression splines (MARS) for the forecasting of firms "bankruptcy," *Account. Manag. Inf. Syst.*, vol. 10(3), pp. 351–374, 2011.

[16] S. Makridakis, S. C. Wheelwright, and V. E. McGEE, *Forecasting: Methods and Applications*. New York: Wiley, 1997.

IOP Conf. Series: Materials Science and Engineering **846** (2020) 012008 doi:10.1088/1757-899X/846/1/012008

Recognition of Image Pattern To Identification Of Braille Characters To Be Audio Signals For Blind Communication Tools

Ramiati*, Siska Aulia, Lifwarda, Nindya Satriani S.Ningrum

Electrical Engineering, State Polytechnic of Padang, Padang, Indonesia

*ramiati76@gmail.com

Abstract. The five senses are a source of information in humans. The sense that is the main source of information is the sense of sight. Some humans are created with limited sense of sight. The blind performs reading and writing activities using Braille characters. Braille characters are writing or printed systems for the visually impaired in the form of codes consisting of six dots in various combinations that are highlighted on the paper so that they can be touched. Those with significant visual impairment need special education or learning services. Information is very important for everyone, including blind people. Submission of information is done through various media. One media that is often used is print media such as books. But books available on the market do not adapt the way for blind people to capture information. This research offers alternative solutions to overcome the above problems, namely communication aids for reading the blind. The implementation is in the form of a scanner and webcame that is equipped with a braille character text to speech system as an alternative to the lack of blind reading media, especially braille print books. The method, the reading of Braille character scripts by studying braille characters from a to z. First, a webcame or scanner captures braille characters. Second, the system will convert Braille characters and translate Braille characters into alphabetical form through Optical Character Recognition (OCR) image processing. Recognition of Braille character patterns in written text using Artificial Neural Networks (ANN). The results of research on braille character testing are in the form of alphabetical texts a to z, and audio signals of alphabet pronunciation. The results of testing the introduction of braille character patterns using a scanner for training data 100% and for testing data 90.38 % and 96.15 %.

1. Introduction

Braille character converter system is a system that can translate Braille characters into latin characters automatically. The system changes the image of braille letters and the identification of alphabetical letters and binary data from Arduino UNO. From the results of testing Braille with the camera using a time from 3 seconds to 5 seconds, the accuracy rate is 92.3% [1]. The system will recognize Braille characters and classify Braille characters into text using the Optical Character Recognition (OCR) method. Then the system will change the braille character that has become a latin character (text) into sound using a diphone system that will chop words into syllables. Diphone system used is using the Text To Speech application [2].

Inspired by the superiority of the human brain, the field of Artificial Neural Network (ANN) or a network of artificial brain cells has been developed to enable the learning process to be implemented in a system [3]. ANN is used to identify braille characters [4]. This new breakthrough makes the intelligence system a step further [3]. Braille research continues to be developed, both hardware and software. Braille research uses computer vision[5]. This research will design a braille character translator system using the Artificial Neural Network (ANN) method. Broadly speaking, this research is braille image processing with sound output with a mini PC, so that braille letters or braille character scripts can be enjoyed by listening. This literature review aims to create a Braille character translator system into an audio signal using ANN through image processing.

2. Methodology

The research method process in this research uses an experimental method which consists of the study of literature, hardware design, and software development.

2.1 Literature study

At this stage the literature discussion activity of a study is carried out by planning a braille characters recognition system in which the results are written text and sound signals, where the writer collects data and learns relevant basic theories from various sources.
The following are the stages of the research methods used in this study as follows:

1. Data Collection

   In this study the data used include:

   a. Sample Data

   Braille characters or letters are letters used by blind people to read and write. Braille characters data with a six-point pattern in the form *.jpg.

Figure 1. Braille Characters[6]

   b. Target data

   This data was obtained from training data which is in the form of targets of letters of the alphabet that identify from braille characters.

   c. Training data and testing data

   This data is used as a training and testing process system.

2. Data Processing

   In data processing, data in the form of braille character images are processed using digital image processing principles to obtain the characteristic features of each braille character. Digital signal processing to identify braille is also applied at this stage with the support of a raspberry pi device[7].

2.2 Hardware Design

At this stage the activity carried out by the writer is to make a hardware design based on Figure 2.

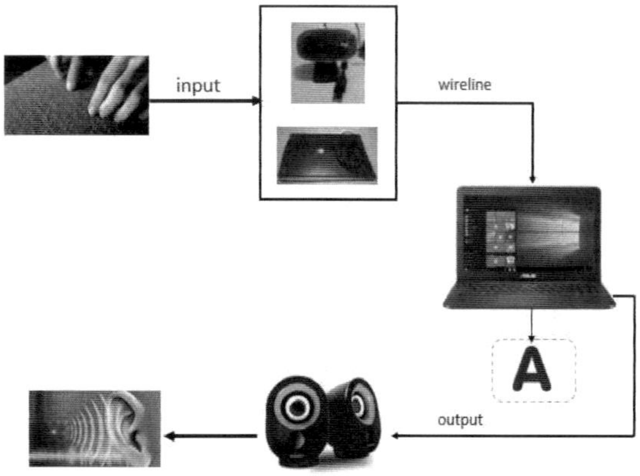

Figure 2. Design of Communication Devices for the Blind Text To Speech

At this stage, the interface display is designed based on the Graphical User Interface (GUI). The purpose of using the interface design based on this GUI is to facilitate the display in the use of the system.

In this study the manufacture of hardware or systems that are built begins with taking pictures of braille characters. The hardware used in this system is a PC (laptop), camera or webcam (Logitech C270 HD), scanner (Canoscan Lide 20) and speakers as sound output. Braille characters are captured by webcame and scanner. Communication between the laptop and webcam / scanner uses Wireline.

2.3 Software Development

At this stage the authors carry out the process of design and manufacture of software for data acquisition, image capture braille characters. Braille image processing using matlab[8][9], block diagram of braille character image processing in figure 3.

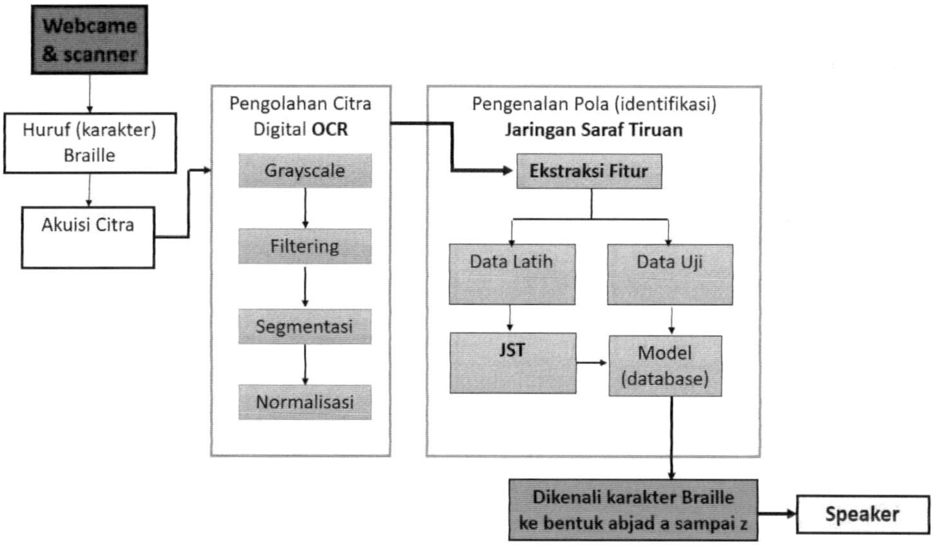

Figure 3. Braille Character Signal Processing Block Diagram

Based on Figure 3, digital image processing is the camera capturing braille letters. The result is a digital image of braille letters analyzed and image acquisition per character[10]. Then the results of image acquisition are processed using digital image OCR to identify the image of letters and numbers that are converted into writing files. The stages of digital image processing are grayscale, filtering, segmentation and normalization. After that, the feature extraction process uses the Multi-level thresholding method in the form of a binary pattern.

Then proceed with the introduction of image patterns. The result of image feature extraction is to obtain certain characteristics of the observed letter characters. Image data that has features is saved and pattern recognition is performed. The process of pattern recognition of an image is divided into two, namely the process of training data and test data. The training data is obtained using the Artificial Neural Network method to produce a training model as a database. The training model is used for the testing process. The test results in the form of identification of braille character identification. Finally audio signal processing. This stage produces sound signals based on identification of recognizable braille letter patterns.

3. Result and Discussion

The review of literature studies in this study is limited to journals published in 2010 to 2018. The results of the implementation of the introduction of braille character patterns into sound signals using matlab. Image processing and pattern recognition of braille images using Artificial Neural Networks and output into voice signals using text to speech. Pattern recognition consists of training and testing. Braille letters are taken from braille books that are captured using a *webcam (Logitech C270 HD Webcame)* and scanner *(Scanner Canon Canoscan Lide 20*) with an image resolution of 165 pixels x 110 pixels.

At the training stage using the Radial Basis Function Network (RBFN) Neural Network. The accuracy of training and testing depends on the training parameters. The training used 48 samples of braille font images where 26 samples used HVS paper and 26 samples used drawing paper. Table 1 below shows the training parameter values.

Table 1. Training Parameters

No	Parameter	Value
1	Max. number of neurons	15
2	Number of neuron to add between display	25
3	Param. goal	1e-06
4	spread	1

➢ Braille Character Image Processing Results

At this stage, the data set is processed using digital image processing techniques to obtain feature extraction from each braille character. The process begins with enhancement techniques, namely the transformation of RGB (color) images into grayscale form. The goal is to determine the distinguishing regions in the image with the condition of only two black and white colors. Figure 4 Data Processing Results of Braille Character Images.

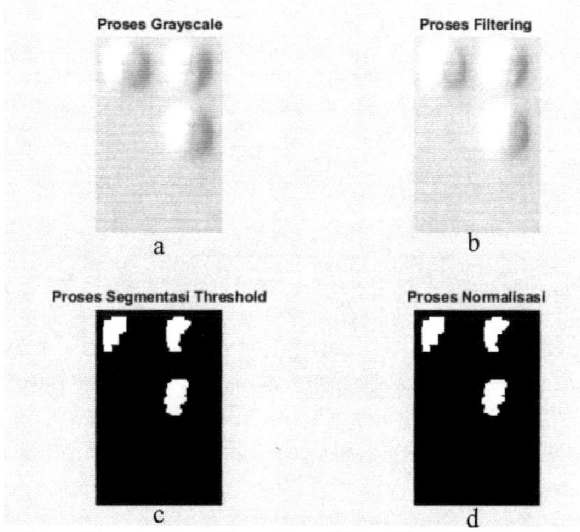

Figure 4. Data Processing Results of Braille Character Images

The second process is filtering to reduce noise, so the resulting image is smoother. Filtering results can be seen in Figure 4b. The third process of segmentation is to divide the image into segments and focus on specific objects. The results of segmentation like Figure 4c. And the final image processing is normalization. Normalization is done before entering the recognition process. This process aims to adjust the input image data with image data in the database. The results of normalization are shown in Figure 4d.

The result of the extraction of the braille letter images. Feature extraction aims to determine the characteristics of each of each braille character. The feature extraction method used in this study is a binary pattern. Multi-level thresholding is an image segmentation method that uses two or more threshold values. Its characteristic extraction technique is converting pixel values to grayscale images into binary or Multi-level thresholding. The results of the extraction of braille characters can be seen in table 2 and figure 5.

Table 2. Data Extraction Results Braille Characteristics

Feature1	Feature2	Feature3	Feature 4	Feature5	Feature6	Target
1	0	0	0	0	0	a
1	0	1	0	0	0	b
1	1	0	0	0	0	c
1	1	0	1	0	0	d
1	0	0	1	0	0	e
1	1	1	0	0	0	f
1	1	1	1	0	0	g
1	0	1	1	0	0	h
0	1	1	0	0	0	i
0	1	1	1	0	0	j
1	0	1	0	0	0	k
1	1	1	0	0	0	l
1	1	0	0	1	0	m
1	1	0	1	1	0	n
1	0	0	1	1	0	o
1	1	1	0	1	0	p
1	1	1	1	1	0	q
1	0	1	1	1	0	r

0	1	1	0	1	0	s
0	1	1	1	1	0	t
1	0	0	0	1	1	u
1	0	1	0	1	1	v
0	1	1	1	0	1	w
1	1	0	0	1	1	x
1	1	0	1	1	1	y
1	0	0	1	1	1	z

The results of identification of braille letter pattern recognition for 100% training data and 100% testing data, can be seen in table 3. For experiments on braille letter pattern recognition can be seen in test 1 (HVS paper) with 90.38 % accuracy results. Testing 2 (drawing paper) with an accuracy of 96.15 %. Testing 1 and testing 2 using a scanner, the results of its implementation can be seen in Figure 7. Testing 3 using a webcam with an accuracy of 57.69 %, can be seen in figure 6. The results of Braille Recognition Experiment can be seen in table 3.

Table 3. Result of Braille Recognition Experiment

Phase	Percentage	Error
Training	100 %	0
Testing	100 %	0
Testing 1	90.38 %	9.62 %
Testing 2	96.15 %	3.85 %
Testing 3	57.69 %	42.31%

In the next stage, an interface is implemented using the Matlab Interface GUI. Figure 5 shows the results of the implementation of the interface in this study. Figure 6 and Figure 7 are the experimental display of braille character identification experiments.

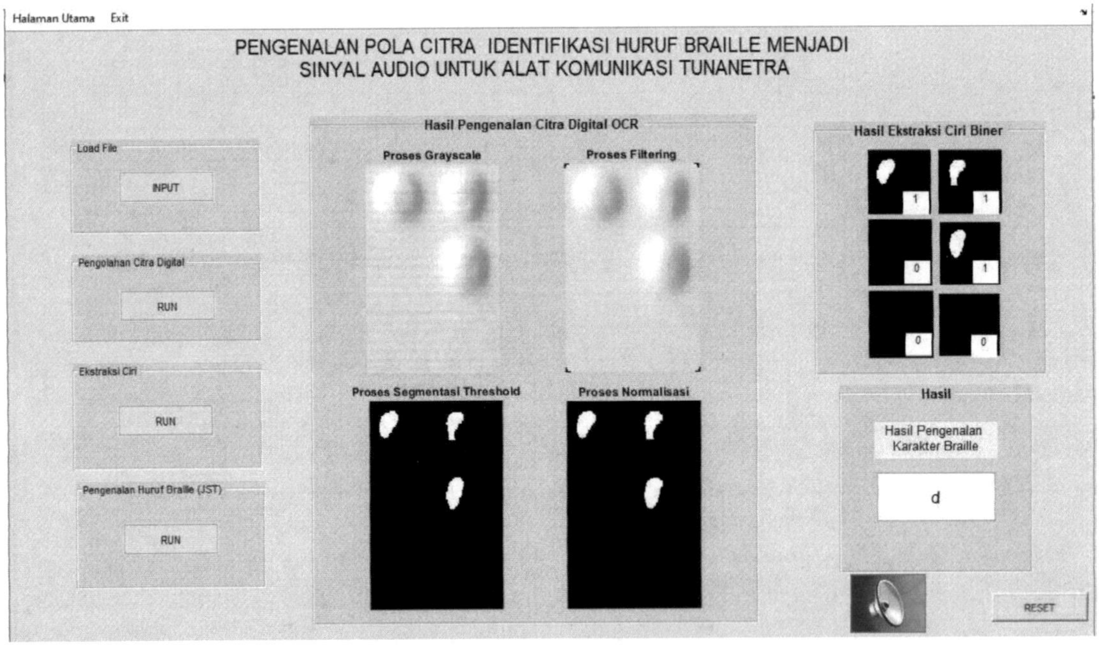

Figure 5. Results of the Implementation of the Pattern Recognition System for Braille Character Images Into Sound Signals

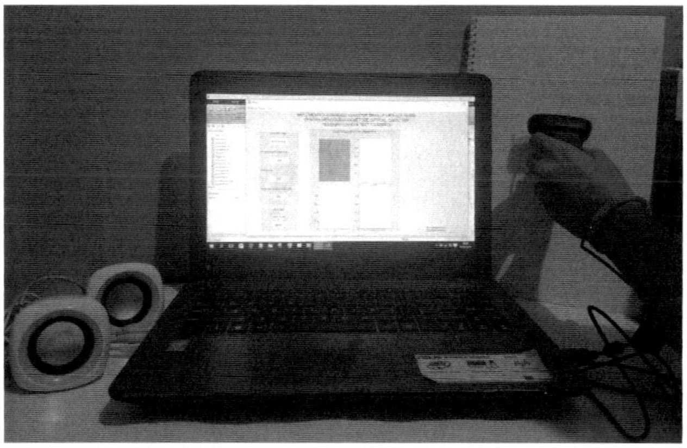

Figure 6. Integration Design of Real Time Braille Character Identification Program using Webcam

Braille character test results using a webcam are far from perfect because of its low accuracy. The system encountered an identification error of 42.31%. Errors due to several reasons, namely the object or sample that is too small, the webcam's focus in recognizing the sample, the webcam's distance from the sample, and the intensity of the light used. The test results can be seen in Figure 6.

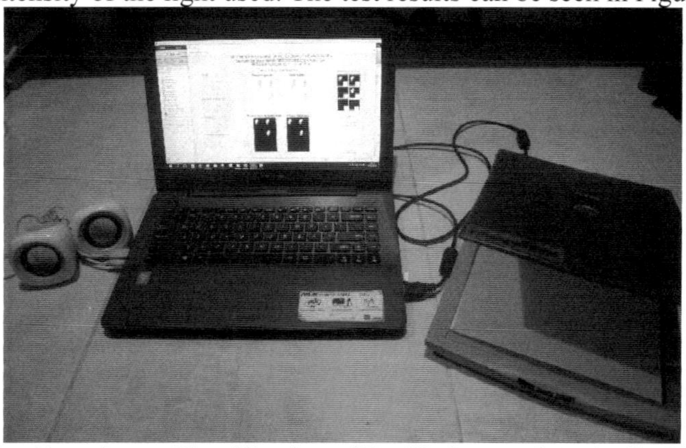

Figure 7. Display Experiment Identification of Braille Letters into Audio Signals for Blind Communication Tools using Scanners

4. Conclusion

Based on testing and evaluation results of this study can be concluded as follows:

1. Artificial Neural Network pattern recognition method with feature extraction of multi-level thresholding binary patterns can recognize Braille characters well with testing accuracy of 100%, 90.38%, 96.15% and 57.69%.
2. This research can be a reference to facilitate communication with the blind. The implementation in this study uses matlab integration with the device, while testing in realtime in the process of using the Raspberry Pi mini PC.

References

[1] D. P. Sari and S. Rasyad, "Identifikasi Huruf Braille Berbasis Image Processing Secara Real Time," no. November, pp. 1–2, 2017.
[2] L. P. Eka Damayanthi, "Pengembangan Aplikasi Text To Speech Dalam Pembuatan Kamus

Untuk Tunanetra," *J. Pendidik. Teknol. dan Kejuru.*, vol. 11, no. 1, pp. 1–10, 2014.

[3] Y. Eninggar *et al.*, "Pengenalan Huruf Braille Berbasis Jaringan Syaraf Tiruan Metode Hebbrule," pp. 1–5.

[4] K. Smelyakov and A. Sakhon, "Braille Character Recognition Based on Neural Networks," *2018 IEEE Second Int. Conf. Data Stream Min. Process.*, pp. 509–513, 2018.

[5] M. P. Arakeri, N. S. Keerthana, M. Madhura, A. Sankar, and T. Munnavar, "Assistive Technology for the Visually Impaired Using Computer Vision," *2018 Int. Conf. Adv. Comput. Commun. Informatics, ICACCI 2018*, pp. 1725–1730, 2018.

[6] S. Hossain, A. A. Raied, A. Rahman, and Z. R. Abdullah, "Text to Braille Scanner with Ultra Low Cost Refreshable Braille Display," *2018 IEEE Glob. Humanit. Technol. Conf.*, pp. 1–6, 2018.

[7] E. Ronando and A. Sudaryanto, "Sistem Pengenalan Pola Huruf Braille Berbasis Audio Menggunakan Metode Naïve Bayes," vol. 3, no. 1, pp. 42–51, 2018.

[8] V. V. Murthy and M. Hanumanthappa, "Improving Optical Braille Recognition in Pre-processing Stage," *ICSNS 2018 - Proc. IEEE Int. Conf. Soft-Computing Netw. Secur.*, pp. 1–3, 2018.

[9] M. Wajid and M. W. Abdullah, "Imprinted Braille-Character Pattern Recognition using Image Processing Techniques," no. Iciip, pp. 0–4, 2011.

[10] S. Isayed and R. Tahboub, "A Review of Optical Braille Recognition," *2015 2nd World Symp. Web Appl. Networking, WSWAN 2015*, pp. 1–6, 2015.

IOP Conf. Series: Materials Science and Engineering **846** (2020) 012009 doi:10.1088/1757-899X/846/1/012009

The Analysis of Fuzzy Logic Method and Multiple Linear Regression in Determining National Rice Production to Improve Food Self-Sufficiency in Indonesia.

Yulia Retno Sari[1], Lidya Rizki Ananda[1], Maha Rani[1]

[1]Computer Science of Department, Universitas Putra Indonesia YPTK Padang

yuliaretnosari2012@gmail.com

Abstract. Indonesia in 2023 targets rice self-sufficiency. To be included in the category of rice self-sufficiency, Indonesia's total rice production should be 82 million tons in accordance with Government Regulation No.2 of 2015. The amount of rice production is uncertain. One way to calculate uncertainty in forecasting is by using the Fuzzy Logic of the Mamdani method and the Linear Regression of the Least Squares method. In the Fuzzy Logic Mamdani method there are 4 steps namely; 1) Fuzzy set formation; 2) Application of Implication functions; 3) Composition of Rules; and 4) Defuzzification. Rice Harvested Area, Rice Production and National Rice Consumption are the input variables and National Rice Production is the output variable. The data used are secondary data from the Central Statistics Agency (BPS) and the Ministry of Agriculture Annual Report 2006 - 2017. The data analysis process also uses matlab and minitab software. From the research results obtained that the Fuzzy Logic of the Mamdani method and the Linear Regression of the Smallest Square method can be used to predict the amount of national rice production. For the results of predictions of Fuzzy Logic the Mamdani method has a relatively large error compared to the Linear Regression of the Smallest Squares method. Furthermore it is also known that there is a significant relationship between harvested area and consumption. These two variables will be the determining factors for realizing rice self-sufficiency in Indonesia.

1. Introduction

Rice self-sufficiency is an effort to meet the country's own food needs. According to Government Regulation No.2 of 2015 concerning the National Medium-Term Development Plan for 2015-2019, to be included in the category of rice self-sufficiency, the number of Indonesia's rice production should be 82 million tons. Since 1984, Indonesia has no longer been capable of being self-sufficiency in rice. The deficit that occurred as a result of large amount of demand caused the government to import rice from the Philippines and Thailand.

There are so many factors that cause rice self-sufficiency cannot be realized. For example, it may be caused by poor agricultural infrastructure in Indonesia, government policies, lack of land for opening new fields and also lack of labors. This is very ironic, considering that Indonesia is a vast tropical region and also has abundant human resources. The government is still trying to find a solution so that the problem of rice self-sufficiency can be overcome and the welfare of the people of Indonesia can be increased. Therefore, the government must re-evaluate every policy decided whether it can succeed in the rice self-sufficiency program.The success of government policy will have an impact on the increasing of the amount of rice production.

The amount of rice production is something that is uncertain, one way to calculate uncertainty in forecasting using the Fuzzy Logic of the Mamdani method [1]. All this time, the forecasting method commonly used is linear regression. Linear regression is used to form an equation of some independent variables (input variables) that are considered to have a relationship with the dependent variable (output variables). The factors that need to be considered in determining the amount of rice production in Indonesia are the Rice Harvested Area, Rice Production and National Rice Consumption.These three factors are the input variables and the output variable is National Rice Production.

Based on the analysis and observations which has been conducted, there are several problems that the writer can formulate as follows ; How is the use of the application of Mamdani Fuzzy Logic method in determining the amount of national rice production. And how is the use of Multiple Linear Regression of The Least Squares method in determining the amount of national rice production. The purpose of this study is as follows ; To find out the use of the application of Mamdani Fuzzy logic method in determining the yield of rice production in Indonesia. And to find out the use of Multiple Linear Regression of the Least Squares method in determining the yield of rice production in Indonesia.

2. Methodology

A term is said to be fuzzy (blurred) if the term cannot be defined explicitly or surely so that it requires affirmation. Fuzzy logic is an appropriate way to map the input space into an output space [2].

Fuzzy logic was first introduced by Prof. LA. Zadeh (1965). In principle, the fuzzy set is the expansion of the crisp set (firm), the set which divides a group of individuals into two categories, namely members and non-members.

In the crisp set (firm), is there are only two possibilities of the membership value, namely 0 or 1. In the fuzzy set, the membership value is in the range of 0 to 1. If x has a fuzzy membership value $\mu_A[x] = 0$, it means that x does not become A set, and so if x has a fuzzy membership value $\mu_A[x] = 1$. It means that x becomes a full member of the A set.Fuzzy set has 2 attributes [3], namely:

a. Linguistics, namely naming a group that represents a certain condition or condition using natural language, such as : YOUNG, MIDDLE-AGED, OLD
b. Numerical, which is a value (number) that indicates the size of a variable, such as : 40, 25, 50, and so on.

The membership function is a curve that shows the mapping of data input points into membership values that have intervals from 0 to 1. One method that can be used to obtain membership value is through a function approach. There are several functions that can be used ; Linear Representation, Triangle Curve Representation, Trapezoid Curve Representation, Shoulder Shape Curve Representation, S-Curve Representation and Bell Curve Representation.

As with any conventional sets, there are several operations that are specifically defined to combine and modify fuzzy sets. Membership values as a result of two sets of operations are often known as fire strength or α-predicate. There are three basic operators created by Zadeh, namely : AND Operators, OR Operators and NOT Operators.

Each rule (proposition) on the fuzzy knowledge base will be associated with a fuzzy relation. The general form of the rules used in the implication function is

$$IF\ x\ is\ A\ THEN\ y\ is\ B$$

where x and y are scalars, A and B are fuzzy sets. Proposition that follow IF is called antecedents, while proposition that follow THEN is called consequent. This proposition can be expanded by using fuzzy operator, such as:

$$IF\ (x_1\ is\ A_1) \circ (x_2\ is\ A_2) \circ (x_3\ is\ A_3) \circ \ldots \circ (x_n\ is\ A_n)\ THEN\ y\ is\ B.$$

where ∘ is an operator (e.g. : OR and AND). Generally, there two implication function which can be used, namely:

a. Min (Minimum) ; This function will cut the fuzzy set output.

b. Dot (product) ; This function will scale the fuzzy set output.

The Mamdani method is often known as the Max-Min Method. This method was introduced by EbrahimMamdani in 1975. To get the output, there are 4 steps required [4]:

a. The formation of a fuzzy set.
   In the Mamdani Method, both input and output variables are divided into one or more fuzzy sets.
b. Application of Implication function
   In Mamdani method, the implication function used is Min.
c. Rule Composition
   Unlike monotonous reasoning, if the system consists of several rules, then inference is obtained from the collection and correlation between rules. Inference is a process of combining many rules based on available data.

There are 3 methods used in fuzzy system inference, namely max, additive and probabilistic OR (**probor**).

i. Max Method (Maximum)
   In this method, the fuzzy set solution is obtained by taking the maximum value of the rule, then using it to modify the fuzzy region and apply it to the output by using the OR (union) operator.If all propositions have been evaluated, the output will contain a fuzzy set that reflects the contribution of each proposition. Generally, it can be written as follow :

$$\mu_{sf}(x_i) = \max\ (\mu_{sf}(x_i), \quad \mu_{kf}(x_i)) \quad \ldots \quad (1)$$

   where :
   $\mu_{sf}(x_i)$ = fuzzy solution membership value up to rule-**i**.
   $\mu_{kf}(x_i)$ = fuzzy consequence membership value i-rule

ii. Additive Method (Sum)
   In this method, the fuzzy set solution is obtained by bounded-sum to all fuzzy area output. It is generally written as follow :

$$\mu_{sf}\ (x_i) = \min(\ 1, \mu_{sf}(x_i) + \mu_{kf}(x_i)) \quad \ldots \quad (2)$$

   where :
   $\mu_{sf}(x_i)$ = fuzzy solution membership value up to rule-**i**.
   $\mu_{kf}(x_i)$ = fuzzy consequence membership value i-rule

iii. Probabilistic Method OR (probor)
   In this method, the fuzzy set solution is obtained by doing *product* to all fuzzy area output. It is generally written as follow :

$$\mu_{sf}(x_i) = \left(\mu_{sf}(x_i) + \mu_{kf}(x_i)\right) + \left(\mu_{sf}(x_i) * \mu_{kf}(x_i)\right) \ldots (3)$$

   where :
   $\mu_{sf}(x_i)$ = fuzzy solution membership value up to rule-**i**.
   $\mu_{kf}(x_i)$ = fuzzy consequence membership value i-rule.

d. Affirmation (*defuzzy*)
   The input of the defuzziffication process is a fuzzy set which is obtained from the composition of fuzzy rules, while the resulting output is a number in the fuzzy set domain. So if it is given a fuzzy set within a certain range, then a certain crisp value must be taken as output.

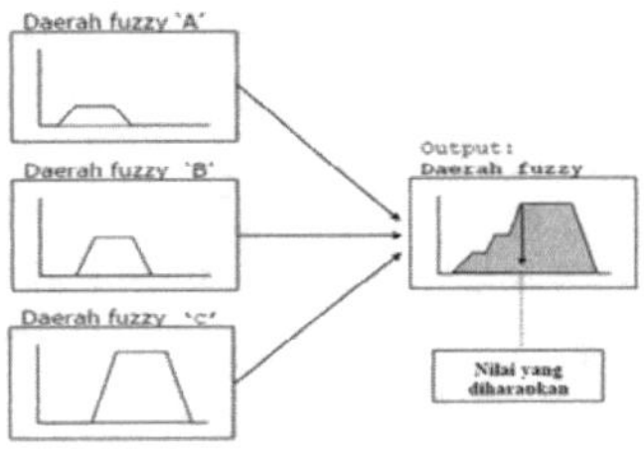

Figure 1. Defuzziffication process

One method of defuzzyfication is the centroid method. Centroid which can be called as *Center of Area (Center of Gravity)* is the most common and most proposed method by many researchers to be used. In this method, a crisp solution is obtained by taking the center point (z*) of the fuzzy region. It is generally formulated as:

$$z^* = \frac{\int_z z\mu(z)\,dz}{\int_z \mu(z)\,dz}, for\ continuous\ variables \ldots (4)$$

$$z^* = \frac{\sum_{j=1}^{n} z_j\mu(z_j)}{\sum_{j=1}^{n} \mu(z_j)}, for\ discrete\ variables \ldots (5)$$

In multiple linear regression, forecasting the value of the dependent variable (Y) is obtained by forming an equation that connects more than one variable, namely $x_1, x_2, \ldots, x_n$.. In general, the multiple regression equation can be formulated as [5]:

$$\hat{Y}_i = b_0 + b_1x_{1i} + b_2x_{2i} + \cdots + b_nx_{ni} + \varepsilon \ldots (6)$$

This research consists of one independent variable Y and three variables X namely x_1, x_2, x_3. Then the multiple regression equation is:

$$\hat{Y}_i = b_0 + b_1x_{1i} + b_2x_{2i} + b_3x_{3i} + \varepsilon \ldots (7)$$

The coefficient values of b_0, b_1, b_2 and b_3 can be obtained by using the least squares method, by first minimizing the remaining squares as follows:

$$S = \sum_{i=1}^{n} e_i^2 = \sum_{i=1}^{n} (Y_i - \hat{Y}_i)^2 = \sum_{i=1}^{n} (Y_i - b_0 - b_1x_{1i} - b_2x_{2i} - b_3x_{3i})^2 = 0 \ldots (8)$$

Then, equation (8) is differentiated to each coefficient b_0, b_1, b_2 and b_3, so the following equations are obtained:

$$\sum_{i=1}^{n} Y_i = nb_0 + b_1\sum_{i=1}^{n} x_{1i} + b_2\sum_{i=1}^{n} x_{2i} + b_3\sum_{i=1}^{n} x_{3i}$$

$$\sum_{i=1}^{n} x_{1i}Y_i = b_0\sum_{i=1}^{n} x_{1i} + b_1\sum_{i=1}^{n} x_{1i}^2 + b_2\sum_{i=1}^{n} x_{1i}x_{2i} + b_3\sum_{i=1}^{n} x_{1i}x_{3i}$$

$$\sum_{i=1}^{n} x_{2i}Y_i = b_0\sum_{i=1}^{n} x_{2i} + b_1\sum_{i=1}^{n} x_{1i}x_{2i} + b_2\sum_{i=1}^{n} x_{2i}^2 + b_3\sum_{i=1}^{n} x_{2i}x_{3i}$$

$$\sum_{i=1}^{n} x_{3i}Y_i = b_0 \sum_{i=1}^{n} x_{3i} + b_1 \sum_{i=1}^{n} x_{1i}x_{3i} + b_2 \sum_{i=1}^{n} x_{2i}x_{3i} + b_3 \sum_{i=1}^{n} x_{3i}^2 \dots (9)$$

Relative error is a measure of error in relation to measurement. Relative error is defined by :

$$e_r = \left|\frac{X_s - X_a}{X_s}\right| \quad \dots (10)$$

where:

e_r= relative error
X_s= true value
X_a= calculation value

To see the average relative errors that occur in a data, can be stated by :

$$Average\ Relative\ Error = \frac{number\ of\ relative\ error}{amount\ of\ data} \dots (11)$$

3. Result and Discussion

In predicting rice production in Indonesia, researchers used two calculations. Firstly, calculation by using the Fuzzy logic of Mamdani method and secondly by using multiple linear regression analysis of the least square method. In the Fuzzy logic Mamdani method, it requires 4 stages namely; Fuzzy set formation, Application of Implication function, Rule Composition and Defuzzification. The data analysis process also uses Matlab and Minitab software.

The data used to predict rice production in Indonesia are secondary data which are obtained from the Central Statistics Agency (BPS) and Ministry of Agriculture Annual Report 2006 – 2017.

Table 1. Rice Production Data

Year	Harvest Area (Ha)	Total Population (million)	Consumption (ton)	Rice Production (ton)
2006	11.786.430	224.179	21.496.757	54.454.937
2007	12.147.637	227.521	20.583.388	57.157.435
2008	12.327.425	230.913	21.576.527	60.325.925
2009	12.883.576	234.356	21.397.141	64.398.890
2010	13.253.450	238.519	21.503.662	66.469.394
2011	13.203.643	241.991	21.652.602	65.756.904
2012	13.445.524	245.425	21.409.667	69.056.126
2013	13.835.252	248.818	21.277.431	71.279.709
2014	13.797.307	252.165	21.340.203	70.846.465
2015	14.115.475	255.462	22.285.201	76.361.248
2016	14.449.372	258.705	22.568.131	77.049.958
2017	14.701.998	261.891	22.846.053	79.175.945

The Fuzzy logic of Mamdani method

The Mamdani method is also called the Max-Min method. To get the output which is a prediction of Rice Production in Indonesia, 4 stages are needed:

a. Formation of Fuzzy Set
This is the first step taken in using the Mamdani method. There are four Fuzzy variables to be modeled, i.e Harvest Area with Fuzzy set ; small, medium and large, Number of Population with Fuzzy set ; few and many, Consumption with Fuzzy set ; decrease and increase. And Rice Production with little, medium and much.

Table 2. Determination of Variable and Speaker Universe

Function	Variable	Universal Speaker (thousand)	Explanation
Input	Harvest Area	11.786 – 14.701	Production Area (Ha)
	Number of Population	224.000 – 262.000	Number of National Population (head)
	Consumption	21.496 – 22.846	Number of Consumption (Ton)
Output	Rice Production	54.454 – 79.175	Number of Rice Production (Ton)

Determine the domain of Harvest Area, Population, Consumption and Rice Production at Fuzzy leadership as follows

Table 3. Fuzzy Set

Function	Variable	Fuzzy Set	Universal Speaker	Domain
Input	Harvest Area	small	11.786 – 14.701	11.786 – 13.244
		medium		12.758 – 13.730
		large		13.244 – 14.701
	Number of Population	few	224.000 – 261.000	224.000 – 242.500
		many		242.500 – 261.000
	Consumption	decrease	21.496 – 22.846	21.496 – 22171
		increase		22.171 – 22.846
Output	Rice Production	low	54.454 – 79.175	54.454 - 66.814
		medium		62.694 – 70.934
		high		66.814 -79.175

i. Harvest Area

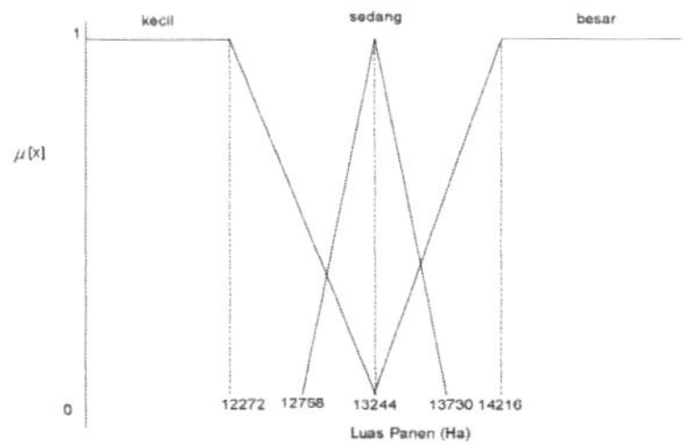

Figure 2. Fuzzy set curve for harvest area variable or Representation of harvest area variable

Membership function:

$$\mu_{LP\ Small}(x) = \begin{cases} 1, x \leq 12.272 \\ \dfrac{13.244 - x}{13.244 - 12.272} = \dfrac{13.244 - x}{972}; 12.272 \leq x \leq 13.244 \\ 0, x \geq 13.244 \end{cases} \ldots (12)$$

$$\mu_{LP\ Medium}(x) = \begin{cases} 0;\ x \leq 12.758\ or\ x \geq 13.730 \\ \dfrac{x - 12.758}{486}; 12..758 \leq x \leq 13.244 \\ \dfrac{13.244 - x}{486}; 13.244 \leq x \leq 13.730 \end{cases} \ldots (13)$$

$$\mu_{LP\ Large}(x) = \begin{cases} 0;\ x \leq 13.244 \\ \dfrac{x - 13.244}{972}; 13.244 \leq x \leq 14.216 \\ 1;\ x \geq 14.316 \end{cases} \ldots (14)$$

If the harvest area is 14.701.998 Ha, then the value of Fuzzy membership in each set is

$$\mu_{LP\ Small}[14.701.998] = 0$$

$$\mu_{LP\ Medium}[14.701.998] = 0$$

$$\mu_{LPLarge}[14.701.998] = 1$$

ii. Number of Population

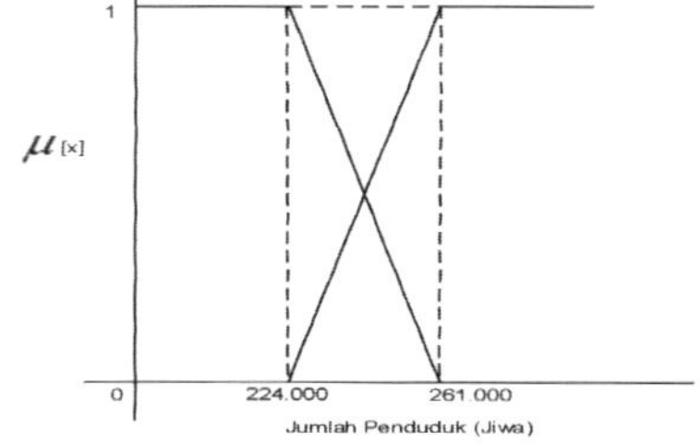

Figure 3. Fuzzy set curve for the number of population's variable or

Representation of number of population's variables

Membership function:

$$\mu_{JP\ Few}(x) = \begin{cases} 1; \ x \leq 244.000 \\ \dfrac{261.000 - x}{37.000}; 244.000 \leq x \leq 261.000 \ldots (15) \\ 0; \ x \geq 261.000 \end{cases}$$

$$\mu_{JP\ Many}(x) = \begin{cases} 0; \ x \leq 244.000 \\ \dfrac{x - 224.000}{37.000}; 244.000 \leq x \leq 261.000 \ldots (16) \\ 1; \ x \geq 261.000 \end{cases}$$

If the total population is 261.891.000, then the value of Fuzzy membership in each set is,

$$\mu_{JP\ Few}[261.891.000] = 0$$
$$\mu_{JP\ Many}[261.891.000] = 1$$

iii. Consumption

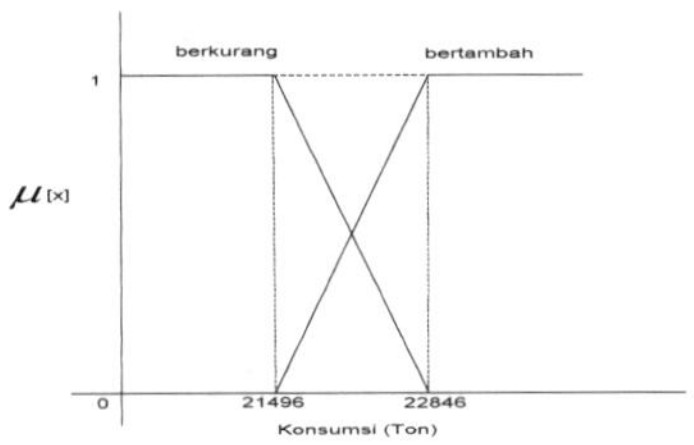

Figure 4. Fuzzy set curves for consumption variables or Representations of consumption variables

Membership Function:

$$\mu_{K\ Decrease}(x) = \begin{cases} 1; \ x \leq 21496 \\ \dfrac{22.846 - x}{1.350}; 21.496 \leq x \leq 22.846 \ldots (17) \\ 0; \ x \geq 22.846 \end{cases}$$

$$\mu_{K\ Increase}(x) == \begin{cases} 0; \ x \leq 21.496 \\ \dfrac{x - 21.496}{1350}; 21.496 \leq x \leq 22.846 \ldots (18) \\ 1; \ x \geq 22.846 \end{cases}$$

If the consumption is 22,846,053 tons, then the value of Fuzzy membership in each set is

$$\mu_{K\ Decrease}[22.846.053] = 0$$
$$\mu_{K\ Increase}[22.846.053] = 1$$

iv. Rice Production

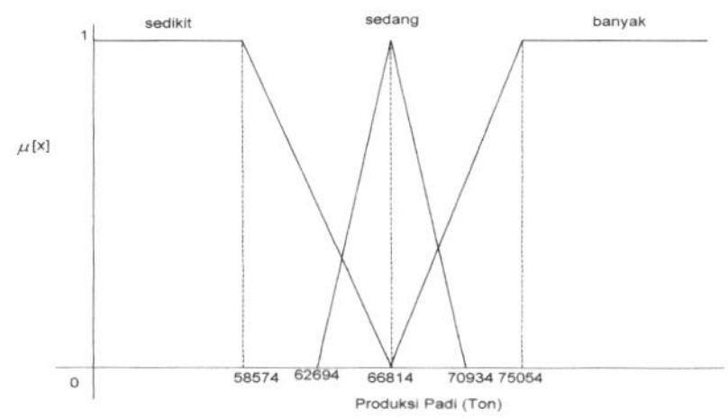

Figure 5. Fuzzy set curves for Rice Production variables or Representations of Rice Production variables

Membership Function:

$$\mu_{PP\ Low}(x) = \begin{cases} 1;\ x \leq 58.574 \\ \dfrac{66.814 - x}{8.240};58.574 \leq x \leq 66.814 \\ 0;\ x \geq 66.814 \end{cases} \quad \ldots(19)$$

$$\mu_{PP\ Medium}(x) = \begin{cases} 0;\ x \leq 62.694\ or\ x \geq 66.814 \\ \dfrac{x - 62.694}{4.120};62694 \leq x \leq 66.814 \\ \dfrac{70934 - x}{4.120};66.814 \leq x \leq 70.934 \end{cases} \quad \ldots(20)$$

$$\mu_{PP\ High}\ (x) = \begin{cases} 0;\ x \leq 66.814 \\ \dfrac{x - 66.814}{8.240};66.814 \leq x \leq 75.054 \\ 1;\ x \geq 75.054 \end{cases} \quad \ldots(21)$$

If rice production is 79.175.945 tons, then the value of Fuzzy membership in each set is

$$\mu_{PP\ Low}[79.175.945] = 0$$

$$\mu_{PP\ Medium}[79.175.945] = 0$$

$$\mu_{PP\ High}[79.175.945] = 1$$

b. Application of Implication function

The rules in the form of qualitative statements which are written in the form of IF **THEN** so that they are easy to understand. From 3 variables and 10 sets, 36 rules are obtained, as below:

[1] **IF** the harvest area is small **AND** the number of population is few **AND** the consumption is reduced **THEN** the rice production will be low

[2] **IF** the harvest area is small **AND** the number of population is few **AND** the consumption is reduced **THEN** the rice production will be medium.

[3] **IF** the harvest is small **AND** the number of population is few **AND** the consumption is reduced **THEN** the rice production will be high.

[4] **IF** the harvest area is small **AND** the number of population is few **AND** the consumption is increased **THEN** the rice production will be low.

[5] **IF** the harvest area is small **AND** the number of population is few **AND** the consumption is increased **THEN** the rice production will be medium.

[6] **IF** the harvest area is small **AND** the number of population is few **AND** the consumption is increased **THEN** the rice production will be high.

[7] **IF** the harvest area is small **AND** the number of population is many **AND** the consumption is decreased **THEN** the rice production will be low.

[8] **IF** the harvest area is small **AND** the number of population is many **AND** the consumption is decreased **THEN** the rice production will be medium.

[9] **IF** the harvest area is small **AND** the number of population is many **AND** the consumption is decreased **THEN** the rice production will be high.

[10] **IF** the harvest area is small **AND** the number of population is many **AND** the consumption is increased **THEN** the rice production will be low.

[11] **IF** the harvest area is small **AND** the number of population is many **AND** the consumption is increased **THEN** the rice production will be medium.

[12] IF the harvest area is small **AND** the number of population is many **AND** the consumption is increased **THEN** the rice production will be high.

[13] **IF** the harvest area is medium **AND** the number of population is few **AND** the consumption is decreased **THEN** the rice production will be low.

[14] **IF** the harvest area is medium **AND** the number of population is few **AND** the consumption is decreased **THEN** the rice production will be medium.

[15] **IF** the harvest area is medium **AND** the number of population is few **AND** the consumption is decreased **THEN** the rice production will be high.

[16] **IF** the harvest area is medium **AND** the number of population is few **AND** the consumption is increased **THEN** the rice production will be low.

[17] **IF** the harvest area is medium **AND** the number of the population is few **AND** the consumption is increased **THEN** the rice production will be medium.

[18] **IF** the harvest area is medium **AND** the number of the population is few **AND** the consumption is increased **THEN** the rice production will be high.

[19] **IF** the harvest area is medium **AND** the number of the population is many **AND** the consumption is decreased **THEN** the rice production will be low.

[20] **IF** the harvest area is medium **AND** the number of the population is many **AND** the consumption is decreased **THEN** the rice production will be medium.

[21] **IF** the harvest area is medium **AND** the number of the population is many **AND** the consumption is decreased **THEN** the rice production will be high.

[22] **IF** the harvest area is medium **AND** the number of the population is many **AND** the consumption is increased **THEN** the rice production will be low.

[23] **IF** the harvest area is medium **AND** the number of the population is many **AND** the consumption is increased **THEN** the rice production will be medium.

[24] **IF** the harvest area is medium **AND** the number of the population is many **AND** the consumption is increased **THEN** the rice production will be high.

[25] **IF** the harvest area is big **AND** the number of the population is few **AND** the consumption is decreased **THEN** the rice production will be low.

[26] **IF** the harvest area is big **AND** the number of the population is few **AND** the consumption is decreased **THEN** the rice production will be medium.

[27] **IF** the harvest area is big **AND** the number of the population is few **AND** the consumption is decreased **THEN** the rice production will be high.

[28] **IF** the harvest area is big **AND** the number of the population is few **AND** the consumption is increased **THEN** the rice production will be low.

[29] **IF** the harvest area is big **AND** the number of the population is few **AND** the consumption is increased **THEN** the rice production will be medium.

[30] **IF** the harvest area is big **AND** the number of the population is few **AND** the consumption is increased **THEN** the rice production will be high.

[31] **IF** the harvest area is big **AND** the number of the population is many **AND** the consumption is decreased **THEN** the rice production will be low.

[32] **IF** the harvest area is big **AND** the number of the population is many **AND** the consumption is decreased **THEN** the rice production will be medium.

[33] **IF** the harvest area is big **AND** the number of the population is many **AND** the consumption is decreased **THEN** the rice production will be high.

[34] **IF** the harvest area is big **AND** the number of the population is many **AND** the consumption is increased **THEN** the rice production will be low.

[35] **IF** the harvest area is big **AND** the number of the population is many **AND** the consumption is increased **THEN** the rice production will be medium.

[36] **IF** the harvest area is big **AND** the number of the population is many **AND** the consumption is increased **THEN** the rice production will be high.

In manually calculating using 2017 data, the rule used is rule 36 [R 36] viz **[36] IF** the harvest area is big **AND** the number of population is many **AND** the consumption is increased **THEN** the rice production will be high. The minimal α-predicate value using the above rules is as follows :

$$\alpha - predicate = (\mu_{LP\ Big} \cap \mu_{JP\ Many} \cap \mu_{K\ Increased} \cap \mu_{PP\ High})$$

$$= \min\ (\mu_{LP\ Big}[14.701.998] \cap \mu_{JPMany}\ [261.891] \cap \mu_{K\ Increased}[22.846.053] \cap \mu_{PP\ High}\ [79.175.945])$$

$$= \min imal(\ 1\ ; 1\ ; 1\ ; 1\ ) = 1$$

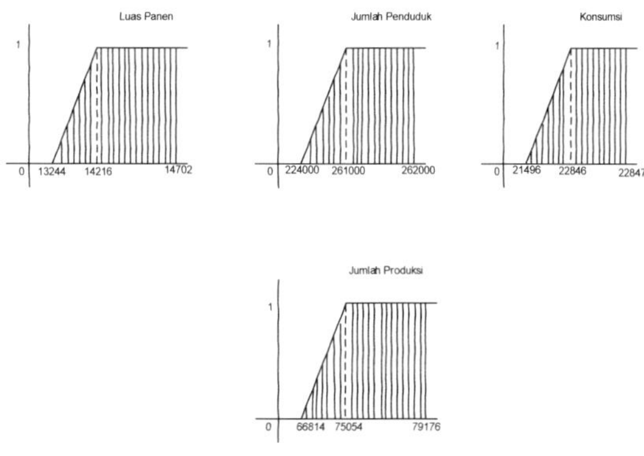

Figure 6. Application of Implication function for Rule [36]

c. Aggregation (Rule Composition)
Formation of the rule composition is done by taking the maximal value on α-predicate.

$$\mu\,(z) = \max\,\{\,R\,36\,\} = \max\,\{\,1\,\} = 1$$

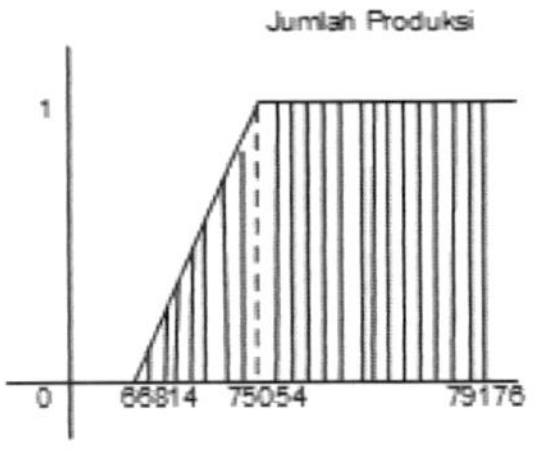

Figure 7. The Area of Composition Result

$$\mu\,(z) = \begin{cases} 0;\ z \leq 66814 \\ \dfrac{z-66814}{8240};66814 \leq x \leq 75054 \ldots (22) \\ 1;\ z \geq 75054 \end{cases}$$

d. Defuzzification

The affirmation method used is the centoid method. Because it's the first calculated moment for each region:

$$m1 = \int_{66814}^{75054} \left(\frac{z-66814}{8240}\right) z\,dz = \int_{66814}^{75054} \frac{z^2-66814\,z}{8240}\,dz = \int_{66814}^{75054} \frac{z^2}{8240} - \frac{66814\,z}{8240}\,dz = \frac{1}{3}\frac{z^3}{(8240)} - \frac{1}{2}\frac{66814\,z^2}{8240}\Big|_{66814}^{75054} = \frac{z^3}{24720} - \frac{33407z^2}{8240}\Big|_{66814}^{75054}$$

$$= [17103030188{,}40873 - 22837993824{,}61310] - [12065739699{,}07540 - 18098609548{,}61310]$$

$$= -5734963636{,}2043 + 6032869849{,}5377 = 297906213{,}3333$$

$$m2 = \int_{75054}^{79176} 1z\,dz = \frac{1}{2}z^2\ \Big|_{75054}^{79176} = \frac{1}{2}[(79176^2) - (75054^2)]$$

$$= \frac{1}{2}[(6268838976) - (5633102916)] = \frac{1}{2}(635736060) = 317868030$$

Then calculate the area of each region:

$$A_1 = \frac{1}{2}(75054 - 66814).1 = \frac{8240}{2}.1 = 4120$$

$$A_2 = (79176 - 75054).1 = 4122$$

The central point can be obtained from:

$$Z = \frac{M1 + M2}{A1 + A2} = \frac{297906213{,}333 + 317868{,}030}{4120 + 4122}$$

$$= \frac{615774243{,}3333}{8242} = 74711{,}74998 = 74.711.750\ ton$$

So, the amount of rice production in Indonesia in 2017 using the Fuzzy count is 74.711.750 tons. By entering the membership function value of each variable into the Fuzzy rules using the Fuzzy application are obtained the forecast result as in Table 11 below. The results obtained between the manual and the application are differences that are far different because the level of accuracy of the parameters results in the manual calculation is not so effective and efficient. So we use forecast results with the Fuzzy application only.

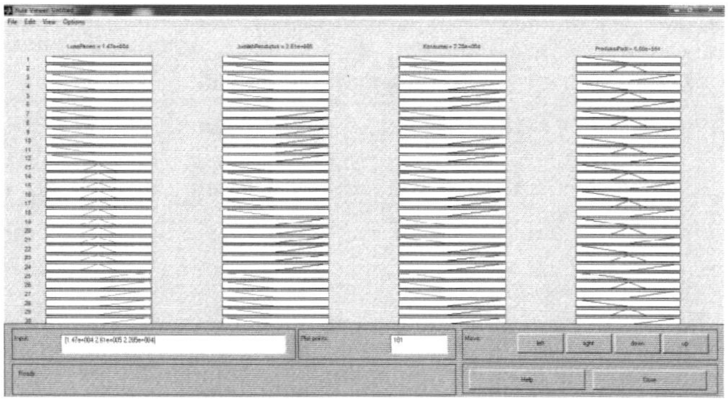

Figure 8. Variable curve $\hat{Y}$ in Matlab

Multiple Linear Regression Analysis

For multiple linear regression, the equation is obtained by eliminating the equations of the least squares method (9). The coefficients of b_0, b_1, b_2 and are b_3 obtained with the help of Minitab software.

```
The regression equation is
Y = - 56669133 + 7.83 X1 + 0.017 X2 + 0.729 X3


Predictor        Coef   SE Coef      T      P
Constant    -56669133  10065662  -5.63  0.000
X1              7.826     1.562   5.01  0.001
X2             0.0172    0.1169   0.15  0.887
X3             0.7292    0.5063   1.44  0.188


S = 706560   R-Sq = 99.4%   R-Sq(adj) = 99.2%
```

Figure 9. Coefficient values using Minitab software

So the equation produced is as follows:

$$\hat{Y}_t = -56669133 + 7{,}83\ x_1 + 0{,}017x_2 + 0{,}729x_3$$

If each of the variables x_1, x_2 and x_3 is substituted with the above equation, then the forecast result of rice production $(\hat{Y}_t)$ using the multiple linear regression method is

Table 4. Results of Rice Production Forecasting with Fuzzy Logic and Multiple Regression

Year	Harvest Area (X1)	Number of Population (X2)	Consumption (X3)	Rice Production (Y)	Fuzzy Rice Production ($\hat{Y}$)	Regression Rice Production ($\hat{Y}_t$)	Fuzzy Relative Errors	Regression Relative Errors
2006	11.786.430	224.179.000	21.496.757	54.454.937	66.800.000	55100792,75	0,226702365	0,011860371
2007	12.147.637	227.521.000	20.583.388	57.157.435	66.800.000	57320011,56	0,16870185	0,002844364
2008	12.327.425	230.913.000	21.576.527	60.325.928	66.800.000	59509413,93	0,107318233	0,013535044
2009	12.883.576	234.356.000	21.397.141	64.398.890	66.800.000	63791834,87	0,037284959	0,009426484
2010	13.253.450	238.519.000	21.503.662	66.469.394	66.800.000	66836373,1	0,004973808	0,005521024
2011	13.203.643	241.991.000	21.652.602	65.756.904	66.800.000	66613985,55	0,015862912	0,013034092
2012	13.445.524	245.425.000	21.409.667	69.056.126	66.800.000	68389192,16	0,032670903	0,009657852
2013	13.835.252	248.818.000	21.277.431	71.279.709	66.800.000	71402043,36	0,062846904	0,001716258
2014	13.797.307	252.165.000	21.340.203	70.846.465	66.800.000	71207593,8	0,057115976	0,005097344
2015	14.115.475	255.462.000	22.285.201	75.361.248	66.800.000	74443801,78	0,11360279	0,012173979
2016	14.449.372	258.705.000	22.568.131	77.049.958	66.800.000	77319602,26	0,133030027	0,003499603
2017	14.701.998	261.891.000	22.846.053	79.175.945	66.800.000	79554430,98	0,156309407	0,004780315

From the above table, it is obtained that the forecast results of Rice Production using Fuzzy Logic for a number of years are 66.800.000 tons, the number is stable for twelve years of rice production in Indonesia. While for the forecast results of Rice Production using Multiple Linear Regression, the resulting numbers are vary and are close to the actual data results.

4. Conclusion

Based on the analysis conducted, it can be concluded that ;

a. Fuzzy logic Mamdani method and Multiple Linear Regression can be used to predict the yield of rice production in Indonesia. Forecasting the amount of rice production using Fuzzy Logic from 2006 - 2017 amounted to 66.8 million tons, while the Multiple Linear Regression results are almost the same as the actual data. That's why the relative error of using Multiple Linear Regression is smaller than that of Fuzzy Logic.
b. It was found that the relationship between the harvest area variable and the consumption variable significantly affected the amount of rice production in Indonesia. So to increase rice production in Indonesia, related parties, in this case the government, must try to open new fields and find alternative staple foods. These are all efforts to realize rice self-sufficiency in 2023.

References

[1] D. L. Rahakbauw, F. J. Rianekuay, and Y. A. Lesnussa, "Penerapan Metode Fuzzy Mamdani Untuk Memprediksi Jumlah Produksi Karet (Studi Kasus: Data Persediaan Dan Permintaan Produksi Karet Pada PTP NUSANTARA XIV (PERSERO) Kebun Awaya, Teluk Elaputih, Maluku-Indonesia)," *J. Ilm. Mat. dan Terap. Vol. 16, No.1*, 2019.

[2] Y. R. Sari, "Aplikasi Logika Fuzzy Metode Mamdani dalam Menentukan Produksi Beras tahun 2018 di Indonesia," *Pros. SISFOTEK*, 2018.

[3] M. Abrori and A. H. Prihamayu, "Aplikasi Metode Fuzzy Mamdani dalam Pengambilan Keputusan Penentuan Jumlah Produksi," *Kaunia*, 2015.

[4] M. Rani, "Logika Fuzzy untuk Menentukan Asupan Kalori pada Terapi Diet terhadap Penderita Obesitas," *J. Inov. Tek. Inform.*, vol. 1, no. 2, 2018.

[5] R. S. Siska Ernida Wati, Dakaria Sebayang, "Perbandingan Metode Fuzzy dengan Regresi Linier Berganda dalam Peramalan Jumlah Produksi (Studi Kasus: Produksi Kelapa Sawit di PT. Perkebunan III (Persero) Medan Tahun 2011-2012)," vol. 1, no. 3, 2013.

THE IMPLEMENTATION OF SOUND OUTPUT IMAGE PROCESSING WITH WIRELESS BASED ON CONTROL TO VEHICLE IDENTIFICATION

Siska Aulia[1], Popy Maria[2], Aprinal A. Asri[3], Ratna Dewi[4]

[1,2,3,4] Electrical Engineering, State Polytechnic of Padang, Indonesia

siska.auliaa@gmail.com

Abstract. Vehicle license plate identification is a system that utilizes image processing techniques to identify vehicles automatically from the vehicle license plate image. This system can be applied in various applications such as tracking stolen cars in homes or parking lots, automatic parking ticket payments, and collecting traffic statistics. Parking space as a facility to park motorists' vehicles to be more organized. Parking space is currently still done conventionally. The development of technology, conventional control systems developed into electronic access systems, one of which is a monitoring system. The monitoring system using CCTV is a monitoring device, but it is expensive at the cost of electrical, network installation and limited mobility of device placement. In this study a Raspberry Pi-based room control system was designed. Raspberry Pi to build inexpensive monitoring systems and electricity-saving technology. This study aims to create a remote-based IP camera control system wirelessly for processing image plates of vehicle sound output with Raspberry Pi. The system starts by capturing the vehicle license plate with a Raspberry Pi-based camera or webcam. The control system is created using a motion program, and plate image processing uses digital image processing techniques and the Optical Character Recognition (OCR) method. The results of the system research can identify a Raspberry Pi server-based vehicle plate. The result is the vehicle license plate or text and voice output with text to speech technology. The results of implementing image processing from nine plate tests are 90.278%.

1. Introduction

Along with the development of information technology allows more practical data access. Currently there are many communication tools to facilitate humans in the communication process [1]. Current technological developments can access images using Internet-based Protcol (IP) cameras. One of the implementation is in the control or monitoring system. Security in the room or location is ideal if the room has direct control. But the limited time and energy that must be ready to control a room at any time makes a person unable to control.

The development of increasingly sophisticated and efficient technology can be utilized to do things that humans cannot do. One of them is controlling using an IP camera and web based that can monitor the room at any time[2]. Problems that often occur in the community are vehicle theft in the garage of a house or campus parking location[3]. Advances in technology have an impact on increasing energy needs. Closed Circuit Television (CCTV) or Surveillance Camera is a monitoring device, but it is expensive in the case of electricity or network installation costs, and the limited mobility of device placement makes the use of this device not optimal [4]. This research offers an alternative solution to overcome the above problems, which is a technology-saving technological device, Raspberry PI (RPi). RPi is a mini computer PCB hardware equipped with CPU and memory[5][6].

The vehicle number plate is the identity number of each vehicle [7]. Of course, every vehicle that has a different number plate for each shows the identity of the vehicle. Apart from that, the vehicle plate image processing uses computer media to detect the location of the plate and the introduction of its character. Vehicle numbers are identified by the Optical Character Recognition (OCR) method [8][9]. Image processing in detecting vehicle plates is very necessary, namely converting the image of character images into writing technology. Based on the above background, the authors make research on the Implementation of Voice Output Image Processing with Wireless Based Control for Vehicle Identification. In this study, a vehicle license plate identification system with image processing and sound output through IP control media on an RPi server integrated with a USB camera passes through a router.

2. Methodology

At this stage the literature discussion activity of a study is carried out, namely planning the vehicle license plate identification system and USB camera control system via Wireless IP (Internet Protocol) network devices and using IPv4 DHCP (Dynamic Host Configuration Protocol), where the authors collect data and study basic theory relevant from various sources such as books, internet, resource persons and research that has been done relating to the research that will be conducted by the author.

At this stage the activity carried out by the author is to make a hardware design planning based on Figure 1.

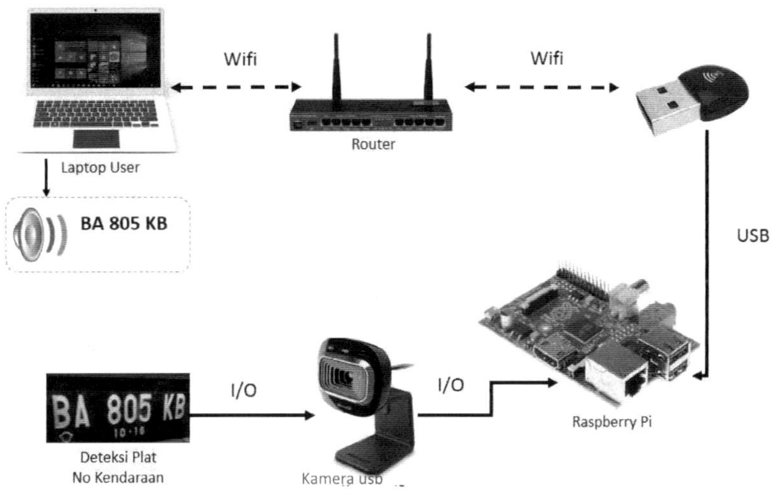

Figure 1. IP Camera Control Design For Vehicle Number Plate Identification

Based on pictures 1 and 2 the hardware used in this system is 1 laptop as a user and Raspberry Pi as a mini PC (server) equipped with a USB camera and USB Wi-Fi dongle, and a wifi extender and 1 unit of the Mikrotik Operating System router. As for the RPi software using the Motion EyeOS operating system and the Motion application for Surveillance Cam The vehicle license plate is captured by a USB camera on the RPi, so the IP function is on the RPi, while communication on the camera is via USB. The mini PC used is the Raspberry pi Model B rev 1. Communication between the RPi to the laptop unit (user) using Wireless LAN (WLAN) passes through the router, while to add wifi signal strength an additional router is needed[10]. Network topology in this system generally uses the star topology.

2.1 Hardware Design

In this research, the manufacture of hardware or a system that is built is divided into two. The first is the image processing system of the vehicle's output sound plate, and the second is the control of the IP camera using a router, where the target is to capture or capture images from the vehicle's license plate. Diagram of research block in general can be seen in Figure 2.

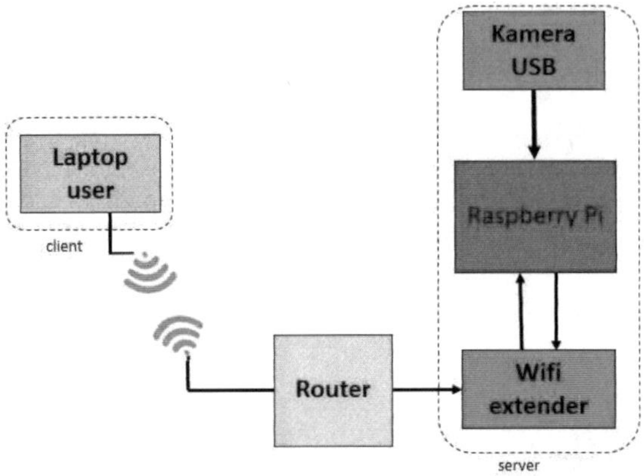

Figure 2. Raspberry Pi Camera Control Block Diagram for Identification of Vehicle Number Plates

The system in RPi where the camera captures images when there is detection of moving objects in general then the data is stored in RPi storage then the data is taken and sent to the user via Wi-Fi dongle that has connected the RPi with the router and PC-Client devices by the user[11]. The design of the hardware design is shown in Figure 1. There are 2 research systems, namely controlling the USB camera from the user to the server via the router shown in Figure 2, and processing the vehicle plate image signal using the pattern recognition method in Figure 4. The image signal from the monitoring results will be processed in the RPi and giving notifications to the user (supervisor). Image signal processing and sound output (text to speech) and character segmentation for identification of vehicle license plates is carried out on a laptop.

2.2 Software Design

At this stage the authors make the process of designing and designing software for data acquisition, retrieval of vehicle license plate image data, and retrieval of control data of the RPi server using a USB camera via a LAN router.

3. Result and Discussion

The test in this study is divided into two systems, namely the testing of a wireless-based control system and the testing of a vehicle's image signal processing. Control testing is to design and build a parking monitoring system in the garage using a Raspberry Pi server that can be controlled automatically. Successful testing of the system shows that the Raspberry Pi can be used as a control system or monitoring server. Testing of vehicle plate image signal processing using Matlab. Image signal processing includes image acquisition, preprocessing and identification of image to text plates using OCR. Voice signal output on this system uses text to speech.

The system will run when the Raspberry Pi server is turned on. Based on the process of taking input data, it was found that each device, namely Raspberry Pi, Router Mikrotik, Wifi Extender and PC Client each has an IPv4 class C address, where the Router Mikrotik as DHCP Server also has an IP network 192.168.88.1, while the device the other wirelessly connected LAN has a changing IP (DHCP Client) where is 192.168.88.xxx, where the xxx variable is automatically determined randomly and the range is determined by the DHCP Server system on the Mikrotik Router.

In taking Raspberry Pi-based camera data, Motion software is needed in the Motion EyeOs Operating System. Configuring the Motion in capturing images is done manually and time triggering the system every 5 seconds or taking video, the captured image is then stored temporarily in memory, then the data is transferred to a PC Client that has been configured via the File Transfer Protocol (FTP) protocol. To view the stream from motion, enter the address 192.168.0.102:8081 in the browser on the side of the client, then the stream will appear as shown in Figure 3.

(a) (b)

Figure 3. Display of Motion Identification Program Results
(a) Identification of "Car" Vehicle Plates with a Raspberry Pi Camera
(b) Identification of "Motor" Vehicle Plates with the Logitech C270 Webcam

Based on Figure 3, the video signal results are processed in Raspberry Pi, motion on the video is detected by the Motion program. Motion detects motion from the vehicle's plate image by monitoring the video signal from one or more cameras. To check the success of the motion program on the OS eye operating system that can only work on a web browser on a PC Client, the web browser uses Google Chrome to view streaming video displayed in real time thirty fps (frames per second) for testing the ability of framing on RPi and USB The camera (webcam) used.

The next process is to check the uploader of the recorded video when motion is detected. Image input for processing the vehicle plate image signal is taken or obtained data by downloading the image from the Raspberry Pi. Following the test of the vehicle plate image signal processing system can be seen in Figure 4.

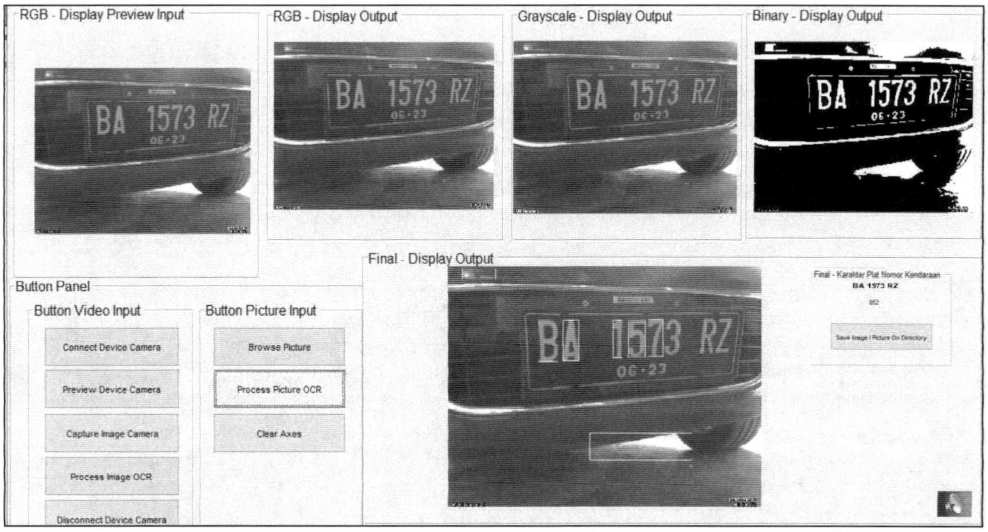

Figure 4. Image Processing Results with Optical Character Recognition Method for Vehicle Number Plate Identification

At the implementation stage of image processing, the vehicle plate image capture is taken from Raspberry Pi data. Then initialize OCR and image preprocessing. The results of the vehicle license plate image that appears through the input display and then cropped to cut only the number plate, can be seen in Figure 5. The process in the system, the first acquisition of the image that is the RGB image plate is converted into grayscale image. The two image preprocessing processes include Imadjust, convert to complement image, convert to binary. Here is also displayed the RGB histogram. Then the OCR process includes segmentation, normalization and correlating the image between the character of the vehicle number plate pattern with the template matching image. The OCR process will simultaneously change the image image into text or text. The final results of the study were identification of license plate characters and pronunciation of vehicle license plate characters. Voice output uses text to speech technology.

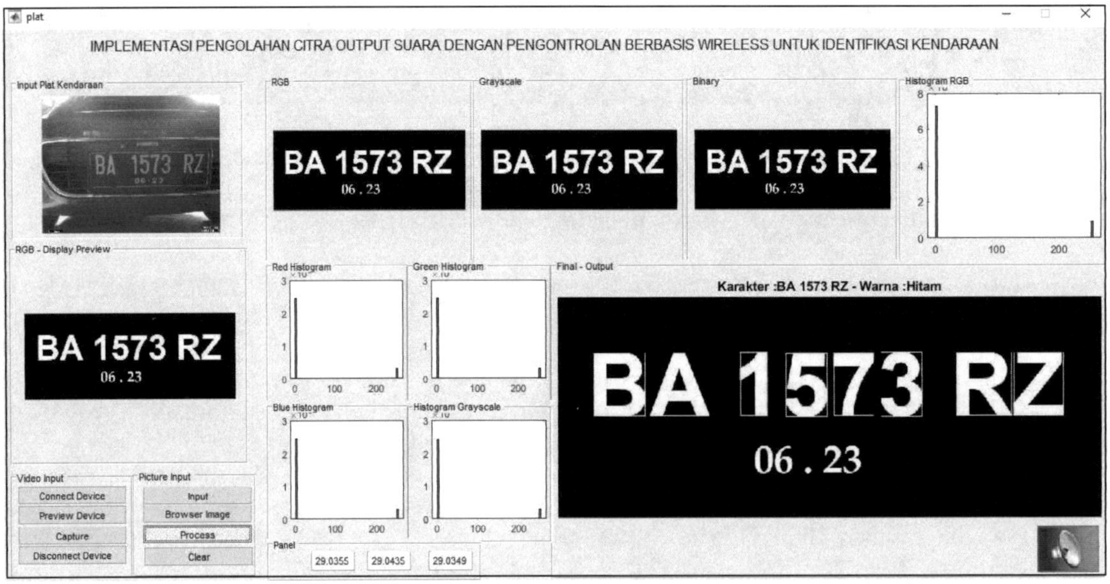

Figure 5. Image Output Results of Voice Output with Text to Speech Technology

Based on Figure 5, the results of the introduction of a car plate image with a black plate type of individual vehicle plates. The percentage of successful identification of vehicle plates based on Figure 5 is 100%. The number plate image in Figure 5 above has the input character pattern "BA 1573 RZ". Research using a wireless USB-based camera, image data obtained from the Raspberry Pi. Data is also collected in real time using the Logitech C270 HD webcam through the matlab program. The vehicle plate that is captured is the 2 wheeled vehicle plate. The real time test results with matlab can be seen in Figure 6 and table 1.

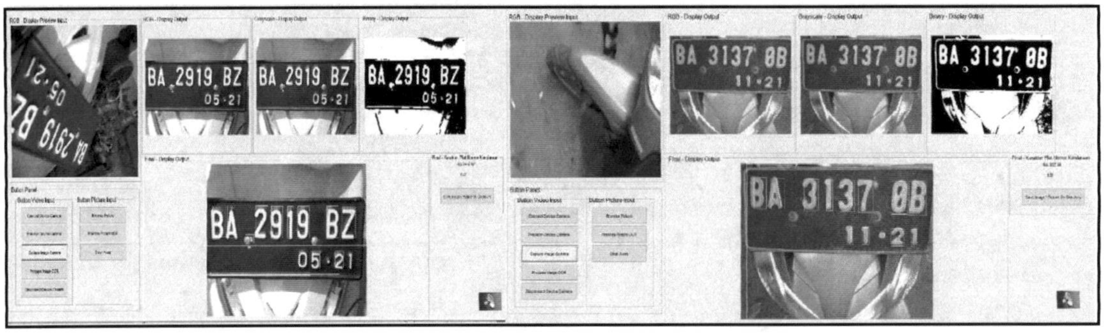

Figure 6. Real time results of Identification of Vehicle Plates with Optical Character Recognition

IOP Conf. Series: Materials Science and Engineering **846** (2020) 012010 doi:10.1088/1757-899X/846/1/012010

Table 1. Results of Identification of Vehicle Number Plates Using the OCR Method

Vehicle license plate	OCR Results	OCR accuracy (%)	Vehicle License Plate Capture Conditions
	Final - Karakter Plat Nomor Kendaraan BA 2919 BZ	100	morning
	Final - Karakter Plat Nomor Kendaraan BA 4020	75	afternoon
	Final - Karakter Plat Nomor Kendaraan BA 3l37 08	87.5	evening
	Final - Karakter Plat Nomor Kendaraan BA 29185	62.5	night
	Final - Karakter Plat Nomor Kendaraan BA 2700 LC	100	morning
	Final - Karakter Plat Nomor Kendaraan BA 2919 BZ	100	afternoon
	Final - Karakter Plat Nomor Kendaraan BA 6975 BZ	100	evening
	Final - Karakter Plat Nomor Kendaraan BA 2599 06	87.5	night (with lamp light)
	Final - Karakter Plat Nomor Kendaraan BA 2919 BZ	100	night (with lamp light)

The test results through the Matlab R2017b GUI, the Logitech C270 HD webcam is configured as an input media with a resolution of 1280 pixel x 720 pixel fixed lens. The lens on the webcame does not have auto focus. Vehicle image processing with RGB color images and vehicle identification testing based on light intensity. Good intensity of light taking conditions in the morning with 100% accuracy. At night a good vehicle identification result with the help of lights. Based on table 1, the average test result of vehicle identification from 9 tests with OCR in realtime is 90.278%. While the form of implementation of testing the vehicle license plate identification output Raspberry Pi server can be seen in Figure 7.

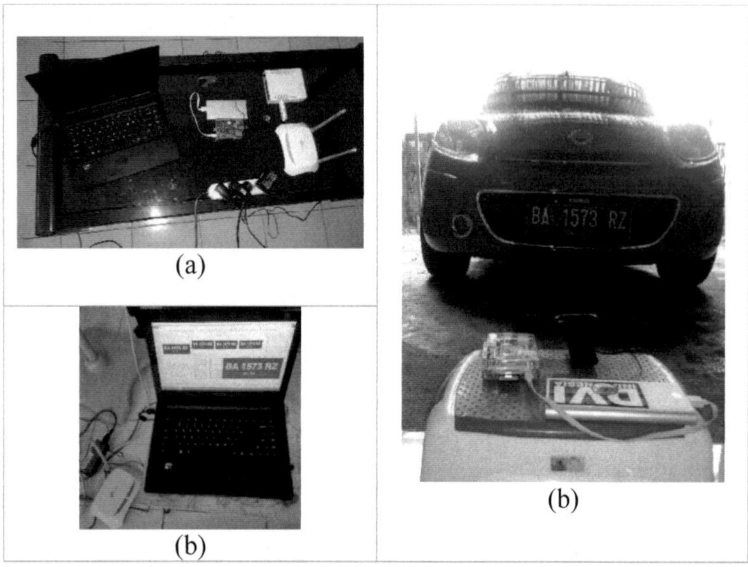

Figure 7. Results of Image Processing Implementation of Voice Output Identification of Vehicle Number Plates
(a) Realization of Equipment, (b) System Testing on PC Users,
(c) Vehicle Number Plate Capture Device using the Raspberry Pi server

4. Conclusion

The results of this study can be concluded that, firstly the system successfully communicates between PC-Client and Raspberry Pi Server which is integrated through a Network Server (Router), secondly has successfully implemented digital image processing using OCR technology to recognize vehicle license plates in a parking system in a home garage and the results of the algorithm are applied with text to speech technology.

References

[1] D. Ashardi, "Rancang Bangun Aplikasi Pemantau Ruangan Melalui Kamera IP Menggunakan Platform Android (Studi Kasus: Laboratorium Teknik Informatika Universitas TanjungPura)."

[2] R. F. Giant, "Perancangan Aplikasi Pemantau Dan Pengendali Piranti Elektronik Pada Ruangan Berbasis Web," vol. 17, no. 2, pp. 70–75, 2015.

[3] K. Kusumawati and D. W. Cahyadi, "Penerapan Teknologi Optical Character Recognition Untuk Mendeteksi Plat Nomor Kendaraan," *Pros. Semin. Nas. Inov. Teknol.*, pp. 12–20, 2017.

[4] B. Yuwono, S. P. Nugroho, P. Studi, and T. Informatika, "Pengembangan Model Public Monitoring System Menggunakan Raspberry Pi," vol. 12, no. 02, pp. 123–133, 2015.

[5] G. Udayana and I. Darmawiguna, "Pengembangan Prototipe Portal Otomatis Dengan Pendeteksian Plat Nomor Kendaraan Berbasis Raspberry Pi," *Karmapati*, vol. 5, no. 2, 2016.

[6] H. M. Shadiq, S. Sudjadi, and D. Darjat, "Perancangan Kamera Pemantau Nirkabel Menggunakan Raspberry Pi Model B," *Transient*, vol. 3, no. 4, pp. 1–6, 2014.

[7] F. A. S. Rusdi Efendi, Endina Putri Purwandari, "Aplikasi Pembacaan Plat Nomor Kendaraan Menggunakan Optical Character Recognition (OCR)," 2017.

[8] A. Solichin and Z. Rahman, "Identifikasi Plat Nomor Kendaraan Berbasis Mobile dengan Metode Learning Vector Quantization," vol. 3, no. 3, pp. 1–7, 2015.

[9] D. P. Sari and S. Rasyad, "Identifikasi Huruf Braille Berbasis Image Processing Secara Real Time," no. November, pp. 1–2, 2017.

[10] E. N. Wijatsongko, A. E. Putra, and B. N. Prastowo, "Sistem Pemantauan Ruangan Dengan Server Raspberry Pi," *IJEIS (Indonesian J. Electron. Instrum. Syst.*, vol. 5, no. 1, p. 65, 2015.
[11] F. R. Hadiputra and T. K. Priyambodo, "Purwarupa Pengendalian Jarak Jauh Pada Mobile Robot Berbasis Web Melalui Jaringan Wireless TCP/IP," *IJEIS (Indonesian J. Electron. Instrum. Syst.*, vol. 6, no. 1, p. 105, 2016.

IOP Conf. Series: Materials Science and Engineering **846** (2020) 012011 doi:10.1088/1757-899X/846/1/012011

The Implementation of IoT (Internet of Things) for Controlling Cow Health

Cipto Prabowo[1] Rafki Mauliadi[2], Afdal[3] , Johan[4], Zurnawita[5,] Alde Alanda[6], Fazrol Rozi[7]
[1,2,,3,6,7]Computer Engineering, Information Technology, State Polytechnic of Padang
[4]Informatics Management, Information Technology, State Polytechnic of Padang
[4]Telecomunication Engineering,, State Polytechnic of Padang

[1]cipto@pnp.ac.id,[2]rafkimauliadi63@gmail.com,[3]afdalmtk@gmail.com,
[4]johandocument2107@gmail.com, [5]zurnawita@pnp.ac.id [6]alde@pnp.ac.id, [7]fazrol@pnp.ac.id

Abstract .Livestock health management is a major factor in the sustainability of cattle farms. However, health care for livestock by veterinarian costs lots of time and money. The purpose of this study was to develop an android IoT device for cow physiological health management through detection of cow temperature, heart rate and breathing rhythm. This device is built based on a microcontroller. By using the concept of IoT to send data, then the information generated on the device will be displayed directly to the android application. Three adult cows were measured using android IoT device compared to conventional measurement methods as controls. This procedure was carried out for morning (06:30-09:30 am), afternoon (11:00-01:00 pm) and evening (04:30-05:30 pm). Independent sample t-test was used to determine whether there were significant differences between Android IoT device and conventional measurement method. The results show that there is no significant difference between the two methods for temperature ($t(58) = -0.191$, $p = 0.849$) and heart rate ($t(58) = -0.459$, $p = 0.648$). Utilization of device makes it easy for cattle farmers to monitor cow health regularly at economical prices and provide remote information. Furthermore, this device needs to be tested on types of ruminants such as goats, sheep and buffaloes.

1. Introduction

Cattle farm in Indonesia is one of renewable natural resources developed for improving the dynamics of adequate economic growth productivity. The cultivation of cattle farm has largely developed in all over regions of Indonesia and it will keep developing because it has some superiorities such as reproduction efficiency, tested data of environmental adaptation, easy to be nurtured for various needs and has an excellent quality of meat and milk.

One of the ways to maintain animals' health is through checking livestock body and physiological condition. Physiological condition used to know the indication of healthy livestocks are body temperature, heartbeat frequency, and respiratory rhythm. This becomes the reference for veterinarian or livestock health staff if abnormality occurs among three of them.

A lot of research about physiological health of cow using microcontroller had been done, however still not interactive with the cattle farmer because it was still website based [[1][2][3].

Based on that problem, cattle farmers need high cost for maintaining livestock health. Therefore, ISEP-I is needed as the tool for conducting control and check on livestock health efficiently.

2. Methodology

2.1. System Analysis

Health problem on cows can possibly occur such as because of not knowing how to check cows' health condition. The success of livestock business is extremely determined by the livestock health. The selling price of healthy cows is certainly more expensive than unhealthy cows.

If seen from the method of checking cows' physiological health done at animal husbandry in Indonesia, still using traditional method, then by designing this system at the tool of temperature check and heartbeat by using temperature sensor and voice sensor will ease the farmers in conducting check on cows' physiological health without doing manual check and can monitor cows' health control and also the cows' health history periodically.

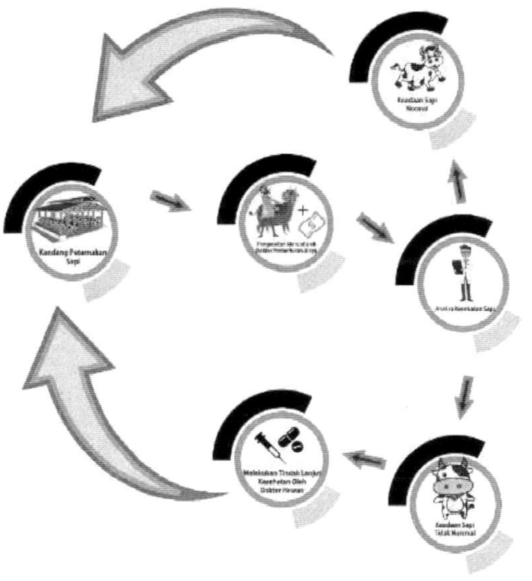

Figure 1. The General Procedure of Cows' Physiological Health Check

2.2. System Design

The system made is in the form of tool and application. The tool is strung of some electronics modules NodeMCU, temperature sensor, voice sensor, battery or power bank, and other electronics components. In the application would be made in the form of data information display sent by the tool. The sotfware used for making the system was arduino IDE using C language, and the wed design PHP based for handling server, CSS, HTML, and bootstrap for interface display from the website to be responsive. Then the information is displayed into android application.[4][5][6][7] [8][9][10]

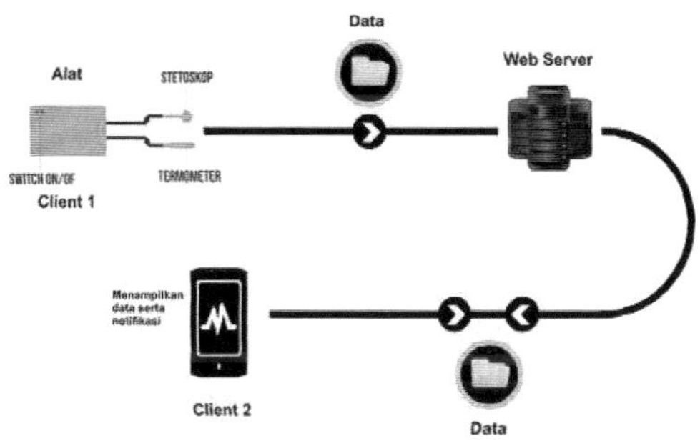

Figure 2. The System Design

Figure 2 shows that the system has two sides; client and server. Client as the user that uses the tool and application. The tool will give cows' health information such as temperature, voice, and number of heartbeats obtained from the combination of some sensors then sent to web server through connection, then the user can see the results of information in android application.

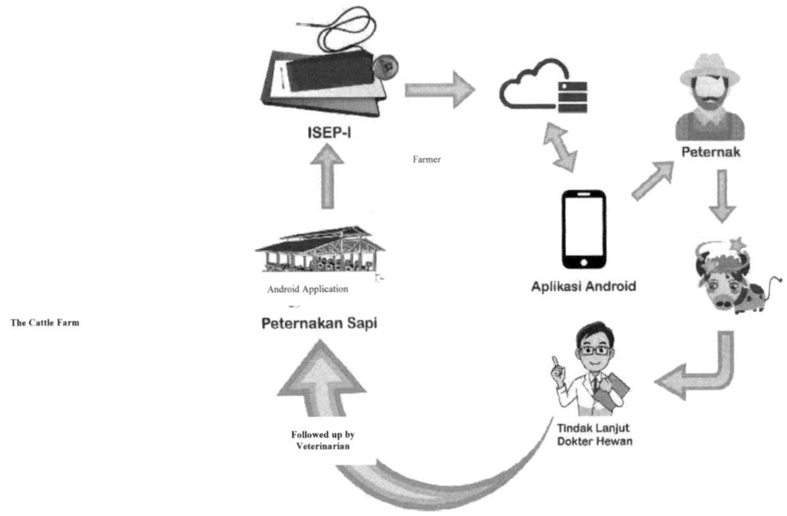

Figure 3. The Cycle of Health Check Using ISEP-I

Figure 3 is the cycle of checking cows' physiological health using ISEP-I. In the usage of this ISEP-I, first turn it on through switch on/off button, then turn on the hotspot from the smartphone with determined username and password. Data will be sent to the server when the tool is connected to the hotspot and also has internet connection. Data sent by the tool can be seen in android application also needs internet connection.

2.3. Test and Data Analysis

ISEP-I sends data in the form of temperature and hearth rhythm every 0,02 second to the server, data for the hearth rhythm will becalculated to be heartbeat per minute (bpm). These data then accepted by the server to be processed and saved by the server. After data are processed by the server, then they are forwarded to the client in the form of android application to monitor the process and result of check. The system workflow can be seen in Figure 4.

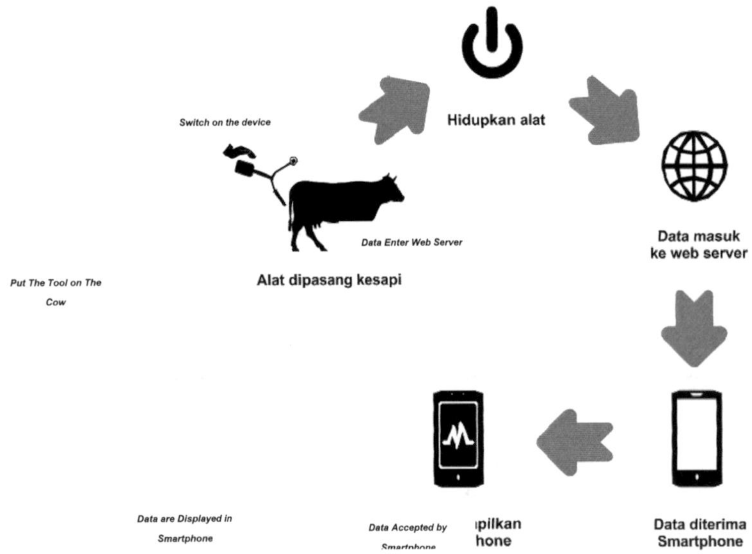

Figure 4. System Workflow

The test is done towards 5 adult cows with conventional measurement method. The measurement procedure is done three different time with morning measurement (06:30-09:30 am), afternoon (11:00-01:00 pm), and evening (04:30-05:30 pm). The test process is started with manual test in which for testing the temperature is done by using thermometer for the cow, then continued with testing using ISEP-I. After that, followed by the heartbeat test at the base of cow tail manually using stethoscope, continued with the examination using ISEP-I to compare the result obtained manually and using ISEP-I tool.

3. Results and Discussion

This part will show the test results of the system had been made. The test is conducted in two stages. First, the temperature sensor test with ESP8266 Wi-Fi module. Second, the heartbeat check test.

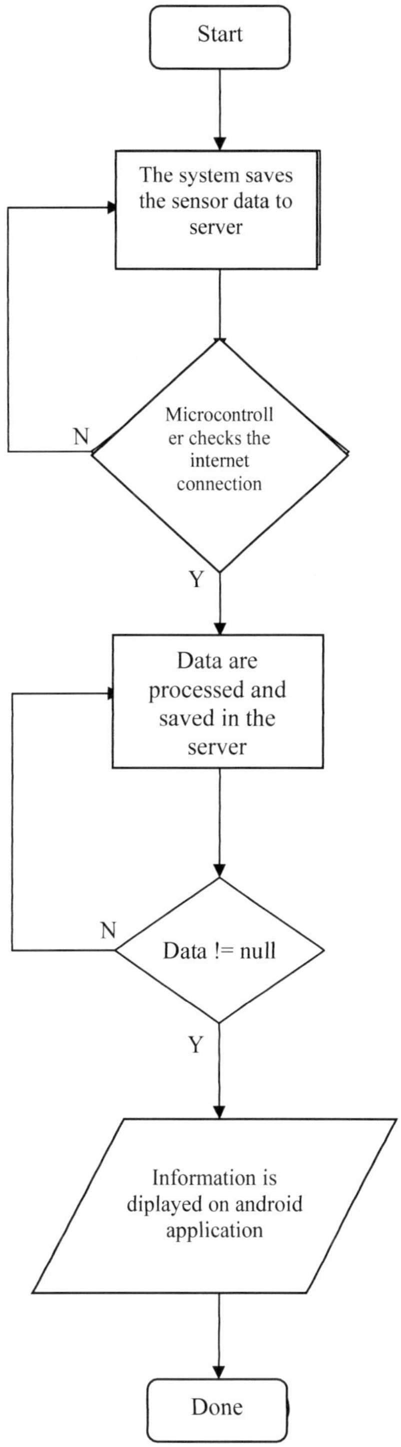

Figure 5. Flowchart of ISEP-I Tool

3.1. The Test of Temperature Sensor

This test is conducted to see if the temperature sensor can obtain the temperature data and save it in server database and then display it on android application for the farmer. Figure 6 shows the prototype of the tool being made, the temperature sensor is put into the cows' butt, the temperature value is displayed on the application, then compared to cows' thermometer tool. Look at figure 6. Cows' thermometer is the tool used to measure the cows' temperature. The way to test and measure the temperature is by putting the temperature sensor into the cows' butt. The test results can be seen in figure 7.

Figure 6. The test of temperature sensor on the cow

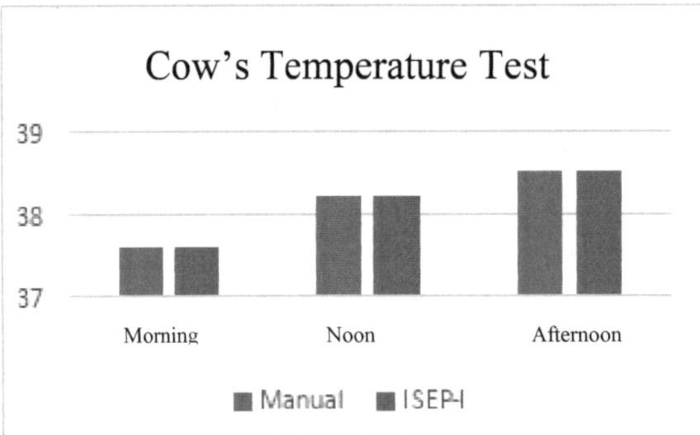

Figure 7. The Result of Morning Test and Measurement on The Temperature Sensor

Based on statistics test independent sample t-test, it obtains that there is no significant difference between manual check (38.06 +- 0.47) and ISEP-I (38.08 +_0.47) (t(58) = -0.191, p = 0.849). The temperature sensor can measure and detect the cows' temperature sent to database through internet connection and displayed on android application with average error 0.002%. The temperature sensor data after client connection process is done then given delay for 500 ms meant so that the sensor can read the data in realtime and then data reading will be started from the temperature sensor. Module ESP8266 works together to process the data after sensor data obtained then sent to database. For saving the temperature data of cows then the new table is made namely "temperature". In order the data can be accessed by the farmers, they must have the account to log into andriod, such as seen in Figure 8.

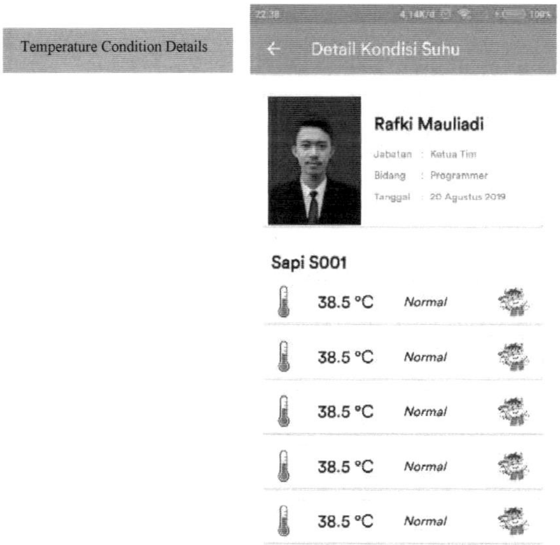

Figure 8. The Diplay of Realtime Temperature Check in Android Application

3.2. The sensor test of heartbeat

The test of heartbeat sensor is done just like temperature measurement, but the testing site is at the base of cows' tail. Figure 9 shows the prototype of the tool being made, heartbeat sensor is put at the base of cows' tail, the score of heartbeat is displayed on android application in the form of BPM then compared to the manual check using stethoscope. The way of testing and measuring the cows' heartbeat using stethoscope is by putting the stethoscope at the base of cows' tail. The test results can be seen in figure 10.

Figure 9. The test of heartbeat sensor on a cow

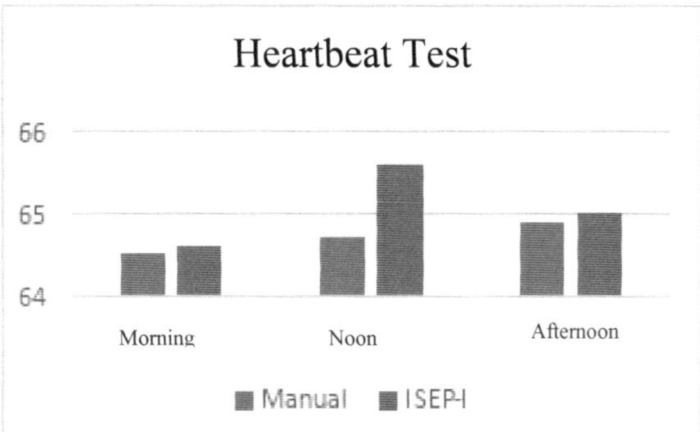

Figure 10. The Result of Morning Test and Measurement of Heartbeat Sensor

In statistics test independent sample t-test, obtained that there is no significant difference between manual check (64.70 +- 2.86) and ISEP-I (65.06 + 3.31) (t(58) = -0.459, p = 0.648). Heartbeat sensor can measure and detect cows' heartbeat which is sent to database through internet connection and displayed in android application with average error 0,05%. Data of heartbeat sensor after client connection process then given delay for 500 ms which is meant to let sensor read data in realtime and then data and the reading of heartbeat sensor data will be started. Module ESP8266 works simultaneously to process the data after data sensor obtained then the data are sent to database. To save the data of cows' heartbeat data then the new table is made namely "heartbeat". In order the data can be accessed by the farmers, they need to have the account to log into android application as seen in figure 11.

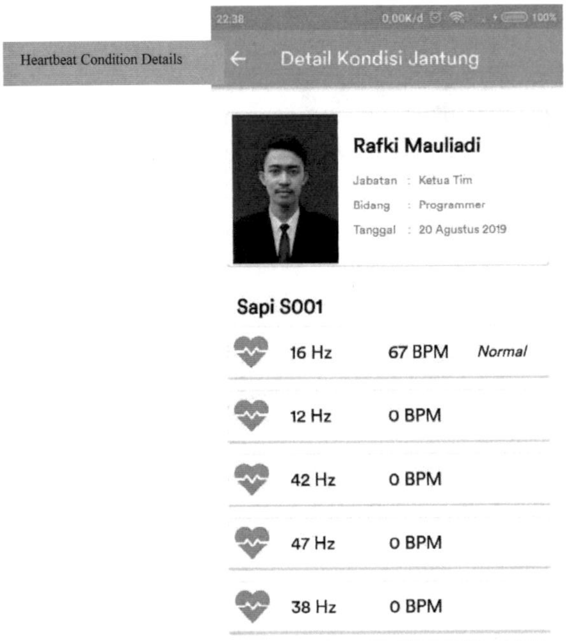

Figure 11. The display of realtime heartbeat check on android application

4. Conclusion

In this research, we propose IoT monitoring system of cows' temperature and heartbeat with the concept of Internet of Things (IoT) to save the data to database. There is efficiency in the check of cows' physiological health android technology based to ease and decrease the cost of livestock check. The farmers can see the information of temperature and heartbeat of cows directly in their android application. Beside that, there is also system of saving data to the server the results of each check. The future research will develop the system which be connected directly to android application without using internet connection.

References

[1] M. E. E. Alahi, S. C. Mukhopadhyay, and L. Burkitt, "Imprinted polymer coated impedimetric nitrate sensor for real- time water quality monitoring," *Sensors Actuators, B Chem.*, vol. 259, pp. 753–761, 2018.

[2] N. Rao and S. Tanweer, "Fast Pattern Discovery in Healthcare Data Using Graphics Processors," *J. Drug Deliv. Ther.*, vol. 9, no. 1-s, pp. 358–360, 2019.

[3] G. Vijay Kumar, A. Bharadwaja, and N. Nikhil Sai, "Temperature and heart beat monitoring system using IOT," *Proc. - Int. Conf. Trends Electron. Informatics, ICEI 2017*, vol. 2018-January, no. i, pp. 692–695, 2018.

[4] M. I. Mahali, "Smart Door Locks Based on Internet of Things Concept with mobile Backend as a Service," vol. 1, no. 3, pp. 171–181, 2016.

[5] A. Rayes and S. Salam, *Internet of Things From Hype to Reality*. 2019.

[6] G. Kortuem, A. K. Bandara, N. Smith, M. Richards, and M. Petre, "Educating the Internet- of-Things Generation," no. February, pp. 53–61, 2013.

[7] K. Bryceson, A. Navas Borrero, and K. Gunasekera, "Internet of Things (Iot) - Smart Agriculture Education At the University of Queensland," *EDULEARN16 Proc.*, vol. 1, pp. 8036–8044, 2016.

[8] J. G.-V. J.M. Barcelo-Ordinas, J.P. Chanet, K.-M. Hou, "A survey of wireless sensor technologies applied to precision agriculture," *ECPA Conf.*, vol. 9, pp. 7–11, 2013.

[9] L. Karim, A. Anpalagan, N. Nasser, and J. Almhana, "Sensor-based M2M Agriculture Monitoring Systems for Developing Countries: State and Challenges," *Netw. Protoc. Algorithms*, vol. 5, no. 3, p. 68, 2013.

[10] A. Kamilaris, F. Gao, F. X. Prenafeta-Boldu, and M. I. Ali, "Agri-IoT: A semantic framework for Internet of Things-enabled smart farming applications," *2016 IEEE 3rd World Forum Internet Things, WF-IoT 2016*, pp. 442–447, 2017.

Measurement of 2.4 GHz Access Point Antenna

Firdaus [1], Afida Nurul Fatimah[1], Yulindon[1], Ratna Dewi[1]

[1]Politeknik Negeri Padang

*firdaus@pnp.ac.id

Abstract. Antenna research on various wireless system applications has been carried out by many R & D telecommunications equipment vendors and academic. In academic researchs, antenna design are not tested on devices but are only based on the requirements and weaknesses of previous studies. In this study we developed an antenna for wireless LAN by first testing a conventional wireless antenna. based on knowledge of conventional antennas then we did a new antenna design. The antenna we tested was the antenna used on the TP LINK WA701ND access point device. In this paper we report the measurement results of the antenna. Measurements were made using antenna measuring devices at the State Polytechnic of Padang to obtain important antenna parameters including radiation pattern, polarization, and frequency response. The measurement results obtained the antenna is omnidirectional, linear polarization and 2.34 - 2.74 GHz frequency band with a center frequency of 2.43 GHz. The minimum return loss is -20.3 dB and the SWR is 1.14.

1. Introduction

Wireless communication systems have become a necessity in a daily life. Not only conventional communication devices, all equipment used in activities, at home, offices, schools are also connected to the internet by wireless. It known as the Internet of Things (IoT)[1]. Communication devices currently use devices manufactured by vendors. To become independent, Padang State Polytechnic began to develop its own communication devices through the study centers. One of them is the Antenna study center which investigates and develops antennas for various telecommunication devices. In the last few years we have been doing antenna research as in [2]–[4] for ultra wideband applications. Because test equipment for ultra wideband systems is not sufficient for antenna testing on ultra wideband devices, the Study Center has begun to develop antennas for WLAN devices. Recent research on WLAN is as in [5]–[11] which refers to standards IEEE 802.11 a/b/g [12]. Substrate material is more popularly used because of the low price and strong material. The antenna is designed in the form of a folded strip/slot [5],cross dipole[6] modified J–type antena[7], combination of two monopole antenna[8], microstrip–based slot antena[9], DRA–based dual frequency[10] and inverted–F antenna [11]. But these antennas are not tested on WLAN devices or compared to conventional WLAN antennas. WLAN antennas is compared with conventional antennas. In this paper we tested the characteristics of antennas used on WLAN access point devices TP-LINK WA-701 ND to find out the important parameters of antennas. The test was carried out using an anechoic chamber antenna measurement supported by a Keysight E5071C Vector Network Analyzer and a computer for controlling measurements. Then the results of this measurement are used as a comparison of WLAN antenna design.

Methodology

TP-Link access point devices are very widely used by users. One of them is TP-LINK WA-701 ND [13]. The antenna of this device is a type of monopole that has a loading coil as shown in the figure 1. Based on the specifications in the device manual in [13] it is omnidirectional and 3 dBi gain. Measurement of frequency response, radiation pattern and polarization plots are not shown in the manual. We conducted tests on the antenna as seen in figure 1 by using the anechoic chamber that available at the Politeknik Negeri Padang. An anechoic chamber and supporting devices are shown in Figure 2. The equipment has 4 units consisting of an anechoic chamber with RF absorber designed for measuring antennas or wireless devices with a frequency range of 700 MHz - 6 GHz, a Vector Network Analyzer Keysight E5071C, a set of control devices (consisting of computers, switches and controller turntable) and a software to operate all equipments.

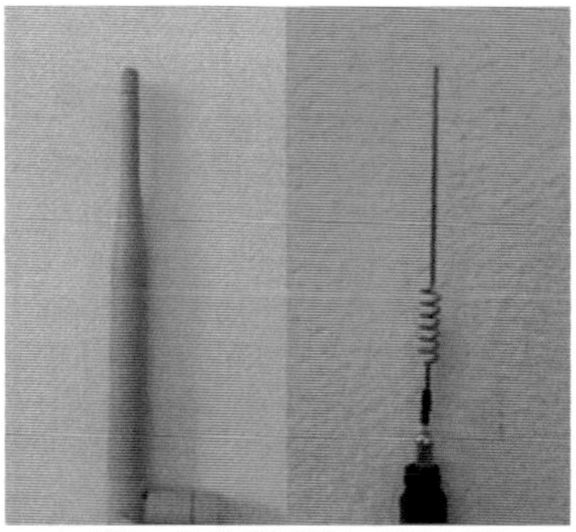

Figure 1. TP – LINK WA 701 ND antenna

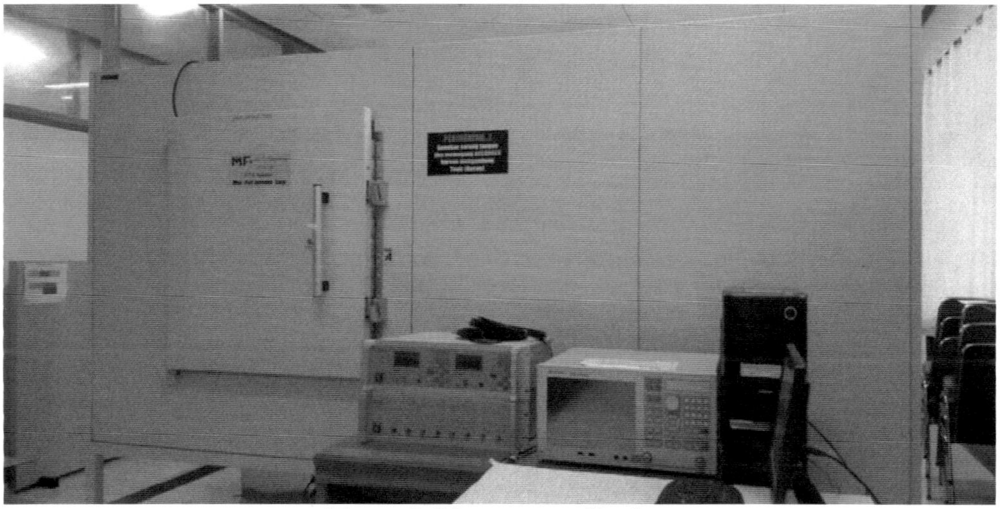

Figure 2. Measurement Devices

2. Result and Discussion

The 3D radiation pattern testing is shown in Figure 3(a) and the 2D is in Figure 3(b). The shape of the radiation pattern obtained is omnidirectional. In figure 3 the maximum power is shown by orange and the minimum is by green colour. The antenna radiation is omnidirectionally distributed in XY plane. Radiation pattern of antenna in XY plane has a circle form emitting the same power in all directions is shown by red line. Meanwhile a small portion of power is emmitted to +z dan –z direction in x – z dan y –z plane is shown by blue line.

Figure 4 shows the frequency response of the antenna. The optimum frequency is working at 2.43 GHz with range from 2.35 GHz until 2.74 GHz. The minimum return loss is -20.53 dB as shown in Figure 4(a). Bandwidth of the antenna at 390 MHz and minimum SWR is 1.13 in frequency at 2.43 GHz as shown in figure 4(b). The measure of the polarization is seen in **Figure** 5. These measurments used the procedure in [14]. Those figures shows that the polarization is linear.

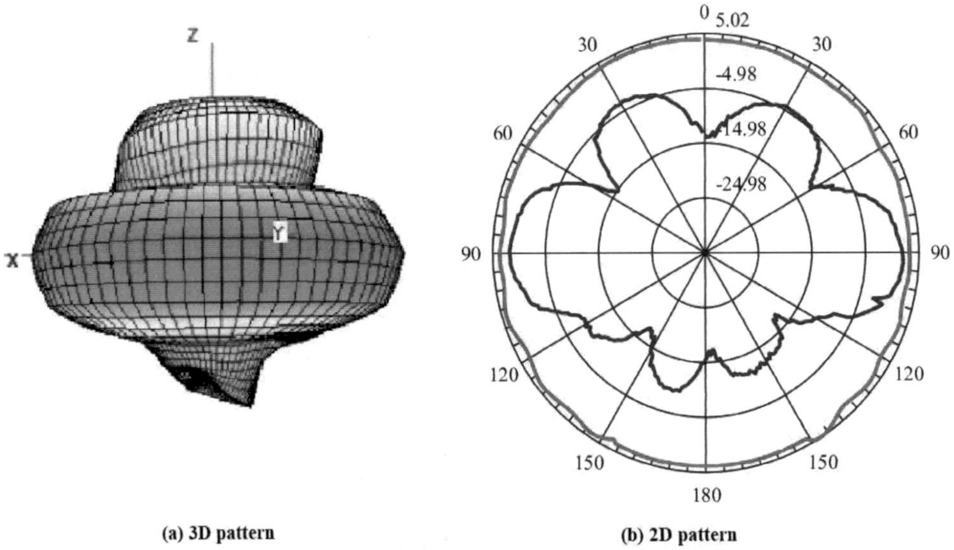

Figure 3. Antenna pattern. (a) 3D pattern. (b) 2D Pattern (x – y plane = red, x – z pattern = blue)

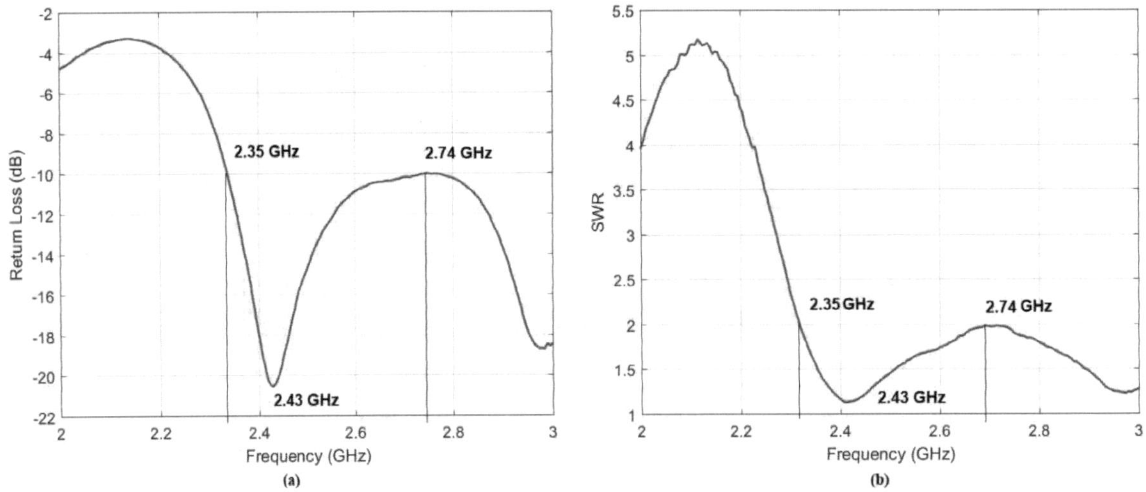

Figure 4. Frequency response. (a) Return Loss, (b) SWR

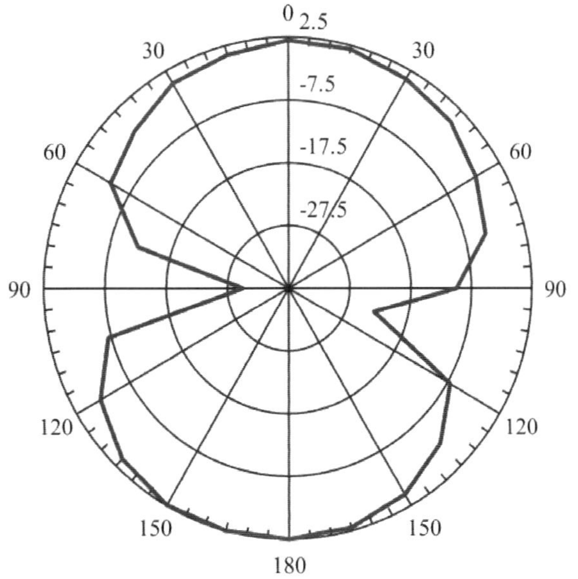

Figure 5. Polarisation measurement of the antenna.

3. Conclusion

Measurement of the antenna TP LINK WD 701 ND has done. Radiation pattern achieved in x – y plane is omnidirectional. Meanwhile in x – z plane is bidirectional that has small portion of the power distributed in +z an –z directions. The optimum frequency works at 2.43GHz. Bandwidth of the antenna is 360 MHz and the SWR is 1.13.

References

[1] A. N. B. Zanella, "Iternet of Things for Smart City," *IEEE Internet Things J.*, vol. 132, no. 2, pp. 265–268, 2014.

[2] U. N. F. Hendri and M. Silvana, "Compact UWB Band Notched with C-Slot Antenna," *Ico-Ascnitech*, no. October, pp. 13–15, 2017.

[3] Firdaus, Yulindon, M. Silvana, Rusfandi, R. Fernandez, and H. Adre, "Study on the Design of UWB Antenna for Omnidirectional Pattem," no. c, pp. 78–81, 2019.

[4] Firdaus, Yulindon, M. Silvana, Asrial, R. Efendi, and N. A. Samsudin, "Optimization of Compact UWB Monopole Antenna with Tapered Connection and Linier Tapered Transformer," *IEEE 2018 Int. Conf. Appl. Inf. Technol. Innov.*, pp. 166–169, 2019.

[5] Y. Dong, J. Choi, and T. Itoh, "Folded Strip/Slot Antenna With Extended Bandwidth for WLAN Application," *IEEE Antennas Wirel. Propag. Lett.*, vol. 16, no. c, pp. 673–676, 2017.

[6] Y. Zhang, Y. Zhang, D. Li, K. Liu, and Y. Fan, "Dual-polarized band-notched antenna without extra circuit for 2.4/5 GHz WLAN Applications," *IEEE Access*, vol. 7, pp. 84890–84896, 2019.

[7] L. Sun and Y. Li, "Slim dual-polarised antenna with omnidirectional patterns for WLAN applications," vol. 55, no. 8, pp. 7–8, 2019.

[8] S. Su, S. Member, C. Lee, and S. Chen, "System for WLAN Notebook Computers," *IEEE Antennas Wirel. Propag. Lett.*, vol. PP, no. c, p. 1, 2018.

[9] S. Nandi, S. Member, and A. Mohan, "A Compact Dual-Band MIMO Slot Antenna for WLAN Applications," vol. 1, no. c, 2017.

[10] M. S. Sharawi, S. K. Podilchak, M. U. Khan, and Y. M. Antar, "Dual-frequency DRA-based MIMO antenna system for wireless access points," 2017.

[11] W. Liao, S. Member, C. Hsieh, B. Dai, and B. Hsiao, "Inverted-F / Slot Integrated Dual-Band Four-Antenna System for WLAN Access Points," vol. 1225, no. c, pp. 11–14, 2014.

[12] IEEE, "IEEE 802.11n Standard for Information Technology-Telecommunikactions and information exchange between systems-Local and metropolitan area networks-Specific

requerements," 2009.
[13] TP-LINK, *TD-W8951ND 150Mbps Wireless N ADSL2 + Modem Router User Guide*. TP-LINK, 2015.
[14] C. A. Balanis, "Antenna Theory." A John Wiley & Son, 2005.
[15] P. Langley and S. Sage, "Induction of Selective Bayesian Classifiers," *Proc. Tenth Int. Conf. Uncertain. Artif. Intell.*, no. 1990, pp. 399–406, 1994.
[16] N. Friedman, D. Geiger, M. Goldszmidt, G. Provan, P. Langley, and P. Smyth, "Bayesian Network Classifiers *," *Mach. Learn.*, vol. 29, pp. 131–163, 1997.
[17] J. Wu, S. Pan, X. Zhu, Z. Cai, P. Zhang, and C. Zhang, "Self-adaptive attribute weighting for Naive Bayes classification," *Expert Syst. Appl.*, vol. 42, no. 3, pp. 1487–1502, 2015.

IOP Conf. Series: Materials Science and Engineering **846** (2020) 012013 doi:10.1088/1757-899X/846/1/012013

Study on Impedance Matching of 2.4 GHz Dipole Antenna

Firdaus[1], Larra Monica[1], Yustini[1]

[1]Electrical Department, Politeknik Negeri Padang

*firdaus@pnp.ac.id

Abstract. Understanding on matching impedance between feedline and load is important in many aplication. Theoritically, matching impedance can be calculated easily if charateristic impedances both load and line is known. Practically, when we do impedance matching at certain frequency we face a problem if the impedance of the feedline and antenna are unknown. by using vector network analyzer and anechoic chamber we tried to study and to approach the impedance match through measurement of return loss, VSWR, and antenna radiation pattern. The results of studies have shown that adjusting dipole antennas to lines can be done by optimizing antenna and lines length and subsequently optimizing the use of balun. In this reasearch we optimized 2.4 GHz dipole antenna and matched it to feedline by using bazooka and balun 1 : 1. To achieve good matching, the antenna length is 0.35 λ and the feedline length multiplies 0.5 λ. The length of the bazooka and balun for the best impedance adjustment are 0.25 λ and 0.2λ.

1. Introduction

The charateristic of impedance feedline and load are very sensitive to frequency. Furthermore, the impedance matching technique needs to be performed accurately in various applications such as antenna design. Theoritically, fundamental design of the antenna can be studied on [1], [2]. On High Frequency (HF) band antenna design and impedance matching can be practically studied on [3] But for higher frequencies base on our knowledge there are no practical references available. In this study we focus on design and matching line and dipole antenna. Some literature about dipole antena can be found as in [4]–[7]. A calculable dipole antenna for frequencies 1-3 GHz using the moment method with the hybrid balun derived in [4] but no empirical testing of the antenna. Optimization of printed dipole antennas with integrated balun is done in [5] by using an algorithm that is able to optimize antenna geometry with a faster time compared to conventional parametric studies. The integrated dual band balun antenna at the 2.4 GHz frequency was designed at [6] but produced a complex antenna design making it more difficult in application. The various crossed dipole applications discussed in [7] are very readable to increase understanding of dipole antenna applications and matching networks. In this paper we conduct a design study and match of the dipole antenna impedance at higher frequencies with practical testing which was not done in the previous literature. In the early part of the study we tested 500 MHz standard dipole antenna to study the concept of impedance adjustment and then tried to design dipole antennas for higher frequencies. In this paper we report the results of a study of dipole antenna design at a frequency of 2.4 GHz. The choice of 2.4 GHz frequency because standard dipole antennas for 2.4 GHz or higher does not available on the laboratorium telecommunications. The results of this study can be used as a reference for designing standard dipole antennas for higher frequencies.

2. Methodology

Dipole antenna is a wire antenna that consists of two conductor rods with a half-wavelength long that is feeded at center. Dipole antenna has omnidirectional patern as shown at **Figure** 1[1]. Matematically pattern at **Figure** 1 can be written as $U = sin\ (\theta)$.

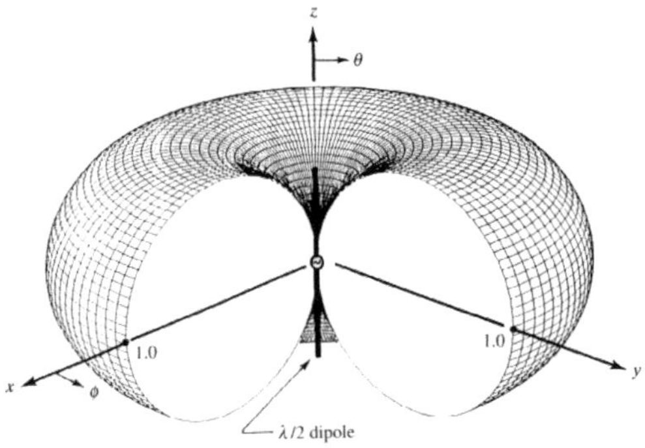

Figure 1. Half wavelength Dipole and pattern[1]

This omnidirectional pattern is formed by the current distribution for a halfwave dipole antenna that maximum at the center tap and minimum at both ends. For balance lines that are flared at both ends, current distribution will be generated as in Figure 2. Standing waves do not eliminate each other on the wire because a current distribution is also formed on the part of the flare (dipole) so that the current flows from one end to the other as in Figure 2(a). Dipole antennas with half wave length the current distribution is as shown in Figure 2(b). maximum current is distributed at the center tap and minimum at both ends. So that the radiation pattern of the signal radiated by the antenna is as in Figure 1.

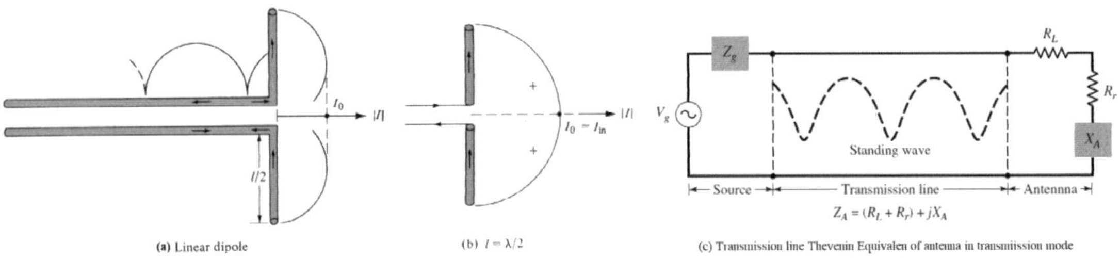

Figure 2. A dipole contruction and it equivalen circuit[1]

The problem in dipole antenna design and antennas in general is to match the antenna impedance with the feedline. Antenna is usually fed with coaxial cable which is unbalanced line while dipole antenna is balance line. For this reason, it is necessary to match the unbalance line with the dipole. There are several ways that we can do to adjust the dipole antenna with the feed line. In this study two methods will be used, namely bazooka balun (**Figure 2**a) and λ / 4 coaxial balun (**Figure 2**b.). Before we design the antenna at the 2.4 GHz frequency, we tested the standard dipole antenna that available in the Antenna and Propagation laboratory of the Padang State Polytechnic Department. The antenna is 500 MHz dipole that has two handle without and with balun. We measured antenna pattern, VSWR and return loss of the antenna for this two balun. By conduct this measurement we can have best knowledge of matching impedance of dipole antenna. Furthermore, we created a 2.4 GHz dipole antenna with length of ½ λ by using conductor with diameter is 1 mm and fed it using a standard 2.4 GHz coaxial cable. The dipole antenna is shown in **Figure 5.**

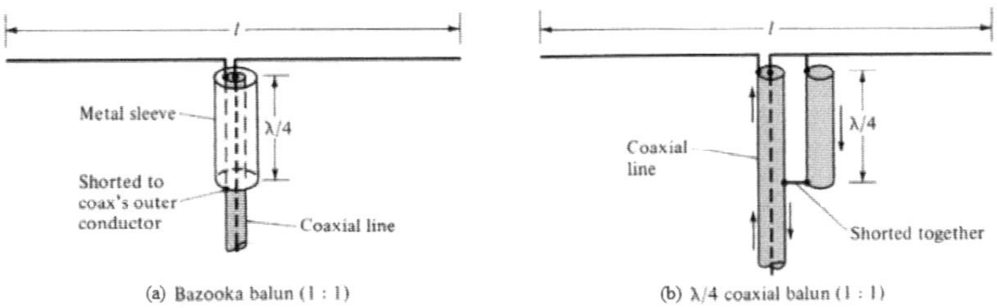

Figure 3. Balun Configuration that used in this study [1]

The methodology of our research is to do a direct measurement of return loss and VSWR antenna using the Vector Network Analyzer to observe whether the 2.4 GHz antenna working frequency has been reached. Antenna was optimized by varying the length of the antenna and feedline and observing its return loss and SWR measurements. The dipole antenna is not exactly 1.5 λ length so the antenna length is reduced slowly to get the 2.4 GHz frequency. If a 2.4 GHz working frequency is achieved then an impedance adjustment is mounted as in Figure 2. Then measurements are taken to optimize the antenna impedance adjustment. Optimization is done by slowly changing the balun length and observing the value of return loss and VSWR and antenna polaradiation so that the antenna with VSWR and minimum return loss is obtained. Antenna pattern is also observed to obtain omnidirectional pattern which are closer to pattern in theory as in Figure 1.

3. Result and Discussion

The measurement of dipole 500 MHz antenna pattern is despicted in Figure 4. The type of balun used on this antenna is 1: 1 balun as in Figure 2b. Pattern of the antenna without balun is not symmetrical as depicted in figure 4a (blue line). this shows the unequal current between the two dipole arms due to unmatching between the dipole and the feedline. Unlike conditions without balun, the antenna pattern looks balanced between the two poles due to match conditions as shown in red line in Figure 4(a). It means that the use of balun on a dipole antenna can effectively increase the antenna impedance matching. The pattern with balun was compared to theoretically omnidirectional pattern $U = \sin(\theta)$ as depicted in figure 4 (b). It shown good agreement betwen measurement and theory. Figure 4(c) is photograph of 500 MHz antenna.

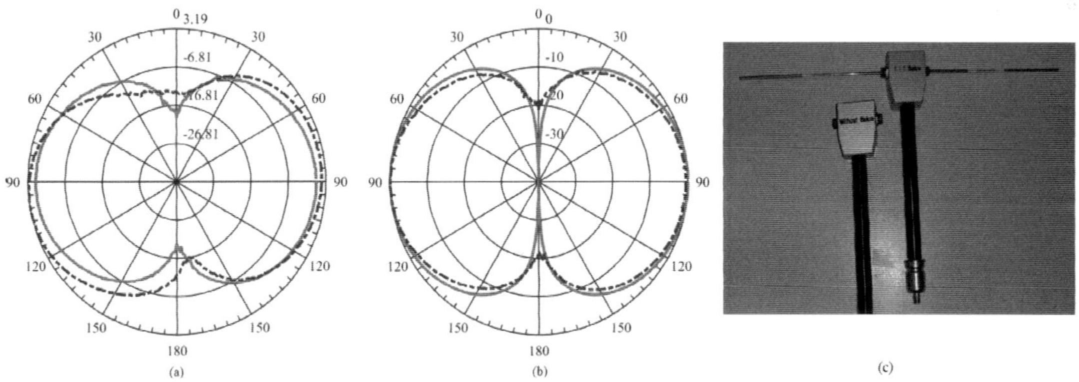

Figure 4. Dipole pattern. (a) blue = Dipole 500 MHz without Balun, red = dipole with balun 1:1, (c) Photograph of dipole 500 MHz antenna.

After testing the 500 MHz dipole antenna, we designed the dipole antenna for the 2.4 GHz. The antenna is made of wire with a diameter of 1 mm as in Figure 5. dipole is 0.5 λ length (3.1 cm) and fed by 1.5 λ coaxial. The antenna length is gradually reduced and its effect is observed on the frequency response. The measurement result is shown in Figure 6 . An antenna with a length of 0.5 λ is very good on the 1.58 GHz. This shows that the actual length of the dipole antenna is shorter than 0.5 λ. The 2.4

frequency is obtained at 0.35 λ dipole length. Figure 4b. Shows the frequency response to VSWR which results are linear to return loss. To observe the effect of the dipole length on the antenna working frequency, the minimum return loss value to dipole length is displayed in Figure 7. by changing the dipole length from 0.5 to 0.35, slowly decreases to -45 dB (0.42 λ). Similar results are shown in the SWR curve of Figure 5b. Using a length of 0.35 λ, we then observe the change in feedline length. The previous line with a length of 1.5 λ was cut by step 0.2 λ and observed it return loss. Changes in feedline length result in slight changes in frequency as in Figure 6a, But for 0.5, 1, and 1.5 λ the antenna still produces the center frequency response at 2.4 GHz as in Figure 6(b). It can be seen that the size of the feder must be a multiple of 0.5 λ. the antenna response for 0.5 λ as good as 1.5 λ length.

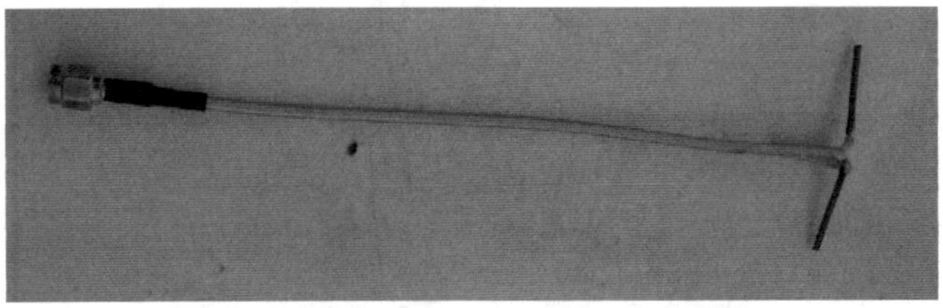

Figure 5 Dipole antenna for 2.4 GHz

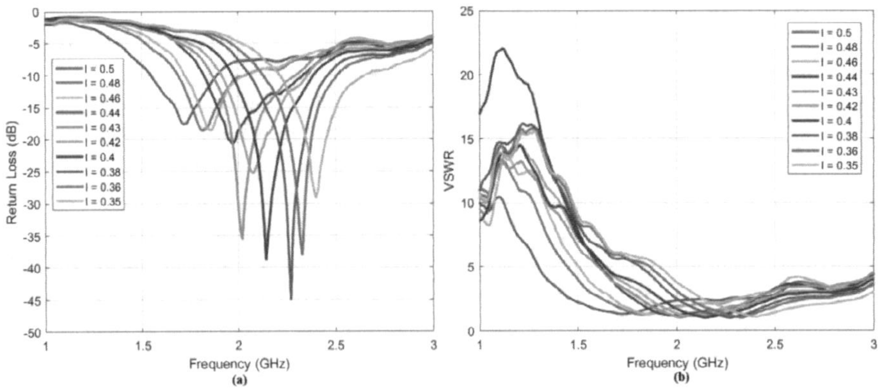

Figure 6. Influence of length dipole on antenna frequency (a) Return Loss (b) VSWR.

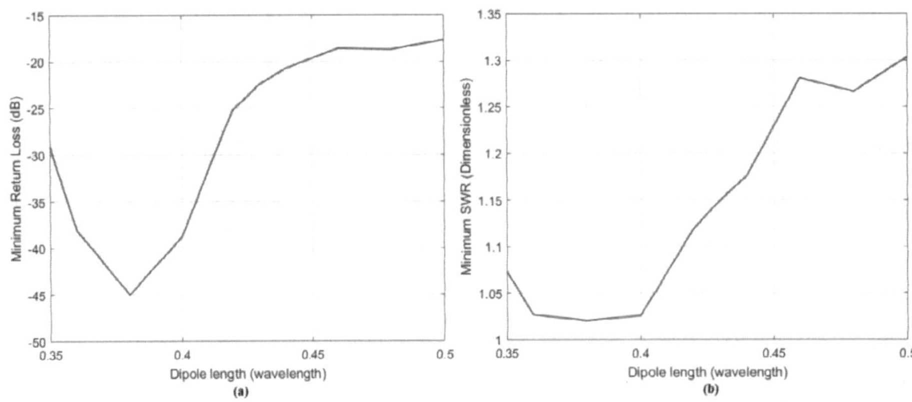

Figure 7. Minimum Return Loss vs dipole length

IOP Conf. Series: Materials Science and Engineering **846** (2020) 012013 doi:10.1088/1757-899X/846/1/012013

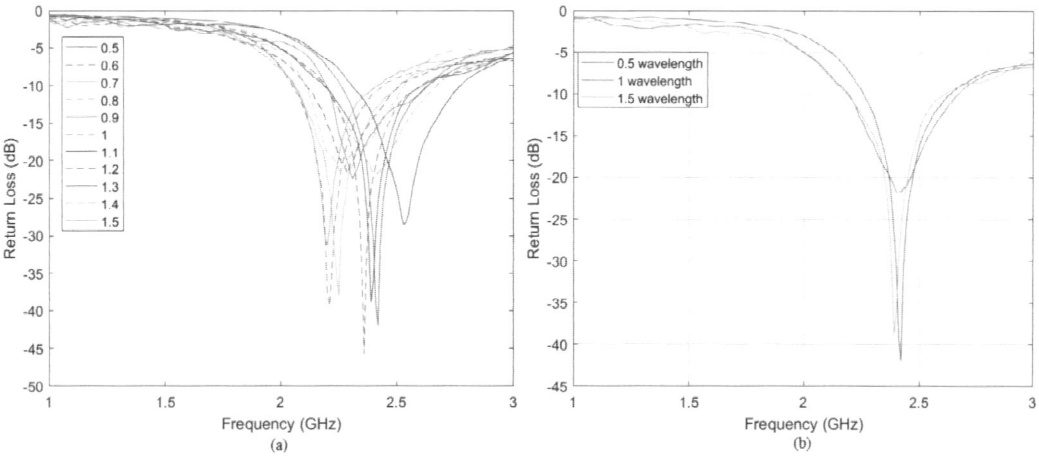

Figure 8. The variation of line length vs antenna impedance

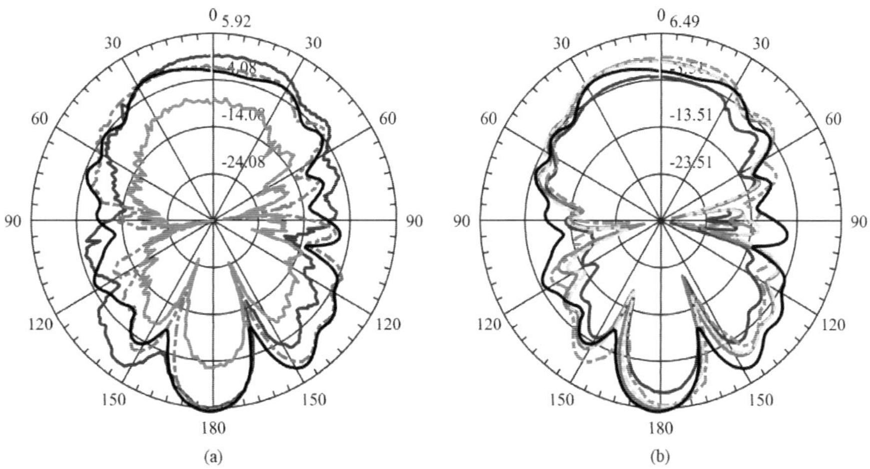

Figure 9. Variation of bazooka and balun length vs antenna pattern. (a). Bazooka (blue line = 0.25 λ, red line = 0.2 λ, magenta line = 0.16 λ .) (b) Balun 1:1. (blue line = 0.5 λ, red line =0.25 λ , magenta line = 0.16 λ. Cyan line = 0.2 λ)

by keeping the cable length of 1.5 λ and a dipole length of 0.35 λ, then the bazooka and balun are mounted to the antenna as in Figure 2. Each balun and bazooka were measured in varying lengths. The pattern is observed to see the impedance matching. It can be seen that the most symmetrical radiation pattern is obtained for the bazooka size ¼ λ. Whereas for balun 1: 1, the most symmetrical pattern is obtained for balun length of 0.16 λ

4. Conclusion

Impedance matching and optimization of dipole antena for 2.4 GHz is presented. In Optimizing the dipole antenna, the antenna length is optimized first to achieve the required frequency. When frequency was reached, the line feed is cut 0.5 λ length or multiple of 0.5 λ. And the last step is to optimize the impedance balancing device. We found that the length of the dipole antenna for the 2.4 GHz frequency is 0.35 λ, and the balun length is 0.25λ for the bazooka and 0.2 λ for the 1: 1 balun.

References

[1] C. A. Balanis, “Antenna Theory.” A John Wiley & Son, 2005.
[2] W. L. Stutzman, *Antenna theory and Design*, 3rd ed. Virginia: John Wiley, 2012.
[3] G. L. Hall, *ARRL Antenna book.pdf*. Connecticut: American Radio Relay league, 2010.
[4] K. K. S. Kim and J. K. T. Kang, “The Design of Calculable Standard Dipole Antennas in the Frequency Range of 1 ～ 3 GHz,” vol. 12, no. 1, pp. 63–69, 2012.
[5] S. Koziel, S. Member, S. Ogurtsov, W. Zieniutycz, and A. Bekasiewicz, “Design of a Planar UWB Dipole Antenna with an Integrated Balun Using Surrogate - Based Optimization,” vol. 1225, no. c, pp. 1–4, 2014.
[6] H. Zhang, H. Xin, and A. B. T. D. Balun, “A Dual Band Dipole Antenna With Integrated-Balun,” vol. 57, no. 3, pp. 786–789, 2009.
[7] S. X. Ta, I. Park, and R. W. Ziolkowski, “Crossed Dipole Antennas: A review,” no. October, 2015.

IoT: Heart Rate Monitoring Tool Using Android with Alert Messanger Telegram System

Yuhefizar[1], Anggara Nasution[2] , Roni Putra[3], Ervan Asri[4] , Deni Satria[5]

[1]Information Technology Department, Politeknik Negeri Padang

*yuhefizar@pnp.ac.id

Abstract. Patients with symptoms of arrhythmias must live a healthy lifestyle and routinely conduct consultations and control the heart rate with the doctor. The most important of these is an integrated health monitoring tool. In this study a portable heart rate monitoring tool is proposed for patients at risk of low-cost IoT-based alert systems using a telegram messager. The ESP8266 Wifi module functions for communication to the web server gateway and pulse sensor to detect heart rates that converted to Bits per Minute (BPM). The heart rate data is then stored in the database server using TCP / IP communication — heartbeat information in realtime with Android display. The system alert will send to the doctor/health worker and family if the heart rate is below 60 BPM and above 100 BPM, the time required for sending system alerts with a telegram messager is around 6 to 7 seconds. Test results show an average error of the heart rate measurement of 4.75% and the entire system is running well. With this tool the patient is at risk of being able to work more safely, and the doctor and family are easier to supervise the patient.

Keywords: IoT, Heart rate, Monitoring, WiFi ESP8266, Pulse sensor, Android, Smartphone, Telegram

1. Introduction

The heart is one of the core organs in the human body, working on a continuous and never stopping mechanism, commonly referred to as heart rate. Normal heart rate in adults ranges from 60 to 100 times per minute [1]. If the heart is not functioning or not operating properly, of course, human life will be disrupted and can even be fatal such as death. Heart disease can occur due to various factors, such as genetic factors and lifestyle patterns. Some heart diseases that have been known by the public are coronary heart disease, heart failure, heart valve abnormalities, and arrhythmias. Arrhythmia is a disorder of the heart rhythm outside the normal sinus rhythm [2]. In Indonesia heart disease reaches 26.4 percent [3]. Arrhythmia is a heart disorder that is characterized by abnormal beats or rhythms, which can be too fast, too slow, or irregular. An arrhythmia occurs when electrical impulses do not function. Symptoms include slow or fast heartbeat, palpitations in the chest, chest pain, shortness of breath, dizziness, if no medical action takes quickly it can cause death. Patients with symptoms of arrhythmia must live a healthy lifestyle and routinely conduct consultations and control heart rate with a doctor/health professional [4]. The development of information technology today marked by the presence of the Internet of Things (IoT). IoT is an internet service that integrated with the use of certain types of sensors [5] - [7]. IoT is now widely used for agriculture [8] - [10], smart city management [11] - [13] and manufacture [14] - [15].

In the health sector IoT is used for monitoring the health condition of patients [16] - [18]. In his research David L et al. measured the patient's heart rate using an infrared sensor and photodiode, the patient's heart rate data is then sent to a smartphone using Bluetooth communication if the patient's heartbeat is abnormal the patient's smartphone will call the doctor or family of the patient who determined. The limited Bluetooth communication distance which is only 10 meters becomes an obstacle besides emergency telephone calls are also limited [19]. Using Bluetooth heart rate module to detect heart rate Muhammad Farouq et al. developed a cellular application for heart patients. This android-based application has a feature to detect heart rate and SMS notifications and the position of the patient if the heart rate is not normal. Limited Bluetooth communication distance is also an obstacle in this study [20]. G.Vijay et al. also developed a monitoring tool for heart rate and blood pressure and body temperature. The data is then saved to the database and displayed on the website. Patient data can be seen by doctors at any time. Data communication using Bluetooth with a limited distance, there are no notifications such as panic buttons, SMS, and E-mail if the patient is in an emergency [21].

Using wireless communication Andrej Škraba et al. [22] developed a prototype tool for heart rate monitoring groups. All patient data for heart rate monitoring groups are displayed on one page using a web browser. There are no notifications such as panic buttons, SMS or e-mail if the patient is in an emergency. They are utilizing the Node-Red cloud application via the MQTT Puneet et al. protocol [23] measuring the heart rate of patients using the PPG technique by implementing an infrared sensor. Patients' heart rate data is saved to the database and displayed on the android application; they also add the feature of the nearest hospital address and distance to the patient. Data communication is still using cable, so it requires infrastructure costs, besides SMS and e-mail notifications are only sent to only one person / certain if the patient is in a dangerous situation. In this paper we develop a heart rate monitoring system for patients at risk based on wireless. This system uses an android application for monitoring the patient's heart rate. For notifications if the patient's condition is abnormal, we use the Telegram messenger application, besides this tool is also equipped with a panic button if the patient is in an emergency condition. With this tool the patient is at risk of being able to work more safely, and the doctor and family are easier to supervise the patient.

2. Methodology

The system block diagram can see in Figure 1, where the system divided into two parts: the first is a heart rate detector and the second is the heart rate monitoring application. The hardware consists of a heart rate sensor, LCD and esp8266 wifi module. Data from the pulse sensor is processed by the 8266 nodemcu microcontroller; then the data is sent to the database with the HTTP protocol. The database used by the Firebase Realtime Database hosted on the cloud. Data is stored as JSON and synchronized in realtime to each connected client. When creating cross-platform applications with the Android SDK, iOS, and JavaScript, all clients will share a Realtime Database instance and receive the latest data updates automatically.

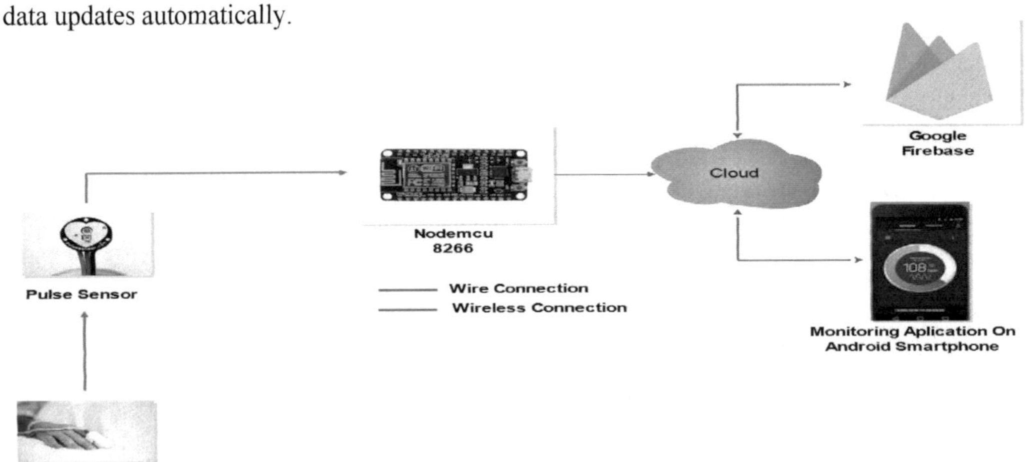

Figure 1. Block diagram of the system

The electronic circuit of the device can see in Figure 2. This tool uses a pulse sensor to detect the patient's heartbeat. Pulse sensors work by utilizing light. When these sensors placed on the surface of the skin, most of the light is absorbed or reflected by organs and tissues (skin, bones, muscles, blood), but some light will pass through body tissues if they are thin enough. If the amount of light intensity that hits the pulse sensor still eats the signal value will be around 512 (the middle value of the 10-bit

ADC range). The greater the light intensity, the higher the ADC value. The signal generated by the sensor produces waves called a photoplethysmogram (PPG) shown in Figure 2. PPG in the medical world is used to measure respiratory rate and heart rate. Pulse sensors have three pins that used differently. Pin 1 used as a data pin connected to pin A0 on the microcontroller. When the heart pumps blood throughout the body, each pulse that occurs accompanied by the emergence of pulse waves such as shock waves that propagate through the arteries to the capillary layer of the hands (fingers) where the pulse sensor installed. The speed of blood flowing slower than pulse waves. Determination of the number of heartbeats per minute (BPM = beat per minute) with this sensor done by dividing 60000 (in milliseconds) the average value of ten IBI (interbeat interval) that passed. IBI is the time difference between a point and a point

Next, the point value is 50% of the P (peak) value minus T (valley) when the graph has risen sharply

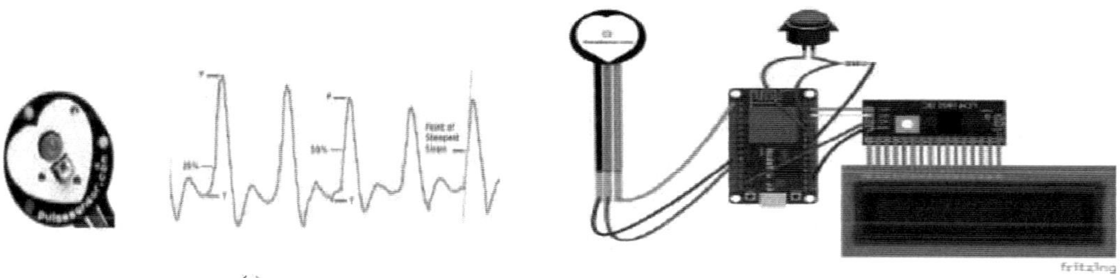

(a) (b)

Figure 2. (a) Pulse sensor, and (b) Hardware design

The heart rate information stored in the database then displayed on the android application on the smartphone. The application will also display the patient's medical history, doctor's data, and patient's family mobile number. The application system flowchart can see in Figure 3 (a). The alert system will function if it is smaller than 60 BPM and greater than 100 BPM, then the SMS will be sent to the doctor. Likewise if the patient presses a panic button, then the system alert will be active, and an SMS will also be sent to the patient's family, as shown in figure 3 (b).

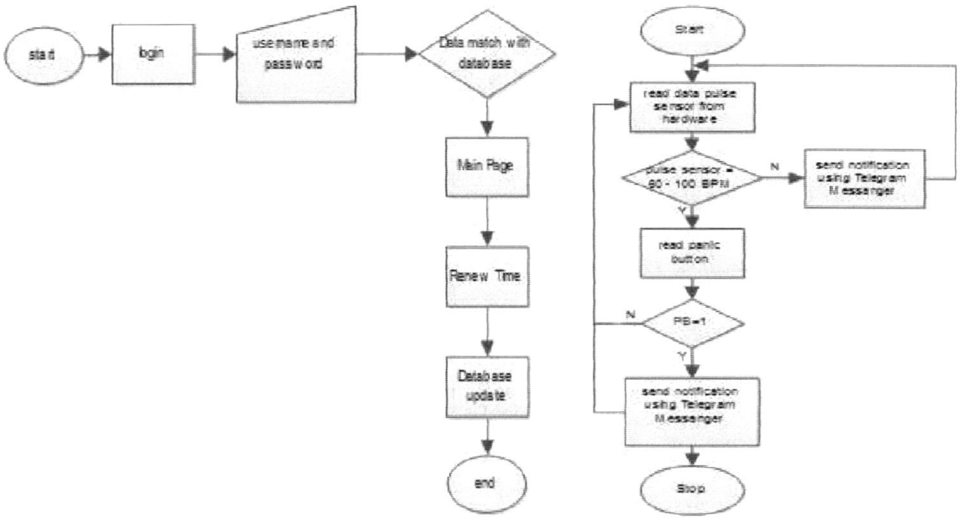

(a) **(b)**

Figure 3. (a) Flowchart system, and (b) The flowchart of Alert system

The process starts from the initiation of UART, SSID and password, pulse sensor and LCD. Heart rate would appear on the LCD if it connected to the network. Furthermore, the patient's heart rate data is sent to the database server to be stored using the TCP / IP protocol. Then the system will check, if the panic button (PB) = '1' then the buzzer will sound and information sent to the database server.

3. Result and Discussion

To observe system performance, several tests performed in each part of the whole system. Pulse sensor testing with ESP8266 Wi-Fi module and system alert testing.

A. Pulse Sensor Testing and Applications for Heartbeat Detection

This test aims to measure the accuracy of the sensor and heart rate information displayed on the smartphone application and observe how accurate the measurement of the human heart rate by comparing the measurement results of the pulse sensor with an oximeter. An oximeter is a device used to measure heart rate (HR = heart rate) and used for premature babies or patients in special conditions. In this experiment, measurements made by placing a pulse sensor on the fingertips of adult human hands, as shown in Figure 4. The test results can see in table 1.

Figure 4. Pulse sensor testing with ESP8266 Wi-Fi module

Tabel 1. Pulse Sensor Measurement and Testing Results

Testing (person number)	Age (Year)	Measurement using Pulse Oximeter (BPM)	Measurement using Pulse Sensor on LCD (BPM)	Error (%)
1	18	76	79	3.94
2	21	65	70	7.69
3	24	69	72	4.34
4	25	70	73	4.28
5	30	76	80	5.26
6	33	72	75	4.16
7	35	79	83	5.06
8	42	80	83	3.75
9	46	77	81	5.19
10	54	79	82	3.79
		Average Measurement Error		4.750

Pulse sensors can measure and detect heart rates, which displayed on the LCD with an average error of 4,750% as seen in table 1. In the heart rate monitoring application, the patient's heart category marked with color. Yellow level indicates the patient's heart rate below 60 BPM, categorized as a patient suffering from bradycardia. Red level if above 100 BPM categorized as tachycardia. While the green level indicates the normal patient's heart rate is in the range of 60 / minute - 100 / minute. The test is carried out to compare the patient's heart category, as shown in Figure 5. The test results show the monitoring application can display the results of the pulse sensor reading placed on the human finger.

B. Testing Alert System Using Telegram Messenger

Making a Telegram Bot requires bot registration by a BotFather account for the bot name and obtaining an API key. The API key used for every API request to the server telegram. Server bot applications are built using the PHP programming language and MySQL database. Communication from Telegram server to bot server utilizes webhook with https protocol. The activating system alerts on this system if the heart rate is below 60 BPM and above 100 BPM is counted ten times; then the system will send a message to the telegram messager application. The test results can see in Figure 6. The time required for sending system alerts is about 6 to 7 seconds based on testing from table 2.

Table 2. Testing the period for sending Telegram Messenger notifications

Trial	Telegram Notification	Time (seconds)
1	exist	5
2	exist	6
3	exist	5
4	exist	5
5	exist	7
6	exist	9
7	exist	10
8	exist	6
9	exist	5
10	exist	8

4. Conclusion

In this study, we implemented an IoT-based heart rate monitoring system with WiFi communication using the ESP8266 wifi module. Heart rate values can display on the LCD and smartphone applications in realtime. Patients and doctors can see heart rate information at any time. Telegram messager alert system will be active if the heart rate is below 60 BPM and above 100 BPM, and there is an emphasis on the panic button.

References

[1] I. B. A. Menown *et al.*, "Resting heart rate and outcomes in patients with cardiovascular disease: Where do we currently stand?" *Cardiovasc. Ther.*, vol. 31, no. 4, pp. 215–223, 2013.
[2] R. Weber, D. Stambach, and E. Jaeggi, "Diagnosis and management of common fetal arrhythmias," *J. Saudi Hear. Assoc.*, vol. 23, no. 2, pp. 61–66, 2011.
[3] https://ugm.ac.id/id/berita/13416-kematian-akibat-kardiovaskuler-masih-tinggi
[4] Yoga Yuniadi, "Mengatasi Aritmia, Mencegah Kematian Mendadak." eJKI Vol. 5, No. 3, December 2017
[5] S. H. Shah, "A Survey : Internet of Things (IoT) Technologies, Applications, and Challenges," *2016 IEEE Smart Energy Grid Eng.*, vol. I, pp. 381–385, 2020.
[6] X. L.D., H. W., and L. S., "Internet of things in industries: A survey," *IEEE Trans. Ind. Informatics*, vol. 10, no. 4, pp. 2233–2243, 2014.
[7] P. V. Paul and R. Saraswathi, "The Internet of Things - A comprehensive survey," *6th Int. Conf. Comput. Power, Energy, Inf. Commun. ICCPEIC, 2017*, vol. 2018–Janua, pp. 421–426, 2018.
[8] J. Bauer and N. Aschenbruck, "Design and implementation of an agricultural monitoring system

for smart farming," *2018 IoT Vert. Top. Summit Agric. - Tuscany, IOT Tuscany 2018*, pp. 1–6, 2018

[9] "IoT-based Intelligent Irrigation Management and Monitoring System using Arduino," *TELKOMNIKA (Telecommunication Comput. Electron. Control.*, vol. 17, no. 5, pp. 2378–2388, 2019.

[10] S. R. Prathibha, A. Hongal, and M. P. Jyothi, "IOT Based Monitoring System in Smart Agriculture," *Proc. - 2017 Int. Conf. Recent Adv. Electron. Commun. Technol. ICRAECT, 2017*, pp. 81–84, 2017.

[11] L. B. Campos, C. E. Cugnasca, A. R. Hirakawa, and J. S. C. Martini, "Towards an IoT-based system for Smart City," *Proc. Int. Symp. Consum. Electron. ISCE*, pp. 129–130, 2016.

[12] M. Rosmiati, M. F. Rizal, F. Susanti, and G. F. Alfisyahrin, "Air pollution monitoring system using LoRa module as transceiver system," *TELKOMNIKA (Telecommunication Comput. Electron. Control.*, vol. 17, no. 2, p. 586, 2019.

[13] R. H. Putra, F. T. Kusuma, T. N. Damayanti, and D. N. Ramadan, "IoT: smart garbage monitoring using android and real-time database," *TELKOMNIKA (Telecommunication Comput. Electron. Control.*, vol. 17, no. 3, p. 1483, 2019.

[14] H. P. Breivold and K. Sandstrom, "Internet of Things for Industrial Automation-Challenges and Technical Solutions," *Proc. - 2015 IEEE Int. Conf. Data Sci. Data-Intensive Syst. 8th IEEE Int. Conf. Cyber, Phys. Soc. Comput. 11th IEEE Int. Conf. Green Comput. Commun. 8th IEEE Inte*, pp. 532–539, 2015.

[15] A. Caputo, G. Marzi, and M. M. Pellegrini, "The Internet of Things in manufacturing innovation processes," *Bus. Process Manag. J.* vol. 22, no. 2, pp. 383–402, 2016.

[16] S. B. Baker, W. Xiang, and I. Atkinson, "Internet of Things for Smart Healthcare: Technologies, Challenges, and Opportunities," *IEEE Access*, vol. 5, no. c. Pp. 26521–26544, 2017.

[17] P. Gupta, D. Agrawal, J. Chhabra, and P. K. Dhir, "IoT based smart healthcare kit," *2016 Int. Conf. Comput. Tech. Inf. Commun. Technol. ICCTICT 2016 - Proc.*, pp. 237–242, 2016.

[18] V. Vippalapalli and S. Ananthula, "Internet of things (IoT) based smart health care system," *Int. Conf. Signal Process. Commun. Power Embed. Syst. SCOPES 2016 - Proc.*, pp. 1229–1233, 2017.

[19] David L Larkai, Ruiheng Wu, "Wireless Heart Rate Monitor in Personal Emergency Response System," *IEEE 18th International Symposium on Design and Diagnostics of Electronic Circuits & Systems 2015*

[20] Muhammad Farouq Bin Mustapha, Toni Anwar, "Mobile Heart Rate Monitor for Myocardial Infarction Patients," *IEEE 6th ICT International Student Project Conference (ICT-ISPC) 2017*

[21] G.Vijay Kumar, A.Bharadwaja, N.Nikhil Sai, "Temperature and Heart Beat Monitoring System Using IOT," *International Conference on Trends in Electronics and Informatics ICEI 2017.*

[22] A. Škraba, A. Koložvari, D. Kofjač, R. Stojanović, V. Stanovov, and E. Semenkin, "Prototype of group heart rate monitoring with NODEMCU ESP8266," *2017 6th Mediterr. Conf. Embed. Comput. MECO 2017 - Incl. ECYPS, 2017; Proc.*, no. June 2017.

[23] P. Bansal, M. Malik, and R. Kundu, "Smart heart rate monitoring system," *2018 IEEMA Eng. Infin. Conf. eTechNxT 2018*, pp. 1–4, 2018.

Implementation of Wireless Sensor Network for Four sensors on Natural Disaster Early Warning System

Suryadi[1], Marta Zuradi[2], Tuti Angraini[3], Laxsmy Devy[4]

[1,2,3,4] Electrical Engineering, Politeknik Negeri Padang

[1]suryadi2708@yahoo.com, [2]adi.zuriadi@gmail.com, [3]tutiangraini@pnp.ac.id, [4]laxmy_d@yahoo.com

Abstract. Wireless Sensor Network (WSN) is likely to be implemented as a long-range transmission network for sensor data to detect potential natural disaster like flood, landslide, and soil liquefaction. The sensor parameters: for example rainfall, soil humidity, vibration, type of soil, and so forth were sent from the location at which natural disaster potentially occurs and were forwarded to data center in real time for further analysis and observation. Therefore, the natural disaster which potentially occurs in the area was easily detected by the parameter of nodes that were gathered in server. When sensor detected the sign of natural disaster, the information would be conveyed through WSN to information center and was broadcasted to society as an early warning of natural disaster such as flood, landslide, soil liquefaction, etc. Such information was automatically broadcasted to society and was adjusted to the impending circumstance e.g. secure, cautious, and warning. By the implementation of WSN technology as an early warning system, it is expected to information concerning natural disaster is broadcasted accurately and rapidly to society, hence society can take quick action and response to natural disaster.

1. Introduction

Wireless Sensor Network (WSN) is a node containing sensors for collecting data in the form of certain parameters and forwarding it to data center or server for further observation[1]. Therefore, WSN technology is likely to be implemented to detect a circumstance in long-range area through sensor data parameters sent from the network.

The landslide, for instance, this natural disaster is commonly caused by several factors such as the geographical area of hillside, which is the main factor of landslide, and the other factors are geological condition, morphology, and human activities. Besides these factors, landslide can also be triggered by high rainfall, earthquake, and volcanic activity[2].

Natural disaster can occur unpredictably, and this is attributable to its unknown signs. To detect what natural disaster may potentially occurs, it requires a few samplings that are measured and tested with several parameters that cause and trigger natural disaster itself. Generally, such process is carried out manually hence this will be difficult and hazardous. By the implementation of Wireless Sensor Network, it enables the long-range data transmission through radio wave. As a result, the location at which natural disaster potentially occurs can be monitored from afar. In addition, it can also convey information for early warning of natural disaster.

A few sensors will be installed at the location which natural disaster like landslide, flood, soil liquefaction, and so on potentially occur to detect the parameters which cause and trigger natural disaster. The parameters are rainfall sensor, soil humidity sensor, sensor that detects type of soil, vibration sensor, and shifting soil sensor. These parameters are later processed by the microcontroller to detect whether natural disaster potentially occurs in the area or not. If there a sign of disaster appears, the microcontroller will convey the information to data center and give a warning of danger.

There will be three warning scales: namely cautious, insecure, and danger. These warning scales are adjusted to the magnitude of impending natural disaster. By means of this early warning system, society in the potentially affected area can beware of and be prepared before the natural disaster actually occurs.

- Wireless Sensor Network (WSN)

Wireless sensor network consists of a bunch of deployed sensors (detectors) and are all capable to cover specific geographical region called sensor area in which many detectable parameters contained. These sensors are designed in such a way that it is capable of sensing, calculating, producing communication elements that enable the administrator to do a measurement, observing, and reacting to an event and phenomenon in an environment, processing the data resulted from the gathered information, as well as capable of communicating either horizontally (with identical sensor), or vertically (with base station) without using cable as the transmission media (wireless)[3]. When this wireless sensor network is connected to the gateway which can access the internet, this wireless sensor network can also access and collaborate with other systems, as shown in Fig 1.

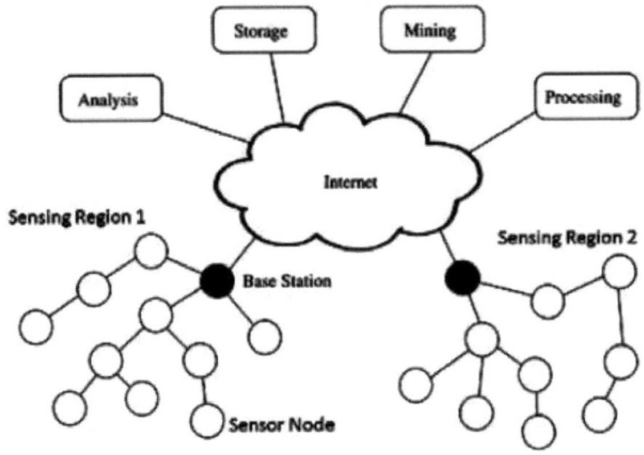

Fig 1. Illustration for the scenario of the usage of wireless sensor network[4].

- SX1278 LoRa Module

Fig 2. SX1278 LoRa Module

IOP Conf. Series: Materials Science and Engineering **846** (2020) 012015 doi:10.1088/1757-899X/846/1/012015

SX1278 is a long-range RF transceiver module (LoRa) which is capable of transmitting data at a distance of up to 10 KM on the open-air area (LoS). This transceiver module works at the frequency of 433 MHz and is equipped with high sensitivity and performance, reliable anti-jamming feature, and extensive-range data transmission. This is highly suitable to be implemented in areomodelling and in unmanned aerial vehicle, robotics, and industry[5].
Specification:

- Chipset: SX1278
- Interface: SPI
- Range: 3 KM (LoS)
- Frequency: 433 MHz
- Power Output: 20 dbm / 100 mW
- Power Supply: 1.7 - 3.8V

- Microcontroller

Arduino UNO is microcontroller board based on the ATmega328. It has 14 digital input/output pins (of which 6 can be used as PWM outputs) 6 analog inputs, a 16MHz crystal oscillator, a USB connection, a power jack, a ICSP header, and a reset button[6].

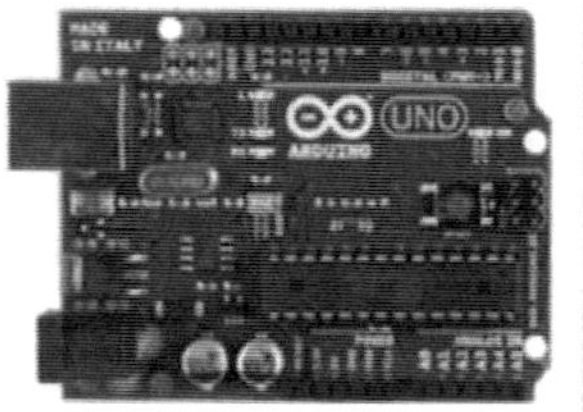

Fig 3. ATmega328 Microcontroller

2. Methodology

- Drafting for System

Drafting for system is made to assist and to facilitate the design process of device. The draft begins with the making of block diagram which each of block has its typical function yet it is interrelated that it build a system. Block diagram system is divided into two blocks i e. client block diagram, server and user block diagram. The detail of the two block diagrams as follow.

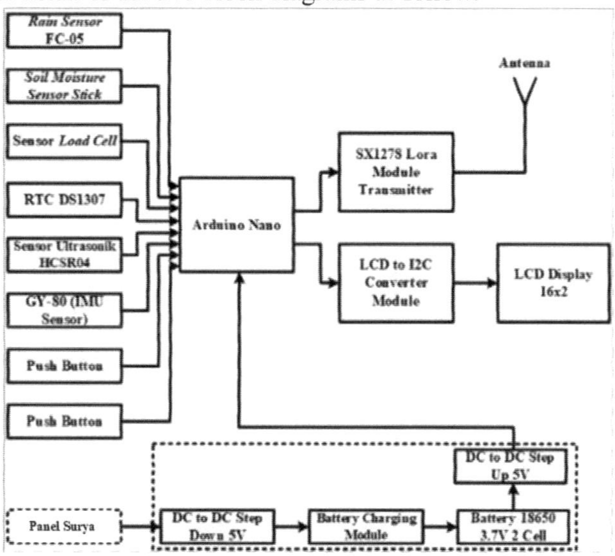

Fig 4. Block diagram of client

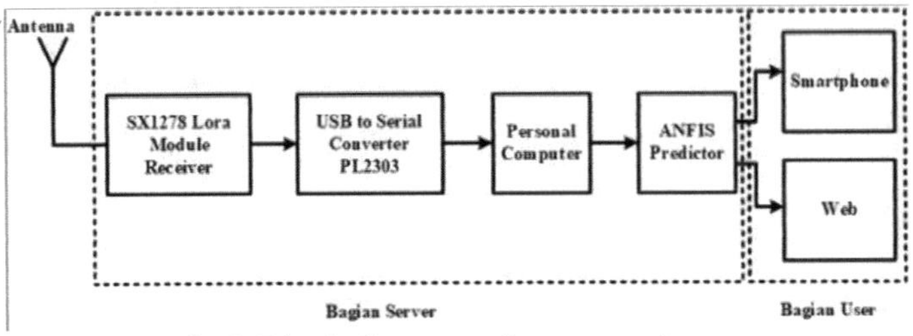

Fig 5. Block diagram of server and user.

- Drafting for Electronics

Based on block diagram in the draft of system, there are two parts that composes this device: part of client, and part of server. Each design of block in each part will be explained in the following.

a) Drafting for Electronics in Client part

In this client part, it is subdivided into 3 main stages: input, process, and output. Each main stage consists of several components. The draft for each part is depicted in the following circuit diagram.

b) Drafting for Electrinics in Server Part

In this server part, it is subdivided into 3 stages, namely client data receiver (lora SX1278 which function is as receiver), USB to serial TTL data converter (USB to TTL converter), and the last is PC as the data processing and decision taking center. The following is block diagram for this server.

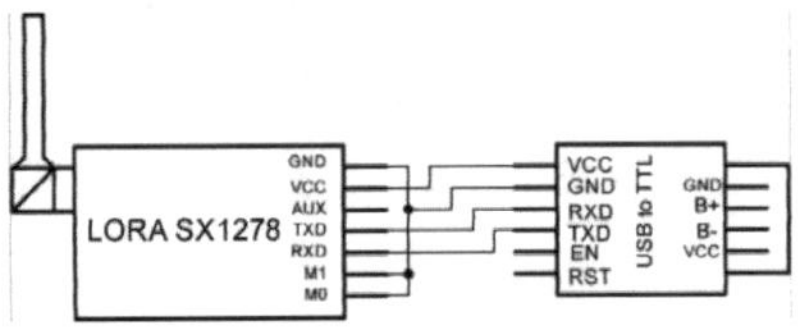

Fig 7. Circuit diagram of server.

- Drafting for Mechanic

The following is the draft for the bodywork of the device. The body is made of acrylic with 3mm thickness. The mechanical motion part is made of iron with 1 cm diameter. Meanwhile, the roof top is made of stainless plate with 2 mm thickness. The bodywork design of the device can be seen in the following figure 8 as detail.

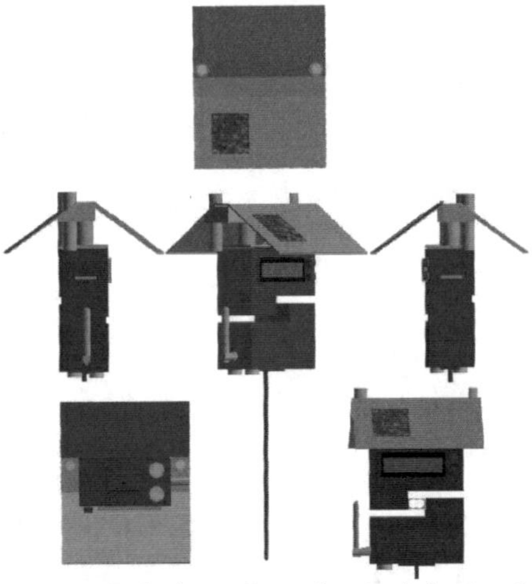

Fig 8. The bodywork design of mechanical device (all side view)

Description :

a) The left side view of bodywork design.
b) The right side view of bodywork design.
c) The top view of bodywork design.
d) The bottom view of bodywork design.
e) The back view of bodywork design.
f) The front view of bodywork design.

- Drafting for Software (Flow Chart)

In the drafting for software, it is explained about the software as a work principle of this device. The flow chart is divided into two parts, the client flow chart and the servee flow chart. The detail explanation for both flow charts are shown in this following figure 9.

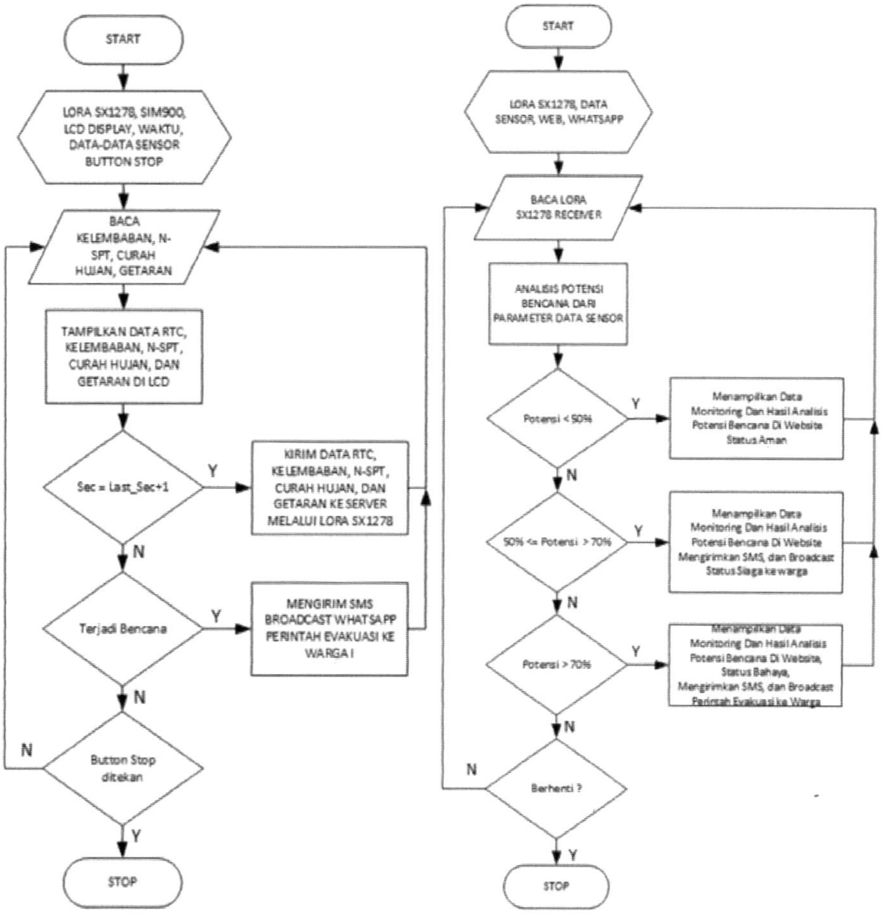

Fig 9. Client flow system (a) dan server flow system (b)

3. Result and Discussion

- The Result of Observation on Humidity Sensor

Table 1: The Result of Observation on Humidity Sensor

Actual Humidity	ADC Values	Humidity Sensor Values	Error (%)
5 %	480	1.93417 %	61.31661
10 %	520	8.60333 %	13.96666
15 %	560	15.2725 %	1.816664
20 %	577	18.10689 %	9.465526
30 %	648	29.94466 %	0.184457
35 %	690	36.94729 %	5.563675
40 %	722	42.28262 %	5.706546
45 %	740	45.28374 %	0.630539
50 %	780	51.95291 %	3.905815
55 %	810	56.95478 %	3.554148
60 %	836	61.28974 %	2.149565
65 %	882	68.95928 %	6.091198
100 %	1023	92.46809 %	7.531914
Average Error			9.37564

Based on table 1 above, it can be concluded that the average score of error for humidity sensor was 9.4%. The score of error was obtained based on the following formula:

$$\text{Error (\%)} = \frac{Actual\ Humidity\ Values - Sensor\ Values}{Actual\ Humidity\ Values} \text{x}100\%$$

$$\text{Average Error (\%)} = \frac{\sum\ \%Error}{nData}$$

The score of humidity sensor was obtained based on the following linearized regression formula:

$$m = \frac{n\sum xy - \sum x \sum y}{n\sum x^2 - (\sum x)^2}$$

$$b = \frac{\sum y}{n} - m\frac{\sum x}{n}$$

$$y = mx + b$$

After being observed and regression linearized, there coefficient and constant of humidity were obtained on ADC score as what was shown in the following formula:

Humidity = 0.016673 * nADC - 7.80958.

The following chart showed the result of humidity after linearization, the blue curve was the real humidity score, and the humidity score after linearization.

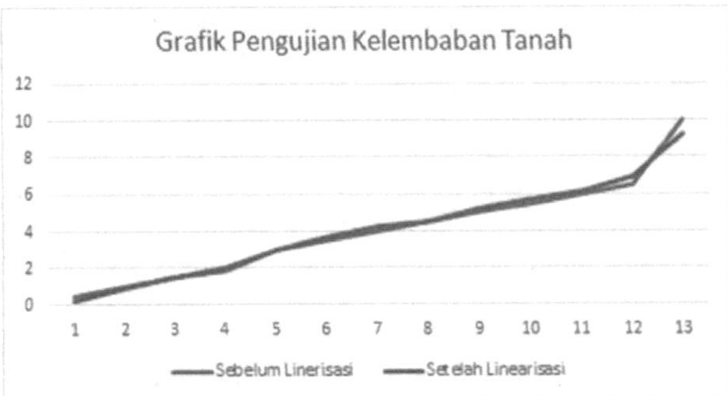

Fig 10. The Comparison Chart on sensor score after being linearized with the real humidity score.

- The Result of Observation on Rainfall Sensor

Table 2. Result of Observation on Rainfall Sensor

ADC	Time	Volume	Rainfall	Classification
660	97 Seconds	200 ml	2.06 ml/s	Drizzle
670	70 Seconds	200 ml	2.86 ml/s	Light rain
680	60 Seconds	200 ml	3.33 ml/s	Heavy rain
700	45 Seconds	200 ml	4.44 ml/s	Downpour

Based on table 2 above, it can be said that the higher score of ADV was the heavier rainfall would become and vice versa. Rainfall was classified into 4 types: drizzle, light rain, heavy rain, and downpour.

- The Result of Observation on Soil Sensor

Table 3. Result of observation on soil sensor

Location	N-SPT	Soil Classification
Seashore	6.04	Light
Rice Field	6.84	Light
Paddy Bund	8.02	Light
Jicama Garden	10.42	Light
House Yard 1	20.56	Medium
House Yard 2	24.34	Medium
House Yard 3	32.33	Medium
Hilly Terrain (Red Soil)	52.56	Dense
Hilly Terrain (Lamay Soil)	84.88	Dense
Hilly Terrain (Rocky Soil)	88.02	Dense

- The Result of Observation on Accelerometer Sensor.

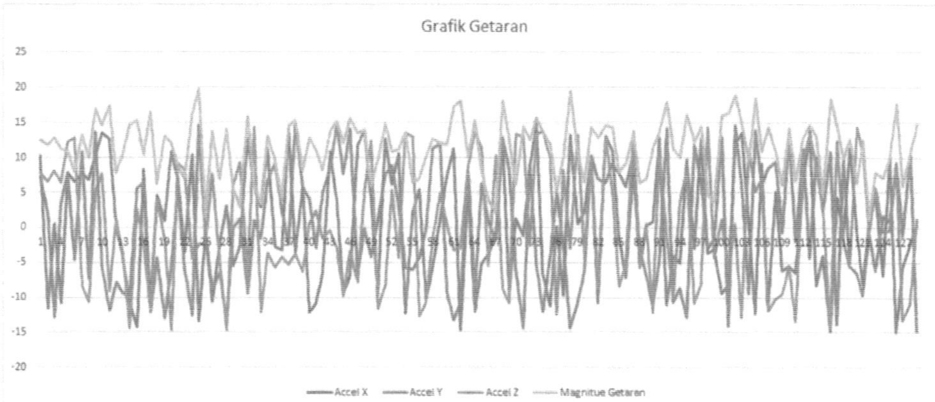

Fig 11. Result chart of observation on Accelerometer sensor

Based on the chart above, there were 4 curves i.e. vibration curve on X-axis which was represented with the blue line, vibration curve on Y-axis which was represented with the red line, and vibration curve on Z-axis which was represented with the grey line, as well as vibration magnitude curve which was represented with the orange line

The magnitude score was affected by the acceleration score on X-axis and Y-axis or in other words was formulated as bellow:

$$\text{Vibration Magnitude} = \sqrt{Accel\ x^2 + Accel\ Y^2}$$

4. The Result of Observation on WSN

The following is a figure result of sensor data conveyed from client to server through WSN, it is seen in figure 12. The data conveyed were time, humidity score, rainfall score, and soil liquefaction detected score, magnitude score, N-SPT score, and type of soil.

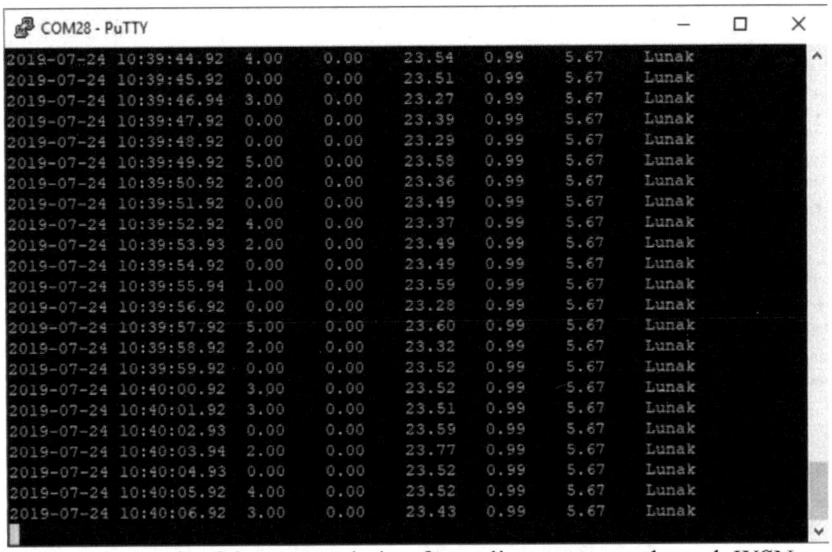

Fig 12. Result of data transmission from client to server through WSN

Table 4. Result of Observation on Lora SX1278

Distances	Baud rate	Obstacle Area	No Obstacle Area	Times
10 Meters	9600	Sent	Sent	1 Sec
100 Meters	9600	Sent	Sent	1 Sec
200 Meters	9600	Sent	Sent	1 Sec
300 Meters	9600	Sent	Sent	1 Sec
400 Meters	9600	Sent	Sent	1 Sec
500 Meters	4800	Sent	Sent	2 Sec
750 Meters	4800	Sent	Sent	2 Sec
1 KMs	2400	Sent	Sent	4 Sec
1.5 KMs	1200	Sent	Sent	6 Sec
2 KMs	600	Sent	Sent	9 Sec
2.1 KMs	300	Sent	Sent	14 Sec
2.5 KMs	300	Not sent	Sent	18 Sec
3 KMs	300	Not sent	Not sent	-

Based on the table of observation above, it can be concluded that the further range between server and client affects the time needed to transmit the data which was longer. The fastest time to send the data from server to client was 1 second, in which range was less than 400 meters and on baudrate of 9600 bps. Additionally, the data transmission was also affected by the baudrate or transfer speed from Lora SX1278. The higher baudrate become, the shorter time and range needed to transmit the data, in this case was the maximum range of baudrate should be at 9600 and at 400 meters. The maximum range Lora SX1278 can reach was at 300 bps baudrate, or at 2.4 km on occupied area (constricted), and at 2.5 km on open-air area (unconstricted).

Table 5. Result of Observation Overall

Humidity	Rainfall	Decay	Vibration Magnitude	Soil classification	Natural Disaster	Status	Action
80 %	High	Detected	M 6.2	14,4 / Soft	Earthquake, landslide and Liquefaction	Hazard (Evacuation)	WhatsApp Broadcast and SMS Sent, information on website updated
70 %	High	-	-	Hard	Flood	Warning	WhatsApp Broadcast and SMS Sent, information on website updated
50 %	Medium	-	M 6.4	Soft	Earthquake and landslide	Warning	WhatsApp Broadcast and SMS Sent, information on website updated
40 %	-	-	M 6.8	Soft	Earthquake and landslide	Warning	WhatsApp Broadcast and SMS Sent, information on website updated
40 %	-	-	M 4.8	Hard	earthquake	Secure	Information on website updated

Then in Figure 4 can be seen a journal that published a paper about the assumption of feature independence in the Naïve Bayes method. For the record, the journal in question is a journal that publishes papers that have been selected.

5. Conclusion

After conducting observation, collecting data and analyzing, it can be concluded:

- The maximum range of transmitting sensors data from client to server of Lora SX1278 100mW on constricted area is 400 meters on baud rate 9600 bps.
- The maximum range of transmitting sensors data from client to server of Lora SX1278 100mW in unconstructed area is 2.5 kilometers and on constricted area is 2.1 Kilometers on baud rate 300 bps.
- The response time of Lora SX1278 100mW in transmitting sensors data from client to server is 1 seconds.
- The fastest response time of early warning system device to broadcast information about impending natural disaster via WhatsApp to society is 30 seconds.

The fastest response time of early warning system device to send short message about impending natural disaster to society is 30 seconds.

Acknowledgment

The authors would like to thank Politeknik Negeri Padang, for sponsoring this work under Penelitian Produk Terapan No : 053/PL9.1.4/LT 2015 Date 3 April 2018.

References

[1] Firdaus, *Pengantar Wireless Sensor Network*. Yogyakarta: Graha Ilmu, 2014.
[2] A. S. Muntohar, *Tanah Longsor*, 1st ed. Yogyakarta, 2008.
[3] E. Imanningtyas, S. R. Akbar, and D. Syauqy, "Implementasi Wireless Sensor Network pada Pemantauan Kondisi Struktur Bangunan Menggunakan Sensor Accelerometer MMA7361," *J. Pengemb. Teknol. Inf. dan Ilmu Komput. Univ. Brawijaya*, vol. 1, no. 7, pp. 545–554, 2017.
[4] W. Dargie and C. Poellabauer, *Fundamentals of Wireless Sensor Networks: Theory and Practice*. Technical University of Dresden, Germany and University of Notre dame, USA, 2010.
[5] "Jual Modul LoRa SX1278 | DigiWare." [Online]. Available: https://digiwarestore.com/id/network-module/sx1278-lora-wireless-transceiver-433mhz-100mw-rf95-432266.html. [Accessed: 13-Dec-2019].
[6] M. Banzi and M. Shiloh, *Make: Getting started with Arduino*. 2014.

MOBILE APPLICATIONS FOR GAS FINDING

Rita Afyenni[1], Rika Idmayanti[2], Refni[3], Intan Pratama Kirana[4], Fitri Nova[5]

[1] Information Technology Department, [Politeknik Negeri Padang]

* afyennirita@gmail.com

Abstract. Current conditions show that almost all Indonesian households use gas as fuel for cooking, but the availability of Liquid Petroleum Gas (LPG) is not proportional to the number of user needs. This condition makes it difficult for the community to get it. Sometimes they have to search directly to each gas station to get information about the availability of gas. Mobile applications for gas finding that have been made, help the community especially in Padang city to find a gas station that still has a stock of gas. This application is equipped with the name of the Gas Station and location map feature. This application can be accessed using an Android smartphone.

1. Introduction

The use of gas stoves in Indonesia as a household need is very common, especially since the Government's success in converting the use of kerosene to the Liquid Petroleum Gas (LPG). Until 2015 Pertamina managed to distribute 57.19 million starter packs of 3 kg LPG. In 2018 based on the Decree of the Minister of Energy and Mineral Resources, Pertamina received a mandate to carry out the supply and distribution of 531131 packages of 3 kg LPG to prospective recipients of the first selected package[1]. These conditions make the community's need for LPG consumption increasing, starting from gas measuring 3 kg, 5.5 kg and 12 kg.

The conversion of kerosene to LPG has both positive and negative impacts on the community. One positive impact of this conversion is increased business opportunities in the distribution of LPG gas. Many people use gas as stove fuel because the cooking process will be easier and faster compared to using firewood as stove fuel. Thus, the sale of gas is a good opportunity for business. While the negative impact is the difficulty of consumers in finding LPG gas, due to the uneven distribution of gas bases in various places while its use has been evenly distributed among the people. Information regarding the availability of LPG gas at each LPG gas selling agent is also difficult to obtain, sometimes the user community must visit directly to the location of the gas base. Sometimes he has to visit several places first to get the gas he needs. This certainly makes it difficult for the gas user community.

Many increasingly sophisticated Android applications that can be used by everyone, this is due to the development of mobile technology[2], [3], [4], [5]. To solve the gas problem as described previously, it is necessary to make an application to overcome the problems that occur. With this Gas Finding Application created, the user community can find out the location of gas sales in the surrounding environment, and the existing gas stock at the base because the gas stock data will always be updated by each gas station owner. This application is equipped with added tracking to assist users in searching for gas. This application can be accessed by android smartphone [6].

2. Methodology

Data collection methods carried out in the form of observation and literature study [7]. Activities carried out in observation are collecting data through field research and requesting data from gas base agents. Data collected are types of gas, gas bases, gas stocks, gas prices, latitude and longitude, photos of gas bases, and other supporting information. Activities carried out in the literature study are collecting data from printed books relating to the application.

Application Development Methods are carried out using stages: 1) analysis, 2) design, 3) making program code using the appropriate programming language, 4) testing and implementation. In the analysis phase, the process of processing data that has been collected is done so that things that become the application needs to be obtained. Activities carried out at the design stage are designing user interfaces for user needs and their databases. At the stage of making program code activities translate the design into program code using the appropriate programming language. The programming languages and tools used are PHP and Android Studio [8], [9], [10]. This activity also creates a database that will be used on the application. At the test and implementation phase is testing the application that has been made to the implementation location, testing the route shown on the map, and submitting the application to the owner of the gas station.

3. Result and Discussion

The application is designed to facilitate use by three types of users, namely 1) people who need gas finding, 2) application administrator and 3) gas station owner. Each user has a different menu feature. Menu features provided for people who need a gas finds are a list of gas bases, information about the gas station, and maps to the location of the gas station. The menu features provided for application administrators are the processing of overall gas data and the data of the application user community. The owner of the gas station is given the right to access the server directly to the web. Gas station owners are also given a special Android application to change their gas stock, but they are not given the right to change the gas stock data at another owner's gas station.

One of the tools for application design used Unified Modeling Language (UML). UML is one of the architectures used to describe object-oriented programming architectures [11], [12]. Figure 1 shows the use case diagram for data processing needs on the server, there are two actors namely 1) the admin as the person who can manage all the data, 2) the gas station is the person who can only access their data station.

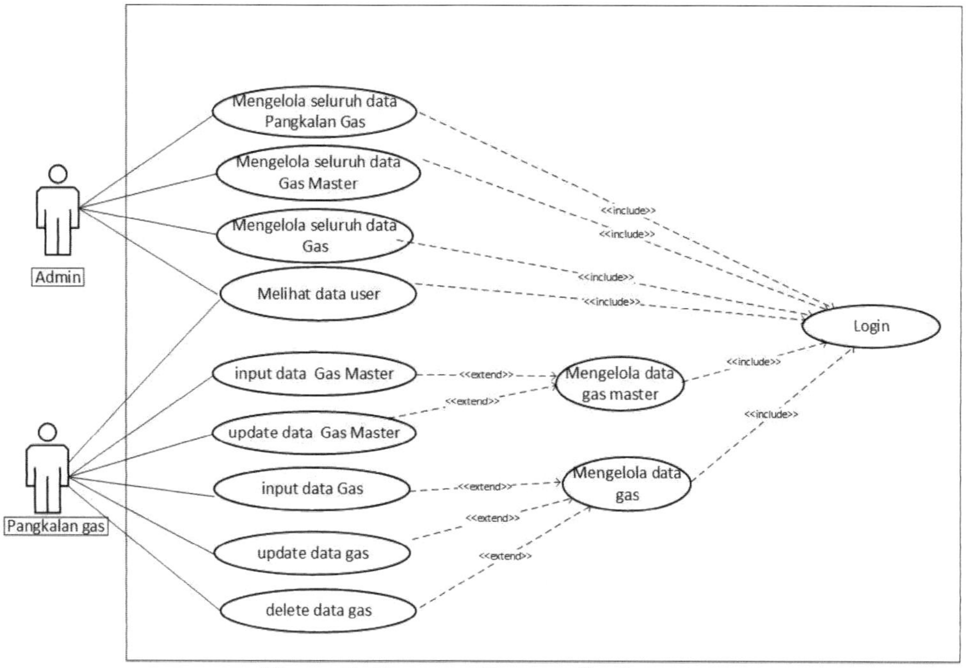

Figure 1. Use case diagrams for the web

Figure 2 shows a use case diagram for the community that uses gas finding applications. Figure 2 explains the two actors, namely 1) the user is the community looking for gas, 2) the gas station is the person who will change the gas stock data.

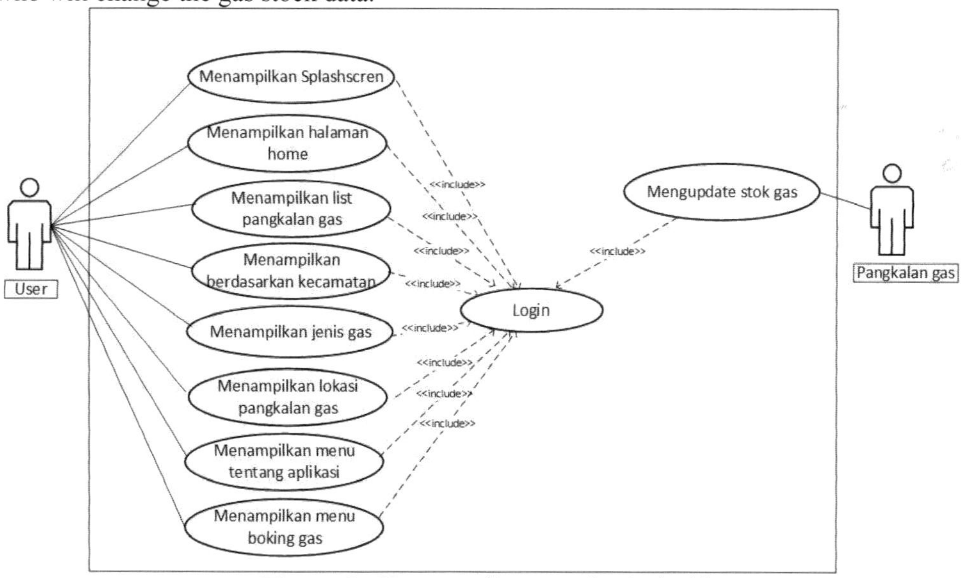

Figure 2. Use case diagrams for Android

Some views for Android applications that have been made, can be seen in Figure 3 and Figure 4.

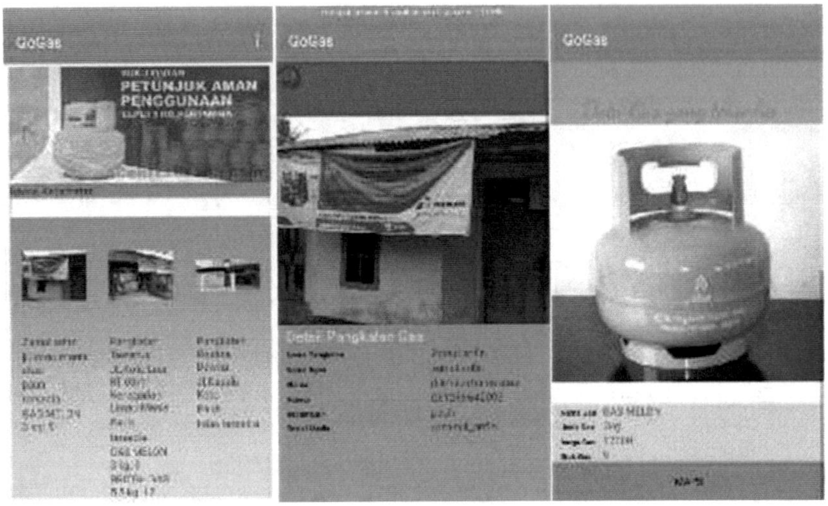

Figure 3. Display a list of gas base and gas details

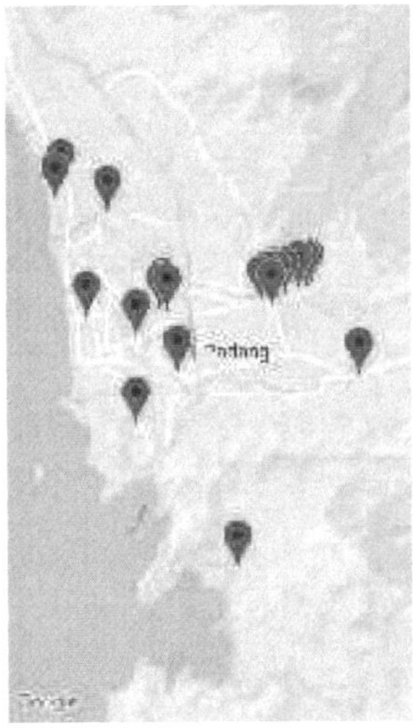

Figure 4. Map location of the gas station.

4. Conclusion

Gas Finding Application using mobile has been successfully created to provide information and tracking the location of the gas station in Padang city. This application uses maps from Google Maps, so that it can provide location information and tracking the location of each gas base in the city of Padang. This application is a means to facilitate users in finding a gas station and find out the gas stock provided. The results of this activity aim to identify the problems that occur and the things needed by the user.

References

[1] Pertamina, "Pertamina Terima Mandat Kelanjutan Program Konversi Minyak Tanah ke LPG," 2018. [Online]. Available: https://www.pertamina.com.

[2] Frans Richard Kodong and Yuli Fauziah, "Pengembangan Aplikasi Smartphone Tracking untuk Monitoring Pergerakan Kurir Pendistribusi Barang pada PT. Synergy First Logistics Yogyakarta," *TELEMATIKA*, vol. 14, no. 1, pp. 38–47, 2017.

[3] A. S. and K. S. S. Elisa, Febi Safitri, "Perancangan Aplikasi Ketersediaan Obat pada Apotek di Kota Medan Berbasis Android," in *SENSASI*, 2019, pp. 369–373.
[4] Y. S. and I. Hafiza Aswin, "Aplikasi Pencarian Lokasi Panti Asuhan di Kota Ambon Berbasis Mobile," *ILKOM*, vol. 10, no. 3, pp. 338–343, 2018.
[5] S. A. and R. E. C. Devi, "Implementasi Global Positioning System (GPS) dan Location Based Service (LSB) pada Sistem Informasi Kereta Api untuk Wilayah Jabodetabek," *SISFOTEK Glob.*, vol. 7, no. 2, pp. 27–33, 2017.
[6] I. N. P. and A. A. K. A. C. W. I Made Widnyana, "Aplikasi Sistem Informasi Geografis Bengkel di Kota Denpasar Berbasis Android," *Merpati*, vol. 3, no. 1, pp. 23–30, 2015.
[7] A. Kristanto, *Perancangan Sistem Informasi dan Aplikasinya*. Yogyakarta: Gava Media, 2018.
[8] M. A. A. and M. K. Angga Hendri Kusuma, "Perancangan dan Implementasi Pencarian Lowongan Industri Menggunakan Metode Simple Additive Weighting (Saw) pada Android," *Jartel*, vol. 8, no. 1, pp. 162–167, 2019.
[9] M. B. K. and M. W. S. Seng Hansun, *Pemrograman Android dengan Android Studio IDE*. Yogyakarta: Andi, 2018.
[10] A. Kadir, *From Zero to A Pro Pemrograman Aplikasi Android*. Yogyakarta: Andi, 2013.
[11] I. and A. F. Sugeng Santoso, "Aplikasi Monitoring Rumah Kos Berbasis Android di Kota Tangerang," *Maklumatika*, vol. 5, no. 2, pp. 129–139, 2019.
[12] L. D. S. and R. Amanda, "Aplikasi Penjualan Barang Tunai dan Kredit pada Toko Komputer Rizcom Madiun," *AKSI (Akuntansi dan Sist. Informasi)*, vol. 3, no. 2, pp. 72–81, 2018.

IOP Conf. Series: Materials Science and Engineering **846** (2020) 012017 doi:10.1088/1757-899X/846/1/012017

The Detecting of Damage Conductor with Guided Image filter

Ismail[1,2], Kamarul Hawari[2], Surfa Yondri[1], Firdaus[1], Yefriadi[1] , Nasrul Harun[1]

[1]Electrical Department, [Politeknik Negeri Padang, Indonesia]
[2]Electrical and Electronics Engineering Department, [Universiti Malaysia Pahang, Malaysia]

*indoismail@gmail.com

Abstract. The Damage of electrical conductor of high voltage transmission system is very common. This problem effected the distribution of electrical power from the sources into the costumer place. The main cause of this problem if come from the damage of electrical conductor. There were many ways to detect damage conductor so far. The most common is through compute the impedance of transmission line. This method is accurate, but it needs to switch of the line for a while. This disruption contribute losses for the costumers and energy provider. In order to handle this problem, our proposed method is to detect the damage conductor with Guided Image Filter. The input image is a visible image and the reference image is invisible image. The processing is very simple. The result obtained has high accuracy. The accuracy is shown on the detecting map.

1. Introduction

In the usage of conductor devices that were commonly made from silver, aluminum and other metal compounds. As time passing by, this conductor fall onto damaged or corrosion gradually as shown in **Figure 1**[1]. It needs very long time to happen. The corrosion on the electrical conductor will causes the rise of temperature around the damaged regions. Finally, the conductivity of conductor fall down[2][3]. Furthermore, the rise of temperature of the conductor also makes the increase of resistive characteristic of electrical conductor [4]. Since the rise of resistive factor and heat of the conductors, they effect the efficiency of electrical energy distribution, secondly they have also possibility to produce disaster such as fire and etc.

Figure 1. The damaged conductor

Since, there are many negative impacts from the existence of damaged region on the electrical conductor, so that it is needed many efforts to prevent the problem occur. One of the method is computing the impedance of the electrical distribution network. This method has high accuracy.

IOP Conf. Series: Materials Science and Engineering **846** (2020) 012017 doi:10.1088/1757-899X/846/1/012017

Otherwise, the network has to be shut down for a while[5][[6]]. This procedure left the disadvantages for the costumer and company side. The losses of supplying energy for a period of time means loses economically and credibility. In order to handle this situation, the authors propose a new method to detect the damaged regions on the conductor through Guided Image Filter with Invisible Image as Reference Image. The invisible image of the damaged region is as shown in Figure 2. The use of visible and invisible image will no need to shut down power system at all[7]. The data of damaged region on the electrical conductor is obtained based on visible and invisible image.

Figure 2. Invisible image of damaged region

2. Methodology

The pair of input images were taken from Thermal Camera FLIR TG167 and Canon EOS 700D. The first one produces invisible image and the other visible image. Both have the same size, format and registered images.

2.1 Guided Image Filter

The proposed method is based on usage of Guided Image Filter with more complete data type. This filter is very common used in visible image processing only. Guided image filter is a linear translation-variant filter that wraps together a guidance image *R*, filter input *t* and an output image *S*. The filtering process according to the following equation,

$$S_i = \sum_j K_{ij}\,(R)_{t_j} \tag{1}$$

The filtering output at pixel is according to Equ. 1. There *i* and *j* are pixel indexes. $\boldsymbol{K}_{ij}$ is a window kernel as a function of guidance image and independent image. The characteristic of this filter is linear with respect to $\boldsymbol{t}$ [8][9].

2.2 Morphological Filter

Mathematical Morphology is a kind of methodology based on object formation generally used for identifying object boundary accurately[10]. It can make a simplification operation based on formation in image data. It makes connection the related and removes the unrelated parts [10].

Dilation

The dilation operator is concerning with the pixel formation modification. In the image processing, the pixel formation modification is implemented in the binary (0, 1) or gray scale (0-1) image. The operation of pixel formation is in the binary level image [10]. The dilation operation is a builds addition dimension of pixel forms. It works to merge two pixel formations using vector addition of each element. **X** and **Y** are pixel formation. The dilation of X and Y symbolized as **X ⊕ Y** [19]. **x ϵ X** is x subset X. And formulated by

$$(X \oplus Y)\, \mathrm{X} = \bigcup_{x \in X} X_y \tag{2}$$

$\mathbf{X_y}$ is a translation of X by y. Since dilation is an operation of pixel formation modification, and it performs a lowest level image, this operation generally used to modify the shape structure image. It can be used to smooth, segment image and so on.

Erosion

Erosion process works at pixel level image, in binary or gray image. It does as a decrement operation. If X and Y is the erosion operation X by Y symbolized as **X ⊖ Y** [14]. y ϵ Y is y subset Y. $\mathbf{X_{-y}}$ is translation erosion X by y. Mathematically, it formulated by following equation:

$$(X \ominus Y) = \bigcap_{y \in Y} X_{-y} \tag{3}$$

The erosion operation presents the discrepancy between two pixels form modification.

Opening operation

The opening operation is combination operation between erosion and dilation. This operation works starting with erosion and follow by dilation operation. Mathematically formulate as follow:

$$X \circ Y = (X \ominus Y) \oplus Y \tag{4}$$

Where X is an image, Y is structuring element. This equation gives smoothing, breaking uncorrelated and removing small objects. Also it sharpens the binary image.

Closing operation

The closing operation of image or pixel form X by the structuring element Y, it symbolized as X • Y. The closing operation presented as following equation:

$$(X \bullet Y) = (X \oplus \mathrm{Y}) \ominus \mathrm{Y} \tag{5}$$

The characteristic of closing operation, mainly are removing the small holes and filling the gaps among part pixel form.

2.3 Algorithm

Our method has several stages to complete. Each stage contains a certain process and output. The stages are as follow:

1. Pre-processing phase, the input images (visible and invisible image) must be in the same size, orientation and data type. The process include making image registration. The output of this process is Initial map.
2. Generating Guided Map phase, in this process the map of damaged region on the conductor has beed detected.
3. Optimization phase, the accurate boundary of damaged region on conductor generated. The fusion the Guided map and visibile image form Damaged Region.

The block diagram of the algorithm as shown in the Figure 3.

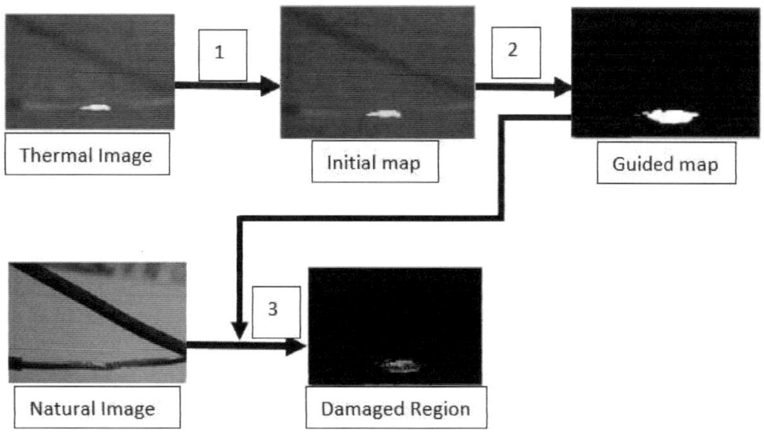

Figure 3. Block diagram of Proposed Method

3. Result and Discussion

Based on our method, the algorithm is able to detect the damage region of electrical conductor accurately. As shown in Figure 4. In Figure 4(a), the input image is a slender surface of conductor. The Figure 4(b), the damaged region detected and Figure 4(c). The detailed of the damaged regions.

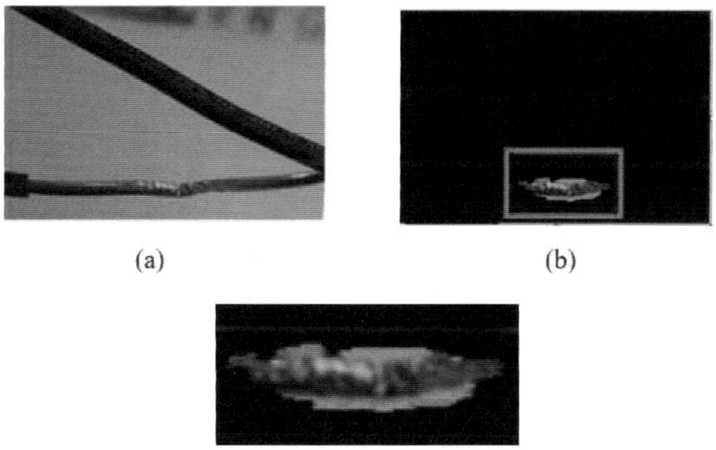

(a) (b)

(c)

Figure 4. The Result of Proposed method. (a) The input image ; (b) Damaged region detected and (c) The Detail damaged region.

From the Figure 4(c), the method show the accuracy of detecting region. It can generate all damaged region.

4. Conclusion

The Proposed method is a kind of new method to implementing the detection of damaged surface on the conductor. This method can be applied easily. The thermal data shortened the processing of defining region map. It helps simplification of algorithm and computing cost lower.

The proposed method gives easiest way to determining the appropriateness prediction of damaged region of the surface conductor. Since this method is supported by a pair of visible and invisible images.

Finally, the utilization of a pair of visible and invisible images help to handle the problem of detecting damaged conductor in electrical fields, especially the transmission line.

References

[1] B. D.Dunn, A.deRooij, andD. S.Collins, "Corrosion of Silver-Plated Copper Conductors.," *ESA J. (European Sp. Agency)*, vol. 8, no. 3, pp. 307–335, 1984.

[2] Leonard, "Corrosion of aluminium and copper in cable conductors – Leonardo Energy - Knowledge Base." [Online]. Available: https://help.leonardo-energy.org/hc/en-us/articles/202771412-Corrosion-of-aluminium-andcopper-in-cable-conductors. Accesses 22 April 2019. [Accessed: 20-Nov-2019].

[3] V. D.A, K.B, W.P.A.A.F, andS.E.F, "Impact of corrosion on the reliability of low voltage cables with Aluminium Conductors," *Proc. IEEE Conf. Conf. Elect. Insul. Dielectr. Phenom.*

[4] A.Abdallah Ahmed Elfaki, "The Effect of Temperature on Conductivity of Conductors and Superconductors," *Am. J. Phys. Appl.*, vol. 5, no. 1, p. 1, 2017.

[5] K. J.Stevens *et al.*, "Conductor damage inspection system for overhead ACSR power cables CDIS on ACSR," *Proc. Int. Conf. Sens. Technol. ICST*, pp. 901–905, 2013.

[6] N. A.Jaffrey andS.Hettiwatte, "Corrosion detection in steel reinforced aluminium conductor cables," *2014 Australas. Univ. Power Eng. Conf. AUPEC 2014 - Proc.*, no. October 2016, 2014.

[7] T.Sosnowski, G.Bieszczad, andH.Madura, "Image Processing in Thermal Cameras."

[8] R. G.Gavaskar andK. N.Chaudhury, "Fast Adaptive Bilateral Filtering," *IEEE Trans. Image Process.*, vol. 28, no. 2, pp. 779–790, 2019.

[9] K.He, J.Sun, andX.Tang, "Guided image filtering," *IEEE Trans. Pattern Anal. Mach. Intell.*, vol. 35, no. 6, pp. 1397–1409, 2013.

[10] S.Rani, D.Bansal, andB.Kaur, "Detection of Edges Using Mathematical Morphological Operators," *Open Trans. Inf. Process.*, vol. 2014, no. 1, pp. 17–26, 2014.

IOP Conf. Series: Materials Science and Engineering **846** (2020) 012018 doi:10.1088/1757-899X/846/1/012018

Android-Based Chicken Development Monitoring Application

Aldo Erianda[1], Putri Huriyati[2], Fanny Sukma[3]

[1]Teknologi Informasi, Politeknik Negeri Padang

aldo@pnp.ac.id

Abstract. Eggs are one of the staples of Indonesian people's food. Based on data from the Central Statistics Agency (BPS), it was found that the total national egg consumption in 2018 was 2,766,760 tons. As for the total production of 2,968,954 tons, so there was an excess of 202,195 tons of national stock. Even though there is excess stock, the current market price is unstable due to seed restrictions, so the chicken is slow to produce. In addition, the weather in the environment around the cage also affects the productivity of laying hens. In linda chicken farms, the problems faced are the same as those above, weather factors affect the health and productivity of chickens. If there is an unstable temperature and weather many chickens will experience pain and may even die. In this case, breeders sometimes do not know how the temperature and the weather are housed, whether or not the chicken needs it. With this Kancer system, breeders can monitor the development of laying hens and cages. Starting from the active or not the chicken is producing, changes in temperature, humidity and notification if there is rain. Microcontroller device that is installed in a cage and connected to a wifi network will send data temperature, humidity and rain detection to the server, then from the server the data will be processed and then sent to the application. And to monitor the productivity of each chicken, the kancer system utilizes the qrcode that has been stored per-chicken data, then through the kancer qrcode application is scanned and the application will display chicken data ranging from age to the number of eggs produced by the chicken.

1. Introduction

The need for national consumption chicken eggs in 2018 was 2,766,760 tons with a total production of 2,968,954 tons, so that an excess national stock of 202,195 tons was obtained. Based on data from the Central Statistics Agency (BPS), in 2017 chicken egg exports were 386 tons, this figure increased by 27.39 percent compared to 2016 [1]. However, although egg production is currently still sufficient for national needs, the price in the market is unstable due to breeding restrictions and drug restrictions which in turn results in slow production of chickens. In addition, extreme weather in a number of areas also affects the productivity of chickens and a decrease in supply to the market makes supply to the market decrease while demand increases.

Linda Chicken Farm is one of laying hens in Padang. In this livestock business, the problem often faced is the problem of chicken disease, this disease can be caused by viruses, weather factors, and temperature factors. Non-conducive weather like continuous rain can cause the chicken to become sick and even die. And temperatures that are too cold or too hot can also affect the health of chickens.
Then in monitoring the development of chickens, the owner of the farm still uses a manual method that is by remembering when these chickens lay eggs, when giving vitamins, and how many chickens

that die. Of course this makes monitoring the development of chickens ineffective because every time more and more data must be processed to see the development of the chicken.

Ease in monitoring the development of chickens is needed, such as ease in monitoring new breeds, chicken productivity in producing eggs, vitamins, feed, dead chickens, and others. So that the development of chickens can be controlled properly. So by developing a monitoring application for the development of poultry by utilizing qrcode technology can help the work of chicken farmers in monitoring the development of chickens and can also help in improving the quality of their chicken farms.

2. Methodology

In conducting this research, the Software Development Life Cycle (SDLC) method was used. SDLC is the stages carried out to build a software system. This stage is done to produce the desired software [2] [2]. The stages of SLDC in building software are as follows:

1). Planning

Planning is a stage that starts from defining the problem, then continues by describing the design of making Android-based Laying Chicken Development Monitoring Application such as project management to be created and the required design documents.

2). Requirement Analysist

Requirements Analysis is a stage that shows the process of analyzing and developing the needs of users of android-based applications and creating functional requirements documents. At this stage, data collection was carried out directly at the Linda Chicken Farm located in Lalang Village, By Pass Padang City. In the process of collecting data, interviews are conducted directly with the owners of chicken farms, so that the system needs to be obtained.

3). Design

Design is a stage that focuses on the design of making information systems based on Android. The system design used is UML in the form of usecase diagrams, sequence diagrams, class diagrams and activity diagrams. 4). Development

Development is the process of changing a design from paper to software. This stage is in the form of preparing files, creating databases, coding programs, and others.

5). Implementation

The implementation phase is the stage where applications that have been made are used by the user. This stage aims to identify mistakes and ensure the application works according to user needs.

3. Results and Discussion

In this discussion consists of analysis and system design, as well as system implementation and testing.

3.1 Use case diagrams

Use Case diagram is a modeling to illustrate the functionality of the software to be made [3]. In the application of monitoring the development of laying hens, functional for the user as a farmer is shown in the use case diagram in Figure 1.

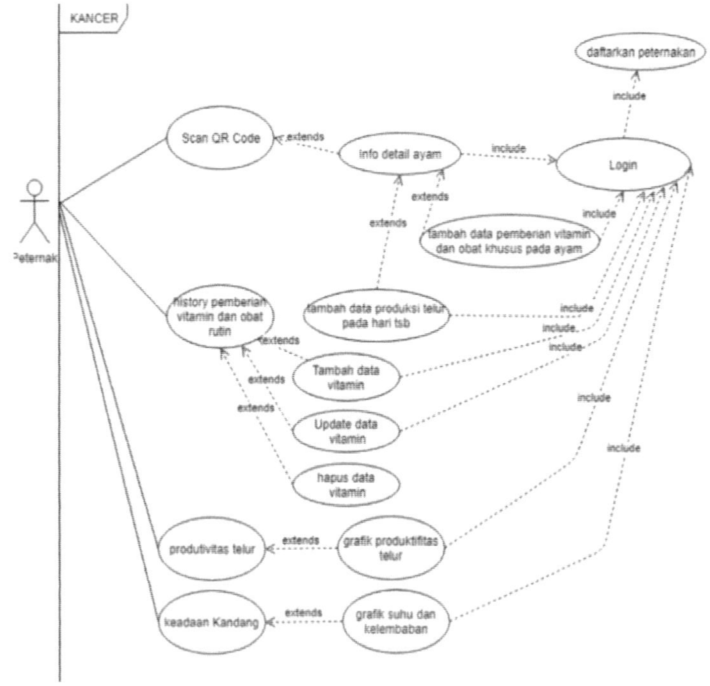

Figure 1. Use Case Diagram

3.2 Sequence Diagram

Sequence Diagram is a diagram that illustrates the behavior of objects in the use case by explaining the life time of objects and messages sent and received between objects [3].

3.2.1 Sequence Registration

Farmers do account registration by entering their email and password, then the email and password will be executed by the createAccount () function, in the validation process the email and password are registered to authenticate. If it works next, the homepage is displayed.

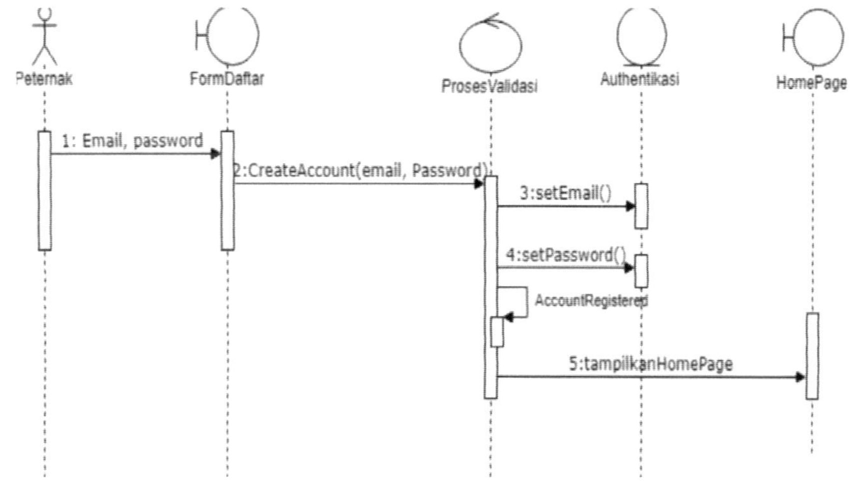

Figure 2. Sequence Registration

3.2.2 Sequence Login

Farmers enter their email and password, then the email and password will be executed by the signIn function in the validation process. The validation process will retrieve the existing email and password on the authentication, if there is a meaning that the account is valid and then display the homepage.

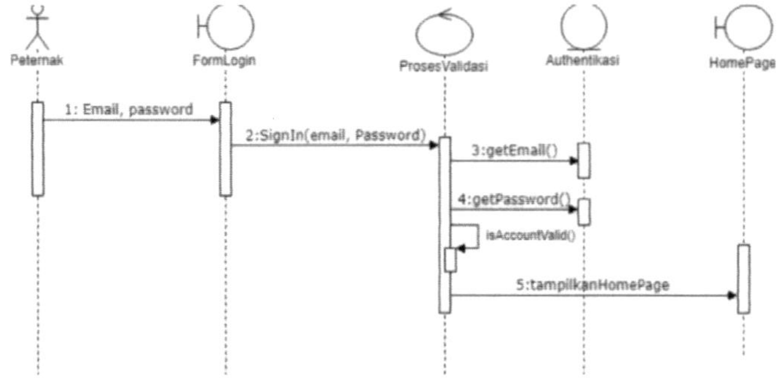

Figure 3. Sequence Login

3.2.3 Sequence Scan

The farmer selects the scan menu, then performs a qrcode scan. The system will process the code that is read in the qrcode, then retrieve data based on the code from the database, then display the detailed chicken data in the detail form.

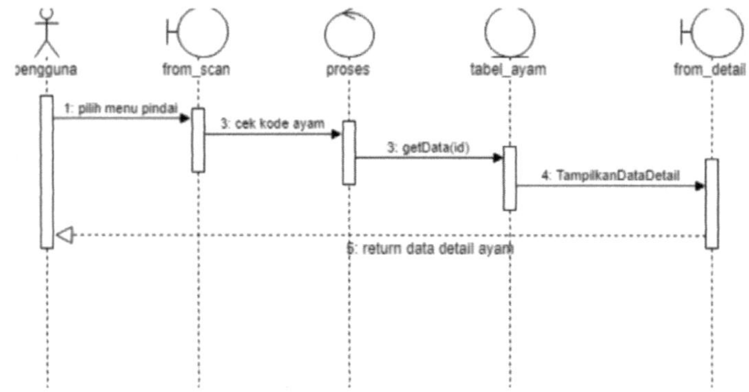

Figure 4. Sequence Menu Pindai

3.2.4 Sequence history of giving vitamins and drugs

The user selects the history menu, then the form will load the existing history data in the database. Users want to add history data, users will be directed to the form added, then the data entered will be processed and sent to the database. If the process is successful, the message will display successfully to the history list form

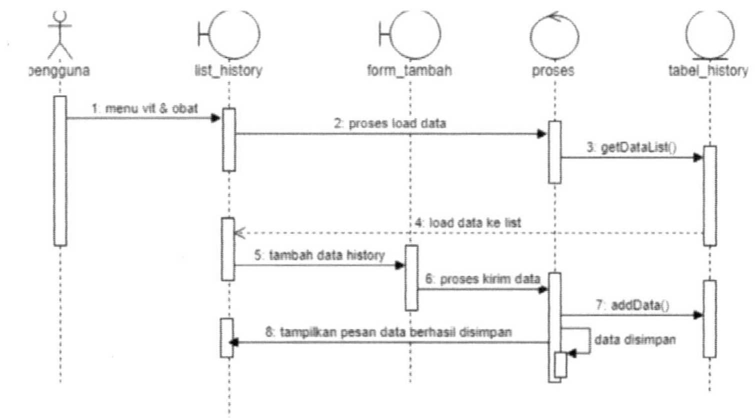

Figure 5. Sequence history of giving vitamins and drugs

3.2.5 Sequence of egg productivity

The user enters the application, then on the homepage will load the egg productivity data taken from the database, then display the data to the homepage.

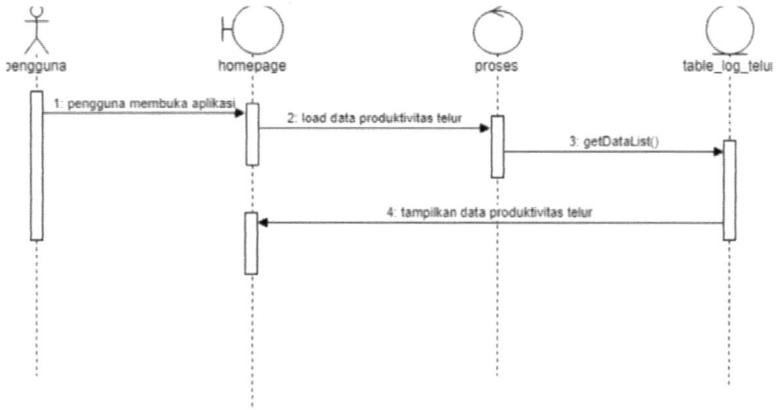

Figure 6. Sequence of egg productivity

3.2.6 Cage State Sequence

The user enters the application, then the homepage will load data on the state of the cage (humidity and temperature) taken from the database, then display the data to the homepage.

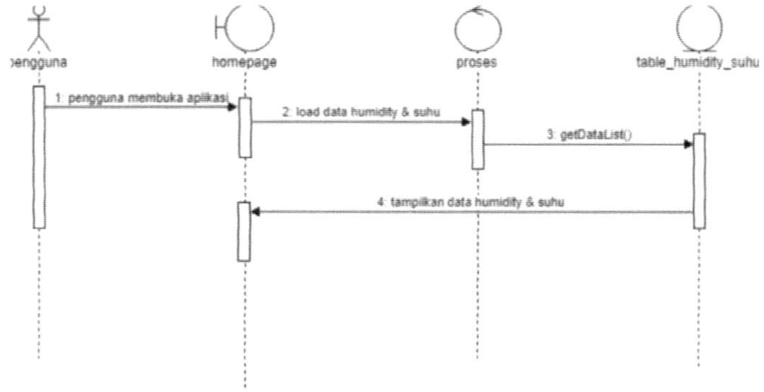

Figure 7. Cage State Sequence

IOP Conf. Series: Materials Science and Engineering **846** (2020) 012018 doi:10.1088/1757-899X/846/1/012018

3.3 User Interface Implementation

The results of the implementation of the Monitoring Application for the development of laying hens consist of the following appearance:

3.3.1 Splash Screen

Splash screen display is the initial appearance of opening the application, this display will emerge for 4 seconds when the application is first opened. The splash screen display can be seen in Figure 8

Figure 8. Splash Screen

3.3.2 Display of Farm Registration Form

Peterkan registration form is useful for registering the user's farm to get access rights to use this application. Data that must be entered when registering livestock is:

a. Farm Name, used to find out the user's farm name.
b. Address, the user's farm address is used.
c. Email, an email used to be used to login to the application.
d. Password, the password used will be used to login to the application.

Display the registration form can be seen in Figure 9.

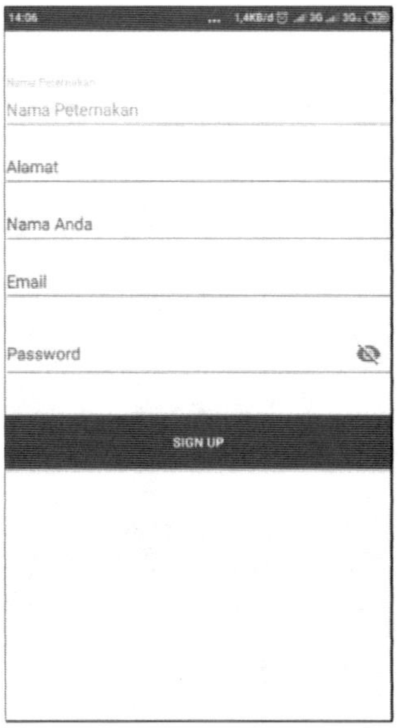

Figure 9. Registration Form

3.3.3 Login Display

The login form is used to start the application. The function of the login form is to limit who can access the application and access data on the system according to the user's access rights.

When entering, the data that must be entered are:

a. Email, functions to open user accounts owned by application users.
b. Password, functions to run the application, so that unauthorized people cannot login.

Display Login Form can be seen in Figure 10.

Figure 10. Login Page

3.3.4 Display Home Page

The successful login will then enter the homepage. On the home page the user can access 2 main menus namely scan and vitamins & medicine. And also, users can see info on the results of temperature sensors, humidity and rain detection on farms, and information on egg productivity that has been produced by farmer's hens. Display the home page can be seen in Figure 11.

3.3.5 Display Scan Qr Code

Figure 12 below is the display when the user presses the scan menu, then the application will open the camera to be able to scan the qr code owned by the chicken.

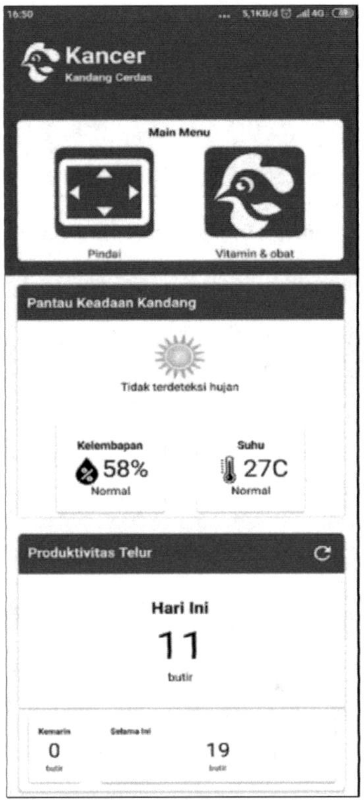

Figure 11. Homepage

IOP Conf. Series: Materials Science and Engineering **846** (2020) 012018 doi:10.1088/1757-899X/846/1/012018

Figure 12. Scan The QR Code

3.3.6 Detailed view of Chicken Information

If, the qrcode scanned by the user is correct and the chicken data is in the database, the application will display detailed data on the chicken information. The detailed data is in the form of gender, age, number of eggs that have been produced.

Then, on this page, the user can add information such as the chicken laying eggs today or not, and also can give special notes to the chicken. Display page detail chicken information can be seen in Figure 13.

Figure 13. Detailed Chicken Information

3.3.7 Display history of chicken productivity

On the chicken information detail page, there is a button to be able to see the history of chicken productivity. If the user wants to know the history of chicken egg productivity, the user can click on the button. Then the application will switch the display to the chicken productivity history page. On this page, users can find out the productivity of chickens and also special records held by chickens. Display history of chicken productivity can be seen in Figure 14.

Figure 14. History of chicken productivity

3.3.8. Display History of Vitamins and Medicines routinely

Figure 15 is a display of the history of vitamins and routine medication given by farmers every month to maintain the quality and health of laying hens. On this page, users can add notes on routine vitamin administration. And on this page the user can find out the last time the vitamin was given.

3.3.9. Display graphs of egg productivity

Figure 16 is a graph display of the productivity of farmed chicken eggs every time. On the graph the user can find out the number of eggs produced by laying hens every day and can compare the number of eggs produced with previous days.

IOP Conf. Series: Materials Science and Engineering **846** (2020) 012018 doi:10.1088/1757-899X/846/1/012018

Figure 15. History of Vitamins and Medications

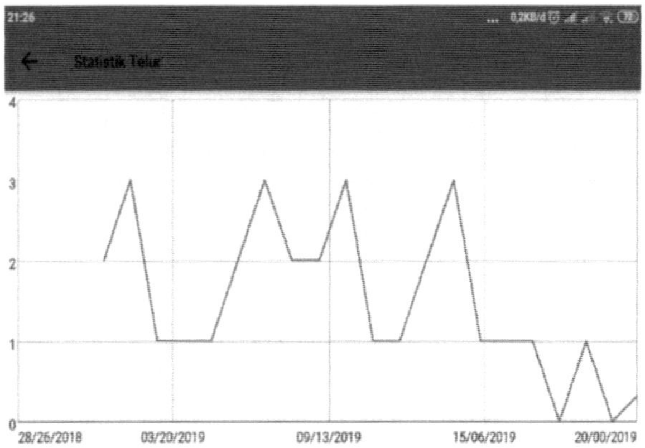

Figure 16. Egg Productivity Graph

3.3.10. Display temperature and humidity charts

The graph display of temperature and humidity serves to provide knowledge to the user how the temperature and humidity that exist in the cage now, if the temperature and humidity rise or fall, the breeders can do something about their chickens. Because temperature and humidity also affect the development and productivity of laying hens.

A graph of temperature and humidity can be seen in Figure 17 and Figure 18.

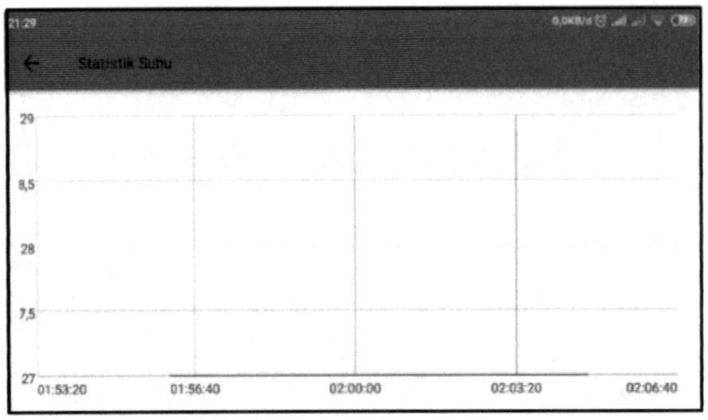

Figure 17. Temperature Chart

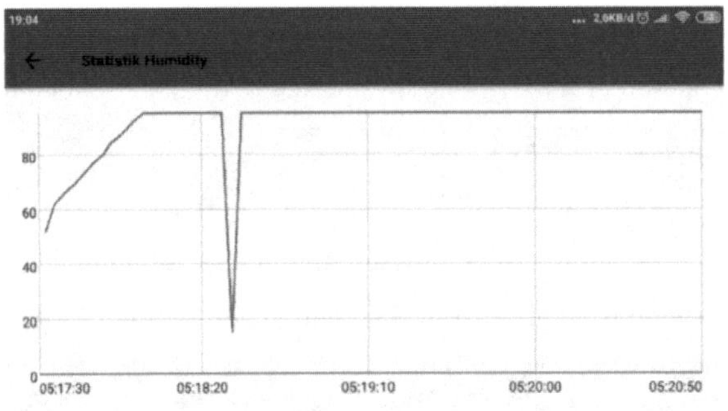

Figure 18. Moisture Graph

3.4 4 Testing Results

After implementing and testing the application monitoring the development of laying hens, it was found that all application features run 100% if the application is connected to the internet. The test results can be seen in the table.

Table 1. Testing Results

No.	The Activity	Informaion	The Result
1.	Animal Husbandry Registration	-when the user enters the farm data and press the save button the farm data will be saved to the Fire base Real time Database	Success
2.	Application Login	-If the user has not logged in, the application will take the user to the login page -If the user has previously logged in, the application will go directly to the main page and load data from the database.	Success
3.	Scan Qr code	-When the application gets the code from the scanned qr-code, the application displays detailed chicken data based on the scanned qr-code. And the application will load data from the database and display it on the application	Success
4.	Add new information about chickens	-The data added on the detail page runs fine and the data is saved to the Fire base Realtime Database	Success

5.	Add Vitamins and Medicines routinely	-This action works fine, and the data entered is stored in the Fire base Real time Database	Success
6.	Edit Vitamins and medication routine	-This action works fine, and the data entered is stored in the Firebase Realtime Database	Success
7.	Egg productivity chart	-The application will display a graph of laying hens productivity every time	Success
8.	Graph of changes	-Tthe application will display a graph of temperature changes each time	Success
9.	Moisture change graph	-The application will display a graph of laying hens productivity every time	Success
10.	User Profile	-The application will display the profile of the user who is logged in.	Success

3.5 Application Testing on Android

Android-based Reimbursement applications are tested on several types of smartphones. Smartphone to be tested on various screens and OS. The following table tests several smartphones, in Table 2.

Tabel 2. Pengujian di beberapa smartphone

Smartphone Name	Screen Size	OS used	Test result
Xiomi Redmi Note 4X	5,5 inch	Nugget	The application is running well
Samsung J3 Pro	5 inch	Oreo	The application is running wellk
Xiomi redmi 3	5 inch	Lolilop	The application cannot be installed
Oppo A37	5 inch	Lolipop	The application cannot be installed
Oppo A57	5,2 inch	Marsmellow	The application cannot be installed

From the above table it can be concluded that the application can be installed on Android with an operating system version of the nugget and above. On the android version of the nugget down the application cannot be installed.

3.6 Comparison after using the application with before using the application

After doing application development and testing application features. Then the application is used by partners and implemented directly to partner farms. After implementing on the farm, the comparison found by the partners before using the application and after using the application, can be seen in Table 3.

Table 3. Comparison Tables

Before	After
- There are no records	- Take notes in the application

- Know the temperature and weather around the cage based on feelings	- Can directly monitor temperature and humidity remotely
- Not knowing exactly how many eggs are produced on the farm	- Can see in the application the amount of production, and data displayed in the form of statistics can see the amount of livestock productivity each time.
- Data is not neatly stored	- Application of all farm data has been recorded and presented with informative
- Manajemen kandang tidak teratur	- Menajemen kandang menjadi teratur

Table 4. Table of User Satisfaction

Partner satisfaction with the Kancer system		
a.	Completeness of Content	satisfied
b.	Accuracy	satisfied
c.	Display	satisfied
d.	Ease	satisfied
e.	Accuracy	satisfied

4. Conclusion

This Laying Chicken Development Monitoring Application is an application that is designed to assist laying hens in monitoring the development of their chickens. This application stores and sends data in realtime. Breeders can monitor the development of their chickens remotely while they are still connected to the internet. In the application, breeders can see detailed information about breeding and laying hens, ranging from age, sex, number of eggs produced, and history of productivity of chicken eggs, and also breeders can see the temperature and humidity in a stable through the application.

References

[1] Badan Pusat Statistik 2019 Produksi Telur Ayam Petelur menurut Provinsi, 2009-2018 *Website BPS*

[2] Tristianto C 2018 Penggunaan Metode Waterfall untuk Pengembangan Sistem Monitoring dan Evaluasi Pembangunan Pedesaan *J. Teknol. Inf. ESIT* **12** 41–56

[3] Hendini A 2016 Pemodelan UML Sistem Informasi Monitoring Penjualan dan Stok Barang (Studi Kasus: Distro Zhezha Pontianak) *J. Khatulistiwa Inform.* **IV** 107–16

IOP Conf. Series: Materials Science and Engineering **846** (2020) 012019 doi:10.1088/1757-899X/846/1/012019

Computational study of action potential initiation and action potential backpropagation in mitral cell of the olfactory bulb

A Yuniati[1], A D Ardiyanti[1]

Department of Physics, UIN Sunan Kalijaga, Jl. Marsda Adisucipto Caturtunggal Depok Sleman Yogyakarta 55281, INDONESIA

*Email : anis.yuniati@uin-suka.ac.id

Abstract. A neuron sends information in the form of electrical activity called an action potential. The action potential propagates along the axon to all parts of the cell body. The opposite phenomena of this mechanism were observed in the neurons from various brain area as action potential backpropagation. We studied computationally the action potential initiation and action potential backpropagation in the model of a mitral cell of the olfactory bulb that consists of soma, primary dendrite, secondary dendrite, tuft dendrite, axon hillock, and initial segment. The neuronal activity was constructed with voltage-dependent sodium channels and potassium channels. In the results, we provide that the action potential initiation occurs in soma with an amplitude of 40 mV and the action potential propagation decrease as the distance gets farther from the soma. By plotting the amplitude as a function of the distance from the soma, we obtain the magnitude of action potential backpropagation amplitude has an average of about 42.7 mV. This result consistent with observations in experiments where the normalization of action potential backpropagation amplitude of the mitral cell as a function of the distance from the soma is constant.

1. Introduction

Action potential (AP) is a way of neurons transmit the information or communicate with one another in the nervous system in the form of an electrical signal. The nerve impulses are generated in the nucleus of nerve cells and travel along an axon that consist of the myelin sheath and the nodes of Ranvier to the axon terminal. The nerve cells are surrounded by a membrane that allows some ions to pass through and block the passage of other ions. The action potential (AP) is the result of ions moving in and out of the cell. The propagation of AP has been widely measured experimentally and studied computationally in various neuronal types. The AP is an all-or-none event with decreasing amplitude as a function of the distance from the soma

A reverse process of AP has been recorded in some experimental methods as action potential backpropagation that occurs along the axons leading to the dendrite caused by active conduction and triggers a second spike on the dendrite originating from the initiation of the action potential of the axon. Dendritic recording electrophysiology using patch-clamp technique, infra-red differential interference contrast (IR-DIC) optics, and fluorescence imaging technique are an example of an experimental procedure that has been used to measure the AP backpropagation[1, 2]. These measurement techniques provide essential data to understand the mechanism and other properties of backpropagating AP, but they also have a few disadvantages. The study of the computational model is an alternative to overcome the limitation from the experimental methods.

The AP backpropagation has been observed in various mammalian neuronal types such as cerebellar Purkinje cells, hippocampal interneurons, dopamine neurons, and mitral cells. From these varieties, only mitral cells of the olfactory bulb that has been measured in direct recordings from both the thick apical and the thin lateral/basal dendrites[3-9]. The functions of backpropagating APs mainly related to synaptic efficacy that affects synaptic plasticity, neuronal firing, and synchronization of developing a network. One of the functions of AP backpropagation is deal with local synaptic feedback, forming an intricate network of dendro-dendritic synapses as observed in the mitral cells of the olfactory bulb and can be concluded that the AP backpropagation may be responsible towards olfactory function[5].

In studying the AP initiation and AP backpropagation phenomena, we construct a model of the mitral cell of the olfactory bulb that consists of soma, primary dendrite, secondary dendrite, tuft dendrite, axon hillock, and initial segment. The amplitude of AP backpropagation formed at different dendrite location is observed as a function of distance from the soma. The model was simulated with the NEURON program using python. The NEURON simulator provides an environment to simulate a biophysical neural network with realistic biological membranes[10-12].

2. Methodology

In this study, we built the model using the following steps:

2.1. Developing the model

The starting point uses the interface in the NEURON main menu is by clicking Build and then Cell Builder. In the top of cell builder will appear an array of buttons and click the Topology where we can set up the model. In this window, we start to model the mitral cell that consists of soma, dendrite, and axon with make session mode. We label each part of the cell with Basename menu and click on the part of the cell. Another option to develop the model cell is writing script with python with the following steps:

a. Connecting python and NEURON simulators by importing the NEURON module into python.
We build the model computationally using the combination of NEURON with Python programming language. There are several ways to use NEURON with python that can be followed. Here we import NEURON in a running python, using NEURON as a module and import it in the first line of the python program list. The first and foremost module is the GUI neuron module. The second modules to be included in the program list are the matplotlib and numpy that are used to display graphics.

b. Construct the model of a mitral cell.
The model of the mitral cell is constructed based on the experimental morphological results with details of topology, segments, and geometry. The morphology specifications of the mitral cell are presented in Table 1.

Table 1. The morphology of mitral cell[13]

	Soma	Pri	Tuft	Sec	Axon hillock	Initial Segmen	Node of Ranvier	Myelin sheath
Diameter (μm)	20	3.5	0.51	3.4	20:1.5	1.5	1	1.5
Length (μm)	25	370	180	500	5	20	1	1000

c. Connecting each part of the cell
All part of the cell is being connected to form a morphology of mitral cell. The dendrites begin with two dendritic tuft branches and then connected with two long primary dendrites, and the last connected to soma are two secondary dendritic branches.
The axon starts with the axon hillock that connects the soma with the initial segment axon. There are two initial axon segments which are connected to five nodes of Ranvier and myelin sheaths that form a long axon.

2.2. *The kinetic models*
In modeling the AP initiation, we have to set the parameter value of each section. The parameters used in the model are initial resistance (R_a) and membrane capacitance (C_m) with the value equal to 100 Ωcm^2 and 1 $\mu f/cm^2$ respectively. These parameters were set to have the same value in every part of the mitral cell. We have to add the parameter of resistance formed after the action potential (R_m) for the process of AP backpropagation and give the value equal to 2800 Ωcm. There are two kinetic models used in the program, active kinetic model and passive kinetic model. An active kinetic model that is used in modeling the AP initiation is Hodgkin-Huxley neuron model to modeling the dynamics of the action potential initiation membrane in soma, axon hillock, and axon initial segment. This model include sodium conductance (g_{Na}), potassium conductance (g_K), leak conductance (g_L) and leak reversal potential (E_L) with the value are 0.12 S/cm^2, 0.036 S/cm^2, 0.0003 S/cm^2, and -54.3 mV respectively. In the AP backpropagation, the chemical mechanism was scripted in NMODL language. NMODL files will be extracted into files with the extension of .c and .o so that they can be recognized. NMODL is used to model a chemical mechanism that occurs in mitral cell that involves potassium ions and sodium ions. In this model, there are two potassium ion channels and three sodium ion channels written in NMODL syntax. The solution of the potential equation for the ion channels use the exponential Euler method. The NMODL files are inserted in each part of the mitral cell with different ion gate compositions according to the specifications of the mitral cell. There are two sodium ion channels in soma and one potassium ion channel, while in dendrites there is one sodium ion channel and three potassium ion channels. A passive kinetic model is used in the tuft dendrite, primary dendrite, and secondary dendrite. The passive kinetic model includes passive conductance (g) and leak reversal potential (E) with the value 0.001 S/cm^2 and -65 mV, respectively. The stimulus is inserted onto the soma to induce the membrane dynamics. In this model, we use IClamp stimulus model with an amplitude of 25 mV, a delay of 20 ms and a duration of 1 ms for AP initiation, while the amplitude of 0.8 mV, a delay of 1 ms and a duration of 1 ms were set for AP backpropagation.

2.3. *Plotting the graphs*
In this model, we use the matplotlib packages feature in python to plot the graph. The graph visualizes the AP initiation in the soma, the AP propagation along an axon, and the AP backpropagation on the soma, secondary dendrite, primary dendrite, and tuft dendrite.

3. Results and Discussion
The model of mitral cell is described in figure 1. The mitral cell consists of two dendritic tuft branches, two secondary dendritic branches, primary dendrite, soma and axon with five myelin and five nodes of Ranvier. The mitral cell that is constructed with nrngui is shown differently from the plotting graph with the python script.

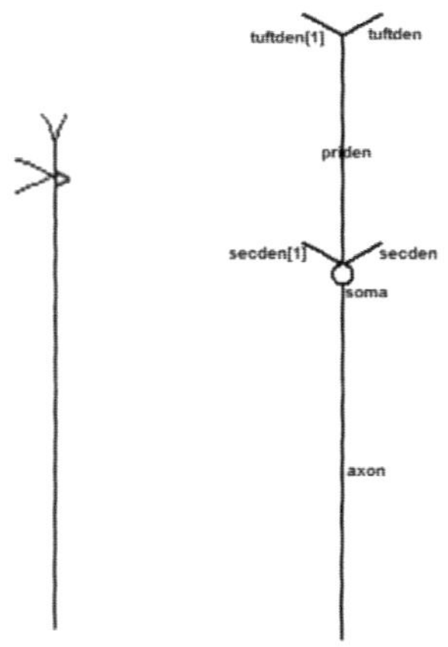

Figure 1. Plotting graph of mitral cell in the program with python (left) and another way to construct a mitral cell using NEURON environment (right)

The results of AP initiation in the soma and AP propagation along an axon are shown in figure 2 and figure 3, respectively. The IClamp stimulus model was applied to provide the initial potential and result of the action potential that propagates along an axon. The AP initiation occurs in soma where the stimulus was applied with the amplitude of 40 mV. The AP initiation is the result of the migration of sodium and potassium ion across the nerve cell membrane, thus triggering the occurrence of depolarization. The action potential then propagates with diminishing amplitude along an axon. The amplitude of AP decrease as the function of distance from the center of the electrode stimulus. The AP amplitude decreases from the first myelin sheath to the fifth node of Ranvier before it reaches the nerve terminal.

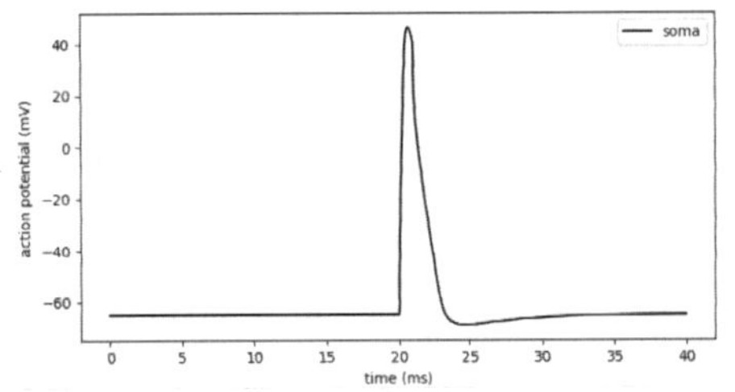

Figure 2. The AP initiation in the soma

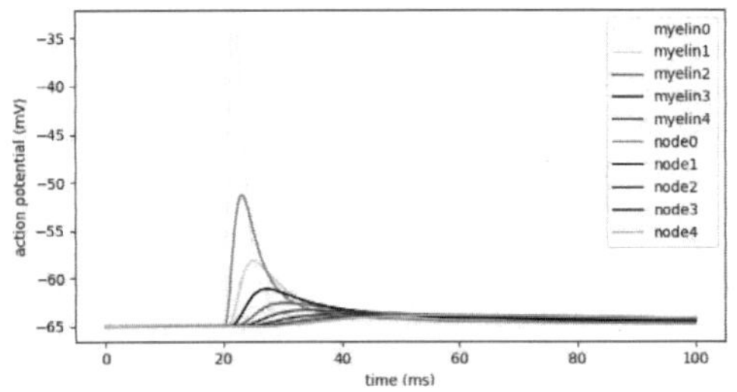

Figure 3. The AP propagation along an axon

The AP backpropagation in the soma, primary dendrite, secondary dendrite, and tuft dendrite of mitral cell is shown in figure 4 with its enlarged peak. The magnitude of the AP backpropagation amplitude in different parts of the mitral cell as a function of distance from the soma is almost the same, as shown in figure 5. The details data of AP backpropagation amplitude and the distance of each part from the soma is described in Table 2.

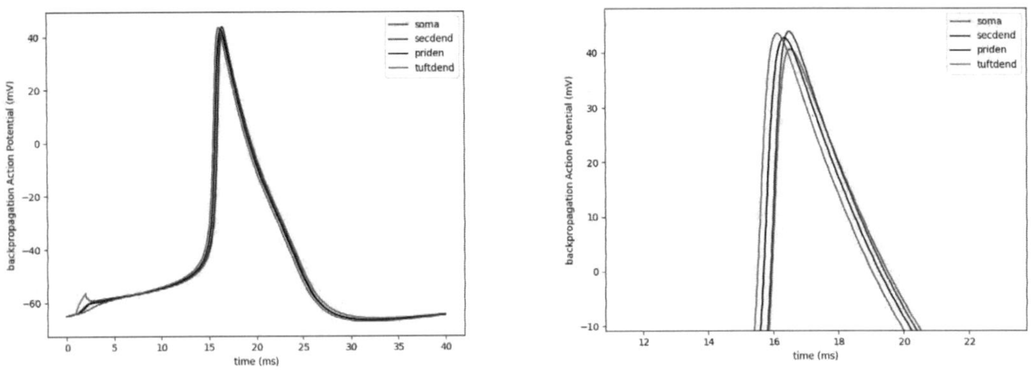

Figure 4. The AP backpropagation of mitral cell along the dendrites (left) and enlarged figure the peak of AP backpropagation (right)

The magnitude of the amplitude of AP backpropagation is influenced by several things such as the density and character of the ion channel, the neuronal morphology and neuronal activity[14]. Moreover, the AP backpropagation is also influenced by the initial resistance (R_a), the resistance formed after the action potential (R_m), the distance from the soma (x) and diameter (d) from the dendrite[15] as described in the following equation

$$V_x = V_0 e^{-\frac{x}{\lambda}} \quad (1)$$

with λ equal to

$$\lambda = \left(\frac{R_m}{R_a}\frac{d}{4}\right)^{\frac{1}{2}} \quad (2)$$

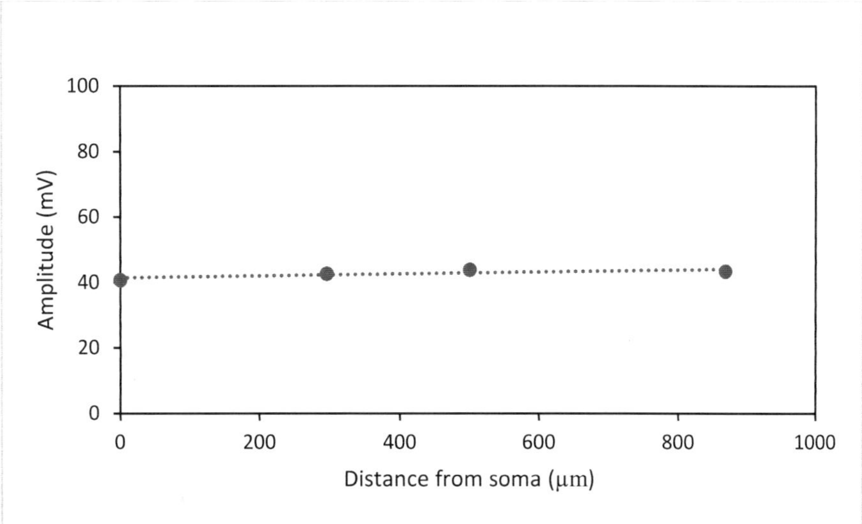

Figure 5. The magnitude AP backpropagation amplitude as a function of distance from soma

The kinetic model parameter of AP backpropagation of this mitral cell is Ra and Rm with the same value in each part. The initial amplitude (V_0) is determined by the amplitude of the stimulus given at 0.8 mV with the different chemical mechanism of sodium ion channels and potassium ion channels in each part. The AP backpropagation amplitude in this model is determined by the initial amplitude (V_0), distance (x), and diameter (d) of the parameters. The distance and diameter are in the order of the micrometer so that the difference of amplitude in each part of the dendrites. These results are consistent with the equation and experimental research conducted by Waters et al., 2005[5], as illustrated in figure 6. The figure shows the summary of AP backpropagation in different cell types. In the mitral cell, the normalized amplitude of AP backpropagation as a function of the distance from the soma is constant. In our simulation, the difference amplitude of the AP backpropagation in various part of the dendrites is very small, almost constant with an average of about 42.7 mV. The soma section which is the closest part to the axon shows the lowest backpropagation amplitude, due to the special attenuation of the sodium ion channel in soma, while the attenuation in the dendrites is only influenced by distance from the source of stimulus injection[15].

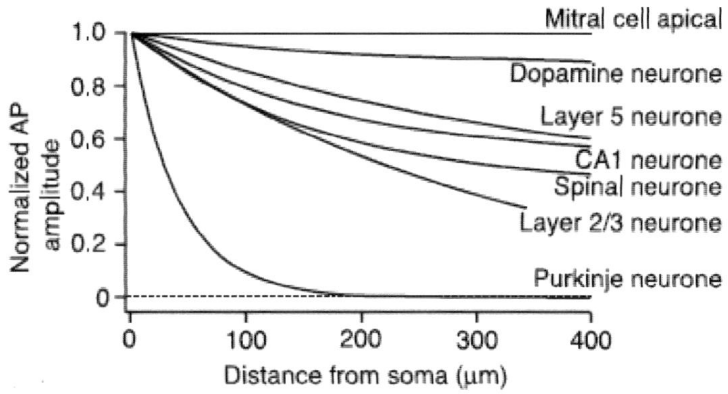

Figure 6. The normalized amplitude of AP backpropagation in different cell types as a function of the distance from the soma. Adapted from [5]

The phenomena of AP backpropagation in different cell types leads to the question on the mechanism of the occurrence of AP backpropagation. The active dendritic conductance and the neuronal morphology of the dendritic tree are considered as the main factor that affects the backpropagation[5]. The synaptic input can modulate the AP backpropagation, and this interaction leads to the enhancement of the backpropagation. In figure 7, we show the modulation AP backpropagation on each part of the dendrites. The modulation was carried out by varying the stimulus with a magnitude of 0.2 mV, 0.4 mV, 0.6 mV, and 0.8 mV. From the variation of these values, the magnitude of the amplitude is obtained equal to the initial modeling stimulus of 0.8 mV.

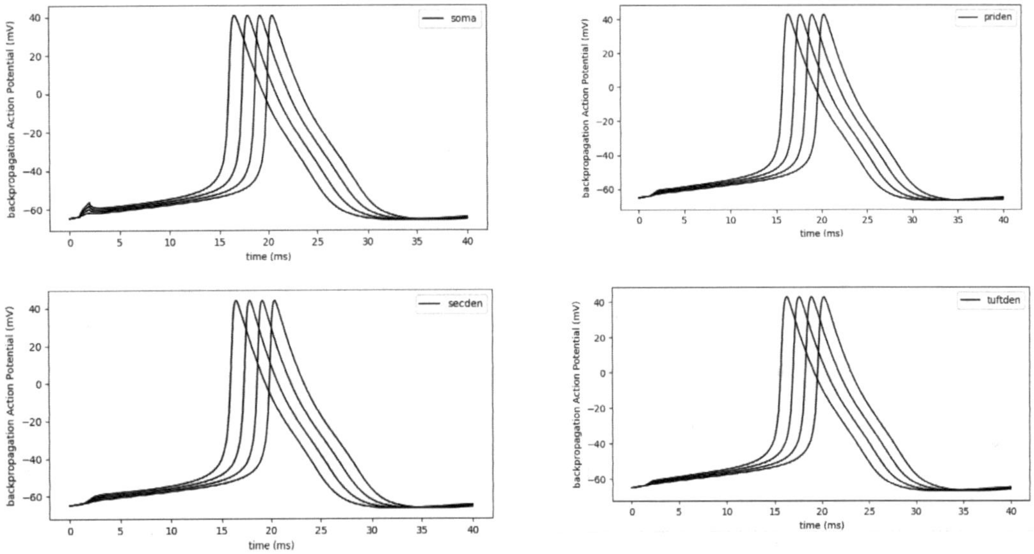

Figure 7. The modulation of AP backpropagation on each part of the dendrites

4. Conclusion

Modeling of the AP initiation and AP backpropagation in the mitral cell of the olfactory bulb has been done using a combination of NEURON and Python simulators. The AP initiation occurs in soma with a magnitude of 40 mV. The AP propagates along an axon with decreasing amplitude as a function of distance from the soma. The AP backpropagation occurs in soma and dendrites with almost the same amplitude in each part of the dendrites about 42.7 mV. The variation of the stimulus was used to modulate the AP backpropagation and result in the amplitude that equal to the initial stimulus about 0.8 mV.

References

[1] G. J. Stuart, Dodt, H.U. & Sakmann, "Patch-clamp recordings from the soma and dendrites of neurons in brain slices using infrared video microscopy, " *B. Pflügers Arch.*, p. 423: 511, 1993.

[2] S. G. Sakmann B., "Patch-Pipette Recordings from the Soma, Dendrites, and Axon of Neurons in Brain Slices. In: Sakmann B., " *In: Sakmann B., Neher E. (eds) Single-Channel Recording,* 1995.

[3] J. Bischofberger and P. Jonas, "Action potential propagation into the presynaptic dendrites of rat mitral cells, " *The Journal of physiology,* vol. 504 (Pt 2), pp. 359-365, 1997.

[4] T. W. Margrie, B. Sakmann, and N. N. Urban, "Action potential propagation in mitral cell lateral dendrites is decremental and controls recurrent and lateral inhibition in the mammalian olfactory bulb, " *Proceedings of the National Academy of Sciences of the United States of America,* vol. 98, pp. 319-324, 2001.

[5] J. Waters, A. Schaefer, and B. Sakmann, "Backpropagating action potentials in neurones: measurement, mechanisms and potential functions," *Progress in Biophysics and Molecular Biology,* vol. 87, pp. 145-170, 2005/01/01/ 2005.

IOP Conf. Series: Materials Science and Engineering **846** (2020) 012019 doi:10.1088/1757-899X/846/1/012019

[6] W. R. Chen, J. Midtgaard, and G. M. Shepherd, "Forward and Backward Propagation of Dendritic Impulses and Their Synaptic Control in Mitral Cells, " *Science,* vol. 278, pp. 463-467, 1997.

[7] J. Christie and G. Westbrook, "Regulation of Backpropagating Action Potentials in Mitral Cell Lateral Dendrites by A-Type Potassium Currents, " *Journal of neurophysiology,* vol. 89, pp. 2466-72, 06/01 2003.

[8] G. Lowe, "Inhibition of Backpropagating Action Potentials in Mitral Cell Secondary Dendrites, " *Journal of Neurophysiology,* vol. 88, pp. 64-85, 2002.

[9] F. Debarbieux, E. Audinat, and S. Charpak, "Action Potential Propagation in Dendrites of Rat Mitral Cells " *The Journal of Neuroscience,* vol. 23, pp. 5553-5560, 2003.

[10] M. L. Hines and N. T. Carnevale, "The NEURON simulation environment, " *Neural Comput.,* vol. 9, 15 Aug 1997.

[11] M. Hines, A. Davison, and E. Muller, "NEURON and Python, " *Frontiers in Neuroinformatics,* vol. 3, 2009-January-28 2009.

[12] M. Hines and T. Carnevale, "NEURON Simulation Environment," in *Encyclopedia of Computational Neuroscience*, D. Jaeger and R. Jung, Eds., ed New York, NY: Springer New York, 2013, pp. 1-8.

[13] G. Y. Shen, W. R. Chen, J. Midtgaard, G. M. Shepherd, and M. L. Hines, "Computational Analysis of Action Potential Initiation in Mitral Cell Soma and Dendrites Based on Dual Patch Recordings, " *Journal of Neurophysiology,* vol. 82, pp. 3006-3020, 1999.

[14] S. Gasparini and M. Migliore, "Action Potential Backpropagation, " ed, 2014.

[15] P. Balbi, S. Martinoia, and P. Massobrio, "Axon-somatic back-propagation in detailed models of spinal alpha motoneurons, " *Frontiers in computational neuroscience,* vol. 9, pp. 15-15, 2015.

IOP Conf. Series: Materials Science and Engineering **846** (2020) 012020 doi:10.1088/1757-899X/846/1/012020

A Virtual Reality Based Surgical Simulation as an Alternative of Halal Surgical Trainings and Better Surgical Planning

Sugeng Rianto

Fisika FMIPA UB

*priantos.ub.ac.id

Abstract. The growing of medical industries need cadaver as a part of full filling better understanding of the components of surgical competency in teaching surgical technical skills in safe and pedagogically efficient environment. However, majority of Muslim scholars argue that using cadaver for surgical trainings are not permissible (haram), thus to find alternative halal procedures replacing cadaver in surgical training are urgent. Moreover, the old approach of 'see one do one, teach one' is no longer acceptable to either the surgical profession or to the well-informed and demanding public. New tools have been developed for teaching and assessing technical skills outside the operating room using virtual reality simulation, which has been applied for many years with great success in many industries including aviation and the military. With software development, the simulators now enable users to perform complete procedures with the added simulation of rare anatomical variations and various pathological conditions. The interface of these high-fidelity systems enables the surgeon to 'feel' the tissue (haptic feedback). Although, the realism of these simulated procedures is still suboptimal and the high cost of virtual reality simulators, the advantages of the high-fidelity, high-cost systems have not yet been demonstrated and our studies have established the potential advantages of procedure-specific simulation. The research using phantom-omny haptic to enable surgeon to "feel, touch, and interact" with tissue, fluid, and bone during surgical procedures using PC based simulator.

1. Introduction

Cadavers or corpes take many important roles in the medical industries. Cadavers can either be used as autopsy material for medical research or as a tool for practicum for novice doctors. Moreover Indonesia, with 280 million population, is actually easy to find corpses and the price can be cheap, even corpses can be obtained without having to buy because there are many unidentified corpses in hospitals. Bodies used for medical practice are usually not identified or are victims of accidents that have been severely damaged, so it is rare in Indonesia to find corpses for surgical practice in perfect condition. In contrast to Indonesia, in developed countries many people are willing to donate bodies for surgical practice even with a very high code of ethics, including keeping corpses in good condition[1].

On the contrary, based on the view of Islamic law, using corpses as research material or practical tools is still an area of dispute between whether or not they are allowed. The stronger opinion (rajih) forbids any types of autopsy, including autopsy in the context of practicum for medical students[2][3]. There are also an authentic traditions that prohibit violating the honor of a corpse, such as chopping, slicing or breaking the bones, particularly forbidden autopsy for Muslim corpses[4].

In addition, medical simulator can be an alternative to replace corpse that is widely used in the medical industry. A remarkable advances in surgical techniques and technology in recent decades, disparities in access to necessary surgical care have dramatically increased [5]. Consistent with other global health care challenges, developing countries and rural regions carry the bulk of the burden. The ability to receive surgical care is dependent on two factors, the accessibility of surgical centers and the availability of qualified healthcare professionals to deliver that care[5]. However, current teqniques have proven to be adequate in meeting the increasing demand for specialized surgical care. In addition to necessary investments to facilities and underlying infrastructure, innovations in surgical education may play a crucial role in ameliorating the world-wide shortage of surgeons, while diminishing gaps in knowledge, and ultimately increase access to care. One of the most promising strategies to achieve these goals derives from an advanced application of virtual reality simulator[6], known as surgical simulator[7].

It is shown that a medical simulator reduce the use of resources. These Emerging technologies can train new procedures, determine the level of competence and create the best planning before operations[8]. Thus, in the near future physicians and surgeons will be trained using simulation, virtual reality, and Web-based electronic learning[9], and It could also be derived as halal surgical procedure.

2. Literatur Review

Virtual reality (VR) is known as a computer simulation of a real or imagined environment that allows the users experience a generated sensory stimulation from the simulated system. For instance, the user can have a visual experience in the three dimensions of width, height, and depth, and may have additional interactive experiences in full real-time motion with sound and possibly with tactile and other forms of feedbacks [10]. Formerly VR could only be run on a special dedicated expensive work station; however, the developments of Graphics Processor Unit (GPU) make it possible to run VR on a standard Personal Computer (PC) with GPU.

In VR the developers often build a computer model to simulate or generate a virtual experience. Sometimes virtual reality is also known as virtual worlds or virtual environments (VEs). These are three dimensional computer generated environments that the user is able to not only experience interactively, but also manipulate in real time [11]. The way humans interact with their physical environments are artificially imitated in VEs. Beside, VEs provide natural interface between humans and computers. The key component of successful VR generation is the provisions of interactions that permit the user to have a sense of touch, as well as the ability to manipulate the virtual world to some extent. To support the interaction between humans and machines, VEs usually include interfacing devices connected through an output of sensory information and input of commands. Several examples of such input and output devices are 3D space mice, data gloves, 3D navigation devices, and Phantom haptic device.

Virtual reality (VR) has been widely used as a training tool in science, engineering, and industry [12][13]. In medical science for instance, VR simulators provide the potential for a realistic, safe, controllable environment for medical students or novice doctors to practice surgical operations, allowing them to make mistakes without serious consequences [13], [14]. Using surgical simulators, a procedure can also be performed for preoperative evaluations and beforehand practices for surgeons. The procedure can be recreated or repeated in a virtual environment without harming patients or placing the patient at risk of trauma or injury.

One of the significant examples is the laparoscopy simulator [15], [16]. This is a virtual reality application in surgery simulation for minimally invasive surgery. The technique provides training systems for surgery that avoid serious damage to the surrounding tissue of the actual area of medical significance. The simulation resembles an operation with a fibre optic camera and surgical tools through a small number of portals in the skin without cutting through muscles and other tissues lying in the way of the organs or joints of interest. The simulator has a synthetic human torso into which can be placed surgical procedures to allow trainees to introduce surgical like devices with force feedback and practice suture skills, gallbladder removal, and other procedures on latex simulated organ, while watching their work on a typical video endoscopic screen.

Despite the recent successful development of many surgical simulators, there are still many research problems remained to be answered in the field in order to provide a truly realistic surgical simulation which includes simulating natural phenomenon involved in real surgery.

3. Methodology

3.1. Volume rendering

Volume rendering is a term that is usually used for a set of techniques to display a series two dimensional (2D) images projection of a 3D discretely sampled data set. Volumetric data are very commonly presented in two ways: direct volume rendering and 3D surface rendering. The direct volume rendering scheme was firstly developed by [10], followed by [11] and [12]. On the other side, researchers such as [13], [14], and [14] have more work on 3D surface rendering. There are also limited research effort that combines both surface and volume rendering together.

In medical visualization, volumetric datasets are gathered from patient images in several ways. Three common devices producing volumetric medical datasets include computed tomography (CT), magnetic resonance imaging (MRI), and single-photon-emission-computed-tomography (SPECT). These devices generate a volume by scanning a series of cross-sections of a human body. Volume graphics is also popularly used in other engineering and science fields for data presentation and visualization. Volume rendering has also been used in haptic visualization especially for exploring scalar and vector fields of various objects [16], [17]

A volume usually consists of many particles that emit and absorb light. Several effects such as iso-surfaces and semi-transparent objects can be achieved by simply manipulating the mapped values of the light emission and absorption coefficients of the original volume data. For haptic visualization, the user will have the potential advantages of the volume rendering when they can naturally view and have sense of touch with the 3D volume in real-time through force feedback of the haptic according to the properties of the real object. However volumetric modelling requires large storage space, and is computationally expensive.

Direct volume rendering does not explicitly extract the geometric of the bounding surface. This method requires every sample value to be mapped to opacity and a color. Geometry extraction usually is undertaken by iso-surfaces visualizations. Volume data can be stored directly in special-purposed graphics memory as a stack of two-dimensional (2D) texture slices or as a single 3D texture object [17], where the texture values represent voxel (volume element) density values of an object. When 3D texture objects are used, the volume rendering process is manageable only by a graphic processor unit (GPU) [18] without the need of assistance from a central processor (CPU). A GPU based volume rendering integrates the value of a continuous volume function reconstructed from discrete sampling points along selected projectors to produce images [19]. Using GPU as a volume processor leads to faster computation and will drastically reduce the computation time.

In this project, a novel combination of surface and direct volume rendering is adapted to take full advantage of all aspects of a volume data representation. Geometry information of objects in immersive virtual environments, depth perception, 3D direct interaction, and manageable collision detection, are all issues that need to be addressed for the objectives of surgical operations and biomedical simulations.

The volumetric object is created by a combination of 2D texture and transfer functions to represent a volumetric entity in a rendering scene and inherits functions to volume's position, orientation and origin. This volumetric data then is stored as 3D texture or is render to the screen.

3.2. Volume Segmentations

A digital image can be split into multiple segments (sets of pixels, also known as super-pixels). This partitioning process usually is defined as Segmentation. The segmentation results a simple representation of an image into something that is more meaningful and easier to analyze [20]. In special case, volume segmentation is typically used to locate scalar field data and boundaries (surface) in volume data set and assign a label to every voxel with the same label share certain visual characteristics.

Although, there are many floating methods on volume segmentation, this project only focuses on two segmentation methods: marching cubes and color transfer functions. Marching cube refer to the work of [21] that is usually utilized for extracting a polygonal mesh of an iso-surface from a three-dimensional scalar field (sometimes called voxels). In two dimensional (2D) data application similar method is called the marching squares algorithm. Marching cubes algorithm proceeds through the scalar field, taking eight neighbor locations at a time (this forming an imaginary cube), then determining the polygon(s) needed to represent the part of the iso-surface that passes through this cube. The individual polygons are then fused into the desired surface.

The gradient of the scalar field at each grid point is also the normal vector of a hypothetical iso-surface passing from that point. When we interpolate to these normal along the edges of each cube, the normal of the generated vertices can be created. These normal are essential for shading the resulting mesh with some illumination model. The Marching cubes application mainly concern with medical visualizations including CT and MRI scan data images, and special effects or 3-D modelling with what is usually called meta-balls or other meta-surfaces.

The color transfer function method on the other hand, has role in emphasizing features in the data mapping values and other data measures to optical properties. The simplest and most widely used transfer functions (TF) are one dimensional (1D). When 1D transfer function is applied, the ranges of data values is mapped into color and opacity and usually is implemented with 1D texture lookup tables. When the lookup table is created, color and opacity are usually assigned separately by the transfer function. For correct rendering, the color components need to be multiplied by the opacity, because the color approximates both the emission and the absorption within a ray segment (opacity-weighted color)[22]. Since in 1D TF utilize only data value measurements to control the assignment of color and opacity, the effectiveness of classifying features in the data may be limited. To generate better visualization and control to the images the gradient magnitude should be included. When we use a combination of data value measurement and gradient magnitude to compose transfer function, the result is two dimensional (2D) transfer functions.

The transfer function design is not simple as in marching cubes method. Beside we need a difficult iterative procedure that requires significant insight into the underlying data set, other information, such as the histogram of data values indicating which ranges of values should be emphasized, should be provided. In the 2D transfer function, the presence of material boundaries can be identified by arches within the value and gradient magnitude distribution. To assign the resulting color and opacity to voxels with the corresponding ranges of data values and gradient magnitudes, a set of brushes are used to paint into the 2D transfer function dependent texture [23].

3.3. Hardware accelerations

Interactive direct volume rendering particularly for haptic rendering has been restricted to high-end graphics workstations and special-purpose hardware, due to the large amount of tri-linear interpolations that are required for high image quality. Implementations using the 2D texture capabilities of standard PC hardware usually render object-aligned slices, in order to substitute tri-linear interpolations by bilinear interpolations. Consequently the resulting images often contain visual artefacts caused by the lack of spatial interpolation.

The advancement of graphics hardware and the creation of its new feature of programmability have inspired people to develop algorithms transferring much of the processing needs from CPU to GPU on common PCs [24][17][23]. The intrinsic parallelism and efficient communication features of a GPU mean it can perform calculations much faster than a CPU. Furthermore, the power of GPUs is currently increasing at a much faster rate compared to that of CPUs, and the algorithm realized on GPU can have great potentials in the future. To achieve fast volume rendering, this project implement an algorithm that takes advantage of FX Quadro. They hold some new characteristics, such as reading texture data in vertex programs, supporting dynamic shift command in pixel programs and even longer lengths for shader programs. A rendering technique is proposed in this project that significantly improves both the performance and the image quality of 2D texture-based rendering. The multi-texturing capabilities of GeForce Quadro are exploited to enable interactive high quality volume visualization to demonstrate hardware based rendering to get efficiently render shaded iso-surfaces and to compute diffuse illumination for semi-transparent volume rendering at interactive frame rates.

The traditional slice-based 3D texture volume rendering is performed by slicing the volume in back-to-front or front-to-back order with planes oriented parallel to the view plane [18]. Each fragment program gets the sampled color from texture by tri-linear interpolation and then blends them with the current value in the color buffer using the proper blending functions, as described in Eq.1 and 2.

$$C_d = (1 - A_s)(C_d + A_s C_s) \quad (1)$$

$$C_d = (1 - A_s)(C_d + A_d C_s) \quad (2)$$

And

$$A_d = A_d + (1 - A_d)A_s$$

C_d, A_s and C_s, are accumulated color from the viewing ray, opacity absorption, and the color opacity weighted emission of the incoming fragment, respectively. In the front-to-back order, the accumulated opacity will be stored in an A-buffer and the A-test should be enabled.

The volume data is stored using 3D texture and resample them in pixel processing. To reduce computing time, the volume bounding cube is not sliced in CPU, but is ray casted in GPU. CPU only produces a polygon parallel to the view plane as the basic ray, and prepares some necessary parameters for GPU. To further speed up the intersection calculations, the slabs are used instead of slices [25], together with pre-integrated classification. A slab is the space between two adjacent slices, which can be rendered as a slice with its immediately neighboring slice, either in the back or in the front, projected onto it.

3.4. Haptic Rendering

Originally *haptic*(Greek) refers to the sense of touch and comes from the Greek verb *haptesthai*, meaning to contact or to touch. Currently, haptic refers to haptic technology that mean a tactile feedback technology that takes advantage of a user's sense of touch by applying forces, vibrations, or motions to the user [26]. This mechanical stimulation may be used to assist in the creation of virtual objects (objects existing only in a computer simulation), the control of such virtual objects, and the enhancement of the remote control for machines and devices (tele-operators). More generally, haptic is normally used today to concern to the science of touch in real and virtual environments. This would include not only the study of touch capabilities in different organisms, including humans, but also the development of engineering systems to create haptic virtual environments.

The concepts of the haptic rendering infrastructure refers to the most common haptic action or rendering surfaces to provides extensive surface rendering functionality in addition to the general force model. Surface rendering restricts itself to the task of producing force renderings of mostly solid objects that have a well-defined surface. This object may be somewhat soft and spongy. We implements a surface-rendering force model at a low level that records state information about the surfaces being touched and the point at which they are being touched. There are two points in space that are important in describing the surface rendering: the proxy and the finger. The proxy is a small sphere that is constrained to remain outside of all surfaces in the scene, whereas the finger is the actual haptic device position which is not as constrained. Visually, proxy and finger are one object that is called as a tip. When the tip touches the surface and the user push further then the proxy and finger will be split out. The proxy remains on the surface of the object and finger a bit immerse from the surface (see Fig.1). Thus the user fell like touches the spongy surface or fell have repulsion from spring. But we see only the proxy on the surface unless we program to show both proxy and finger with transparent surface to show the finger. The stylus is visual representation of the haptic in virtual environment. The stylus is composed of handle and the tip (see Fig.1). In the normal haptic, stylus handle have no collision detection to the surface, so that it could be immersed inside the object.

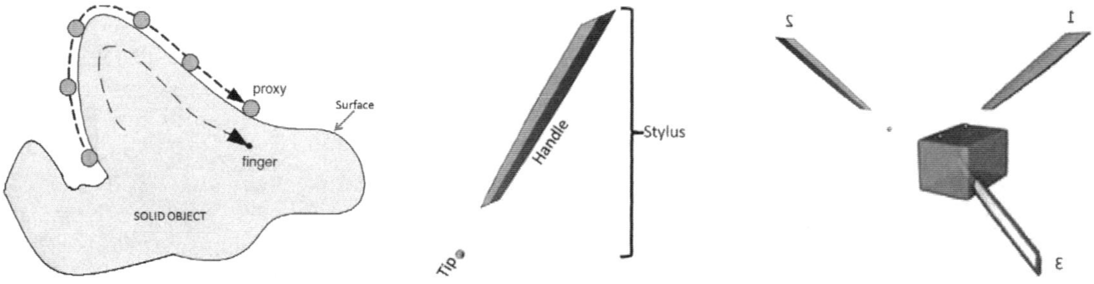

(a) (b) (c)

Figure 1. (a) The proxy and finger representing haptic tip when touching virtual surface. One model of visually haptic representations. (a) Haptic with visual stylus, handle, and Tip. (b) Haptic no.1 touch the surface from the top object, Haptic no. 2 touch from the side with handle immerse in the object, and haptic no.3 in space with no touch.

A haptic rendering is the process of how a virtual environment system allows user to touch and sense virtual objects in a simulated environment. The implementation of this process is designing force computations, so that we can decide what force should be exerted and how we will deliver the forces to users. A force computation generating feedback into a haptic is known as a force model. This force model has a means of mapping from the haptic device position to a force vector [27]. It should also deliver constrain or reaction forces that are sensed by the user through the haptic device. Although a force model is generally updated in a scene-graph loop at about 60 Hz, this force needs to be used extensively by the real-time loop at about 1000 Hz by repeating the call to its evaluate function. To keeping the stability and quality of the haptic rendering the loop is maintained in the high frequency processing, whereas the visual aspects can have frame rates only in the order of 30 Hz [28], [29].

In this project haptic rendering algorithm refers to *Reachin API engine* that execute two parallel event loops. One loop is execute once per millisecond and is used to interface with the haptic rendering interface and hardware. This loop is called as real-time loop and is keeping in millisecond interval to preserve the stability and quality of the haptic rendering. The real-time loop does not interface directly to the scene graph or field network. It works with temporary, simplified and localized objects which are derived from the scene graph representation. At the concurrent time, in a separate loop is running the scene graph loop. The scene graph loop simply runs as fast as possible given the CPU constraints of the real-time loop. This loop responsible for interfacing with, and interpreting, the scene graph data structures (via traversals).

Haptic rendering apply a collision traversal that refer to detecting collisions of the haptic device with various object in the scene. The collision traversal and visual rendering that occurs in scene graph loop are detected in real-time loop for the calculation of the force reaction and this force then is delivered back to the user in real-time loop. Although the haptic and graphics rendering are split into different loops, they remain synchronized by sharing the same data set. When each loop has separate data set, then, manual synchronization should be done. The synchronization is particularly useful to avoid issues concerned with artefacts that come up when the haptic rendering loop appear faster or slower than the real-time loop. For example, when simulating tissue cutting, changes in geometry require frequent updates in both graphic and haptic dataset in real-time. While haptic loop is writing to the shared memory, graphic loop is reading, and then the artefact may appear. To overcome this situation the synchronization to be done to avoid conflict between loops.

The haptic rendering goal is to create a 3D model that will be recognized by the device (haptic). The model may be similar to the visual object but the may have different resolution to keep high frame rate interaction with the device. The object that is established on the display is not directly generating object that can be recognized by haptic unless the haptic rendering is undertaken in a separate thread. More detail explanation regarding force model design and its interaction with object in virtual environment (VE) will be explained in the next sections.

3.5. Surface and volume Haptics

The establishment of a surgical simulator requires developing and integrating several components consist of defining the virtual three-dimensional (3D) environment, determining the properties of the objects in the VE, managing the interface to the 3D interaction devices (haptic, space mouse) including force feedback, and design of the visual displays. Traditionally force model algorithms rely on surface modelling [30],[31], and deal with the problem of interacting force with various surface representations. The surface can be a form of geometric roughness, object collisions, an incremental surface tracing, and probe-surface contact forces interactions (shears, torques, viscosities etc.). There two main visual renderings to be considered during force model design include surface and volume rendering. Surface renderings are believe to be computationally intensive particularly during triangulation or polygon build-up process from the vertices [32]. The other important issue in interactive environment is collision detection between surface and device to generate force. The number of contact points in surface modelling is related to the number of vertices on the contacting surfaces, in the range of tens to hundreds contact points per collision instance. On the other hand volumetric objects can have the thousands for relatively small contact surfaces of volume elements [33], [34].

The iso-surface that has been discussed in previous section means nothing for haptic force-feedback unless we define the force model to generate feedback during touching haptic over the surface. The force model that is applicable for surface rendering usually is defined as surface haptic. The surface haptic should provide several properties of different forces generating feedback such as surface roughness, plastic or membrane like surfaces, deformed surface, magnetic or sticky surface, or slippery surface. These types of surface haptic can be rendered based on texture bump mapping or scalars based values and are used to send the force feedback through haptic device to user hand according the real physical properties of the object. For instance, bone should be feel hard, fat is soft and sticky, and skin is flexible as a membrane.

The surface haptic is not too complex to be calculated, however, every data that is not a part of the surface is unrepresented in the surface haptic. On the other hand, the volume data is defined as a vector valued function to produce force feedback solely from the data around the haptic tip and the velocity of the tip. The possible force could be generated based on local gradient vector depicts the orientation as well as the magnitude changes in the scalar data to generate viscosities like forces. This force may applicable to represent fluidic contents in the volume.

Force models for volume haptic usually are approximated from a proxy point that is executed at every scene graph loop in the servo motor. The volume datasets are measured based on the gradient or curvature information from scalar values of a volume object. In volume haptic, scalar value and gradient magnitude of the images have analogy into material (type of objects), surface penetrability and surface distinctness (see Table 1).

Table 1. Conversion of Transfer Functions to Visual and Haptic Volume Rendering

Visual Volume Rendering	Haptic Volume Rendering
Scalar value→colour	Scalar value→material
Scalar value→opacity	Scalar value→surface penetrability
Gradient magnitude→opacity	Gradient magnitude→surface distinctness

3.6. Deformations

A deformation are created when the body suffering stress. Stress is internal forces acting within deformable body and measures the average forces per unit area of a surface within the body on which internal forces act. These internal forces are a reaction to external forces applied on the body. In the case of haptic rendering, the deformation happens as a surface reaction to a colliding device/tool. The pre-deformation state of an organ surface exhibits the viscoelastic behaviors of the tissues on and underneath the surface [35]. This kind of behaviors is controlled by elastic deformation, where each layer is considered as a linear, isotropic elastic material. By using Hooke's law, strains are measured along the principal axes; the principal axes are found by using only a few geometric operations aligning the original and deformed triangle and computing the oriented bounding box or ellipse to the triangle. The major and the minor axes therefore provide the two principal axes. By measuring strains along the principal axes, the shear-strains do not need to be explicitly computed. The principal stresses are computed using Hooke's law [36] as in Eq.3.

$$\sigma_1 = \frac{E}{(1-v^2)}(\varepsilon_1 + v\varepsilon_2) \text{ and } \sigma_2 = \frac{E}{(1-v^2)}(\varepsilon_2 + v\varepsilon_1) \quad (3)$$

where E, v and ε are the elastic modulus or the Young's modulus, the Poisson ratio, and the principal strains respectively. Once the in-plane stress has been computed, it is then converted to the forces at the vertices (nodes) of the triangle. Assume that mass is constant and distributed over the triangles. The vertices acceleration is determined by implicit integration method to obtain their new position of the triangle vertices.

3.7. Force Cuttings

The force-feedback uses a method of local model based on reaction force as the blade is pressed into skin and constrained motion in a volume. There are two issues to consider governing the force, motion, and cutting. Firstly, reaction forces (force model) that correspond to the tissue resistance felt as the scalpel pressed deeper into the cut (it is called as axial force) and the blade is constrained by the surrounding tissue to move in the cut direction. Secondly is visual modelling during blade's motion.

There are two thread were applied during a cutting process during surgical cutting simulation. When the tissue is cut the surface topology are change and the material on the cutting path to be removed. The first issue related to graphical modelling using mesh triangulation and the second using volume removal.

In 3D case, elastic behavior was simulated with damped mass-spring systems instead of the FEM. Assuming a line-shaped scalpel were used, the re-meshing process around an active node can be done for curved instead of flat surfaces,. Then the scalpel can intersect a curved 3D surface in multiple points, so a consistent model of cutting allows multiple incisions, each with an active node. These active nodes can interact: incisions may meet, leading to annihilations, or incisions may hit folds, leading branches. We assume that the surface is given as a triangle mesh with a boundary, and that no further information on the surface shape is known. The virtual scalpel is a line segment, and its movement is given by sampled positions of its endpoints. The endpoints are assumed to move with constant velocity between the samples. During a cut, active nodes are part of the boundary of the surface. An incision is defined by the two boundary edges incident with an active node again are called cut edges. This process is useful to reflect the topology change of the surface triangle mesh after cutting.

Material removal, on the other hand, is also approximated according to the trajectory of the cutting tool. A voxel will be removed from the volume of the tissue with two conditions: when the contact force is bigger than the threshold of the tissue cutting resistance force and when the voxel exist over the grooving triangle. According to the properties of brittle material, material removal will only occur within the local area around the action point of the contact force. Therefore, the material is removed along the negative direction of the normal vector of each related voxel in the tissue volume.

4. Implementation and Result

This section describes the implementation of 3D modelling and haptic rendering, and shows some of the experimental results. The results presented here concentrate on visual and haptic renderings, and the discussions are split into human organ modelling and surgical tool modellings.

4.1. Human Organ Modelling

This section presents three human organ models generated from a CT scanner according to algorithms as presented in previous sections. The raw images were collected from Visible Human data set (VHD) [37]. The established models can be modified in certain ways such as deformation, volume clipping, and volume and surface cutting. These models are results of image reconstruction and segmentations.

The objects are segmented using two dimensional (2D) color transfer functions. This color transfer function can segment more than seven distinct objects. For instance, the head can be segmented into skin, connective tissue, ape-neurosis, loose areolar tissue, brain, pericardium and bone. The iso-surfaces can be generated by choosing one of the segmented values from the color transfer function. Besides using the color transfer function, the segmentation of the organ can also be determined according to the CT number. A Normalized CT number that ranges from 0 to 2000 is used in the project, where 0 value is associated to -1000 Hounsfield Unit (HU) of original CT Number and 2000 is 1000 HU. CT number assign the density of a voxel in CT scan on an arbitrary scale on which air has density -1000, water 0, and compact bone +1000. Similarly, Hounsfield unit is the numeric information contained in each pixel of a CT image. HU is related to the composition and nature of the tissue imaged and is used to represent the density of tissue. CT number and HU can be interchanged each other. The original image slices are coded based on gray scale values. The gray scale is also associated with the HU or CT Number. In this project, two or more colors and two transfer functions are used to segment the objects of interest.

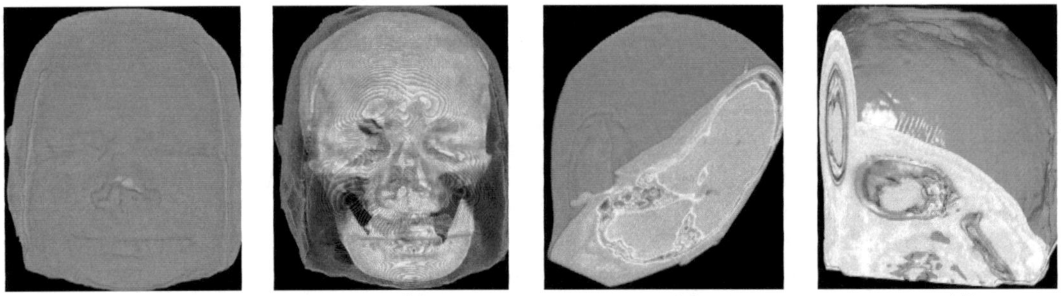

(a) (b) (c) (d)

Figure 2. Volume and isosurfaces rendering for the human head from visible human data set; (a) One 2D transfer function of the volume rendering; (b) Two 2D transfer function of the volume rendering; (c) Three 2D transfer function of the volume rendering with volume clipping; (d) Volume clipping to show the combination of the volume and isosurfaces rendering.

The results shown in Fig.2 show various manipulations of the volume rendering techniques. The first image shows the full volume when only one two dimensional (2D) color transfer function (CTF) is used. In this image all the gray scale images that represent the original CT number are coded based on the combination of two user defined colors mapping (green, orange). In the second image two 2D CTFs are used to show the separation between the skin and the bone. In the third image three 2D CTFs are employed to describe a more detailed separation between organs of interest. The last image shows the combination of volume and iso-surfaces rendering. The volume is represented by the CTF and the iso-surfaces by indexed face set based polygonal surface that is generated from segmentation using 2D transfer function.

4.2. Deformation and Cutting

Several examples of volume haptic and their interactions with organ models are shown in Fig.3. The skin may be deformed and penetrated by a tip, when the user applies a force greater than the boundary stiffness threshold of the skin. However, a bone cannot be penetrated unless the skin as a boundary layer is occluded (see Fig.3).

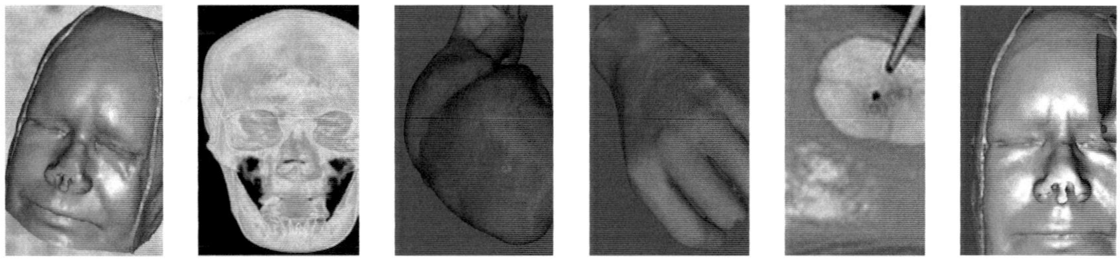

(a) (b) (c) (d) (e) (f)

Figure 3. Segmented objects for organ of interest: (a) Head; (b) Skull; (c) Heart; (d) Hand; (e) Boundary occlusion when the pressing force is applied greater than the surface stiffness; (f). An example of an object model deformation when the haptic tip interacts with the model

The model deformation examples are shown in Fig.3ef. The deformed coordinates are used for graphics rendering whereas the un-deformed coordinates are used for haptic rendering. The haptic interaction is used to determine the extent of the deformation. Since the graphic and haptic renderings are separated, the different geometry substitution for the purposes of haptic rendering to be develop to keep the high frame rate. For example a high-resolution mesh should be used to provide smooth looking graphical deformation while a coarse mesh can be used to calculate the haptic interaction and object deformation.

The quality of deformation is controlled by the stiffness field. This field controls the extent of the deformation as a function of the distance from the point of contact. The deformation is generated nicely when the triangles in the object surface are not larger than the size of the proxy. Larger triangles often cause the haptic tip to end up on the wrong side of the surface.

Two types of deformations are studied in this project: elastic and clay deformations. The elastic deformation is a local geometry deformation that only occurs when the force is applied to the object surface. When the forces no longer exist on the surface the geometry will return to its initial condition. The clay deformations, on the other hand, have properties like clay where the geometry will remain in the last form when the forces are applied.

The cutting algorithm is an extension of the deformation algorithms. While deformation only changes the geometries of the meshes and the volume to some extents, cutting can result into surface and volume removal and new surface additions on top of geometric changes.

5. Discussion

Several techniques to approximate the formulation of constrained and reaction forces with 3D volumetric model have been developed. There are three main haptic interactions that have been developed in this project include: surface haptic, volume haptic, and proxy-based volume haptic. Each type of the surface haptic calculate forces to touch virtual objects based on explicit surface descriptions that represent iso-values of scalar data. The explicit surfaces can be polygons that prevent the haptic tip from penetrating the object. The surfaces will return force feedback into the user's hand and push the tip away from the object surface.

Volume-based haptic rendering, on the other hand, has no surface representation, thus the standard proxy-based method cannot be implemented. There are procedures to create the direct volume haptic have been explained in [38][39]. These methods propose a haptic feedback as a computation of vector function. The computation is formulated using the scalar data around the tip and the current tip velocity. A force to feel the boundary surface is obtained by the gradient vector representing the variation of the orientation and the magnitude in the scalar data. In this project, we employed local or global surfaces that is adapted from [40] to compute proxy-based surface haptic. Moreover, to get a haptic interaction with a volume data, a direct volume haptic has been generated. This work refer to [41] and [16] with some modifications including force model design and hardware acceleration in computation and visualization. The residual force and the new proxy position were not calculated using residual force minimization based on adaptive step-length numerical solver as in [16], in this project implement our adapted CIP that I consider have more efficient computation than its predecessor, especially when predicting forces through the closest of two boundary surfaces stiffness values.

Furthermore, along with haptic penetration, the boundary surface is removed when a pressure is forced over a surface greater than its predefined threshold. Adaptive surface re-meshing method is employed and the voxels underneath are removed. This approach splits different layers not purely by visibility of objects of interest, careful analysis of ray profiles and boundary values need to be carried out at possible feature transition areas.

6. Conclusion

A method to effectively calculate the forces for interactive haptic-volume rendering in immersive virtual environment for surgical simulation is developed. Overall, this study has achieved high quality visualizations and force feedbacks for haptic on a reasonable frame rate. The volume size of 512x512x220 and 256 x 256 x 128 can be visualized at about 20 fps without haptic jerking. The speed slows down to about 5 fps during large deformations and generally will drop to below 10 fps when the pressure force is applied over the boundary surface for an un-penetrable object (bone), the frame rates keep at above 15 fps when the GPU computations were employed.

In Addition, since all surgical related simulations in this project are not related or used any organ, skin or bone of the human, it could be simplified that the surgical procedure derived from this simulation can be revealed as hallal surgical procedures.

References

[1] Prasetya Online, “Sugeng Prianto: Lebih Presisi dan Efisien Dengan Bedah Virtual,” *Prasetya*, 2011. [Online]. Available: https://prasetya.ub.ac.id/berita/Sugeng-Prianto-Lebih-Presisi-dan-Efisien-Dengan-Bedah-Virtual-3606-id.html. [Accessed: 01-Oct-2019].

[2] Farid Ma’ruf, “Hukum Praktikum Bedah Mayat (Otopsi) « Konsultasi Islam,” *Konsultasi.com*, 2009. [Online]. Available: https://konsultasi.wordpress.com/2009/04/24/hukum-praktikum-bedah-mayat-otopsi/. [Accessed: 22-Sep-2019].

[3] B. S. Atiyeh, M. Kadry, S. N. Hayek, and R. S. Musharafieh, “Aesthetic Surgery and Religion: Islamic Law Perspective,” *Aesthetic Plast. Surg.*, vol. 32, no. 1, pp. 1–10, Jan. 2008.

[4] Shiddiq Al Jawi, “Otopsi Dalam Pandangan Syariah Islam - mozaik www.inilah.com,” *Mozaik.com*, 2014. [Online]. Available: https://mozaik.inilah.com/read/detail/2165268/otopsi-dalam-pandangan-syariah-islam. [Accessed: 23-Sep-2019].

[5] J. Dankelman and N. Di Lorenzo, “Surgical training and simulation,” *Minim. Invasive Ther. Allied Technol.*, vol. 14, no. 4–5, pp. 211–213, Jan. 2005.

[6] Sugeng Rianto and Rachel Li, “Fluid dynamic visualisations of cuttings-bleeding for virtual reality heart beating surgery simulation,” in *Conference: Computer Science 2010, Thirty-Third Australasian Computer Science Conference (ACSC 2010)*, 2010.

[7] Tempo Interaktif, “Dosen Brawijaya Temukan Alat Pengganti Mayat - Nasional Tempo.co,” *Tempo* , 2011. [Online]. Available: https://nasional.tempo.co/read/327151/dosen-brawijaya-temukan-alat-pengganti-mayat/full&view=ok. [Accessed: 01-Oct-2019].

[8] Nicolaks Davis, “Cutting-edge theatre: world’s first virtual reality operation goes live | Science | The Guardian,” *The Guradian*, 2016. [Online]. Available: https://www.theguardian.com/technology/2016/apr/14/cutting-edge-theatre-worlds-first-virtual-

reality-operation-goes-live. [Accessed: 01-Oct-2019].
[9] Sam Daley, "10 Companies Using AR And VR In Surgery | Built In," *Buildin*, 2019. [Online]. Available: https://builtin.com/healthcare-technology/augmented-virtual-reality-surgery. [Accessed: 01-Oct-2019].
[10] C. J. Morris and D. Ebert, "Direct Volume Rendering of Photographic Volumes Using Multi-Dimensional Color-Based Transfer Functions," in *Proceedings of the symposium on Data Visualisation 2002*, Barcelona, Spain: Eurographics Association, 2002, p. 115–ff.
[11] Z. Wang, C.-K. Chui, Y. Cai, and C.-H. Ang, "Multidimensional volume visualization for PC-based microsurgical simulation system," in *Proceedings of the 2004 ACM SIGGRAPH international conference on Virtual Reality continuum and its applications in industry - VRCAI '04*, 2004, p. 309.
[12] L. Kim and S. H. Park, "A Haptic Sculpting Technique Based on Volumetric Representation," in *Lecture Notes in Computer Science*, Perales FJ and D. BA, Eds. Springer, Berlin, Heidelberg, 2004, pp. 14–25.
[13] H. Zhang, S. Payandeh, and J. Dill, "Simulation of Progressive Cutting on Surface Mesh Model." DRAFT6-08 Sept02, 2002.
[14] S. Choi and H. Z. Tan, "Toward realistic haptic rendering of surface textures," in *ACM SIGGRAPH 2005 Courses on - SIGGRAPH '05*, 2005, p. 125.
[15] C. Basdogan, C. Basdogan, and A. A. Srinivasan, "Haptic rendering in virtual environments," *STANNEY, K. (ED.), Handb. VIRTUAL Environ. LAWRENCE ERLBAUM, INC*, vol. 2002, pp. 117--134, 2002.
[16] K. Lundin, B. Gudmundsson, and A. Ynnerman, "General Proxy-Based Haptics for Volume Visualization," in *First Joint Eurohaptics Conference and Symposium on Haptic Interfaces for Virtual Environment and Teleoperator Systems*, pp. 557–560.
[17] Wei Li, K. Mueller, and A. Kaufman, "Empty space skipping and occlusion clipping for texture-based volume rendering," in *IEEE Transactions on Ultrasonics, Ferroelectrics and Frequency Control*, pp. 317–324.
[18] J. Kruger and R. Westermann, "Acceleration Techniques for GPU-based Volume Rendering," *Proc. 14th IEEE Vis. 2003*, p. 38, 2003.
[19] M. Ikits, J. Kniss, A. Lefohn, and C. Hansen, "Volume Rendering Techniques," in *GPU Gems*, R. Fernando, Ed. Addison-Wesley, 2004, pp. 667–732.
[20] S. De, S. Bhattacharyya, and S. Chakraborty, "True color image segmentation by an optimized multilevel activation function," *2010 IEEE Int. Conf. Comput. Intell. Comput. Res.*, pp. 1–4, Dec. 2010.
[21] W. E. Lorensen, H. E. Cline, W. E. Lorensen, and H. E. Cline, "Marching cubes: A high resolution 3D surface construction algorithm," in *Proceedings of the 14th annual conference on Computer graphics and interactive techniques - SIGGRAPH '87*, 1987, vol. 21, no. 4, pp. 163–169.
[22] C. M. Wittenbrink, T. Malzbender, and M. E. Goss, "Opacity-weighted color interpolation for volume sampling," in *IEEE Symposium on Volume Visualization (Cat. No.989EX300)*, p. 135–142,.
[23] Milan Ikits, Joe Kniss, University of Utah, Aaron Lefohn, and Charles Hansen, "Volume Rendering Techniques," in *GPU Gems* - , 2004.
[24] Moreland and K. Dean, "Fast high accuracy volume rendering," 2004.
[25] Christof Rezk Salama, "Volume Rendering Techniques for General Purpose Graphics Hardware (Deutsche Kurzfassung) | Request PDF," 2003.
[26] S. D. Laycock1 and A. M. Day1, "The Haptic Rendering of Polygonal Models involving Deformable Tools."
[27] Anonimous, *ReachinAPI 4 Programmer's Guide*, Revision N. Reachin Technology AB, 2005.
[28] A. Liu, F. Tendick, K. Cleary, and C. Kaufmann, "A Survey of Surgical Simulation: Applications, Technology, and Education," *Presence Teleoperators Virtual Environ.*, vol. 12, no. 6, pp. 599–614, 2003.
[29] M. Agus, A. Giachetti, E. Gobbetti, G. Zanetti, and A. Zorcolo, "Real-time haptic and visual

simulation of bone dissection," *PRESENCE Teleoperators Virtual Environ.*, vol. 12, no. February, pp. 110–122, 2003.

[30] S. Choi and H. Z. Tan, "Toward realistic haptic rendering of surface textures," *Comput. Graph. Appl. IEEE*, vol. 24, no. 2, pp. 40–47, 2004.

[31] K.-W. Chen, P.-A. Heng, and H. Sun, "Direct haptic rendering of isosurface by intermediate representation," in *Proceedings of the ACM symposium on Virtual reality software and technology*, Seoul, Korea: ACM Press, 2000, pp. 188–194.

[32] J. Kim, S. De, and M. A. Srinivasan, "Computationally efficient techniques for real time surgical simulation with force feedback," in *Proceedings. 10th Symposium on Haptic Interfaces for Virtual Environment and Teleoperator Systems, 2002. HAPTICS 2002*, 2002, pp. 51–57.

[33] D. Marchal, F. Aubert, and C. Chaillou, "Collision between deformable objects using fast-marching on tetrahedral models," in *Proceedings of the 2004 ACM SIGGRAPH/Eurographics symposium on Computer animation*, Grenoble, France: ACM Press, 2004, pp. 121–129.

[34] C. Wagner, M. A. Schill, and R. Manner, "Collision detection and tissue modeling in a VR-simulator for eye surgery," in *Proceedings of the workshop on Virtual environments 2002*, Barcelona, Spain: Eurographics Association, 2002, pp. 27–36.

[35] F. Khatyr, C. Imberdis, P. Vescovo, D. Varchon, and J.-M. Lagarde, "Model of the viscoelastic behaviour of skin in vivo and study of anisotropy," *Ski. Res. Technol.*, vol. 10, no. 2, pp. 96–103, 2004.

[36] N. Magnenat-Thalmann, P. Kalra, J. L. Lévêque, R. Bazin, D. Batisse, and B. Querleux, "A computational skin model: fold and wrinkle formation.," *IEEE Trans. Inf. Technol. Biomed.*, vol. 6, no. 4, pp. 317–23, Dec. 2002.

[37] G. A. H. A. J. Bordner, "Boundary element solution of the linear Poisson-Boltzmann equation and a multipole method for the rapid calculation of forces on macromolecules in solution," *J. Comput. Chem.*, vol. 24, no. 3, pp. 353–367, 2003.

[38] S. Gibson *et al.*, "Simulating Arthroscopic Knee Surgery using Volumetric Object Representations, Real-Time Volume Rendering and Haptic Feedback," in *Lecture Notes in Computer Science*, Springer, Berlin, 1997, pp. 369–378.

[39] R. S. Avila and L. M. Sobierajski, "A haptic interaction method for volume visualization," in *Visualization '96. Proceedings.*, 1996, p. 197–204,485.

[40] Y. J. Kim, M. A. Otaduy, M. C. Lin, and D. Manocha, "Six-Degree-of-Freedom Haptic Display Using Localized Contact Computations," 2002.

[41] K. Lundin, A. Ynnerman, and B. Gudmundsson, "Proxy-based Haptic Feedback from Volumetric Density Data," in *Proceedings of Eurohaptics 2002*, 2002, pp. 104–109.

Implementation of Average Filter and Median Filter for OCR Pre Processing of Incoming Letters Image

[1]I Made Avendias Mahawan, [2]I Putu Agus Eka Darma Udayana

[1,2]Department of Informatic Engineering, STMIK STIKOM Indonesia, Indonesia

[1]avendias@stiki-indonesia.ac.id, [2]agus.ekadarma @stiki-indonesia.ac.id

Abstract. Letters are one of the tools to communicate through writing and are useful for conveying information. STMIK STIKOM Indonesia campus (called STIKI Indonesia) receives letters from other institutions reaching an average of 50 letters a month, an automatic system is needed that can help manage the incoming letters and can determine the purpose of the letter. OCR (Optical Character Recognition) is one of the ways that can be used to find a text contained in an image, but to get good results in the OCR process requires a pre-process in order to reduce noise in letter images, combining average filters and median filters be an option because it succeeded in reducing noise and smoothing the image of the organization's stamp and signature that piled on the name of the sender of the letter. The results obtained from OCR after going through pre-processing are an array of words from letters. The accuracy achieved based on the date of the letter, letter number, subject of the letter and the sender of the letters from 50 test images reached 90%.

1. Introduction

STMIK STIKOM Indonesia is a campus that currently has 3,953 active students and 1,400 new students entering the academic year 2019/2020, with the large number of students making this campus popular in Eastern Indonesia especially in Bali. As an educational institution under Kementerian Riset Teknologi dan Pendidikan Tinggi this campus has good cooperation between educational institutions or other non-educational institutions in Indonesia and especially in Bali, such as research collaboration with the government of Badung Regency in 2018-2019. One example of good collaboration is invitations to attend or participate in activities organized by other institutions. At present, incoming letters or invitations received by STMIK STIKOM Indonesia reach an average of 50 letters a month, to assist front office staff (FO) in managing incoming letters, STMIK STIKOM Indonesia is starting to make improvements by developing a letter filing system so that it can makes it easy to record, search and report letters every year. The system developed for mail management has not been able to run automatically, such as tracing letter numbers, letter dates, senders and letter destinations. In this case it need to add features that can automatically search letter details based on Optical Character Recognition (OCR) technology. To get good results in the OCR process, we need a way that can be done to reduce noise in the letter image, in this study the author tries to combine two methods, namely averaging filter and median filter in the OCR preprocess.

Several studies have been carried out to conduct character recognition in print media on digital media, OCR is an effective solution for the process of converting print documents to digital documents. In research [1]a word input system was designed in a text to speech application by utilizing digital image processing. OCR technology is also applied to research conducted by [2], in this study proposed a feature extraction method using Centroid to Boundary and classification using Backpropagation Neural Network that can improve system performance better including time and accuracy of computational processing. In research [3] evaluating OCR performance on Arabic letters data, the results obtained are low accuracy, this is due to the absence of performance evaluation metrics and tools for proper evaluation to evaluate Arabic letters. Based on the three previous studies, OCR is an appropriate technology for managing incoming letters automatically, the researcher considers that an appropriate preprocess is needed so that it can produce good output on OCR, in this study the researcher tries to combine average filter and median filter to reduce noise created during the data acquisition process.

2. Methodology

2.1. Digital Image Processing

According to [4] digital image processing can be interpreted as processing an image on the two dimensions using a digital computer, image processing is also interpreted as a technique used to process, manipulate and modify images. Digital image is a function of two-dimensional light intensity f (x, y), where x and y show spatial coordinates. The value of "f" at each point (x, y) shows the level (value) of the color of the image at that point [5].

2.2. Pattern Recognition

Pattern recognition is a study of science that can describe something based on the characteristics of an object and then classify it based on the characteristics of the object. The pattern itself is an entity that is defined and can be identified and given a name. Pattern is a collection of measurement results and expressed in vector or matrix notation [4]. Pattern recognition can also be one way to preserve culture in Indonesia, especially in Bali, such as the classification of Balinese letters or also as a classification of types of Balinese carvings [6].

2.3. System Overview

The process flow in this study is explained in Figure 1.

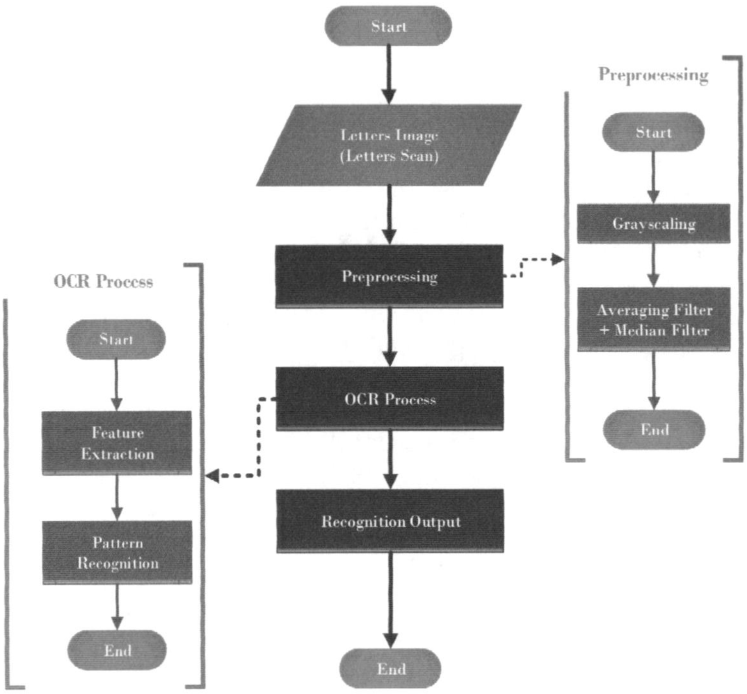

Figure 1. System Overview

IOP Conf. Series: Materials Science and Engineering **846** (2020) 012021 doi:10.1088/1757-899X/846/1/012021

The main process in this study starts from data acquisition, followed by preprocessing, OCR process and character output obtained from the image of the letter. Pre-processing consists of grayscaling, filtering consists of a combination of averaging filter and median filter. The OCR process consists of a feature extraction process followed by a pattern recognition process.

2.4. Data Acquisition
The method used to collect data is observation and interviews, the results obtained are 50 incoming letters used as test data. The obtained letter is then scanned in the .jpg file format.

2.5. Process Flow
Before the OCR process, the previous process was the grayscaling process which was calculated using the following equation [6]

$$I\,(x, y) = \alpha.R + \beta.G + \gamma.B \qquad (1)$$

Information :
I (x, y) = Graysclale Image
R = Red value in the color image
G = Green value in the color image
B = Blue value in the color image
With a constant $\alpha = 0.2989$; $\beta = 0.5870$; $\gamma = 0.1140$

The results of the grayscaling process are shown in Figure 2

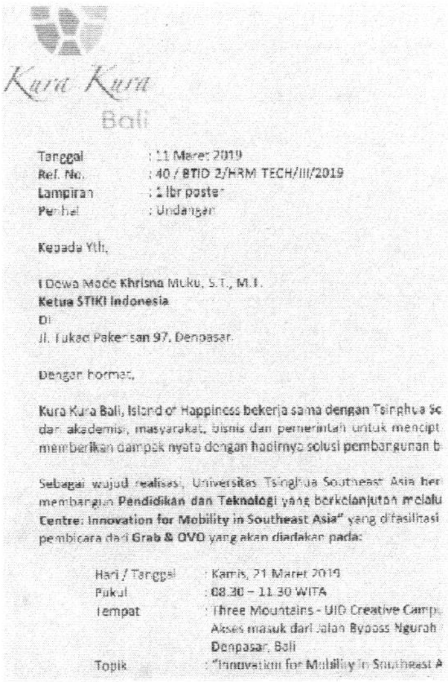
Kura Kura
Bali

Tanggal : 11 Maret 2019
Ref. No. : 40 / BTID 2/HRM TECH/III/2019
Lampiran : 1 lbr poster
Perihal : Undangan

Kepada Yth,

I Dewa Made Khrisna Muku, S.T., M.T.
Ketua STIKI Indonesia
Di
Jl. Tukad Pakerisan 97, Denpasar

Dengan hormat,

Kura Kura Bali, Island of Happiness bekerja sama dengan Tsinghua Sc
dan akademisi, masyarakat, bisnis dan pemerintah untuk mencipt
memberikan dampak nyata dengan hadirnya solusi pembangunan b

Sebagai wujud realisasi, Universitas Tsinghua Southeast Asia ber
membangun **Pendidikan dan Teknologi** yang berkelanjutan melalu
Centre: Innovation for Mobility in Southeast Asia" yang difasilitasi
pembicara dari **Grab & OVO yang akan diadakan pada:**

Hari / Tanggal : Kamis, 21 Maret 2019
Pukul : 08.30 – 11.30 WITA
Tempat : Three Mountains - UID Creative Camp
Akses masuk dari Jalan Bypass Ngurah
Denpasar, Bali
Topik : "Innovation for Mobility in Southeast A

Figure 2. Grayscaling Image

The next process is filtering, the filter used is the average filter and the median filter. Averaging filters can be calculated by the example applied to the following pixel image.

65	66	56
68	70	68
65	64	64

Figure 3. Example Pixel f(x,y)

The average filter results of the replacement pixel for g (y, x) are calculated as follows:
g (y, x) = 1/9 x (65 + 66 + 56 + 68 + 70 + 68 + 65 + 64 + 64) = 65,111 ≅65, Based on the calculation of g (y, x), the pixel in image 3 is replaced by a value of 65 [7]

Median filter works using filters and is calculated by sorting the intensity of a group of pixels, then the pixel value is replaced by a certain value, the filter on the median filter contains odd pixels, the filter is

IOP Conf. Series: Materials Science and Engineering **846** (2020) 012021 doi:10.1088/1757-899X/846/1/012021

shifted point by point reaching the entire image area, the pixel values are sorted ascending then the median value is calculated. The value obtained is used to replace the value at the center of the filter [7]. An example of a filter median filter is shown in Figure 4.

	123	125	126	130	140	
	122	**124**	**126**	**127**	135	
	118	**120**	**150**	**125**	134	
	119	**115**	**119**	**123**	133	
	111	116	110	120	130	

Figure 4. Median Filter

The feature extraction process uses the template matching method, where the pixels of each image are matched with the template provided, the character that has the highest match is considered as the character.

3. Result

This study experienced problems when doing OCR on the image of letters that have noise. Through literature studies and trials, researchers find solutions to overcome these problems. The researcher combines two filters, the average filter and the median filter in the OCR preprocess. The results obtained by the system can recognize more words, compared to before using additional filters. Figure 5 shows the results before and after using the filter.

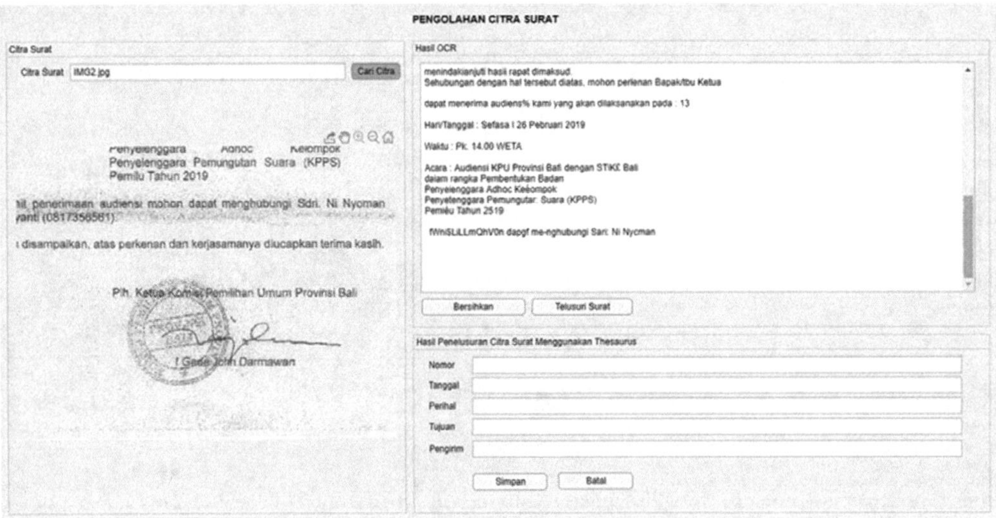

Figure 5a. Result Before Using Filter

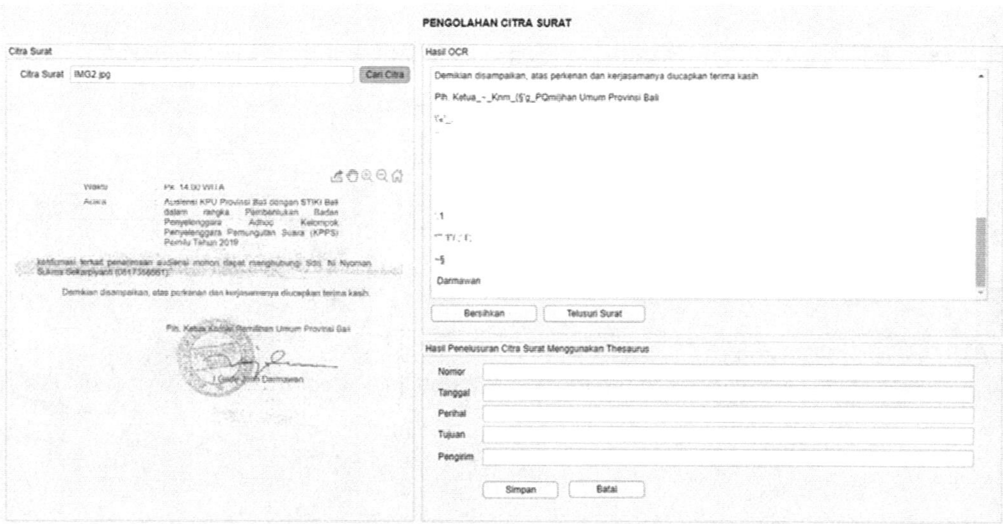

Figure 5b. Result After Using Filter

Figure 5a shows the OCR results have not been able to find the sender of the letter, but after going through the filtering process the results obtained in Figure 5b can recognize some characters that are not readable in Figure 5a. The characters that are read in Figure 5b show the sender of the letter, based on the testing process carried out on 50 images, the results obtained 45 images of the letter can indicate the existence of a number, date, subject, destination and sender of the letter. There are 5 letters that cannot indicate the sender of the letter, it is caused by the stamp of endorsement piling up in the name of the sender of the letter.

4. Conclusion

Based on the analysis of 50 test data, it can be concluded that the level of accuracy resulting from the merging of the averaging filter and the median filter in the OCR preprocess is 90%. 45 OCR image images can indicate the number, date, subject, destination and sender of the letter, 5 images cannot show the sender of the letter because the endorsement stamp merges with the sender of the letter. The researcher suggests that for further research a segmentation method is added to separate the endorsement stamp and the sender of the letter.

References

[1] K. Apriyanti and T. Wahyu Widodo, "Implementasi Optical Character Recognition Berbasis Backpropagation untuk Text to Speech Perangkat Android," *IJEIS (Indonesian J. Electron. Instrum. Syst.*, vol. 6, no. 1, p. 13, 2016.

[2] B. A. Pratama, W. T. A. Budi, R. K. Nur, and F. Sthevanie, "Mobile OCR using centroid to boundary and Backpropagation Neural Network," *2015 3rd Int. Conf. Inf. Commun. Technol. ICoICT 2015*, pp. 364–368, 2015.

[3] M. A. Alghamdi, I. S. Alkhazi, and W. J. Teahan, "Arabic OCR evaluation tool," *Proc. - CSIT 2016 2016 7th Int. Conf. Comput. Sci. Inf. Technol.*, 2016.

[4] D. Putra, *Pengolahan Citra Digital*. Yogyakarta: Andi, 2010.

[5] R. C. Gonzalez and R. E. Woods, *Digital Image Processing*. USA: Pearson/Prentice Hall, 2008.

[6] I. M. A. Mahawan and A. Harjoko, *Pattern recognition of balinese carving motif using learning vector quantization (LVQ)*, vol. 788. 2017.

[7] A. Kadir and A. Susanto, *Teori dan Aplikasi Pengolahan Citra Digital*. Yogyakarta: Andi, 2013.

IOP Conf. Series: Materials Science and Engineering **846** (2020) 012022 doi:10.1088/1757-899X/846/1/012022

Tajweed Segmentation Using Pattern Recognition, Extraction and SURF descriptor Algorithms

T A Zuraiyah[1], S Madenda[2], R A Salim[3], R Noviana[4]

[1]Department of Computer Science Universitas Pakuan

[2, 3, 4] Departement of Computer Science and Information of Technology Universitas Gunadarma

tjut.awaliyah@unpak.ac.id, sarifuddinmadenda@gmail.com, ravisalim@gmail.com, rinanov1169@gmail.com

Abstract. This paper proposes a method of detection and recognition algorithms and recitation of the Qur'an. The methods and algorithms are packaged into a software system application to allow a person to learn recitation and how pronunciation is good and right and real-time. Any recitation of the Qur'an written in a different color of the letters and punctuation instead of recitation and tajweed each have different shapes. Variable color is used as a reference for the segmentation process forms of recitation and SURF algorithm used for feature extraction process forms. Feature any form of recitation is stored in the data base (knowledge base) accompanied by explanatory text recitation of data and audio pronunciation files. In the process of recognition when the user enters a query tajweed image, Euclidian distance is used to measure the similarity between the query and the shape feature tajweed recitation shape features that exist in the database.

1. Introduction

Al-Qur'an holy book that must be practiced by every Muslim. One is the regular practice reading it with good pronunciation and correct. Tajweed is the main part of the Qur'an which regulate short length strains and humming sound when read. Currently available Qur'an Read Pen Digital utilizing barcode technology. The Qur'an is in the form of printed books that are equipped with a Read Pen Digital and barcode reading system. Each recitation in the Qur'an was given a barcode indexed in the database Read Pen Digital. Tajweed barcode unreadable when touching readings digital pen tajweed in the Qur'an and automatically sound the tajweed recitation accordance with role.

In fact, the Quran tajweed digital has been available on the online media and can be accessed easily by anyone who wants to learn the rules of recitation and how to read it [1][2][3][4], But until now has not provided the information and communication technology tools that can facilitate a person to learn the visual form of recitation and listening to audio how well and properly pronouncing[4].

This paper proposes a method of recitation in an image recognition Quran recitation based on the features digital form. When viewed from the side of its characteristics, any recitation of the Qur'an recitation can be identified by the color and the number of objects (the number of letters and punctuation) and the shape of each object. The recitation form. Figure 1 shows an example of the image from the Qur'an where tajweed written in a different color (red box marked dashed) of letters and punctuation marks that are not recitation (black) and white background. In the image shown that recitation is always consisting of two or more objects, so it takes a feature extraction algorithm that is able to define the shape of the object into one unified collection.

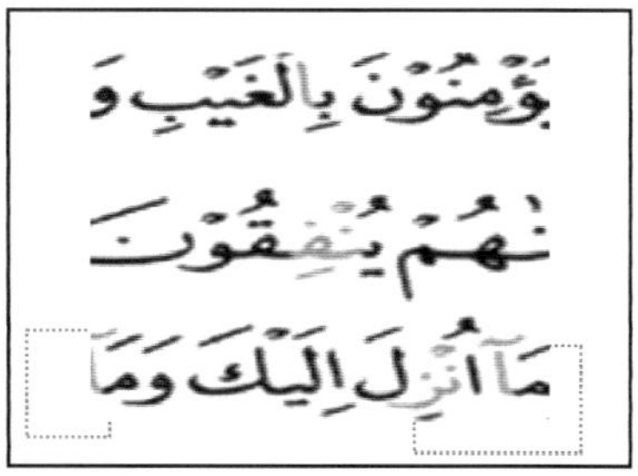

Figure 1. Example of Image Alqur'an Tajweed

In [5] using convolution techniques and Curtis Bray Distance to detect recitation in a sheet image of the Qur'an, but the introduction of forms and recitation has not been done. Tajweed detection techniques (especially in the recitation ikhfa) has been developed in [6] by using HSV segmentation algorithm and canny edge detection and SVM classifier. Tajweed recognition technique has not been used so that the shape feature's accuracy is still low. In addition, the handwriting recognition method based on similarity of features Arabic letter forms have been developed in [7]. Features shape extracted using methods freeman or chain code. The method developed is only able to extract a single character and can not be applied to the recitation because it has some letters and punctuation is a unity to recognize it. other than that[5]Arabic handwriting recognition using SVM classifier confined to recognize characters. One algorithm that can mengektraksi feature some form of object and represent in a unified way tajweed is speeded-Up Robust Feature (SURF). SURF algorithm produces features such main points or more dominant keypoint represent object shape[6][7],

2. Research Method

Methods of recognition and learning Qur'an recitation proposed by the scheme shown in Figure 2. This method is divided into two parts: the first is the stage of making a knowledge base of recitation of the Qur'an and the second is the stage of introduction of recitation. The knowledge base contains data tajweed shape feature, a text explanation of the recitation, Tajweed and recitation audio recitation image itself. Features tajweed form obtainable by the process of segmentation, cropping, counting the number of objects and object feature extraction form (recitation). Tajweed introduction stage query based on common features tajweed form, together with the four first process at the stage of formation of a knowledge base. The number of objects forming tajweed query matched to the number of objects on-Tajweed recitation that exist in the knowledge base. Furthermore, tajweed have the same number of objects will be calculated degree of similarity to the query recitation based on shape features (extracted using SURF algorithm). Tajweed which has the highest degree of similarity is expressed as a recitation identified. Then tajweed recognizable image is displayed along with explanatory text and audio examples tajweed law correct pronunciation.

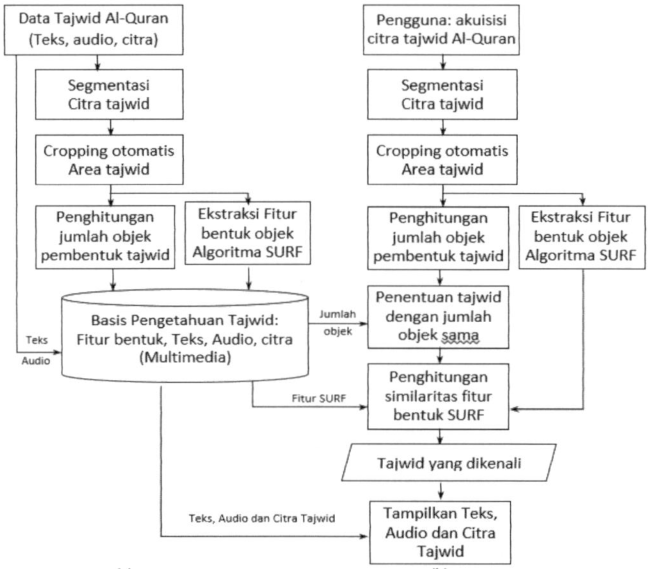

Figure 2. Flowchart Research
(a) Stages of Development Knowledge Base (b) Stages Introduction

2.1. Image Acquisition

Tajweed acquired image of each page of Surat Al Baqorah Qur'anic Tajweed Digital publisher of Al-Kalam. There are 150 images recitation of three recitation (Idgham Bighunnah, Idgham Bilaghunnah and Ikhfa haqiqi) used in the creation of a knowledge base. Each image is included with the text data Tajweed recitation[15] and audio pronunciation examples[15], Acquisition, recitation destination is the focus point of the camera, so that the recitation is always located at the midpoint of the image. It is intended to facilitate the processing. Each image acquisition is automatically normalized on the size of 500x 500 pixels through rescale process. This size was chosen for reasons of speed and size of the object recitation process visually quite representative to be recognizable.

In the image acquisition process of recitation to prepare test data is done in the same manner as in the image acquisition is used to input image data to the knowledge base. Test data set is divided into two categories namely:

1. Image data derived from the Al-Quran Al-Kalam Surat Al-Baqoroh to perform the acquisition process is different for the same recitation
2. Image data obtained from the Al-Quran Qordoba with the acquisition of the same process and the same recitation anyway.

Data acquisition This test is used to test the introduction of forms of recitation of interest whether it remains recognizable in case of translation (shift) because it is done with a different image acquisition and whether identified when using the Qur'an with different prints (color is also different). Table 1 shows the results of the test data acquisition process.

Table 1. Image Acquisition Process Test

No.	law Tajweed	Test the image of Al-Kalam	Test image Qur'an Qordoba
1	Idghom Bighunnah Mim4.PNG	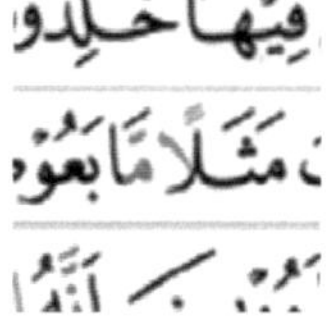	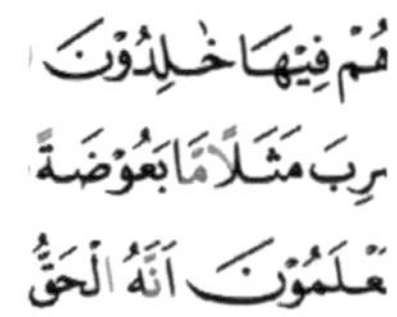

No.	law Tajweed	Test the image of Al-Kalam	Test image Qur'an Qordoba
2.	Idghom Bilaghunnah Lam4.PNG	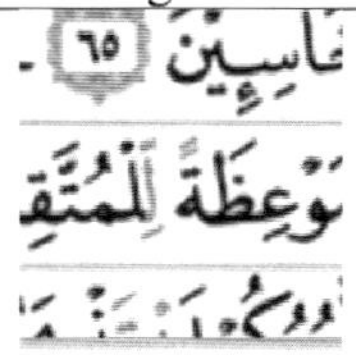	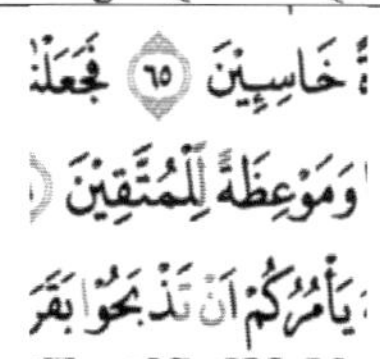
3.	IkhfaHaqiqi Fa1.PNG		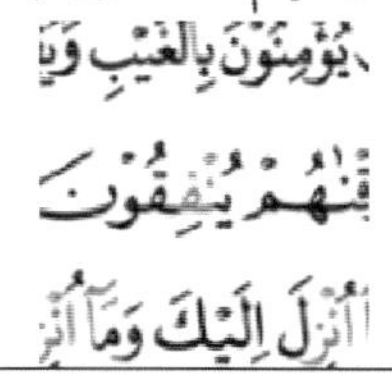

2.2. Segmentation Tajweed

Image in Figure 1 and Figure 3 shows that the recitation of letters and punctuation marks are printed with different colors of letters and punctuation marks instead of Tajweed (colors Hitan) and also different from the background color (white). This means to separate objects from non-Tajweed recitation and from the backside to do with the background segmentation algorithm based on the color feature. The author of the study results on the colors used for coloring recitation of the Qur'an is concluded that there is no standardization of color in perwanaan recitation. This means that the same recitation can be printed with different colors in two of the Qur'an from different publishers. This conclusion indicates that the variable color Hue can not be used as a reference in the segmentation process. Meanwhile, it can be seen that all in all a recitation of the Qur'an, both from the same publisher wants a different sort of publisher, printed using bright colors and a high saturation level. The brightness and saturation levels of the colors of recitation located at coordinates value (V)> 0.5 and saturation (S)> 0.5 for all kinds of color hue (H) in the HSV color space 3D. Two threshold value S (THS) and V (THV) is used in this study as a reference in the process of segmentation and extraction tajweed form. Figure 3 shows an example of image segmentation process results. On the left is the image of recitation as an input and that right is tajweed segmentation results. The brightness and saturation levels of the colors of recitation located at coordinates value (V)> 0.5 and saturation (S)> 0.5 for all kinds of color hue (H) in the HSV color space 3D. Two threshold value S (THS) and V (THV) is used in this study as a reference in the process of segmentation and extraction tajweed form. Figure 3 shows an example of image segmentation process results. On the left is the image of recitation as an input and that right is tajweed segmentation results. The brightness and saturation levels of the colors of recitation located at coordinates value (V)> 0.5 and saturation (S)> 0.5 for all kinds of color hue (H) in the HSV color space 3D. Two threshold value S (THS) and V (THV) is used in this study as a reference in the process of segmentation and extraction tajweed form. Figure 3 shows an example of image segmentation process results. On the left is the image of recitation as an input and that right is tajweed segmentation results

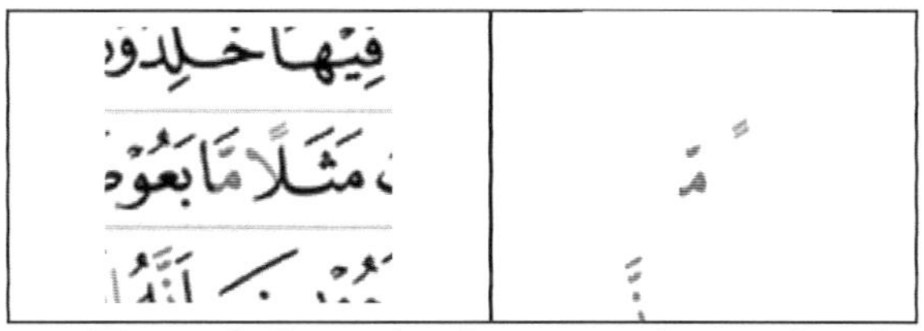

Figure 3. Example of segmentation results tajweed

2.3. Automatic cropping

Image segmentation results in Figure 4 show two segmented recitation. Tajweed as an object of interest you want to identify is always located at the center of the image. This means that if there are other areas of recitation at participating segmented, the recitation has to be removed. To eliminate cropping algorithm is used. Based on the results of manual analysis of the imagery used tajweed 500x500 pixels, the maximum size is 180x180 pixels recitation. This means automatic cropping process can be set in the middle of the image area in the four-point coordinates (row, column): (160.160), (160, 340), (340, 160) and (340.340). Figure 4 shows the results of cropping image (middle) and a binary image of the results of cropping (right).

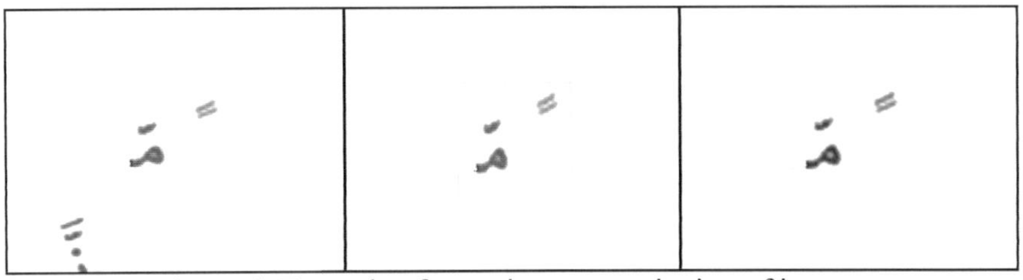

Figure 4. Example of cropping area recitation of interest

2.4. Feature Extraction Tajweed

As has been stated above that a recitation composed of two or more objects (letters and punctuation) that shape it. Therefore in this paper we use two types of recitation feature that the number of objects and features of the object shape. Object shape features extracted using SURF algorithm. This algorithm is known for reliable against geometric transformations (translation, scaling, rotation) and lighting, where it is often the case at the time of image acquisition using mobile devices. SURF algorithm determines the main points or keypoints dominant object and extracting represent its shape. Figure 5. Shows the result of the determination of SURF keypoints

Figure 5. Example of cropping area recitation of interest

2.5. Making the Knowledge Base

Each recitation is processed into a shape feature of recitation is accompanied by text files, audio files and image files, so that the unity of the information into a knowledge that will be used in the learning process of recitation of the Quran. There are 150 tajweed used in this paper as an example of the establishment of a knowledge base.

2.6. Determination Tajweed with Same Object

Determination Tajweed with the same object made through proses selection is the number of objects based on analysis of the characteristics of each recitation of the various sources of different Qur'an has the same number of objects. Besides the selection order is based on the similarity number matrix extraction process results SURF. In the SURF algorithm after Haar wavelet filtering process and obtained a description of the vector generating matrix of different orders on each image extracted. This Strategy is done automatically, especially to support the introduction process.

2.7. Recognition and Learning Tajweed Qur'an

Introduction stage is the stage where a user enters into the system tajweed image with the aim to identify recitation, an explanation recitation and listening to examples of correct pronunciation. After going through the selection process the data to be recognized are fewer in number. In the process of recognition is tested with a test images and test images results were compared with the results database SURF feature selection by calculating the shortest distance using the Euclidean distance. Euclidean distance formula using the following equation:

$$d_{ij} = \sqrt{\sum_{k=1}^{n} (X_{ik} - X_{jk})^2} \quad (1)$$

3. Results And Analysis

In the process of introduction of recitation is done testing include: segmentation and automatic cropping, SURF feature extraction, selection tajweed determination based on the number of objects and the order matrix SURF extraction and recognition result.

3.1.Segmentation Process Testing and Automated Cropping

Tajweed image segmentation process is done by using color segmentation in HSV space. Cropping the image segmentation results aim to take part image that contains only a recitation you wish to identify. Referring to the image acquisition technique in which the focal point is the image of the object intended Tajweed recitation then surely it is at the midpoint of the image. In addition, from the results of manual analysis of the imagery used size 500x500 pixels, the maximum size is 180x180 pixels recitation. This means automatic cropping process can be set on a four-point coordinates (row, column): (160.160), (160, 340), (340, 160) and (340.340). Tajweed image cropping process results are then converted to a binary image. Table 2 below shows the results of image segmentation, cropping and binary image.

Table 2. Image Segmentation process results, cropping and image Binary

No.	law Tajweed	First image	Cropping image segmentation	Binner image
1	Idgham Bighunnah (Mim1)			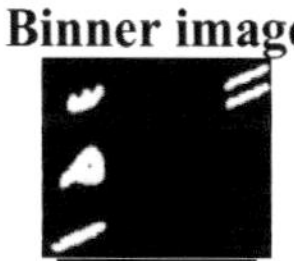
2	Idgham Bilaghunnah (Lam 1)	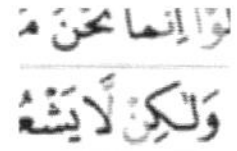		
3	Ikhfa haqiqi (Fa 1)	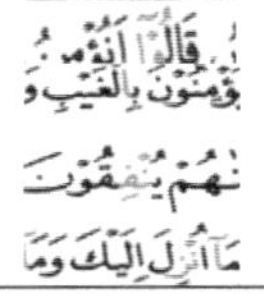		

3.2. Feature Extraction Process Testing Using SURF

SURF algorithm steps[8][9][10][11][12][13][14] through several stages as in Figure 6 is a reading a reference image and the image of the target, the detection of the feature points, the process of feature extraction, find points of matching between a reference image and the image of the target, the process of determining the location of an object by using matching point and display objects detected

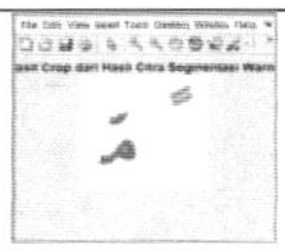	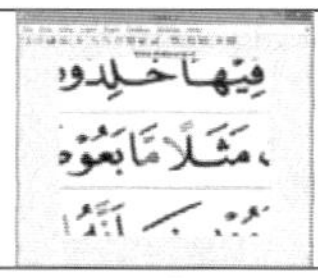
Target image	Figure 16. Reference Image
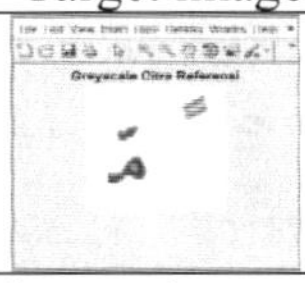	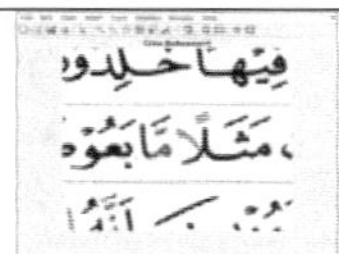
Results Conversion to Graycale Citra Target	Results Conversion to Graycale Reference Image
	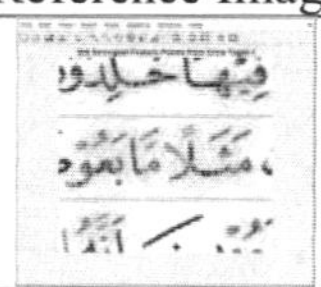
The detection process results Dots Image Features Target	The detection process results Dots Image Features Reference

Figure 6. Results Feature Extraction process steps SURF

3.3. Results Introduction Process Image

The results of image recognition performed on the test images 150 images comprising 39 idghom bighunnah images, 20 images and 91 idghom bilaghunnah haqiqi ikhfa image. Rated accuracy of this recitation detection process by 98% from 3 150 test data found no data was spotted in appendix 3. The amount of data that is recognizable senanyak means 147 (147/150 * 100%) = 98%. Figure 7 shows an example of the recognition of three recitation in breadfruit and tanwin nun.

Figure 7. Sample Results Introduction Process Three Laws of Tajweed Nun Breadfruit And Tanwin which is Recognized

As for the test image data derived from the Al-Quran Qordoba not been identified dikarena number of objects and the order matrix resulting from the SURF feature extraction process is different. Figure showing Qordoba image recognition process analysis.

Table 3. Recognition Process Unrecognized

Alphabet	Test image	Objects and Matrix Order Number Trains	Number of Objects and the Order Matrix Test	Citra Detection Results	Results Detection	Information
Dal 4	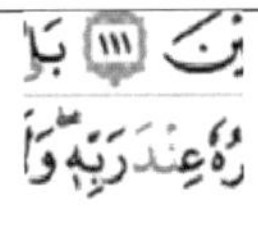	4 26 x 64	4 26 x 64		Unrecognized Sound does not come	Not found the ideal spots
Kaf 1	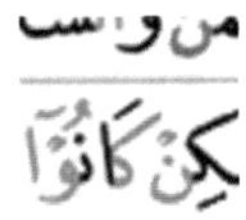	4 38 x 64	4 38 x 64		Unrecognized Sound does not come	Not found the ideal spots
Zai 1	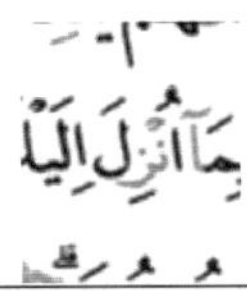	6 32 x 64	6 32 x 64		Unrecognized Sound does not come	Not found the ideal spots

4. Conclusion

In general, the application of methods and implementation of an algorithm developed was made into a prototype software that can automatically extract the image of the recitation of the Qur'an so as to facilitate the learning process in the Qur'an recitation. In particular the conclusions of this study were adapted to the formulation of the problem and research objectives are:

1. From the analysis of the most significant feature is used for the introduction of recitation in an image of the Quran recitation is a feature of forms. Features the form in question is the number of objects forming letters and punctuation marks recitation and features the shape of any object. Features color is only used for the segmentation process of recitation and is not used in the

recognition process. This is because some of the printing of the Qur'an uses different colors to mark the recitation of the same.

2. Making the knowledge base as a dictionary for information storage recitation recitation of the Qur'an which is equipped with multimedia data such as feature recitation, descriptive text, sound and image pronunciation Tajweed recitation. The knowledge base is used as information to facilitate the process of image recognition recitation.
3. From the test results as many as 150 images in 39 image tajweed idghom bighunnah recitation, Tajweed 20 idghom bilaghunnah law and legal recitation ikhfa haqiqi 91 values obtained 98% accuracy as there are three images that are not recognizable. From the analysis of the three images that are not recognizable due to the process of matching the feature points between the test images with trainer image was not found.
4. Learning system recitation of the Qur'an has been able to provide convenience in the process of image recognition recitation, so the detection results in the form of recitation, pronunciation and pronunciation examples in the form of audio files can be displayed properly.

Acknowledgments
Thanks are given to Kemenristek Higher Education which has provided support in the form of grants Doctoral Research Grant.

References

[1] S. Hidayat, "Al-Quran Digital (Variety, Problems and Future)," vol. 1, no. 1, pp. 1-40.
[2] MJ Aqel and NM Zaitoun, "Tajweed: An Expert System for Proficiency Recitation of the Holy Qur'an," in *Procedia Computer Science*, 2015, vol. 65, pp. 807-812.
[3] I. Ahsiah, NM Noor, and MYI Idris, "checking system to support Tajweed recitation," in *2013 International Conference on Advanced Computer Science and Information Systems, ICACSIS 2013*, 2013, pp. 189-193.
[4] A. Ismail, Idris MYI, NM Noor, Z. Razak, and ZM Yusoff, "MFCC-VQ approach for Qalqalah Tajweed rule checking," *Malaysian J. Comput. Sci.*, Vol. 27, no. 4, pp. 275-293, 2014.
[5] B. El Kessab, C. Daoui, B. Bouikhalene, and R. Salouan, "Isolated Handwritten Eastern Arabic Numerals Recognition Using Support Vectors Machines," *TELKOMNIKA Indones. J. Electr.*, Vol. 15, no. 2, pp. 346-351, 2015.
[6] SC Kan *et al.*"SURF binarization and fast codebook construction for image retrieval," J. Vis. Commun. Image Represent., Vol. 49, pp. 104-114, 2017.
[7] JT Pedersen, "SURF: Feature detection and description," *Comput. Vis. Image underst.*, Vol. 110, no. 3, pp. 346-359, 2011.
[8] U. Singh, SS Singh, and MK Srivastava, "Object Detection and Localization Using SURF Supported By K-NN" *Int. J. Comput. Sci. Trends Technol.*, Vol. 3, no. 2, pp. 88-93, 2015.
[9] M. Lopez-De-La-Calleja, T. Nagai, M. Attamimi, M. Nakano-Miyatake, and H. Perez-Meana, "Object Detection Using SURF and Superpixels *," *J. SOFTW. Eng. Appl.*, Vol. 6, no. September, pp. 511-518, 2013.
[10] H. Bay, A. Ess, T. Tuytelaars, and L. Van Gool, "speeded-Up Robust Features (SURF)," *Comput. Vis. Image underst.*, Vol. 110, no. 3, pp. 346-359, 2008.
[11] C. Wilson *et al.*"A power-efficient real-time architecture for SURF feature extraction," in 2014 International Conference on Reconfigurable Computing and FPGAs, reconfig 2014, 2014.
[12] AM Ahsan and D. Bin Mohamad, "Features Extraction for Object Detection Based on Interest Point," *TELKOMNIKA Indones. J. Electr.*, Vol. 11, no. 5, pp. 2716-2722, 2013.
[13] G. Wang, Z. Wang, Y. Chen, and W. Zhao, "Robust point matching method for multimodal retinal image registration," *Biomed. Signal Process. Control*, Vol. 19, pp. 68-76, 2015.
[14] Z. Gu, L. Cai, Y. Yin, Y. Ding, and H. Kan, "Registration of Brain Medical Images Based on SURF Algorithm and R-RANSAC Algorithm," *TELKOMNIKA Indones. J. Electr.*, Vol. 12, no. 3, pp. 2290-2297, 2014.
[15] AAA Rauf, *Guidelines Dauroah Qur'an: study Tajweed Sciences, arranged Applicative*Revised cet. Jakarta, Indonesia: Alfin Press, 2006.

Implementation of Convolutional Neural Networks to Recognize Images of Common Indonesian Food

[1]I Putu Agus Eka Darma Udayana, [2]Made Sudarma, [3]Putu Gede Surya Cipta Nugraha

[1,3]Department of Informatic Engineering STMIK STIKOM Indonesia, [2]Department of Electrical and Computer Engineering, Faculty of Engineering Udayana University

[1]agus.ekadarma@stiki-indonesia.ac.id, [2]sudarma@ee.unud.ac.id, [3]surya.cipta@stiki-indonesia.ac.id

Abstract. One of the applications of computer vision in the popular culture is food recognition which is popularized in the internet with a "hotdog and not hotdog" problem. Food recognition is also useful in many popular lifestyle apps such as calorie counter app or any diet related app. In this paper is proposed a CNN aided technique for recognizing food that is common in Indonesia. The technique is consist of 3 main phases, one is pre-processing normalize the data one is model formation and training which is known as the common binary classifier template, boosted with pooling and evaluated by cross-entropy technique in this paper the model used is the pre-trained model to test the testing data afterward, which is show a promising result with a relatively short training time. The experiments focused on how CNN can be used as a component to recognize food so that in the future it can be used to develop better calorie counter applications. In this experiment 10,000 data were used for training and 50 tests for each food category with a total of 500 food image data used for testing with the best accuracy reaching 88% for one of the categories.

1. Introduction

Apply human intelligence to computer systems is a very popular study conducted by researchers, for example research on computer vision. In the implementation of problems encountered in computer vision research is the classification of general objects, for example mimicking the ability of humans to interpret information from pieces of images. To achieve this people usually apply feature engineering to overcome this problem, but technically this approach will not give the best results. Because the scope of the problem is limited, they can be addressed or bound only to certain dataset sets without optimization for general solutions. This problem occurs because of different points of view in terms of scale, brightness, object deformation and so on. Academics have struggled with this problem for a long time; this problem usually falls into the scope of case studies of artificial neural networks, solutions inspired by the neural network of human ideas carried out until the era of deep learning. At that time in 1989, Y Lecun et al. managed to make a classification of postal image codes using a sophisticated neural bait model called the Convolution Neural Network (CNN) [1]. But hardware constraints came that made this research not continued, until early 2009 when Jurgen came up with the idea of the Repeated Neural Network (RNN) which helped him recognize human handwriting. Since then, along with the development of GPU based computational neural networks, in mid 2012 convolutional neural network experiments were able to rival humans in certain data sets [2]. In another study in 2016 by Suartika et al. managed to classify animals categorized as poultry using the CNN method with an accuracy of 20% - 50%. In conclusion, it is said to be able to increase the value of system accuracy needed for optimal training data [3]. Deep learning is a hot topic in research, because of its ability to model, a series of very complex problems such as pictures and sounds. One method of registering for learning in image processing/image recognition is Convolutional Neural Network (CNN) which comes not without logical reasons, it turns out the main idea of CNN stems from efforts to imitate networks in the changing human visual cortex to make CNN effective for processing large image data. But because this method also has its own weaknesses, namely CNN training time which is considered very time consuming. But once again the GPU developed into a savior with the presence of the Versatile Graphics processing unit and the emergence of the limitations of cloud computing is not as big a problem as before. [3] Based on this background, in this study the authors developed a system that is able to recognize the image of popular foods in Indonesia using the CNN method. The use of the CNN method in this study is motivated by the ability of the CNN method in public objects and needs to be deeply identified. Besides the problems that arose in the past, namely the long training time can be solved by the existence of cloud computing. To test the system developed, the authors tested the accuracy of food image recognition. By using accuracy testing, it will be known how optimal the CNN method and training data are to get the accuracy of food image recognition [4].

2. Fundamental Theory

2.1. Convolutional Neural Network

Convolutional neural network or CNN is a level up of a very popular artificial neural network technique called multilayer perceptron which is designed to tackle 2 dimensional data problem like images, CNN itself is falls into a category of Deep neural network because of the added depth of the network. the CNN is popular among researcher who tackle image recognition problem, the problem which MLP fails to solve effectively due to the fact that MLP doesn't store the information of image spatial data and assume that every pixel is an independent feature which limits the ability of MLP to solve the image recognition problem.

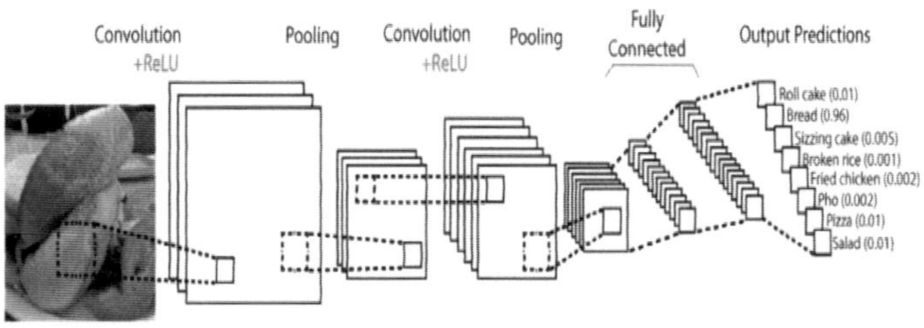

Figure 1. CNN Method Process

CNN is a term coined orignally by a researcher Kunishiko Fukushima from NHK broadcasting science research laboratory Kinuta Setagaya in Tokyo Japan with the name of Neocognitron [5]. The concept which Kunishiko coined later was perfected by Lichen a researcher from AT&T Bell Laboratories in New Jersey USA, Lecunn implement a full-pledged CNN model he named Le net that he trained to recognize a written digit [1]. Flies a couple of years a man named Alex Krizhevsky with his implementation of CNN is able to win an ImageNet Large Scale Visual Recognition Challenge 2012. A contest that later propels the popularity of deep learning to the mainstream scene as a powerful solution. more powerful than SVM (Support Vector Machine) a popular method at that time.

2.2. Architech of Convolutional Neural Network

Artificial Neural Network consist of various layers and a bunch of neurons for each layer. Both of that thing would not be possible if we stick to the old procedural programming paradigm as it would be different one case to another [6]. In the case of a multilayer perceptron, a network without a hidden layer could map a linear equation whatever the variable is, compared to a network with one or two hidden layers that could map almost of the equation with a simple data. But with the case with more complex data multilayer perceptron have a limit, the limit of MLP would be problem that needs more than 3 layers although there is a method to decided a number of neuron on each layer to get an optimal

IOP Conf. Series: Materials Science and Engineering **846** (2020) 012023 doi:10.1088/1757-899X/846/1/012023

result many scholars would not recommend a uses of more than 2 layer cause the overfitting problem and the diminishing return of the network capabilities to solve problem. With the revolution of deep learning, it can backup an MLP in its most weak are treating complex data. There is a function that can transform input data to data that would be easier to feed into a neural network, which is the main revolutionary aspect of deep learning. The layer that transforms the data before its trained to be a classifier class. That thing alone makes an effective neural network above 3 layers possible. But because of the flexibility of this transforming layer which is mainly used for feature extraction, there is no basic rule on behalf of how many layers that we should use in a deep learning network.

2.2.1. Convolution Layer. Convolution layer does all the convolution operation in the network that is coming from the previous layer that layer is the main process that is a fundamental of a convolutional neural network a convolution meaning a recurrent process to apply a function to an output of another function in the case of image processing it means to apply a kernel to every possible offset (as illustrated down below) [7].

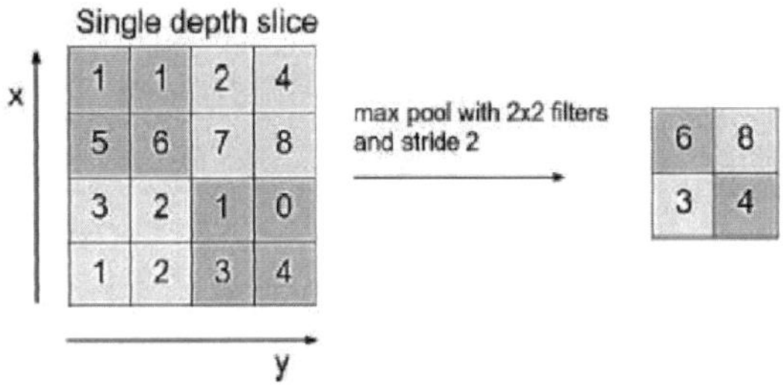

Figure 2. Convolution Operation

2.2.2. Subsampling Layer. Subsampling is a process of size reduction of an image. In the realm of image processing subsampling could be also to get more invariance of position and feature of an image in almost of all CNN. Subsampling is most common technique is a max pooling max pooling is to divide every output of the convolution layer to various smaller grid and then get the maximum of each grid to get the reduced format of the input image (which is illustrated in the image down below) [8].

2.2.3. Fully Connection Layer. Fully connected layer is a layer that we usually find in a multi layer perceptron application that one layer that has a main goal of transforming a dimension data so that the data could be classified as a linear data. Each neuron of a convolution layer needs to be transformed to a one dimensional data at first before it can be faded to a full they connected layer [2]. But it should be implemented in the last layer of the network to avoid a data loss which makes this layer is not as often used as the other kind of layer.

2.2.4. Activation Layer. Activation function refer to a function that transform data to N+1 dimension that makes the data can be cutted in the hyperlane space that makes the classification of the data point possible in the realm of CNN the activation function that are commonly used is a sigmoid function[2]. A sigmoid function transform an input of x to an either 1 of 2 value 0 and 1 with a form of distribution like presented in figure 3.

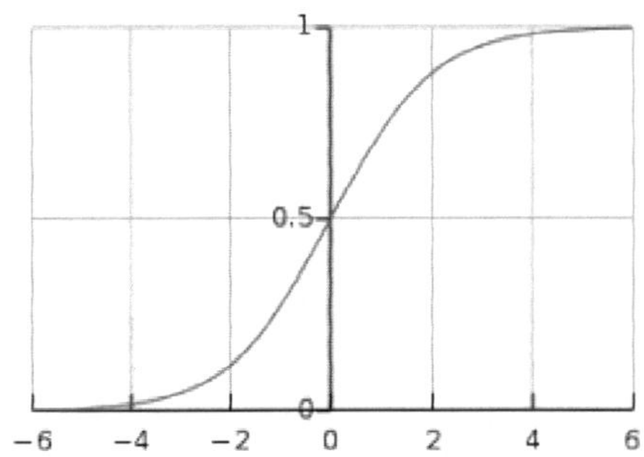

Figure 3. Sigmoid Function

A sigmoid function is basically best to classify a binary problem because its output tendency to be between zero and and one.

3. Methodology

3.1. Image Preprocessing

In the process of preprocessing food images are processed by wrapping and cropping. The wrapping process aims to determine the edge of the input image, after determining the maximum edge of the input image the cropping process will then be carried out so that the system can focus on recognizing the main objects of the food image.

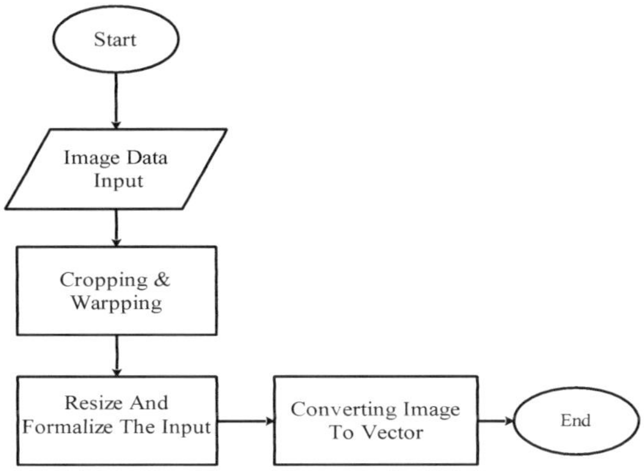

Figure 4. Flowcahart Preprocessing CNN

Figure 5. Ilustration Wrapping And Cropping

In accordance with figure 5, the image processing process begins by changing the size of any food image size to 128 x 128 pixel and then changing it to a gray scale for the training process.

3.2. Training Phase

The training process is used to obtain high accuracy from food image classification. This training phase consists of the feedforward process and the backpropagation process. To start the feedforward process, it takes the number and size of layers to be formed, the size of the subsampling and vector images. The feedforward process in vector images will go through a process of convolution and Max Pooling to reduce the size of the image and multiply the neurons.

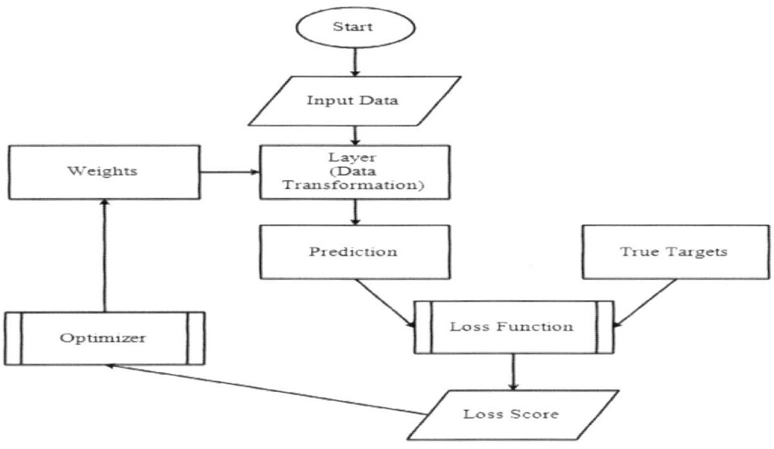

Figure 6. Flowchar Training Process CNN

20/20 [==============================] - 41s 2s/step - loss: 0.0975 - acc: 0.9750 - val_loss:

0.0428 - val_acc: 0.9920
Epoch 796/800
20/20 [==============================] - 40s 2s/step - loss: 0.0760 - acc: 0.9740 - val_loss: 0.0531 - val_acc: 0.9837
Epoch 797/800
20/20 [==============================] - 40s 2s/step - loss: 0.0894 - acc: 0.9650 - val_loss: 0.0604 - val_acc: 0.9820
Epoch 798/800
20/20 [==============================] - 41s 2s/step - loss: 0.1327 - acc: 0.9600 - val_loss: 0.0736 - val_acc: 0.9740
Epoch 799/800
20/20 [==============================] - 40s 2s/step - loss: 0.0848 - acc: 0.9710 - val_loss: 0.0920 - val_acc: 0.9620
Epoch 800/800
20/20 [==============================] - 41s 2s/step - loss: 0.1137 - acc: 0.9630 - val_loss: 0.1151 - val_acc: 0.9660

Figure 7. Contoh Hasil Proses Training

In the training process, the author uses data as much as 10,000 food image data. The following is an example of the results of the training conducted by the author. As shown in Figure 8 the training process in this study is set up to 800 epochs with a loss of 0.1137 and 0.9630 acc using 1,000 food image data for each class. The accuracy of 0.9630 shows the system has been able to classify food types with an accuracy of 96.30% with misclassification of 11.37% in the case study data training. To test the system's ability to classify images it is necessary to test the system using food images outside of the training data.

4. Result

4.1. Implementation

The implementation of Convolutional Neural Networks (CNN) method in the classification of food image types in this study uses the python programming language. In its application, the CNN method uses several steps to predict the process of food image recognition. In the initial stage, food images will be included in the Convolution process with sizes Conv2D (32, (3, 3)), Conv2D (64, (3, 3)), Conv2D (128, (3, 3)) and input form (128, 128, 3). To reduce the dimensions of each food input image feature but still retain the most important information from the image, the Pooling process is used. In this study the pooling method used is Max Pooling with the pool_size = (2, 2) movement, followed by the 'Flattening' and 'Fully Connected' process to classify the features obtained in the previous process into various classes.

Figure 8. Food Image Recognition Results (Fried Rice)

Figure 9. Food Image Recognition Results (Donuts)

In this step, to be able to classify food images the user must enter food images in the system and then the system tries to classify the types of food inputs from the user. Figure 8 and 9 are simulations of food type classification using the CNN method which results in the classification of fried rice and donuts.

4.2. Testing
In the research that has been carried out the application of the Convolutional Neural Networks (CNN) method has been able to classify food images that are inputted to the system. In its implementation to be able to classify food image types quite well, with the prediction of system errors in classifying food images by 11.37% in the image of training data. In testing the accuracy of the system this time the author will focus testing on 10 food image classes using 50 new food images for each test class outside the training data images with the provisions of using sigmoid activation for each type of food. The following are the results of tests that have been done by the author:

Table 1. Result of Accuracy Testing

Food Name	True		False		Total of Image Testing
	Amount of Image	Percentage (%)	Amount of Image	Percentage (%)	
Pizza	42	84	8	16	50
Steak	41	82	9	18	50
Sushi	44	88	6	12	50
Omelette	41	82	9	18	50
Noodles	49	98	1	2	50
Ice Cream	44	88	6	12	50
Hamburger	41	82	9	18	50
Fried Rice	48	96	2	4	50
Donat	48	96	2	4	50
Ceke	42	84	8	16	50
Min	41	82	1	2	50
Max	49	98	9	18	50
Average	44	88	6	12	50

Based on the test results of the classification of food images in table 1 using 50 food image test data outside the food image used in the training data with sigmoid activation produces good test results. In the test results, the system can classify food with an average error is 12% and average acc is 88%.

5. Conclution
Preprocessing methods and classification methods using Convolutional Neural Networks are reliable enough to determine the truth of the classification of object images. This is proven by the accuracy of 88%. The difference in input image size does not significantly affect the results of the system accuracy, because the input image has previously been resized in the preprocessing process. Based on the results of research conducted by the CNN method is very reliable in dealing with changes in input parameters. The more optimal the use of training data, the higher the accuracy of the system will result.

Acknowledgment
The author would like to thank the research assistant team and the ISETcon editorial team who provided input for writing this paper.

References

[1] LeCun Y 1990 *Handwritten Digit Recognition with a Back Propagation Network.*

[2] Coates A, Lee H and Ng A Y 2011 *An Analysis of Single-Layer Networks in Unsupervised Feature Learning* Qiche Gongcheng/Automotive Engineering Vol. **24** Issue 2.

[3] Suartika E P I P, Wijaya A W and Soelaiman R 2016 *Klasifikasi Citra Menggunakan Convolutional Neural Network (Cnn) pada Caltech 101* Jurnal Teknik ITS Vol **5** Issue 1.

[4] Udayana, I P A E D, Sudarma M, Kumara I N S 2017 *Balinese Latin Text Becomes Aksara Bali Using Rule Base Method* International Journal of Research in IT, Management and Engineering Vol **07**, Issue 5.

[5] Karpathy A *CS231n Convolutional Neural Network for Visual Recognition* [Stanford University: cs231m.github.io].

[6] Zufar M and Setiyono B 2016 *Convolutional Neural Networks untuk Pengenalan Wajah Secara Real - Time* J. Sains Dan Seni Its Vol. **5** Issue 2.

[7] Abroyan N 2017 *Convolutional And Recurrent Neural Networks For Real-Time Data Classification* Seventh Int. Conf. Innov. Comput. Technol., no. Intech, pp. 42–45.

[8] Nurfita R and Ariyanto G 2018 *Implementasi Deep Learning Berbasis Tensorflow Untuk Pengenalan Sidik Jari* Emitor: Jurnal Teknik Elektro Vol. **18** Issue 01.

Hierarchy of Grid Partition (HGP) Integrating data in Software Engineering and databases

Verdi Yasin*, Muhammad Zarlis, Opim Salim Sitompul, Poltak Sihombing.

Universitas Sumatera Utara, Medan, Indonesia.

*Verdiyasin29@gmail.com

Abstract. Hierarchy is a method of forming a system into grouping data in a database, while to integrate some data into an integrated data system in software engineering, there needs to be an appropriate method so that the data does not occur in an integrated system. In addition to making it easier to know the object attributes of a data, it is necessary to create a partition method, after we can do data integration with the grid method, then the combination of several methods becomes a new method, the Hierarchy of Grid Partition (HGP) Method. The purpose of the Hierarchy of Grid Partition (HGP) method is to integrate the data of an engineering software and database, in order to facilitate the user in finding the data needed and not duplicate the data. To develop this method, there needs to be design tools such as object-oriented modeling which is a unified modeling language (UML). The function of the Hierarchy of Grid Partition (HGP) method makes it easy to find data, because all data has been formed in a data partition grid system, the conclusion of this method is that the data access system is more effective and efficient.

Keywords: Hierarchy of Grid Partition (HGP) Method, Data Integration, Object Oriented Modeling, Data Access Systems, effective and efficient.

1. Introduction

The amount of data stored on the server database management system at this time, of course with greatly increased the amount of data, then the opportunity for duplicate data, will be very possible, [2], [9] therefore researchers feel interested in developing a model that can overcome several obstacles related to very large data nowadays. In developing the concept of developing this data control system, the researcher develops the results of his own ideas and observes several issues related to the amount of data that is difficult to determine the certainty of data accuracy. Especially the identity of the data distribution system object. As a solution that will be developed by researchers to achieve certainty of the data in controlling the processing system. So researchers developed a research methodology model by measuring the similarity of data through data integration to facilitate the integrated control of data in its distribution and access systems. [2] The above method facilitates the processing of large amounts of integrated data, in the view of researchers this method is very appropriate to be used as one method to measure the certainty of data such as Population data. There are several problems or problems that often occur from the data is that it is possible to duplicate data, even there is difficulty detecting duplicate data or the difficulty of finding out the amount of data there is more than one entity. Data control with intelligent methods, it is necessary to design an integrated data processing network system with several other database servers. In addition there needs to be an appropriate model in synchronizing data with system integration. So that way the data control system is fast, precise and effective will be implemented to the maximum. [1], [2], [4]

The research is expected to produce a new way of working to integrate data with smart through a network of database processing systems integrated with a database management system (DBMS) server, this modeling method can enable a process of data integration to be carried out maximally. Because all data items are stored or recorded by a digital system, it will be easy to integrate. So that on this occasion the researchers conceptualized the design of smart systems for controlling and processing data in an integrated manner through a network cloud data system. [1], [2], [9]

2. Methodology

To develop a concept of developing a data control system that has a very large chance of difficulty determining the certainty of the accuracy of the data entity, the researcher gave birth to a new methodology, the method of measuring data similarity with the integration of data stored in a database, in order to find accurate data. [2], [8] The following is a method in the process of integrating and testing data, can be seen in Figureure 1.

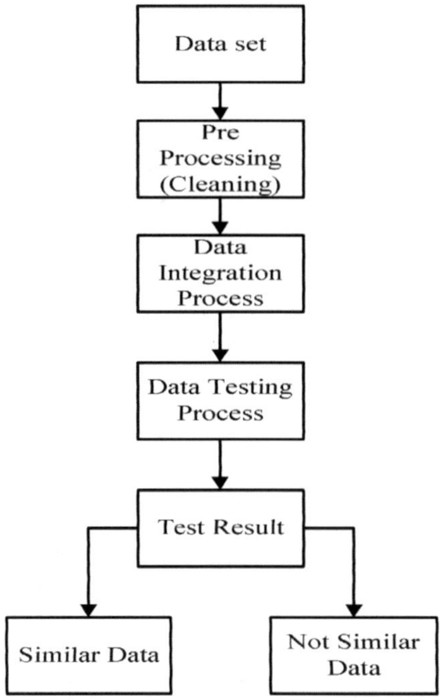

Figure. 1 Research Methods in data integration and testing

3. Discussion and results

In the concept of developing an integrated data control system, it is necessary to carry out an analysis process of an integrated data processing model, and then make a formula to be able to search for data stored on an integrated data network system, including data in the database server. So that, the system can easily find it. [9], [10] The following is an architectural image of an integrated smart data processing system model.

IOP Conf. Series: Materials Science and Engineering **846** (2020) 012024 doi:10.1088/1757-899X/846/1/012024

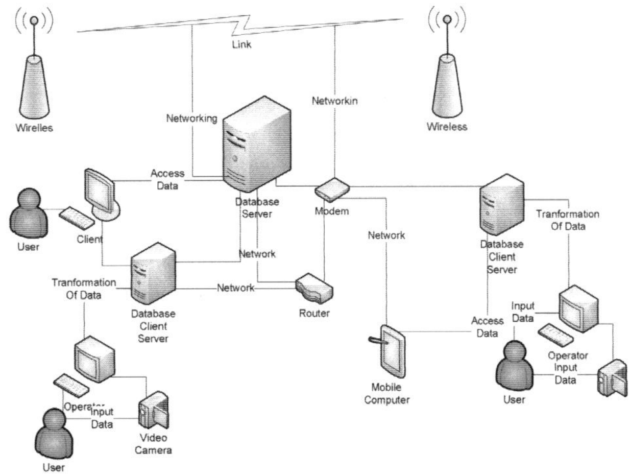

Figure. 2 Integrated Data Control System Network

Figure. 2 network integrated data control system, so easily, if you do data processing and do not need to be difficult in accessing data that has been stored, anywhere can always be accessed effectively and quickly controlled. [2], [5]

To measure the data processing in order to be able to analyze properly, the researcher makes an appropriate method to facilitate data access quickly and is not difficult and has no doubt about the results of the search, assisted by data processing tools that use SQL databases. [2], [10]

In integrating this data, researchers used an object-oriented software engineering system model, with the use case diagram of system integration and data similarity testing, as follows:

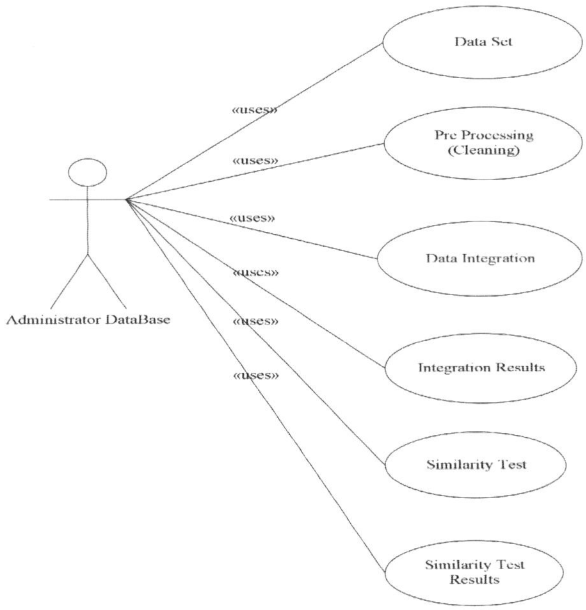

Figure. 3 Data Process Integration and Similarity Testing Diagram

Figure. 3 Data Integration and Similarity Testing Process Diagram, is a design model of the process of testing the available data sets, performed by the database administrator, in order to find data that has similarities, so that it can produce accurate and valid data.

In addition, researchers developed a model, reducing duplicate data. And the process flow method of the integrated smart data control system is to be shown in the following Figureure 4:

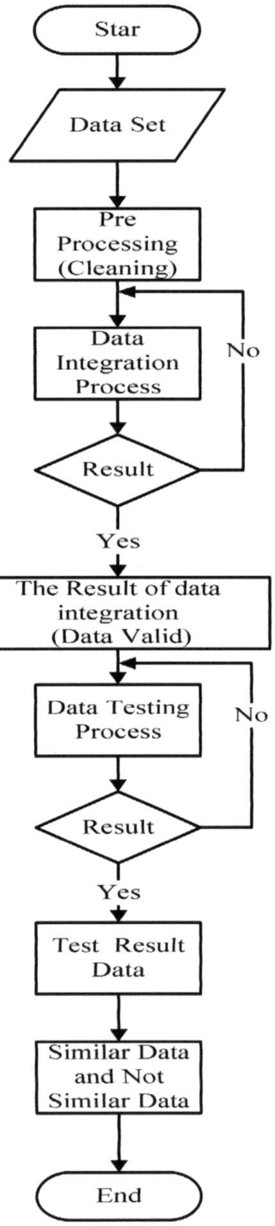

Figure 4 Flow Process Integration and Data Similarity Testing

Figure.4 Process Flow Integration and Testing Similarity of data, researchers will use data sets, using the main data from population data to be integrated processing, because the population data is very appropriate to be tested in this method. Residents have attribute entities to be able to measure the effectiveness of the data control model effectively. In this case, the researcher limited only to primary data namely name, gender, date of birth. Thus researchers can use mathematical formulas to find out the similarity of data in a population entity. So as to reduce the occurrence of duplicate population data.

The formula model created by researchers to determine valid data is as follows:

Namely:

Va: attributs
Va: Name, Date of Birth, Gender
K: Similarity

The formula for measuring data similarity in a database integration system is as follows:

Formula Model: ***K: Va***

K: (name, date of birth, gender)

And the following researchers make a model of attribute relations into integrated entities, namely as follows:

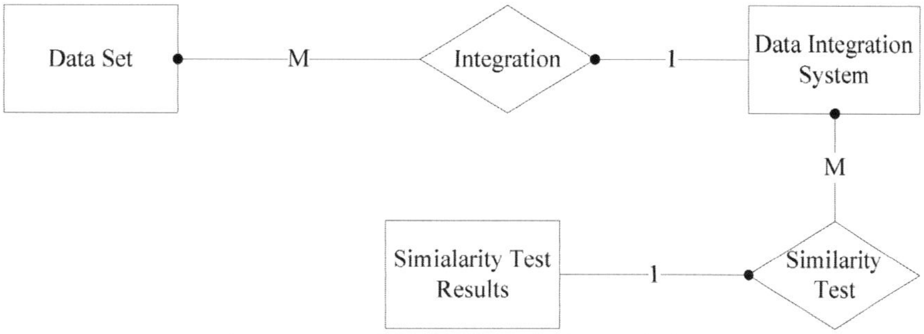

Figure. 5 Model Data Integration Relations

The following is an image of a data table layout in an SQL database that will be tested Data similarity with the aim of producing accurate data (valid data), inaccurate data (invalid data), data is not feasible or data allows problems (Fraud data), namely as follows:

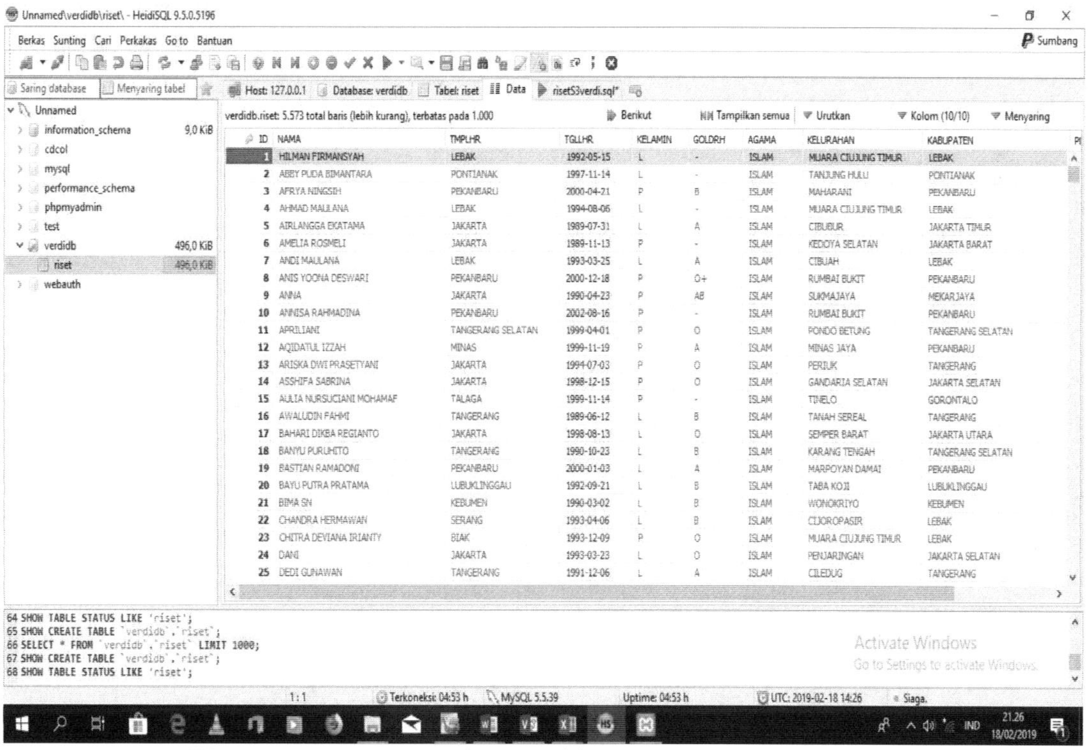

ID	NAMA	TMPLHR	TGLLHR	KELAMIN	GOLDRH	AGAMA	KELURAHAN	KABUPATEN
1	HILMAN FIRMANSYAH	LEBAK	1992-05-15	L	-	ISLAM	MUARA CIUJUNG TIMUR	LEBAK
2	ABBY PUDA BIMANTARA	PONTIANAK	1997-11-14	L	-	ISLAM	TANJUNG HULU	PONTIANAK
3	AFRYA NINGSIH	PEKANBARU	2000-04-21	P	B	ISLAM	MAHARANI	PEKANBARU
4	AHMAD MAULANA	LEBAK	1994-08-06	L	-	ISLAM	MUARA CIUJUNG TIMUR	LEBAK
5	AIRLANGGA EKATAMA	JAKARTA	1989-07-31	L	A	ISLAM	CIBUBUR	JAKARTA TIMUR
6	AMELIA ROSMELI	JAKARTA	1989-11-13	P	-	ISLAM	KEDOYA SELATAN	JAKARTA BARAT
7	ANDI MAULANA	LEBAK	1993-03-25	L	A	ISLAM	CIBUAH	LEBAK
8	ANIS YOONA DESWARI	PEKANBARU	2000-12-18	P	O+	ISLAM	RUMBAI BUKIT	PEKANBARU
9	ANNA	JAKARTA	1990-04-23	P	AB	ISLAM	SUKMAJAYA	MEKARJAYA
10	ANNISA RAHMADINA	PEKANBARU	2002-08-16	P	-	ISLAM	RUMBAI BUKIT	PEKANBARU
11	APRILIANI	TANGERANG SELATAN	1999-04-01	P	O	ISLAM	PONDO BETUNG	TANGERANG SELATAN
12	AQIDATUL IZZAH	MINAS	1999-11-19	P	A	ISLAM	MINAS JAYA	PEKANBARU
13	ARISKA DWI PRASETYANI	JAKARTA	1994-07-03	P	O	ISLAM	PERIUK	TANGERANG
14	ASSHIFA SABRINA	JAKARTA	1998-12-15	P	O	ISLAM	GANDARIA SELATAN	JAKARTA SELATAN
15	AULIA NURSUCIANI MOHAMAF	TALAGA	1999-11-14	P	-	ISLAM	TINELO	GORONTALO
16	AWALUDIN FAHMI	TANGERANG	1989-06-12	L	B	ISLAM	TANAH SEREAL	TANGERANG
17	BAHARI DIKBA REGIANTO	JAKARTA	1998-08-13	L	O	ISLAM	SEMPER BARAT	JAKARTA UTARA
18	BANYU PURUHITO	TANGERANG	1990-10-23	L	B	ISLAM	KARANG TENGAH	TANGERANG SELATAN
19	BASTIAN RAMADONI	PEKANBARU	2000-01-03	L	A	ISLAM	MARPOYAN DAMAI	PEKANBARU
20	BAYU PUTRA PRATAMA	LUBUKLINGGAU	1992-09-21	L	B	ISLAM	TABA KOJI	LUBUKLINGGAU
21	BIMA SN	KEBUMEN	1990-03-02	L	B	ISLAM	WONOKRIYO	KEBUMEN
22	CHANDRA HERMAWAN	SERANG	1993-04-06	L	B	ISLAM	CIJOROPASIR	LEBAK
23	CHITRA DEVIANA IRIANTY	BIAK	1993-12-09	P	O	ISLAM	MUARA CIUJUNG TIMUR	LEBAK
24	DANI	JAKARTA	1993-03-23	L	O	ISLAM	PENJARINGAN	JAKARTA SELATAN
25	DEDI GUNAWAN	TANGERANG	1991-12-06	L	A	ISLAM	CILEDUG	TANGERANG

Figure. 6 Layout Data Selector Table (DPS) uses SQL DataBase

The following is an object attribute database table layout that will be tested via SQL database, to measure the similarity of object attributes from the data set of population data that will be used for permanent voter data (DPT) in general elections by the National Election Commission (KPU) of the Republic of Indonesia , which is as follows:

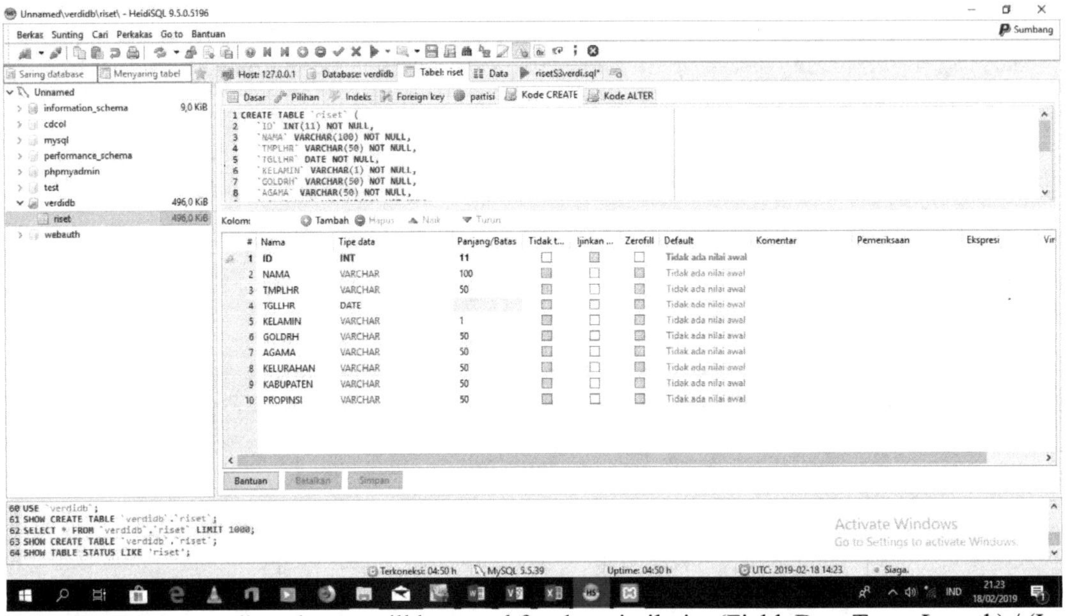

Figure.7 Object attribute layout will be tested for data similarity (Field, Data Type, Length) / (In Indonesian)

The following is a picture of the Data Similarity Test Layout Layout table, referring to the formula made by the researcher, based on the formula that has been created or found in data integration namely the K: Va Formula Model as follows:

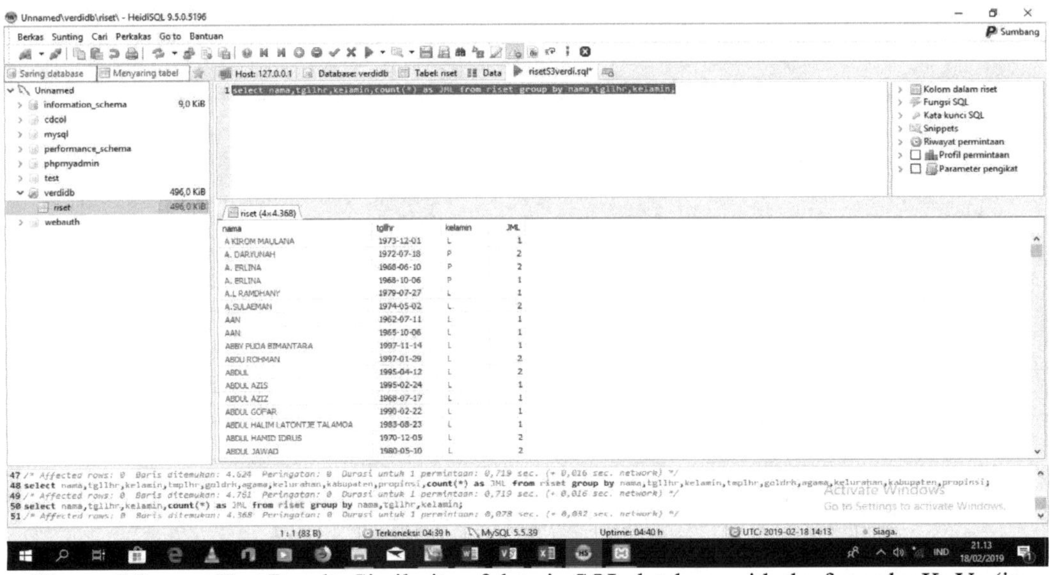

Figure.8 Layout Test Results Similarity of data in SQL database with the formula: K: Va (in Indonesian)

IOP Conf. Series: Materials Science and Engineering **846** (2020) 012024 doi:10.1088/1757-899X/846/1/012024

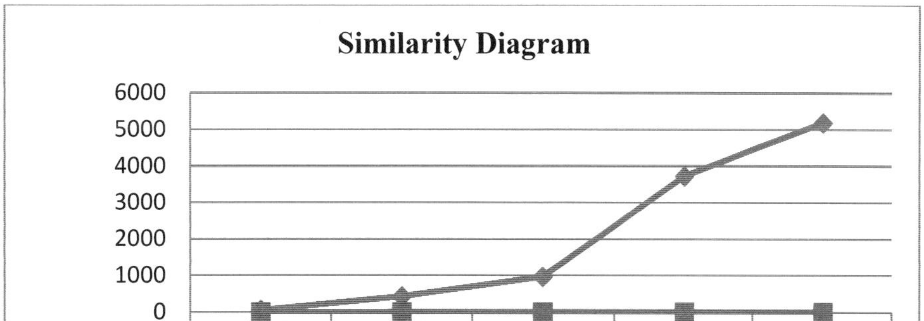

	Ki = 4	Ki = 3	Ki = 2	Ki = 1	Total Data
Series1	60	429	972	3724	5185
Series2	1.16%	8.27%	18.75%	71.82%	100.00%

Figure.9 Similarity Index Diagrams for Data Test Results (in Indonesian)

4. Conclusion

This research paper can produce similar data and not similar data, therefore the Hierarchy of Grid Partition (HGP) method is a model that can be used to measure the similarity of data, so as to produce similar data and not similar data, so as to optimize the data tested in measuring similarity.

References

[1] Abraham Silberschatz and Henry F. Korth and S. Sudarshan, 2011, Database System Concepts, sixth edition, McGraw-Hill Companies, Inc., ISBN 978-0-07-352332-3

[2] David L. Olson and Dursun Delen, 2008, Advanced Data Mining Techniques, Springer-Verlag Berlin Heidelberg, ISBN: 978-3-540-76916-3 e-ISBN: 978-3-540-76917-0

[3] David L. Olsen, 2007, Big Data, http://www.gartner.com/it-glossary/big-data (access June 10, 2018)

[4] Djoni Setiawan K., and Hendra Bunyamin, 2018, Utilization of inverted index on the process of tracking the similarity of the contents of pdf document files for final assignments of students, National Seminar on Information and Communication Technology 2018 (SENTIKA 2018) ISSN: 2089-9815, Yogyakarta, March 23 2018 page 356 -365

[5] Felipe Gómez-Cuba and Rafael Asorey-Cacheda and Francisco J. González-Castañ, 2013, Last-Mile Smart Grid Communications Model and Its Application to Study of Wired-Access Leased Broadband, IEEE Transactions on Smart Grid, Vol. 4, No. 1, March 2013

[6] Gehlot et al., 2012, International Journal of Modern Engineering Research (IJMER) www.ijmer.com Vol.2, Issue.4, July-Aug 2012 pp-1923-1928 ISSN: 2249-6645.

[7] González-Castaño. IEEE Transactions on smart grid, vol. 4, no. 1, March 2013

[8] Humasak Simanjuntak, Rosni Lumbantoruan, Wiwin Banjarnahor, Erisha Sitorus, Magdalena Panjaitan, Sintong Panjaitan, 2017, Evaluation of the Similarity of Entity Relationship Diagrams with Tree Edit Distance Algorithm, JNTETI, Vol. 6, No. 1, February 2017

[9] Verdi Yasin, 2012, Object Oriented Software Engineering: Modeling, Architecture and Designing (Modeling, Architecture and Design), Mitra Wacana Media Publisher, Jakarta.(in Indonesian)

[10] Xing He, Robert Caiming Qiu, Wentao Huang, Longjian Piao, and Haichun Liu, 2013, A Big Data Architecture Design for Smart Grids Based on Random Matrix Theory, IEEE.

IOP Conf. Series: Materials Science and Engineering **846** (2020) 012025 doi:10.1088/1757-899X/846/1/012025

Cold Smoking Equipment Design Of Smoked Fish Products With Closed Circulation Using Temperature And Concentration Monitoring System Based On Arduino Uno

Titik Ayu Dwi Jayanti, Agus Sudarmanto*, Muhammad Izzatul Faqih

Walisongo Islamic State University of Semarang, Semarang, Indonesia

*agussudarmanto@walisongo.ac.id

Abstract. This study aims to develop a cold smoking equipment by improving the air circulation system in a fish smoking equipment with the position of the fish hanging vertically. The components used are DS18b20 sensor, MQ-2 sensor, arduino microcontroller, LCD, micro SD. Cold smoke testing is done by monitoring the temperature and concentration of smoke. The type of fish used in this study is the type of flying fish (*Decapterus*). The analysis used to determine the outcome of cold-smoked fish is water content. The results showed that the average temperature in the smoking room was (23.18-44.81)°C. While the smoke concentration obtained is (230-252) ppm Temperature and smoke concentration as desired and the results of the sensor data recording are stored in micro SD. The average value of the water content of smoked fish is (25.53% -32.45)%. The lowest water content of smoked flying fish is 25.53% at 6 hours fogging duration and the highest water content of floating fish is 32.45% at 3 hours fogging duration.

1. Introduction

One of the processed products favored by the people of Indonesia and abroad is smoked fish. Fish is included in the category of fast decay so that the value of the quality of fish freshness is an important factor to consider. Therefore, the decline in the value of the quality of fish freshness will reduce the value of nutrition and the value of fish selling power.

The growth of the fisheries subsector becomes an important role as a contributor to production equipment in various parts of Indonesia, but not all regions of Indonesia can be fulfilled from the development of existing technology. As a result of uneven fish distribution, proper processing is needed for fish distribution in Indonesia.

Processing can make fish durable and allow it to be distributed from production centers to consumers. The existence of high temperatures, fish meat becomes cooked without being processed first [7]. Fish production in the last 20 years has only been processed (around 23-47)%. Processing in the form of salted fish, smoked fish, and boiled fish. The amount is mostly processed traditionally like conventional smoking. Isamu (2012) defines smoking is the process of penetration of volatile compounds in fish which produced from burning wood and producing products with specific taste and aroma [6]. Flyfish composition after smoked has phenol content depending on the process applied to the smoking, duration, and temperature in the smoking chamber will affect the smoked fish yield [10].

The technique of smoking fish is generally known by the public there are two, namely hot smoking and cold smoking. Hot smoking process usually often called fish baking [2]. According to Syam, on hot smoking, the storage temperature is around (70-100) ° C [8], while cold smoking is around (15-50) ° C [1]. The hot smoking for drying process takes place quickly, while the cold smoking for drying takes place slowly. The water content produced by cold smoking is relatively low, so cold smoking can be applied for long-lasting preservation [1].

Central Java is the center of fishermen and fishing both small and medium scale which has the potential to produce processed fish products. Tegal, one of the cities that is directly adjacent to the Java sea, has the potential to produce fisheries that are worth developing. Fishery product processing is still done traditionally in the form of the home industry [5]. Fish smoking in Tegal City currently still uses smoking houses. Judging from these conditions, of course, the smoking equipment used has not been equipped with a temperature monitoring system and the concentration of smoke, the smoke produced cannot be managed properly so that it impacts on health and the environment. Swastawati explains that the smoke emitted irregularly can also affect the level of fish maturity, and the resulting texture becomes less attractive. This happens because the smoke that comes outbreaks down more in the environment than is absorbed in the body of the fish. Then the need for tools that can help for the processing of smoked fish [14].

Progress in Science and Technology is currently experiencing a very rapid increase. The technology certainly cannot be separated from the use of a sensor and transducer microcontroller. The advantage of this microcontroller technology is that it is a digital measurement and automation tool with an LCD display [12]. Sensors and transducers are currently made into innovation and creativity that can be applied to a tool for experimentation. The instrumentation system used to make measurements has an input in the form of the actual value of the variable being measured, and output in the form of a measured variable value as shown in Figure 1 [14].

Figure 1. Measurement System Diagram

The results of advances in science and technology have provided convenience and benefits for the community. As research conducted by Fansuri (2011) has designed microcontroller-based cold smoking equipment for fish products. The results of trials on this cold smoking equipment showed a stable temperature concentration value with an average (29.71-39.29) ° C, the voltage released from the smoke sensor to measure smoke concentrations ranged from 4.11 to 4.33V. The time needed to smoke the tuna until cooked is 31 hours. The fuel used is a mixture of rambutan wood and coconut shells. The organoleptic test results of smoked fish that the taste is good and not tasty. Smoked fish can last one week. The water content of smoked fish is 51.91%, the water content of smoked fish that has been produced has fulfilled SNI with a maximum value of 60% [3].

The development of tools still has advantages and disadvantages, the results of the development of these tools now still have significant shortcomings, such as there are still gaps in the fumigation chamber causing some smoke to come out, the position of the smoked fish is not hung so it needs to be flipped manually, then there needs to be developed so that the fogging process evenly and data storage are still connected directly using a laptop or computer.

2. Research Methods

2.1. Research Time and Place

This research was conducted on April 23 - November 24 2019. Software and hardware and tool design were carried out at the Physics Laboratory of the Walisongo State Islamic University, Semarang. The field test was carried out in Muarareja Village, Kec. Tegal Sari, Tegal City, Central Java Province.

2.2. Hardware System Design

The system designed on this device is divided into 3 main parts. Namely the input consisting of a sensor DS18b20 and sensor MQ-2, the processor consists of Arduino uno and the output portion of Liquid Crystal Display (LCD) and Micro SD.

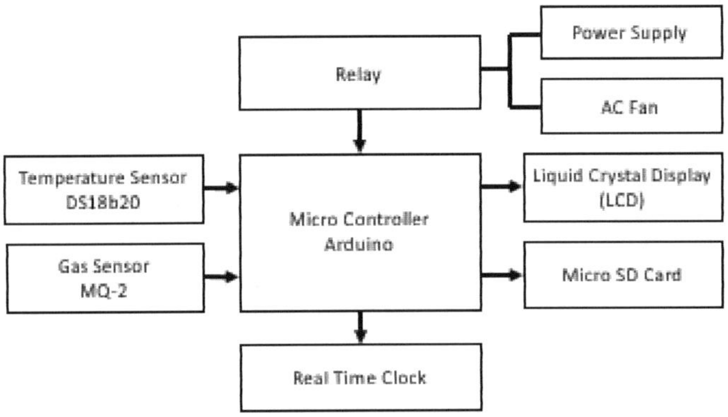

Figure 2. Hardware System Design.

2.3. Software System Design

Software design using Arduino IDE (Integrated Development Environment). Arduino IDE is software that functions to configure the arduino uno microcontroller board which contains a text editor for writing code.

2.4. Research Procedure

The cold smoking equipment for smoked fish products is designed to use Arduino Uno as a microcontroller, devices connected to Arduino Uno are DS18b20 sensor, MQ-2 sensor, relay, real time clock, liquid crystal display (LCD) and micro SD. Each has a different function in the cold smoking equipment working system.

DS18b20 sensor and MQ-2 sensor as a detector for temperature and smoke concentration. The sensor will read the temperature of the smoking room and send digital data to Arduino Uno for processing after it is forwarded by a 12 volt AC fan.

The 12 volt fan is a temperature control device and the smoke concentration is controlled by Arduino as a microcontroller. Two fans in the smoking room will work optimally if the temperature in the smoking room is <15 ° C and will work minimally when the temperature in the smoking room> 45 ° C. The 12 volt AC fan is controlled by a relay that works according to the temperature data that is input to Arduino Uno via the display.

The display is a temperature and smoke concentration monitoring device that is read by the DS18b20 sensor and the MQ-2 sensor. This display uses a Liquid Crystal Display (LCD) and the data will be stored on a micro SD..

3. Results And Discussion

3.1. Testing of the Smoking Equipment

Equipment testing is done by looking at the workings of the DS18b20 temperature sensor and MQ-2 sensor, whether the temperature value of the DS18b20 sensor is vulnerable according to the desired specifications. MQ-2 sensor testing can be done by looking at the output value of the MQ-2 sensor.

This value can be converted to smoke density in ppm. The equation used to convert digital values to parts per million (ppm) values.

3.2. Design Results of Cold Smoking Equipment
The design of cold smoking equipment for smoked fish products with a temperature monitoring system and the concentration of smoke is divided into 3 parts, namely the kiln (smoke making), smoke distribution pipes, and the smoking room. The following design of cold smoking can be seen in Figure 3.

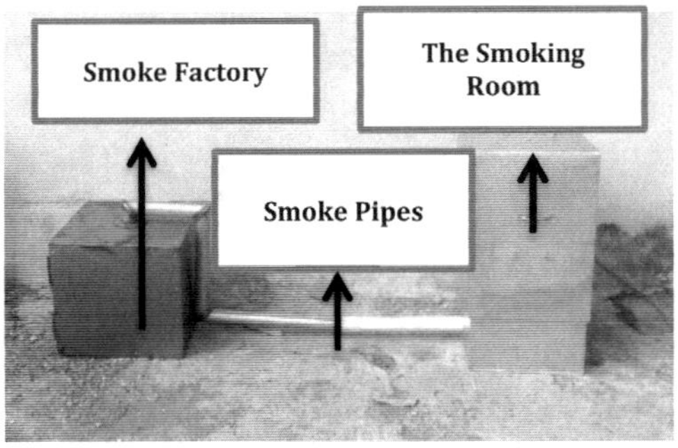

Figure 3. Cold Smoking Equipment

The smoking room functions as a container for the material to be smoked. Made of galfalum with a size (70x40x40) cm. There is a glass in front of the smoking area which aims to make it easier to see the fish in the smoking room. Then in the smoking room there are 2 shelves to hang the fish with a distance between rack 1 and 2 is ± 20 cm. The smoking room has two 12 volt AC fans, a fan used (5x5x2.5) cm [12]. Between shelves one and two there is a DS18b20 temperature sensor and a MQ-2 sensor. The right outside of smoking there is a component box that functions to place components and electrical circuits, such as Arduino Uno, RTC, LED, LCD, micro SD, and relay. The size of the component box (15x 10) cm. Display smoke chamber can be seen in Figure 4.

(a) (b) (c) (d)

Figure 4. (a) Fish Smoking Glass (b) Component Box (c) Sensor (d) Shelves and Fans in the Fish Smoking Room

The smoking process of this equipment is a coconut shell that is burned at the incinerator, produces smoke, then the smoke enters the pipeline connecting the smoke chamber and the incinerator. The smoke goes to the smoking room because it is inhaled by a 12 volt AC fan. The bottom 12 volt AC fan, which amounts to 2 pieces, will turn on when the temperature is less than 15 ° C. if the temperature in the smoking room is above 45 ° C, one of the fans will turn off. Smoke entering the smoke chamber is then recorded by the DS18b20 sensor and the MQ-2 sensor. Data recorded on the temperature and smoke sensors are then stored in micro SD. The position of the fish to be smoked by hanging on a rack in the smoking room. The position of the fish can be seen in Figure 5.

Figure 5. Position of Flying Fish Before Smoked

Viewed from Figure 5, the purpose of the fish is hung vertically, so that fish during the smoking process does not need to be alternated. Smoked fish evenly cooked. This research was conducted 4 times with variations in smoking time. The duration of smoking is 3 hours, 4 hours, 5 hours and 6 hours. The following is a display of smoked fish results can be seen in Figure 6.

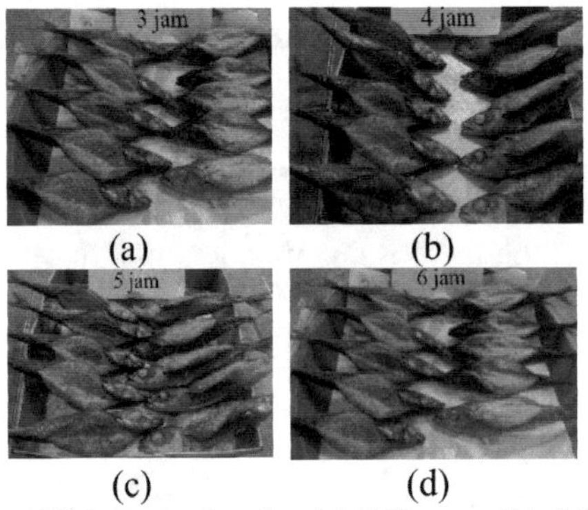

Figure 6. Smoked Fish Results With Long Smoke (a) 3 Hours, (b) 4 Hours, (c) 5 hours (d) 6 hours

Figure 6 shows the results of smoked fish using cold smoking equipment. Judging from the color texture of brownish smoked fish. Taste and aroma of concentrated smoked fish and tender fish meat.

3.3. Results of Cold Smoking Equipment Test

Testing cold smoking equipment is done by controlling the smoke chamber. Testing the temperature and concentration of smoke carried out to find out how long the process of smoking fish and to find out how much the temperature and concentration of smoke in the smoke chamber. The following results of testing the temperature in the smoking room can be seen in Figure 7.

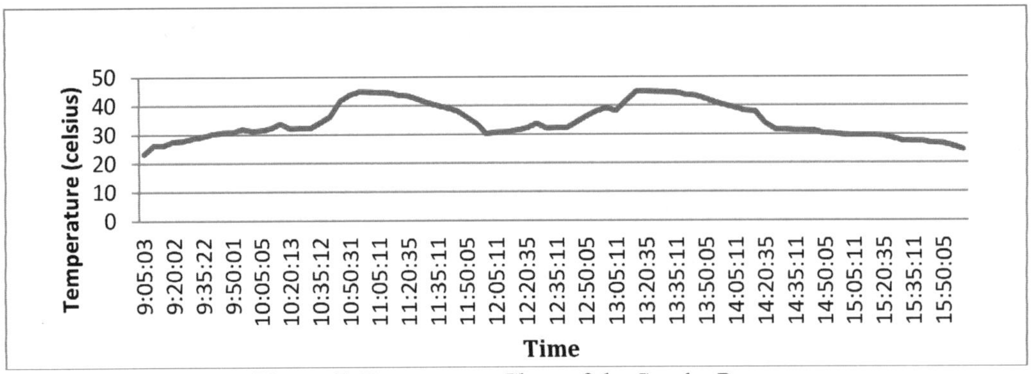

Figure 7. Temperature Chart of the Smoke Room

The x-axis in Figure 7. is the time of data collection and the y axis is the temperature in the smoking room. The temperature in the smoking room ranged between (23.18 ° C-44.81 ° C). The results of the lowest temperature observation were 23.38 ° C and the highest temperature was 44.81 ° C. The increase in temperature is due to the addition of charcoal fuel in the combustion place, so the distribution of heat and smoke is more so that the temperature in the smoking room increases. The decrease in temperature is because the charcoal in the incinerator is not sufficient to distribute heat and smoke into the smoking room. The temperature sensor test results can be said that the DS18b20 temperature sensor works well according to the desired specifications. In addition to testing the temperature sensor MQ-2 sensor testing is also performed to detect the concentration of smoke in the smoke chamber. Smoke concentration and smoke velocity in the smoke chamber will determine the quality of smoked fish [10]. The following results of monitoring temperatures in the smoking room can be seen in Figure 7.

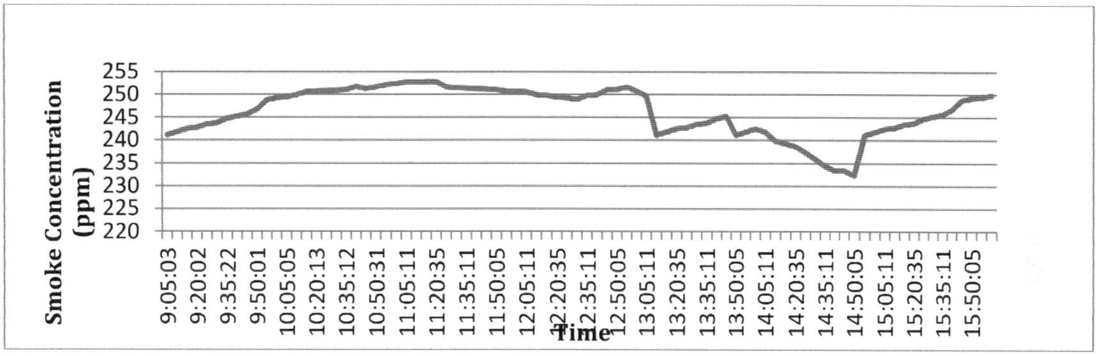

Figure 8. Graph of Smoke Concentration Test Results

The x-axis in Figure 8 is the time of data collection, while the y-axis is the Analog Digital Converter (ADC) value which is changed in units of parts per million (ppm). Based on observations during smoking for 3 hours, 4 hours, 5 hours and 6 hours, it can be seen that the concentration of smoke in the smoking room is (230-252) ppm. To get optimal results, the smoke concentration is expected to be above 100 ppm in the smoking process.

3.4. Water Content Analysis of Smoked Fish

The analysis used in this study was the water content in smoked fish. Food durability has a close relationship with water content. The lower the water content in food is expected to be more durable food ingredients. The content of water content in food affects the resistance of food ingredients to microbial attack. Water is a suggestion for microorganisms to develop so that the smoking process has the aim of reducing the levels of water in fish [14]. The lower the water content the slower the growth of microbes so that food is durable [16]. The results of the cold smoking water content of flying fish are presented in Figure 9

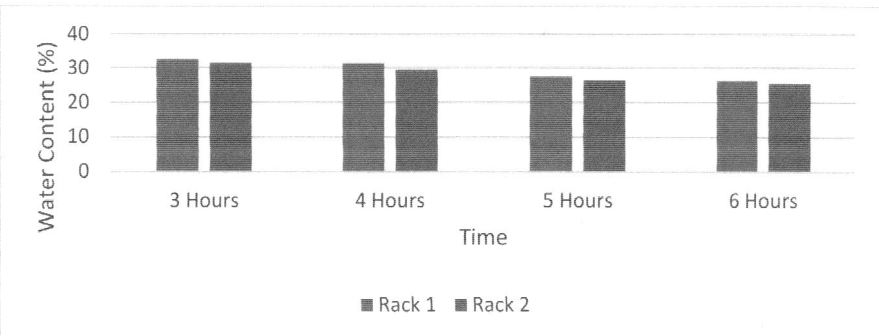

Figure 9. Water Content Graph of Smoked Fish

Based on Figure 9 the average water content of smoked fish is (25.53% -32.45)%. The average length of a fish is 16.7 cm and the average weight of a fish is 192.3gram. There are 10 fish in the smoking room. The lowest water content of smoky fish is 25.53% which occurs at 6 hours of smoking and the highest water content of smoked fish is 32.45% which occurs at 3 hours of smoke fogging. The efficiency of the time needed to bloat the flying fish to the water content is 6 hours compared to 3 hours of smoking. This means that the longer the smoking, the lower the water content in smoked fish.

Previous research by Fansuri (2011) used tuna with 31 hours of smoking. The value of water content of smoked tuna is 51.91% The difference in body shape of the two different types of fish can affect the decrease in water content. The decreased water content of swift fish is faster because of flying fish have a slim body [3]. This was confirmed by Fauzi (2011) who stated that the type, shape, and size of the fish, the temperature of the ignition and the duration of smoking greatly influenced the smoked fish yield [4].

4. Conclusions

Cold smoking equipment for smoked fish products with closed circulation using an Arduino Uno based temperature and smoke monitoring system has been created. The design of cold smoking equipment is divided into three parts, namely the incinerator, smoke distribution pipes, and smoking chamber. The result of the temperature in the smoking room is the lowest temperature of 23.18 °C and the highest temperature is 44.81 °C. Smoke concentration in the smoking room is (230-252) ppm. Temperature recording data is stored on a micro SD card. The average results of the lowest water content value of smoked flying fish are 25.53% occurred at 6 hours smoking duration and the highest water content of smoked fish is 32.45% occurred at 3 hours fogging duration. The efficiency of the time needed to bloat the flying fish to the water content is 6 hours compared to 3 hours of smoking. This means that the longer the smoking, the lower the water content in smoked fish.

Suggestion

Suggestions, it is necessary to develop further about cold smoking equipment and the need for additional alarms to find out the fuel that has run out and used different types of fish and can be developed with ozone technology for preserving fish with longer shelf life.

References

[1] Adawyah. (2007). "*Pengolahan Dan Pengawetan Ikan"*. Jakarta: Bumi Aksara
[2] Afrianto, dan Liviawaty. (2015). "*Pengawetan Dan Pengolahan Ikan"*, Yogyakarta: Kanisius
[3] Fansuri, Aldo Saragih. (2011). "*Rancang Bangun Alat Pengasapan Dingin Berbasis Mikrokontroller"*, Skripsi Departemen Ilmu dan Teknologi Kelautan. Fakultas Perikanan Dan Ilmu Kelautan : Institut Pertanian Bogor
[4] Fauzi dan Muchtar, (2011). "*Pengasapan Ikan Menggunakan Asap Skala Rumah Tangga"*. Jurnal perikanan dan kelautan vol.16, No.1, Hal 103-116
[5] Harris, dan Karmas. 1989. "*Evaluasi Pada Pengolahan Bahan Pangan". Bandung* : ITB
[6] Isamu, Kobajashi dkk.(2012). "*Karakteristik Fisik, Kimia, Dan Organoleptik Ikan Cakalang (Katsuwonus Pelamis) Asap Di Kendari"*. Jurnal Teknologi Pertanian. Vol.13 no.2 (agustus 2-12) 105-110
[7] Kadir, (2004). "*Pengaruh Suhu Dan Lama Penyimpanan Terhadap Jumlah Kandungan Bakteri Dan Kualitas Fisik Ikan Tongkol Asap (Euthynnus Afinis)". Journal of Biological Physics*. Vol.6, No.2 Mei 2004.
[8] Marenta TM, and Arwani SN. (2008). *Pengawetan Ikan Bawal Dengan Pengasapan Dan Pemanggangan*. Jurnal Ilmu-ilmu Pertanian. Vol.7 no.2:33-47
[9] Sebayang .(2002). "*Penerapan Teknologi Pengasapan Ikan Bagi Masyarakat Nelayan"*. Jurnal Pengabdian Masyarakat Vol.8 no.28: 25-34
[10] Baron, dkk. *Effect Of Smoking Processes On The Contens Of 10 Major Phenolic Compound In Smoked Fillets Of Herring (Clupea Harengus)* Food Chemistry, 85, 111-120

IOP Conf. Series: Materials Science and Engineering **846** (2020) 012025 doi:10.1088/1757-899X/846/1/012025

[11] Sudarmanto, Agus. (2013) "*Pembuatan Alat Laboratorium Fisika Fakultas Ilmu Tarbiyah Dan Keguruan IAIN Walisongo Untuk Praktikum Osilasi Cairan Dan Ayunan Matematis Secara Digital*", Phenomenon Jurnal Pendidikan MIPA UIN Walisongo Semarang. Vol. 3 no.1 :5-16
[12] Suharningsih. (2012). *"Rancang Bangun Jemuran Otomatis Berbasis Mikrokontroller"*. Jurnal Teknik elektro industri politeknik elektro negeri Surabaya. Avaibel at http://respostory.uksw.edu/handle/123456789/2772
[13] Swastawati, (2013)."*Karakteristik Kualitas Ikan Asap yang Diproses Menggunakan Metode dan Jenis Ikan Berbeda"*. Jurnal Aplikasi Teknologi Pangan. Vol.3 no.3:126-132
[14] Syam, Rafiuddin. (2013). " Dasar-dasar Teknik Sensor". Fakultas Teknik Universitas Hasanuddin, Makassar
[15] Winanmo. 1997. *Kimia pangan dan Gizi* . Jakarta: PT Gramedia

IOP Conf. Series: Materials Science and Engineering **846** (2020) 012026 doi:10.1088/1757-899X/846/1/012026

Expert system on core competency detection and student achievement based on Fuzzy TOPSIS method

Sutiah[1], Supriyono[2], Indah Aminatuz Zuhriyah[3], Zainul Arifin[4]*
Universitas Islam Negeri Malang, Kota Malang, Indonesia
Email: sutiah@pai.uin-malang.ac.id, priyono@ti.uin-malang.ac.id, indahaz@gmail.com, zainularifin65@gmail.com

Abstract. The improvement of information technology is experiencing rapid development. The expert system experienced a rapid increase so that it is helpful to human needs in resolving existing problems. The following research has built intelligent systems based on the fuzzy TOPSIS method intending to help to detect core competencies and student achievements. Testing system on the expert system based on the Fuzzy TOPSIS method the conditions in the Ministry of Education Islamic Religion Faculty of Tarbiyah and Islamic State University Maulana Malik Ibrahim Malang. The results show quite a good result that can help in making process decisions. From testing showed that the fuzzy method of TOPSIS resulted in a pretty good level of accuracy in expert systems on core competency detection and student learning achievements.

1. Introduction

State Islamic religious college is a strategic unit of education in preparing outstanding human resources and excellence. Global competition demands that universities should be able to deliver their students to compete globally both nationally and internationally. The quality of higher education graduates, including the state Islamic religious college, can be measured from how the level of achievement of student learning and the level of graduates in the field of employment as well as their field of expertise. Because of that, the quality of the college can be measured. One of them is from how the state Islamic religious college prepares students to be able to master the core competencies and produce graduates who achieve the best. Student learning achievement is demonstrated through the performance of academic, intellectual, skill, and moral skills, personality attitudes. The success of achieving high learning achievements in lectures in higher education becomes the ideals and expectations of every student. High learning achievement is expected to have implications for the ease of obtaining decent work and being able to compete openly in global competition. However, each student faces various obstacles, challenges, and various difficulties so that during the lecture, it is not certain that every student can face challenges and succeed in achieving the aspirations.

The technique for Order Performance by Similarity to Ideal Solution (TOPSIS) is a well-known method Multiple Criteria Decision Making (MCDM) problem. Some fuzzy criteria in decision making have an important role in solving problems in the decision-making process. Among the popular methods used in solving such problems is the Fuzzy TOPSIS, where solutions are produced based on the shortest distance from the ideal positive solution and the farthest distance from the ideal solution that is negative [1]. Related research uses the implementation of the evaluation and selection of projects that propose to project assessment and assist in the decision-making process of choosing the wrong One of the best results for oil companies in Iran [2]. The research uses six criteria to compare alternative investments as criteria in the utilization of AHP methods and Fuzzy TOPSIS methods. The AHP method is used to analyze the problem structure of the project selection and to determine the weight of a criterion, while the Fuzzy TOPSIS method is used to perform the final rating. The final results of the study were conducted to illustrate the use of models in project selection issues. The research shows the result of calculating the weight of criteria in the implementation using the Fuzzy TOPSIS method by changing the criterion rating.

Decision making can use a combination of different weights according to the criteria rating priority. Fuzzy TOPSIS can be used to solve decision-making problems by doing a summation of multiplication operations on the triangular fuzzy number (TFN). Based on the background discussion, the following studies have researched the development of the Neural Network-based Fuzzy Topsis optimization applied to an expert system in detecting student learning in measuring Instrument indicators development of core competencies and internal and external factors affecting the success of learning achievement achieved by students. The implementation of expert systems in detecting student learning can help students and lecturers in decision making to support the improvement of student learning achievement. This development was done to produce expert system applications by optimizing the Fuzzy TOPSIS method on the expert system of detection of core competencies and learning factors affecting student learning achievement. The research is integrated with the use of Fuzzy TOPSIS based on neural networks (NN) because it is very effective for the digging process. Fuzzy logic is often used to solve complex problems. The Neural Network is a representation of a human brain performance that seeks to mimic the way the human brain works to solve existing problems.

2. Literature review

Another study implemented the Fuzzy TOPSIS method on the Supply Chain Management (SCM) problem. The research object is in one of the companies in India by applying many criteria [3]. Research related to the combination of statistical analysis, fuzzy theory, and TOPSIS to evaluate the service quality of rail transit. Analysis results show the stableness of the fuzzy TOPSIS method by the ranking change of service quality for a line from different comparison sets of metro lines and provide suggestions and guidance for the optimization of rail transit infrastructure and investment [4].

Research related to making the best selection on the recycling method involves complex decision variables is considered to be a multiple criteria decision-making (MCDM) problem. Research for taproot an evaluation model based on the Fuzzy Analytic Hierarchy Process (AHP) and the technique for order performance by similarity to ideal solution (TOPSIS) to enable the industry practitioners to perform performance evaluation in a fuzzy environment. The results of the study are to determine the best method for recycling plastics among the various plastic recycling processes and identify that the mechanical recycling process is found to be the best plastic recycling process using the integrated approach [5].

3. Problem definition

The use of sections to divide the text of the paper is optional and left as a decision for the author. Where the author wishes to divide the paper into sections the formatting shown in table 2 should be used.

4. Research methodology

Following is the stage in research with the R & G approach.

IOP Conf. Series: Materials Science and Engineering **846** (2020) 012026 doi:10.1088/1757-899X/846/1/012026

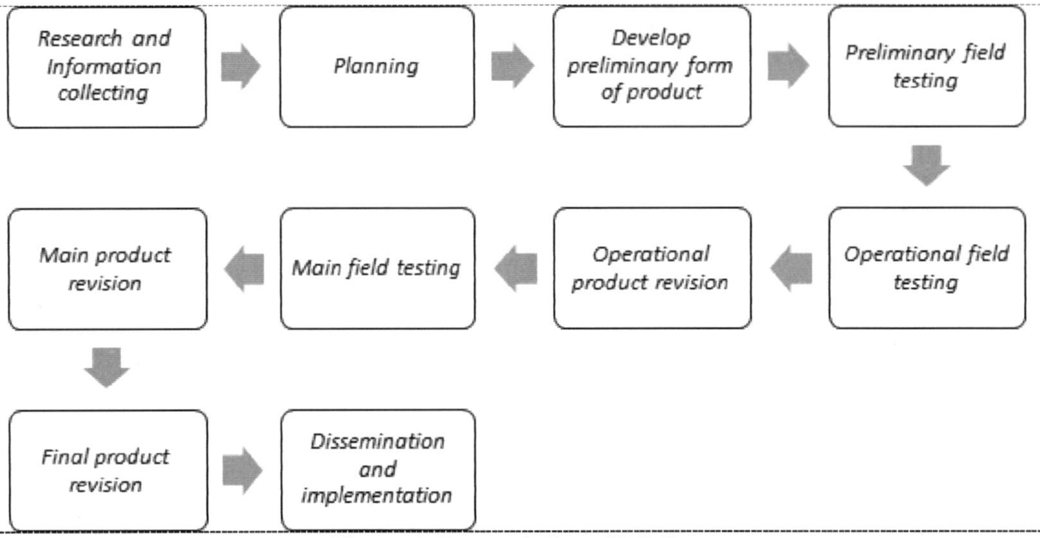

Figure 1. Research Steps R & D

1. Potential and problems.
Research always begins with potential or problems. Likely is everything that if you are going to have added value. Questions can also be turned into potential if researchers can power the problem. The problem will occur when there are irregularities, among which are expected with the circumstances occurring. Solving this problem is through R&D by learning it so that it is capable of modeling effective, integrated handling systems, or a useful pattern for solving problems. The potential and issues expressed in a study should be with empirical data. Data about potential and difficulties should not be sought by themselves. Still, it can also be based on other people's research reports as well as from documentation of activity reports originating from individuals or certain agencies that are always up to date.

2. Collecting information
Approving potentials and problems factually and up to date, the next step is to receive a variety of information and literature studies that can be used as a material to plan to ensure the product to address this problem. This study aims to find concepts and theoretical foundations that can strengthen a product, especially concerning educational outcomes, such as program-shaped products, models, systems, software, approaches, etc. On the other party through this literature review will examine the scope of a product, the breadth of use, supporting conditions so that the product can optimally, as well as its limitations and benefits. Literature studies are also needed to determine the most appropriate measures in developing the product.

3. Product Design
The existing product of this R&D study has a lot of types. One of the goals is to generate a new system of work. With the system, it should be made a new design work based on the assessment of the old work system, so that there can be weaknesses to the system. Besides, research needs to be done to other units that are considered to be a sound work system. Besides, it should be done with the assessment of cutting-edge references related to the modern working system, along with a good indicator of the working system. The end result of this activity is usually a new product design that has been completed with specifications. This design is still hypothetic, because its effectiveness is still not proven, and is only known after passing tests. Product design must be realized in the form of an image or chart, so it can be used as a handle to assess and make it and will make it easier for others to understand it.

4. Design validation
Design validation is an activity process aimed at assessing whether the design of the product, in which case the new working system is rationally more effective than the old one or not. It is said to be rational because validation at this stage is still based on logical thinking, not yet based on the fact of the field. Product validation can be carried out by presenting some experts or experienced experts to provide an assessment of the new product that was designed. Each expert is asked to give the value of the new design so that the next step can be known for its strengths and weaknesses. Design validation can be executed on a discussion forum. Before discussing, researchers presented the research process until found the design, along with its excellence.

5. Design Improvements
Once the product design is finished, it is validated through discussions with experts and other experts. Then it can be known his weaknesses. The weakness was then tried to be reduced by the way to improve the design. The task of improving the design is the researcher who will produce the product.

6. Product Trial
The design of the products that have been made can not be immediately tested first. However, it must be made in advance to produce products, and that product is tested. Testing can be Dilaksankan through experiments, comparing the effectiveness and efficiency of the old working system with the new working system.

7. Product Revision
Product testing of the limited samples could indicate that the performance of the new work system turned out to be better when compared to the old system. The difference is very significant so that the new system of work can be applied or enforced.

8. Trial usage.
After testing for a successful product, and there may be a less important revision, the next step is that the new working system is enforced or applied to real conditions for a wide scope. In the operation of the new work system, it remains to be assessed barriers or shortcomings that arise for further repairs.

9. Final Product Revisions
There are advantages and disadvantages to reviewing a product if, in the repair, it has a real condition. In testing the use of the product, researchers recommend that product makers as researchers always evaluate how the product is done in this case is the business system.

10. Dissemination and implementation
In the stages of the manufacture of the product in bulk will be executed if the tested product can be produced effectively and is worth mass-produced. As an example of making machines that can convert waste into useful materials, the mass will be produced when taking feasibility studies both from the economic, technological, and environmental aspects. Entrepreneurs and researchers must work together to produce a product.

5. Result and discussion
In this research, data were obtained from a questionnaire that was distributed to 70 respondents among students Islamic religious education in UIN Maulana Malik Ibrahim Malang being the object of research.

Development trials cover Three steps of expert testing, limited trials, and a more extensive trial. In expert tests involved experts in the development of the Neural Network-based Fuzzy Topsis optimization application product applied to the expert systems of competency application detection and learning achievement by building Fuzzy Topsis structures Neural Network-based networks. The development of the method is emphasized on the optimization of the Fuzzy TOPSIS based Neural Network on the expert system of detection of core competencies, learning factors, and student learning achievements.

Table 1. Instrument Research

Code	Core Competencies Academic Characters	Level of Practices					Level of Importance				
H	**Honesty**	1	2	3	4	5	1	2	3	4	5
H-1	Telling what I can do and can not to group member in a project.										
H-2	Admit friends' strength in class activities.										
H-3	Confessing my weaknesses to my academic supervisors										
H-4	Telling true resources in oral and writing presentation										
H-5	Controlling myself for not to present and report a fictive data										
H-6	Copying and pasting to work on my assignments										
H-7	Pretending to understand when lecturers explain a topic of material.										
H-8	Giving appraisal to friends thought it not like as it is.										
I	**Appreciating**										
I-1	Honouring the improvement of a friend in a lesson.										
I-2	Listening to friend who is asking questions or giving an idea.										
I-3	Paying attention to every words of friends' presentation										
I-4	Respecting friends with low achievement										
I-5	Encouraging less active friend to be more active.										
I-6	Prioritizing harmony in giving different ideas for the aim of seeking agreement by using unhatred words.										
I-7	Giving appraisal to friends' effort and work.										
J.	**Tolerance**										
J-1	Appreciating differences of ideas of friends who are different from themselves										
J-2	Appreciating the attitude of others who are different from myself										
J-3	Accepting the different ways of a group member in presenting and asking questions.										
J-4	Appreciating the different ways of class members in completing assignment										
J-5	Accepting diversity in a group, for instance difference gender, academic competence, race, religion and interest										
K.	**Discipline**										
K-1	Following academic rules and regulations at my faculty										
K-2	Coming to a class before the lecture begin.										
K-3	Completing and submitting my assignment by the deadline										
K-4	Organizing my learning activities daily.										
K-5	Scheduling, timing and prioritizing my learning activities.										
K-6	Targeting 70% and above to obtain my target of learning output										
K-7	Following rules set by classroom agreement or lecturer in learning activities										
K-8	Following the template, style or format given by lecturer in completing an assignment										
L	**Patient (SABAR)**										
L-1	Devoting myself to achieve my goal of learning										
L-2	Willing to hear long explanation and argumant from a friend.										
L-3	Accepting the final result though it is quite disappointing										
L-4	Controlling my emotion in debating or discussion.										
L-5	Staying motivated though I get unexpected result of my learning										
L-6	Controlling my emotion when my ideas against or rejected strongly by others										
L-7	Staying in an effort though I failed many times until I obtain a result.										
L-8	Making myself enjoy under a group leader or lecturer pressure in a task.										
M.	**Confident**										
M-1	Pushing down my nervousesness when I am trying to perform or present. (e.g assignment presentation).										
M-2	Encouraging myself while I am presenting my assignment										

Code	**Core Competencies Academic Characters**	**Level of Practices**					**Level of Importance**				
M-3	Being confident to ask question or share my ideas without afraid by making mistakes.										
M-4	Encouraging myself to participate more in classroom activities										
M-5	Encouraging myself to be more confident to perform in the classroom.										
M-6	Assuring myself on my own ability in completing a task and assignment.										
N.	**Responsible**										
N-1	Completing my own part as group member in group project and discussion										
N-2	Totally involving my own in group discussion.										
N-3	Taking a part as moderator in classroom discussion.										
N-4	Checking some errors and mistakes by my own..										
N-5	Revising my report, task or assignment by myself as suggested by lecturers.										
N-6	Taking a role of group leader in group work and discussion										
N-7	Working and completing individual assignment by myself.										

The Fuzzy TOPSIS Algorithm is a troubleshooting solution with process sequencing to get the value of a user's preference as a final value to get a ranking of all the alternatives and criteria that have been calculated. There are several working steps of TOPSIS:

A. Create a normalized decision matrix

B. Create an ideal definite solution matrix and an ideal negative solution

C. Determine the model of the perfect positive answer and the ideal negative solution

D. Determines the distance between the values of each alternative with the matrix of the ideal positive solution and the ideal negative matrix solution

E. Specifying the preference value for each alternative

Pengujian aplikasi ditunjukkan seperti pada gambar berikut ini.

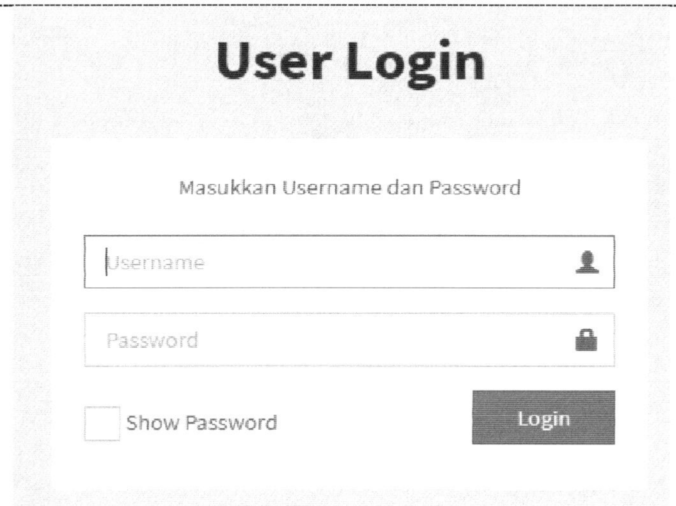

Figure 2. Login page

IOP Conf. Series: Materials Science and Engineering **846** (2020) 012026 doi:10.1088/1757-899X/846/1/012026

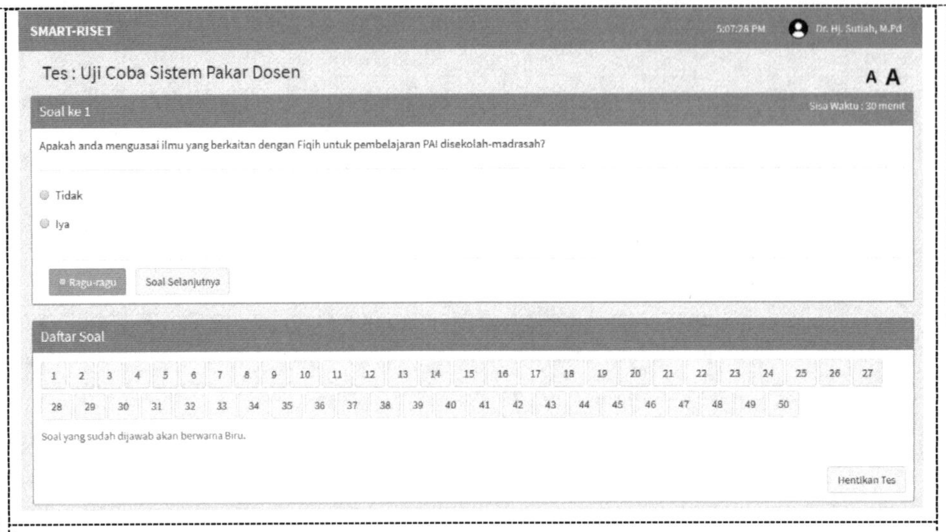

Figure 3. Expert Testing System

6. Conclusion

On the results of the implementation of several trials conducted on the application of expert systems in enhancing core competencies and detecting student achievement using the Fuzzy TOPSIS method at the Maulana Malik Ibrahim State Islamic University of Malang, it can be concluded that the resulting level of use of fuzzy TOPSIS achieves better results.

7. References

[1] A. W. Nizami and A. Mahmudi, "Problem-solving-based learning to improve students' learning interest," *AIP Conf. Proc.*, vol. 2014, no. September, 2018.

[2] V. A. Bhosale and R. Kant, "Examining the solutions to overcome the SCKFBS using fuzzy AHP and fuzzy TOPSIS method," *IEEE Int. Conf. Ind. Eng. Eng. Manag.*, vol. 2017-Decem, pp. 403–407, 2018.

[3] S. Kumar, S. Kumar, and A. G. Barman, "Supplier selection using fuzzy TOPSIS multi criteria model for a small scale steel manufacturing unit," *Procedia Comput. Sci.*, vol. 133, pp. 905–912, 2018.

[4] J. Li, X. Xu, Z. Yao, and Y. Lu, "Improving Service Quality With the Fuzzy TOPSIS Method: A Case Study of the Beijing Rail Transit System," *IEEE Access*, vol. 7, pp. 114271–114284, 2019.

[5] S. Vinodh, M. Prasanna, and N. Hari Prakash, "Integrated Fuzzy AHP-TOPSIS for selecting the best plastic recycling method: A case study," *Appl. Math. Model.*, vol. 38, no. 19–20, pp. 4662–4672, 2014.

Application of Monitoring Database for Accreditation Instrument UKI PAULUS

Lande Sudianto[1], Petrus Simon[2]

[1] Study Program of Electrical Engineering Study Program, Faculty of Engineering, [Universitas Kristen Indonesia Paulus]
[2] Study Program of Accounting, Faculty of Economics and Business, [Universitas Kristen Indonesia Paulus]

* Corresponding author: sudianto@ukipaulus.ac.id

Abstract. Accreditation is a form of quality and feasibility evaluation of higher education institutions or study programs conducted by independent organizations or bodies outside of tertiary institutions. To achieve the target of accreditation scores, universities and study programs must adjust their internal conditions in accordance with the standards given by the National Accreditation Board for Higher Education (BAN-PT). In accordance with Permenristekdikti No 32/2016, BAN-PT develops accreditation instruments that are relevant to the development of the higher education sector in Indonesia and follows the global developments. Study Program Accreditation Instrument (IAPS 4.0) BAN-PT uses 9 criteria. The Accreditation Instrument of Monitoring Database of the Study Program UKI Paulus was built using a PHP interface and MySql database that can integrate the entire contents of the data and other supporting documents so that during the evaluation processes, the information of achievement determined by BAN-PT from each standard is obtained. The system development steps, uses the Waterfall method which starts from the analysis phase, system design, program code writing, system testing and system maintenance. From 30 respondents who joined the test using black box method, the user accessibility level reached 95.52%. It shows that the system run optimally. This system is able to provide easy access and management of study program accreditation documents.

1. Introduction

Accreditation of higher education institutions is a process of comprehensive evaluation and assessment of the commitment of tertiary institutions to the quality and capacity of the implementation of tertiary education programs, to determine the feasibility of programs and educational units. Evaluation and evaluation in the context of institutional accreditation is carried out by a team of assessors consisting of peer experts and / or experts who understand the nature of higher education management. Decisions regarding quality are based on evaluations and assessments of various evidence related to established standards and based on the reasoning and consideration of peer experts [1] [2][3].

Evidence required includes a written report prepared by an accredited tertiary institution, verified and validated through a field assessment team visit or assessment to the university location. BAN-PT is an institution that has the authority to evaluate and assess, and determine the status and quality ratings of tertiary institutions based on established quality standards. Thus, the aims and benefits of accreditation of tertiary institutions are as follows

a. Providing guarantees that accredited tertiary institutions meet the quality standards set by BAN-PT, so that they are able to provide protection for the public from the implementation of tertiary institutions that do not meet the standards.
b. Encourage universities to continuously make improvements and maintain high quality.
c. Accreditation results can be used as a basis for consideration in the transfer of college credit, the provision of assistance and allocation of funds, as well as recognition from other agencies or agencies.

The quality of a tertiary institution is a reflection of the totality of conditions and characteristics of the institution inputs, processes and outputs or services measured by a number of standards set by BAN-PT. Higher education accreditation standards include standards on higher education commitment to institutional capacity and commitment to the effectiveness of educational programs, which are packaged in seven accreditation standards.

Higher education performance assessments are based on meeting the demands of accreditation standards. Process able accreditation documents for tertiary institutions must have fulfilled the initial requirements marked by valid and valid permits in the administration of tertiary institutions from authorized officials; has a statute and statutes and strategic plan documents or master plan for development that clearly shows the vision, mission, goals and objectives of higher education; basic values adopted and various aspects regarding the organization and management of tertiary institutions, the decision making process of program administration; quality assurance system; and has a number of accredited study programs.

Colleges describe and analyze all the indicators in the overall context of the accreditation standards with regard to the dimension of quality, namely: relevance, academic atmosphere, the internal management and organization, sustainability, efficiency including productivity. Is the additional dimension of leadership, equity, and governance [2].

The accreditation status is a reflection of the performance of a Study Program in describing the quality, efficiency, and relevance of a study program at the UKI Paulus Makassar in particular the Electrical Engineering Study Program. Accreditation management has not been done by using a special system to handle the data needed in filling the accreditation instruments. Accreditation instrument data filing system of the Electrical Engineering Study Program UKI Paulus Makassar is still in the form of manual. The data required for accreditation are still stored in various sources, so as to perform the accreditation must collect the data that makes the manager seek accreditation data. Study programs often lose accreditation-supporting data in the form of attachments, and cause the lengthy process of compiling and filling tables on the accreditation instrument. In line with the development of information and communication technology that offers convenience and speed in conveying access to information to meet the needs of accreditation more quickly and easily. Then developed an accreditation forms database information system that can manage and store accreditation forms data, which at the same time can be a means of measuring the performance of the Study Program based on standards set by BAN-PT.

2. Methodology

The stages of the development of the Accreditation Form Database system in the Electrical Engineering Study Program UKI Paulus uses the SDLC (System Development Life Cycle) paradigm approach [3] shown in Figure 1 which consists of 5 stages:

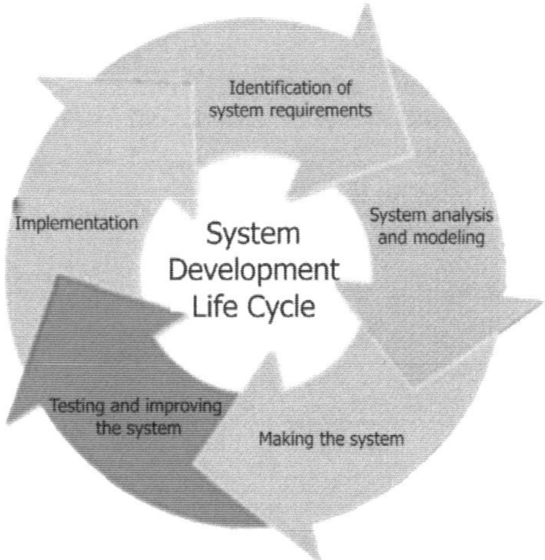

Figure 1. System Development Life Cycle

1. Identification of system requirements, namely determining the devices needed in making the system both primary and secondary data, hardware and software.
2. System analysis and modeling, system analysis and modeling are carried out with a flow map to find out the functions or activities that will be applied in the system. Flow map system is made in the form of context diagrams, data flow diagrams, entity relationship diagrams, relationships between files, file structure, program structure, menu structure, input format and output format. Analysis and modeling of the system is done based on UML (Unified Modeling Language) [4].
3. Making the system, the process of making the software The Accreditation Instrument of Monitoring Database of the Study Program UKI Paulus is complete using UML techniques, namely the selection of program packages that are appropriately analysis suitable for building systems. Making the system starting from system design, database design, and system display design. The design of the system display serves as a reference to create a user interface in system implementation.
4. Testing and improving the system, carried out testing of The Accreditation Instrument of Monitoring Database of the Study Program UKI Paulus was based on the whitebox and blackbox testing methods. Whitebox testing is used to test the accuracy of the algorithm flow and data structure of the system created. And blackbox testing is used to test the truth of system functionality based on the results of needs analysis [5][6][7]. Testing is done in the evaluation year / academic year 2018/2019.
5. The system implementation is carried out on server computers and intranet networks of the Electrical Engineering Study Program UKI Paulus

3. Result and Discussion

Accreditation is carried out by BAN-PT on the basis of open criteria. Accreditation is a process undertaken to determine the feasibility of education programs and units in the formal and non-formal education channels at each level and type of education. Decisions regarding quality are based on evaluations and assessments of various evidence related to established standards. The purpose and benefits of study program accreditation is to guarantee that accredited study programs meet the quality standards set, encourage study programs to continuously make improvements and maintain high quality, and the results of accreditation can be used as a basis for consideration in transferring college credit, giving assistance and allocation of funds, as well as recognition from other agencies or agencies [5].

There are 9 (nine) standards that explain the commitment of undergraduate study programs to institutional capacity and commitment to the effectiveness of educational programs [1].

1. Vision, mission, objectives, and strategy
2. Civil service, governance, and cooperation
3. Student
4. Human resource
5. Finance, Facilities and Infrastructure
6. Education
7. Research
8. Community Service
9. Tridharma's Outcomes and Achievements

The results of the analysis of system requirements in the form of minimum specifications of hardware and software are used as follows:

Computer hardware, with specifications:

- Intel Duel Core Processor
- 32 GB hard drive
- 2GB RAM
- Mouse and keyboard

Computer software, consisting of:

- Microsoft Windows 7 Operating System.
- Macromedia Dreamweaver application.
- ApacheFriends XAMPP application.
- Browser: Mozilla Firefox, Google Chrome

The system flow map is made in the form of context diagrams, data flow diagrams, entity relationship diagrams, relationships between files, file structures, program structures, menu structures, input formats and output formats. This flow map illustrates data flow, data storage, and data output. It also displays the data flow and the processes that exist in the system that become a reference in making the user interface in system implementation.

Making the system using several applications as follows:

a. Macromedia Dreamweaver application that functions as a professional HTML editor to visually design and manage websites and web pages .
b. The ApacheFriends XAMPP application functions as a stand-alone server, which consists of the Apache HTTP Server program, MySQL database, and language translator written in the PHP programming language (KYE Aryanto et al, 2016) .
c. PHP (Hypertext Preprocessor) is a scripting language that can be embedded or inserted into HTML which serves to make the *website* display dynamic and interactive.
d. MySQL is a *multithread* , and multi-user SQL database management system (DBMS) software . which functions as a data storage or *database server* [8][9].

Figure 2 shows the start page of the Information System Accreditation Form Database of the Electrical Engineering Study Program UKI Paulus. Each user must fill in their *username* and *password* to enter the system where each *user* has different access rights

Figure 2. System login page

Figure 3 displays the *database* structure presented in the *Physical Data Model* (PDM) format explaining how data is stored on a computer by presenting information such as *record formats*, *record orderings*, and access *paths*. The Accreditation Instrument of Monitoring Database of the Study Program UKI Paulus created using the programming language PHP and *database* MySQL. The PHP programming language was chosen with the aim of *web*-based and multi-platform. The program is made by paying attention to the UML design that has been made.

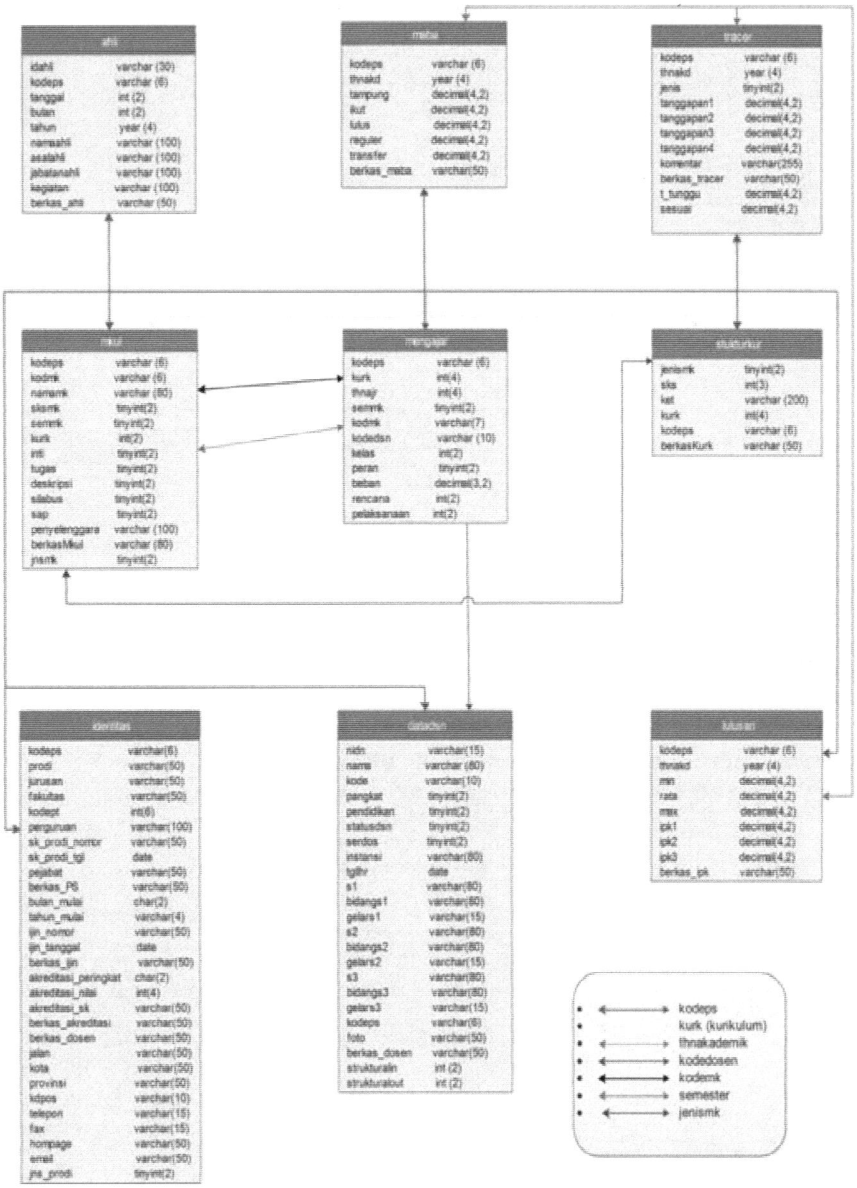

Figure 3. *Physical data model*

To run The Accreditation Instrument of Monitoring Database of the Study Program UKI Paulus for intranet networks by enabling the *ApacheFriends XAMPP* application on a *server* computer . By using a *browser* (Mozilla Firefox, Google Chrome) the system can be accessed through the site address http: Through the URL the application is run to play the program, namely Home. By filling in the user login correctly, it will be able to use the accreditation forms according to the appearance. Here are some program screen displays of the available menu options, as follows:

- Page menu_input_data.php were winnowed l right in Figure 4 , a menu of options for selecting the link inputting data borang among others: new students, student status, ipk graduate, tracer studies, courses, teaching, other research, community services, energy experts / experts, study assignments, scientific / professional organizations, staff, academic guidance, lecturer achievements, finance.

Figure 4. *Menu_input_data.php page*

- The menu_lihat_data.php page shown in Figure 5 is an option menu for selecting links to view the accreditation forms that correspond to the BAN-PT accreditation instrument table.

Figure 5. *The menu_view_data.php page*

IOP Conf. Series: Materials Science and Engineering **846** (2020) 012027 doi:10.1088/1757-899X/846/1/012027

- The input_dosen.php page shown in Figure 6 is a form for filling in the lecturer identity.

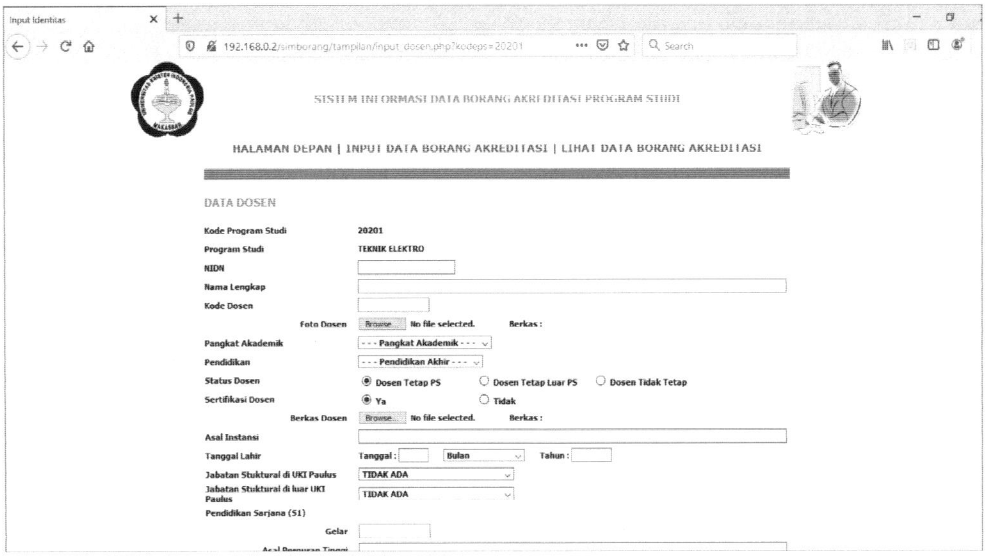

Figure 6. *Input_dosen.php page*

- The lihat_data_dosen_tetapPS.php page shown in Figure 7 is a form for filling the permanent lecturer identity of the study program.

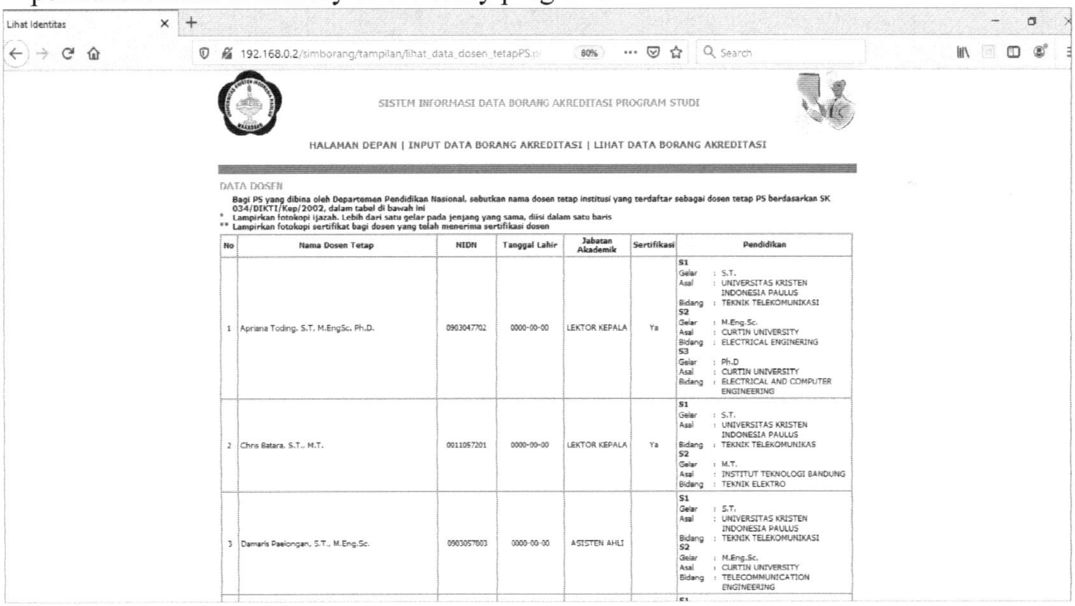

Figure 7. *Input_dosen.php page*

Testing is done in the evaluation year / academic year 2018/2019. The results of testing with the *whitebox* and *blackbox testing* methods, obtained that the system is running according to the flow algorithm and the desired data structure and each. Table 2 shows the results of system testing. Based on Table 2, it can be concluded that all functional aspects of The Accreditation Instrument of Monitoring Database of the Study Program UKI Paulus have been functioning properly and in accordance with the system design.

Table 1. System test results

Function	Desired Results	Testing	Test result
Login	Each user (user) must fill in the form username and password correctly	Fill in the correct username and password	Users can enter the main menu page
		Fill in the wrong username and / or password	The message "Fill in your User Name and Password". Users cannot enter the main menu page
Fill in form data	Fill in data forms and verified and saved to the database by the system	Fill in complete data	Data is stored in a *database*
		Fill in incomplete data or Fill in the same data	Data is not stored in the *database,* there is a message display "data incomplete" or "data already registered"
Changing d ata	Data changes according to the code of each data	Fill in or click the data code of the form to be changed	Form data changes in the *database*
Wipe data	Data is deleted according to the code of each data	Fill in or click the form data code to delete	Data forms are deleted in the *databas*

Testing the level of user accessibility by measuring the level of user satisfaction with the system built. The test is performed by the admin istrator , dean, head of study program , faculty, administrative staff, and students involved in the process of accreditation. Responden provides an assessment of the 5 questions given, namely:

a. Display of The Accreditation Instrument of Monitoring Database of the Study Program UKI Paulus.
b. Complete menu / form on the accreditation form database system in the UKI Paulus Electrical Engineering Study Program
c. Speed of response and access to the database system Accreditation Forms in the UKI Paulus Electrical Engineering Study Program.
d. The accuracy of the data displayed in the Accreditation Form Database system of the UKI Paulus Electrical Engineering Study Program.
e. Accreditation Form Database System UKI Paulus Electrical Engineering Study Program meets the needs of the accreditation forms table

Each question was rated using a scale of 5 (1 / No; 2 / Poor; 3 / Enough; 4 / Good; 5 / Very Good). From 30 respondents, the measurement results obtained from the user's accessibility level obtained by 95.52% for the scale (Good - Very Good)

4. Conclusion

This research has resulted in the UKI Paulus Electrical Engineering Study Program Accreditation Database information system, which is able to present accurate data in accordance with the needs of the BAN-PT accreditation form. Presentation of web-based data forms provides convenience and speed in filling accreditation instrument tables with user accessibility levels obtained at 95.52%

References

[1] BAN-PT, "Panduan Penyusunan Laporan Evaluasi Diri", Jakarta, 2018.
[2] Direktur Jenderal Pendidikan Tinggi, "Sistem Penjaminan Mutu Perguruan Tinggi (SPM-PT)". Jakarta, 2010.
[3] Menteri Pendidikan Nasional, "Peraturan Menteri Pendidikan Nasional Nomor 63 Tahun 2009 Tentang Sistem Penjaminan Mutu Pendidikan", Biro Hukum dan Organisasi Departemen Pendidikan Nasional Indonesia. Jakarta, 2009.
[4] L. Sudianto, P. Simon. "The Encyclopedia of Virtual Art Carving Toraja". MATEC Web of Conferences, 2018.
[5] Rully Indra Laelaningrum, dkk. "Sistem Informasi Visitasi Akreditasi Program Studi Teknik Informatika". Transmisi, 20, JULI 2018, p-ISSN 1411-0814 e-ISSN 2407–6422.
[6] Ahmad Fauzi, M.Kom, Oman Komarudin, S.Si., M.Kom, "Perancangan Perangkat Lunak Borang Akreditasi Program Studi", SYSTEMATICS, 2012
[7] K.Y.E. Aryanto, I.K.R Arthana, "Sistem Informasi Penyusunan Borang Akreditasi Daring Untuk Program Studi Diploma, Sarjana Dan Fakultas", Seminar Nasional Vokasi dan Teknologi (SEMNASVOKTEK). Denpasar-Bali, 22 Oktober 2016
[8] Anggriani Profita dkk, "Pengembangan Sistem Informasi Manajemen Borang Akreditasi (SIMBA) Berbasis Website", Seminar Nasional Teknik Industri. Surabaya, 2017.
[9] Fitriana Nurhayatun Nuphus dkk, "Sistem Informasi Akreditasi Perguruan Tinggi (SIAP) untuk Pengisian Borang Standar 3 BAN-PT". JUSTIN (Jurnal Sistem dan Teknologi Informasi) Vol. 7 , No.2 , April 2019.

Smart City Recommendations Using the TOPSIS Method

Nurliana Nasution [1], Gita Widi Bhawika [2], Anjar Wanto [3], Ni Luh Wiwik Sri Rahayu Ginantra [4], Teuku Afriliansyah [5]

[1] Universitas Lancang Kuning, Pekanbaru - Indonesia
[2] Institut Teknologi Sepuluh Nopember, Surabaya - Indonesia
[3] STIKOM Tunas Bangsa, Pematangsiantar - Indonesia
[4] STMIK STIKOM Indonesia, Denpasar - Indonesia
[5] STKIP Bumi Persada Lhokseumawe, Aceh - Indonesia

* nurliananst@unilak.ac.id

Abstract. This study aims to recommend a city that is suitable as a Smart City on the island of Sumatera. This study uses a Decision Support System with the TOPSIS (Technique For Others Reference by Similarity to Ideal Solution) Method. The TOPSIS method is one method that is often used for ranking problems. The research data used are data from 10 major cities on the island of Sumatera. with the largest population in 2019 obtained from the Wikipedia Website (https://www.wikipedia.org/wiki/Sumatra) and the Central Statistics Agency website (https://www.bps.go.id/) as a reference in determining the assessment of each Alternative later. The 10 major cities are Medan, Palembang, Bandar Lampung, Pekanbaru, Batam, Padang, Jambi, Bengkulu, Banda Aceh, and Pematangsiantar. To be able to determine and recommend cities in the islands of Sumatera that are eligible to become Smart Cities, there are 4 criteria as an assessment of each Alternative, namely, Infrastructure, Population, Area, and Economic Level. Based on calculations using the TOPSIS method, the results obtained that the city of Medan has the highest value in the eligibility to become a Smart City.

1. Introducing

Smart City is an urban area environment that is made smart and smart by using ICT technology so that it can improve the quality of life of residents and the efficiency of urban infrastructure, and provide the best and ever-increasing services to its citizens. A Smart City usually has a physical cyber system that can improve urban behavior and capabilities by providing ICT-based facilities. Infrastructure for Smart City must be geographically and functionally expandable, this is needed to grow the shared physical environment and to meet the increasing needs and demands of users/residents of the city [1]–[4]. The term "Smart City" has spread and spread throughout the world and influenced government programs and strategies in city development. Many ideas and breakthroughs by the government are trying to create various services, ranging from smart transportation and smart energy to smart and educated people [5]. Future cities like that are expected to be able to use ICT technology effectively and efficiently in the structure of the urban environment that aims to improve and rationalize public services in the future [6]. However, things like this, however, often fail to guarantee safety and security as the main focus. This is important because it is not only one of the most fundamental principles in urban planning and management but also human welfare. Thus, safety and security are factors that are an integral part of human well-being and also of every Smart City design [7]. Smart City promotes a modern and advanced lifestyle. Infrastructure, innovation, and technology are components that make smart cities efficient and independent [8]. The level of smart city sustainability is measured based on a series of parameters, including public management, social cohesion, governance, technology, urban planning, environment, mobility and transportation, international projections, human capital, and the economy. Smart City has tons of applications to monitor and control each of these parameters [9].

In Indonesia, the central government helps prepare a strategic data storage facility and general applications to assist regions in building smart cities. The central government will provide a smart city development master plan for local governments. Therefore, the government does not arbitrarily choose cities that will be used as smart cities, special studies and assessments are needed, one of them is by building a system that can recommend cities that are worthy of being Smart Cities in Indonesia, especially on the island. Sumatera, the results of this system can later be used as a reference for the government in determining which regions are eligible to be selected as the next smart city. There are many branches in the field of computer science that can solve complex problems, this is evident from the many studies that have been done, such as research in the field of data mining [10]–[14], the field of artificial neural networks [15]–[25], and the field of decision support systems [26]–[33]. Based on this explanation, researchers used a decision support system to resolve the above problem. In this case, researchers use the TOPSIS method, because this method is one of the methods of decision support systems that are often used for ranking based on predetermined variables and criteria. Many previous studies discuss the smart city [34]–[36]. Based on this background, the research was conducted. It is expected that by using the TOPSIS method, the results will be obtained in the form of a Smart City recommendation on the island of Sumatera that can be used as a government reference in determining which cities in Sumatera are eligible to be used as Smart Cities.

2. Methodology

2.1. Method of collecting data

Data collection was carried out by observation contained on the Wikipedia Website (https://www.wikipedia.org/wiki/Sumatra) and the Central Statistics Agency website (https://www.bps.go.id/), not conduct research or surveys directly to cities that are alternative to obtain data.

2.2. Stages of the TOPSIS Method

At this stage, the workflow of the TOPSIS (Technique For Others Reference by Similarity to Ideal Solution) method used in recommending a city worthy of being a Smart City in Sumatera is based on alternatives and criteria that have been analyzed.

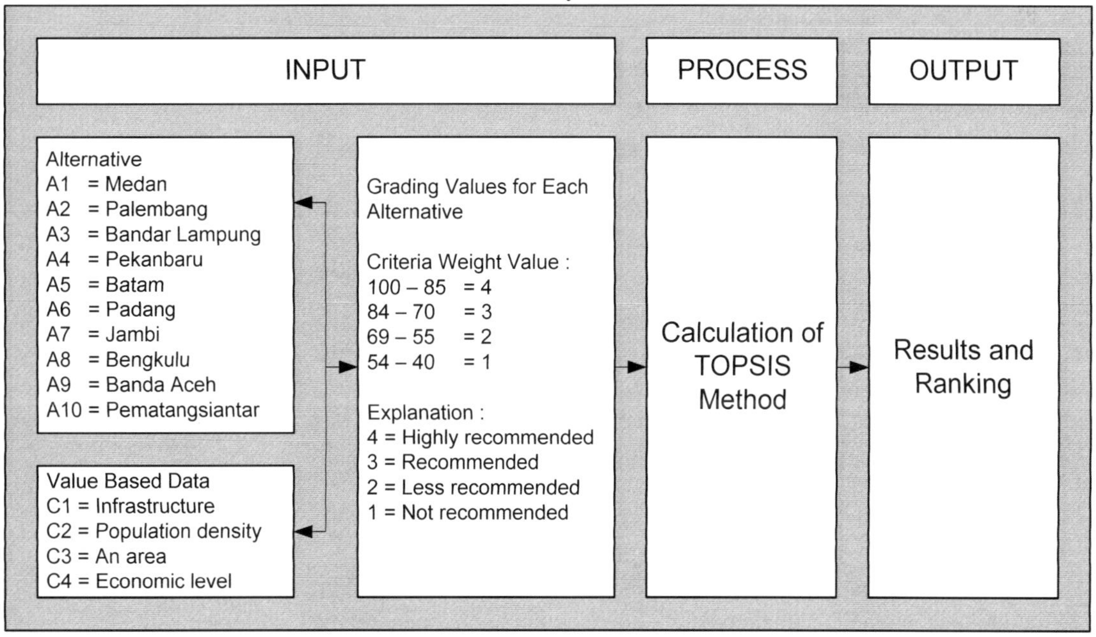

Figure 1. Stages of the TOPSIS Method

3. Results and Discussion

Manual calculations using the TOPSIS method are based on assessments that have been carried out with data analysis based on 10 major cities in Sumatera island obtained from the Website of Wikipedia and the Website of the Indonesian Central Statistics Agency. Alternative Criteria value is given based on the Infrastructure, Population, Area and Economic Level criteria can be seen in the following table.

Table 1. Alternative Value Criteria

No	Alternative	Criteria			
		C1	**C2**	**C3**	**C4**
1	Medan	90	90	70	85
2	Palembang	40	85	75	45
3	Bandar Lampung	75	80	60	60
4	Pekan Baru	85	75	80	70
5	Batam	45	70	90	65
6	Padang	80	65	85	80
7	Jambi	70	60	65	50
8	Bengkulu	65	45	60	90
9	Banda Aceh	55	55	40	40
10	Pematangsiantar	60	40	45	75

Furthermore, the valuation of each Alternative based on the value of the weight that has been determined based on the assessment of the criteria of Infrastructure, Population, Area and Economic Level can be seen in the following table.

Table 2. Alternative Criteria Weight Value

No	Alternative	Criteria			
		C1	**C2**	**C3**	**C4**
1	Medan	4	4	3	4
2	Palembang	1	4	3	1
3	Bandar Lampung	3	3	2	2
4	Pekan Baru	4	3	3	3
5	Batam	1	3	4	2
6	Padang	3	2	4	3
7	Jambi	3	2	2	1
8	Bengkulu	2	1	1	4
9	Banda Aceh	2	2	1	1
10	Pematangsiantar	2	1	1	3

The following is a manual calculation of the TOPSIS (Technique For Others Reference by Similarity to Ideal Solution) method in recommending cities that are worthy of being Smart cities in Sumatera.

a. Normalized decision matrix

$$X_1 = \sqrt{4^2 + 1^2 + 3^2 + 4^2 + 1^2 + 3^2 + 3^2 + 2^2 + 2^2 + 2^2} = \sqrt{73} = \mathbf{8{,}5440}$$

$$R_{11} = \frac{4}{8{,}5440} = \mathbf{0{,}4682} \qquad R_{61} = \frac{3}{8{,}5440} = \mathbf{0{,}3511}$$

$$R_{21} = \frac{1}{8{,}5440} = \mathbf{0{,}1170} \qquad R_{71} = \frac{3}{8{,}5440} = \mathbf{0{,}3511}$$

$$R_{31} = \frac{3}{8{,}5440} = \mathbf{0{,}3511} \qquad R_{81} = \frac{2}{8{,}5440} = \mathbf{0{,}2341}$$

$$R_{41} = \frac{4}{8{,}5440} = \mathbf{0{,}4682} \qquad R_{91} = \frac{2}{8{,}5440} = \mathbf{0{,}2341}$$

$$R_{51} = \frac{1}{8{,}5440} = \mathbf{0{,}1170} \qquad R_{10\,1} = \frac{2}{8{,}5440} = \mathbf{0{,}2341}$$

For X_2, X_3 and X_4 done in the same way based on the values of C2, C3 and C4. From the normalization calculation, the value of R is obtained :

0,4682	0,4682	0,3586	0,4781
0,1170	0,4682	0,3586	0,1195
0,3511	0,3511	0,2390	0,2390
0,4682	0,3511	0,3586	0,3586
0,1170	0,3511	0,4781	0,2390
0,3511	0,2341	0,4781	0,3586
0,3511	0,2341	0,2390	0,1195

0,2341	0,1170	0,1195	0,4781
0,2341	0,2341	0,1195	0,1195
0,2341	0,1170	0,1195	0,3586

b. Normalized weight value. Matrix R Normalized W = 4, 3, 2, 3

$Y_{11} = 0{,}4682 \times 4 =$ **1,8727** $\quad Y_{61} = 0{,}3511 \times 4 =$ **1,4045**
$Y_{21} = 0{,}1170 \times 4 =$ **04682** $\quad Y_{71} = 0{,}3511 \times 4 =$ **1,4045**
$Y_{31} = 0{,}3511 \times 4 =$ **1,4045** $\quad Y_{81} = 0{,}2341 \times 4 =$ **0,9363**
$Y_{41} = 0{,}4682 \times 4 =$ **1,8727** $\quad Y_{91} = 0{,}2341 \times 4 =$ **0,9363**
$Y_{51} = 0{,}1170 \times 4 =$ **04682** $\quad Y_{101} = 0{,}2341 \times 4 =$ **0,9363**

Next, the calculation is done in the same way to get the results from $Y_{ij} = W_i \,.\, R_{ij}$

1,8727	1,4045	0,7171	1,4343
0,4682	1,4045	0,7171	0,3586
1,4045	1,0534	0,4781	0,7171
1,8727	1,0534	0,7171	1,0757
0,4682	1,0534	0,9562	0,7171
1,4045	0,7022	0,9562	1,0757
1,4045	0,7022	0,4781	0,3586
0,9363	0,3511	0,2390	1,4343
0,9363	0,7022	0,2390	0,3586
0,9363	0,3511	0,2390	1,0757

c. The Positive Ideal Solution (A^+)

Y_1^+ = Max { 1,8727, 0,4682, 1,4045, 1,8727, 0,4682, 1,4045, 1,4045, 0.9363, 0.9363, 0.9363 }
= Max {1,8727} (And so on until Y_4^+)

Solusi Ideal Positif (A^-)

Y_1^- = Min { 1,8727, 0,4682, 1,4045, 1,8727, 0,4682, 1,4045, 1,4045, 0.9363, 0.9363, 0.9363 }
= Min {0,4682} (And so on until Y_4^-)

d. The alternative distance is weighted with a positive ideal solution.

$$D_1^+ = \sqrt{(1{,}8727 - 1{,}8727)^2 + (1{,}4045 - 1{,}4045)^2 + (0{,}7171 - 0{,}9562)^2 + (1{,}4343 - 1{,}4343)^2}$$
$$= \sqrt{(-0{,}2391)^2}$$
$$= \sqrt{0{,}0571}$$

= 0,2391 (And so on until getting results D_{10}^+ **= 1,6214**)

e. The alternative distance is weighted with a negative ideal solution.

$$D_1^- = \sqrt{(1{,}8727 - 0{,}4682)^2 + (1{,}4045 - 0{,}3511)^2 + (0{,}7171 - 0{,}2390)^2 + (1{,}4343 - 0{,}3586)^2}$$
$$= \sqrt{(1{,}4045)^2 + (1{,}0534)^2 + (0{,}4782)^2 + (1{,}0757)^2}$$
$$= \sqrt{1{,}9726 + 1{,}1096 + 0{,}2285 + 1{,}1571}$$
$$= \sqrt{4{,}4678}$$

= 2, 1137 (And so on until getting results D_{10}^- **= 0,8563**)

Result of D_i^+ : 0,2391; 1,7851; 1,0417; 0,5551; 1,6153; 0,9170; 1,4484; 1,5813; 1,7439; 1,6214
Result of D_i^- : 2,1137; 1,1567; 1,2472; 1,7912; 1,0658; 1,4241; 1,0280; 1,1731; 0,5850; 0,8563

f. Preference value for each alternative

Alternative 1 $\quad V_1 = \frac{2{,}1137}{2{,}1137 + 0{,}2391} = \frac{2{,}1137}{2{,}3528} = \mathbf{0,8983}$

Alternative 2 $\quad V_2 = \frac{1{,}1567}{1{,}1567 + 1{,}7851} = \frac{1{,}1567}{2{,}9418} = \mathbf{0,3931}$

Alternative 3 $\quad V_3 = \frac{1{,}2472}{1{,}2472 + 1{,}0417} = \frac{1{,}2472}{2{,}2889} = \mathbf{0,5448}$

Alternative 4 $\quad V_4 = \frac{1{,}7912}{1{,}7912 + 0{,}5551} = \frac{1{,}7912}{2{,}3463} = \mathbf{0,7634}$

Alternative 5 $\quad V_5 = \frac{1{,}0658}{1{,}0658 + 1{,}6153} = \frac{1{,}0658}{2{,}6811} = \mathbf{0,3975}$

Alternative 6 $\quad V_6 = \frac{1{,}4241}{1{,}4241 + 0{,}9170} = \frac{1{,}4241}{2{,}3411} = \mathbf{0,6083}$

Alternative 7 $\quad V_7 = \frac{1{,}0280}{1{,}0280 + 1{,}4484} = \frac{1{,}0280}{2{,}4764} = \mathbf{0,4151}$

Alternative 8 $\quad V_8 = \frac{1{,}1731}{1{,}1731 + 1{,}5813} = \frac{1{,}1731}{2{,}7544} = \mathbf{0,4259}$

Alternative 9 $V_9 = \frac{0,5850}{0,5850+1,7439} = \frac{0,5850}{2,3289} = \mathbf{0,2511}$

Alternative 10 $V_{10} = \frac{0,8563}{0,8563+1,6214} = \frac{0,8563}{2,4777} = \mathbf{0,3456}$

4. Conclusion

The TOPSIS method can recommend cities that are worthy of being Smart City. This study results that the city of Medan (Alternative 1) has the highest value in the feasibility of being a Smart City based city that is equal to 0.8983 or the highest compared to other alternatives.

References

[1] A. Monzon, "Smart Cities Concept and Challenges," *2015 International Conference on Smart Cities and Green ICT Systems (SMARTGREENS)*, pp. 1–11, 2015.

[2] K. R. Kunzmann, "Smart cities: A new paradigm of urban development," *Crios*, vol. 7, pp. 9–19, 2014.

[3] F. Cicirelli, A. Guerrieri, G. Spezzano, and A. Vinci, "An edge-based platform for dynamic Smart City applications," *Future Generation Computer Systems*, vol. 76, pp. 106–118, 2017.

[4] A. Sharifi, *A typology of smart city assessment tools and indicator sets*, vol. 53. Elsevier B.V., 2020.

[5] J. Laufs, H. Borrion, and B. Bradford, "Security and the smart city: A systematic review," *Sustainable Cities and Society*, vol. 55, p. 102023, 2020.

[6] M. Berry, "Technology and organised crime in the smart city: an ethnographic study of the illicit drug trade," *City, Territory and Architecture*, vol. 5, no. 1, pp. 1–11, 2018.

[7] A. G. Reddy, D. Suresh, K. Phaneendra, J. S. Shin, and V. Odelu, "Provably Secure Pseudo-Identity Based Device Authentication for Smart Cities Environment," *Sustainable Cities and Society*, vol. 41, pp. 878–885, 2018.

[8] P. Chamoso, A. González-Briones, F. De La Prieta, G. K. Venyagamoorthy, and J. M. Corchado, "Smart city as a distributed platform: Toward a system for citizen-oriented management," *Computer Communications*, vol. 152, no. December 2019, pp. 323–332, 2020.

[9] C. Richter, S. Kraus, and P. Syrjä, "The Smart City as an opportunity for entrepreneurship," *International Journal of Entrepreneurial Venturing*, vol. 7, no. 3, pp. 211–226, 2015.

[10] R. Rahim *et al.*, "C4.5 Classification Data Mining for Inventory Control," *International Journal of Engineering & Technology*, vol. 7, pp. 68–72, 2018.

[11] I. Parlina *et al.*, "Naive Bayes Algorithm Analysis to Determine the Percentage Level of visitors the Most Dominant Zoo Visit by Age Category," *Journal of Physics: Conference Series*, vol. 1255, no. 1, 2019.

[12] I. S. Damanik, A. P. Windarto, A. Wanto, Poningsih, S. R. Andani, and W. Saputra, "Decision Tree Optimization in C4.5 Algorithm Using Genetic Algorithm," *Journal of Physics: Conference Series*, vol. 1255, no. 1, pp. 1–7, 2019.

[13] D. Hartama, A. Perdana Windarto, and A. Wanto, "The Application of Data Mining in Determining Patterns of Interest of High School Graduates," *Journal of Physics: Conference Series*, vol. 1339, no. 1, pp. 1–6, 2019.

[14] M. Widyastuti, A. G. Fepdiani Simanjuntak, D. Hartama, A. P. Windarto, and A. Wanto, "Classification Model C.45 on Determining the Quality of Custumer Service in Bank BTN Pematangsiantar Branch," *Journal of Physics: Conference Series*, vol. 1255, no. 012002, pp. 1–6, 2019.

[15] E. Siregar, H. Mawengkang, E. B. Nababan, and A. Wanto, "Analysis of Backpropagation Method with Sigmoid Bipolar and Linear Function in Prediction of Population Growth," *Journal of Physics: Conference Series*, vol. 1255, no. 1, pp. 1–6, 2019.

[16] G. W. Bhawika *et al.*, "Implementation of ANN for Predicting the Percentage of Illiteracy in Indonesia by Age Group," *Journal of Physics: Conference Series*, vol. 1255, no. 1, pp. 1–6, 2019.

[17] W. Saputra, J. T. Hardinata, and A. Wanto, "Resilient method in determining the best architectural model for predicting open unemployment in Indonesia," *IOP Conference Series: Materials Science and Engineering*, vol. 725, no. 1, pp. 1–7, 2020.

[18] A. Wanto and J. T. Hardinata, "Estimations of Indonesian poor people as poverty reduction efforts

facing industrial revolution 4 . 0," *IOP Conference Series: Materials Science and Engineering*, vol. 725, no. 1, pp. 1–8, 2020.

[19] A. Wanto *et al.*, "Model of Artificial Neural Networks in Predictions of Corn Productivity in an Effort to Overcome Imports in Indonesia," *Journal of Physics: Conference Series*, vol. 1339, no. 1, pp. 1–6, 2019.

[20] A. Wanto *et al.*, "Analysis of the Accuracy Batch Training Method in Viewing Indonesian Fisheries Cultivation Company Development," *Journal of Physics: Conference Series*, vol. 1255, no. 1, pp. 1–6, 2019.

[21] I. S. Purba *et al.*, "Accuracy Level of Backpropagation Algorithm to Predict Livestock Population of Simalungun Regency in Indonesia Accuracy Level of Backpropagation Algorithm to Predict Livestock Population of Simalungun Regency in Indonesia," *Journal of Physics: Conference Series*, vol. 1255, no. 1, pp. 1–6, 2019.

[22] A. Wanto *et al.*, "Analysis of the Backpropagation Algorithm in Viewing Import Value Development Levels Based on Main Country of Origin," *Journal of Physics: Conference Series*, vol. 1255, no. 1, pp. 1–6, 2019.

[23] S. Setti, A. Wanto, M. Syafiq, A. Andriano, and B. K. Sihotang, "Analysis of Backpropagation Algorithms in Predicting World Internet Users," *Journal of Physics: Conference Series*, vol. 1255, no. 1, pp. 1–6, 2019.

[24] M. K. Z. Sormin, P. Sihombing, A. Amalia, A. Wanto, D. Hartama, and D. M. Chan, "Predictions of World Population Life Expectancy Using Cyclical Order Weight / Bias," *Journal of Physics: Conference Series*, vol. 1255, no. 1, pp. 1–6, 2019.

[25] A. Wanto *et al.*, "Forecasting the Export and Import Volume of Crude Oil , Oil Products and Gas Using ANN," *Journal of Physics: Conference Series*, vol. 1255, no. 1, pp. 1–6, 2019.

[26] P. P. P. A. N. . F. I. R.H Zer, Masitha, A. P. Windarto, and A. Wanto, "Analysis of the ELECTRE Method on the Selection of Student Creativity Program Proposals," *Journal of Physics: Conference Series*, vol. 1255, no. 1, pp. 1–7, 2019.

[27] A. P. W. Budiharjo and A. Muhammad, "Comparison of Weighted Sum Model and Multi Attribute Decision Making Weighted Product Methods in Selecting the Best Elementary School in Indonesia," *International Journal of Software Engineering and Its Applications*, vol. 11, no. 4, pp. 69–90, 2017.

[28] D. R. Sari, N. Rofiqo, D. Hartama, A. P. Windarto, and A. Wanto, "Analysis of the Factors Causing Lazy Students to Study Using the ELECTRE II Algorithm," *Journal of Physics: Conference Series*, vol. 1255, no. 012007, pp. 1–6, 2019.

[29] K. Fatmawati *et al.*, "Analysis of Promothee II Method in the Selection of the Best Formula for Infants Under Three Years," *Journal of Physics: Conference Series*, vol. 1255, no. 1, pp. 1–7, 2019.

[30] P. Alkhairi, L. P. Purba, A. Eryzha, A. P. Windarto, and A. Wanto, "The Analysis of the ELECTREE II Algorithm in Determining the Doubts of the Community Doing Business Online," *Journal of Physics: Conference Series*, vol. 1255, no. 1, pp. 1–7, 2019.

[31] S. Sundari, Karmila, M. N. Fadli, D. Hartama, A. P. Windarto, and A. Wanto, "Decision Support System on Selection of Lecturer Research Grant Proposals using Preferences Selection Index," *Journal of Physics: Conference Series*, vol. 1255, no. 1, pp. 1–7, 2019.

[32] S. R. Ningsih, R. Wulansari, D. Hartama, A. P. Windarto, and A. Wanto, "Analysis of PROMETHEE II Method on Selection of Lecturer Community Service Grant Proposals," *Journal of Physics: Conference Series*, vol. 1255, no. 1, pp. 1–7, 2019.

[33] T. Imandasari, M. G. Sadewo, A. P. Windarto, A. Wanto, H. O. Lingga Wijaya, and R. Kurniawan, "Analysis of the Selection Factor of Online Transportation in the VIKOR Method in Pematangsiantar City," *Journal of Physics: Conference Series*, vol. 1255, no. 1, pp. 1–7, 2019.

[34] C. Zheng, J. Yuan, L. Zhu, Y. Zhang, and Q. Shao, "From Digital to Sustainable : A Scientometric Review of Smart City Literature between 1990 and 2019," *Journal of Cleaner Production*, pp. 1–51, 2020.

[35] M. W. Wathne and H. Haarstad, "The smart city as mobile policy: Insights on contemporary urbanism," *Geoforum*, vol. 108, no. February 2019, pp. 130–138, 2020.

[36] M. Lom and O. Pribyl, "International Journal of Information Management Smart city model based on systems theory," *International Journal of Information Management*, no. February 2019, pp. 1–11, 2020.

E-Green Jukebox: The Development of an Electronic Jukebox That Accept Aluminium Can as Token

Wan Mohd Rumaizi Wan Taib[1], Nor Haizat Omar[1], Nur Anizah Jalani[2], Zakiah Mohd Yusoff[2], Amar Faiz Zainal Abidin[3]

[1]Department of Electrical Engineering, Politeknik Ibrahim Sultan, Malaysia
[2]Faculty of Electrical Engineering, Universiti Teknologi MARA Cawangan Johor, Kampus Pasir Gudang, Malaysia
[3]Faculty of Electrical and Electronics Engineering Technology, Universiti Teknikal Malaysia Melaka, Malaysia

*wanrumaizi.poli@gmail.com, norhaizatomarpis@gmail.com

Abstract. Jukebox has been part of the outside dining experience in the western family restaurant in the year 1940s to 1990s. The progression of technology in entertainment industry makes the jukebox become obsolete due to the relatively expensive price to play a song from the jukebox. At the same time, most of the shopping mall food court in Malaysia required the customers to self-service and self-clean after dinning. One of the problem encountered is the customers do not usually separate the waste when throwing the waste. This project aims to revitalize the jukebox by reintroducing the song machine with a slight twist, the token used is aluminum can. This is to encourage the customers who purchase can drink in the shopping mall food court to recycle the aluminum can and in the same time to enjoy the retro feeling of dining experience with song played by jukebox. This paper presented the development of this jukebox called Green Jukebox. The result shows the potential of the prototype for the real application.

1. Introduction

Jukebox, or originally known as nickelodeon, is a device that accepts token from the user to play the song selected by the user. Jukebox is commonly found in the western family restaurant especially in North America in 1940s to 1970s. The trend also catching up in Malaysia in the early 1980s to the late 1990s. The jukebox can commonly found in the western-themed restaurant in Malaysia such as Kentucky Fried Chicken chains, and Mc Donald chains. As the technology progresses, the cost of accessing a music song is relatively cheap compared to the golden ages of jukebox, this makes the jukebox become obsolete. By early 2000s, jukebox no longer can be seen in the market due to the relatively expensive price to play a song from the jukebox.

Nowadays, most of food court in the shopping mall in Malaysia required the customers to self-service and self-clean after dinning. One of the noticeable problem surfaced is the customers do not usually separate the waste when throwing the waste. Thus, this project attempts to address this issue at the same time repurposing the jukebox. This is done by reintroducing the song machine with a slight twist where the aluminum can is used, instead of the typical coin. The project aims to create a prototype encourage the customers who purchase can drink in the shopping mall food court to recycle the aluminum can and in the same time to enjoy the retro feeling of dining experience with song played by jukebox. This project is not something outside the box as there are several recycle machines that had been introduced in the market. Based on literature survey done, there are four prominent recycle machines that reward user when recycling the recyclable items. These recycle machine are called reverse vending solutions. The first reverse vending machine is called Tomra T-53 Plastic and Can Recycling Machine which refund the money to user when the user insert related plastic or aluminum can [1]. The reverse vending machine fundamentally used the infrared (IR) technology which customize to the product. Tomra patented this technology and named the technology as Tomra Sure Return [1]. Instead limiting itself refunding coin for the recycled items [2], the second reverse vending machine offers several options: 1) food voucher; 2) bus ticket; 3) 10 cent donation to charity group [3]. The second reverse vending machines were installed in Sydney, Australia. Third reverse vending machine produced by Pugedon [4] where each recycled items and the machine will dispense food for the street cats and dogs. Another reverse vending machine is offers a reward points through RFID card instead of coins given [5].

The remaining of this project will explain the methodology of the project, the result and discussion of the project, and conclusion of this project.

2. Methodology

Figure 1 is the block diagram of the proposed prototype which consists of three main parts: 1) inputs; 2) outputs; 3) controller. The controller used in this project is Arduino Mega which have more input and output pins accessible for the programmer to use. Arduino microcontroller is known to be programmer friendly and cost effective where the success of the use of Arduino-based microcontroller can be seen in several applications: educational kit [6][7][8], electronic game board [9][10], can crusher [11][12], etc. There are two inputs of the proposed prototype which are push button, and inductive proximity sensor. Four outputs used in this proposed prototype which are light emitted diode (LED), servo motor, liquid crystal display (LCD), and speaker. The LCD used in this project is 20X4 LCD which has 20 columns and 4 rows of characters. A 100 Watt stereo speakers with external power source are used to play the music chosen by the user. These speakers are connected to the music player 3 (MP3) shield produced by Cytron called C-Sheild-EZMP3. This shield receives instructions from Arduino Mega and select relevant music file stored inside the Micro SD card and translate the file to relevant acoustic voice that to be transmitted via speakers. Several push buttons employed in the product prototype with the purpose to detect user action in navigating and selecting the menu and song. The servo motor is used as a mechanical actuator that move the recycled items into the recycle container. The function of the inductive proximity sensor is to detect whether the recycled items either a metal or a non-metal. Lastly, the LED is used as indicator for the user to inform the user whether the jukebox is ready to accept new recycled item.

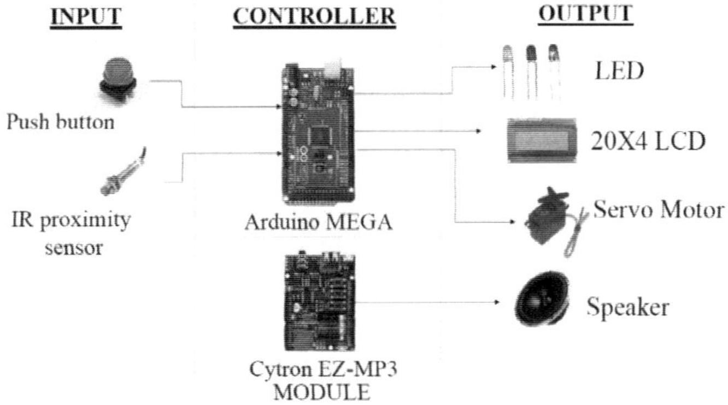

Figure 1. Block Diagram of the proposed prototype

In figure 2 shows the schematic diagram of the proposed prototype using Proteus software. Notice that there are several difference between the schematic diagram and actual implementation as there are several components are not available in the Proteus software. A simple buzzer is replacing the speaker and the MP3 shield. Instead using the 20X4 LCD with I2C, the LCD used here is 16X2 without I2C module. A pushbutton also being used to replace the IR proximity sensor. The remaining circuitry remained the same. This schematic is essential in order to simulate the prototype before developing the hardware version of it.

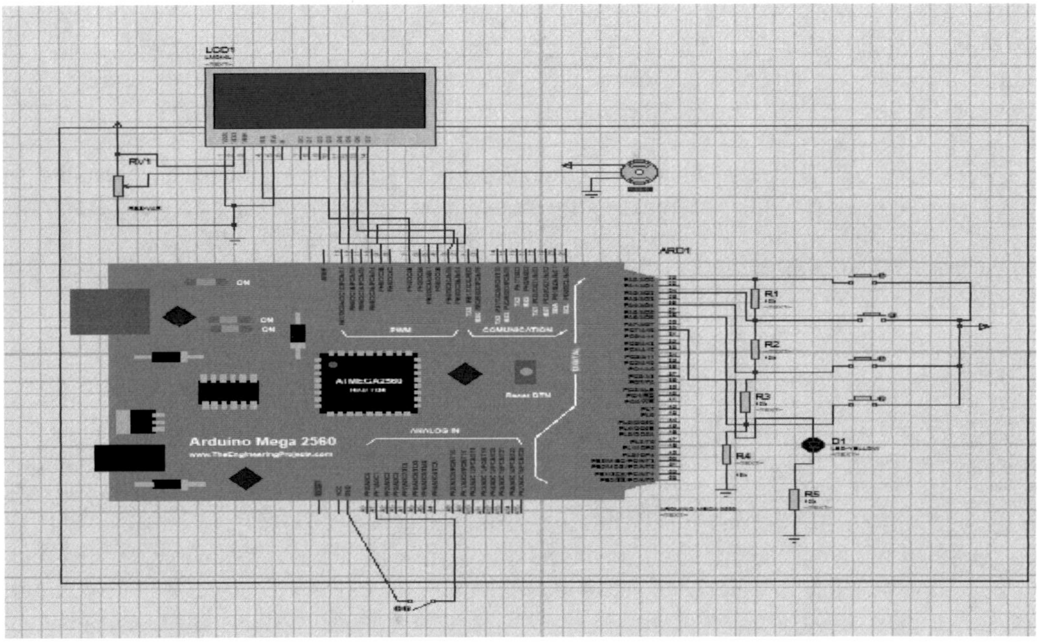

Figure 2. Schematic diagram of the proposed prototype

Figure 3 shows the actual image of the construction of the proposed prototype. The LCD and the push buttons being display on the top of the jukebox. The area for the user to place the aluminum are also located at the top of the jukebox. The containers are place at the bottom of the jukebox where the left container is for the non-aluminum and the right container is for the aluminum.

Figure 4 shows the flowchart of the proposed prototype. After the prototype being turned ON, the LCD will display the welcome menu, the EZ-MP3 shield will play a voice that read out the menu displayed at the LCD and require the user to press push button A. User is expected to insert the recycled item and pressed button A. Once the Arduino able to detect the user pressed push button A, the IR proximity sensor will detect the item recycled in the container. If the item is empty or non-aluminum item inserted, Arduino will prompt the LCD to notify the user that item is a non-aluminum item and EZ-MP3 shield will play a voice that read out the notification. After five second delay, the servo motor will be activated to rotate to the left or -90° which displace the item into non-aluminum container. Then, Arduino will prompt the LCD to display the main menu the EZ-MP3 shield will play a voice that read out the menu displayed at the LCD and require the user to press push button A. On the other hand, if the IR proximity sensor detects an aluminum-based item, Arduino will prompt the LCD to notify the user that item is an aluminum item and EZ-MP3 shield will play a voice that read out the notification. Then, the servo motor will be activated to rotate to the right or +90° which displace the item into aluminum container. Soon after that the user is requested to choose song to be played by pressing push button 1 to go the next song, and push button 2 to return to previous song. The LCD will update the corresponding song to the action by the user and the EZ-MP3 shield will read out the song and the artiste name. Once the user, pressed button A again, Arduino will prompt the EZ-MP3 shield to play the song at a greater volume that can be heard in the targeted area. Once the song is finished, Arduino will prompt the LCD to display the main menu the EZ-MP3 shield will play a voice that read out the menu displayed at the LCD and require the user to press push button A.

Figure 3. Construction of the Proposed Prototype

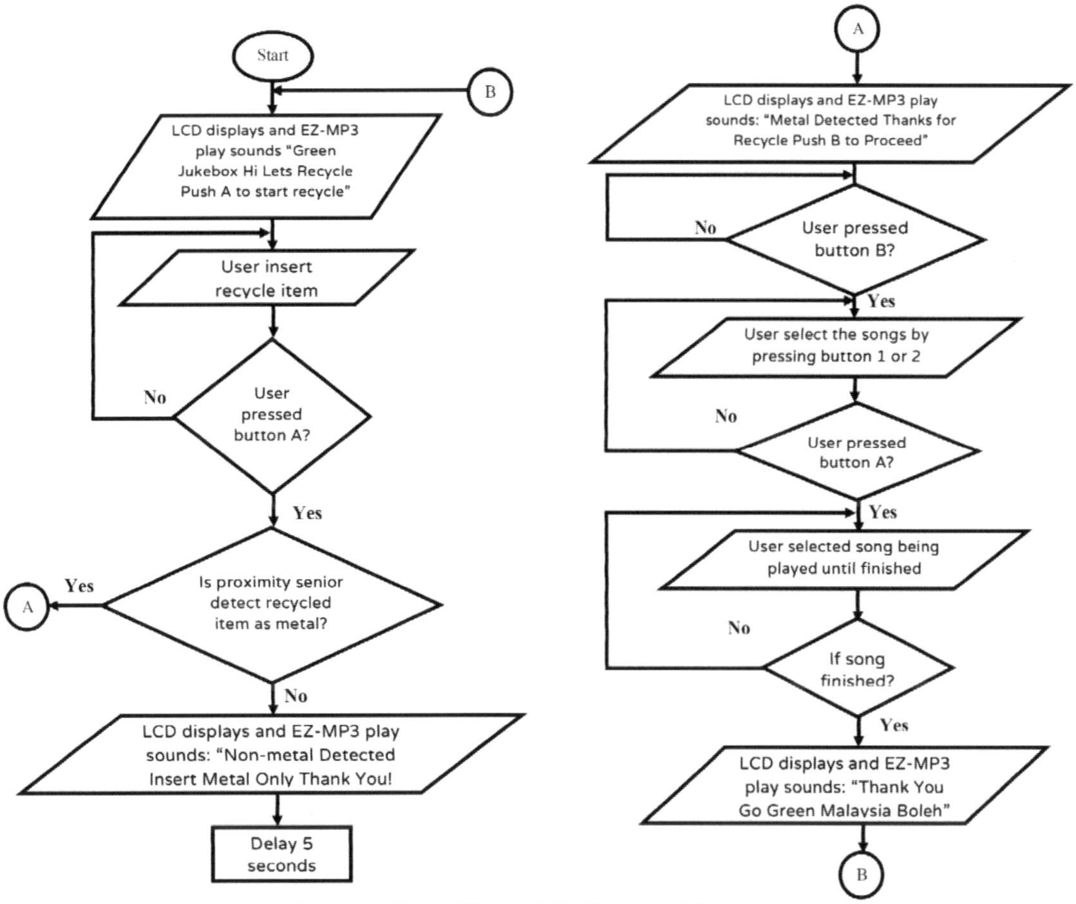

Figure 4. Flow Chart of the Proposed Prototype

3. Result and Discussion

Table 1 shows the scenario tested in order to verify the functionality of the proposed prototype. All possible scenario is tested which consists of 16 steps which are listed in Table 1. The result shows that the actual outcome is similar to the expected outcome. This shows that the proposed prototype function according to the requirement.

Table 1. Result of the testing of all scenario for the proposed prototype

Step	Description	Image
1	At the beginning, the LCD will display "GREEN JUKEBOX Hi Let's Recycle Push A start recycle" and follow with the sound of the Green Jukebox speaking based on the text appear on the LCD.	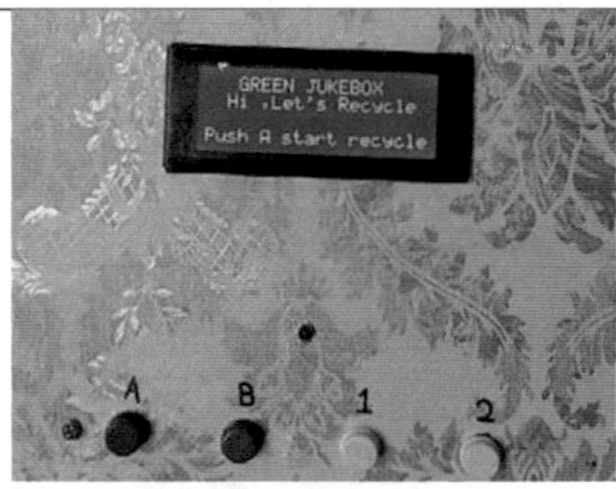
2	If the user place a plastic bottle (non-metal) in the platform.	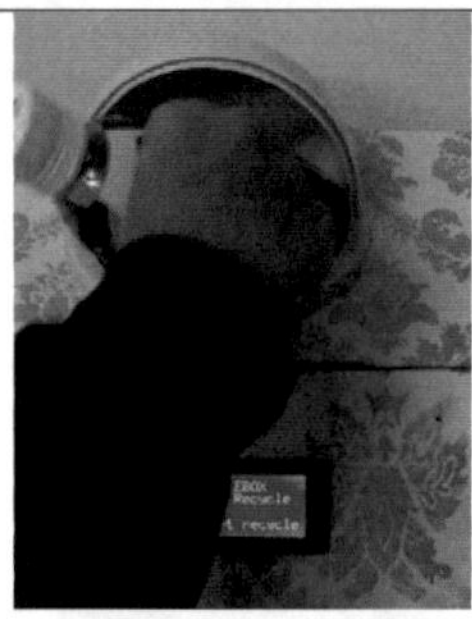
3	Button A is pushed to activate the proximity sensor. The sensor will recognize the item thrown based on the magnetic field concept.	
4	When the non-metal item detected, the servo will send the item to the left compartment of the Green Jukebox. Followed by the sound of the E-Green Jukebox speaking based on the text shown on the LCD.	

<table>
<tr><td>5</td><td>The program will return to main menu. Step 1 is repeated.</td><td></td></tr>
<tr><td>6</td><td>Now, the aluminum can is placed in the platform.</td><td>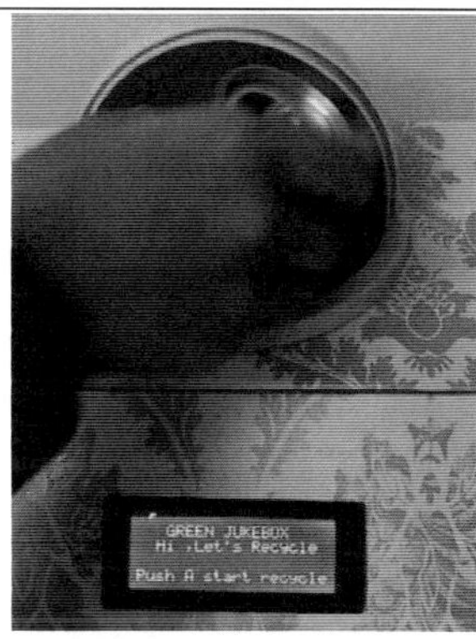</td></tr>
<tr><td>7</td><td>Button A is pushed to activate the proximity sensor. The sensor will recognize the item thrown based on the magnetic field concept.</td><td></td></tr>
<tr><td>8</td><td>The sensor will detect the metal based on monitoring the reading of proximity sensor. The servo motor will turn to right and the aluminum can will fall to metal compartment. Followed by the sound of the E-Green Jukebox speaking based on the text shown on LCD.</td><td></td></tr>
</table>

9	Then, button B is pressed to proceed	
10	LCD will display the first song title and artist name.	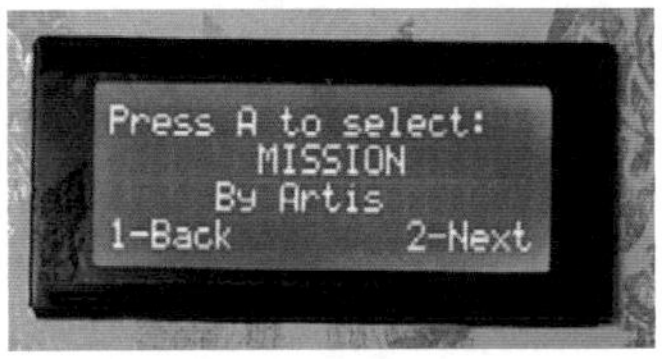
11	If the user want to change the song, the user will press button 2 to view the next song in the list	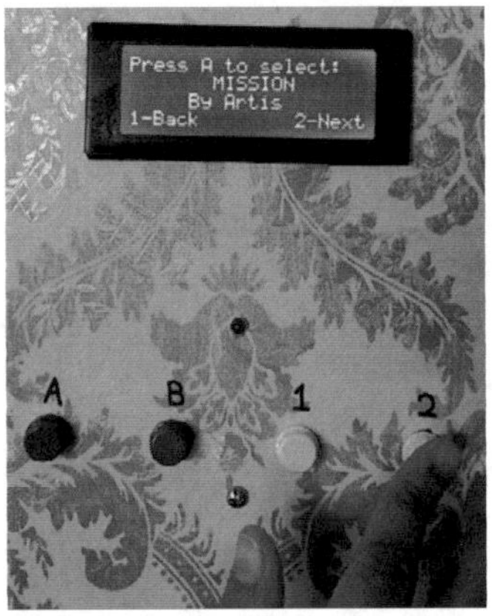
12	LCD will display the second song title and artist name. Led will light up when the button being pressed.	

IOP Conf. Series: Materials Science and Engineering **846** (2020) 012029 doi:10.1088/1757-899X/846/1/012029

13	If the user want to return to previous song, the user will press button 1 to view the previous song in the list.	
14	If the user want to play the song, the user need to press button A.	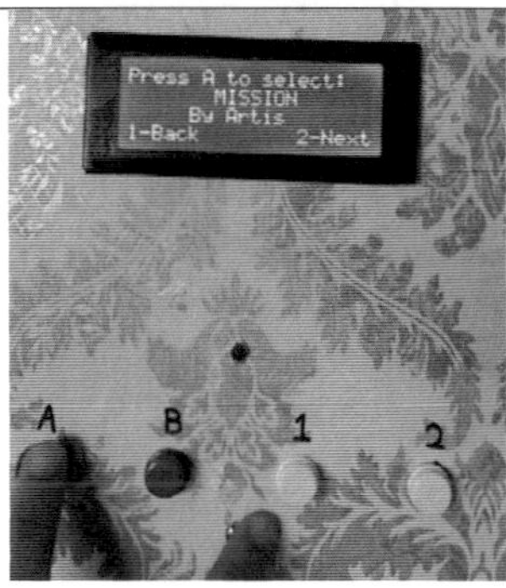
15	The song will be played. LED will light up when playing song.	
16	When the song completed, LCD display “Thank You Go green!! Malaysia Boleh” and follow with the sound of the E-Green Jukebox speaking based on LCD display. A delay of five seconds, then it goes to Step 1 back.	

4. Conclusion

This paper presented the prototype of the proposed electronic jukebox that accept aluminum can as token. The proposed project is operates using Arduino Mega as controller, four push buttons as inputs, IR proximity sensor as sensor, EZ-MP3 module, and several outputs such as the servo motor, LCD, LED, and speaker. The proposed prototype tested using two materials: 1) plastic; 2) aluminum and the result shows it able to detect all the beverage correctly.

References

[1] "T-53s is a TOMRA InPac™ reverse vending solution with a clean and robust design for versatile placement — ideal for low to medium return volumes of non-refillable containers," 2019. [Online]. Available: https://www.tomra.com/en/coolection/reverse-vending/reverse-vending-machines/inpact/t-53s.

[2] S. Sharma and A. Monga, "Implementation of Reverse Vending Machine Based on FPGA," *Int. J. Emerg. Technol. Comput. Appl. Sci.*, vol. 14, no. 153, pp. 265–271, 2014.

[3] "Sydney installs machines that offer bus tickets in exchange for recycling," *sciencealert*, 2019. [Online]. Available: https://www.sciencealert.com/sydney-installs-machines-that-offer-bus-tickets-in-exchange-for-recycling.

[4] "Smart Recycle Bins," *Pugedon*, 2019. [Online]. Available: https://www.pugedon.com/pugedon.

[5] R. Tomari, A. A. Kadir, W. N. W. Zakaria, M. F. Zakaria, M. H. A. Wahab, and M. H. Jabbar, "Development of Reverse Vending Machine (RVM) Framework for Implementation to a Standard Recycle Bin," *Procedia Comput. Sci.*, vol. 105, no. Iris 2016, pp. 75–80, 2017.

[6] M. F. Mohd Ab Halim, N. A. I. Idrus, A. F. Z. Abidin, S. F. Sulaiman, R. Rifin, and M. S. Saelal, "Development of an Electronic Educational Kit with Android Application that Test Student Knowledge in Series and Parallel Resistor for Electrical Circuit (Res-Circuit Quiz Board)," *Int. J. Innov. Technol. Explor. Eng.*, vol. 8, no. 9, pp. 1–5, 2019.

[7] M. Z. Muhammad Farhan Zuhaili, C. A. Siti Asma, Z. A. Amar Faiz, A. Mohd Anuar, R. Norain, and W. H. Wan Haszerila, "Development of An Educational Quiz Kit that Test Student Knowledge on Control System ' s Second Order Transient Response by using DC Motor Speed Control as Application (Transient Response Quiz Box)," *ARPN J. Eng. Appl. Sciemces*, vol. 13, no. 13, pp. 4079–4082, 2017.

[8] M. F. M. A. Halim *et al.*, "Resducational Kit the Development of an Electronic Quiz Board that Test Engineering Students Knowledge on Resistors Concept in Electrical Circuit," *Int. J. Recent Technol. Eng.*, vol. 8, no. 1, pp. 202–207, 2019.

[9] I. A. Rozani *et al.*, "E-Othello: The Development of an Electronic-Hardware Version of Traditional Othello Board Game," *ARPN J. Eng. Appl. Sci.*, vol. 14, no. 3, pp. 694–699, 2019.

[10] A. F. Hafizan *et al.*, "E-Congkak: The Development of an Electronic Congkak Board Game to Promote Traditional Board Game to Younger Malaysian Generation," *ARPN J. Eng. Appl. Sci.*, vol. 13, no. 24, pp. 9624–9630, 2018.

[11] N. A. A. Hadi *et al.*, "Development of an Automatic Can Crusher Using Programmable Logic Controller," *Int. J. Electr. Comput. Eng.*, vol. 9, no. 3, pp. 1795–1804, 2019.

[12] A. Patel, A. Wagh, N. Tilekar, and S. Thakur, "Cam Follower Can Crusher Vending Machine," *Int. J. Res. Eng. Appl. Manag.*, vol. 2, no. 06, pp. 12–15, 2016.

IOP Conf. Series: Materials Science and Engineering **846** (2020) 012030 doi:10.1088/1757-899X/846/1/012030

Improved of Smart Vacuum Flask with Beverage Identifier and Temperature Indicator

Mario Anak Apok[1], Wilzy Anak James[1], Amar Faiz Zainal Abidin[2], Saipol Hadi Hasim[3], Nor Haizat Omar[3], Tengku Nadzion Tengku Ibrahim[3]

[1]Faculty of Electrical Engineering, Universiti Teknologi MARA Cawangan Johor, Kampus Pasir Gudang, Malaysia
[2]Faculty of Electrical & Electronics Engineering Technology, Universiti Teknikal Malaysia Melaka, Malaysia
[3]Department of Electrical Engineering, Politeknik Ibrahim Sultan, Malaysia

*saipolhadi@pis.edu.my, norhaizatomarpis@gmail.com

Abstract. Vacuum flask has been an essential part of the everyday life of a working person. Nowadays, vacuum flask is not the only function as a liquid carrier but also has been equipped with technology to diversify its function from the essential function. This paper proposes to create a smart vacuum flask that able to classify the beverage type and indicate the temperature range of the beverage. This is attempted to address the user difficulty in remembering the type and temperature of the beverage inside the vacuum flask. This project is built using Arduino as the microcontroller, while LM35 is used to read the temperature sensor. A simple concept of tomography using LDRs and LEDs is used to determine the beverage type. The flask is tested to test nine scenarios: 1) cold and clear beverage; 2) warm and clear beverage; 3) hot and clear beverage; 4) cold and juice beverage; 5) warm and juice beverage; 6) hot and juice beverage; 7) cold and milk beverage; 8) warm and milk beverage; 9) hot and milk beverage. The result obtained indicates that the prototype able to classify all nine scenarios correctly.

1. Introduction

Basically, vacuum flask is a container that has a vacuum space which we called as thermal insulator to prevent thermal losses. As we want to keep the temperature remains in the container, thermal insulation is the most ideal insulator. The vacuum part normally made by the industry nearly vacuum and this shows that industries are mostly agree that the heat cannot travel through vacuum. For the structure of the vacuum flask, mostly vacuum flask has an inner chamber and an outer plastic or metal case which will separate the glass with a vacuum in between. Sometimes, the layer it coated with reflective metal and nowadays lot of durable vacuum flask created which is using two stainless steels with vacuum layer and a reflective layer. For the stopper part, usually the industries prefer to use screw down stopper as it will close the outlet part tightly.

In our daily life, the application of vacuum flask has been implemented in many types of products which make our life more convenient. Example of application of vacuum flask is vacuum flask with heater, ice box, hot milk container, etc. There are lot of companies such as Bakelite (Thermos 1925 Ltd),

British Vacuum Flask Company and many more. Recently, there is a shift in the vacuum flask manufacturing where the vacuum flask created is no longer serve the function as beverage carrier but also equipped with technology in order to differentiate the product. The introduction of additional feature like temperature indicator [1], automatic stirrer [2], braille [3], tea thermos timer [4], and drink remainder [5] that encourages the existing consumer to purchase additional vacuum flask for these new features.

For this project, the authors attempt to include two new features: 1) temperature indicator; 2) beverage indicator. The temperature indicator proposed in the prototype should able to classify the three ranges of temperatures which are hot, warm and cold. The reason behind this idea is to avoid user accidentally burns their lips when they drink their beverage by knowing the temperature of the beverage in advance. The second feature proposed in this project is a simple beverage identifier that able to classify three type of beverage: 1) clear beverage; 2) juice beverage; 3) milky beverage.

The remaining of this project will explain the methodology of the project, the result and discussion of the project, and conclusion of this project.

2. Methodology

Figure 1 is the block diagram of the proposed prototype which have three inputs which are 1) Light Dependent Resistor (LDR); 2) push button; 3) temperature sensor LM35. The LDR is used to capture the light produce by the light emitted diode (LED) inside the vacuum flask. Using basic tomography concept where it is commonly known that liquid with greater substance will block more light through it compared to liquid with less substance. The temperature sensor is used to determine the temperature of the beverage. This project use LM35 sensor which are installed at the inner metal container (in the vacuum area). The push button is used in this project is as feedback for Arduino to determine the user request. When the user wants to know the temperature and the beverage type, the user is required to press the pushbutton, then the Arduino will process the user request and provide the suitable feedback based on the two sensors mentioned.

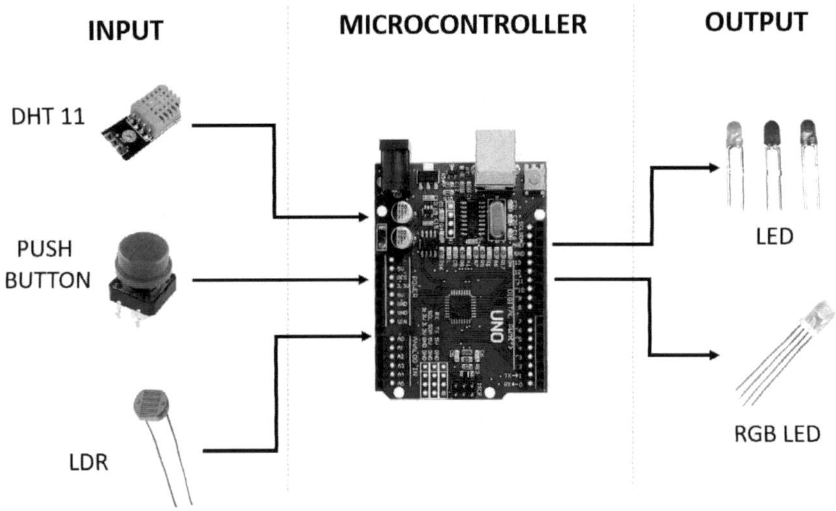

Figure 1: Block diagram of the proposed prototype

Arduino UNO is used as the microcontroller for this project. Arduino microcontroller is known to be programmer friendly and cost effective where the success of the use of Arduino-based microcontroller can be seen in several applications: educational kit [6] [7] [8], electronic game board [9] [10], can crusher [11], etc. The use of Arduino UNO in this project rather than the smaller Arduino Nano because the library of Arduino Nano is not available in the Proteus simulator during the project development. There are two electronic components used in this project which are:1) LED; and 2) Red-Green-Blue (RGB) LEDs. As stated earlier, the LED which is monochrome color is used to provide light for the tomography process in the vacuum flask in order to identify the type of the beverage. While the RGB LEDs are used as indicator for both features proposed by this project.

Figure 2 shows the flow chart of the project, after Arduino detect the user pressed the push button, the LED inside the vacuum flask will turn ON and the LDR will takes several readings until the reading stabilize within the margin of error of 5%. This operation usually takes less than a second. Then, Arduino will decide the respective RGB LED that will be activated which depending on the reading value obtained by the LDR. Next, the LM35 value will start to read the temperature value of the inner metal container of the vacuum flask for several times until the reading stabilize to +/- 1°C. Based on that information, the activated RGB LED just now will turn ON appropriate color based on the temperature range.

3. Result and Discussion

As the proposed prototype is non-invasive sensor where the temperature sensor and LDR sensor is place in the vacuum area of the inner side of the vacuum flask, calibration of the LM11 is deemed necessary. This calibration needs to be done as it necessary to find the linear correlation between the temperature of the liquid and the relationship with the temperature of the inner container. Table 1 shows the relationship between these two parameters and clearly it can be seen that the difference between these two parameters are =/- 5°C and this has been taken into account in the program algorithm.

Table 1: Temperature Testing Result

Condition	Inside Container (°C)	Outside Container (°C)	Differences (°C)
Cold	6	11	5
	5	10	5
	4	9	5
Warm	29	25	4
	28	24	4
	29	24	5
Hot	50	44	4
	49	44	5
	41	36	5

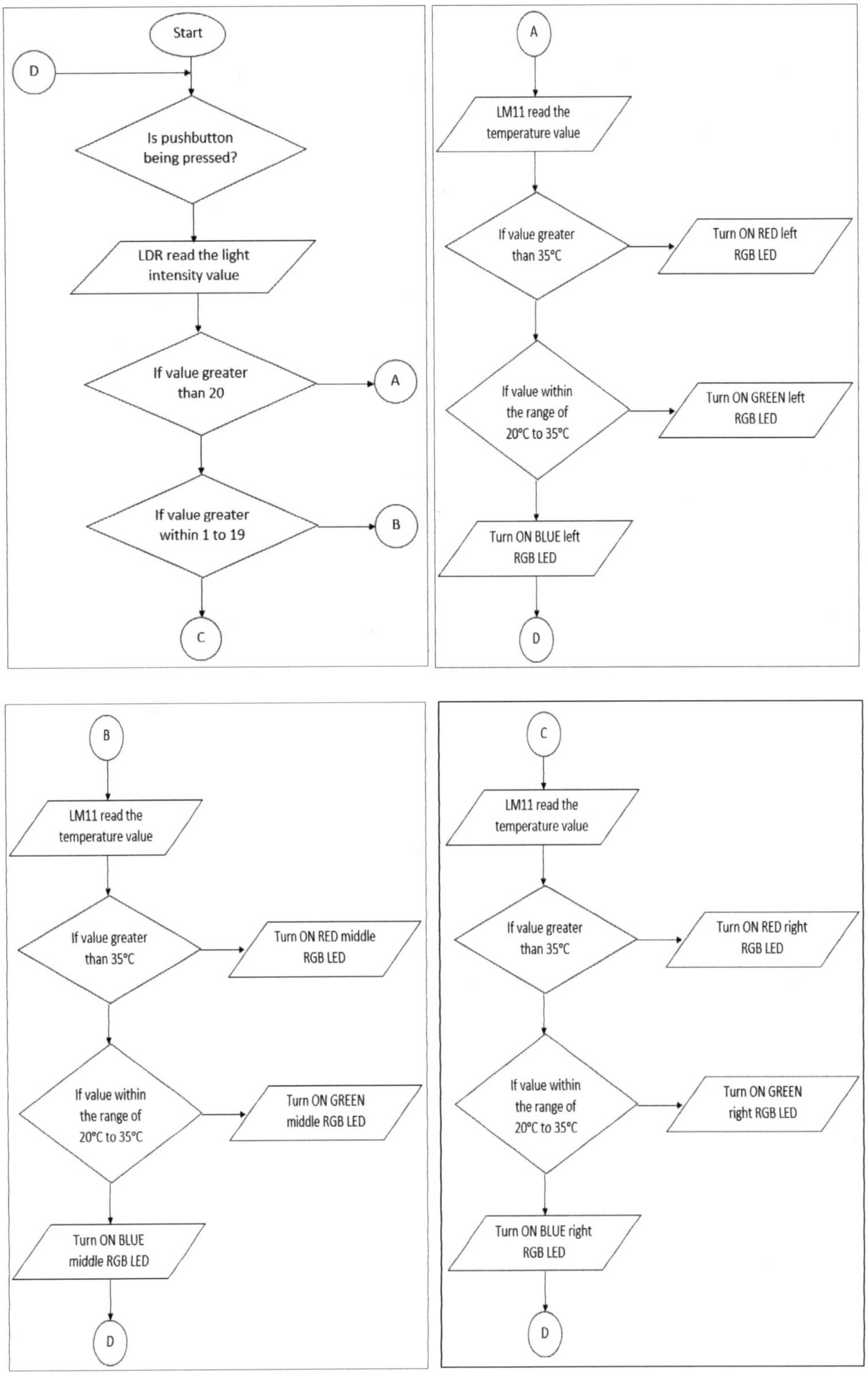

Figure 2: Flowchart of the proposed prototype

Table 2: Tomography Testing Result

	Beverage Type	Reading 1	Reading 2	Reading 3	Average
Clean Beverage	Mineral water	96	97	96	96.33
	Reverse Osmosis water	104	102	103	103
	Pipe water	102	102	103	101.67
Juices	Miranda Strawberry	5	4	6	5
	Fanta Oren	1	2	1	1.3
	Sunquick Blackcurrant	4	5	6	5
Milky Base/Dark Beverage	Nestum	0	0	0	0
	Coffee	0	0	0	0
	Milk	0	0	0	0

Table 2 shows the readings taken from several types of beverage for each category to be classified in the vacuum flask. This result used as prove to the range to be set in the vacuum flask is suitable. The calibration shows that the value of these three categories differ a lot which is easy for this project to classify them accordingly. This justify the selection of the value of 0 to represent milk-based beverage, 1 to 19 to represent juice beverage, and remaining as the clear water.

Figure 3 shows three types of beverage from different category which are Cold Plain Water (Clear Beverage), Mountain Dew (Juice) and Milo (Milk Base/Dark Beverage). These testers will be tested in three conditions which are hot, cold and warm. The result will appear as the LED which represents the beverage type will light up with the temperature condition color indicator.

(a) (b) (c)

Figure 3: Beverage used for testing (a) Dasani mineral water (b) Mountain Dew Blue Spark (c) Milo

Figure 4 shows the result of the testing with plain water which light up the correct LED that represents clear beverage. The clear beverage is in three conditions which are hot, warm and cold. The red LED indicates that the vacuum flask contains clear beverage is hot, while blue LED indicates that the clear beverage is cold, and the green LED indicates warm clear beverage.

Figure 5 shows the result of the testing with Mountain Dew Blue Sparkle which light up the middle LED that represents juice beverage. The juice beverage is in three conditions which are hot, warm and cold. The red LED indicates that the vacuum flask contains hot beverage, while blue LED indicates that the juice beverage is cold, and the green LED indicates warm juice beverage.

(a) (b) (c)

Figure 4: Result for Dasani Mineral Water in (a) cold condition (b) hot condition (c) warm condition

(a) (b) (c)

Figure 5: Result for Mountain Dew Blue Spark in (a) cold condition (b) hot condition (c) warm condition

Figure 6 shows the result of the testing with plain water which light up the left LED that represents milk-based beverage. The milk-based beverage totally blocks the light to pass through beverage. The clear beverage is in three conditions which are hot, warm and cold. The red LED indicates that the vacuum flask contains hot milk base beverage, while blue LED indicates that the milk base/dark beverage is colds and the green LED indicates warm milk base/dark beverage.

(a) (b) (c)
Figure 6: Result for Milo in (a) cold condition (b) hot condition (c) warm condition

Table 3: Overall Result for the proposed prototype

Scenarios	Temperature Value (°C)	LDR Reacting (light intensity)	Expected Results		Actual Results	
			LED Reacting (position)	LED Color	LED Reacting (position)	LED Color
Cold Plain Water	x < 20	y > 15	Right	Blue	Right	Blue
Warm Plain Water	21 < x < 34	y > 15	Right	Green	Right	Green
Hot Plain Water	x > 35	y > 15	Right	Red	Right	Red
Cold Mountain Dew	x < 20	1 < y < 14	Middle	Blue	Middle	Blue
Warm Mountain Dew	21 < x < 34	1 < y < 14	Middle	Green	Middle	Green
Hot Mountain Dew	x > 35	1 < y < 14	Middle	Red	Middle	Red
Cold Milo	x < 20	y = 0	Left	Blue	Left	Blue
Warm Milo	21 < x < 34	y = 0	Left	Green	Left	Green
Hot Milo	x > 35	y = 0	Left	Red	Left	Red

Table 3 summarizes the overall result shown in the previous three figures. The prototype tested on three types of beverages which are Milo (milk base/dark beverage), Mountain Dew (juice beverage) and Plain Water (clear beverage). These beverages tested in three conditions which are hot, warm and cold.

4. Conclusion

This paper presented the prototype of the proposed smart vacuum flask that able to determine the beverage type and beverage temperature. The proposed project is operates using Arduino UNO as controller, a push button as input, two sensors: LDR and LM35, and the RGB LEDs as the outputs. For the temperature this will be +/- 5^0C of temperature value as it due to material loss. The proposed prototype tested using three kind of beverage and the result shows it able to detect all the beverage correctly.

References

[1] R. Seth, "Yanko Design," Electronic Productions, 2009. [Online]. Available: http://www.yankodesign.com/2009/03/26/lets-talk-about-camping-essentials/. [Accessed 2009].

[2] Rechargeable USB Electric Heating Automatic Stirring Insulated Coffee Milk Tea Travel Mug Thermos Cup & Lid Warmer.," Ali Express, 2009. [Online]. Available: http://www.aliexpress.com/item/New-Rechargeable-USB-Electric-Heating-Automatic-Stirring-Insulated-Coffee-Milk-Tea-Travel-Mug. [Accessed 2009].

[3] A. Roblin, "Braille Tea-Brewing Thermoses.," Tuvie, 2010. [Online]. Available: http://www.tuvie.com/special-teatime-a-tea-thermos-concept-for-visually-impaired-people/. [Accessed 2010].

[4] E. Gandelman, "Tea Thermos with Timer.," Tuvie, 2010. [Online]. Available: http://www.tuvie.com/tea-thermos-concept-by-eddie-gandelman/. [Accessed 2010].

[5] T. Turner, "Hydration Thermos with Touch Sensor.," Yanko Design, 2014. [Online]. Available: http://www.yankodesign.com/2014/01/22/ehydration/. [Accessed 2014].

[6] M. F. A. Halim, N. A. I. Idrus and A. F. Z. Abidin, "Development of an Electronic Educational Kit with Android Application that Test Student Knowledge in Series and Parallel Resistor for Electrical Circuit (Res-Circuit Quiz Board)," *International Journal of Innovative Technology and Exploring,* vol. 8, no. 9, pp. 1-5, 2019.

[7] M. F. Z. M. Zakaria, S. A. C. Aziz, A. F. Z. Abidin, M. A. Adip, N. Rahim and W. H. W. Hassan, "The Development Of An Electronic Educational Quiz Board That Test Student Knowledge On Control Principle's Second Order Transient Response by Using DC Motor Speed Control as Application," *ARPN Journal of Engineering and Applied Sciences,* vol. 13, no. 13, 2018.

[8] M. F. M. A. Halim, A. M. M. Nor, A. A. Badari, A. F. Z. Abidin, K. A. Kadiran, A. Adam, S. H. Mohammad and M. H. Harun, "Resducational Kit The Development Of An Electronic Quiz Board That Test Engineering Students Knowledge on Resistors Concept in Electrical Circuit," *International Journal of Recent Technology and Engineering (IJRTE),* vol. 8, no. 1, 2019.

[9] I. A. Rozani, A. F. Z. Abidin, M. S. Karis, M. N. M. NIzam, A. H. Azahar, M. H. Harun, Z. M. Yusoff, K. K. Hassan and B. N. Shah, "E-OTHELLO: The Development Of An Electronic-Hardware Version Of Traditional Othello Board Game," *ARPN Journal of Engineering and Applied Sciences,* vol. 14, no. 3, 2019.

[10] A. F. Hafizan, A. F. Z. Abidin, N. Z. N. Suhaimi, M. M. Mustam, F. Sukarman, K. A. Kadiran, S. A. Saleh and R. Rifin, "E-CONGKAK: The Development Of An Electronic Congkak Board Game To Promote Traditional Board Game To Younger Malaysian Generation," *ARPN Journal of Engineering and Applied Sciences,* vol. 13, no. 24, 2018.

[11] N. A. A. Hadi, L. J. Uee, A. M. Annuar, Z. Zaidi, Z. A. Ghani, M. F. M. A. Halim, A. F. Z. Abidin and M. B. N. Shah, "Development of an automatic can crusher using Programmable Logic Controller," *International Journal of Electric and Computer Engineering,* vol. 9, no. 3, pp. 1795-1804, 2019.

IOP Conf. Series: Materials Science and Engineering **846** (2020) 012031 doi:10.1088/1757-899X/846/1/012031

Technical Design and Financial Feasibility Analysis of Off-Grid Photovoltaic Power Supply System for Residential Load

Muhammad Imran Hamid[1*], Afifah[2]

[1]Electrical Engineering Dept. Andalas University - Indonesia
[2]Business Administration Dept. Politeknik Negeri Padang - Indonesia

*imran@eng.unand.ac.id.

Abstract. Due to intensive promotion to use renewable energy, by the various benefits obtained by utilizing, and then supported by advances in generation technology: causing various parties to divert the type of fuel for their energy use activities. The same thing happens to residential electricity consumers, driven by the advantageous of electricity power generation that can be done on the consumer side, independent operation, reliability and quality of power, many consumers try to install renewable energy based generation such as photovoltaic system for their electricity supply. However, looking at the tariff of utility electricity and considering the investment needed: building a photovoltaic-based supply system is not necessarily the right choice. This paper presents a technical design for electricity supply of a residential load through an off-grid photovoltaic system. The technical basis for designing and the capacity of its constituent components is explained, as well as the analytical techniques carried out to assess the financial aspects of the development and operation of the system. The results of this analysis can be used for deciding whether it is appropriate to divert the source of electrical energy to the photovoltaic system. To show the steps of the analysis, a case study was carried out by designing an off-grid photovoltaic system for a residential load characterized by its daily load curve. The capacity and costs were calculated and financial analysis is carried out by looking at the extent to which the designed system can provide comparable benefits if the power is supplied from the utility at the prevailing price, as well as how long the system must be operated to obtain a breakeven return on investment. Based on the daily load curve in this case study, it is known the total daily energy consumption of the load is 6.885 kWh. This amount of energy must be provided from photovoltaic module with 5 hour effective per day, so the minimum capacity of PV-module is 1463.65 W. For an autonomy time of 1 day, a minimum energy storage capacity of 8.965 kWh is required. Investment costs which include the cost of procurement, system development, and the tax are Rp. 150,008,037. The financial benefits are calculated based on prices of the PLN 2016 non-subsidized tariff. It was assummed that the energy used that should be supplied from PLN, is substituted by photovoltaic generation system. NPV analysis and Payback period shows that the off-grid photovoltaic generation system will be feasible if the electricity price is 2.3 times the current price, and the discount factor (DF) for construction is 2.5%, in such electricity prices and DF factor, the payback period is reached if generation system is operated at least until the end of the 25th year. Financial analysis showed the inability of the off-grid electricity supply scheme based on the designed photovoltaic generation for current conditions.

1. Introduction

The need for energy today continues to increase along with the civilization development, economic, and population growth. Unfortunately at this time, energy consumption is still dependent on the non-renewed fossil energy. According to Energy Economic Statistics of Indonesia 2016, the three main sources of energy supply in Indonesia are still held by fossil energy such as petroleum, coal, natural gas and their products. The coal supplies around 24.8%, petroleum 30.2% and 19.03% by natural gas [1]. Meanwhile, fossil energy reserves in Indonesia are increasingly depleting. Based on the 2014 Reserve/Production (R/P), if it is assumed that no new reserves of fossil energy found in Indonesia, it was predicted that the petroleum will run out in 12 years, natural gas 37 years, and coal 70 years [2]. Thus it is very necessary to change the dependency of fossil to the renewable energy. It reasonable, considering that renewable sources such as wind power, hydropower, geothermal, sea water waves, solar power, bio-energy, and others renewable sources are very promising in Indonesia.

Encouraged by the above conditions, various parties devoted attention to efforts in the context of diverting energy sources. Various studies have been carried out by researchers to find new energy sources, optimize the exploration of existing renewable energy and try to create efficient technologies for energy consumption. Likewise, various sectors of energy use try to divert their energy use from initially relying on fossil-based fuels to fuels from renewable sources. On the side of policy makers, it is seen that many government in various countries around the world emphasize the use of renewable energy in their policy.

Utilization of energy is carried out in a variety of human activities: household, transportation, industry and also to produce or convert to other form of energy such as electrical energy. In the household sector, the diverting of energy use can be seen as the replacement of fuel from the use of kerosene to gas and electricity, as well as in transportation and industry, the diverting of energy is carried out in the form of using bio-fuel as previously using diesel.

Specifically for electrical energy, in addition to the diverting at the generation-side level -where energy for generating the electricity is diverted from coal and petroleum to renewable energy such as water, wind, photovoltaic (PV), and other renewable sources- diverting is also done at the consumer-side level. This can be done by utilizes the advantage of electrical energy that is, it can be generated at the consumer installations. In electricity user installations, in addition to being connected with electrical loads, it can also be connected to electric power generation devices. This can be found as in *home-based photovoltaic generation* applications. This is inseparable from the progress in the field of electric power generation technology which is achieved today.

Electric energy conversion technology using photovoltaic systems has enabled the electric generation can be carried out on the side of consumers. The capacity of energy that can be generated stretches from hundreds to several kilowatts. This technology also increases the reliability of energy supply for consumers because it can be used as a back-up power supply. Even with certain schemes, consumers not only consume electricity but can also supply the generated electricity by selling it to the utility grid. Furthermore, with the off-grid system, consumers can provide, regulate, plan electricity consumption independently. Consumers are also protected or isolated from various problems of fault, interruptions as well as electrical power quality problems that often occur in the electricity grid.

The reliability, security and independence of electricity supply made possible by using the photovoltaic system caused many parties and consumers interested and decided to realize and install it on their electricity supply system. However, realizing a photovoltaic system requires no small investment, considering the components used in this system are still relatively expensive due to some factors. In this regard, a number of questions have not been fully answered, is it more efficient to use electricity from own photovoltaic generation than from utility? Maybe in a certain country this is efficient because it is supported by various factors such as high electricity utility prices and the renewable generation energy incentives, however in other countries it is not profitable related to subsidies and other factors.

This paper presents the financial analysis of photovoltaic generation system applied to residential applications. The components that make up the system and its functions are described, as well as the patterns of electricity usage that result in the design of the photovoltaic system. Based on the design, the initial investment value of the system is calculated. Financial assessment is done based on investment criteria such as net present value (NPV) and payback period (PBP). The energy produced is assessed by the price of electricity sold by state electric company. This is done with the assumption that the energy generated is used as a substitution when the energy needed by the load must be supplied from state electric company. A residential load with the characteristics of electricity usage in form of load curve is used as a case study.

2. Methodology

2.1 Residential Photovoltaic System Design

The design of a photovoltaic generation system in a location, including for residential use, is aimed basically to adapt the designed system to specific energetic and geographical condition, as well as to energy behavior of the load. In addition to the feasibility of the geographical position in relation to the availability and fluctuations of photovoltaic energy sources, several things related to operation need to be determined from the beginning, because this will result in the selection of components and technology used. Whether the system will be operated off-grid or on-grid, whether the system will be built in single or three-phase system [3], and whether the coupling component is done in DC or AC side and others considerations.

The *off-grid or stand-alone system*, is completely isolated from the existing electricity system (grid). To be able to meet the availability of electricity supply for a full day, in the absence of solar energy sources at night, the presence of energy storage components such as batteries is absolutely necessary. While the *on-grid system* is a system built using components that allow energy generated in consumer installations to be connected to the grid. Power-supply for consumer installations can be done from two directions, from own photovoltaic system, and in the absence of a solar energy at night; the power-supply for the load is taken from the grid. In the condition that the energy generated exceeds consumer consumption, the excess energy can be fed into the grid. In such a system, the consumer and grid must continue to be connected because the inverter component uses the grid voltage as a reference for generating power. This causes the reliability of consumer supply decreases and is affected by the condition of the utility grid. To improve system reliability and prevent the system from being affected by the distribution system in the event of a fault and poor power quality, on-grid systems can be equipped with energy storage components (*hybrid system*). When the grid supplied are is disconnected and in the absence of solar energy, the power-supply of consumer installations can be taken from the storage unit. Such a system requires that the inverter used has the ability to work on-grid or off-grid by having its own voltage reference for power generation. **Figure 1** describes the systems.

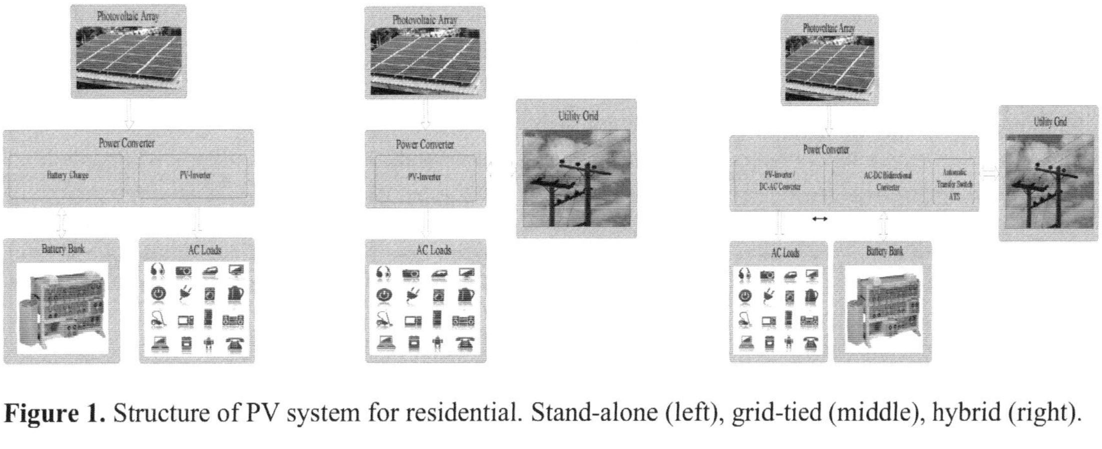

Figure 1. Structure of PV system for residential. Stand-alone (left), grid-tied (middle), hybrid (right).

Related to geographical location, the data needed is the amount of irradiance at the location where the PV system will be built. To get this data, direct measurements followed by statistical analysis must be carried out. The data can also be obtained through data provided by various institutions related to atmospheric conditions and irradiation monitoring. These institutions provide specific data on the amount of irradiation at a location, and also provide a database on the amount of radiation for a certain time.

The PV system design for a residential load begins by recognizing the characteristics of the load to be supplied. This is shown in the form of load curves describes the variation of consumed energy in full day duration (24 hours). From the curve, load parameters such as average, maximum and minimum load and when they occur, load and loss factors, and also the amount of energy used for one day can be calculated. The information is then used to determine the photovoltaic system capacity and the amount of battery capacity related to the planned autonomy time duration. Autonomy time is the time for which the storage device will support and ensure the load still supplied by the electrical power. The next step is determining the capacity and features of each component that makes up the photovoltaic system. Considering the operating patterns (on-grid or off-grid) of the system must be involved. The following is an overview of the planning of each component making up photovoltaic system for residential:

2.2 Photovoltaic Module

This component functions to change the amount of photon energy contained in solar sun into DC electricity in its terminals. In addition to the level of irradiation and temperature, a number factors influence the amount of power that can be generated, in general the relation of these factors to the output power -presented by DC current- is written by the following relation [4]:

$$I_{pv} = I_{ph} - I_d - I_{sh} \tag{1}$$

$$I_{ph} = \left(I_{SCR} + K_1(T-25)\right) - \frac{G_a}{100}$$

$$I_d = I_o\left[e^{\frac{q(V_{pv}+I_{pv}.R_{se})}{A_o kT}} - 1\right]$$

$$I_{sh} = \frac{\left(V_{pv} - R_{sh}\right)}{R_{sh}}$$

$$I_o = I_{or}\left[\frac{T}{T_r}\right]^3 \exp\left[\frac{qE_{Go}}{A_o k}\left(\frac{1}{T_r} - \frac{1}{T}\right)\right]$$

Where:

I_{pv}, V_{pv}	: current and voltage of the PV cell.
I_{ph}	: light generating current,
I_o	: diode dark saturation current.
K_1	: short circuit temperature coefficient at ISCR.
T and G_a	: cell temperature and irradiance on cell surface.
I_{or}	: cell saturation current at reference temperature Tr.
E_{GO}	: band gap energy.
k and q	: Boltzmann cons. = 1.3807×10^{-23} JK^{-1} and electric charge = 1.6022×10^{-19} C
R_{se} and R_{sh}	: parasitic series and shunt resistances, associated with cells in operation condition.
A_o	: diode ideality (quality) factor; its value is taken between 1 and 2
I_{SCR}	: short circuit current of the cell on 1000 W/m^2 and 25°C of temperature.

In the market, PV modules are available in the form of modules which are series and parallel configuration of PV cell. The important thing for each module is the availability of the relationship *vi-curve* which shows the conditions of the output voltage and current for various levels of test irradiance and temperature. From this curve, the condition of short-circuit current and no-load voltage, and the point where the maximum power operation happened (MPP) can be known. Other important aspects of PV modules are lifetime and efficiency. Referring to the glass as the basic material of PV modules, theoretically the life time of PV module is predicted to be up to 50 years. Over time, the performance of PV modules decreases [5]; in general the manufactures provide a guarantee of operation for up to 25 years for their module will still within the permissible level of performance [6]. From the aspect of efficiency, studies shown that mono-crystalline and concentrated photovoltaic are photovoltaic module with good efficiency [7]. **Table 1** shows the performance comparison of several types of PV modules available on the market.

Table 1. Characteristic of Some PV-Module Types

Solar Cell Type	Efficiency-Rate	Advantages	Disadvantages
Monocrystalline Solar Panels (Mono-SI)	~20%	High efficiency rate; optimized for commercial use; high life-time value	Expensive
Polycrystalline Solar Panels (p-Si)	~15%	Lower price	Sensitive to high temperatures; lower lifespan & slightly less space efficiency
Thin-Film: Amorphous Silicon Solar Panels (A-SI)	~7-10%	Relatively low costs; easy to produce & flexible	shorter warranties & lifespan
Concentrated PV Cell (CVP)	~41%	Very high performance & efficiency rate	Solar tracker & cooling system needed (to reach high efficiency rate)

2.3 Photovoltaic Inverter

PV-Inverters are characterized by the number of phases and their capacity in kilowatt, working voltage, frequency, and efficiency. In the market, single-phase inverters are available in small capacities, usually in transformer-less topologies while three-phase inverters are designed in a higher power rating and use a topology with a transformer (multi-stage inverter). Voltage rating, frequency of the inverter is usually determined and ordered before the purchase is made. A number of manufacturers produce PV inverters specifically for off-grid and on-grid use, but other manufacturers distinguish these functions only on the software and hardware settings. The efficiency of a PV inverter is defined as:

$$\eta_{inv} = \frac{P_{AC}}{P_{DC}} = \frac{P_{DC}-P_{loss}}{P_{DC}} \tag{2}$$

Where:

P_{DC} : instantaneous DC power
P_{AC} : instantaneous AC power
P_{loss} : power losses

The efficiency values of inverter products from various manufacturers on the market vary from 94 - 98% [8]. Each inverter has efficiency level as function of its load factor, and presented as efficiency curves. Generally efficiency of the inverter shows a different value at each level of power conversion, higher efficiency is shown when operating below 50% of its rating. Due to the varying conditions of this efficiency, to assess the overall efficiency of an inverter, European Central Climate, arrange the formula as:

IOP Conf. Series: Materials Science and Engineering **846** (2020) 012031 doi:10.1088/1757-899X/846/1/012031

$$\eta_{euro} = 0.03.\eta_{5\%} + 0.06.\eta_{10\%} + 0.13.\eta_{20\%} + 0.10\eta_{10\%} + 0.48.\eta_{50\%} + 0.2.\eta_{100\%} \quad (3)$$

The level of inverter capacity for an installation can be calculated energetically by knowing the total energy to be consumed in one year (annual) and deriving the equation of the inverter efficiency curve available from manufacturer [8]. Furthermore, by using equations (4) and (5), the optimum capacity can be obtained by differentiate the energy equation respect to inverters power rating and equate it with zero:

$$\frac{dE}{dP_{Dinv,N}} = 0\,,$$

Then, for $P_{inv,N} > P_{max}$:

$$P^{opt}_{inv,N} = P_{max}\sqrt{\frac{B}{3C}} \quad (4)$$

and for $P_{inv,N} \leq P_{max}$:

$$P^{opt}_{inv,N} = P_{max}\frac{A+B+C}{A+\frac{4}{3}B} \quad (5)$$

Where:

P_{max}	: Maximum DC power from the installed PV module
A, B and C	: A number obtained by modeling (making the function of efficiency vs. load factor) the characteristics of the inverter efficiency.

The equation for optimum inverter capacity for a certain PV plant is arranged with two conditions: the condition when the P inverter is made larger and smaller than the module capacity. Designing a PV inverter smaller than a PV module will cause most of the operating time, the inverter will work at a 100% rating and vice versa, this working mode will affect the inverter's life time. Optimum size of an inverter for a given installation can vary from1.11 to 2.02. [9].

2.4 Battery

In a photovoltaic-based and other intermittent energy supply systems, energy storage is very important. This is even more important if the supply system is off grid, it will function as a main supply in the absence of energy sources. In the on-grid system, energy storage functions as a useful back-up in the case of grid supply are interrupted or experiencing a fault. The most practical form of energy storage today is the battery storage. The battery is characterized by its output voltage and capacity measured in Ah or kWh.

In general applications, there are a number of battery types that can be used, but the selection of batteries must take into account the price, capacity, voltage and cycle-life. Cycle life needs to be considered because it involves battery capacity in terms of the number of discharge/charge cycles the battery can provide before capacity drops to a specified percentage of rated capacity. Batteries are considered to be at the end of their lifespan when 20% of their original capacity is gone. The selection must also consider what is called a deep-cycle, i.e. how far the battery can be discharged relative to its capacity [10]. A comparison of several types of batteries is shown in **Table 2**. Note that two types of batteries dominate the use in photovoltaic systems, namely Lead-Acid based batteries and Lithium-Ion based batteries.

The amount of battery storage is designed based on the energy usage that can be derived from the load curve. After the amount of energy is known, the next step is to determine the days of autonomy, i.e. how many days are worth of energy to be stored in the battery bank. And the next is to choose the type of battery by considering the output voltage of the solar panel, the rate of discharge, the depth of discharge we need, and the operation temperature. According to following formula:

$$Ah_a = \frac{Ah_d \times TC \times DA \times DM}{DoD} \quad (6)$$

Where:

Ah_a : Ampere-hours adjusted
Ah_d : Ampere-hours day
TC : temperature correction factor, according to type of battery
DA : autonomy time (day, 1 – 4 days
DM : design factor, 1 – 1.25
DoD : depth of discharge, 0.2 – 0.8

Battery storage capacity can also be designed based on the total energy (kWh) that must be stored. Noted that Ah value in formula (6) is the total energy needed by the load, to calculate how many batteries are needed, a datasheet from the battery manufacturer is needed which shows the output voltage, the capacity of the unit so that the battery configuration can be arranged according to the current and operating voltage of the system.

Table 2. Characteristic of some Battery Storage Types

Battery Type	Advantage	Disadvantage
Lead-Acid, - VRLA/SLA (Valve regulated LA, Sealed LA) consist of AbsGlasMat and Gel - Flooded LA 1 kW to 10 MW Application	Better in subtropical (low temperature) operation, relative ECO friendly	low specific energy, Limited Cycle Life 400-500 cycles, 50 - 55% DoD, low charge rate, 80-85% efficiency,
NiCd (Nickel Cadmium) 1 kW to 0.5 MW	NiCd batteries have superior performance characteristic to lead-acid with respect to their tolerance of full charge and discharge	lower energy density, present some environmental hazards
NIFE (Nickel Iron), 1 kW to 1 MW.	Better energy density,	high self-discharge rate
Lithium-Ion, - Iron phosphate - Nickel, Cobalt, Manganese 1 kW to 1 MW	Low discharge rate, high energy density (lighter and smaller). 80 - 100% DoD, ECO friendly, no maintenance or venting, higher charge rate (refilled much faster), 95% efficient , life cycle: 2000 to 5000 cycles, better operate in tropical (hot temperature)	The internal resistance can heat, thus causing failure of the battery; Higher cost.
Sodium–Sulfur (NaS) Battery, 0.8 to more than 10 MW	high power density, high efficiency, up to 90% and long lifetime (over 10 years)	Contain many hazardous elements, operate at high temperatures of approximately 270°C–350°C, cycles 250
Flow Batteries (FBs);VRB, Iron–chromium (FeCr), zinc–bromine (ZnBr2), Zinc–air. 10 kW to more than 10 MW	Cost and storage of the electrolyte more convenient, the electrolyte does not degrade with time, and so, the life of VRB can theoretically reach ten years,	when a VRB is deactivated, the membrane can be highly acidic or alkaline requiring special measures of disposal.
Sodium–Nickel–Chloride (NaNiCl2) Batteries, 100 kW to less than 10MW	Wide range operating temperature	Operate at high temperatures of approximately 270°C–350°C

IOP Conf. Series: Materials Science and Engineering **846** (2020) 012031 doi:10.1088/1757-899X/846/1/012031

2.5 Charge Controller

Charge controller is used to protect batteries from over charging and/or excessive discharge. This is achieved by regulating the current for charging to the battery, avoiding overcharging and overvoltage, regulating the current released/taken from the battery so that the battery does not over-charged, and monitor battery temperature. Charge controllers are characterized by their working voltage, amperage, full charge level and low voltage cut.

2.6 Financial Analysis of Residential PV System

Internal Rate of Return (IRR) and Pay Back Period (PBP) are the indicators used to assess the feasibility and profitability of an engineering project. In this method, a cash flow analysis using the NPV is carried out to be able to evaluate the indicators. NPV is used based on the present value of cash inflows and outflows over a period of time, from this, the Internal Rate of Return (IRR) which the discount rate that makes the NPV of the project equal to zero can be found. NPV is written as [11]:

$$NPV = \sum_{t-1}^{T} \frac{C_t}{(1+r)^t} - C_o \tag{7}$$

Where:

C_t : net cash inflow for the period
C_o : initial investment
r : discount rate
t : number of period

Cash outflows consist of maintenance and replacement costs of the project, while cash inflow typically consists of financial benefits by selling the energy produced by the PV plant. Capital investments (C_o) are the initial costs incurred for investment projects carried out. They are calculated considering the cost of all the components that make up the PV system, including the cost of procuring solar modules, inverters, batteries, charge controllers, electrical cables and civil infrastructure. Information regarding the cost obtained from the supplier websites or from any literature.

In the Payback Period Analysis, we used the payback period to assess the feasibility and profitability of the project. Payback period is the period needed to return the investment expenditure (initial cash investment).

In this financial analysis, the financial benefits of the project are obtained from the amount of electricity generated, assessed by the tariff of electricity released by the electricity company, in this case, PT. Perusahaan Listrik Negara (PLN). This is done by assuming the energy generated is used as a substitute if the energy supply is fed from PLN. Thus, through the financial analysis conducted, it can be assessed whether it is better to realize/built the PV generation system for supplying the connected load or connect it to the PLN grid.

Electricity tariffs in Indonesia are built on the basis of two cost components: connection and electricity usage cost. By type of use, the tariff is divided into main groups: tariffs for social services, household, business, industrial, electricity tariffs for government offices and street lighting, tariff for traction purposes, for electricity power in the medium voltage, and tariffs for special services on low, medium and high voltage which are not included in the tariff groups mentioned previously [12] . Each main tariff group consists of several groups to form totally 12 tariff groups. In each of these groups there are groups that receive subsidies from the government. The payments for the tariffs can be made in prepaid or postpaid scheme. The tariffs for low voltage (single or three-phases) households are classified into three tariff groups: the tariff for small households with power up to 2200 VA, tariff for medium households with power 3500 - 5500VA, and tariff for large households with power up to 6600 VA or above. Subsidies for these household tariffs are only for small household with a maximum power connection of 900 VA. The electricity tariffs as mentioned above are generally dynamic, a tariff adjustments is made regularly to adapt the factors that affect the cost of electricity production.

3. Result and Discussion: The Case Study

This section will present a case study of a photovoltaic power plant design to supply a residential load with load-curve characteristic as illustrated in **Figure 2 (a).** For one year, energy consumption is obtained by multiplying the average daily energy consumption and number of days in a year. The power supply system for this load is designed as an off-grid photovoltaic system equipped with battery storage and charge controller with autonomy duration for one day. PV-inverter for this purpose is the type that has references for voltage generation independently (off-grid inverter). The scheme of the system is presented in **Figure 2 (b).**

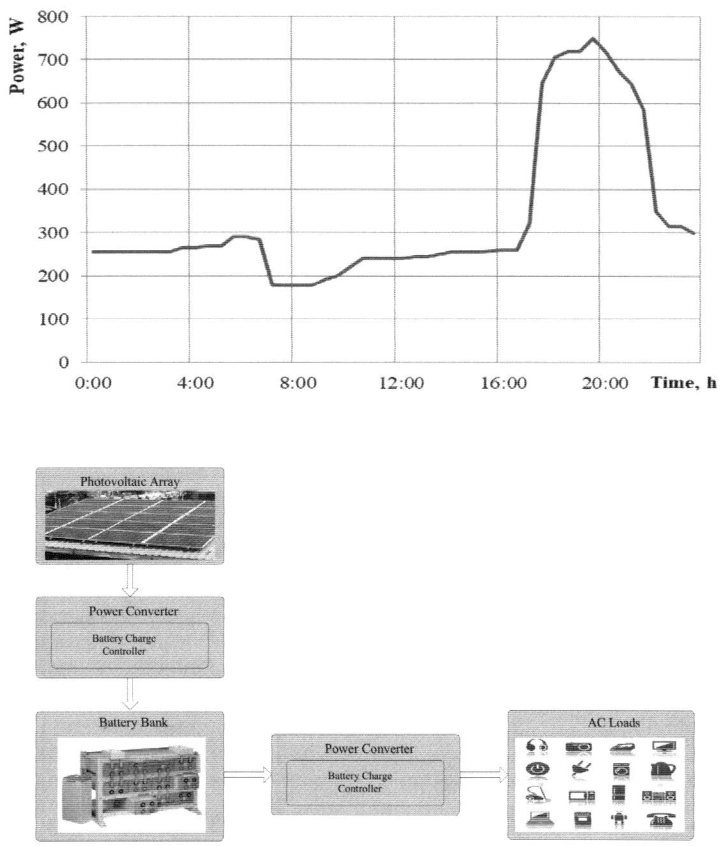

Figure 2. Load curve for the case study (upper); Schematic for the PV system (lower)

From this daily load-curve, it is known that the peak load occurs at 19.30 pm of 690 Watt, while the lowest load occurs at 08.00 am at 155 Watt level. With further analysis, the average electricity consumption is 143.44 Watt with a total daily energy consumption of 6.885 kWh. This amount of energy must be provided from photovoltaic modules via inverter, it is assumed the inverters efficiency is 96%, so the input power to the inverter is 6.885/0.96 = 7.171 kWh. With this condition, and assuming that effective irradiance at the location is 5 hour, efficiency of wiring is 98%, then minimum capacity of the total photovoltaic module that must be provided is 7.171 kWh/0.98/5h = 1463.65 W. For this needs, taking into account the uneven variations in the sun's intensity for one day, 6 modules of Panasonic VBHN245SJ25 are chosen with a capacity of 245 Watt each. These modules are arranged in 3 series configuration (3×2) to obtain maximum 100 Volt voltage during no load condition on their terminals.

Based on the obtained daily energy consumption, then an energy storage component capacity with autonomy time of 1 day, depth of charge 80%, minimum energy storage capacity of (7.171kWh×1d)/0.8 = 8.965 kWh is required. The Lithium-Ion battery from LG Chem with capacity of 9.8 kWh was chosen for this purpose. To charge the battery for 5 hours duration in a DC bus voltage of 48 volts, a charge controller with a minimum current capability of 7.171 kWh/5h/48 Volts = 29.88 Amp is required. For this purpose, the Steca Solarix MPPT 5020 solar charge controller is chosen. The DC output of the charge controller is fed to fill the battery; other output can be used as supply for existing DC loads. To supply the AC loads, an inverter that with DC side connected to battery terminal/ DC output of charge controller is needed. Considering the maximum load at 650 W, the Steca XTS 1400-48 inverter with DC input range of 38 - 68 Volt, AC output of 230 Volt and power capacity of 900 VA is selected for this purpose.

The financial cash flow for this 9.8 kWh power supply and storage system is shown in the **Table 3**. The system is planned to operate in normal time of 25 years with three scenario of discount factor, they are 2.5%, 5%, and 7.5%. Investment costs which include the cost of procurement, system development, and the tax is Rp. 150,008,037. Operating and maintenance is planned to ensure no degradation on energy producing, the costs are taken at 1% of investment which is done every 5 years. The benefits obtained are ascribed to the production of energy produced, the price of which is calculated based on the non-subsidiary energy price of PLN: Rp. 1,467.26 /kWh.

Table 3. Financials analysis of the system based on current energy price of PT. PLN

Year	Capital Investment	Annual Return			Cash Flow (Rp.)	Discounted Factor, DF 2.5%		Discounted Factor, DF 5%		Discounted Factor, DF 7.5%	
		Produced Energy	Energy Price (Rp.)	Benefit (Rp.)		Disc. Cash Flow	Disc. Net Cash Flow	Disc. Cash Flow	Disc. Net Cash Flow	Disc. Cash Flow	Disc. Net Cash Flow
0	**-150,008,037**	0	0	0	-150,008,037	-150,008,037	-150,008,037	-150,008,037	-150,008,037	-150,008,037	-150,008,037
1		2,513	1,467.28	3,687,311	3,687,311	3,597,377	-146,410,660	3,511,725	-146,496,312	3,430,057	-146,577,980
2		2,513	1,467.28	3,687,311	3,687,311	3,509,636	-142,901,024	3,344,500	-143,151,812	3,190,751	-143,387,229
3		2,513	1,467.28	3,687,311	3,687,311	3,424,035	-139,476,989	3,185,238	-139,966,573	2,968,140	-140,419,089
4		2,513	1,467.28	3,687,311	3,687,311	3,340,522	-136,136,467	3,033,560	-136,933,013	2,761,061	-137,658,028
5	-1,500,080	2,513	1,467.28	3,687,311	2,187,231	1,933,193	-134,203,273	1,713,753	-135,219,261	1,523,535	-136,134,493
6		2,513	1,467.28	3,687,311	3,687,311	3,179,557	-131,023,716	2,751,528	-132,467,732	2,389,236	-133,745,258
7		2,513	1,467.28	3,687,311	3,687,311	3,102,007	-127,921,709	2,620,503	-129,847,229	2,222,545	-131,522,713
8		2,513	1,467.28	3,687,311	3,687,311	3,026,348	-124,895,361	2,495,717	-127,351,511	2,067,484	-129,455,229
9		2,513	1,467.28	3,687,311	3,687,311	2,952,535	-121,942,827	2,376,874	-124,974,638	1,923,241	-127,531,988
10	-1,500,080	2,513	1,467.28	3,687,311	2,187,231	1,708,661	-120,234,165	1,342,770	-123,631,868	1,061,231	-126,470,757
11		2,513	1,467.28	3,687,311	3,687,311	2,810,265	-117,423,900	2,155,895	-121,475,973	1,664,243	-124,806,514
12		2,513	1,467.28	3,687,311	3,687,311	2,741,722	-114,682,178	2,053,233	-119,422,740	1,548,133	-123,258,381
13		2,513	1,467.28	3,687,311	3,687,311	2,674,851	-112,007,327	1,955,460	-117,467,280	1,440,124	-121,818,258
14		2,513	1,467.28	3,687,311	3,687,311	2,609,611	-109,397,717	1,862,343	-115,604,937	1,339,650	-120,478,608
15	-1,500,080	2,513	1,467.28	3,687,311	2,187,231	1,510,208	-107,887,509	1,052,095	-114,552,842	739,210	-119,739,398
16		2,513	1,467.28	3,687,311	3,687,311	2,483,865	-105,403,644	1,689,200	-112,863,642	1,159,243	-118,580,155
17		2,513	1,467.28	3,687,311	3,687,311	2,423,283	-102,980,362	1,608,762	-111,254,880	1,078,365	-117,501,790
18		2,513	1,467.28	3,687,311	3,687,311	2,364,178	-100,616,183	1,532,154	-109,722,726	1,003,131	-116,498,660
19		2,513	1,467.28	3,687,311	3,687,311	2,306,515	-98,309,668	1,459,194	-108,263,532	933,145	-115,565,515
20	-1,500,080	2,513	1,467.28	3,687,311	2,187,231	1,334,803	-96,974,864	824,344	-107,439,188	514,903	-115,050,612
21		2,513	1,467.28	3,687,311	3,687,311	2,195,375	-94,779,490	1,323,532	-106,115,655	807,481	-114,243,131
22		2,513	1,467.28	3,687,311	3,687,311	2,141,829	-92,637,661	1,260,507	-104,855,149	751,145	-113,491,987
23		2,513	1,467.28	3,687,311	3,687,311	2,089,589	-90,548,072	1,200,483	-103,654,666	698,739	-112,793,248
24		2,513	1,467.28	3,687,311	3,687,311	2,038,624	-88,509,448	1,143,317	-102,511,349	649,990	-112,143,258
25		2,513	1,467.28	3,687,311	3,687,311	1,988,901	-86,520,547	1,088,873	-101,422,476	604,642	-111,538,616
		Total			**-63,825,575**	**-86,520,547**		**-101,422,476**		**-111,538,616**	

Financial analysis is carried out to determine at what level the price of PLN energy will provide favorable conditions for deciding to build an own energy supply installation from photovoltaic independently at a calculated cost. From the analysis conducted as in **Table 3**, at the current level of energy prices from PLN at Rp. 1467.28; there is no discount factor that gives a positive NPV. A positive NPV system is obtained when the PLN energy price is Rp. 3337.00 obtained at a 2.5% DF rate. This shows that financially, for the technical specifications and planned lifetime of the system, the development of an independent PV-based supply and storage system is only feasible if the PLN energy price level is 2.3 times the current price, also with a maximum discount rate of 2.5%. The payback period obtained from the payback period analysis for this condition is reached at the end of the 25th year, where the value of the discounted net cash flow starts to be positive, **Table 4**.

Table 4. Financials analysis of the system based on positive NPV of energy price of PT. PLN

Year	Capital Investment	Annual Return			Cash Flow (Rp.)	Discounted Factor, DF 2.5%		Discounted Factor, DF 5%		Discounted Factor, DF 7.5%	
		Produced Energy	Energy Price (Rp.)	Benefit (Rp.)		Disc. Cash Flow	Disc. Net Cash Flow	Disc. Cash Flow	Disc. Net Cash Flow	Disc. Cash Flow	Disc. Net Cash Flow
0	**-150,008,037**	0	0	0	-150,008,037	-150,008,037	-150,008,037	-150,008,037	-150,008,037	-150,008,037	-150,008,037
1		2,513	3,337.00	8,385,964	8,385,964	8,181,429	-141,826,608	7,986,633	-142,021,404	7,800,897	-142,207,140
2		2,513	3,337.00	8,385,964	8,385,964	7,981,882	-133,844,726	7,606,317	-134,415,087	7,256,649	-134,950,491
3		2,513	3,337.00	8,385,964	8,385,964	7,787,202	-126,057,525	7,244,111	-127,170,976	6,750,371	-128,200,120
4		2,513	3,337.00	8,385,964	8,385,964	7,597,270	-118,460,255	6,899,154	-120,271,822	6,279,415	-121,920,706
5	-1,500,080	2,513	3,337.00	8,385,964	6,885,884	6,086,118	-112,374,137	5,395,270	-114,876,552	4,796,422	-117,124,284
6		2,513	3,337.00	8,385,964	8,385,964	7,231,191	-105,142,946	6,257,736	-108,618,816	5,433,782	-111,690,502
7		2,513	3,337.00	8,385,964	8,385,964	7,054,820	-98,088,126	5,959,748	-102,659,068	5,054,681	-106,635,820
8		2,513	3,337.00	8,385,964	8,385,964	6,882,752	-91,205,374	5,675,951	-96,983,117	4,702,029	-101,933,791
9		2,513	3,337.00	8,385,964	8,385,964	6,714,880	-84,490,494	5,405,667	-91,577,449	4,373,980	-97,559,811
10	-1,500,080	2,513	3,337.00	8,385,964	6,885,884	5,379,242	-79,111,253	4,227,335	-87,350,114	3,340,989	-94,218,822
11		2,513	3,337.00	8,385,964	8,385,964	6,391,319	-72,719,934	4,903,100	-82,447,014	3,784,948	-90,433,874
12		2,513	3,337.00	8,385,964	8,385,964	6,235,433	-66,484,501	4,669,619	-77,777,395	3,520,882	-86,912,992
13		2,513	3,337.00	8,385,964	8,385,964	6,083,349	-60,401,151	4,447,256	-73,330,139	3,275,239	-83,637,753
14		2,513	3,337.00	8,385,964	8,385,964	5,934,975	-54,466,176	4,235,482	-69,094,657	3,046,734	-80,591,019
15	-1,500,080	2,513	3,337.00	8,385,964	6,885,884	4,754,466	-49,711,710	3,312,228	-65,782,429	2,327,195	-78,263,825
16		2,513	3,337.00	8,385,964	8,385,964	5,648,995	-44,062,716	3,841,707	-61,940,723	2,636,438	-75,627,386
17		2,513	3,337.00	8,385,964	8,385,964	5,511,214	-38,551,501	3,658,769	-58,281,954	2,452,501	-73,174,886
18		2,513	3,337.00	8,385,964	8,385,964	5,376,795	-33,174,707	3,484,541	-54,797,413	2,281,396	-70,893,490
19		2,513	3,337.00	8,385,964	8,385,964	5,245,653	-27,929,053	3,318,611	-51,478,802	2,122,229	-68,771,261
20	-1,500,080	2,513	3,337.00	8,385,964	6,885,884	4,202,255	-23,726,799	2,595,217	-48,883,584	1,621,028	-67,150,234
21		2,513	3,337.00	8,385,964	8,385,964	4,992,888	-18,733,910	3,010,078	-45,873,507	1,836,434	-65,313,800
22		2,513	3,337.00	8,385,964	8,385,964	4,871,110	-13,862,800	2,866,741	-43,006,766	1,708,310	-63,605,489
23		2,513	3,337.00	8,385,964	8,385,964	4,752,303	-9,110,497	2,730,229	-40,276,536	1,589,126	-62,016,363
24		2,513	3,337.00	8,385,964	8,385,964	4,636,393	-4,474,104	2,600,218	-37,676,318	1,478,257	-60,538,107
25		2,513	3,337.00	8,385,964	8,385,964	4,523,310	49,206	2,476,399	-35,199,919	1,375,123	-59,162,984
		Total			**53,640,752**	**49,206**		**-35,199,919**		**-59,162,984**	

4. Conclusion

Design and financial analysis of an off-grid photovoltaic generation system for supplying a residential load have been presented. The load is characterized by a daily load curve that shows the average power consumption of 143.44 watts, maximum load power of 155 Watts and a daily energy usage of 6,885 kWh. To supply this load in an off-grid system with solar-based generation with autonomy time for one day, it requires a photovoltaic module with a total capacity of at least 1463.65 Watts, and a minimum battery storage capacity of 8,965 kWh which is controlled using a battery controller of 29.88 Amperes. The supply for AC loads is carried out through an off-grid inverter with a capacity of 900 VA, 230 V. For realization, it is necessary to optimize and adjust the capacity obtained from the design and availability of component capacities produced by the manufacturer in real conditions.

Financial analysis for this DC-Coupling off-grid system is carried out to assess the feasibility in the environment of the existence of other energy supply, i.e. state electrical company, PLN. For cash flow analysis, the investment cost consists of cost for component procurement and construction of facilities, including taxes, where prices are obtained from the website of the supplier/manufacturer. The financial benefits obtained are calculated based on electricity prices according to the non-subsidized electricity tariff of PLN 2016, which is done with the assumption that the energy used that should be supplied from PLN, is substituted with the energy generated through own photovoltaic generation system. NPV analysis and payback period that has been done shows that in supplying electricity through off-grid photovoltaic generation system it will be feasible if the electricity price is 2.3 times the current price, and the discount factor (DF) for construction (if assumed to be built using a loan) is 2.5%, in such a condition of electricity prices and DF factor, the payback period is reached if generation system is operated at least until the end of the 25th year.

Financial analysis carried out showed the inability operations of the independent electricity supply scheme based on the photovoltaic generation that have been designed for current conditions. This was due to the high investment that had to be provided because most of the photovoltaic generation system components were still imported equipment. A number of schemes can be carried out by policy makers to encourage the use of renewable energy, for example by incentive treatment, reduction of interest rates for the construction of renewable plants, reduction or elimination of import duty for renewable generation components and the most important is a policy to encourage the availability and production of renewable generation components domestically.

References

[1] Ministry of Energy and Mineral Resources Republic of Indonesia, "2016 - Handbook of Energy & Economic Statistics of Indonesia.," vol. 45, no. 12. pp. 48–50, 2005.

[2] S. Nugraha, "Indonesia Energy Outlook 2010," p. 139, 2016.

[3] M. Anwari, M. I. Hamid, and Taufik, "Photovoltaic energy conversion system using single-phase grid-tied inverter," in *INTELEC, International Telecommunications Energy Conference (Proceedings)*, 2009.

[4] E. Koutroulis, K. Kalaitzakis, and N. C. Voulgaris, "Development of a Microcontroller-Based , Photovoltaic Maximum Power Point Tracking Control System," *IEEE Trans. Power Electron.*, vol. 16, no. 1, pp. 46–54, 2001.

[5] D. C. Jordan and S. R. Kurtz, "Photovoltaic degradation rates - An Analytical Review," *Prog. Photovoltaics Res. Appl.*, vol. 21, no. 1, pp. 12–29, 2013.

[6] J. Weaver, "Solar module lifetime predictions are getting better. PV magazine -Photovoltaics Markets and Technology 2018 Available at: <https://www.pv-magazine.com/2018/12/10/solar-module- lifetimepredictions-are-getting-better/>," p. 2018, 2018.

[7] M. F. Nayan, S. M. S. Ullah, and S. N. Saif, "Comparative analysis of PV module efficiency for different types of silicon materials considering the effects of environmental parameters," *2016 3rd Int. Conf. Electr. Eng. Inf. Commun. Technol. iCEEiCT 2016*, no. Dc, 2017.

[8] T. Khatib, A. Yasin, A. A. Mohammad, and I. A. Ibrahim, "On the effectiveness of optimally sizing an inverter in a grid-connected photovoltaic power system," in *2017 14th International Conference on Smart Cities: Improving Quality of Life Using ICT and IoT, HONET-ICT 2017*, 2017, vol. 2017-Janua, pp. 48–52.

[9] C. Demoulias, "A new simple analytical method for calculating the optimum inverter size in grid-connected PV plants," *Electr. Power Syst. Res.*, vol. 80, no. 10, pp. 1197–1204, 2010.

[10] I. Alsaidan, A. Khodaei, and W. Gao, "Determination of optimal size and depth of discharge for battery energy storage in standalone microgrids," *NAPS 2016 - 48th North Am. Power Symp. Proc.*, 2016.

[11] Gujari A. S, "A Technoeconomic Feasibility of Inverter Selection for MegaWatt(MW) Scale Grid Connected Solar Photovoltaic Power Plant," in *International Conference on Energy, Communication, Data Analytics and Soft Computing (ICECDS-2017)*, 2011, vol. 1, no. June, pp. 6–8.

[12] Ministry of Energy and Mineral Resources Republic of Indonesia, "Peraturan Menteri Energi dan Sumber Daya Mineral Republik Indonesia Nomor 28 Tahun 2016 tentang Tarif Tenaga Listrik yang Disediakan oleh PT Perusahaan Listrik Negara (Persero)." 2016.

THE DESIGN OF WHEELCHAIR SYSTEMS WITH RASPBERRY PI 3-BASED JOYSTICK ANALOG AND VOICE CONTROL

Putri Madona[1], Husna Khairun Nisa[2], Yusmar Palapa Wijaya[3], Amnur Akhyan[4]

[#] Electronics Engineering, Politeknik Caltex Riau, Rumbai, Pekanbaru, 28265, Indonesia

E-mail: dhona@pcr.ac.id, husnakn09@gmail.com, yusmar@pcr.ac.id, akhyan@pcr.ac.id

Abstract, In this study, an electric wheelchair that combines two controls: joystick analog and voice control is designed. IC MCP3008 is used to navigate wheelchairs by using Josytick, where joystick analog data will be converted into digital data. The movements resulted from the joystick analog on the xAxis axis (horizontally) are the right turn and left turn, and on the yAxis axis (vertically) are forward and backward. The movements on the yAxis and xAxis axes set by the user affects the speed of the wheelchair. Meanwhile, the AMR-Voice application on Android is used to navigate wheelchairs by using sound. There are five commands in this voice control: "Forward", "backward", "left", "right", "stop". The order will be sent to Raspberry Pi 3 via the HC-06 module to then be recognized for the command. If the voice commands are received accordingly, Raspberry Pi 3 will provide an activation signal to the motor driver to move the wheelchair in the direction corresponding to the command given by the user. Voice control testing on wheelchairs is tested in quiet rooms and noisy rooms. The results of the wheelchair control testing with sound indicate that the accuracy and speed of the wheelchair response rely heavily on Internet connection and room conditions. The average response when the condition of the room is quiet is 0.16 s and when the condition of the room is noisy is 5.18 s. Wheelchairs with joystick control and the voice made can be used for the disabled, whether for those who can move their fingers or not, at a low cost so that they can be an alternative in developing countries.

1. Introduction

Paralysis is a major issue that requires someone to use a wheelchair. The usefulness of this tool is often interpreted as accommodation for people with paralysis. Wheelchair assistance for these patients is widely approved to provide the proper running therapy for a patient who is paralyzed. The large benefits of wheelchairs can help patients in their daily activities. Unfortunately, wheelchairs generally still require the help of others to push them. But there is a time when the disabled must be able to move their wheelchair independently. Wheelchairs are needed in Indonesia because of the many disabilities in the country. Based on the data of Sakernas 2017 survey institution, the number of people with disabilities in Indonesia was 21 million. Currently wheelchairs are still widely used manually,

IOP Conf. Series: Materials Science and Engineering **846** (2020) 012032 doi:10.1088/1757-899X/846/1/012032

that is with the help of others. Psychally, a person with disabilities does not want to trouble the surrounding people. A lot of wheelchair development has been done, one of which is a wheelchair capable of being controlled with voice by using module STM32F103C6T6 (STM32) [1]. On another study, brain signal-based wheelchairs are developed with various methods of brain signal processing [2], [3], [4]. This research develops a wheelchair-control system using joystick analog and voice. Speed settings can be performed on joystick control mode. The relatively inexpensive use of components is expected to be a solution to the availability of wheelchairs with special controls for the disabled in developing countries such as Indonesia.

2. Methodology

A. Raspberry Pi 3

Raspberry Pi 3 has 1GB RAM and Broadcom VideoCore IV graphics at a higher clock frequency than before that run at 250MHz. There are 40 pins or 26 pins of GPIOs on Raspberry Pi 3. In the GPIO, sensors are generally connected or used to drive the relay. GPIO is used as an alternative to Raspberry communication to a device outside, exactly like a USB port or Ethernet. The GPIO pins consist of:

1. Power supply (3.3 V dan 5 V, 2 sets each)
2. UART (Universal asynchronous receiver transmitter, 1 set)
3. SPI (Seriall Peripheral Interface)
4. I2C (Inter-Integrataed Circuit) –EEPROM
5. GPIO (General Purpose Input Output)
6. Raspberry Pi 3 Block Diagram

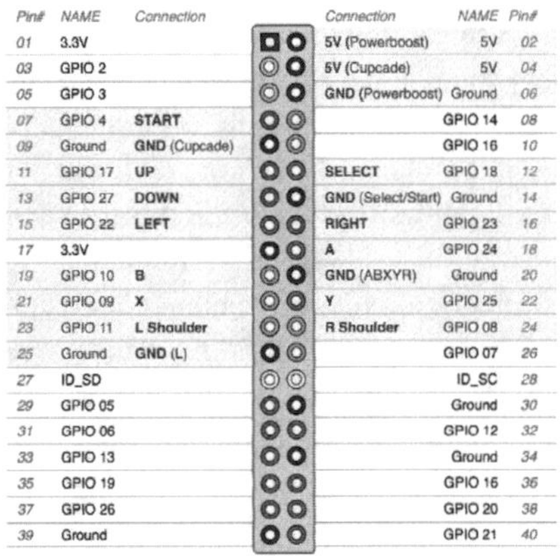

Figure 1. Data sheet GPIO Raspberry Pi 3

B. Joystick Analog

Joystick Analog is two potentiometers combined for one vertical movement (yAxis) and another for horizontal movement (xAxis). The Joystick also has a select switch. This Joystick generates 2.5 V output of X and Y when the position is not moved. Adjusting the direction of the joystick will cause the output to vary from 0V to 5V depending on the direction of the movement.

Figure 2.Analog Joystick

Configuration pin:
1. GND : Ground
2. 5V : 5V DC
3. VRx : X-axis proportional voltage
4. VRy : Y-axis proportional volatge
5. SW : *Switch pushbutton*

C. IC MCP3008

This IC MCP3008 is a 10-bit Analog to-Digital (A/D0 Converter) device. There are 4 pairs of false differential inputs or 8 single inputs. Differential Nonlinear (DNL) and Integral Nonlinearity (INL) are determined at ± 1 LSB. Communication with the device is performed by using a simple serial interface that is compatible with the SPI protocol. The IC MCP30008 has a conversion rate of up to 200 ksps. The MCP3004/3008 device operates at a wide voltage range (2.7 V-5.5 V).

Figure 3. IC MCP 3008

D. AMR-Voice Control

AMR-Voice is an Android application available in the Google Play store. When it is impossible for the user to type with the touch keyboard on Android, it can be replaced by saying a sentence, then Google Voice will analyze the sound it takes. Therefore, this developed voice control requires an Internet connection when used.

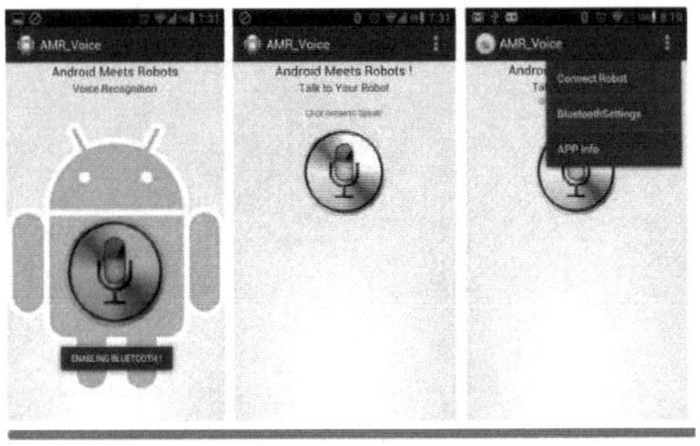

Figure 4. AMR-Voice Application

E. HC-06 Bluetooth Module

Bluetooth is one form of data communication wirelessly, based on radio frequencies. This Bluetooth module is replacing serial communication using cables. Bluetooth consists of two types of devices, namely Master (data sender) and Slave (receiver).

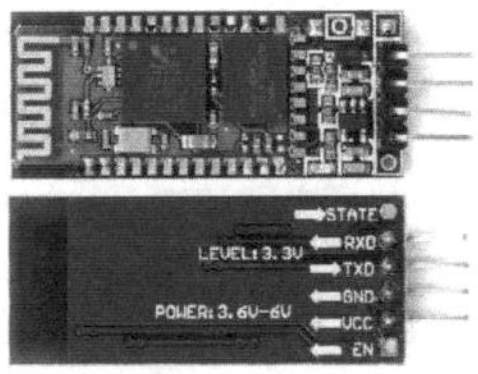

Figure 5. HC-06

Communication can be done immediately after both Bluetooth modules pair the connection. Connection via Bluetooth is serial communication, namely the presence of TXD communication pins and RXD communication pins.

F. System Block Diagram

From the two controls used will be integrated into a system capable of navigating the direction of the wheelchair. Figure 6 is a block diagram of wheelchair control using joystick and voice control. And Figure 7 shows a control system flowchart using a joystick.

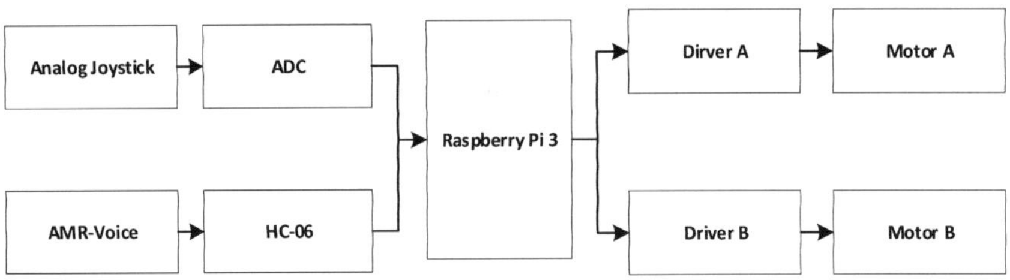

Figure 6. Block Diagram of Wheelchair Control Using Joystick and Voice Control

Analog Joystick produces analog data which will be converted into ADC digital data . Figure 7 shows that to move the wheelchair forward, the user must move the yAxis axis on the joystick analog until the value is larger than 600. Users can increase the speed on the wheelchair by moving the yAxis axis on a joystick analog up to a maximum value of 1024. As for the backward movement, the user must move the yAxis axis on the joystick analog until the value is smaller than 470. The backward speed can also be adjusted by moving the yAxis axis on the joystick analog up to a maximum value of 0. As with the forward and backward movements, for the right turn the user must move the xAxis axis on the joystick analog until the value is arger than 600. Users can increase the speed on the wheelchair by moving the xAxis axis on an analog joystick up to a maximum value of 10-24. As for the left Turn movement, the user must move the xAxis axis on the joystick analog until the value is smaller than 470. The left turn speed can also be set up by moving the xAxis axis on the joystick analog up to a maximum value of 0.

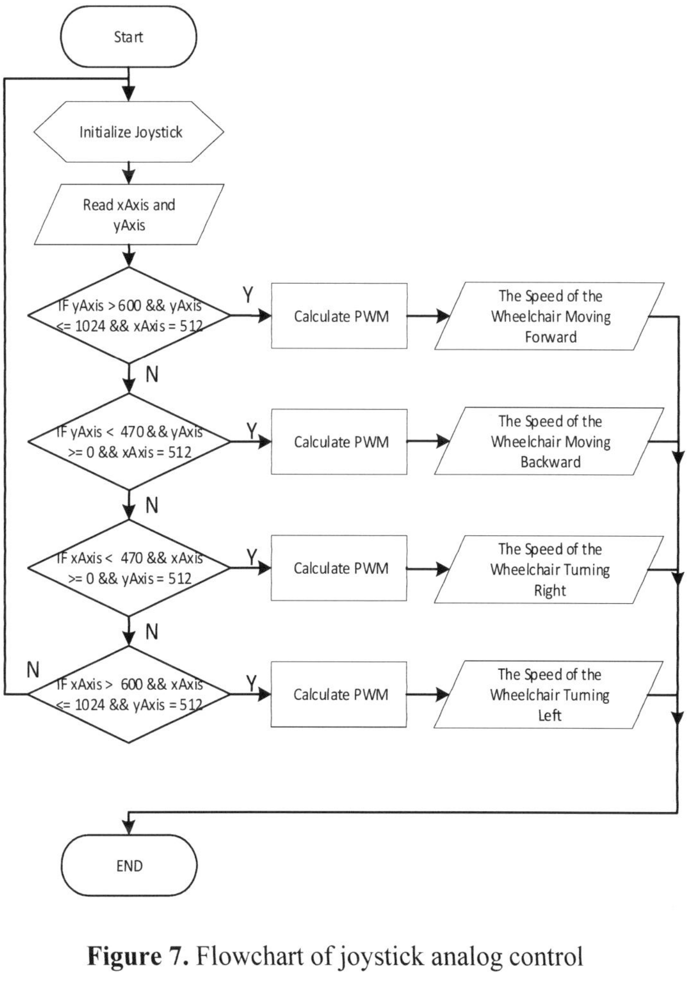

Figure 7. Flowchart of joystick analog control

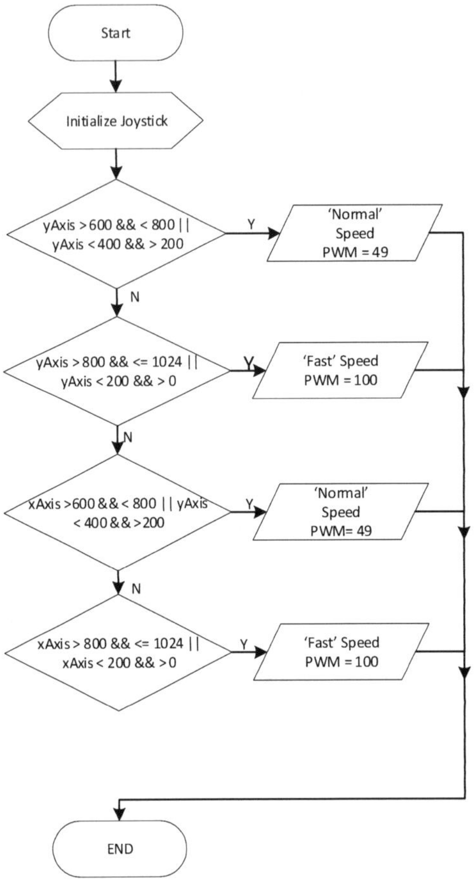

Figure 8. Flowchart of Speed Control

In this system, the speed of the wheelchair movement is divided into two: ' normal ' speed mode and ' fast ' speed mode. Normal speed mode has been determined with PWM 49 and the values on xAxis and yAxis are > 600 and < = 800 or xAxis and yAxis are < 400 and > 200, while fast speed mode uses PWM 100 and the values on xAxis and yAxis are > 800 and < = 1024 or < 200 and > 0. As shown in Figure 8. These two speed modes can be used to all directions from joystick control.

While controlling the wheelchair with voice control uses an Android application: AMR-Voice. This application makes use of Google Voice feature, the data read from Google Voice will be sent via Bluetooth Android and will be received by HC-06 module. The commands that can be executed by raspberry are only five commands; namely, forward, backward, right, left, and stop. When the user gives a command other than the five predefined commands, Raspbery PI will not execute the command and the wheelchair will not respond to anything. Figure 9 is a wheelchair-controlled flowchart using voice control.

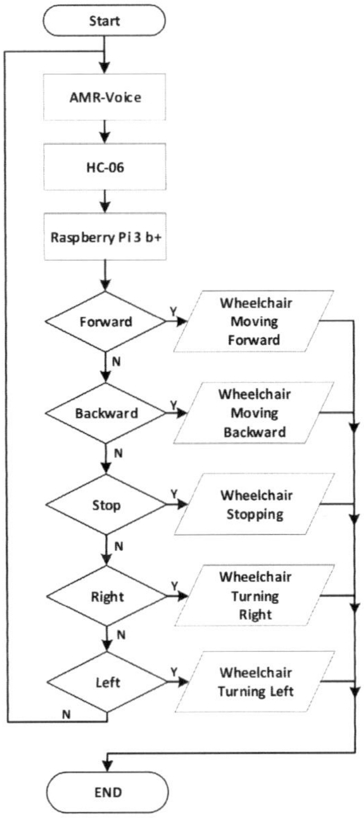

Figure 9. Flowchart of Voice kontrol

3. Result and Discussion

A. Wheelchair Control Testing Using Joystick Analog

Table 1. Motor Movement

No.	Joystick Output ADC Value		Movement		Movement Direction
	xAxis	yAxis	Motor 1	Motor 2	
1	< 470 >= 0	-	Stop	CW	Left
2	> 600 and <= 1024	-	CW	Stop	Right
3	-	< 470 >= 0	CCW	CCW	Backward
4	-	> 600 and <= 1024	CW	CW	Forward

This testing is done to determine the accuracy of the joystick output value to the movement of the two wheelchair motors, which represents the direction of the wheelchair movement. At the time of testing, the wheelchair uses joystick analog to turn left, then the user must shift the analog lever left so that the angle on the xAxis axis becomes < 470 and > = 0. Then, Motor 2 (M2) is moving clockwise (CW), while Motorl (M1) is in the state of Stop. On the other hand, if the wheelchair moves right then the user must shift the analog lever right so that the angle on the xAxis axis becomes > 600 and < = 1024 then Motorl (M1) rotates clockwise (CW) while the Motor2 (M2) is in the state of stop. For the backward movement, the user must move the analog lever backwards, so that the angle on the yAxis axis becomes < 470 and > = 0. Then both motors move counterclockwise (CCW), while for forward the movement, the user must drive Analog lever forward so that the angle on the yAxis axis becomes > 600 and < = 1024, then the motor will move clockwise (CW).

Table 2. Movement Direction And Speed

Joystick Output	**ADC**	**Movement Direction based on XY axis input**	**Wheelchair Movement Direction**	**PWM**	**Duty Cycle (%)**	**Speed Mode**	**Accurate (%)**
XAxis	**yAxis**						
498	690	Forward	Forward	49	50%	Normal	100
495	887	Forward	Forward	100	100%	Fast	100
495	388	Backward	Backward	49	50%	Normal	100
497	0	Backward	Backward	100	100%	Fast	100
700	510	Right Turn	Right Turn	49	50%	Normal	100
910	509	Right Turn	Right Turn	100	100%	Fast	100
382	510	Left Turn	Left Turn	49	50%	Normal	100
99	512	Left Turn	Left Turn	100	100%	Fast	100

The test in Table 2 was done to see the suitability of the joystick output data to the direction and speed of the wheelchair motor movements. The test results show that each joystick analog output data has precise accuracy of the movement direction and of the wheelchair motor speed.

B. Wheelchair Control Testing Using Voice Control

Figure 10. AMR-Voice Display Receiving Command from User

User-spoken voice commands are accepted by AMR-Voice so that the command is sent to Raspberry Pi 3 via the HC-06 module, after which the command will be executed. If appropriate, the wheelchair will move according to the given command. If not appropriate, the wheelchair will not respond to the command. Table III is a response time of AMR Voice that is performed in two conditions; namely, ' quiet ' condition and ' noisy ' atmosphere condition.

From table 3, it is seen that the time response of the AMRVoice application is different in quiet and noisy conditions, where the respond time when the condition is ' quiet ' will be better than the ' noisy ' condition. It will also affect the response of the wheelchair movement.

Table 3. Data Of Wheelchair's Movemoent Comparasion Result Againts 'Noisy' And 'Quiet' Condition

No	*User Input*	Raspberry Data	*Raspond Time (s)* 'Noisy' Condition	'Quiet' Condition	Wheelchair Movement
1	Forward	*Forward#	4.12	0.12	Forward
2	Backward	*Backward#	6.03	0.20	Backward
3	Right	*Right#	4.60	0.19	Right
4	Left	*Left#	5.02	0.15	Left
5	Stop	*Stop#	6.15	0.18	Stop

4. Conclusion

The purpose of this research is to make a wheelchair control system that is cheap and easy to use for people with disabilities who have limited limbs or cannot use their fingers. This wheelchair control using a joystick and voice successfully moves the wheelchair in the desired direction. Joystick controls are for users who can still use their hands to move the joystick, while voice controls are for users who are unable to use their fingers to push or direct the joystick. Voice control has a very good response time in quiet noise situations with a good internet connection.

Acknowledment

The research was funded by the The Higher Education Applied Research Grant, Minister of Research Technology and Higher Education, Republic of Indonesia, in accordance with the research contract of the 2019 budget year.

References

[1] Z. Raiyan, M. S. Nawaz, A. K. M. A. Adnan, and M. H. Imam, "Design of an arduino based voice-controlled automated wheelchair," *5th IEEE Reg. 10 Humanit. Technol. Conf. 2017, R10-HTC 2017*, vol. 2018-January, pp. 267–270, 2018.

[2] S. U. Upase, "Speech recognition based robotic system of wheelchair for disable people," *Proc. Int. Conf. Commun. Electron. Syst. ICCES 2016*, 2016.

[3] H. Wang, T. Li, F. Zheng, and Y. Yan, "A Wheelchair Platform Controlled by a Multimodal Interface," *Proc. - 2015 2nd Int. Conf. Inf. Sci. Control Eng. ICISCE 2015*, pp. 587–590, 2015.

[4] S. Priyanayana, A. G. Buddhika, and P. Jayasekara, "Developing a voice controlled wheelchair with enhanced safety through multimodal approach," *IEEE Reg. 10 Humanit. Technol. Conf. R10-HTC*, vol. 2018-December, pp. 1–6, 2019.

[5] S. He *et al.*, "A P300-Based Threshold-Free Brain Switch and Its Application in Wheelchair Control," *IEEE Trans. Neural Syst. Rehabil. Eng.*, vol. 25, no. 6, pp. 715–725, 2017.

[6] P. Madona, R. Mujiono, Y. Palapa, "Controlling The Direction Of Wheelchair Movement Using Raspberry-Pi Based Brain Signals.", *ICAE 2019*

[7] B. He, B. Baxter, B. J. Edelman, C. C. Cline, and W. W. Ye, "Noninvasive brain-computer interfaces based on sensorimotor rhythms," *Proc. IEEE*, vol. 103, no. 6, pp. 907–925, 2015.

[8] O. R. Pinheiro, L. R. G. Alves, and J. R. D. Souza, "EEG Signals Classification: Motor Imagery for Driving an Intelligent Wheelchair," *IEEE Lat. Am. Trans.*, vol. 16, no. 1, pp. 254–259, 2018.

[9] H. G. M. T. Yashoda, A. M. S. Piumal, P. G. S. P. Polgahapitiya, M. M. M. Mubeen, M. A. V. J. Muthugala, and A. G. B. P. Jayasekara, "Design and development of a smart wheelchair with multiple control interfaces," *MERCon 2018 - 4th Int. Multidiscip. Moratuwa Eng. Res. Conf.*, pp. 324–329, 2018.

A Wideband Rectangular Patch Microstrip Antenna using Quad-Slotted Ground Plane

Rudy Fernandez[1*], Muhammad Aulia Ilham[1], Hanalde Andre[1], Firdaus[2]

[1]Electrical Engineering Dept., Universitas Andalas [2]Electrical Engineering Dept., Politeknik Negeri Padang

*rfernandez@eng.unand.ac.id

Abstract. This research proposes the use of a quad-slotted ground plane in enhancing the bandwidth of rectangular patch microstrip antenna. A feeding line is used for exciting the antenna patch. On the ground plane, there are 4 slots which are designed in a rectangular form with the same dimension. Each slot has a height of 22.5 mm and a width of 15.4 mm. The antenna is designed on FR4 epoxy which has a dielectric constant of 4.4 and a thickness of 1.67 mm. The result shows that the return loss of -14.51 dB in the operating frequency of 1,810 MHz in the LTE Band 3 Frequency, VSWR is 2.19 and the gain is 3.2 dBi. The achieved bandwidth of 200 MHz can cover the frequency range of the LTE Band 3.

1. Introduction

In Indonesia, Long Term Evolution (LTE) operating frequencies have been regulated by the Indonesian Government in Peraturan Menteri Komunikasi dan Informatika Republik Indonesia nomor 27 Tahun 2015. The LTE Frequencies comprise of 5 bands which are band 1,3,5,8 and 40. Band 3 has a frequency range for the uplink from 1710 MHz to 1785 MHz and the downlink from 1805 MHz to 1880 MHz [1]. Band 3 provides uplink and downlink capacity of at least 150 MHz in total. This technology is designed to improve network capacity and speed [2], uses an antenna such as a microstrip antenna. Unfortunately, the antenna has a disadvantage which is a narrow band.

Diverse methods have been developed to broaden the bandwidth of the microstrip antenna. In [3], using a dual-layer substrate is applied to increase the bandwidth on band 40 LTE. An additional patch layer is put on top of a single layer antenna. The additional layer has the same dimensions as a single layer antenna layer but with no ground plane. The increased bandwidth requires a thicker antenna dimension. Research in [4] designs an array antenna with a dual-layer substrate for bandwidth enhancement. It can cover the band 3 LTE frequency range but with a relatively larger antenna dimension. Moreover, applying a rectangular ring slot etched on the patch is used for enhancing the bandwidth of an LTE antenna [5]. This antenna structure does not satisfy the required bandwidth. The three separate radiating elements are used in [6] resulting in a complexity design. Furthermore, the use of metamaterial with negative refractive index is chosen but requiring a challenge in analyzing [7]. Given the design and result of previous researches, this paper proposes a simple antenna design with four quad slots on the ground plane to get a wider bandwidth.

IOP Conf. Series: Materials Science and Engineering **846** (2020) 012033 doi:10.1088/1757-899X/846/1/012033

2. Methodology

A simple and conventional rectangular patch microstrip antenna is used as a basic design to operate in the center frequency of Band 3 LTE. As a substrate for the antenna, FR4 epoxy with a dielectric constant of 4.4 and a thickness of 1.67 mm is chosen. The substrate is economical and easy to find in the market.

The design of the antenna follows the outline and procedures as in [8] as seen in figure 1(a). It is optimized using an EM simulator software to meet a required parameter for a good antenna which is return loss ≤ -10 dB. The result shows that the return loss is achieved but with a narrow bandwidth as in figure 1 (b).

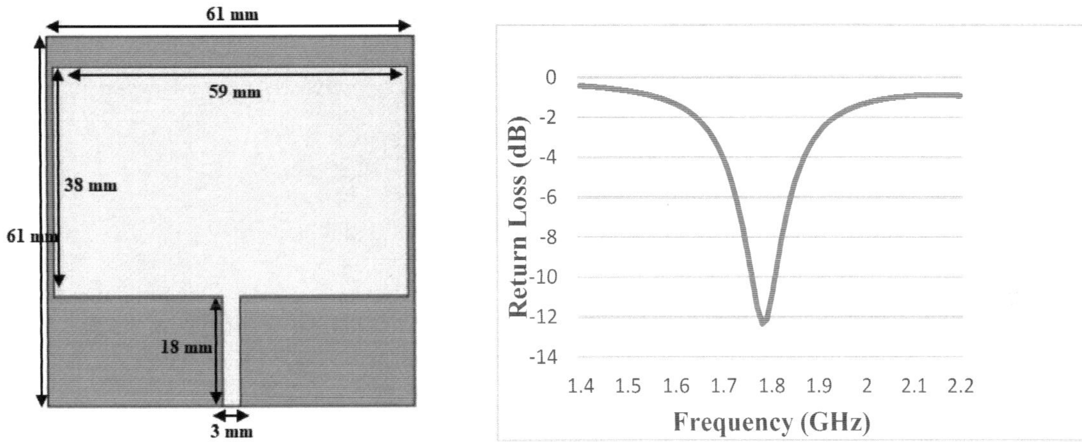

Figure 1. The design of a conventional antenna (a) and simulated result (b)

To cover the uplink and the downlink of Band 3 LTE, the antenna requires the bandwidth of 170 MHz. Regarding this requirement, this research proposes a quad slot over the ground plane. Each slot has the same dimension in a rectangular form with a fixed length of the gap between the slots.

In the initial design, the slot has a width of 5 mm and a length of 20 mm. After simulated and optimized, the width and the length get increased, becoming 15.4 mm and 22.5 mm respectively as shown in figure 2.

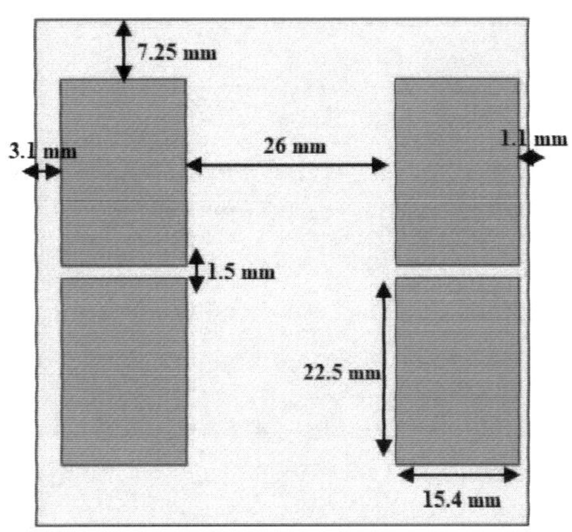

Figure 2. The geometry of the proposed slot in the ground plane

IOP Conf. Series: Materials Science and Engineering **846** (2020) 012033 doi:10.1088/1757-899X/846/1/012033

The increasement followed by slightly adjustment in the patch which is the length reduced as in figure 3.

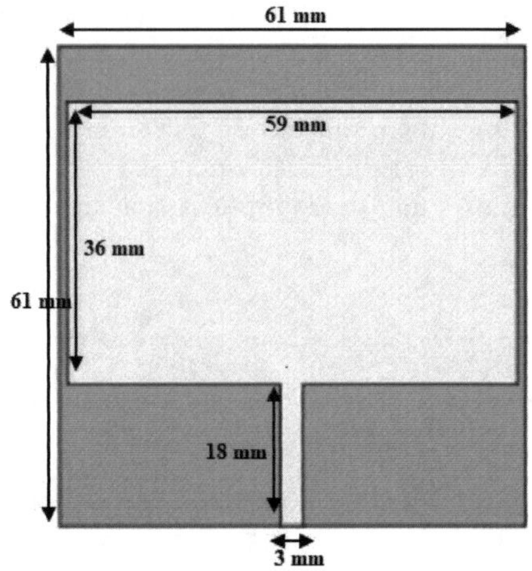

Figure 3. The geometry of the proposed antenna

3. Result and Discussion

The proposed antenna is simulated using an EM simulator to show the performance of the antenna. It is based on the antenna parameters which are return loss, VSWR, gain, and radiation pattern.

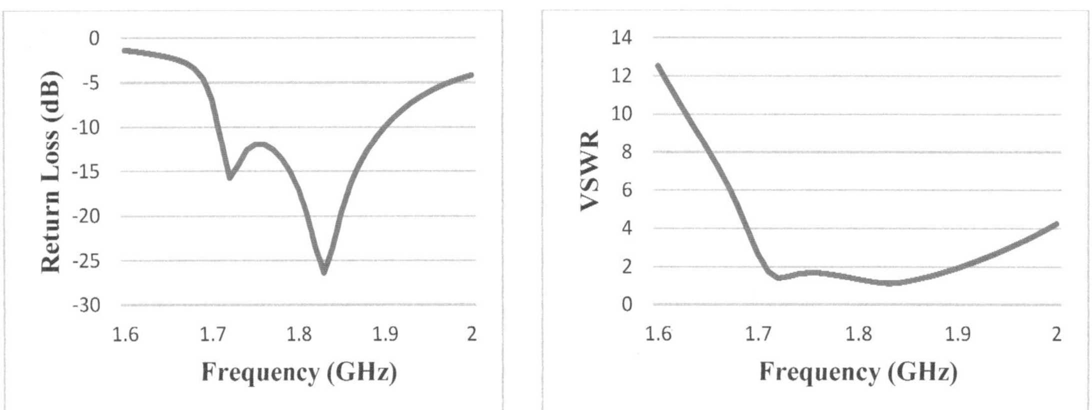

Figure 4. The simulated return loss (a) and VSWR (b)

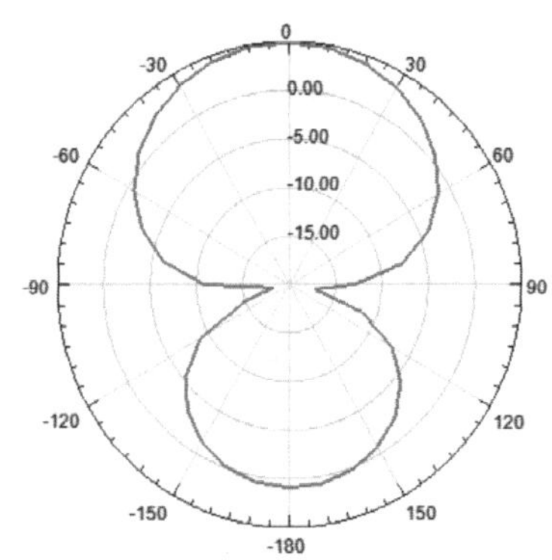

Figure 5. The simulated radiation pattern of antenna

IOP Conf. Series: Materials Science and Engineering **846** (2020) 012033 doi:10.1088/1757-899X/846/1/012033

Figure 4 (a) and (b) shows that the return loss of the antenna in operating frequency is around -17.91 dB and VSWR is 1.296. The bandwidth is 192.8 MHz in the frequency range of 1706.9 – 1899.7 MHz, whereas the gain of the antenna is 3.2 dBi. The simulated radiation pattern of antenna is shown in figure 5.

The hardware realization of the rectangular microstrip patch antenna with quad-slots is carried out by fabricating the simulated antenna. The fabricated antenna is shown in figure 5.

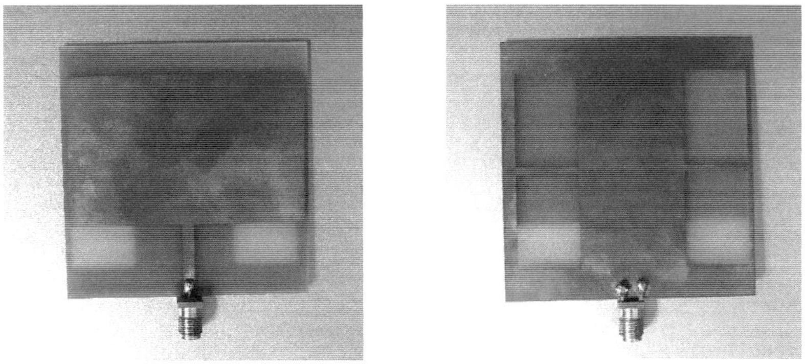

Figure 6. The fabricated antenna in front view (a) and back view (b)

The performance of the fabricated antenna is measured by connecting to a Network Analyzer based on the antenna parameters. The return loss and VSWR of both simulated and fabricated antenna are compared in figure 6.

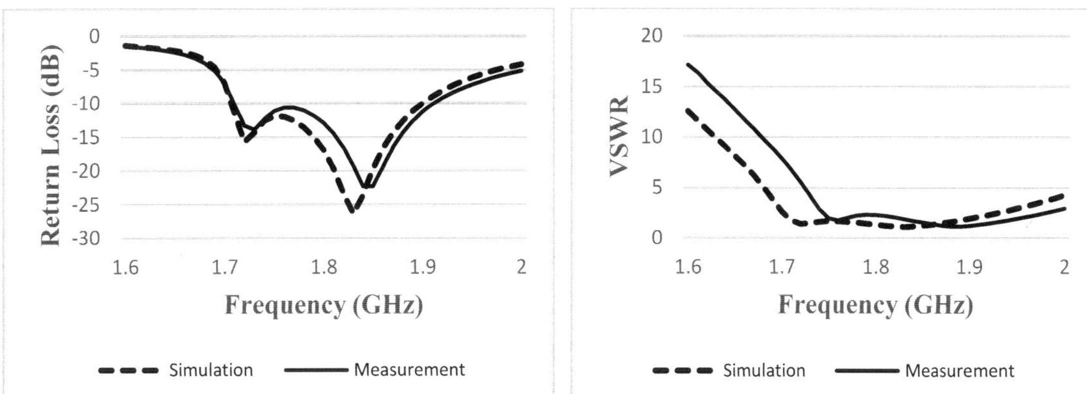

Figure 7. The comparison of simulation (dashed line) and measurement (solid line) result in return loss (a) and VSWR (b)

The result data of the antenna parameter for simulation and measurement are compared and presented in table 1.

Table 1. The comparation of antenna parameter for both result

Antenna Parameter	Simulation	Measurement
Return Loss (dB)	-17.91	-14.51
VSWR	1.29	2.19
Bandwidth (MHz) (Frequency Range)	192.8 (1706.9 – 1899.7)	200 (1710 – 1910)

From table 1 shows that even the return loss and VSWR of simulation results has slightly better value than measurement results. In terms of bandwidth, both results can cover the frequency range of Band 3 LTE while the measurement bandwidth is wider. Overall, the simulation result shows a good agreement with the measurement result.

4. Conclusion

The proposed antenna can cover the operating frequency range of band 3 LTE. The simulated antenna shows a good agreement with the fabricated antenna. The measurement result shows that the return loss of -14.51 dB in the operating frequency of 1,810 MHz in the LTE Band 3 Frequency, VSWR is 2.19 and the gain is 3.2 dBi. The quad-slotted in the ground plane is proven to enhance the bandwidth which is 200 MHz.

Acknowledgment

This research is funded and supported by a research grant for Dosen Jurusan Teknik Elektro Unand and by an International Conference Grant program from Universitas Andalas Tahun Anggaran 2019

References

[1] Menkominfo, "Peraturan Menteri Komunikasi dan Informatika Republik Indonesia Nomor 27 Tahun 2015 tentang Persyaratan Teknis Alat dan atau Perangkat Perangkat Telekomunikasi Berbasis Standar Teknologi Long Term Evolution," in *Kementerian Komunikasi dan Informatika Republik Indonesia*, 2015, pp. 1–11.

[2] 3GPP, "TS 136 101 - V8.3.0 - LTE; Evolved Universal Terrestrial Radio Access (E-UTRA); User Equipment (UE) radio transmission and reception (3GPP TS 36.101 version 8.3.0 Release 8)," 2008.

[3] R. Fernandez and F. A. Pratama, "Penggunaan Dual-Layer Substrate untuk Meningkatkan Bandwidth Antena Mikrostrip pada Frekuensi LTE," *J. Nas. Tek. Elektro*, vol. 7, no. 2, pp. 117–121, 2018.

[4] R. P. Hariyadi, Tommi. Megantara, "Directional 2x2 MIMO Microstrip Antenna Design and Optimization for LTE Band-3 Application," in *2018 5th International Conference on Information Technology, Computer and Electrical Engineering (ICITACEE)*, 2018, vol. 5, pp. 161–166.

[5] C. Chiang, P. Liu, C. Tseng, and C. Hsu, "Planar Microstrip-fed Rectangular Antenna for LTE Applications," in *2018 International Conference on Microwave and Millimeter Wave Technology (ICMMT)*, 2018, vol. 7, no. 2, pp. 1–4.

[6] U. Kumar, D. K. Upadhyay, and B. L. Shahu, "Improvement of performance parameters of rectangular patch antenna using metamaterial," in *2016 IEEE International Conference on Recent Trends in Electronics, Information and Communication Technology, RTEICT 2016 - Proceedings*, 2017, pp. 1011–1015.

[7] Y. Zhao, "Dual-wideband microstrip antenna for LTE indoor base stations," *Electron. Lett.*, vol. 52, no. 8, pp. 576–578, 2016.

[8] C. A. Balanis, *ANTENNA THEORY*, 3rd ed. New Jersey: John Wiley and Sons, Inc., Publication, 2005.

IOP Conf. Series: Materials Science and Engineering **846** (2020) 012034 doi:10.1088/1757-899X/846/1/012034

Twin loop induction sensor for detection Partial discharge

Eka Putra Waldi[1 *], Asri Indah Lestari[1], Rudy Fernandez[1], Saifa Mulyadi[1], Darmawan[1], Heru Dibyo Laksono[1], Andre H[1], Aulia[1], Ariadi Hazmi[1], Yoshinobu Murakami[2], Naohiro Hozumi[2]

[1]Department of Electrical Engineering, Faculty of Engineering, Universitas Andalas, Kampus Limau Manis, Padang, West Sumatra 25163INDONESIA.
[2]Electrical and Electronic Information Engineering, Toyohashi University of Technology, 1-1 Hibarigaoka, Tempaku, Toyohashi, Aichi Prefecture 441-8580, JAPAN.

*Corresponden Email: ekawaldi@eng.unand.ac.id

Abstract. Partial-discharge is a symptom of high voltage insulation aging, which is a small voltage pulse with high frequency and superposition with high voltage sources that have low frequencies. To find out the existence of this signal, it needs a reliable measurement technique to avoid misinterpretation. This paper presents a technique for measuring partial discharge with an induction method. The sensor used is circular twin loops. The first loop passes through a partial discharge signal, and the second induction receiver produces an induced voltage signal. Various sizes of loop diameters are designed with dimensions 3.2, 4.7, 8.9, and 11.4 cm to look for sensor sensitivity. The performance sensor is analyzed by reading the sensitivity test to the input given with 50-ohm oscilloscope impedance. From the measurement results, the higher the diameter of the sensor loop has good sensitivity.

1. Introduction

Partial discharge is a symptom of insulating aging. The energy produced by partial discharge is not too large, but it can burn the surface of insulation slowly. This combustion process can propagate like a tree; therefore, that at a particular time, then there is a short circuit between high voltage and ground. This incident makes electrical energy unusable accordingly that the activity of partial discharge requires periodic detection.

Partial discharge produces a momentary pulse of electric current that passes through an impedance that changes shape to a pulse voltage. The voltage caused by partial discharge has high frequency and low amplitude, which blends with the source voltage wave, which has high magnitude and low frequency. One measurement technique is known as the conventional method using a low voltage sensor. This sensor can detect the series with a coupling capacitor (CCa), which also functions to pass high-frequency waves (high pass filter) [1]–[4]. Digital techniques capable of changing analog signals to digital have been carried out. Therefore, the data storage is done digitally and using a sampling rate between 10 MS/s to 100 MS/s [3][5]–[8]. Digital observation can speed up analysis and be more reliable than physical analysis, and data can be stored and analyzed again without retaking measurements. The advancement of the digital technology era, the sampling rate reaches hundreds of MHz, which allows reading of the discharge signal as high frequency with induction pattern [9]–[16]. One of the advantages of this sensor is naturally protected when the measurement of insulation failure occurs.

In this study, a partial discharge sensor with an induction pattern was designed with simple and economical manufacturing. The sensor used is a twin loop separated from each other by an air gap, one side (one of the loops) functions as an input, and the other side functions as an output. The working principle of this sensor is the same as an air-core transformer with a coil ratio of one. The accuracy of the sensor readings must be following the capabilities of the digital equipment required. Accordingly, it needs to be examined. The electrical characteristics of the sensor to the digital reading with the 50-ohm oscilloscope input impedance and is expected to produce the same sensor capabilities as the tools used in the laboratory. Sensor materials are made from recycled goods, making it easy to present at a lower cost.

2. Methodology

In general, twin loop induction sensors work according to the laws of electromagnetic induction, where the instantaneous current through the primary conductor produces a momentary magnetic field in the form of flux. This flux field induces a secondary conductor in front of it. The induced secondary conductor produces an opposing flux marked by the potential difference between the conductor ends. This voltage is utilized as a result of the instantaneous current response to be detected. Where the voltage produced is proportional to the change in current produced over time. Mathematically, the resulting induced voltage can be expressed by [9], [13], [17]:

$$E = M \frac{di_0}{dt} \tag{1}$$

Where i_o is the seat current of the PD pulse that passes through the primary side, which changes with time function, and M is the mutual inductance between the secondary sensor loop with its primary part. Where M can be used according to the equation:

$$M = \varepsilon_r \pi \left(\frac{d}{4}\right) \tag{2}$$

Substitution M in equation 1 with equation 2, then equation 1 becomes to equation 3:

$$E = \varepsilon_r \pi \left(\frac{d}{4}\right) \frac{di_0}{dt} \tag{3}$$

The sensor output voltage that follows equation 3 can increase if the sensor diameter increases. Alphabet of d is the diameter of the circle, the name of the sensor. This study fully uses equation 3, which is to increase reading sensitivity.

The sensor used, as shown in Figure 1.a, two separate loops (loops) are formed following the shape of a PVC pipe circle. The first part is the primary loop whose edges are connected to the red connector as an input of the partial discharge signal, and to the black connector as a connector to ground. The second part is the secondary loop as an induction receiver generated by the primary loop, the end of the loop is connected to the BNC connector which can later be connected to the measuring instrument, in this study the measuring instrument used is a digital oscilloscope. This sensor can be said as a transformer with a core of air with a ratio of primary-secondary winding is 1: 1. Finally, the information above can easily be illustrated in Figure 1b.

In this study, there are 4 types of sensors that will be tested. The types of sensors are variations in the diameter of the circle. The diameter of the circumference used is 3.2, 4.8, 8.9, and 11.4 cm. The conductor miners used as coils are 0.5 mm. The explanation above can be seen in Figure 2.

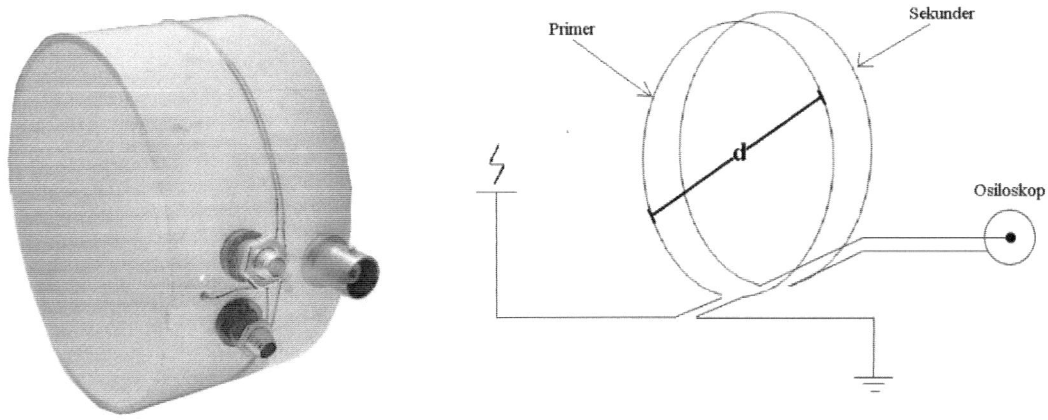

a b

Figure 1.a. The shape of the sensor used. b. Sensor equivalent circuit.

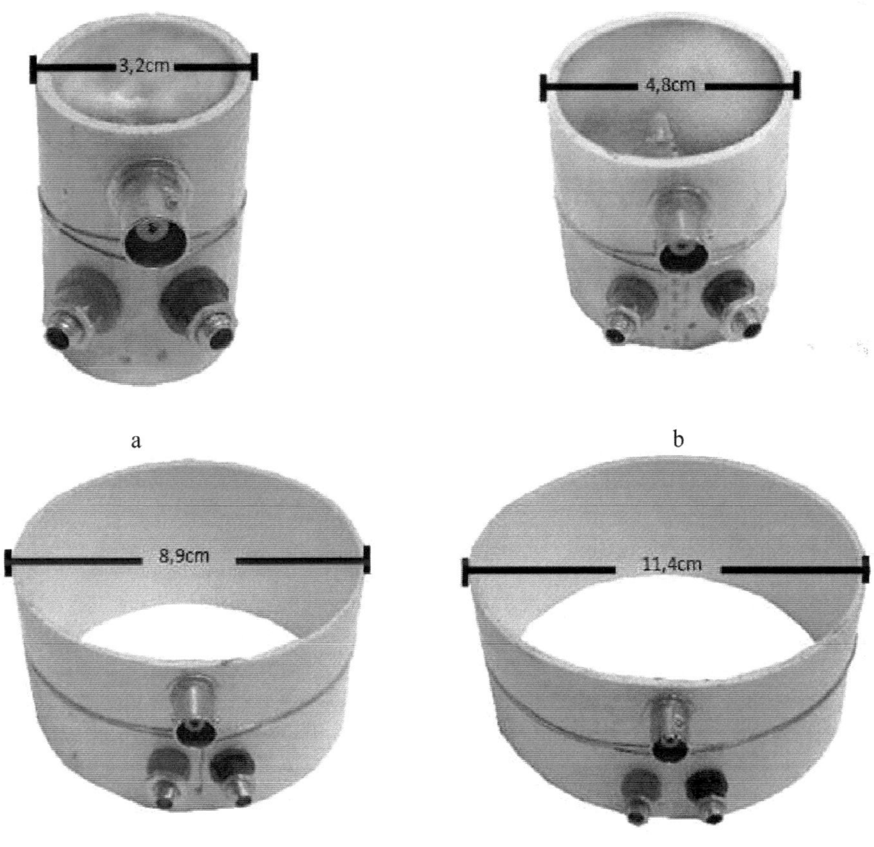

a b

c d

Figure 2. Sensor loop with diameter (a) 3.2 cm (b) 4.8cm (c) 8.9cm (d) 11.4cm

The linearity test sensor sensitivity is done by increasing the input provided and reading the output in the form of an increase as well. As shown in Figure 3. The function generator (FG) as the source voltage and charge regulator for the input calibrator, where the product of the charge calibrator capacitance (C) with the input voltage from the source FG (V) is the simulation charge magnitude of the motions partial discharge. The formula can be written as follows.

The order of testing is as follows. First, the charge value is set through the generator function in the form of a pulse voltage and inputted to the Charge calibrator (Haefly Type 9218 signal calibrator), the charge value can be calculated using formula 3. One of the output calibrator (CC) outputs is connected to the upper CCa connector (Haefely Couling Capacitor Type 9230). The lower part of the CCa connector is connected with a commercial sensor (CS) made by Haefly with the name of the PD detector Type 9231. The CS ground side is connected to one of the Twin loop induction sensor (TLIS) inputs, and the other side is charged to the unconnected CC, which this circuit forms a closed circuit.

Momentary voltage (high frequency) that enters the circuit will be responded by CS, where the output is connected to channel 1 oscilloscope. Likewise, TLIS responds according to formula 1, whose output is connected to channel 2 oscilloscopes. Optimal results in reading the signal used a sampling rate of 200 Mega Sampling per second (MS/s). All equipment used can be controlled automatically or manually using the Labview program, and measurement results can also be stored with digital data.

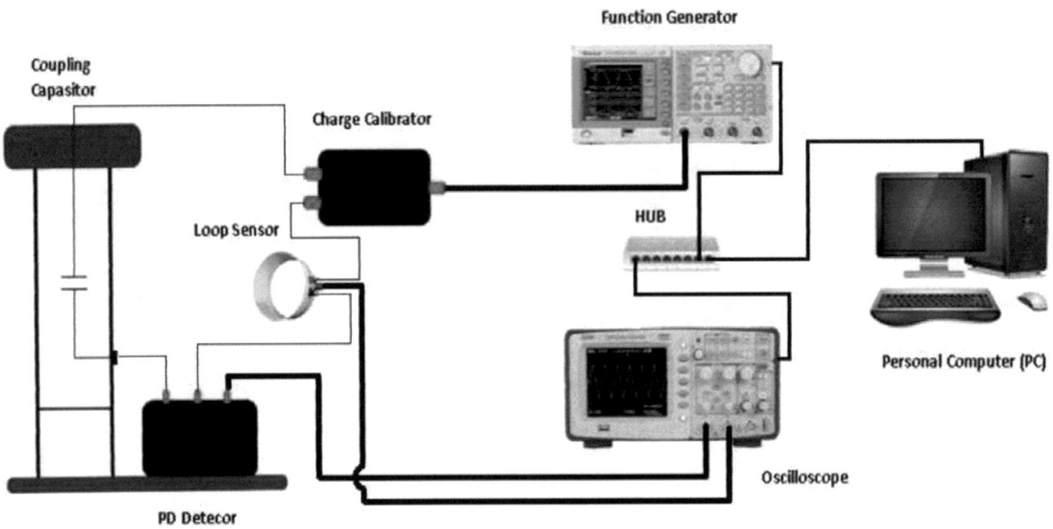

Figure 3. Circuit test loop sensor for magnitude.

3. Result and Discussion

In Figure 4 illustrates the results of commercial sensor (CS) measurements arranged in series with the TLIS sensor. The amount of charge injected to the circuit is 705 pC. From the measurement results, there is no change in the magnitude of the CS sensor even though the diameter of the TLIS sensor changes. The CS sensor waveform has the property of critical response because this wave does not oscillate; the results show a peak value, which means it has a clear polarity. TLIS sensor measurement results are marked with rectangles with dotted lines. The maximum TLIS sensor readings on the sensor diameter are 11.4 cm; the TLIS magnitude reading is 36% smaller than the CS sensor. TLIS response is earlier than CS sensor. However, TLIS pulse width is shorter compared to CS sensor. The oscillating waveform allows the magnitude value to be represented by the peak or valley value.

Figure 5 illustrates the characteristics of the linearity of commercial sensors and TLIS sensors. This graph draws an input function in the form of a charge value that is given to the sensor and produces an output response in the form of a voltage. The sensor response can convert to a charge form, but the shape of the readings or the accuracy of the measuring instrument (oscilloscope) is not visible. The results illustrate that the Commercial sensor satisfies the input given, so does the TLIS sensor has excellent linearity in taking PD values, namely at the peak or valley. The commercial sensitivity of the sensor to the reading of the given input is better than that of the TLIS sensor, but measurements are carried out using a digital oscilloscope that has high sensitivity so that this problem can resolve.

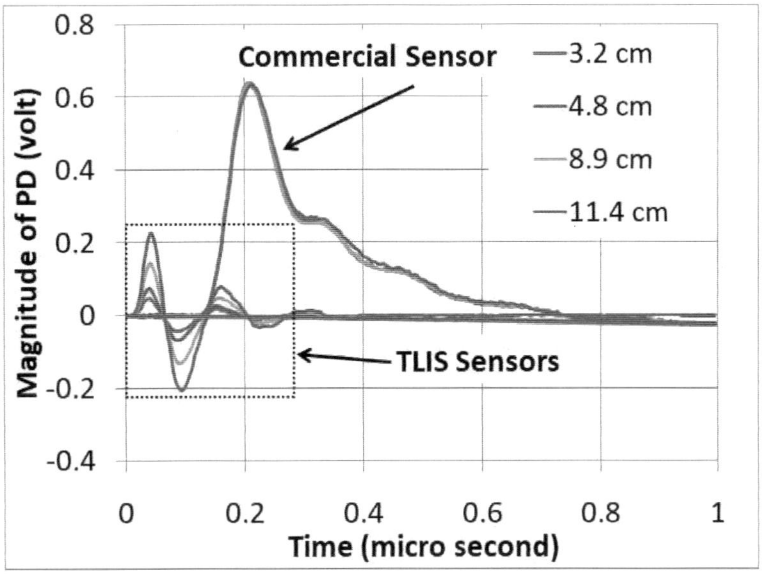

Figure 4. Respond to commercial sensors and TLIS sensors

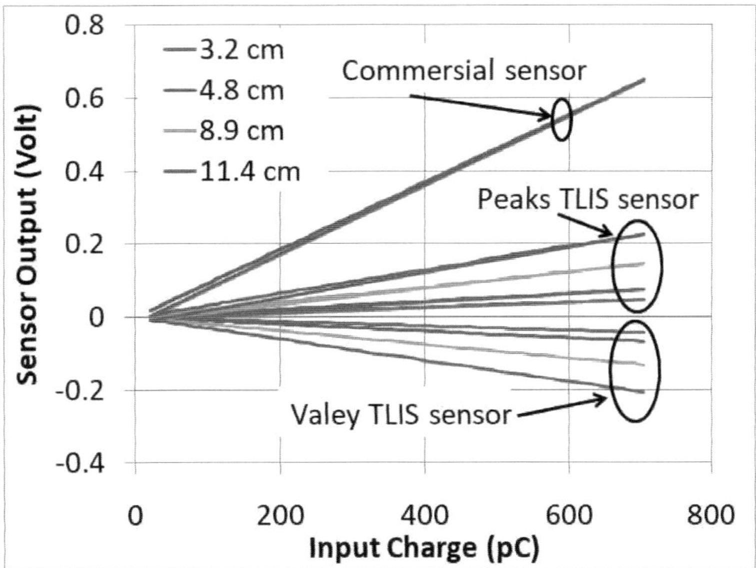

Figure 5. Linearity test of CS sensor and TLIS sensor

Figure 6 illustrates the combination of two different types of graph functions. The graph of the first type of function is a function, which is the response of the TLSI sensor output to the circle diameter owned by the sensor. The first function is marked by a small circle with the direction of the function in the form of an arrow, which is numbered 1, where the direction is to the magnitude of PD. A simple way to compare the magnitude of the peak and valley values visually (in graphical form), must be placed at the same value, the valley value has a value that is opposite to the peak value, then the value of the valley that is negative is positive so that it is easily seen that the magnitude Peak ratio is greater than the absolute magnitude Valley. The second function is indicated by a small circle with the direction of the function in the form of an arrow, which is numbered 2, where the direction is to the percentage error magnitude function. The PD magnitude value is determined at the valley value on an oscillating PD wave. The error value is calculated by dividing the difference in magnitude peak and valley and divided by the magnitude peak multiplied by 100%. From the graph, it can be said that even though the sensor diameter is magnified, an error occurs of approximately 10%. Following formula 3, it is the result of a substitution of formulas 1 and 2, to see the effect of the sensor diameter. The results show that there is conformity with the measurement results where the sensor response increases with increasing sensor diameter for both polarities.

Figure 7 shows the calculated sampling rate values at the peaks and valleys of a PD wave [5]. There is no change in the value of the sampling rate even though the diameter of the sensor is enlarged in the selection of the peak or valley to determine the magnitude of the PD. However, there is a comparison between the peak and valley sampling rates. From the figure, it can conclude that the detection of magnitude PD using the valley wave value has a lower sampling rate compared to the determination using the peak wave value.

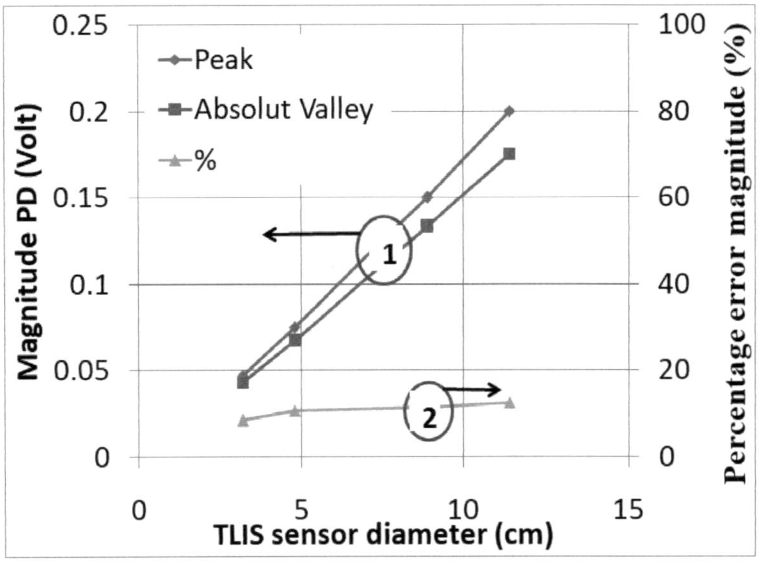

Figure 6. Magnitude PD is a function of the sensor diameter and the percentage of errors in the reading of magnitude in the magnitude valley.

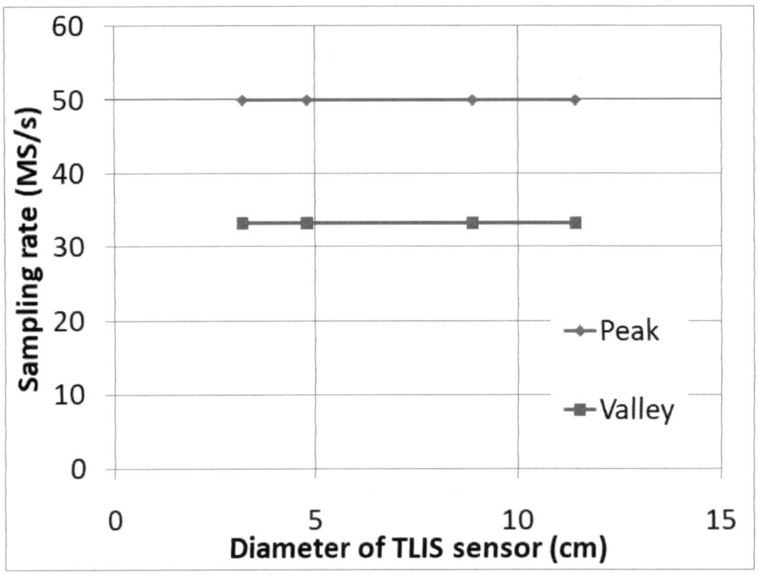

Figure 7. The sampling rate for magnitude and valley detection

4. Conclusion

TLIS sensors have magnitude sensitivity properties if the sensor diameter is enlarged, while the digitizing process is more efficient using valley waves because it has a low sampling rate.

Acknowledgments

This research was supported by "Direktorat Riset dan Pengabdian Masyarakat direktorat Jendral penguatan Riset dan Pengembangan Kementrian Riset, Teknologi dan Pendidikan Tinggi." Under the research contract number. Number: 051/SP2H/LT/DRPM/2019. The fiscal year 2019

References

[1] A. Cavallini, G. C. Montanari, and M. Tozzi, "PD apparent charge estimation and calibration: A critical review," *IEEE Trans. Dielectr. Electr. Insul.*, vol. 17, no. 1, pp. 198–205, 2010.

[2] J. Veen and P. C. J. M. van der Wiellen, "The application of matched filters to PD detection and localization," *IEEE Electr. Insul. Mag.*, vol. 19, no. 5, pp. 20–26, Sep. 2003.

[3] E. Gulski, "Computer-aided Measurement of Partial Discharges in HV Equipment," *IEEE Trans. Electr. Insul.*, vol. 28, no. 6, pp. 969–983, 1993.

[4] M. Hikita, K. Yamada, A. Nakamura, T. Mizutani, A. Oohasi, and M. leda, "Measurements of Partial Discharges by Computer and Analysis of Partial Discharge Distribution by the Monte Carlo Method," *IEEE Trans. Electr. Insul.*, vol. 25, no. 3, pp. 453–468, 1990.

[5] E. P. Waldi, A. Aulia, A. Hazmi, H. Abral, S. Arief, and M. H. Ahmad, "An optimized method of partial discharge data retrieval technique for phase resolved pattern," *TELKOMNIKA (Telecommunication Comput. Electron. Control.*, vol. 14, no. 1, p. 21, 2016.

[6] R. E. James and B. T. Phung, "Development of Computer-based Measurements and their Application to PD Pattern Analysis," *IEEE Trans. Dielectr. Electr. Insul.*, vol. 2, no. 5, pp. 838–856, 1995.

[7] A. Cavallini, A. Contin, G. C. Montanari, and F. Puletti, "Advanced PD Inference in On-Field Measurements. Part I: Noise Rejection," *IEEE Trans. Dielectr. Electr. Insul.*, vol. 10, no. 3, pp. 216–224, 2003.

[8] A. Cavallini, M. Conti, A. Contin, and G. C. Montanari, "Advanced pd inference in on-field measurements. part 2: identification of defects in solid insulation systems," *IEEE Trans. Dielectr. Electr. Insul.*, vol. 10, no. 3, pp. 528–538, 2003.

[9] J. A. Ardila-Rey, M. V. Rojas-Moreno, J. M. Martínez-Tarifa, and G. Robles, "Inductive sensor performance in partial discharges and noise separation by means of spectral power ratios," *Sensors (Switzerland)*, vol. 14, no. 2, pp. 3408–3427, 2014.

[10] J. A. Ardila-Rey *et al.*, "Behavior of an inductive loop sensor in the measurement of partial discharge pulses with variations in its separation from the primary conductor," *Sensors (Switzerland)*, vol. 18, no. 7, Jul. 2018.

[11] G. Robles, J. M. Martínez, M. Rojas, and J. Sanz, "Inductively coupled probe for the measurement of partial discharges," *Rev. Sci. Instrum.*, vol. 79, no. 5, pp. 1–5, 2008.

[12] J. M. Martínez-Tarifa, G. Robles, M. V. Rojas-Moreno, and J. Sanz-Feito, "Partial discharge pulse shape recognition using an inductive loop sensor," *Meas. Sci. Technol.*, vol. 21, no. 10, 2010.

[13] B. A. Siddiqui, P. Pakonen, and P. Verho, "Novel inductive sensor solutions for on-line partial discharge and power quality monitoring," *IEEE Trans. Dielectr. Electr. Insul.*, vol. 24, no. 1, pp. 209–216, 2017.

[14] M. Kunicki, "Comparison of capacitive and inductive sensors designed for partial discharges measurements in electrical power apparatus," *E3S Web Conf.*, vol. 19, p. 01035, 2017.

[15] M. V. R. Moreno, G. Robles, R. Albarracín, J. A. Rey, and J. M. M. Tarifa, "Study on the self-integration of a Rogowski coil used in the measurement of partial discharges pulses," *Electr. Eng.*, vol. 99, no. 3, pp. 817–826, 2017.

[16] G. F. C. Veloso *et al.*, "Detection of partial discharge in power transformers using Rogowski coil

and multiresolution analysis," *2009 Brazilian Power Electron. Conf. COBEP2009*, pp. 1006–1010, 2009.

[17] J. A. Ardila-Rey, R. Albarracín, F. Álvarez, and A. Barrueto, "A validation of the spectral power clustering technique (SPCT) by using a rogowski coil in partial discharge measurements," *Sensors (Switzerland)*, vol. 15, no. 10, pp. 25898–25918, 2015.

IOP Conf. Series: Materials Science and Engineering **846** (2020) 012035 doi:10.1088/1757-899X/846/1/012035

Accessibility Analysis of Potential Vertical Evacuation Site (PVES) from Tsunami of West Sumatra Governor's Office Escape Building in Padang.

Andi Syukri[1], Desnila Sari[1]

[1]Civil Engineering Department, Politeknik Negeri Padang

*andisyukri@pnp.ac.id

Abstract. Potential Vertical Evacuation Site (PVES) from tsunamis, namely as shelter, is an alternative evacuation solution for the coastal communities in Padang City to people from tsunami despite off to using horizontal evacuation routes. PVES building as evacuation shelter is not identified yet as a proper shelter regarding as structural, management and social impact for safety and it needs recommendation and technical data for better routes and evacuation. The accessibility of PVES's does not technically measure from the community who will evacuate to it. They just acclaimed as shelter just designed resistant from the earthquake and the location just based on the Tsunami Evacuation Map in 2011 and Stanford University research evaluation for vertical evacuation buildings. The adequacy of evacuation routes to PVES need to be confirmed and have a prior recommendation for surround community how to get there safely. Based on traffic Google Map data create a real-time simulation for people act with their activity and it will project the how they reaction if they need to evacuation to PVES promptly. This study will obtain level of travel time vulnerability, capacity and evacuation ratio for West Sumatra Governors Escape Building as a first building had been built as PVES in Padang City, West Sumatra.

1. Introduction

Padang as the the most populous population in the province of West Sumatra is known as one of the cities that has predicted as the highest risk of tsunami impact in the world. The city of Padang has two tsunami vulnerabilities namely its proximity to the Sunda fault and in particular the high level of vulnerability in this place [1]. In 2004, the Indian Ocean destroyed northern Sumatra and specifically the northernmost city, Banda Aceh. Padang faced a similar danger to the fault that caused the 2004 Tsunami. Tsunamis will change their geographical shape [2]. Padang had four times the population of Aceh before the tsunami. PVES as an aternative solution will give significant inpact for people evacuation fro tsunami. It will change the paradigm and utilize Padang's people evacuation behavior during emergency response. Many tremors is generate people evacuate massively to inland but it is not effective beacuse traffic jam and bottle neck in intersection will increase remaining time for evacuation horizontally [3]. Lack of routes as main factor cause this problem and drive vehicle during evacuation. PVES have been built in Padang to counter this problem but it is not give big contribution to community [4]. This reasearch will seek the condition on the acessibiltyof it. So, the scope of this research is

a. Number of Potential Vertical Evacuation Sites (PVES) that have been built in the city of Padang
b. The VES building is still potential in nature, yet a comprehensive study of the feasibility of access, function and strength of the structure has not been conducted
c. Many of these VES's are overlapped by not thinking about the optimum capacity to evacuate to tsunamis.
d. Utilization of real time traffic data technology by using Google Map on the traffic lane in Padang City and the path to potential VES.
e. There is no clear path and travel time for evacuation to nearby VES.

The basic contribution of this research is that it is part of a series of road maps of research on the Vertical Evacuation Infrastructure from Tsunami [5]. The results are expected to be able to assess and justify the role of potential buildings as vertical evacuation sites in Padang, both for the ratio of evacuation contributions to tsunamis, access, capacity and feasibility of the evacuation buildings themselves. This will synchronize with schemes related to disaster management technology and management.

2. Methodology

This research is a part of research road maps on Vertical Evacuation of Tsunamis. Previous studies have reviewed preliminary evacuation studies for tsunamis which have resulted in several existing buildings that can be used as evacuation sites. To measure how feasible, the evacuation route is, it is necessary to conduct a parametric study of this evacuation which then decides whether the evacuation infrastructure development uses horizontal or vertical lines. Vertical evacuation options that can be used can be in the form of TEREP (Tsunami Evacuation Raised Earth Park) in the form of artificial hills, then Evacuation Buildings (Escape Building) or better known as shelters and POBET (Pedestrian Overpasses Bridge for Evacuation from Tsunami), namely by utilizing pedestrian crossings in such a way. By designing concepts that adapt to the abilities and needs of the community and proceed into the detailed planning stages of construction and budgeting.

The proposed research stage is still in the stages of the parametric study of the study of evacuation sites in the city of Padang. Two types of evacuations, namely horizontal and vertical in the city of Padang, need to be assessed for the vulnerability and effectiveness and safety of their use. Vertical Evacuation of Tsunamis is an effective and safe alternative for communities where horizontal evacuation routes are not sufficient to evacuate to a place higher than sea level. To measure the effectiveness and safety of potential vertical evacuation sites (VES) in Padang, a feasibility analysis of the accessibility needs of potential vertical evacuations in Padang City, West Sumatra is needed. The research method for analyzing the accessibility of potential vertical evacuation sites for tsunamis is,

a. Conduct a feasibility study of all VESs in Padang based on function, ergonomics and building accessibility.
b. Sampling community data around the PVES environment to prepare community movement patterns.
c. Record data of Padang City real time traffic conditions using Google map data traffic.
d. Overlaying the feasibility of evacuation site data, community movement patterns and traffic data.
e. Conduct analysis, capacity, evacuation routes and evacuation scenarios of VES in Padang City

In this study is an analysis of potential vertical evacuation sites in the city of Padang. The sub districts covered by the study area that have vulnerability to tsunami evacuation [5] are seven districts out of 11 in Padang, including:

a. Kecamatan Koto Tangah
b. Kecamatan Nanggalo
c. Kecamatan Padang Utara
d. Kecamatan Padang Barat
e. Kecamatan Padang Selatan
f. Kecamatan Lubuk Begalung
g. Kecamatan Bungus Teluk Kabung

In the northern region is an area that has a fairly flat topography and has not enough buildings. In contrast to the city center which has many tall buildings and evacuation center and in the southern area already has a hill as a place to evacuate naturally from tsunamis. The study area has a population density of 127 inhabitants / km2 to 34072 inhabitants / km2 with a population impacting tsunami hazard 345906 inhabitants out of 927,168 inhabitants of the entire city of Padang or about 37% of the population will have an impact on the tsunami hazard as shown in the following table

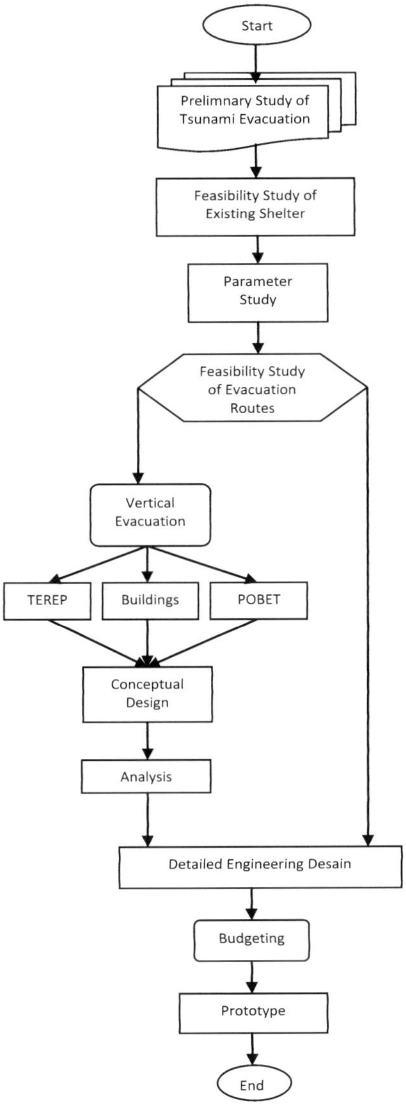

Figure 1. Reasearch Road Map of Vertical Evacuation

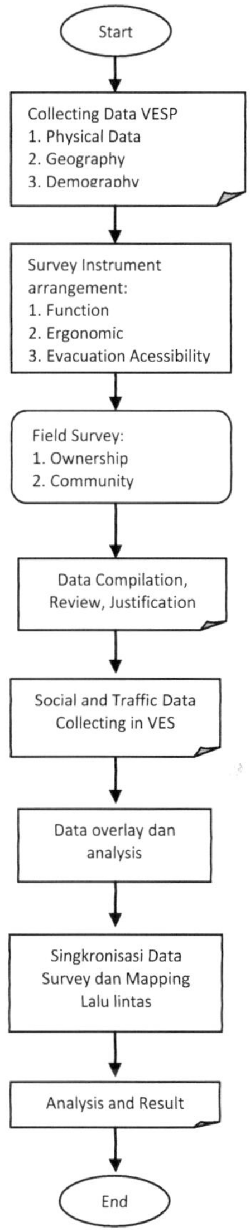

Figure 2. VES Accessibilty research flow diagram

Table 1. Potential Vertical Evacuation Site (PVES) in Padang

No.	PVES	Address	Distance from shoreline (meter)	Capcity (peoples)
1.	Escape Building Ktr. Gubernur	Jl. Jend. Sudirman No.51 Padang	600	3500
2.	Mesjid Raya Sumbar	Jl. Khatib Sulaiman	750	4000
3.	Pasar Inpres	Jl. M.Yamin, Pasar Raya	500	3750
4.	SMPN 4 Padang	Jl. Pulau Karam No.82 Padang	700	1200
5.	SMPN 25 Padang	Jl. Beringin, Belanti Timur,	800	2300
6.	Sekolah Al Azhar 32 Padang	Jl. Khatib Sulaiman	750	1000
			Total	30.550

Source: Mercy Corp Indonesia (NGO-2014)

Vertical evacuation is a function to save themselves in higher area [6]. It is not only hills (natural shelters) or multi-story buildings. Because of the study area, there are no hilly or very sloping areas then the choice is to use several buildings that have more than three story's floors or use the existing multi-story building as a vertical evacuation building. In order to reach a safe area to evacuate, the speed of people evacuating is a key factor before the tsunami comes. The Japan Institute for Fire Safety and Disaster Preparedness (1987) [7] provides an explanation that the running conditions and average walking speed in disaster evacuations are shown in the table

Table 2. Average Evacuation for Walking

Walking Condition	Average Velocity (m/s)
Accompanying Baby	1.07
Accompanying Children	1.02
Single Resident	0.948
Amount of Resident	0.751

Source: Institute for fire and average and Disaster Preparedness, 1987

To analyze the distance of refugees in this study area, use the lowest value from the table above, which has a speed of 0.751 m / s. According to Sing et al, 2010, the tsunami in the city of Padang came within 20-30 minutes and also in refugees could evacuate for 10 minutes. Then, there are three long groups of time to evacuate, 10 minutes, 20 minutes, and 30 minutes or 600 seconds, 1200 seconds, and 1800 seconds. If the speed of this group is multiplied by the refugee walking speed of 0.751 m / s, the distance of each group is 450 meters, 900 meters and 1350 meters.

3. Result and Discussion

Potential Vertical Evacuation Site (PVES) is effectively accommodating the evacuees if the site is more than 500 meter far from the shore line. It prevents uncertainty impact of tsunami effect like debris, overtopping of the waves and gigantic first wave of tsunami. Based on the simple calculation based on the distance on the Google map, It shows PVES 3 (Pasar Inpres, Pasar Raya Padang) is the closest PVES from the shore line and PVES 5 (SMP 25 Padang) as the the farthest site from the shorelines. Figure 1 shows the condition of designated PVES distance from the shorelines.

Estimate the range allowing the Capacity [8] ,which has a population equivalent to the number of people who will be accommodated, as calculated the distance allowing accommodation (L2) (semicircle). Capacity of governor escape building is about 3500 people will accommodate 6573 people/km^2 [9] in Padang Barat District. Based on the calculation, it will have the effective radius PVES 1 is about 0.81 km or 81 meter

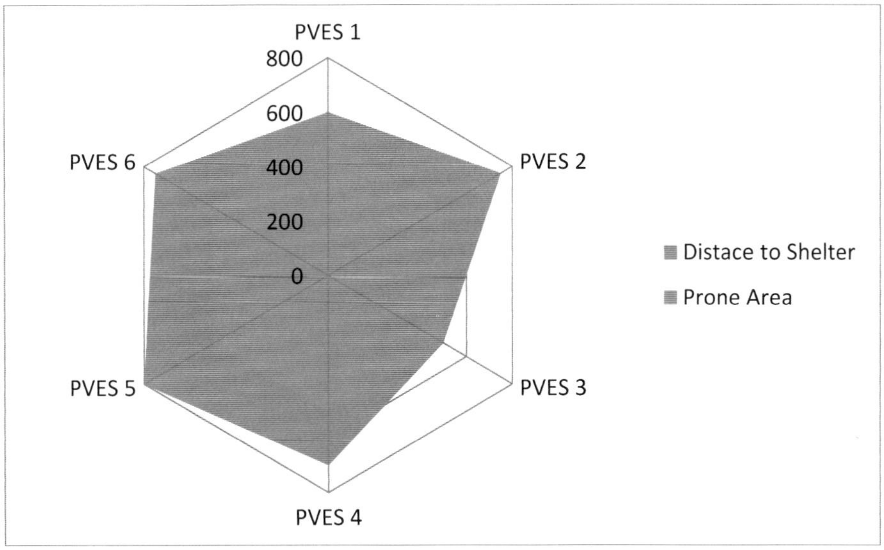

Figure 3. Distance PVES in Padang from the Shorelines

$$L2 = \sqrt{\frac{Capcity(person)/Population\_Density}{3.14} \times 4}$$
$$= \sqrt{\frac{3500/6573}{3.14} \times 4}$$
$$= 0.81km$$

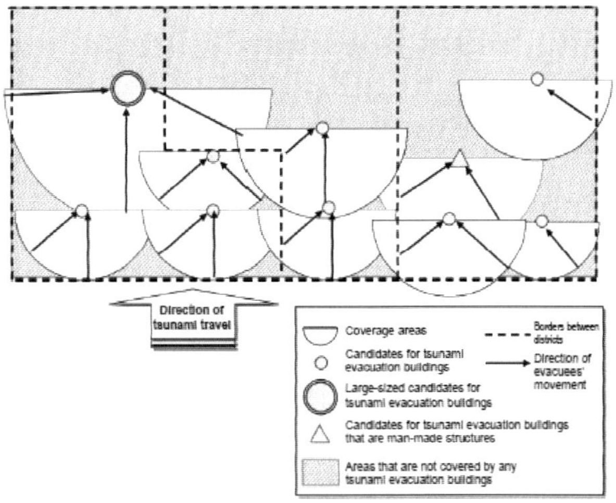

Figure 4 Image of the selection of candidates for tsunami evacuation buildings

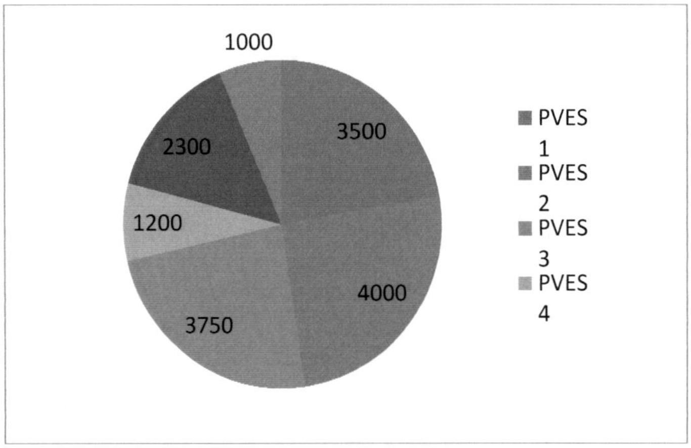

Figure 5 PVES Capacity in person in Padang

Based on figure 5, PVES in Padang have an approximate travel time 15 minute to evacuate by walking the shortest route have been chosen based on the Google Map Traffic to the site. Generally, PVES2 (SMP 25 Padang) have the longest travel time to reach the site. It is longer than remaining time tsunami reach to the shoreline of Padang. It will not cover any evacuees from the shoreline able to reach the site. Otherwise, PVES 4 (SMPN 4 Padang) has the shortest travel time for the people who can reach the site immediately.

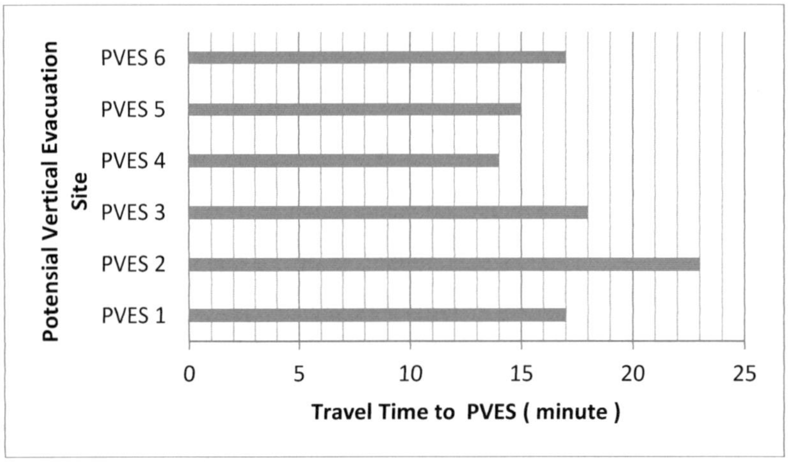

Figure 6. Travel Time Evacuation by walking to PVES in Padang

This paper will explain especially about the analysis of the evacuation route to PVES 1 (Governor's Office Escape Building) who has distance from the shoreline about 600 meters far with 3500 people capacity in the site. Based on this fact, it needs to be investigating how effective the accessibility people near by the shelter can reach the site promptly and safely to PVES 1. PVES 1 have two evacuation routes from the shoreline to the site. The Routes 1 is started from Samudra Street and continues to Ahmad Yani Street. It has 2.3 kilometers far from the shoreline and the Route 2 passes Ujung Gurun Street and it has 1.9 kilometers far from the shorelines.

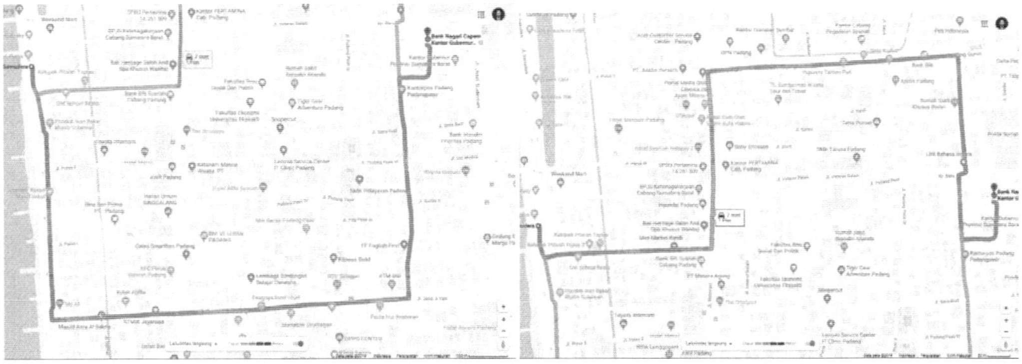

Figure 7. Route 1 and 2 for Evacuation to PVES 1

Based on analysis and data from 12 hours in 7 days it will take analysis on travel time, velocity and capacity ratio of traffic. Travel time for evacuation to PVES 1 in between 6 until 14 minute. It is still in the remaining time from tsunami arrival time to reach the shoreline of Padang. It has increasing the travel time as linear from 06.00 am until 14.00 pm but in the afternoon travel time for the people it takes 12 minute or more. Then in 21.00 am travel time the people will decreasing into normal around 8 minute for travel. Then, if we analyze based on the velocity for evacuation Route 1 may have decreasing velocity gradually start from 06.00 am until 10.00 am in the morning but have same velocity in between of 10.00 am until 22.00 pm with velocity is around 10 km/h to 12 km/h. Regarding to traffic capacity, Route 1 have the same flow for any day based on the research in 7 days. It has same condition in every day with the maximum capacity is around 17.00 pm until 21.00 pm with traffic capacity ratio around 0.6 to 0.4.

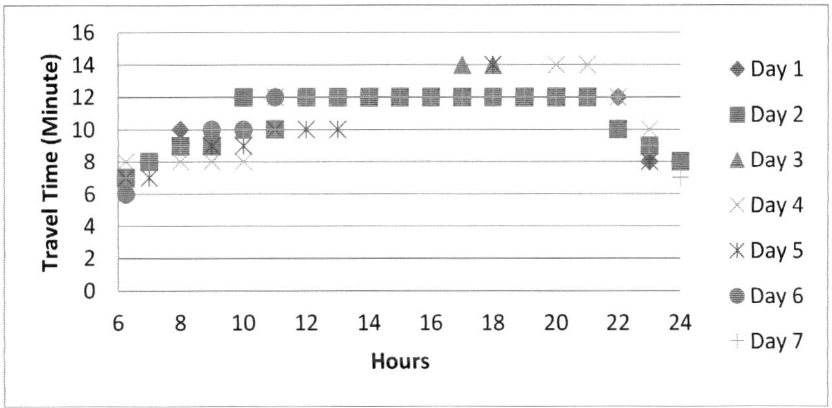

Figure 8. Route 1 Travel Time to PVES 1

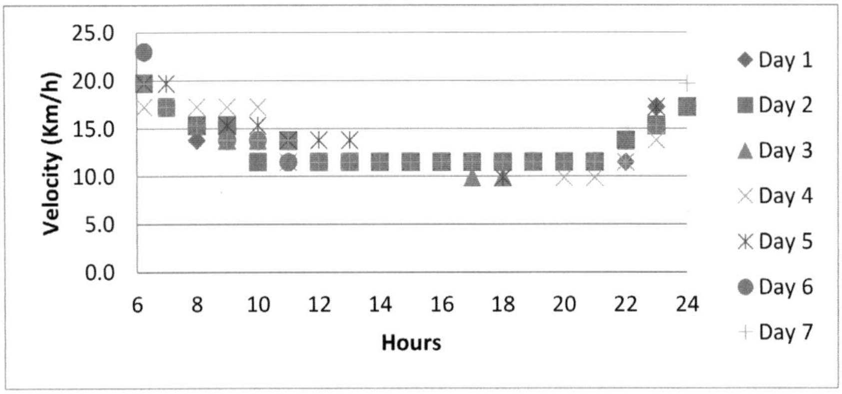

Figure 9. Route 1 Evacuation Velocity to PVES 1

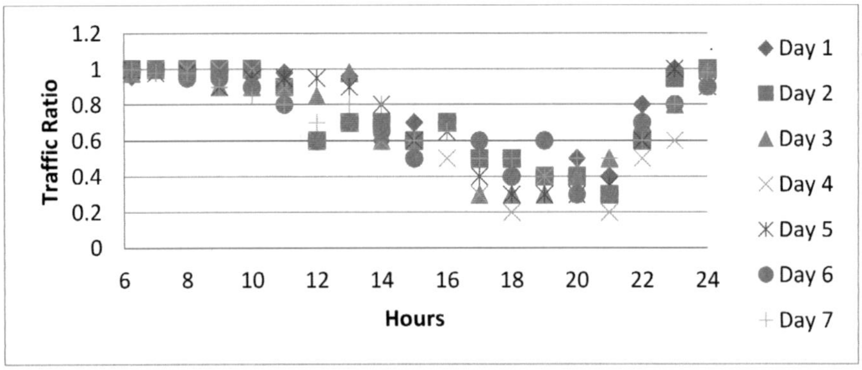

Figure 10. Route 1 Traffic Capacity Ratio to PVES 1

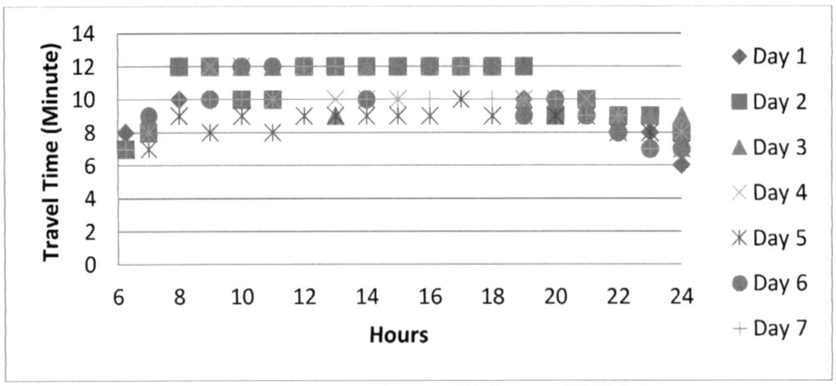

Figure 11. Route 2 Travel Time to PVES 1

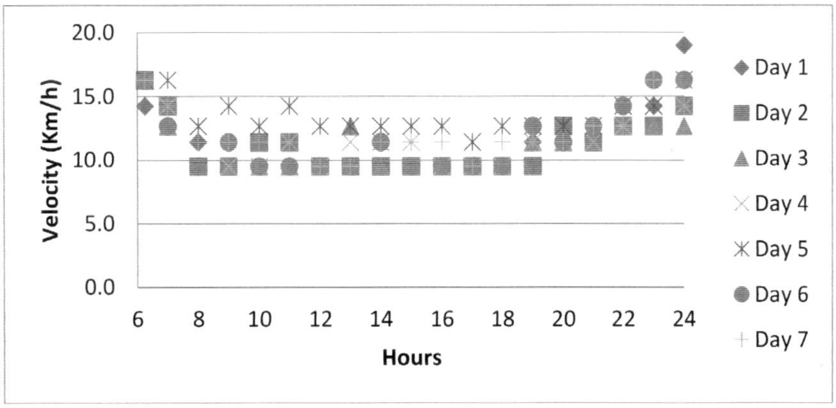

Figure 12. Route 2 Velocity Evacuation to PVES 1

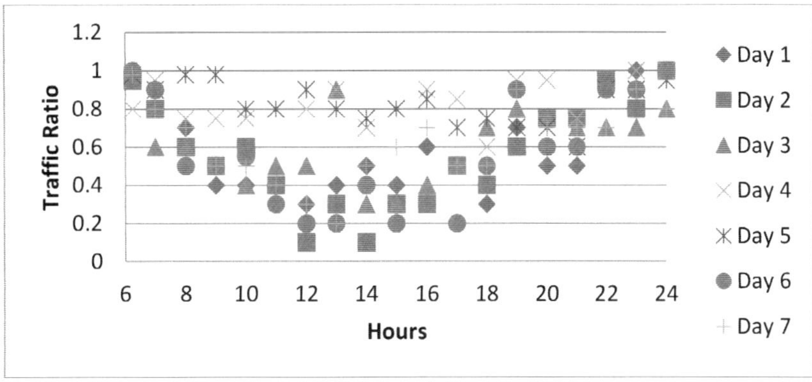

Figure 13. Route 2 Traffic Ratio

Based on analysis and data from 12 hours in 7 days it will take analysis on travel time, velocity and capacity ratio of traffic. Travel time for evacuation route 1 to PVES 2 in between 6 until 12 minute. It is still in the remaining time from tsunami arrival time to reach the shoreline of Padang. It has increasing the travel time as linear from 06.00 am until 08.00 pm but in the afternoon travel time for the people it takes less than 12 minute. Then in 19.00 am travel time the people will decreasing into normal around 6 minute for travel. Then, if we analyze based on the velocity for evacuation route 2 may have decreasing velocity gradually start from 06.00 am until 08.00 am in the morning but have same velocity in between of 08.00 am until 20.00 pm with velocity is around 10 km/h to 12 km/h. Regarding to traffic capacity, Route 1 have the discreet flow for any day based on the research in 7 days. It has same condition in every day with the maximum capacity is around 17.00 pm until 21.00 pm with traffic capacity ratio around 1 to 0.2. It means, there is a lot uncertainty traffic condition for evacuation in this route.

4. Conclusion

This analysis is a one of accessibility analysis that has been doing for PVES in Padang, West Sumatra. It presents how analysis works for the eligibility of route for evacuation to designated PVES. Governor's Escape Building is one of the representative first escape buildings that have been built in Padang. It will give a lot contribution to the community surrounding. Base on the research (Syukri A. , 2018), PVES 1 is only able to accommodate 25% of the people who will evacuate from tsunami. So, analysis intervention to seek the ability of route can evacuate people safely should be determined. Route 1 base on the travel time to evacuation takes longer than Route 2 but and also in velocity but in traffic capacity ratio Route 1 is more adequate and safe for evacuation to PVES 1.

References

[1] V. Cedillos *et al.*, "An evaluation of infrastructure for tsunami evacuation in Padang, West Sumatra, Indonesia," in *9th US National and 10th Canadian Conference on Earthquake Engineering 2010, Including Papers from the 4th International Tsunami Symposium*, 2010, vol. 6.

[2] F. Ashar, D. Amaratunga, and R. Haigh, "The Analysis of Tsunami Vertical Shelter in Padang City," *Procedia Econ. Financ.*, vol. 18, no. September, pp. 916–923, 2014.

[3] A. Syukri and Mukhlis, "Studi Jalur Evakuasi Tsunami Horizontal Di Kabupaten Padang Pariaman," *Rekayasa Sipil*, vol. 8, no. 2, pp. 1–12, 2016.

[4] M. Di Mauro, "Tsunami risk reduction for densely populated Southeast Asian cities: analysis of vehicular and pedestrian evacuation for the city of Padang, Indonesia, and assessment of interventions," *Nat. Hazards*, vol. 68, pp. 373–404, 2013.

[5] A. Syukri, "Study of Intersection Coverage Area for Pedestrian Overpasses as Vertical Evacuation from Tsunami in Padang , West Sumatra," vol. XV, no. April, pp. 1–10, 2018.

[6] F. Ashar, D. Amaratunga, and R. Haigh, "Tsunami Evacuation Routes Using Network Analysis: A case study in Padang," *Procedia Eng.*, vol. 212, no. 2017, pp. 109–116, 2018.

[7] F. E. M. A. FEMA, "Guidelines for Design of Structures for Vertical Evacuation from Tsunamis. Second Edition (FEMA P-646)," *FEMA P-646 Publ.*, no. August, 2012.

[8] A. Syukri, G. Yaldi, D. Hamid, L. Murdiansyah, A. Rozaan, and A. Roza, "Conceptual Design of Pedestrian Overpasses Bridge for Vertical Evacuation from Tsunami (POBET) in Padang City – West Sumatra," *J. Ilm. Rekayasa Sipil*, vol. XIV, pp. 1–10, 2017.

[9] B. P. Statistik, "Padang Barat dalam Angka," 2019.

IOP Conf. Series: Materials Science and Engineering **846** (2020) 012036 doi:10.1088/1757-899X/846/1/012036

Mobile Application Security Penetration Testing Based on OWASP

Alde Alanda[1], Deni Satria[2], H.A Mooduto[3], Bobby Kurniawan[4],

[123]Information Technology Department, Politeknik Negeri Padang[1]alde@pnp.ac.id, [2]deni@pnp.ac.id, [3]hamodutoo@gmail.com, [4]bobbykurniawan@gmail.com

Abstract. Evolution of smartphones and smart devices affected one of the most used operating systems on smartphones and smart device is Android. Android growth with fast and affected the growth of applications used by that operating system. That application developed by many developers and can be downloadable on the play store. besides the benefits and features that operating systems are given and support from the application can affect security to users . The purpose of this research is to know vulnerability and technics that used to find a vulnerability in Android operating systems and Android applications. In addition, to give recommendations and prevention from the vulnerability. Technics and methods that used based on research from the OWASP Foundation consisting of 10 main vulnerability in Android application that is improper platform usage, insecure data storage, insecure communication, insecure authentication, insufficient cryptography, insecure authorization, client code quality, code tampering, reverse engineering, and extraneous functionality. The results from testing of five applications downloaded from Play Store. 4 application have vulnerability based on OWASP Mobile Top Ten documentation. The OWASP documentation can give an illustration of the vulnerability that most found in Android applications from the market.

1. Introduction

Android is an operating system that is widely used for mobile or smartphone. The development of Android can be said to be very fast so that a lot of updates from the previous operating system version. Android has applications that are used to do various things, such as word or data processing, image processing, sound processing, video processing, and various other application features.

The application was developed using Java and Kotlin programming languages. The Kotlin programming language is a new, more practical programming language. However, at this time many developers have developed a framework for the creation and development of other mobile applications, both Android and IOS. The framework was developed using various programming languages and various technologies. Examples of such frameworks are React Native, Flutter, and Kivy. The rapid development of the Android operating system resulted in this operating system being widely used for mobile or smartphone platforms. This development ultimately makes the application

that comes from the developer only concerned with the function without regard to the security of the application.
Tests conducted on mobile applications since 2012, found several general categories from the client side that cause weaknesses in mobile applications, these weaknesses include insecure data storage with a percentage of 63%, insecure transmission of data with a percentage of 57%, lack of binary protection with a percentage of 92%, client-side injection with a percentage of 40%, hard-coded password / keys with a percentage of 23%, and leakage of sensitive data with a percentage of 69%.[1]

2. Literature Review

A. OWASP

OWASP stands for Open Web Application Security Project is a non-profit organization located in the United States and was established on April 21, 2004. The OWASP Foundation itself was online on December 1, 2001, and has become an international organization and supports all OWASP activities in the world. OWASP is an open community that focuses on understanding, developing, obtaining, operating and maintaining applications that can be trusted. All OWASP tools, documents, and forums are free and open to anyone interested in improving and learning application security.[2]
One product of OWASP on mobile application security is documentation of the top 10 weaknesses that exist in mobile applications under the name OWASP Mobile Top Ten. OWASP Mobile Top Ten consists of the 10 most common weaknesses found during the development of mobile applications.

Table 1. OWASP Top 10 Mobile Application Vulnerability

No	Vulnerabilities
1	Improper Platform Usage
2	Insecure Data Storage
3	Insecure Communication
4	Insecure Authentication
5	Insufficient Cryptography
6	Insecure Authorization
7	Client Code Quality
8	Code Tampering
9	Reverse Engineering
10	Extraneous Functionality

B. Penetration Testing

Penetration testing is a very useful measurement tool for finding and finding weaknesses in network infrastructure, showing how vulnerable the network is when attacked . Penetration testing has proven effective in helping to deal with security issues on the network. Penetration testing techniques are not only aimed at applications, but can also be applied to networks, and operating systems, where the main purpose is to find and then try to exploit vulnerabilities that are known or detected in previous evaluations contained in certain technologies [3].Mobile devices such as smartphones and tablets are widely used for personal and business purposes. A mobile device may carry sensitive data and becomes an easy target for cyber criminals.[4]
Penetration testing, also referred to as pentest or white hat hacking, is the process of a company hiring computer security professionals to try to break into their IT infrastructure with the intent to find where

IOP Conf. Series: Materials Science and Engineering **846** (2020) 012036 doi:10.1088/1757-899X/846/1/012036

the greatest vulnerabilities lie. Basically, a company hires security professionals to evaluate and hack into their network, servers, and services before the malicious users can do the same thing.[5] most penetration tests tend to focus on the exploitation of networks and the extraction of sensitive data, the loss of service and the inability of an organization to utilize its own wireless network can also have a significant impact on their productivity.[6]

3. Methodology

this research uses internal network penetration testing method. The steps that will be carried out in this research can be seen in the picture below

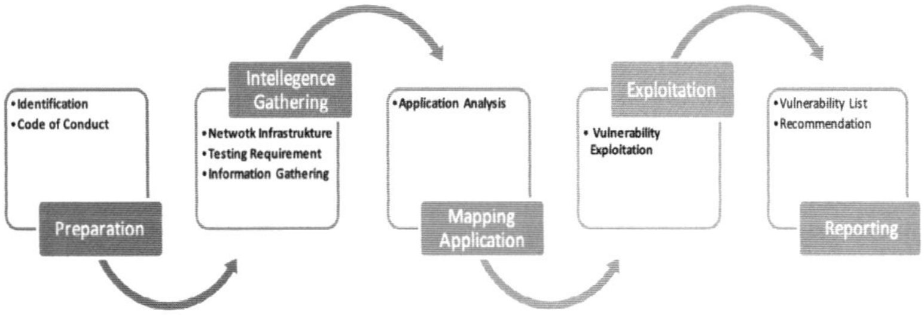

Figure 1. Methodology

1. Preparation, which is the stage that defines the scope of security testing, which includes the identification of security controls used, testing objectives, and sensitive data. In general, the preparation stage is all synchronization with the client in which there is an agreement between the examiner and the client to protect the examiner from legal actions[7]
2. Intelligence Gathering is a stage to analyze the scope and architecture of the application to get a general understanding of the application.
3. Mapping the Application, is the next step to complete the previous stage by scanning automatically and manually exploring applications. Mapping can provide a deeper understanding of the application to be tested such as entry points, data held, and other potential major weaknesses.
4. Exploitation is the stage where security testers penetrate an application by exploiting the weaknesses that have been identified during the previous process. At this stage also the determination of real weaknesses and true positives.
   At this stage, using the attack reference found in OWASP top 10 to get the type of attack most often faced by mobile applications. At this stage several attack techniques will be carried out that are likely to be carried out on mobile applications.
   The OWASP Security Knowledge Framework is intended to be a tool that is used as a guide for building and verifying secure software. It can also be used to train developers about application security. Education is the first step in the Secure Software Development Lifecycle.
5. Reporting, an important stage for the client, security testers provide weakness reports regarding the applications found and can explain the adverse effects of application weaknesses.

IOP Conf. Series: Materials Science and Engineering **846** (2020) 012036 doi:10.1088/1757-899X/846/1/012036

4. Result and Discussion

Application testing is based on documentation and methods contained in OWASP, which consists of 10 security issues in mobile applications that have been researched and are often encountered in the development of mobile applications.

Table 2. Vulnerability testing

Code	Vulnerabilities
M1	Improper Platform Usage
M2	Insecure Data Storage
M3	Insecure Communication
M4	Insecure Authentication
M5	Insufficient Cryptography
M6	Insecure Authorization
M7	Client Code Quality
M8	Code Tampering
M9	Reverse Engineering
M10	Extraneous Functionality

A. Improper Platform Usage

Testing conducted for this weakness is an abuse of one of the android features, Intent. The intent is used to move from one activity to another. Testing to be done is to do sniffing or tapping the application to get data from the intent that is processed when there is change inactivity. This can happen if the developer implements BroadcastReceiver which is used to communicate with other systems or applications, so that broadcast messages are sent through interprocess-communication and can be tapped, so developers shouldn't use BroadcastReceiver to send sensitive or confidential data.

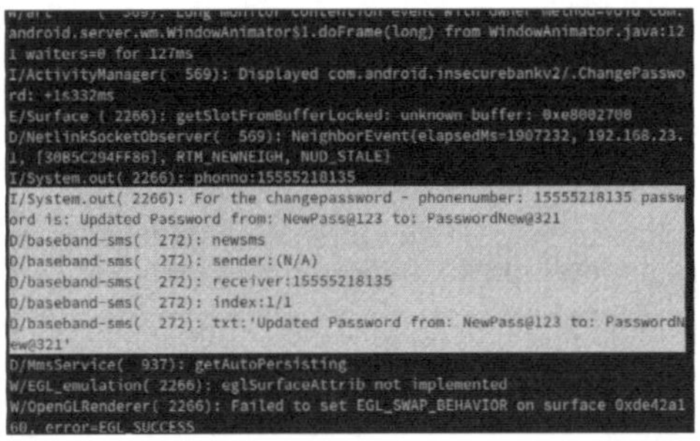

Figure 2. M1 Testing

The results of the test are in the form of data sent to the system log. The data sent can also be known by the application, so it can cause danger if the data contains sensitive information. The summary table of the test is listed in Table 3.

IOP Conf. Series: Materials Science and Engineering **846** (2020) 012036 doi:10.1088/1757-899X/846/1/012036

Table 3. Testing Result for M1

Code	M1
Vulnerability	Improper Platform Usage
Risk Level	High
Impact	Theft of sensitive information and control of application logic
Recommedation	Implement secure coding and pay attention to application development security advice from the official documentation.

B. M2-Insecure Data Storage

Testing of unsafe data storage is done in many ways, namely by testing the security of internal storage, external storage, content service providers, log files, XML data, binary data, cookies, SQL database (SQLite). The stored data can be used by attackers and malicious applications to obtain sensitive and confidential data that can result in misuse of data. One that will be discussed is regarding internal storage, which is shared preference. Developers often use MODE_WORLD_READBALE & MODE_WORLD_WRITABLE for storing data in applications that can be accessed by other applications or attackers and also do not use encryption to store sensitive information.

Figure 3. M2 Testing

The results of the above test allow a user's login account to be stolen and used by other applications for certain purposes. As for the summary of the above test described in Table 4.

Table. 4 Testing Result for M2

Code	M2
Vulnerability	Insecure Data Storage
Risk Level	High
Impact	Theft of sensitive information, privacy violations, fraud, reputation damage, and violations of external policies.
Recommendation	1. Pay attention to the data stored. 2. Use strong encryption for data storage.

C. M3-Insecure Communication

Security testing is done by analyzing the use of communication in sending data between thebackend and the client-side.

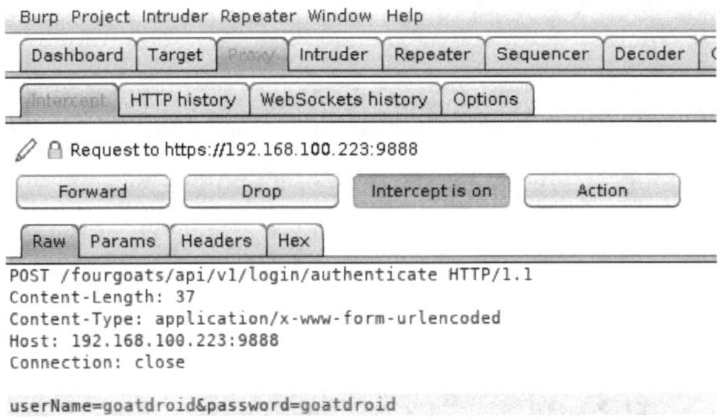

Figure 4. M3 Testing

The results of the above test are in the form of login account information that is sent in the form of the original text so that the information could be stolen and can harm the user. The summary of the tests above is described in table 5

Table 5. Testing Result M3

Code	M3
Vulnerability	Insecure Communication
Risk Level	High
Impact	Theft of sensitive information, violation of privacy, violation of user rights.
Recommendation	1. Use strong encryption on the data sent. 2. Use SSL / TLS in communication between server and client.

D. M4-Insecure Authentication

Insecure authentication is the weakness of the application regarding the user management of the application. Weaknesses that occur caused by users who are not entitled to access the application features without having to perform authentication. This can be done by bypassing the application and exploiting weaknesses in the application authentication process.

Figure 5. M4 Testing

The results of testing obtained are bypass logins that can be performed so that users can enter the main page of the application without authentification. The summary of the tests is described in table 6

Table 6. Testing Result M4

Code	M4
Vulnerability	Insecure Authentication
Risk Level	High
Impact	Identity theft, violation of data access rights
Recommendation	1. Make sure use of true on android: exported. 2. Make sure all authentication is done on the server side and not locally on the application.

E. M5-Insufficient Cryptography

The use of cryptography on data or information contained in the application is highly recommended so that the data can be safer and cannot be easily obtained by irresponsible users. Poor use of cryptography will not be able to protect the data or confidential information, because irresponsible users can easily obtain the data or information. Tests on M5 are performed to provide a description of poor cryptographic examples.

Figure 6. M5 Testing

The results obtained in testing the weaknesses of the use of cryptography are poor cryptographic implementation so it is easy to decrypt. The summary of the test is described in Table 7

Table 7 ***Testing Result*** **M5**

Code	M5
Vulnerability	Kriptografi yang tidak cukup
Risk Level	High
Impact	Identity theft, violation of data access rights
Recommendat ion	1. Avoid storing sensitive data on internal storage. 2. Use cryptographic standards the past 10 years.

	3. Use a strong cryptographic algorithm in accordance with NIST's recommendations.

F. M6-Insecure Authorization

The issue of authorization has a close relationship with the problem of authentication and is sometimes combined. Managing the wrong authorization can result in unauthorized use of the feature and can cause problems for other users. Tests on M6 are carried out to give an idea of the application of bad authorization.

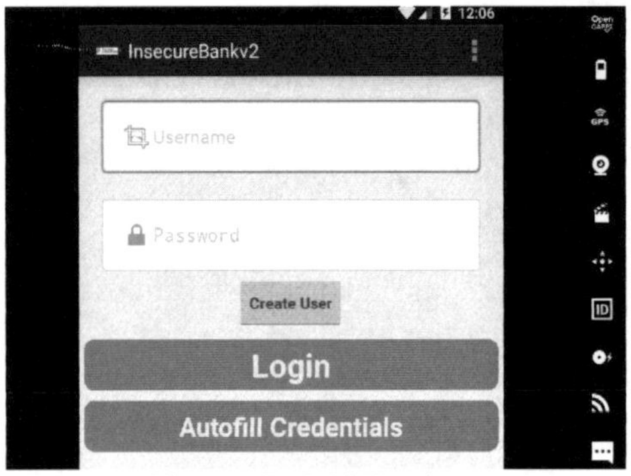

Figure 7. M6 Testing

A summary of the test results regarding authorization issues is explained in Table 8

Table 8. Testing Result for M6

Code	M6
Vulnerability	Insecure Authorization
Risk Level	High
Impact	Identity theft, violation of data access rights
Recommendation	1. Verify that access rights can only be obtained from the backend system. 2. The backend system must be able to verify every request that comes. 3. Does not use storage or source code to set permissions.

G. M7-Poor Code Quality

Poor code quality can cause errors in the application so that the application does not run as it should even stop working. Applications that cannot filter input can cause irresponsible parties to run services without the need for authentication or authorization, change the application workflow and do what they want.

Figure 8. M7 Testing

The results of testing regarding poor code quality are application services that can be run through applications or other sources aside from the application itself and do not need to authenticate even run applications. A summary of the test results regarding authorization issues is explained in Table 9

Table 9. Testing Result M7

Code	M7
Vulnerability	Poor Code Quality
Risk Level	Medium
Impact	Identity theft, violation of data access rights
Recommendation	1. Menulis kode yang baik sesuai dokumentasi dan mudah dimengerti. 2. Gunakan penyaringan terhadap semua input pada aplikasi. 3. Memperhatikan penggunaan *permission* pada aplikasi.

H. M8-Code Tampering

Code tampering is a problem that occurs with an application due to changing and tampering with the application code by another party. This can cause changes to the logic and running of the application so that unauthorized parties can benefit from changing the application code.

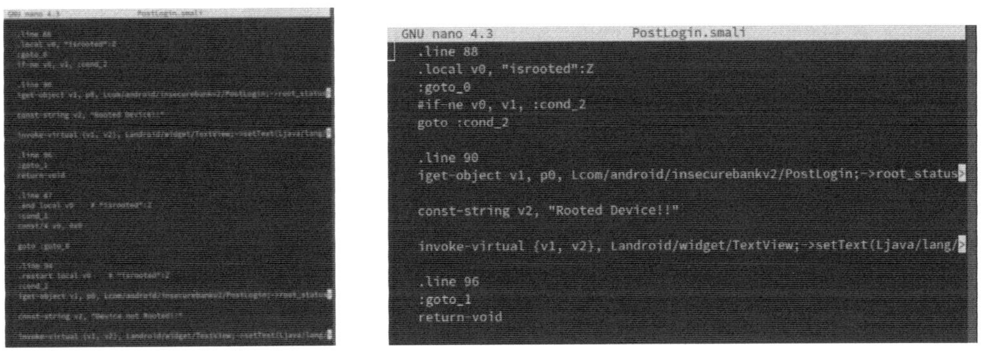

Figure 9. M8 Testing

The result of testing regarding bad code tampering is that the process and logic in the application can be changed with the desire so that it can make the application no longer detect the root of the device. A summary of the test results regarding the problem of code tampering is explained in table 10

Table 10. Testing Result M8

Code	M8
Vulnerability	Code Tampering
Risk Level	Tinggi (Parah)
Impact	Loss of income by piracy, damage to reputation.

Recommendation	1. The application must be able to check changes or additions to the code in the application. 2. Using anti-tampering.

I. M9-Reverse Engineering

Reverse engineering is a technique that is widely used to make changes to behavior and analysis of source code in the field of application development. Using reverse engineering techniques each party can get important information about the application starting from the application flow, application algorithms, cryptography used, information about backend server and other valuable information.

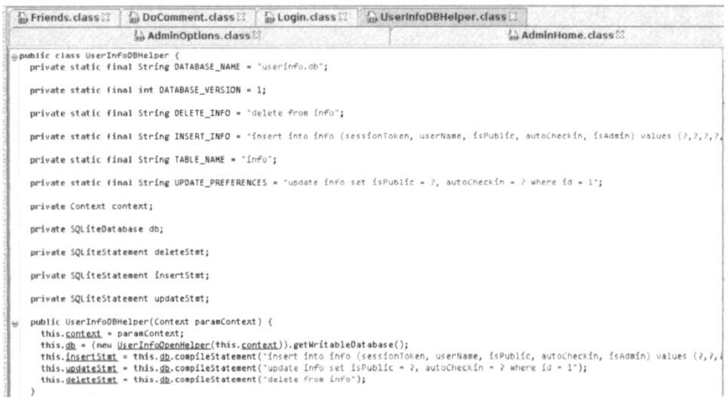

Figure 10. M9 Testing

The results of tests regarding reverse engineering are important information from the application, namely the user database management process. A summary of the test results regarding authorization issues is explained in Table 11

Table 11. Testing Result M9

Code	M9
Vulnerability	Reverse Engineering
Risk Level	Medium
Impact	Identity theft, intellectual property theft, and destruction of the backend.
Recommendation	using obfucation tool

J. M10-Extraneous Functionality

Extraneous Functionality that is usually found in applications is a backdoor. This backdoor planted by the developer on the application is developed, it could aim to facilitate the developers to improve the application when it was damaged, but can also be used to break into systems or applications for malicious purposes.

```
blic void postData(String paramlString) throws ClientProtocolException, IOException, JSONException, InvalidKeyException, NoSuchA
  HttpResponse httpResponse = new DefaultHttpClient();
  HttpPost httpPost1 = new HttpPost(DoLogin.this.protocol + DoLogin.this.serverip + ":" + DoLogin.this.serverport + "/login");
  HttpPost httpPost2 = new HttpPost(DoLogin.this.protocol + DoLogin.this.serverip + ":" + DoLogin.this.serverport + "/devlogin");
  ArrayList arrayList = new ArrayList(2);
  arrayList.add(new BasicNameValuePair("username", DoLogin.this.username));
  arrayList.add(new BasicNameValuePair("password", DoLogin.this.password));
  if (DoLogin.this.username.equals("devadmin")) {
    httpPost2.setEntity(new UrlEncodedFormEntity(arrayList));
    httpResponse = httpResponse.execute(httpPost2);
  } else {
    httpPost1.setEntity(new UrlEncodedFormEntity(arrayList));
    httpResponse = httpResponse.execute(httpPost1);
  }
```

Figure 11. M10 Testing

The test results regarding foreign functions are in the form of a backdoor in the application in the application login function so that the developer or those who know it can access the main page of the application by only using the username "devadmin" without requiring a password. A summary of the test results regarding authorization issues is explained in table 12

Table 12. Testing Result M10

Code	M10
Vulnerability	Extraneous Functionality
Risk Level	High
Impact	Unauthorized access to sensitive functions and theft of intellectual property.
Recommendation	1. Check the application configuration to find hidden functions. 2. Did not enter a trial code in the production application. 3. Check the application log so that it does not display sensitive data from the backend.

From the 10 weaknesses discussed earlier in this study also tested several applications that can be downloaded at Playstore. The results of security testing on 5 applications in PlayStore can be seen in Table 13

Table 13. Penetration Testing Result

Applications	Vulnerability									
	M1	M2	M3	M4	M5	M6	M7	M8	M9	M10
App A	-	√	-	-	-	-	-	-	-	-
App B	√	-	-	-	-	-	√	-	-	-
App C	-	-	-	-	-	-	-	-	-	-
App D	-	√	-	-	-	-	-	-	-	-
App E	-	-	-	√	-	-	-	-	-	-

5. Conclusion

In this research do some assessments for the weaknesses in mobile applications based on OWASP Mobile top 10. From the results of testing several android applications on Playstore applications have vulnerabilities. 80% vulnerability the same as on OWASP Mobile to 10. In addition to testing, it also provides recommendations for various vulnerabilities of mobile applications so it can increase application security

References

[1] D. Uttamchandani, *Handbook of Mems for Wireless and Mobile Applications*. 2013.
[2] Owasp, *OWASP Top 10 - 2013*. 2013.
[3] D. Satria, A. Alanda, A. Erianda, and D. Prayama, "Network Security Assessment Using Internal Network Penetration Testing Methodology," *JOIV Int. J. Informatics Vis.*, 2018.
[4] Y. Wang and Y. Alshboul, "Mobile security testing approaches and challenges," in *2015 1st Conference on Mobile and Secure Services, MOBISECSERV 2015*, 2015.
[5] L. Epling, B. Hinkel, and Y. Hu, "Penetration testing in a box," in *Proceedings of the 2015 Information Security Curriculum Development Conference on - InfoSec '15*, 2015, pp. 1–4.
[6] J. R. R. Brian Sak, *Mastering Kali Linux Wireless Pentesting*. 2016.
[7] R. Rahim, E. F. Armay, D. Susilo, R. F. Marta, and A. Alanda, "Cloud computing security issues and possibilities," *Int. J. Eng. Adv. Technol.*, 2019.

IOP Conf. Series: Materials Science and Engineering **846** (2020) 012037 doi:10.1088/1757-899X/846/1/012037

Bioethanol Producing Equipment Prototype Using Cassava Peel Using Microcontroller-Based Destilation Principles

Yefriadi, Nadia Alfitri, Anton Hidayat, Efendi, Nasrul, Yohanda Putera Maulana

Electro Department, [Politeknik Negeri Padang]

Abstract. Bioethanol is a biochemical liquid in the process of fermentation of sugars from carbohydrate sources using the help of microorganisms followed by the distillation process. Bioethanol can be used as a substitute for fuel oil depending on its purity. Bioethanol with 95-99% content can be used as premium substitution (gasoline), while 40% content is used as a substitute for kerosene.Cassava peel obtained from cassava plant products (Manihot esculenta Cranz or Manihot utilissima Pohl) is the main food waste in developing countries. Each kilogram of cassava can usually produce 15-20% of tuber skin. The cassava skin starch content is quite high, allowing it to be used as a source of bioethanol feedstock. In this research, the distillation process to produce bioethanol is carried out by controlling the temperature and pressure in the distillator so that the bioethanol produced can be maintained.

1. Introduction

Bioethanol is a biochemical liquid in the process of fermentation of sugars from carbohydrate sources using the help of microorganisms followed by the distillation process[1][2]. Bioethanol can be used as a substitute for fuel oil depending on its purity. The use of bioethanol is possible as a substitute fuel for gasoline because of ethanol-like characteristics of gasoline (gasoline)[3]. In addition, bioethanol is environmentally friendly because it has a relatively clean combustion result. This is because in the chemical structure there is an inherent oxygen atom in it. This will help improve the combustion reaction between ethanol and air.

Bioethanol is so popular in many countries, in the US for example that currently uses more than 52 billion liters of bioethanol as a fuel mixture annually[4]. however, the cost of bioethanol production is high compared to petroleum-based fossil fuels[5][6]. Currently, large scale fuel ethanol production is mainly based on sugar containing raw materials (i.e. sugarcane) and starch grains (i.e. corn, wheat and cassava) which are not desirable due to their food and feed value[7]. Lignocellulosic biomasses such as agriculture residues, woody biomass, and algae, industrial and municipal solid waste are the potential resources for bioethanol production worldwide[8]. They are not only renewable, but also low cost, abundant and sustainable. Rice straw, wheat straw, cassava peel, corn straw, and sugarcane bagasse are the major agricultural residues, which are suitable for large scale bioethanol production in terms of quantity of biomass available [9][10]. Moreover, starchy industrial bio- masses such as waste from starch processing factories, potato food factories, and beverage and brewery factories are promising feedstock for high yield large scale production in tropical locations [11][12].

Ethanol is traditionally produced from feedstock high in sugar and/or starch content. A third possible feedstock is lingo-cellulose. These obsolete feedstocks are attractive, but the technology to convert cellulosic material to ethanol (sometimes referred to as 'second generation') is not yet commercially available. Most used feedstocks for fuel ethanol are wheat, corn, sugar cane and sugar beet. The sugars can be fermented to ethanol, while starch first has to be hydrolyzed to obtain free sugars. Next, the sugars are fermented to ethanol which is followed by a purification step yielding pure ethanol[13].

The process of extracting starch from cassava is a well-known technology. Cassava has been used as source of starch for decades. Cassava is high in starch content (70 – 85%, dry base / 28 – 35% wet base) and the starch from cassava is of a high quality compared to other starch sources. Cassava starch is used as raw material in many indust ries, among which paper-, food- and texti le indust ries. Also the technology of producing ethanol from starch is internationally well –developed[13].

Cassava peel obtained from cassava plant products (Manihot esculenta Cranz or Manihot utilissima Pohl) is the main food waste in developing countries. Each kilogram of cassava can usually produce 15-20% of tuber skin. The cassava skin starch content is quite high, allowing it to be used as a source of bioethanol feedstock[14]. In this research, the distillation process to produce bioethanol is carried out by controlling the temperature and pressure in the distillator so that the bioethanol produced can be maintained.

The project BEST, Bioethanol for Sustainable Transport, deals with the introduction and market penetration of bioethanol as a vehicle fuel, and the introduction and wider use of ethanol vehicles and filling stations with ethanol on the market. During the project more than 70,000 ethanol vehicles and ethanol buses have been put in operation, evaluated and demonstrated. More than 300 pumps for high blend ethanol, E85 and ED95, have been opened at filling stations and low blends with petrol and diesel have been developed and tested. Through BEST the participating cities and regions aimed to prepare a market breakthrough for ethanol vehicles and for bioethanol as a fuel and also to inspire and obtain followers. During the project several incentives promoting bioethanol vehicles and bioethanol fuels have been introduced locally and sometimes also nationally. Also several sever barriers to further market penetration was faced and in some, but not all, cases overcome. The participating cities/regions are: BioFuel Region (SE), Brandenburg (DE), Somerset(UK), Rotterdam (NL), Basque Country and Madrid (ES), La Spezia (IT) Nanyang (China) Sao Paolo (Brazil) Co-ordinating City is Stockholm (SE). The project is co-financed within the 6th framework; Sustainable Energy Systems/Alternative Motor Fuels: Biofuel Cities.The project started in January 2006 and end of 2009[15].

E85 is a fuel blend that consists of 86 weight per cent anhydrous bioethanol, 11,6 weight per cent petrol and the remaining percentages being two types of denaturants and colouring. Flexifuel vehicles running on E85 can make a significant contribution towards the achievement of EU climate targets whilst supporting economic development in agriculture. Energy security and the resilience of transportation and the economy will be increased by diversifying the range of fuels on offer in the EU. E85 is a liquid fuel that can be used in dedicated or converted flexifuel vehicles, making it an ideal fuel type for a transport system based on the supply of liquid fuels. As flexifuel vehicles can operate on blend of bioethanol and petrol, as mentioned above, they are appropriate during the transition to a new transport system as users can change their fuel choice according to the availability of fuel. Depending on how bioethanol is produced, use of E85 in converted vehicles can significantly reduce emissions of fossil carbon dioxide (CO2), further details in BEST D9.21 Report on life cycle greenhouse gas impacts of ethanol supply chains at BEST sites, (2009) and the report BioEthanol for Sustainable Transport – Results and recommendations from the European BEST project, Final report, (2009)[15].

2. Methodology

The purpose of this study was to obtain bioethanol from cassava peel. By maintaining the stability of temperature and pressure in the distillation chamber during the distillation process, it is expected that the bioethanol produced has high levels. The equipment used is temperature sensor, microcontroller, pressure sensor, level sensor and so on, as can be seen in Figure 1.
Bioethanol feedstock is made using cassava peel which is dried and finely ground into flour to be fermented to obtain feedstock of bioethanol. The feedstock is then distilled, during the distillation process the temperature and pressure in the distillation chamber is controlled.

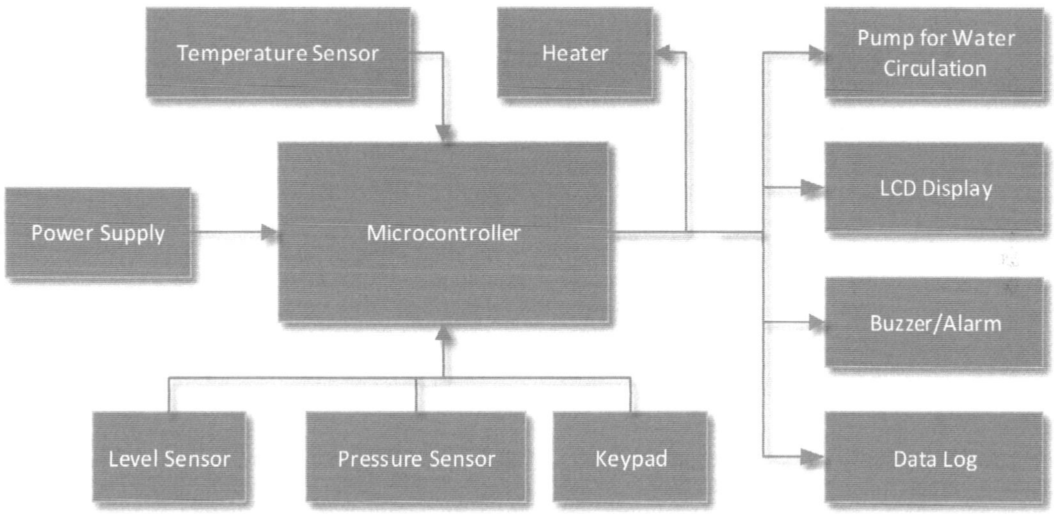

Fig 1. Block system diagram.

The distillation room is heated using an electric heater. Microcontroller accepts input from temperature sensor, level sensor and pressure sensor. Temperature sensor functions to detect the temperature in the distillation room. The level sensor functions to detect the lowest level of ethanol raw material in the distillation chamber, so that it can be known whether the raw material is still there or it is almost gone, while the pressure sensor is to determine the pressure in the distillation chamber. Keypad is a human machine interface, with this keypad the user can enter the desired temperature data in the distillation room. Pump for water circulation serves to cool the ethanol vapor in the pipe coming out of the distillation chamber so that the ethanol vapor becomes ethanol liquid. LCD display functions to display temperature and pressure data in the distillation chamber and the time the device is used. Buzzer or alarm serves to give a warning if an unexpected rise in temperature occurs, the buzzer will also sound if the raw material in the distillation chamber runs out. Data log functions to store time, temperature and pressure data in the distillation chamber.

The prototype of the distillation apparatus made in this study can be seen in Figure 2. The description of the picture is as follows; 1. control panel, 2. distillation chamber, 3. Cooling water, 4. Water cooling fan, 5. Bioethanol output hose.

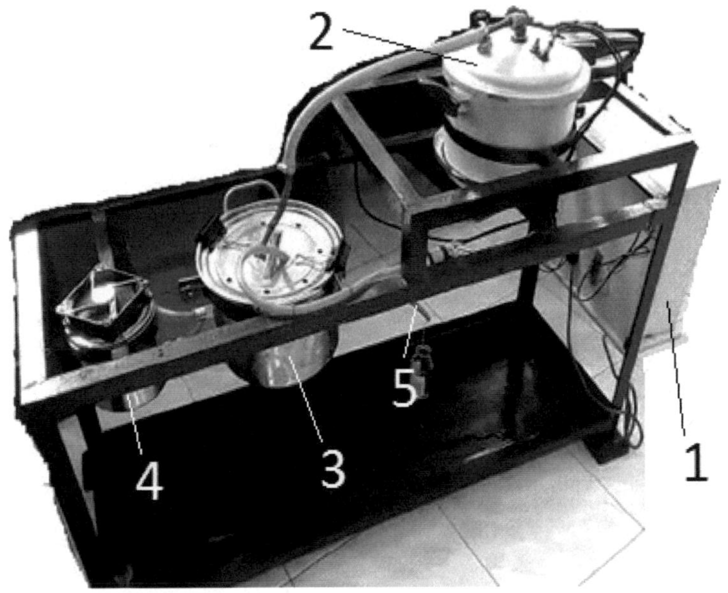

Fig.2 The prototype of the distillation apparatus

3. Result and Discussion

Distillation equipment testing was carried out with two samples of 100ml and 200ml fermented cassava skin waste liquid. From the distillation process, 40 ml and 60 ml of bioethanol were obtained. The test is carried out for 10 minutes and 16 minutes. In table 1, the test results can be seen.

Table 1. Testing results of the bioethanol

No	Feedstock of bioethanol from cassava peel (ml)	bioethanol results by the distillation process (ml)	temperature (^{0}C)	duration (menit)
1	100	40	90	10
2	200	60	90	16
3	500	260	90	30

As can be seen in table 1, using this tool takes 30 minutes to get a quarter liter of bioethanol, it can be estimated that it will take around 1 hour to get one liter of bioethanol if using this tool.

Visually, the results of bioethanol distillation can be seen in Figure 2a and 2b, Figure 2a when bioethanol has not gone through the distillation process, while Figure 2b is bioethanol which has passed the distillation process.

IOP Conf. Series: Materials Science and Engineering **846** (2020) 012037 doi:10.1088/1757-899X/846/1/012037

(a) (b)

Fig.2 a. Feedstock of bioethanol b.Bioethanol

Figure 3, shows the relationship between time and temperature rise, from the graph shows that it takes about 25 minutes for the distillation device to reach 90^0 C, then the temperature in the distillation chamber is kept at 90^0C. The distillation has a capacity of 4 liters and the heater used to heat it is with a power of 350 watts.

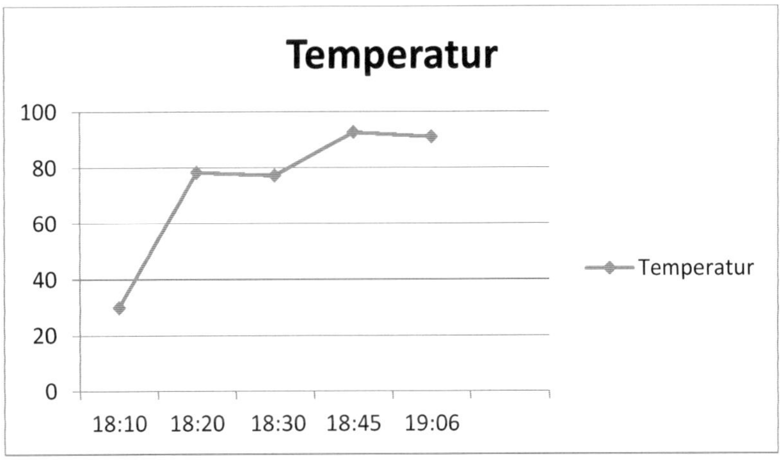

Fig 3. Relationship between time and temperature of distillation process

Furthermore, by using a alcohol meter, measurements of bioethanol grade produced from the distillation tool, and the results obtained by 60%.

4. Conclusion

From the tests conducted, conclusions can be made as follows :

1. The increase in temperature in the distillation chamber takes 35 minutes to reach 90^0C.
2. The temperature control system works quite well, the temperature can be kept constant at 90^0C

3. The level of bioethanol produced is quite good at 60%, but cannot be used as fuel because to be mixed with gasoline the ethanol purity grade must reach 99.6% (fuel grade ethanol). This tool will be developed further in order to produce bioethanol with grade above 90%. by varying the temperature of distillation, high ethanol grades will be obtained.
4. By using this tool. it is estimated that it will take around 1 hour to get one liter of bioethanol.

References

[1] K. Waldron, *Bioalcohol production Edited by*, no. 3. 2010.

[2] F. Applications, *The Biotechnology of Ethanol*. 2000.

[3] B. Murachman, D. Pranantyo, and E. Sandjaya, "Study of Gasohol as Alternative Fuel for Gasoline Substitution : Characteristics and Performances," vol. 3, no. 3, pp. 175–183, 2014.

[4] R. K. Niven, "Ethanol in gasoline: Environmental impacts and sustainability review article," *Renew. Sustain. Energy Rev.*, vol. 9, no. 6, pp. 535–555, 2005.

[5] R. F. A, *Accelerating Industry Innovation - 2012 Ethanol Industry Outlook*, 0306 ed. 2012.

[6] et al. Banerjee S, Mudliar S, Sen R, Giri B, Satpute D, Chakrabarti T, "Com- mercializing lignocellulosic bioethanol: technology bottlenecks and possible remedies," *Biofuels Bioprod Biorefining*, vol. 4:7793, 2010.

[7] Ó. J. Sánchez and C. A. Cardona, "Trends in biotechnological production of fuel ethanol from different feedstocks," *Bioresour. Technol.*, vol. 99, no. 13, pp. 5270–5295, 2008.

[8] A. Limayem and S. C. Ricke, "Lignocellulosic biomass for bioethanol production: Current perspectives, potential issues and future prospects," *Prog. Energy Combust. Sci.*, vol. 38, no. 4, pp. 449–467, 2012.

[9] N. Sarkar, S. K. Ghosh, S. Bannerjee, and K. Aikat, "Bioethanol production from agricultural wastes: An overview," *Renew. Energy*, vol. 37, no. 1, pp. 19–27, 2012.

[10] C. A. Cardona, J. A. Quintero, and I. C. Paz, "Production of bioethanol from sugarcane bagasse: Status and perspectives," *Bioresour. Technol.*, vol. 101, no. 13, pp. 4754–4766, 2010.

[11] et al. Kosugi A, Kondo A, Ueda M, Murata Y, Vaithanomsat P, Thanapase W, "Production of ethanol from cassava pulp via fermentation with a surface engineered yeast strain displaying glucoamylase," *Renew Energy*, vol. 34:1354–8, 2009.

[12] S. Linggang, L. Y. Phang, M. H. Wasoh, and S. Abd-Aziz, "Sago pith residue as an alternative cheap substrate for fermentable sugars production," *Appl. Biochem. Biotechnol.*, vol. 167, no. 1, pp. 122–131, 2012.

[13] K. Kuiper, L., Ekmekci, B., Hamelinck, C., Hettinga, W., Meyer, S., Koop, "Bio-ethanol from Cassava," *Proj. number PBIONL062937*, no. November, pp. 1–13, 2007.

[14] Z. Abidin, E. Saraswati, and T. Naid, "Bioethanol Production from Waste of the Cassava Peel (Manihot esculenta) by Acid Hydrolysis and Fermentation Process," no. March, pp. 2–6, 2016.

[15] S. by the European and European, *The BEST experiences with distribution of bioethanol for vehicles*, no. 4. 2010.

IOP Conf. Series: Materials Science and Engineering **846** (2020) 012038 doi:10.1088/1757-899X/846/1/012038

Analysing Motorcyclist Characteristics and Parking Behaviours towards Different Parking Schemes

Gusri Yaldi[1*], Imelda M. Nur[2], Apwiddhal[1], B. Army[1], Syaiful Amri[1], Syaifullah Ali[1], Wisafri[1]

[1]Civil Engineering Department, Politeknik Negeri Padang
[2] Administration Businesses Department, Politeknik Negeri Padang

*Corresponding Email

Abstract. Unlike developed countries, traffic characteristics in developing countries are dominated by passenger cars and also motorcycles. For example is Indonesia where more than 70% of transport mode share is contributed by motorcycles and a number of the riders are below age indicating poor law enforcement. This figure is suspected to worsen the transport negative impacts like road accidents as well as congestion level due to poor behaviour of motorcycle riders. Therefore, any efforts to reduce the trip number by mode of motorcycles on the road as well as improving their rider behaviours are necessary. For example is by managing motorcycle parking facilities like in this research. The results of this study suggest motorcycles were used as the primary transport mode for low income people mainly for studying and working. Further, motorcyclists tend to park for long time period with parking areas located about 50 meters from the activity centres. Off-street parking facilities are preferred than on-street parking. These findings are expected could be a considerable contribution for policy makers in minimizing the negative impacts of transport, in particular related to motorcycle usage.

1. Introduction

One of the emerging issues from transport sector in developing countries especially in Asia is the high number of road accident involving motorcycles like in Malaysia [1], Cambodia [2], Thailand [3] as well as in Indonesia [4]. According to World Health Organization (WHO), more than 1.2 million road users die annually due to road accidents. It resulted in a massive impact towards nation economic building where the estimated financial losses contributed by road accident could reach six per cent of GDP as reported by Gwilliam [5] in Ma et al. [6]. It also negatively impacts the health sector where road accidents are suspected as the main cause of death to the community within age of 15-29 years [7]. Number of death on road was reported beyond 90% occurred in low and mid income level countries, including Indonesia were the fatality rate is as double as developed countries. Particularly in Indonesia, more than 70% of road traffic accidents involved motorcycles [8] and the highest victims were at age range (15-24) years [9].

Road accident number in Indonesia has been showing an increasing trend since 2001 as illustrated by Figure 1. Based on its casualty, number of road users die on road is showing a more obvious increasing trend. In average, about 18% was killed while seriously and slightly injured were about

21% and 61% respectively. The highest recorded accident number occurred in 2012 where nearly 30000 people die on road or in average more than 80 people die on road on the daily basis [10]. The latest data in 2016 suggest the pattern remain the same (see Figure 1).

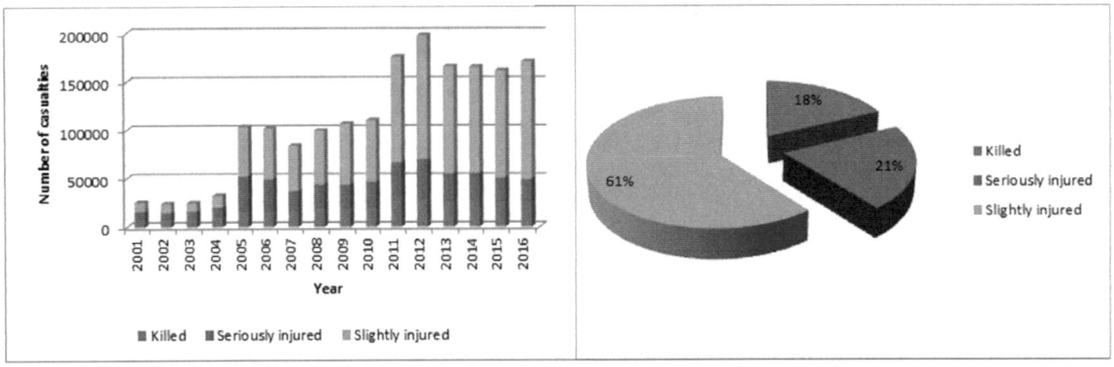

Figure 1. Road accident trend in Indonesia

Based on transport modes, the highest percentage belongs to motorcycles [7, 8]. Figure 2 suggests number of road accidents involving motorcycles is in average about 73% followed by car and truck with 15% and 9% consecutively. This tendency may be related to the motorization trend in Indonesia where all transport modes such as Motorcycle, Car, Bus and Truck are experiencing a growing number. Motorcycles contribute to the transport market share about 81% and remain the highest one. It is followed by cars and trucks about 11% and 6% each while buses are the smallest one with 2% [10].

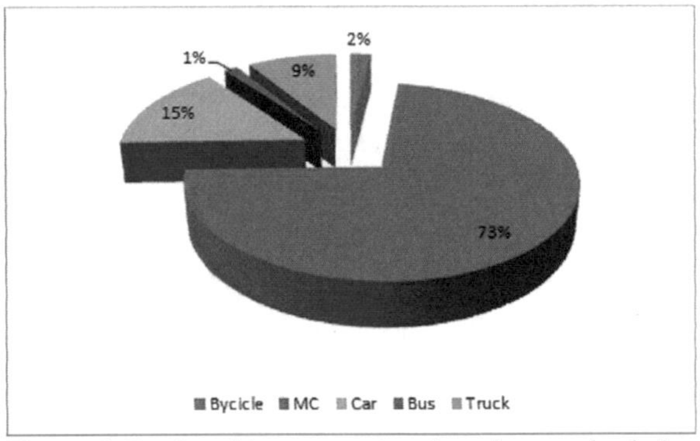

Figure 2. Road accident percentage based on mode choice

In case of urban area and road space usage, private cars are often blamed for the congestion occurrence, fossil fuel crisis of as well as air pollution and hence many researches have been undertaken in order to determine the solution for example is by increasing transport supply facilities as well as controlling its demand [11]. Including in this strategy is managing parking supply as well as increasing public transport usages. However, it becomes unique in case of negative impacts from transport sector due to motorcycle usages. It requires much less road, consumes insignificant petrol and produces less pollutant than private cars. In term of transport cost, it is much cheaper compared to private cars.

These situations tend to increase trip number conducted by using motorcycles and hence seem to be unstoppable. In contrast, motorcycles were found to involve in majority of road accidents and suspects as the main cause once accidents occurred [12]. Therefore, major efforts must be devoted in managing motorcycle usages so that less trip number conducted by this transport mode. Consequently, number of motor cycles on road could be minimized and number of traffic accident involving motorcycles could be gradually reduced. This research is aimed at analysing the behaviour of motorcycle users towards different virtual parking schemes and expected could help the transport authority in developing effective motorcycle parking schemes in order to reduce trip number undertaken by motorcycles.

2. Methodology

This research was conducted in Padang city, West Sumatra where the population of motorcycles contributed about 70% of total motor vehicles [13] and road accident number was dominated by motorcycles. The current practice of on-street parking for motorcycle is still poorly managed, and to certain extends it is illegally operated and causes disadvantages to both society and local government. Figure 3 illustrates the current practices of motorcycle parking in Padang city.

Figure 3. Existing motorcycle parking portraits in Padang city

Meanwhile, the behaviour of motorcyclists towards different virtual parking schemes was obtained by using SP survey. There were three virtual parking schemes were proposed to the motorcyclists as depicted by Figure 4. This scheme is the same one as used by Yaldi et al. [14] in studying the behaviour of private car users related to different parking facility options. The first scheme offers on-street parking with unlimited time parking duration and the parking cost is fixed. Parking area is equipped with clear parking signs as well as automatic parking meter machine.

The second scheme is also on-street parking with similar facilities; however it uses progressive parking cost with additional parking fee after the first hour. This scheme is also used parking inspectors. The last one is off-street parking scheme and the cost is progressive with better security facilities. More details can be seen in Figure 4.

The data was collected from the motorcyclist by means of home interview survey using Revealed Preference (RP) and Stated Preference (SP) survey forms. The respondents are limited to the motorcycle users only. RP was used in order to gather information regarding the road user socio-economic and socio-demographic data as well as daily trip characteristics such as age, occupation, education, monthly income, household size, motorcycle ownership, trip purpose, trip mode, parking location, parking duration, and parking cost.

Regular parking cost in Padang city varies based on parking locations, motor vehicles types, and peak & off-peak periods. For motorcycles, the minimum parking fee is Rp.2000 or about 0.13USD. However, the minimum motorcycle parking fee used in this research is Rp.3500 (0.23USD) since the existing parking fee for motorcycles is considered very cheap especially for on-street parking where the parking duration is unlimited.

IOP Conf. Series: Materials Science and Engineering **846** (2020) 012038 doi:10.1088/1757-899X/846/1/012038

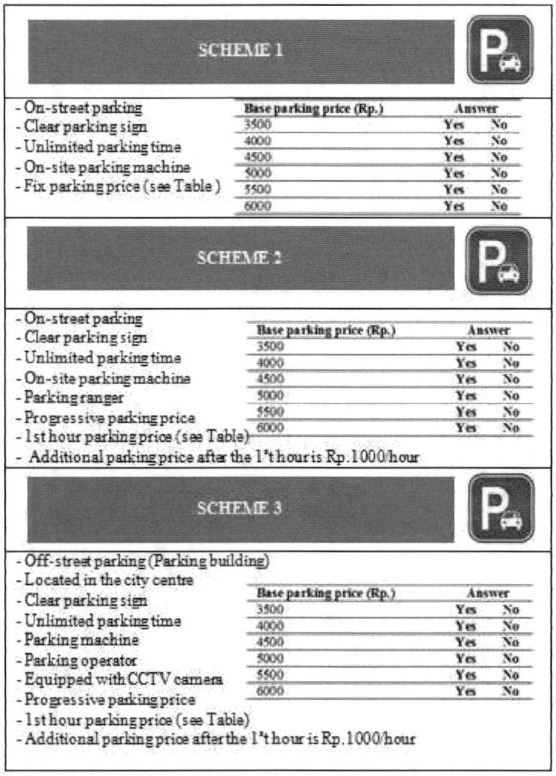

Figure 4. Proposed virtual parking schemes

3. Result and Discussion

Based on the RP data, the percentage of male and female respondent is about 54% and 46% respectively and hence it can be assumed that the data represent the relatively equal numbers of respondent based on gender. This is important since the motorcycle is not only used by men but also by women for different purposes. In addition, a number of motorcycles accidents involved female riders [15] and there are different behaviours in using motorcycles between male and female riders [16]. Figure 5 suggests that more than 90% of respondent ages are above 18 years old where the minimum age allowed to ride a motorcycle in Indonesia is 16 years [17]. Therefore, almost all of motorcyclists involved in this researched were assumed to have fulfilled the requirement to ride a motorcycle.

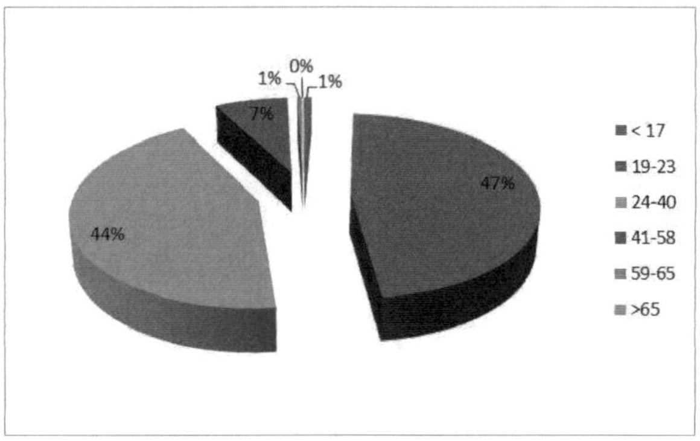

Figure 5. Respondent percentages based on ages

Furthermore, motorcycle usages were mainly for study and work trip purposes. The average monthly income was found nearly Rp.2 million or about 133USD (sees Figure 6). This number suggests the motorcycles were used by mostly low income people [18]. Yet, about 13% of motorcycles were used by medium and high income respondents. It can be seen, although majority of motorcycles were used by poor people, it is still used by medium and rich people with likely different reasons. Based on these characteristics, it could be assumed that motorcycle users were mainly poor people with the main usage is for working and studying purposes. Hence, the related stakeholder needs to seriously address this information in minimizing trip number by mode of motorcycle. For example is by providing school or worker buses.

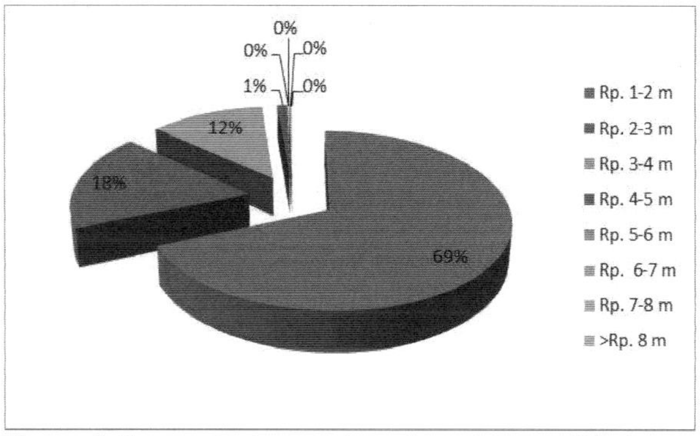

Figure 6. Respondent percentages based on monthly income

The average household size is 4 people and the motorcycle ownership is less than two for each household. It was found that there were two trips conducted by the respondent every day, and the frequency for parking also twice a day. More than 80% motorcyclists parked their motorcycles in off-street parking facilities with an average parking duration of nearly 100 minutes (see Figures 7 & 8, and Table 1). Nearly 70% of motorcyclists parked the vehicles longer than one hour and only about 11% motorcyclists parked for a maximum duration of 30 minutes. These data suggest that majority of motorcyclists parked for a long duration and one third of them parked longer than three hours.

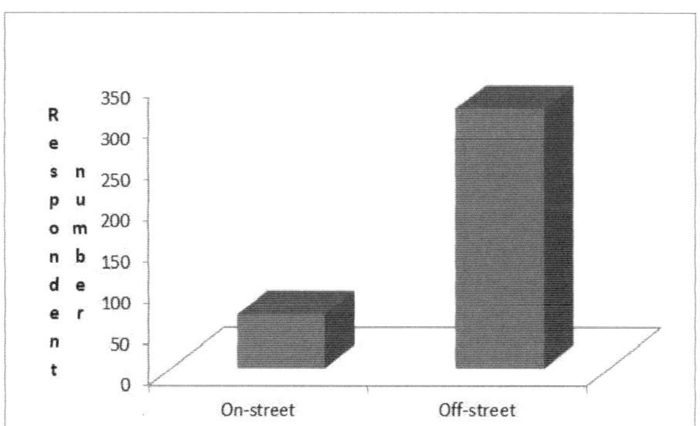

Figure 7. Respondent number based on existing parking locations

Majority of motorcyclists paid more than Rp.2000 for each parking, and nearly 30% of motorcyclists parked their vehicle along unpaid parking areas. In average, each motorist spent nearly 11% of their monthly income just for parking or equals to Rp.213.000 (14.2USD) (see Tables 2 & 3). For rich people, this amount of spending may be insignificant. Yet, that amount is likely a considerable expense for those who live below poverty line. Thus, increasing parking cost could

IOP Conf. Series: Materials Science and Engineering **846** (2020) 012038 doi:10.1088/1757-899X/846/1/012038

prevent motorcycle usages; however it must be compensated by providing reliable and affordable public transports.

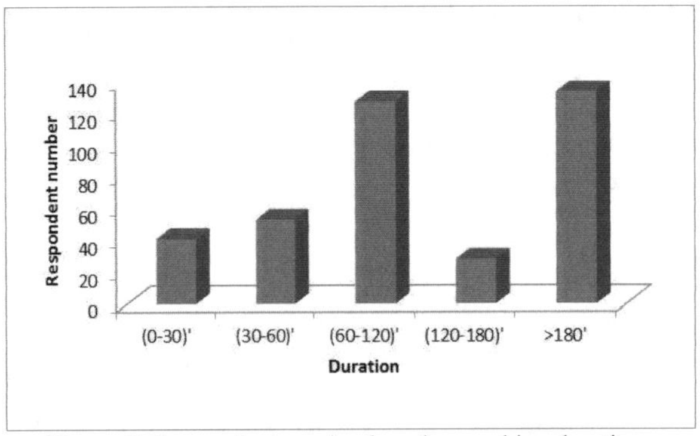

Figure 8. Respondent number based on parking duration

Table 1. Percentage of motorcyclist based on parking duration

Parking duration (minutes)	% motorcyclist
0-30	11
30-60	14
60-120	33
120-180	7
>180	35
Total	100

Table 2. Percentage of motorcyclist based on parking cost

Parking cost (Rp.)	% motorcyclist
0	29
1000	1
2000	59
3000	10
>3000	1
Total	100

Table 3. Percentage of motorcyclist based on parking cost-monthly income ratio

Parking cost-monthly income ratio (%)	% motorcyclist
<10	88
10-20	9
20-30	3
Total	100

The motorcyclists usually parked within a distance of less than 50m from their final destinations as indicated by about 72% of respondent (see Figure 9). Further, off-street parking is preferred by the motorcyclist as confirmed by Figure 10. It can be assumed that motorcyclists prefer an off-street parking location very close to their destination. Based on this statistics, different parking fees could be promoted based on the distance of parking area to the activity centres.

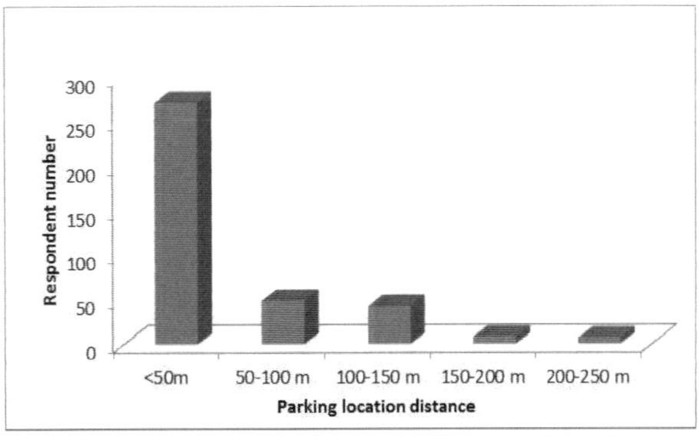

Figure 9. Respondent number based on distance of parking location

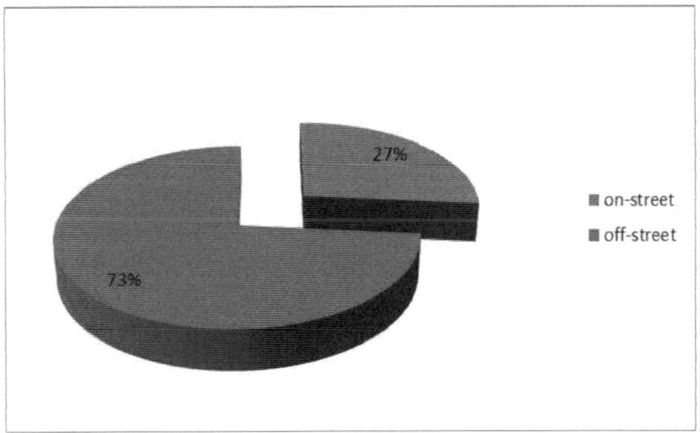

Figure 10. Respondent percentages based on parking location preferences

Figure 11 depicts the behaviour of motorcyclist towards three virtual parking schemes based on parking location, cost, duration and security facilities as illustrated by Figure 1. It can be seen that scheme 1 is preferred by the motorists when the parking cost is not more than RP.3500. Once it is increased, the preference is shifted to scheme 3. It can be drawn that the motorist considers parking cost more important than the parking safety and convenience facilities provided in off-street parking scheme. However, it becomes more important than parking cost when the fee is increased.

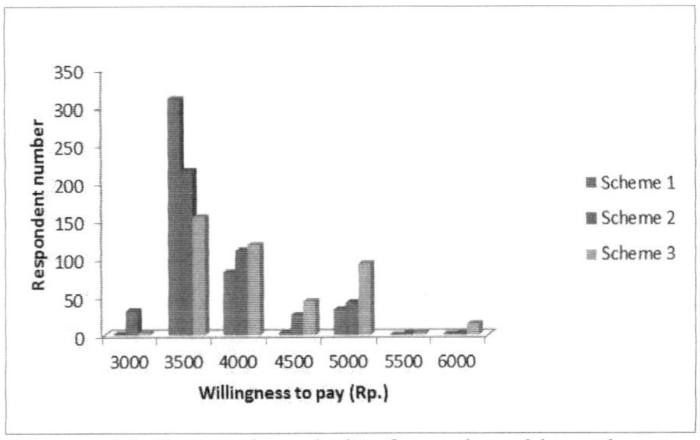

Figure 11. Respondent choice for each parking schemes

Meanwhile, the Willingness to Pay (WTP) for each parking schemes is showed by Figure 12. The motorcyclist still tends to choose on-street parking with an average parking fee about Rp.4000 fixed for unlimited time. If the parking fee is increased, they tend to select off-street parking. Figure 12 also depicts the Ability to Pay (ATP) of the motorcyclist for parking their vehicles. The ATP constantly decreases once the parking fee higher than Rp.2000. It is sharply decreased once it reaches more than Rp.3000. However, more than72% of motorists has the same WTP and ATP value which is Rp.4000. About 50% of motorists have ATP up to Rp.5000, yet the WTP remains Rp.4000. It can be assumed that the motorist would be able to pay more for parking once the parking service is increased based on their preferences.

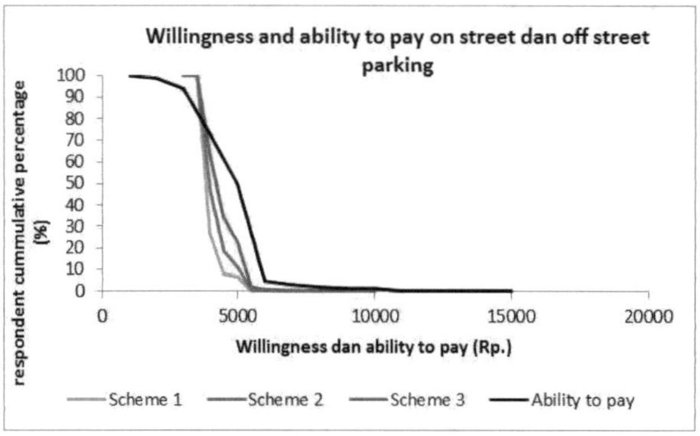

Figure 12. WTP and ATP

4. Conclusion

Findings from this research suggest that there is motorcycle dependency for people who live under poverty line. Motorcycles were mainly used for studying and working. Majority of motorcyclist parked for long time period closed to the activity centres. Since majority of motorcycle users are considered poor people, parking fee is considered more important than the parking service. However, the motorists tend to willingly select a better parking facility like off-street parking-in this case scheme 3 once the parking fee is increased up to double of existing parking cost. Based on the promoted parking schemes, the motorcyclists seem to have a higher ability in paying the parking fee than the parking tariff for each scheme. This could imply that the motorcyclist expects better parking facilities close to their final destinations. Yet, providing more parking facilities might trigger a higher trip number undertaken by motorcycles. Therefore, the relevant stakeholder is suggested to rigorously manage existing and future motorcycle parking facilities; for an example is by providing parking area where its distance is above the motorist preference. For parking fee, it should beyond their affordability. In the meantime, school and working bus policies could be promoted as well as strong law enforcement.

5. Acknowledgment

The authors would like to say thank to Ministry of Research, Technology and Higher Education as well as Politeknik Negeri Padang for supporting this research financially. A great appreciation is devoted for the reviewer whose comments and suggestions are valuable in improving the quality of this paper.

References

[1] M. G. Masuria, *et al.*, "Children, Youth and Road Environment: Road Traffic Accident," in *Asia Pacific International Conference on Environment-Behaviour Studies*, Sarawak, Malaysia, 2010.

[2] S. A. Sarm and K. Kanitpong, "Analysis of Factors Affecting the Severity of Motorcycles Casualties in Phonm Penh Usinga a Bayesian Approach " *Asian Transport Studies,* vol. 4, pp. 430-443, 2016.

[3] S. Baral and K. Kanitpong, "Factors Affecting the Severity of Motorcycles Accidents and Casualties in Thailand by Using Probit and Logit Model," *Journal of the Eastern Asia Society for Transportation Studies,* vol. 11, pp. 2175-2188, 2015.

[4] G. Sugiyanto, "The Effect of Congestion Pricing Scheme on the Generalized Cost and Speed of a Motorcycle," *Walailak Journal,* vol. 15, pp. 95-106, 2017.
[5] K. Gwilliam, "Cities on the Move: A World Bank Urban Transport Strategy Review," World Bank, Washington, DC2002.
[6] H. Ma, *et al.*, "Motorization Process and Management in Big Cities in China," *IATSS Research,* vol. 31, pp. 42-47, 2007.
[7] WHO, "Global Status Report on Road Safety 2015," 2015.
[8] Kemenhub. (2012, 6 Juli). *72 Per cent road accident involving motorcycle (in Bahasa)*. Available: http://www.dephub.go.id/read/berita/direktorat-jenderal-perhubungan-darat/13119
[9] K. Polri. (2013, September 6). *Road traffic accident data*. Available: http://www.korlantas-irsms.info/graph/ageInjuryData
[10] BPS, "Road transport Statitistics," Indonesia Statistics2017.
[11] W. Yan-ling, *et al.*, "Current Situation and Analysis of Parking Problem in Beijing," *Procedia Engineering* pp. 777 – 785, 29 September – 1 October 2004 2016.
[12] M. G. Masuri, *et al.*, "Public Participation in Shaping Better Road Users in Malaysia," in *Asia Pacific International Conference on Environment-Behaviour Studies* Berlin, 2014.
[13] BPS, "Padang city census report 2016," ed. Padang: Statistics Padang, 2018.
[14] G. Yaldi, *et al.*, "Defining Suitable Parking Controls to Minimize Negative Impacts of Road Traffic: A Case Study in Padang City," *International Journal on Advanced Science, Engineering and Information Technology,* vol. 6, pp. 600-606, 2016.
[15] M. M. A. Manan and A. Várhelyi, "Motorcycle fatalities in Malaysia," *International Association of Traffic and Safety Sciences,* vol. 36, pp. 30-39, 2012.
[16] Fitroh, *et al.*, "Analysis of Motorcyclists Driving Behavior in Bandung City," *Journal of Eastern Asia Society for Transportation Studies,* vol. 11, 2015.
[17] Menhumkam, "UU Lalulintas dan Angkutan Jalan," vol. 22 Th 2009, S. RI, Ed., ed. Jakarta: Setneg RI, 2009.
[18] BPS. (2018). *National poverly line*. Available: https://www.bps.go.id/pencarian.html?keywordforsearching=garis+kemiskinan&yt1=Cari

Designing Hygienic and Energy Saving of Water Dispenser Machine

Dodon Yendri[1*], Hadzimah Rizza[2], Budi Rahmadya[3], Derisma[4]

[1,2,3,4] Computer System Department, Faculty of Information Technology, Universitas Andalas, Padang, 25231, Indonesia

Email: dodon@fti.unand.ac.id

Abstract. The use of a dispenser at the office is not as frequent as at home, because it is only used during working hours or at certain times. The use of dispensers must be in accordance with established health standards, so the quality of water produced by the dispensers must also be hygienic and free of germs or bacteria. Generally, water dispensers are operated for 24 hours, meaning that the dispensers use up a lot of electrical energy. In addition, the use of dispensers must also comply with established health standards, so the quality of the water produced by the dispensers must also be hygienic and free of germs or bacteria. Meanwhile the quality of the water produced by the device cannot be monitored. To overcome this problem, a study was conducted by designing a hygienic and energy efficient water dispenser design by applying a forward scheduling approach. The design is carried out through several components, namely ultraviolet (UV), dispenser heaters, Arduino Uno sensors, DS18B20 sensors, HC SR-04 sensors, DC pumps, 2 relays, and RTCDS1307. The results of this study indicate that the dispenser heating system can work based on a set schedule by applying an advanced scheduling approach referring to the RTC DS1307 time reading. The UV lamp can be lit for 10 minutes to sterilize the main container of the dispenser with an estimated suppression of E. Coli bacteria about 87.5 colonies / minute. This system is able to suppress the growth of bacteria about 875 colonies for once lighting. This system can control the quality of water that is safe for consumption and saves electricity energy up to 79.87 kWh per month.

1. Introduction

In line with the development of technology, automation of electronic equipment has become a demand for users. One of them is the provision of drinking water using a dispenser. Dispenser is widely used in households and offices now because it is more practical for getting hot and cold water. A dispenser is supported by components like heat component for heating dan cool component for cooling. The heater component requires more energy than the cooling one. A dispenser absorbs 330-450 watts of power for heating components and 60-80 watts for cooling components.

The dispenser is equipped with a switch to activate and deactivate the heater, but in practice this switch is often ignored so the heater continues to work all day. This is a waste of energy. Another side, many dispensers in general do not have a system to suppress bacterial growth and also there are not periodic cleaning of the container. It makes bacteria grow unnotice. There are eight types of bacteria develop in the dispenser container [1].

The quality of the water dispenser must be hygienic. Parameters of hygienic is if the water have total coliforms and Escherichia Coli (E. Coli) with units / units of colony forming units in 100 ml water samples and the maximum content is 0 [2]. The decree of the Minister of Health of the Republic of Indonesia No. 715 of 2003 regulates the requirements for tableware containers, where a container is hygienic if the number of bacteria does not exceed 100 colonies / cm2.

Research on dispenser operation has been carried out by researchers before using the fuzzy method [3]. The results of the study can heat the water temperature with a success percentage of 89%. But in its use the research has not been able to save energy because the heating system stays on all day. The next study [4], to maintain the quality of drinking water can be done by the disinfection process by providing ultraviolet (UV) radiation in the main container. The results showed that ultra violet rays were able to kill E. coli bacteria in drinking water. Furthermore, ultraviolet disinfection for 10 minutes can reduce the bacteria as much as 875 colonies [5].

Based on the research above, the writer combines research by designing energy-efficient and hygienic dispenser heating systems using the scheduling method. The scheduling method is used to set the active or non-active heater scheduling based on the DS18B20 sensor temperature reading. To ensure the quality of water, a disinfection process is carried out by providing ultraviolet (UV) radiation to the container while to save energy, a relay is used for circuit breakers and time readers use RTCDS1307.

Some studies related to the use of ultraviolet radiation disinfection include purification of drinking water [6], evaluation of the application and contribution of ultraviolet germ irradiation relative to air disinfection [7], efficiency of ultraviolet radiation in reducing cell counts of Staphylococcus aureus ATCC 25923 and Escherichia coli K-12 [8], increasing the process of water disinfection with ultraviolet [9], investigating the effectiveness of ultraviolet light for disinfection of wastewater [10], evaluating the effects of ultraviolet disinfection on water quality [11].

2. Methodology

Basically this research is carried out with a methodology based on the reading time of the RTCDS1307 system which is arranged using the scheduling method as shown in the following Figure 1.

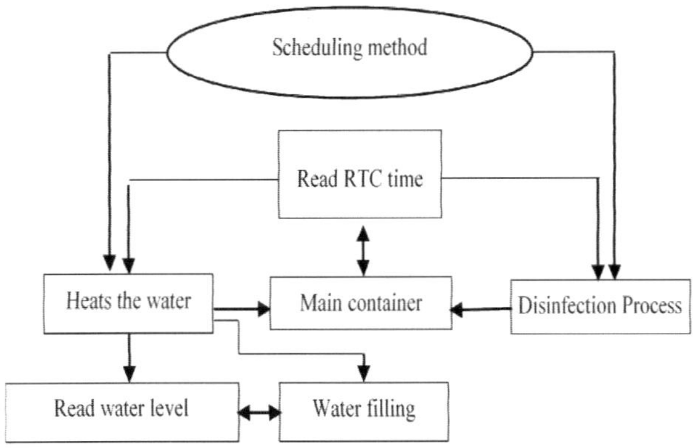

Figure 1. Methodology

The operation of the system is regulated at a schedule of 8:15 a.m. to 8:25 a.m. for disinfection or cleaning, from 10:00 to 11:00 a.m., 00: 00-13: 00 and 15:00-16:00 for heating water. The first time the system is activated, the system will read the RTCDS1307 time. If the time is read according to the scheduling time for example at 8:15 a.m. the system will activate the main disinfection or cleaning function until 8:25 by activating the UV lamp. If the RTCDS1307 time has passed the 8:25 mark, the system will standby until 09:59. Furthermore, when the RTCDS1307 time enters 10:00 a.m. the system will carry out the function of the heating system until 11:00 a.m., and likewise at 12-13.00 and 15:00-16:00. To accommodate and heat water, electric pot is used which is equipped with a heater.

IOP Conf. Series: Materials Science and Engineering **846** (2020) 012039 doi:10.1088/1757-899X/846/1/012039

This energy-efficient and hygienic dispenser heating system research is carried out by blocking the design process in Figure 2.

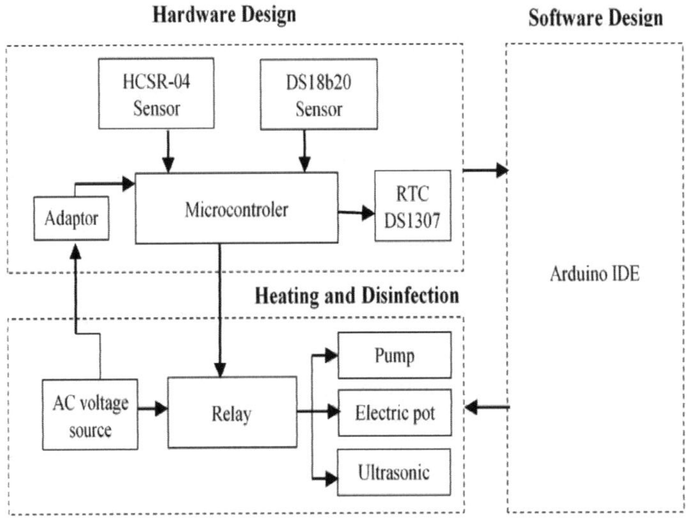

Figure 2. Block design process system

From Figure 2, it can be explained that the design of energy-efficient and hygienic dispenser heating systems consists of three main activities which are hardware design, software design and heating disinfection.

3. System Design

3.1 Hardware Design and Parametric Stud

The design of hardware consists of heating components (electric pots and heaters) which are used as storage containers and heating water as an object whose operation schedule will be set, ultraviolet (UV) lights [11],[12],[13] for disinfection in containers main reservoir, DS18B20 sensor [14],[15] to measure temperature, sensor HC SR-04 [16],[14] to read the water level in a heating tank, dc pump as a valve to control the flow of water from the main container to a heating tank with a capacity of 1.6 liter. Relay 1 to activate and deactivate the system according to the instructions of the microcontroller, RTCDS1307 [16] as a system time reader and Arduino Uno as the control center that will process the sensor reading data. Relay 2 for regulating UV lights, pumps and heaters. Energy-efficient and hygienic dispenser heating system hardware designs can be seen in Figure 3.

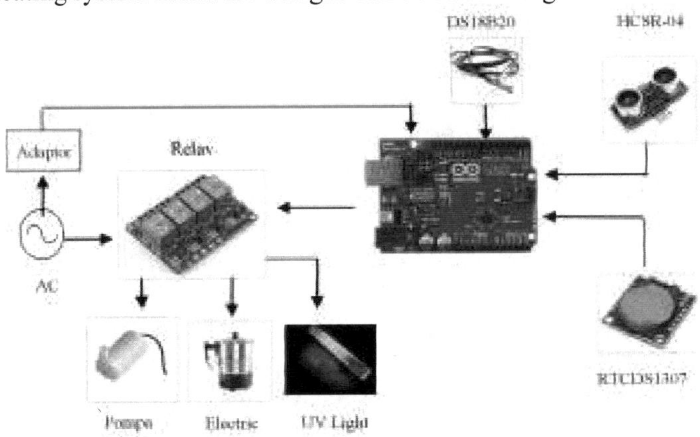

Figure 3. The design of the hardware system

3.2 Software Design

The software designing was conducted for the hardware to work in accordance with its functions to activate the built hardware components, The software application using C programming through Arduino IDE for running system. In this system, a scheduling method was used during the operation to acquire the hygiene and energy saving heating system of the dispenser. The scheduling method using forward scheduling approach is shown in Figure 4.

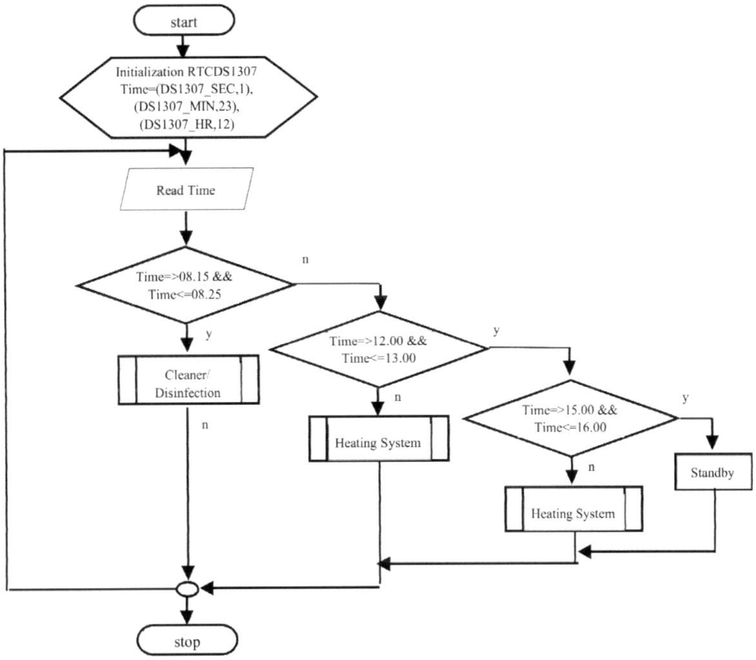

Figure 4. System flowchart of dispenser heating

The value of the RTCDS1307 sensor is reading by the system and adjusted to the predetermined schedule in reference to the scheduling method, namely during 08.15-08.25 where the system conducts "disinfection" process, the system runs the function of "water heating" during 10.00-11.00, 12.00-13.00, and 15.00-16.00. If the time read by the RTCDS1307 sensor is currently not in those time limits, then, the system will go off (standby). In that order, the current will not flow on the system and will surely save energy during the operation.

3.3 Process Design

3.3.1 Cleaner/Disinfection

The disinfection process is performed daily by activating the UV light for 10 minutes from 08.15-08.25 with an assumption that during this time, the employees have just started their activities, thus, none of them requires hot water. After the disinfection process is conducted, the UV light is switched off and the system is back on standby for the next schedules.

3.3.2 The Process of Water Filling and Dispenser Heating

The filling and heating of the dispenser are initiated by reading RTCDS1307's time and detecting temperature through the DS18B20 temperature sensor. If the time on RTC DS1307 has been in accordance with the scheduling system (10.00-11.00, 12.00-13.00 and 15.0016.00), the HCSR-04 sensor will be activated to measure the distance of water height inside the electric pot by reading the floating object which has been re-established. If the reading distance between the floating object and

HCSR-04 sensor is valued as low or equal to 6 cm, then, the water container is determined as full and the water flowing will be stopped, on the contrary, if the reading distance is higher or equal to 10 cm, then, the container is determined as empty and the pump will be activated by flowing the water to fill the electric pot.

The first water heating is performed when the container has been determined as full (the reading distance of HCSR-04 sensor is 6 cm) and low temperature or equal 42°C, then, the heating system will be activated until the temperature reaches its maximum limit (90°C) or before the end of the schedule. If the schedule has not over, thus, the system will be standby for the next change of temperature. The heater will be active again when the temperature is at 86oC. However, if the schedule has ended, thus, the automatic heating system will shut down and the system will go off.

4. Result and Discussions

4.1 The Implementation of Hardware

This system was built with several components, namely the main container which consists of UV light and DC pump placed on the upper part of toolbox, DS18B20 temperature sensor installed on the electric pot equipped with a heater and functioned as the heating component, HCSR-04 ultrasonic sensor installed on middle part of tool box's ceiling right across the surface of floating object installed on the electric pot. While Arduino UNO, RTCDS1307 sensor, relay 1, and relay 2, as well as other components, were arranged on the lowest bottom of the plywood-based dispenser boxes with 40 cm x 60 cm x 20 cm dimension, such as seen in Figure 5.

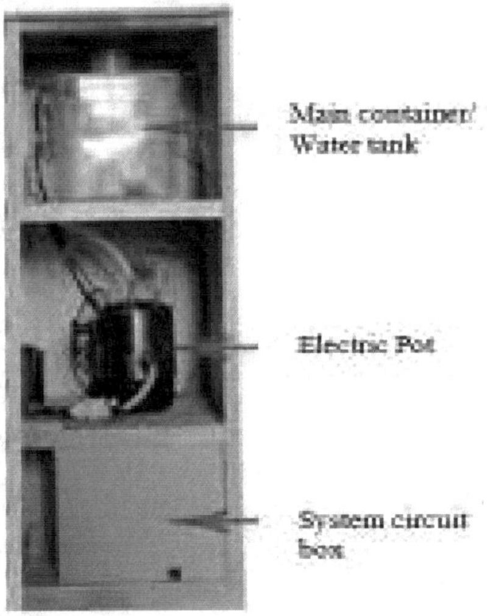

Figure 5. Overall dispenser heating system

4.2 Implementation

The software in this system was using the Arduino IDE. The programming was performed to read the components and implement the scheduling method for the water heating process. The scheduling method was used through the application of a backward scheduling approach on the operation of dispenser's heater and UV light in accordance with the pre-determined schedule.

4.3 Testing

4.3.1 The UV Light testing

This testing was performed to ensure whether the UV light could light up based on the schedule of the scheduling method. This testing was functioned as the sterilization of the main container on the dispenser for 10 minutes. The result of UV light's testing for 5 days from October 25th to 29th 2018 is shown in Table 1.

Table 1. The testing of UV Light

No	Date	Time on RTC	Condition of UV Light
1	10/05/2018	08.15-08.25	ON
2	10/06/2018	08.15-08.25	ON
3	10/27/2018	08.15-08.25	ON
4	10/28/2018	08.15-08.25	ON
5	10/29/2018	08.15-08.25	ON

From the results of the tests in table 1 above, it can be seen that the UV lights can be lit well and successfully operate according to the given schedule and the predetermined duration of time is 10 minutes every day. So based on the results of the study [6] that the UV light is turned on for 10 minutes. This means that the system can suppress bacteria as much as 87.5 x 10 = 875 colonies.

4.3.2 The DS18B20 Temperature Sensor Testing

This testing was used to discover the temperature read by the sensor (in °C unit) by taking the data every five minutes and then the result was compared to the analog thermometer. Comparison of temperature testing results with DS18B20 sensor and analog thermometer can be seen in Table 2.

Table 2. Comparison of temperature testing with DS18B20 sensor and analog thermometer

No	Time	DS18b20 sensor (°C)	Analog Thermometer (°C)	Difference
1	22:00	24.75	25.41	0.66
2	22:05	43.88	44.63	0.75
3	22:10	65.75	66.37	0.62
4	22:15	87.25	88.24	0.99
5	22:16	90.03	90.55	0.52

Based on Table 2, it seen that the temperature measurement results by using a DS18B20 sensor are almost similar to the analog thermometer. There is only very little difference between the two measurements. The difference was due to the specification and accuracy of DS18B20 sensor which amounted to ±0.5°C to -10°C, and -10°C to +85°C. However, in general, the reading of the DS18B20 temperature sensor was categorized as good with a small difference.

4.3.3 The Heater testing

This testing was performed to test the capability of the tool in heating the water inside the electric pot. The energy used by heater was 350 watt. The testing was begun by heating the water in the heater along with the heater of the conventional dispenser which commonly used by the community with the same energy. Firstly, both tanks filled with 1 liter of water volume respectively, then the 1-hour change of temperature was observed (from 10.00 to 11.00) with the temperature set from 42oC until it reaches the set point (90oC) based on the timing of both heaters. The testing results of both heating systems can be seen in Table 3 as follows.

Table 3. The testing of temperature change time and heater status

Time	Heating with the conventional			Heating with the experiment dispenser		
	Tempera ture	Status	Time (Minute)	Tempe rature	Status	Time (Minute)
10.00	42°C	ON		42 °C	ON	
10.07	68 °C	OFF	7			
10.12	63 °C	ON				
10.13				90 °C	OFF	13
10.14	69 °C	OFF	2			
10.28	64 °C	ON				
10.30	71 °C	OFF	2			
10.31				86 °C	ON	
10.33				90 °C	OFF	2
10.43	62 °C	ON				
10.45	78 °C	OFF	2			
10.51				86 °C	ON	
10.53	70 °C	ON		90 °C	OFF	2
10.55	77 °C	OFF	2			
11.00	73 °C	OFF		88 °C	OFF	
Total Heating Time			**15**			**17**

Based on Table 2, it also seen that the cycle of water heating in the conventional dispenser is the difference with the experiment dispenser. In 1 hour (from 10.00-11.00), the conventional dispenser was turning on or active for five times with 15 minutes of total time and 7 minutes of maximum time, namely during the first activation which started from 42oC up to 68oC. While on the experiment dispenser, in an hour operation, the heater was turned on for three times with 17 minutes as the total active time and 13 minutes as the longest active time during the first activation with 42oC as the initial temperature until the 90oC of maximum temperature was achieved.

Seen from the heat resistance, the conventional dispenser was having a big difference with the experiment dispenser. On the conventional dispenser, the decrease of temperature was 5oC (from 68oC to 63oC) which occurred for four minutes (from 10.07-10.12) while on the experiment dispenser, the decrease of temperature was 4oC (from 90oC to 86oC) which occurred for 17 minutes (from 10.14-10.31). The same condition happened on the next temperature decrease from 69oC to 65oC which occurred for 15 minutes (from 10.14-10.28), while on the experiment dispenser, the temperature decreased from 90oC to 86oC which occurred for 17 minutes (from 10.34-10.51). The difference of these changes was caused by the different set point which determined for both dispensers. On the conventional dispenser, the highest temperature was set at +/- 68°C and the lowest temperature was set at +/- 60°C. While the lowest temperature on experiment dispenser was set at 86oC and the highest temperature was set at 90oC. In addition, the material and the impermeability of the heater tank also affect the temperature resistance. From the testing, it can be seen that the heater of experiment dispenser was better compared to the conventional dispenser.

4.4 System Testing and Analysis

4.4.1 The Testing and Analysis of Disinfection Process

The disinfection process in the system is done by irradiating the main container using a UV lamp for 10 minutes. The test results in table 1 show that UV lights are always on for 10 minutes every day for five days. Based on previous research [11], irradiation of UV lamps can suppress bacterial colonies by 87.5 per minute. Thus, every 10 minutes of irradiation every day can suppress bacterial colonies of 87.5 x 10 = 875 colonies every disinfection.

4.4.2 The System Testing and Analysis Based on the Scheduling

The schedule of the system operation was set at 08.15-08.25, 10.00-11.00, 12.00-13.00 and 15.00-16.00. The system testing which based on the scheduling was aimed to ensure that the disinfection process and heating system could control the filling and heating the water on the heater tank in accordance with the pre-determined schedules in which the average operating time in a day was ± 45.5 minutes to acquire the set point value of temperature (90°C) while it took ± 10.08 minutes of average time for operating the conventional dispenser.

4.4.3 The Analysis of Scheduling Based on Backward Scheduling Appriach

The analysis of scheduling based on a backward scheduling approach was initiated by setting a schedule of which a required activity has to be finished, then counting backward to determine the proper time for the first time the activity begins. The implementation of backward scheduling is assuming that the finished date is known and the start date is desired. The disinfection process has to be done by 08.25 with a length of 10 minutes, therefore, it has to be started from 08.15 with an assumption that the employees are already in the office. In similar to the first heating process (10.00-11.00). Employees who yet to have their coffee or tea at their homes will be able to have it after the water is heated. Then, the second heating is adjusted to the break and lunchtime which is around 12.00-13.00. In that order, the filling and heating of water in the dispenser have to be conducted. While the third heating is performed before the employees go home namely at 15.00-16.00, thus, the hot water in the dispenser has already available, in that order, the employees who want to make coffee or tea before coming home are able to do so.

4.5 The Measurement Analysis of Energy Use

After the overall testing of the system can run properly, then, the analysis and measurement of energy use on experiment dispenser were performed which followed by comparing them with the energy measurement of the conventional dispenser. The analysis of the energy measurement of both dispensers can be seen in Table 4 as follows.

Table 4. The analysis of energy measurement

Experiment dispenser	Conventional Dispenser
a. The energy used in kW = kW = 350 /1000 = 0.35 kWh b. Total usage time in a day = 45.5 minutes (0.75 hour), kWh = 0.35 x 0.75 = 0.256 kWh/day c. Arduino's Energy = 0.0005 x 24 = 0.012 kWh/day d. UV Light's Energy = 0.005 x 0.16 = 0.0008 kWh/day e. Pump Energy = 0.0015 x 0.016 = 0.000024 kWh/day f. Total in 1 day = 0.256 + 0.012 +0.0008 + 0.000024 = 0.20 kWh/day g. Total monthly energy usage: kWh in 1 day x 24 days = 0.20kWh x 24 = 4.8 kWh	a. The energy used in kW = 350/1000 = 0.35kWh b. Total operation time of heater in the first hour = 17 minutes = 0.28 hour c. Total operation time of heater for the next 1 hour = 8.5 minutes = 0.14 hour d. Total usage in a day = (0.28 + 0.14) x 24 hours = 10.08 hour e. Total monthly energy usage = 10.08 x 0.35 x 24 days = 84.67 kWh

According to the measurement results in Table 3, it can be seen that the energy usages of both dispensers are very much different. The experiment dispenser consumed 4.8 kWh/month energy while the conventional dispenser consumed 84.67 kWh/month. It means that the experiment dispenser can save 79.87 kWh energy per month. This significant difference was due to the testing of the conventional dispenser which required 17 minutes of time for heating in the first hour and 8.5 minutes for the following hours or equal to 3 hours 53 minutes in a day. In addition, the conventional dispenser was active for 24 hours while for the experiment dispenser, total heater operation in a day was 45.5 minutes on average.

5. Conclusion

There are several conclusions is taken which described as follows:

1. The heating system of the experiment dispenser could work in accordance with the scheduling based on the backward scheduling approach which has been determined in reference to the time reading of RTC DS1307.
2. The UV light could active to perform 10 minutes-disinfection on the main container of the dispenser with an estimation of bacteria suppression amounted to 87.5 colonies/minute and 875 colonies for 1 radiation time.
3. The heating system of the dispenser through the application of forwarding scheduling method could save the energy usage up to 79.87 kWh per month.

Acknowledgement

This research was performed to see into what extent the experiment dispenser can produce hygienic water quality and save energy during the usage of electrical power compared to the conventional dispenser which commonly used in the community. In this research, the author has received massive support from colleagues in providing ideas and recommendations for the success of the research. Due to that matter, the author would like to thank Hadzimah Rizza, Budi Rahmadya, and Derisma as well as other friends who cannot be mentioned individually. Special gratitude toward the rector of Andalas University who has given an opportunity for the author to publish this paper.

References

[1] Hed Series Peltier-Type Temperature Control System for Chemical Liquids, Chemical Thermo-con *SMC Corp. Am.* 356–72

[2] Permenkes R 2017 Peraturan Menteri Kesehatan Republik Indonesia Nomor 32 Tahun 2017 Tentang Standar Baku Mutu Kesehatan Lingkungan Dan Persyaratan Kesehatan Air Untuk Keperluan Higiene Sanitasi, Kolam Renang, Solus Per Aqua dan Pemandian Umum *Peratur. Menteri Kesehat. Republik Indones.* 17–20

[3] Baumgartner A and Grand M 2006 Bacteriological quality of drinking water from dispensers (coolers) and possible control measures *J. Food Prot.* **69** 3043–6

[4] Wandrivel R, Suharti N and Lestari Y 2012 Kualitas Air Minum Yang Diproduksi Depot Air Minum Isi Ulang Di Kecamatan Bungus Padang Berdasarkan Persyaratan Mikrobiologi *J. Kesehat. Andalas* **1** 129–33

[5] A. S, As Z A and Setiadi G 2014 Efektivitas "Portable UV Disinfection" dalam Menurunkan Angka Bakteri (Escherichia Coli Spp) pada Air Minum *J. Kesehat. Lingkung. J. dan Apl. Tek. Kesehat. Lingkung.* **11** 223

[6] Johnson K M, Anil Kumar M R, Ponmurugan P and Gananamangai B M 2010 Ultraviolet Radiation and Its Germicidal Effect in Drinking Water Purification *J. Phytol.* **2010** 12–9

[7] Memarzadeh F, Olmsted R N and Bartley J M 2010 Applications of ultraviolet germicidal irradiation disinfection in health care facilities: Effective adjunct, but not stand-alone technology *Am. J. Infect. Control* **38** S13–24

[8] De Sá Silva C A, De Andrade N J, Soares N D F F and Ferreira S O 2003 Evaluation of ultraviolet radiation to control microorganisms adhering to low-density polyethylene films *Brazilian J. Microbiol.* **34** 175–8

[9] Said M B E N and Otaki M 2013 Enhancement of ultraviolet water disinfection process *African J. Biotechnol.* **12** 2932–8

[10] Oparaku N F, Mgbenka B O and Ibeto C N 2011 Waste Water Disinfection Utilizing Ultraviolet Light *J. Environ. Technol.* **4** 73–8

[11] Choi Y and Choi Y june 2010 The effects of UV disinfection on drinking water quality in distribution systems *Water Res.* **44** 115–22

[12] Mofidi A A, Rochelle P A, Chou C I, Mehta H M, Verne L and Linden K G 2002 Bacterial

Survival After Ultraviolet Light Disinfection : Resistance , Regrowth and Repair *Am. Water Work. Assoc. Annu. Conf. Exhib.* 1–11

[13] Charnas C L, Bell K Y, Adams D, Chambers J, Cohrs D, Hammer D and Schmidt P 2012 Wastewater UV Disinfection Systems– Lessons Learned During Performance Testing *IUVA News* **14** 14–9

[14] Nie S, Cheng Y and Dai Y 2013 Characteristic Analysis of DS18B20 Temperature Sensor in the High-voltage Transmission Lines' Dynamic Capacity Increase *Energy Power Eng.* **05** 557–60

[15] Fezari M and Al Dahoud A 2019 Exploring One-wire Temperature sensor "DS18B20" with Microcontrollers *Univ. Al-Zaytoonah Fac. IT* 1–9

[16] Sidam P, Mate M, Nagarare P, Katle S, Dhapodkar S and Nasare R 2018 IOT Based Smart Garbage Bin Alert System using HC-SR04 Ultrasonic Sensor *Int. J. Innov. Sci. Res. Technol.* **3** 277–80

IOP Conf. Series: Materials Science and Engineering **846** (2020) 012040 doi:10.1088/1757-899X/846/1/012040

Ad8232 Sensor Implementation to Detect Fatigue

Milda Yuliza[1], Tuti Angraini[2], Roza Susanti[3], Effendi[4], Teguh Wijaksana Isma[5]

[1,2,3,4,5] Electronics Engineering Department, Politeknik Negeri Padang

Abstract. Fatigue is a protection mechanism to avoid further damage, so that recovery occurs after a break. Fatigue (fatigue) is a subjective feeling. Work fatigue is a condition that is accompanied by a decrease in efficiency and the need to work. the effects of excessive activity result in fatigue. lack of knowledge results in excessive effects of fatigue. the way to overcome this problem is by utilizing an electrocardiograph (ECG) sensor. The use of electrocardiograph sensors (ECG) to determine the fatigue of smartphone users based on heart rate rhythm. The sensor used as a fatigue detection is the AD8232 type. The result of this process is safe or unsafe when using a smartphone. When BPM is worth 60 <BPM <120, the condition of smartphone users is safe, whereas when BPM <60 or BPM> 120, the condition of smartphone users is unsafe. Based on the results that have been tested on four samples there was a change in heart rate rhythm (BPM) during one hour and fifteen minutes of testing.

1. Introduction

Fatigue is a protection mechanism to avoid further damage, so that recovery occurs after a break. Fatigue (fatigue) is a subjective feeling. Work fatigue is a condition that is accompanied by a decrease in efficiency and the need to work [1]. one of the measurement points that can be measured to determine fatigue is the heart [2]. the value obtained from the measurement of the heart is the value of the heart rate, heart rate is the speed of the heart rate as measured by the number of contractions (heart rate) per minute (bpm) [3]. One of the sensors that can be used to detect heart signals is an electrocardiograph (ECG), electrocardiogram (ECG) or electrocardiogram (ECG) is a graph produced by an electrocardiograph to detect heart abnormalities by measuring the electrical activity produced by the heart, as the heart contracts. ECG can help diagnose various heart health conditions [4].

2. Methodology

Heart rate values are calculated using the haskell and fox formula, can be seen in Figure 1. Based on the haskell and fox formula, the normal range of heart rate is 60-120 bpm [3].

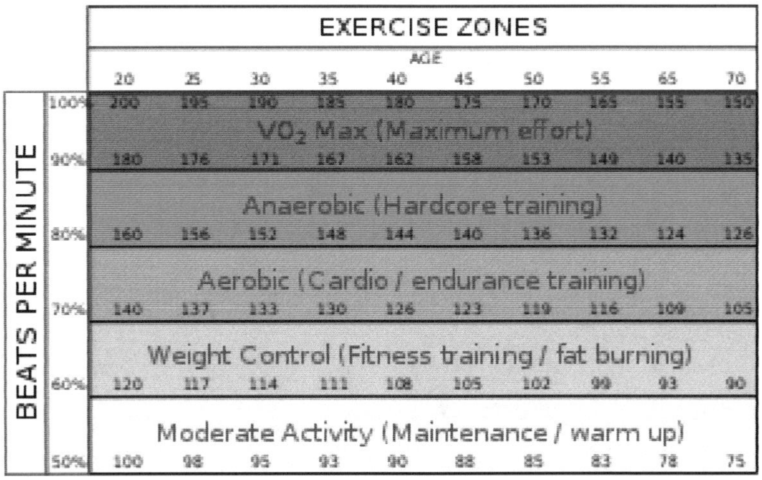

Figure 1. Formula Fox and Haskell

The tool can work when button 1 is pressed. whereas if button 2 is pressed, the tool cannot work. If the heart rate is in unsafe condition, the red LED will be active and the buzzer is active. whereas if the heart rate is in safe condition, the green LED will be active and the buzzer will be inactive. the work process can be seen in figure 2.

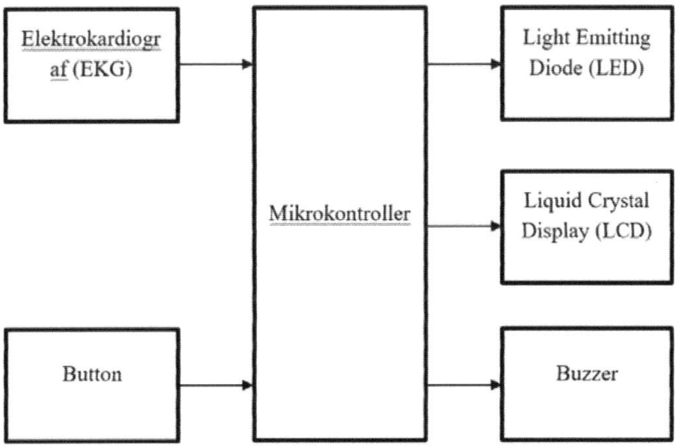

Figure 2. Block diagram of the AD8232 sensor implementation to detect fatigue

3. Result and Discussion

Tests carried out are to compare BPM values on ECG sensors with ECG devices. The ECG tool used is the Philips ECG type G30. Tests carried out as many as 10 by comparing the results of the electrocardiograph sensor output with the instrument. Value reading is done simultaneously. The average value of the error percentage of 10x experiments was 2.39. The results of the comparison of BPM values using sensors with BPM values using Philips ECG tools can be seen in table 1 below:

Table 1. Table of Comparison of BPM Values Using Sensors and BPM Values Using Philips ECG Tools

Experiment	BPM Values with the AD8232 Electrocardiograph Sensor	BPM Value with a Philips Type ECG Tool	Percentage of error (%)	Percentage of error (%)
1	116	112	-3,57	3,57
2	105	106	0,94	0,94
3	99	99	0,00	0
4	103	104	0,96	0,96
5	95	100	5,00	5
6	89	91	2,20	2,2
7	98	97	-1,03	1,03
8	102	105	2,86	2,86
9	94	97	3,09	3,09
10	103	107	3,74	3,74
Average Percentage of errors			2,34	

The formula for finding the percentage error value is:

$$\%\, Error = \frac{|\text{True Value} - \text{Value Obtained}|}{\text{True Value}} \times 100\%..$$

$$\%Error = \frac{100 - 95}{100} \times 100\ \%$$

$$\%\ Error = 5\%$$

While the formula to look for the average error is:

$$Average\ Error = \frac{Error\ Value\ 1 + Error\ Value\ 2 + \cdots + Error\ Value\ 10}{10}$$

$$Average\ Error = \frac{23,4}{10}$$

$$Average\ Error = 2,34$$

A graph of the comparison of ECG sensor values with a Philips ECG tool can be seen in the following figure 3:

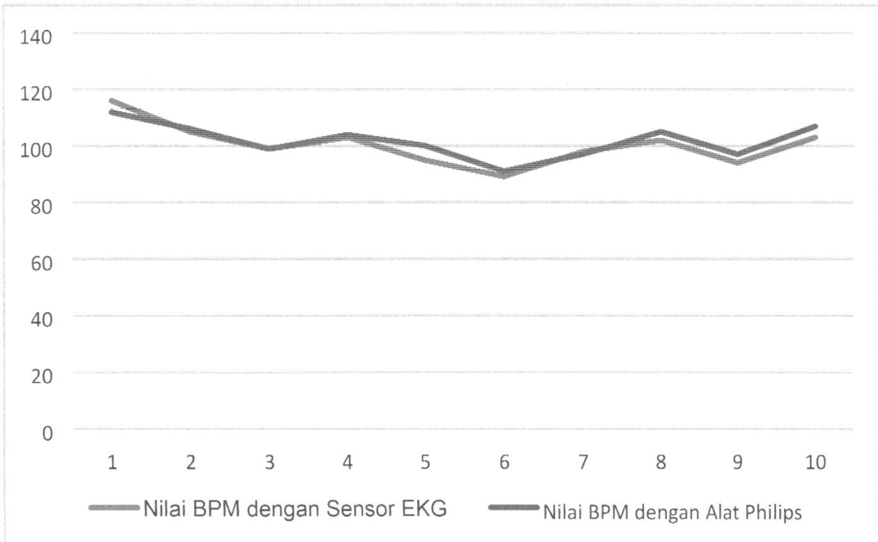

Figure 3. Comparison Graph of ECG Sensor Values with Philips ECG Equipment

Based on Figure x, the y-axis is the Beat Per Minute (BPM) value while the x-axis is an experiment conducted. Based on the graph the value of the ECG sensor with the Philips ECG is not much different.

4. Conclusion

Electrocardiograph Sensor (ECG) AD8232 is used to detect fatigue when using a smartphone based on heart rhythm. The error value of the Electrocardiograph sensor compared to the Philips type ECG is 2, 34%.

References

[1] Z. Hidayat, "Pengaruh Stres dan Kelelahan Kerja terhadap Kinerja Guru Smpn 2 Sukodono Kabupaten Lumajang," *J. Penelit. Ilmu Ekon. WIGA*, vol. 6, no. 1, pp. 36–44, 2016.

[2] Y. Suryana and R. Aziz, "Sistem Pemonitor Detak Jantung Portable Menggunakan Tiga Sensor Elektroda," *J. Al-AZHAR Indones. SERI SAINS DAN Teknol.*, vol. 4, no. 1, pp. 14–17, 2017.

[3] E. Wahyudi and S. Hartati, "Case-Based Reasoning untuk Diagnosis Penyakit Jantung," *IJCCS (Indonesian J. Comput. Cybern. Syst.*, vol. 11, no. 1, pp. 1–10, 2017.

[4] Wikipedia, "Heart Rate," *Encyclopedia of Stress*, 2007. [Online]. Available: https://en.wikipedia.org/wiki/Heart_rate. [Accessed: 18-Jul-2019].

IOP Conf. Series: Materials Science and Engineering **846** (2020) 012041 doi:10.1088/1757-899X/846/1/012041

Design of PLC-based Control System for 18650 Lithium-ion Battery Dismantling Machine

Suryo Hadiwibowo [1], Satria Mahardhika Wisambodhi [1], Muslim Mahardika [1], Indra Perdana [1]

[1] Faculty of Engineering, Universitas Gadjah Mada, Yogyakarta - Indonesia

* suryohad@gmail.com

Abstract. The dismantling process of 18650 lithium-ion battery aims to separate the shell from its contents that contain lithium. The dismantling process is carried out using pneumatic components that automated by using a programmable logic controller (PLC). This research was conducted through 5 stages of the process, namely: machine designing and manufacturing, pneumatic system designing, PLC control designing, assembly, and testing. The design of pneumatic systems and their control is carried out to get the easiest and most effective working system.

1. Introduction

The 18650 lithium-ion (li-ion) battery is the most common used battery type on the market today [1]. The battery has some of the most important building blocks, one of which is lithium. Lithium is an element that is scarce and difficult to obtain [2]. To overcome this, a recycling process is needed on the used li-on battery to recover the lithium [3]. The recycling process requires a device / machine to separate the battery shell from its contents containing lithium [4]. Based on this case, a research need to be conducted to examine and make a machine that can facilitate the disassembly of the 18650 lithium-ion battery.

2. Machine Designing and Manufacturing

18650 lithium-ion battery dismantling machine are designed using Autodesk Inventor Professional 2019, resulting final design which is shown on the Figure 1 and its components are shown on Table 1. Manufacturing process are using conventional tools and electric welding machine are used to weld the frame.

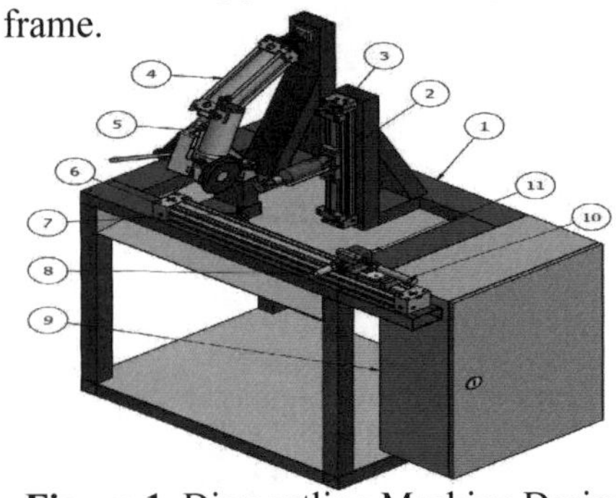

Figure 1. Dismantling Machine Design

Explanation :
1. Main Frame
2. Mini Grinder
3. Linear Actuator Cylinder
4. Cylinder Pneumatic Grinder
5. Hand Grinder
6. Linear Actuator Cylinder Mini Grinder
7. Solenoid Stopper
8. Cylinder Pneumatic Clamp
9. Panel Box
10. Cylinder Pneumatic Ejector
11. Assembly Vise

3. Pneumatic System Design

3.1. Pneumatic System Design Requirements
Machine could split the battery shell crossways and cut off both ends of the battery. It works in the following order Clamp's pneumatic locks the vise → Linear actuator moves forward to split the battery's shell → Linear actuator stops on the first cutting position → Grinder's pneumatic moves down → Grinder's pneumatic moves up after it cuts the battery's first end → Linear actuator moves forward → Linear actuator stops on the second cutting position → Grinder's pneumatic moves down → Grinder's pneumatic moves up after it cuts the battery's second end → Clamp's pneumatic opens the vise → Ejector pneumatic pushes the battery → Ejector pneumatic shortens → Linear actuator return to the initial position

3.2. Pneumatic System Diagram
Pneumatic system diagram is shown in Figure 2. Clamp's pneumatic controlled with 5/2 (S1 and S2) solenoid valve, linear actuator controlled with 5/3 (S3 and S4) solenoid valve, grinder's pneumatic controlled with 5/2 (S5 & S6), ejector pneumatic controlled with 5/2 (S7 and S8) solenoid valve, and linear actuator for mini grinder controlled with 5/3 (S9 and S10) solenoid valve. Besides, a speed control devices are installed in the clamp's pneumatic S1 and S2 output, linear actuator in S3 output, grinder's pneumatic in S5 and S6 output, and linear actuator for mini grinder in S9 and S10 output. For the pressure generation, an electric compressor with 8 bar of maximum pressure, 2 HP power, and debit of 195 liter/minutes are used to drive this machine. Before the air from the compressor enters into the system, the air pass through an air filter, regulator, and a safety valve.

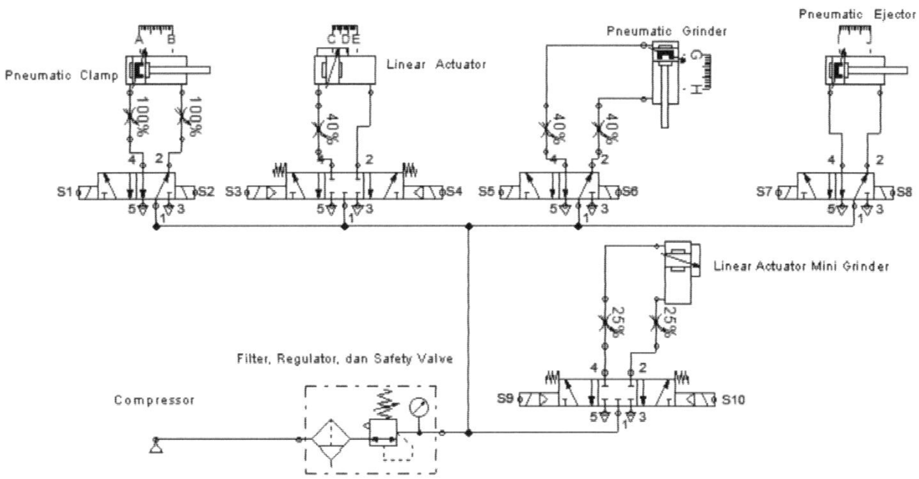

Figure 2. Pneumatic System Diagram

4. PLC Control Design

4.1. PLC Input and Output
As the design of the pneumatic system that has been made, it is determined that the PLC used is the Omron CP1E-E30SDR-A. This PLC has 18 input, 12 output with 100-240V AC power supply and 24 VDC power output. Allocation of input and output adresss on the PLC can be seen on Table 1 and Table 2.

Table 1. Input Address

ADDRESS	COMPONENT	INPUT SIGNAL
0.00	EMG	Emergency pushbutton

ADDRESS	COMPONENT	INPUT SIGNAL
0.01	START	Pushbutton Start
0.02	B	Reed switch pneumatic clamp long
0.03	D	Reed switch linear actuator position 2
0.04	H	Reed switch pneumatic grinder long
0.05	G	Reed switch pneumatic grinder short
0.06	E	Reed switch linear actuator position 3 (last)
0.08	A	Reed switch pneumatic clamp short
0.09	J	Reed switch pneumatic ejector long
0.10	I	Reed switch pneumatic ejector short
0.11	C	Reed switch linear actuator position 1 (start)

Table 2. Output Address

ADDRESS	COMPONENT	OUTPUT SIGNAL
100.00	S11	Solenoid stopper shortens
100.01	S1	Solenoid valve pneumatic clamp extending
100.02	S2	Solenoid valve pneumatic clamp shortens
100.03	S3	Solenoid valve linear actuator forward
100.04	S4	Solenoid valve linear actuator backward
100.05	S5	Solenoid valve pneumatic grinder extending
100.06	S6	Solenoid valve pneumatic grinder shortens
100.07	S7	Solenoid valve pneumatic grinder extending
101.00	S8	Solenoid valve pneumatic ejector shortens
101.01	S12	Solenoid relay grinder rotates
101.03	S13	Solenoid relay mini grinder rotates

4.2. System Software Design

A CX Programmer software was used to program the system control process, the flow chart was shown in Figure 3.

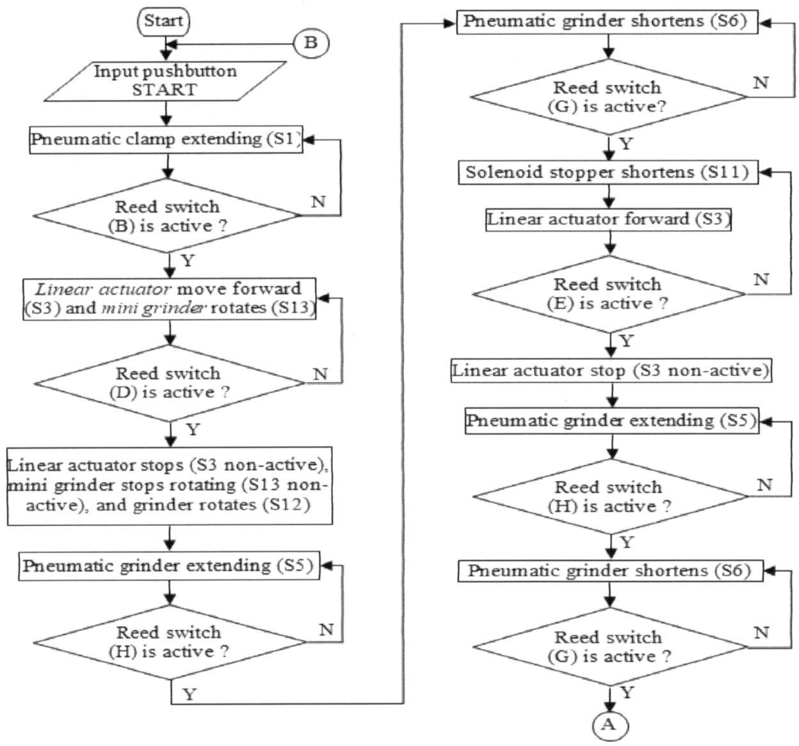

IOP Conf. Series: Materials Science and Engineering **846** (2020) 012041 doi:10.1088/1757-899X/846/1/012041

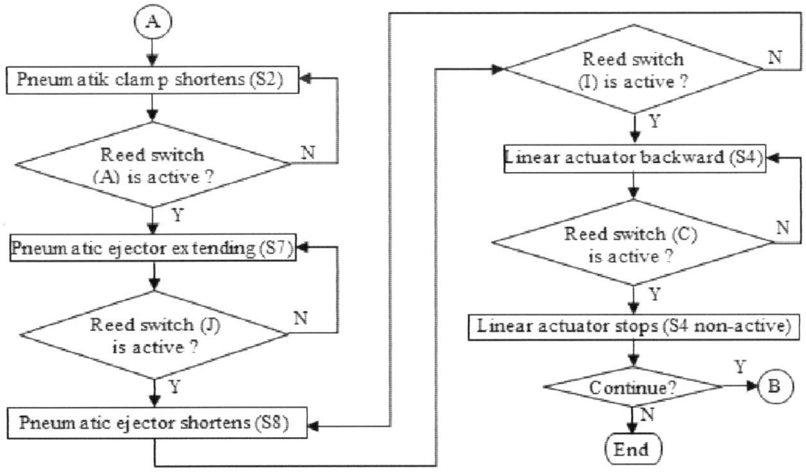

Figure 3. Pneumatic control system program flowchart

5. Assembly

Assembly process is performed on the pneumatic component and electrical component toward the dismantling machine's frame. To ease the installation, a wiring diagram is made as seen in Figure 4.

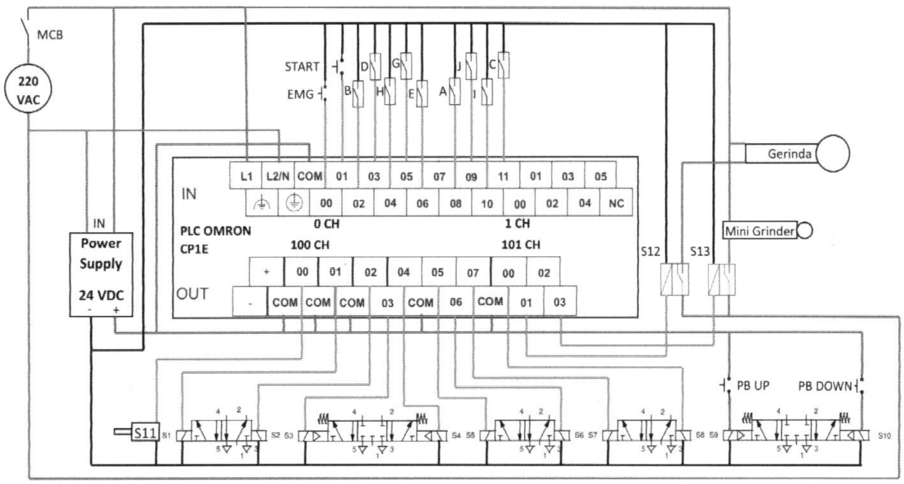

Figure 4. Wiring Diagram

6. Testing

A total of 10 lithium-ion model 18650 battery were tested on the dismantling machine shown in Figure 5 and recorded the time in each cycle in Table 4. While dismantled battery result is shown in Figure 6.

Figure 5. Dismantling machine

Figure 6. Dismantled battery results

Table 3. Time Recording Results

Test	Time (seconds)
1	44,30
2	44,16
3	43,70
4	46,11
5	46,22
6	45,61
7	44,60
8	45,94
9	46,35
10	45,85

From Table 3, the average time for the 18650 lithium-ion battery dismantling process is 45.28 seconds. If the process is done manually it will require an average time of 80.54 seconds, so this dismantling machine could reduce the dismantling time to 35.26 seconds or 43.78%.

7. Summary

Battery dismantling machine able to be manufactured with a design that has been made and can be used for the process of dismantling a lithium-ion battery model 18650 by using the pneumatic control for the mechanism movement. The application of PLC to control the pneumatics ease the machine control, machine function flexibility, and simplify the electronic system. Based on test results, this machine could run as the PLC programmed and need a dismantling time with an average of 45,28 seconds.

8. Acknowledgments

This work was fully supported by Indonesia Endowment Fund for Education (LPDP) Ministry of Finance Republic of Indonesia.

References

[1] Muenzel, V., Hollenkamp, A. F., Bhatt, A. I., Hoog, J. D., Brazil, M., Thomas, D. A., and Mareels, I.: A Comparative Testing Study of Commercial 18650-Format Lithium-Ion Battery Cells. Journal of The Electrochemical Society. Vol. 162 (8), pp. A1592-A1600 (2015)
[2] U.S. Geological Survey.: Lithium. pp. 98-99. Mineral Commodity Summaries (2019)
[3] Huang, B., Pan, Z., Su, X., and An, L.: Recycling of lithium-ion batteries: Recent advances and perspectives. Journal of Power Sources. Vol. 399, pp. 274-286 (2018)
[4] Mahardika, M., Perdana, I., Hadiwibowo, S., and Wisambodhi, S. M.: Method of 18650 Model Lithium-Ion Battery Dismantling. Indonesia Patent P00201907450 (2019)

IOP Conf. Series: Materials Science and Engineering **846** (2020) 012042 doi:10.1088/1757-899X/846/1/012042

A Review of Automated Testing Approach for Software Regression Testing

Fransiskus Anindita Kristiawan Pramana Gentur Sutapa[1], Sri Suning Kusumawardani [*1], Adhistya Erna Permanasari [1]

[1] Faculty of Engineering, Universitas Gadjah Mada, Yogyakarta - Indonesia.

* f.a.kristiawan@mail.ugm.ac.id

Abstract. One of the important parts of software development life-cycle is software testing and one of which is regression testing. Regression testing is done to verify that previously detected errors have been corrected correctly and no new errors arise as a result of incorrect corrections so that the quality of the software being tested is maintained. It is important and must be done. However, the conventional method is not efficient as it is time-consuming, not reusable, and prone to error. In this paper, a comprehensive review of the automated testing approach is presented to be used by other researchers in this field of study. The review result shows that the automated testing approach is suitable to enhance the regression testing with some plausible options of tools e.g. Selenium, SAHI, and Robot-framework. Furthermore, some plausible options of execution methods are also discussed in this paper. This review further concludes that the parallel execution method is considered as a promising choice to conduct the most efficient regression testing process.

1. Introduction

One of the important stages of software development life-cycle is software testing. According to Galin [1], software testing is a formal process carried out by a special testing team that will examine a software unit, several integrated software units or an entire software package by running programs on a computer. All related tests are carried out in accordance with approved test procedures in approved test cases. One type of software testing that is important to conduct is regression testing. Regression testing is carried out to verify that previously detected errors have been corrected correctly and no new errors arise as a result of incorrect correction [2]. Regression testing is important and must be done.

However, the conventional method which is the manual regression test is not efficient as this method is very time consuming, the test is not reusable, and the process is prone to tester error. In this paper we explain and describe several previous studies about automated testing approaches on the regression testing. The review result shows that the regression test can be done based on automated testing approach by employing several tools namely Selenium, SAHI, and Robot framework. This paper also discusses some plausible option of methods to execute the automated test to be used by other researchers in this field of study which is consist of serial execution method and parallel execution method.

2. Regression Testing

2.1. Regression Testing
When a change is made to the software, several aspects of the configuration of the software (program, documentation, or data) will change [3]. Regression testing is carried out to verify that previously detected errors have been corrected correctly and no new errors arise as a result of incorrect corrections [2]. Regression testing helps to ensure that the changes made to the software do not lead to unwanted behavior or additional errors in the software[3]. According to [1], a regression test suite consists of 3 different test case class, which are:

a. A representative sample regarding the test that will run all software functions.
b. Additional test that focuses on software functions that will likely be affected by the changes that have been made.
c. Test that will focus on the changed software components.

2.2. Regression Testing with Conventional Method (Manual Testing)
The conventional method to conduct regression testing is manual testing using a human tester. It is considered not efficient as it is very time consuming, the test is not reusable, and the process is prone to tester error. Manual testing is a software testing without any automatic software testing tools or any script test. In manual testing, a tester positioned him/herself as a user and utilizes the tested software features to ensure that said software is bug-free [4]. In manual testing, test scenario or test case is written first and then followed by various steps in the testing execution[5]. To execute manual testing in the same section of software whenever a change is made is a relatively weary activity, and time-consuming. Another disadvantage of manual testing is a smaller extent of testing scope, a higher number of errors, and a higher number of testing staff requirements. Furthermore, several challenges in manual testing are tiring activities and dependent on tester skills such as patience level, speculative level, creativity, and competencies to consider every possibility of test data inputs[4].

3. Regression Testing with Automated Testing Approach

3.1. Automated Testing Approach
Regression testing can be done with the automated testing approach. Automated testing is a method in software testing that utilizes special software to execute a test and compares the expected execution result to the actual execution result. Automated testing could be done by utilizing software that specifically provides automated testing features. Automated testing enables an automated test execution, automated test execution scheduling, and automated test report generation [6]. According to [6] there are advantages that can be acquired by employing automated testing:

a. Able to run the existing automated test so that the effort to execute tests every time there are changes becomes easier
b. Able to run testing more often due to faster execution duration thus bigger assurance of the system can be acquired
c. Able to obtain the same level of consistency repeatedly, which is hard to obtain in manual testing.

Various research regarding automated testing has been done by several researchers. Those research has proved and concluded fundamental things, such as the advantages of automated testing execution. Previous research by Peethambaran [7] and Shivatri [8] proves that automated testing is faster than manual testing. Next, research conducted by Amannejad also stated that the utilization of testing tools is an advantage of automated testing. The coding of automated testing cases takes a lot of time if done one by one for every test case. Moreover, if the tester who is involved in automated testing does not have sufficient knowledge regarding the programming language to make automated testing source code[9].

Another research conducted by Kaur and Gagan Gupta also elaborates that if testing is done manually, the shortage of testing time will be bigger due to one problem occurring if done with manual testing will take time and even if there is a lot of time spent on it, not all errors can be found[10]. Whereas follow-through research about automated testing by Naidu elaborates that compared to manual testing, automation testing can escalate testing coverage, accuracy, time savings, and also costs [11]. The same statement also mentioned by Singh and Bindia[12].

3.2. Automated Testing Tools

In the field of automated software testing research, many software to supports the testing process is available to be employed. Various software that has been researched are Selenium WebDriver, SAHI, and Robot framework.

3.2.1. Selenium. In his research de Castro et al. stated that Selenium is a tool that he considered to be the most popular tool to carry out an automated test on a web-based software [13]. Subsequentially, Leotta stated that Selenium has 2 approaches in test script development, which are Capture and Replay and Programmable Test[14]. Both have their own unique advantages but Programmable test offers a cheaper cost in maintenance test case than Capture and Replay. Aside from the cost, Selenium also possesses more advantages such as co-evolve on the test script. In test script development with Capture and Replay approach there exists a tool known as Selenium IDE. Whereas in test script development with Programmable Test there exists a tool named Selenium WebDriver. Selenium WebDriver is a programming library that can be used to control a browser during an automated testing process. Selenium WebDriver supports the programming languages Java, C #, Ruby, Python, Perl, PHP, and Javascript[15].

Naidu stated that Selenium has advantages in test case creation perspective, execution time on a single thread, logging and reporting, integrated script editor, and easier installation compared to SAHI. Leotta attempts to compare between Capture & Replay test using Selenium IDE with Programmable Test using Selenium WebDriver, to observe which one is more effective. Leotta stated that the initial cost to create a test case using Capture & Replay is cheaper. However over time when the tested application is in reparation then the created test case must also be repaired, Programmable Test offers a cheaper cost, but this is also affected by the technical skills owned by the test case scripter [11].

3.2.2. SAHI. SAHI is a competitor to Selenium. In contrast to Selenium, SAHI has a feature to execute test cases in parallel, thus enabling a shorter time of execution if the feature is used. However, according to Naidu, SAHI's performance is thwarted by selenium in the perspective of test case creation, execution time on a single thread, logging and reporting, and integrated script editor. Moreover, according to Naidu, the installation of SAHI is more difficult compared to Selenium. If Selenium is targeted towards programmers, then SAHI is targeted towards non-programmer testers [11].

3.2.3. Robot Framework. Research related to automated testing has been done by Peethambaran by using a tool called Robot framework. Peethambaran developed an automated test for the Metinfo application, an application in the field of Forest Data Collection and Analysis. The research used Robot framework which is equipped with Apium plugin. In Peethambaran's research, it is proven that the execution process carried out by automated testing can be done faster than manual testing[7].

4. Execution Method

4.1. Serial Execution Method

In executing an automated test, two methods could be used. One of the methods of execution in automated tests is serial execution. In serial execution, each scenario in an automated test will be executed alternately. An automated test scenario will be executed after the execution of the previous automated test scenario has been completed. This makes the automated test serial execution process inefficient in terms of time. However, in some automation tools, serial execution is the default execution method. Hence currently researchers are trying to develop the parallel execution capability on some of the automation tool which is explained in the next section.

4.2. Parallel Execution Method

Another execution method to execute an automated test is the parallel execution method. Research related to the execution of automated tests in parallel has been conducted by Naidu who tried to execute automated tests in parallel with Selenium without the use of certain testing frameworks. The results of the study show that the process of executing automated tests in parallel cannot be carried out if Selenium IDE is used, but if SAHI is used it can be done [11]. Another research related to the execution of automated tests in parallel has subsequently been conducted by Abdull Razak and Fahrurazi who tried to carry out automated test execution in parallel on the PyTestFramework framework. In that research Selenium is used as an automation tool and Python as a platform. In this research, automated test execution can be carried out in parallel but Selenium Hub or Selenium Server is needed as a link to forward the test script to the browser driver. In that study, it was known that the existence of the hub caused a delay in the automated test execution process [16].

Another research related to the execution of automated tests in parallel has subsequently been conducted by Fajar who developed a platform so that Selenium users can carry out automated test execution in parallel without the use of Selenium Server. In his research, Selenium Server was eliminated and was replaced by calling the Webdriver API directly by developing a platform independently. The platform will take time investment to build and maintain. Furthermore, given that the platform is a non-open source platform, there is no assurance of continue support from its developers, which is important in the continuity of the regression testing process with an automated testing approach [17].

Research by Sutapa et al. presented a parallelization approach for the execution of Serenity framework based automated test. In his study between the test script with the browser driver will be connected directly without the need to use a hub so that there is no delay in the execution of automated tests. In his study, a parallelization tool based on open source is used. The tool is the Maven Failsafe Plugin. In his study, the speed of execution of the automated test can increase by 118.14% so that the time required for the test execution is highly reduced [18].

5. Conclusions

Regression testing can be conducted either by manual testing or automated testing approach. Based on the literature review, the automated testing approach is providing a faster execution time and the minimum possibility of human error. The review result shows that the automated testing approach is suitable to enhance the software regression testing with some plausible options of tools such as Selenium, SAHI, and Robot framework. The use of automated testing approach could also make a lesser cost due to the advantages of reusing the test.

Furthermore, some plausible options of execution methods which consist of serial execution method and parallel execution method are also available. Based on the literature review, we conclude that the parallel execution method is considered as a promising choice to conduct the most efficient regression testing with an automated testing approach due to the reduced time required for the test execution than the serial one.

Acknowledgments

Authors would like to thanks to Directorate of Research of Universitas Gadjah Mada (UGM) for their research grant support, Department of Electrical and Information Engineering UGM and Center for Academic Innovation and Studies UGM for their support during the research.

References

[1] D. Galin, *Software Quality Assurance From Theory to Implementation*. 2004.

[2] S. Ulewicz, B. Vogel-heuser, and S. Member, "Industrially Applicable System Regression Test Prioritization in Production Automation," *IEEE Trans. Autom. Sci. Eng.*, vol. 15, no. 4, pp. 1839–1851, 2018.

[3] R. S. Pressman, *Software Engineering A Practitioner's Approach*, 8th ed. New York: McGraw-Hill, 2015.

[4] M. Qasim, "Shortening Testing Time of a Web-based Business Application in Scrum using Automated Testing," University of Tampere Faculty, 2017.

[5] V. Kumar, "Comparison of Manual and Automation Testing." International Journal of Research in Science and Technology, 2012.

[6] M. F. and D. Graham., *Software Test Automation: Effective Use of Test Execution Tools*. Harlow, Essex, U.K: Addison-Wesley, 1999.

[7] A. Peethambaran, "Automated Functional Testing Using Keyword- driven Framework," no. May, 2015.

[8] A. Shivatri, "Selection and Implementation of Test Framework For Automated System Test of Mobile Application." Tesis, Faculty of Computer Science Departement of Computer Engineering, Technise Universitat Chemnitz, Chemnitz., 2016.

[9] Y. Amannejad, V. Garousi, R. Irving, and Z. Sahaf, "A search-based approach for cost-effective software test automation decision support and an industrial case study," *Proc. - IEEE 7th Int. Conf. Softw. Testing, Verif. Valid. Work. ICSTW 2014*, pp. 302–311, 2014.

[10] H. Kaur and G. Gupta, "Comparative Study of Automated Testing Tools: Selenium, Quick Test Professional and TestComplete," *J. Eng. Res. Appl.*, vol. 3, no. 5, pp. 1739–1743, 2013.

[11] T. J. Naidu, N. A. Basri, and S. Nagenthram, "SAHI vs. Selenium: A comparative analysis," *Proc. 2014 Int. Conf. Contemp. Comput. Informatics, IC3I 2014*, pp. 967–970, 2014.

[12] I. Singh and B. Tarika, "Comparative Analysis of Open Source Automated Software Testing Tools : Selenium , Sikuli and Watir," *Int. J. Inf. Comput. Technol.*, vol. 4, no. April, pp. 1507–1518, 2016.

[13] A. C. de Castro, A.M.F.V., Macedo, G. A., Collins, E. F. dan Dias-Neto, "Extension of Selenium RC Tool to Perform Automated Testing with Databases in Web Applications." 8th International Workshop on Automation of Software Test.

[14] M. Leotta, D. Clerissi, F. Ricca, and P. Tonella, "Capture-replay vs. programmable web testing: An empirical assessment during test case evolution," *Proc. - Work. Conf. Reverse Eng. WCRE*, pp. 272–281, 2013.

[15] S. Stewart, "Selenium WebDriver," *Selenium Documentation*, 2012. [Online]. Available: http://docs.seleniumhq.org/docs/03_webdriver.jsp. [Accessed: 05-May-2018].

[16] R. dan F. F. R. Abdull Razak, "Agile Testing with Selenium." Agile Testing with Selenium, 5th Malaysian Conference in Software Engineering.

[17] K. Mustofa and S. P. Fajar, "Selenium-Based Multithreading Functional Testing," *IJCCS (Indonesian J. Comput. Cybern. Syst.*, vol. 12, no. 1, p. 63, 2018.

[18] F. A. K. P. G. Sutapa, S. S. Kusumawardani, and A. E. Permanasari, "Parallelized Execution of Serenity Framework Based Automated Test: A Case Study," in *Proceeding of 2019 11th International Conference on Information Technology and Electrical Engineering (ICITEE)*, 2019.

IOP Conf. Series: Materials Science and Engineering **846** (2020) 012043 doi:10.1088/1757-899X/846/1/012043

Intratextual Factors in Source Text Trimming for Indonesian-English Machine Translation

Yan Mujiyanto [1]

[1] Universitas Negeri Semarang, Semarang – Indonesia

* yanmujiyanto@mail.unnes.ac.id

Abstract. Source text analysis is a crucial phase in the translation process. Several issues are proposed by experts in this particular field of study. Nord [1], for instance, proposes that this phase may be constrained by two types of factors, i.e. extra-textual and intra-textual ones, the latter consisting of eight factors which need consideration, i.e. subject matter, content, presuppositions, text composition, non-verbal elements, lexis, sentence structure, and suprasegmental features. This paper aims to consider the importance of each of the eight factors as the basis of trimming academic manuscripts in Indonesian before their mechanical rendering to English for international publication. The object of this study was 50 manuscripts in Indonesian prepared for the publication. The manuscripts were analyzed taking the eight factors into account. The trimming process, which refers to a process of making a text tidier for inter-lingual rendering, is applied to the texts considering the issues so as to be more ready for the translation. The effectiveness of the process is shown by means of comparing the results of rendering the original texts and the trimmed ones assuming that the latter is more qualified than the former. Given that the intra-textual factors are relatively influential in source text analysis for trimming purposes, it is suggestible to consider them as a supplementary procedure for the mechanical rendering of academic manuscripts in Indonesian prior to their translation to English.

1. Introduction

One of the three major stages of the practice of translation, source text analysis, has gained considerable attention from translation experts. Whatever the translation model they are proposing, this special stage will not escape the discussion. As two pioneers of the theory and practice of modern translation, Nida & Taber [2] proposed three stages in the translation process. They include source text analysis, transfer, and restructuring in the target language to produce target text that is equivalent to its counterpart in the source language. Furthermore, experts such as Snell-Hornby [3], Newmark [4], Larson [5], Mujiyanto [6] have strived to modify the three stages with their respective justifications.

Hatim & Mason [7] analyze source text from the perspective of register analysis which focuses attention on the text as discourse, text as action, and text as a sign. They also emphasize the importance of inter-textuality and intentionality in the text that is intended to be rendered into a target language. Hatim & Mason [8] propose text analysis in translation focusing their attention on assumptions based on the standard of textuality, intertextuality, and intentionality. Hewson & Martin [9] redefine translation by introducing an inter-lingual translation model which they call the variational model. In the model, source text analysis is referred to as the "production of inter-linguistic homologies", which includes a discussion on the nature of the source text along with its socio-cultural content and potential for the creation of target text forms.

Bell [10] introduces a translation model by dividing the source text analysis into three parts, namely syntactic, semantic, and pragmatic analyses based on the basic concepts of systemic-functional linguistics. Furthermore, Gentzler [11] summarizes several theories concerning contemporary translation, while Nord [12] emphasizing the importance of translating as a purposeful activity. In line with such models, Venuti [13] edits some articles on translation studies; Grego [14] considers specialized translation from theoretical issues and operational perspectives. Munday [15] introduces translation studies from the perspective of theories and their applications followed by a discussion of the relationship between text analysis and translation Munday [16]. Colina [17] examines the principles of translation; Tyler [18] invites readers to analyze the source text before translating it into the target language. Furthermore, Mujiyanto [19] offers strategies in translation practices that include source text analysis as well while Mujiyanto [20] introduces the model of clause restructuring in English-Indonesian translation. Discussing more source text analysis, Nord [1] focuses his attention to theory, methodology, and didactic application of a model for translation-oriented text analysis. In his proposed model, he shows two categories of significant factors, namely intra-textual and extra-textual factors. The former include subject matter, content, presuppositions, text composition, non-verbal elements, lexis, sentence structure, and suprasegmental features while the latter includes sender or writer of the source text, his/her intention of writing the text, target recipient, medium of communication, place of communication, time of communication, motives for communicating the message, and text functions.

Since Nord's [1] introduction of the function of text analysis in translation, there have been many studies on the implementation of this type of analysis in translation from and to various languages in the world. Research on text analysis as the basis for the practice of translation considers the source text from various perspectives, such as philosophical, theoretical, and pedagogical ones. Zlateva [21], for example, uses text analysis as a tool in translation training. He questions the reasons, processes, and achievements of the analysis in handling current issues. Munday [16] explains the link between text analysis and translation. Meanwhile, Ayupova [22] examines text analysis as part of the steps in the pretranslation rendering process. Laviosa, Pagano, Kemppanen, and Ji (Eds.) [23] analyze text and its context in the study of translation practices. Furthermore, when reviewing news translation, Troqe and Marchan [24] examine a text concerning fieldwork through various surveys. Rossetti and Gaspari [25] conduct a pilot study of trainee translators' perceptions concerning difficulties and time effectiveness in translation to create a model of text analysis in translation memory use and post-editing of raw machine-translation output. Meanwhile, Laviosa, et al. [23] edit topics which are related to textual and contextual analysis in empirical translation studies. Hansen-Schirra, Czulo, and Hofmann (2017) edit works related to empirical modeling in translation and interpreting processes.

Dicerto [26], [27], [28] in tandem introduces a new model for source text analysis in translation, analyzes multimodal source text for translation, and applies multimodal source text analysis as a model in pretranslation process. Meanwhile, Boukhaffa [29] examines source text analysis to create a model for enhancing the translation memory outcome. The research studies mentioned above generally focus on intra-textual factors which include subject matter, text organization, use of non-verbal elements, use of vocabulary, sentence structure, stylistic features, and suprasegmental features as they are suggested by Nord [1]. Those studies mainly focus on the use of grammar and vocabulary in the source language that will be translated into a target language.

The implementation of machine translation has been studied for more than five decades ago. Nevertheless, the advancement of information technology has triggered translation experts to proceed to improve this type of translation for the betterment of rendering results. Nirenburg, Somers, & Wilks [30] summarize essays on theories and practices in the use of machine translation in a professional rendering. Furthermore, Birch, Osborne, & Koehn [31] suggest a model of predicting the success of (or failure in) implementing machine translation. Kastberg [32] sees the use of machine translation tools for a professional translator, categorizing it into (1) fully automated machine translation

(FAMT), (2) human aided machine translation (HAMT), and machine-aided human translation (MAHT). Meanwhile, Lennon [33] considers the use of machine translation from the perspective of different cultures, while Rozmyslowicz [34] puts forward issues that are related to the use of machine translation constraining the construction of translation theories.

No matter what translation model is applied in translation practices, the implementation of machine translation is currently unavoidable. Several tools that are popular among translators include such online machines as DeepLTranslator, Google Translate, SDLTrados, and SYSTRAN. Language aids that are also popular among them are grammar checkers, spell checkers, dictionaries, thesauruses, plagiarism checker, etc. Although the use of such devices facilitates the translators in undertaking their work, relying on them without taking manual work into account may lead to result in translations that do not fulfill the intention. The use of such devices requires the source text to be written in standardized language that is commonly used in academic writings such as research articles, reference books, academic essays, etc. To produce standard English texts, Indonesian essays that are written for that purpose should be trimmed before their translation to English.

This study focuses on the analysis of texts written in Indonesian as the source language prior to their machine translation into English as the target language. It involves the occurrence of intratextual factors as proposed by Nord [1]. The results of the analysis are used as the basis to trim the source texts before their rendering into English so that it results in quality translation fulfilling the requirement as standard texts for internationally accredited publication. Besides, this study aims to show differences in readability levels between the source texts (ST) and their respective trimmed results (TR).

2. Methodology

This study was designed based on a descriptive qualitative approach. Fifty academic manuscripts written in Indonesian were chosen as the object of this study. The manuscripts aimed to be submitted for internationally reputable indexed journal publications. They were analyzed in terms of their intra-textual factors including (1) subject matter or main topic of the texts; (2) text organization or content of the text; (3) pragmatic presuppositions including the use of non-verbal elements; (4) text composition including the structuring of the texts; (5) uses of non-verbal elements including visual elements such as tables, graphs and diagram; (6) uses of lexis including its particular features; (7) uses of various types of sentence structure; (8) suprasegmental features commonly used in writing including, among other things, italics, punctuations,, dashes, parentheses, etc., as suggested by Nord [1].

The results of the analysis were used as the bases for the trimming process. Then, the trimmed texts along with their originals were rendered to English employing *Google Translate*. The data for this research were two sets of linguistic elements covered in the intra-textual domains found in the translations of both the original texts and the trimmed ones. A comparison of the two translation results would show whether the impact of the trimming process is significant in causing the manuscripts to be more readable and, thus, viable for publication in internationally reputable journals.

3. Findings and Discussion

3.1. Findings

Figure 1 shows a model of translation process. In the model, Trimming is positioned between Source Text Analysis and Machine Translation, which is an inevitable effort in bilingual rendering. This particular phase executes the results of the ST analysis by taking necessary actions to make the text viable for the next phase, Machine Translation, which may or may not envolve human aid.

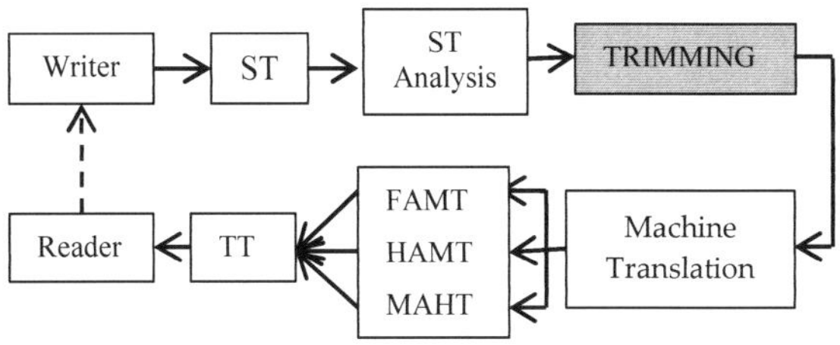

Figure 1. Stages of Translation Process

As stated in the introduction, the trimming process includes eight intra-textual factors that can be presented in the following sub-subsections.

3.1.1. Subject Matter

The analysis of the ST in terms of its subject matter is important due to the reasons proposed by Nord [1], pp. 85-86). According to Nord, there are at least six indicators that can be used to obtain information about the existence of subject matter in a text. In ST trimming, the Subject Matter provides the following results.

Table 1. Results of Trimming the Subject Matter

Indicators	Source Text	Trimming Results
Text coherence	Needs improvement	Improved
Hierarchy of subjects	At random	Deductive
Fulfillment of expectation	Personal	Conventional
Verbalization of subject matter	Implicit	Explicit
Culture bound	Yes	Improved
Relevance to convention	Irrelevant	Relevant

Source: Nord [1]

Table 1 presents the summary of the trimming results involving subject matter based on the six indicators as seen in Column 1. The table shows that the trimming process has yielded several improvements. While the text coherence and the culture-bound in the ST are improved through the process, the hierarchy of the subject is made deductive, the fulfillment of the reader's expectation is made explicit, and the text as a whole is made more relevant to the prospective reader's needs.

Excerpt 1

ST: Analysis of Soft Skills, Work Culture and Service Quality on Student Satisfaction of Anesthesia Specialist Doctor Education Program.

TR: The Influence of Soft Skills, Work Culture, and Service Quality on Student Satisfaction Level of Anesthesia Specialist Doctor Education Program.

In **Excerpt 1**, for instance, the term "Analysis" has been replaced with "Effect" and the term "Level" has been added to make this topic relevant to its contents. This is because the source text intends to show the influence of independent variables mentioned in the text towards certain phenomena. Therefore, the replacement of the term "Analysis" with "Effect", is thus done to make the the topicexplicitly related to the contents of the text.

3.1.2. Contents

According to Nord [1] content is "the reference of the text to objects and phenomena in an extra-linguistic reality" (p. 90) It includes linking devices as well as other logical connections, theme-rheme relationship, functional sentence perspective, sentence patterns, tense, mood, etc. (p. 91).

IOP Conf. Series: Materials Science and Engineering **846** (2020) 012043 doi:10.1088/1757-899X/846/1/012043

Table 2. Results of Trimming the Contents

Indicators	Source Text	Trimming Results
Verbalization of extratextual factors	Explicit	Explicit
Spread of information units	Random	Systemic
Differences between external and internal situation	Ok	Ok
Gaps of cohesion/ coherence	Gaps	Deviced
Availability of conclusion	Available	Improved
Paraphrase	Available	Improved

Table 2 presents the summary of the trimming results involving the contents of the articles under analysis based on the six indicators as seen in Column 1. The table shows that the trimming process has yielded some improvement. While the availability of conclusion and the paraphrase are improved through the process, the Verbalization of extratextual factors is made explicit, the Spread of information units is made systemic, and the Gaps of cohesion/coherence is devised in order to make the whole contents of the draft more relevant to the prospective reader's needs.

Excerpt 2

ST: Family support, support for the family is not only aimed at children with autism but also to families. Parents and siblings of children with autism experience more stress and depression than families who have children with normal development.

TR: Support is not only aimed at autistic people but also their families. Parents and siblings with autism suffer more depression than families with normal-growing children.

In **Excerpt 2**, the spread of information with "Support" as Theme followed by the rest of the clause as Rheme is expressed circularly in the ST. As a result, it uses 42 words merely to convey the meaning that can be expressed in 30 words. It is seen in the excerpt that several sentence elements have been dropped during the trimming process, resulting in a compact expression

3.1.3. Presuppositions

A pragmatic presupposition refers to one that is implicitly assumed by the source text writer, taking it for granted that this will also be the case with the target text reader ([1], p. 95); it usually refers to objects and phenomena of the source culture when the target reader is not fully aware of the aspects of the source culture in the ST and transferred into the TT. Therefore, adjustment of the explicitness level to the background knowledge of the intended TT recipient is a must (p. 98). The existence of presuppositions in translation-oriented source text analysis is indicated by the items presented in Table 3.

Table 3. Results of Trimming the Presuppositions

Indicators	Source Text	Trimming Results
Reference to reality model	Vague	Made clear
Verbalization of reference to reality	Available	Available
Availability of implicit allusion	Available	Available
Existence of redundancies	Exists	Minimized
Verbalization of known/unknown presupposed information	Available	Made clearer

Table 3 presents the summary of the trimming results involving the use of pragmatic presuppositions in the articles under analysis based on the five indicators as seen in Column 1. The table shows that the trimming process has yielded some changes. While the availability of the reference to reality model and the verbalization of known/unknown presupposed information are made clearer by changing them to become explicit through the process, the existence of redundancies is minimized. The rest rfemains as it is to make the whole contents of the draft fulfill the prospective reader's needs.

Excerpt 3

ST: According to Berthal (Muqowim, 2012: 5) soft skills are defined as personal and interpersonal behaviors that develop and maximize human performance (including military lecturers). Sharma in Utama et al, (2010: 3), stated that soft skills are all aspects of generic skills which also include cognitive elements related to non-academic skills.

TR: According to Berthal (as cited in Muqowim, 2012: 5), soft skills are defined as personal and interpersonal behavior that develops and maximizes human performance (including military lecturers). Sharma as cited in Utama et al (2010: 3) states that soft skills are all aspects of generic skills including cognitive elements related to non-academic skills.

In **Excerpt 3**, the ST presupposes the availability of the names "Berthal" and "Sharma" along with their claims on "soft skills", "generic skills", and "non-academic skills". However, the presence of the names "Muqowim" and "Utama, et al" needs clarification because their presence can potentially cancel the presupposition. The addition of the prepositional phrase "as cited in" during the trimming process has supposedly improved the readability level of the TR

3.1.4. Text Composition

Text composition refers to text structuring that may consist of several shorter texts or parts of a bigger text where the macrostructure of the text is marked by chapters and paragraphs and the microstructure by syntactic structures, lexical devices, or suprasegmental features. Both the micro- and macrostructure are important for the source text analysis because, (1) a text can be comprised of smaller elements with different functions ; (2) the beginning and the end of the text may play specific roles in its comprehension; (3) some text types may be subject to conventions; (4) if a text is complex or incoherent, the microstructure analysis may result in information about the subject matter (Nord [1], p. 101). Based on the notions mentioned above, the source texts are analyzed employing the indicators presented in **Table 4** before their trimming process.

Table 4. Results of Trimming the Text Composition

Indicators	Source Text	Trimming Results
Text dependence on larger unit	Independent	Dependent
Signals of macrostructure	Available	Improved
Availability of conventional composition	ST- oriented	TT-oriented
Realization of certain thematic progression	Redundancy	Compact

Table 4 presents the summary of the trimming results involving the realization of the text composition of the articles based on the four indicators as seen in Column 1. The table shows that the trimming process has yielded some changes. While parts of the larger units are made interdependent to one another, the signals of macrostructure are improved, the availability of conventional composition is made target text-oriented, and the realization of certain thematic progression is made more compact. Again, these aim to make the whole contents of the draft fulfill the prospective reader's needs.

Excerpt 4

ST: Interviews and observations regarding: how the emotional and social abilities, how the range of concentration and attention, how the ability and general knowledge according to age such as understanding the limbs, the concept of color, shape, communication skills, fine motor and gross motor, ability to care for themselves. severity of autism using CARS (Childhood Autism Rating Scale).

TR: Interviews and observations about emotional and social abilities, their concentration and attention span, age-specific abilities and general knowledge such as understanding limbs, color concepts, shapes, communication skills, fine motor and gross motor skills, self-care ability, and severity autism uses Childhood Autism Rating Scale (CARS).

In **Excerpt 4**, The ST is composed of clausal elements that potentially relate to one another. However, the elements had been arranged without observing the strict principles of text composition, generally employing the macro- and microstructure of academic texts. As a result, besides structuring an incomplete sentence and offering inconsistent use of clause elements, the readability level of the TS is supposed to be relatively low. The TR shows that a number of sentence elements have been restructured in such a way that it supposedly improves the readability level.

3.1.5. Non-Verbal Elements

Non-verbal elements are various signs which do not belong to any linguistic code and which are used as supplements to them. It aims to illustrate, disambiguate, or even intensify the message contained in a text (Nord [1], p. 108). They may include photos, illustrations, emblems, special types of print, etc. They should not be mistaken for suprasegmental features.

Table 5. Results of Trimming the Non-Verbal Elements

Indicators	Source Text	Trimming Results
Inclusion of non-verbal elements	Available	Available
Functions of non-verbal elements	Illustration	Clarification
Conventionality bound to text type	Clarification	Clarification
Determination by medium	Unfocused	Focused
Relations to source culture	Related	Related

Table 5 presents the summary of the trimming results involving the use of non-verbal elements in the articles under analysis based on the five indicators as seen in Column 1. The table shows that the trimming process has yielded several improvements. While the inclusion of non-verbal elements remains the same, the functions of non-verbal elements and the conventionality bound to text type are clarified, the determination by the medium is focused and the relations of the elements to the source culture are made explicit. These aim to make the whole contents of the draft fulfill the prospective reader's needs.

Excerpt 5

Table 6. Variable test results that affect student satisfaction

Exogenous Variables	Score t	α- 5%	Explanation
Soft Skills Lecturer	4.77	1,96	Significant
Work Culture	3.12	1,96	Significant
Service quality	3.82	1,96	Significant

$R^2 = 0.72$

ST: Based on the results of calculations with the Lisrel program 8.54 using a significance level of 5% (1.96) as in Table 4.26 above, the soft skills variable of the lecturer obtained a score of t 4.77> 1.96, the work culture variable gained a score of t 3.12> 1.96 and the variable of service quality obtained t score 3.82> 1.96. The t value of each exogenous variable is greater than the t score at the significance level of 5%.

TR: Table 4 shows that with a significance level of 5% (1.96), the soft skills variable of lecturers gave a score of t 4.47> 1.96; work culture variable t 3.07> 1.96. The t value of each exogenous variable is greater than the t value at the significance level of 5%.

In **Excerpt 5**, the table is followed by a lengthy sentence describing it. While the ST strives to elaborate the meaning of each of the figures in the table to clarify it, the use of unnecessary elements, as well as the repetitions, has caused the description to be more than required by its prospective reader. If the description is meant to make the table easier to understand, then it should have been simpler than the table itself. The TR shows it clearly.

3.1.6. Lexical Items

According to Nord [1], lexical items refer to the affiliation of words to stylistic levels and registers, word formation, connotations, rhetorical figures, parts of speech, morphological aspects, collocations, idioms, addressing, word choices, degree of originality, etc. Nord also states that "the choice of lexis in a particular text is determined by extratextual and intratextual factors" (p. 112).

Table 7. Results of Trimming the Lexical Items

Indicators	Source Text	Trimming Results
Reflection of extratextual factors in using dialects, varieties, register, convention, functions, etc.	Available	Improved
Indication of writer's attitude through lexical items (stylistic markers, connotations, figures of speech, coinages, etc.)	Minimal	Maximiz-ed
Representation of fields of lexis (terminologies and metalanguage)	Common	Specified
Dominant use of parts of speech and patterns of word formation	Nominalization	Verbal-ization
Dominant use of rhetorical style	Available	Varied

Table 7 presents the summary of the trimming results involving the use of lexical items in the articles under analysis based on the five indicators as seen in Column 1. The table shows that the trimming process has yielded several improvements. While the reflection of extratextual factors in using dialects, varieties, register, convention, functions, etc. is improved, the indication of writer's attitude through lexical items (stylistic markers, connotations, figures of speech, coinages, etc.) is maximized, the representation of fields of lexis (terminologies and metalanguage) is made more specific, parts of speech and patterns of word formation are verbalized, and the use of rhetorical style is made varied. These aim to make the whole contents of the draft fulfill the prospective reader's needs.

Excerpt 6

ST: Based on the results of the study showed that the influence of soft skills of lecturers on service quality has a coefficient of 0.79 this means that the influence of soft skills of school lecturers on service quality is $(0.79)^2$, ie 0.6241 or 62.41% changes that occur in the quality of service to students of anesthesia specialist doctor UNDIP Semarang is caused by the soft skills of lecturers, the remaining 37.59% service quality is influenced by other variables.

TR: The results of this study indicate that the influence of soft skills of lecturers on service quality gives a coefficient of 0.79. That is, soft skills of lecturers improve the quality of service by $(0.79)^2$ or 62.41% to students of anesthesia specialist doctor UNDIP Semarang; the rest (37.59%) is influenced by other variables.

In **Excerpt 6**, the ST has been made simpler by trimming it employing the similar lexical items to construct a new sentence showing close relations among its elements. Using a complex sentence, whose elements are arranged, the TR promises a high level of readability compared with its counterpart in the ST.

3.1.7. Sentence Structure

Sentence structure, among other things, includes questions whether (1) it is mainly paratactic or hypotactic; (2) it contains simplex or complex sentences; (3) there are deviations from functional sentence perspective; (4) the text flow with syntactic figures of speech, parallelism, chiasm, rhetorical question, parenthesis, ellipsis, etc. (Nord [1], pp. 118-120); it has specific functions in the source text.

Table 8. Results of Trimming the Sentence Structure

Indicators	Source Text	Trimming Results
Sentence length including coordination and subordination	Lengthy	Rationalized
Dominant occurence of sentence types	Personal-ized	Impersonal-ized
Theme-rheme (marked-unmarked)	Marked	Unmarked
Occurence of text 'relief'	Simple	Modalized
Functions of syntactic figures of speech (parallelism, chiasm, rhetorical question, parenthesis, ellipsis, etc.)	Inconsistent	Consistent
Syntactic features which are recipient-oriented including adoption and adaptation	Writer-oriented	Reader-oriented

Table 8 presents the summary of the trimming results involving sentence structure based on the six indicators as seen in Column 1. The table shows that the trimming process has yielded the following improvement: the text length is rationalized, the sentences are impersonalized, the theme-rheme relations are made unmarked, the sentences are modalized whenever necessary, the functions of syntactic figures of speech are made consistent, and the syntactic features are made reader-oriented. These are made for the benefits of the prospective reader's needs.

Excerpt 7

ST: Speech therapy is very helpful because people with autism have a lack of communication and must be done during the development of speech.

TR: Speech therapy for people with autism is very important and must be given to him during the development of speech because he faces problems in communication.

In **Excerpt 7**, the ST is trimmed by means of replaceing the adjective "helpful" with "urgent". The subject which is a noun phrase "Speech therapy" is extended by modifying the noun phrase "penyandang autis" to the prepositional phrase "bagi penyandang autis" to function as the post modifier of the subject. The passive verb "to be done", a generalized verb in Indonesian, is replaced with a particular verb "diberikan" ("to be given"), which more common in English. The replacement, modification, and word arrangement have made the sentence more readable.

3.1.8. Suprasegmental Feautures

Suprasegmental features function "to highlight or focus certain parts of the text and to push others to background" (Nord [1], p. 80), and they possess both an informative (i.e. denotative) and a stylistics (i.e. connotative) function. In writing, they involve such elements as italics, spacing, bold type, quotation marks, dashes, parentheses, underlining, affirmative words (actually, in fact), emphatic evaluations (fantastic, great), clefts (It was John who...), ellipsis, etc.

Table 9. Results of Trimming the Suprasegmental Features

Indicators	Source Text	Trimming Results
Representation of suprasegmental features	Inconsist-ent	Made consistent
Specification of suprasegmetal features	General-ized	Specified
Provision of clues to emotional state of the sender	Implicit	Explicit
Division of text to prosodic units for clarity, stress, or focus	Available	Specified
Correspondence of suprasegmental features to theme-rheme structure	Inconsist-ent	Made consistent
Adaption of suprasegmental features in TL patterns	Minimal	Improved

Table 9 presents the summary of the trimming results involving sentence structure based on the six indicators as seen in Column 1. The table shows that the trimming process has yielded the following improvement: the suprasegmental features and the division of text to prosodic units are specified for clarity, stress, or focus, the representation of suprasegmental features is made consistent, the clues to emotional state of the sender are made explicit, the correspondence of suprasegmental features to theme-rheme structure is made consistent, and the adaption of suprasegmental features in TL patterns is improved. These are made for the benefits of the prospective reader's needs.

Excerpt 8

ST: Research and Development research methods, through 4 stages, namely (i) the preliminary research stage, (ii) the model development stage, (iii) the validation stage and (iv) the final model.

TR: This research applies the Research and Development approach through four stages, namely (1) preliminary research, (2) model development, (3) validation, and (4) final model.

Excerpt 8 shows the misuse of the comma (,), numbering, and repetition of the word "tahap" (step"). The suprasegmental features are then trimmed by correcting how the features are commonly used in English texts except for the italicization capitalization of the words "Research and Development" which has been adopted from the English language. The trimming process, along with the correct use of lexis and sentence structure, has improved the sentence.

3.1.9. Text Readability

As shown in Mujiyanto [6], the readability of the source texts and their trimmed ones' translations was assessed using the seven readability formulas available in http://www.readabilityformulas.com. They were employed to help determine the grade levels of the two translations. The followings are the seven formulas along with their uses. (1) The Flesch Reading Ease formula provides numbers ranging from 0 to 100: A lower score indicates harder reading and vice versa; *(2)* The Flesch-Kincaid Grade Level indicates the average reader of the texts; (3*)* The Fog Scale compares syllables and sentence lengths, a higher score indicates easier reading and the other way round; (4) The SMOG Index *i*ndicates the average reader capable of reading the text with minimal difficulties; (5) The Coleman-Liau Index relies on characters: A 10.6 grade implies the text appropriateness for 10-11th graders; (6) Automated Readability Index shows the grade level index needed to comprehend the text: Index 3 implies 3rd graders ability to comprehend the texts; (7) Linsear Write Formula is designed to calculate the grade level of a text-based on the sentence length and the number of words containing three or more syllables.

Table 10. Readability Indices

No	Readability Formulas	TL Manuscripts		Quality
		Original	Trimmed	
1	Flesch Read-ing Ease score	48.3	51.2	(Fairly) difficult to read
2	Gunning Fog	11.9	11	Hard to read
3	Flesch-Kincaid Grade Level	9.4	8.5	Ninth Grade
4	Coleman-Liau Index	9	10	Nine (Tenth) Grade
5	The SMOG Index	9.1	8.3	Ninth (Eight) Grade
6	Automated Readability Index	6.1	5.5	Fifth to Sixth Graders
7	Linsear Write Formula	7.3	5.8	Seventh (Sixth) Grade
8	Readability Consensus	8	8	Reading Level: (Fairly) difficult to read; 12-14 yrs. old (Seventh and Eighth graders)

Table 10 shows that, in general, the trimmed text gives lower scores than the original one, except the Coleman-Liau Index. As for the Flesch Reading Ease score, is noted that the higher the score of the text, the easier it is to understand. Therefore, by and large, the trimmed text is relatively more readable than the original one meaning that it is more appropriate for college graduates and beyond who are the target readers of the texts. In other words, the trimmed text is more readable than the original one, implying that the trimming of a text prior to its machine translation is needed for ease of understanding.

Table 11. Word Statistics

No	Characters, Syllables, Words	TL Manuscripts	
		Original	Trimmed
1	Total # of words	3,292	3,191
2	Total # of unique words	27%	35%
3	Average # of words per sentence	12	10
4	Average # of characters per word	4.6	4.7
5	Average # of syllables per word	2	2
6	Percent of double syllables in text	18%	17%
7	Percent of single syllables in text	62%	62%
8	Percent of 3+ syllables in text	21%	21%

The use of these measures also results in word statistics as they are presented in Table 11 showing the situation in which the translation of the original text contains 3,292 words and it's trimmed one 3,191 words. The table also shows that the average number of words per sentence and percent of double syllables in the trimmed text is lower than that of the original one. Besides, the total number of unique words and the average number of characters per word in the trimmed text is higher that that of the original one implying that the trimming process contributes to the standadization of the text. In other words, the trimming process.

4. Discussion

Nord's [1] introduction to text analysis in translation has triggered many researchers to undertake studies that are focused on the functions as well as the significance of text analysis in translation context. Zlateva's [21] questions about the reasons, processes, and achievements of the analysis have been answered comprehensively, while. Munday's [16] explanation about the relations between text analysis and translation has also been answered. Meanwhile, Ayupova's [22] assessment of text analysis as a step in the pretranslation rendering process and Laviosa's et al. [23] examination of the source text and its context in translation practices have yielded urgent findings on the importance of source text analysis which is oriented to translation purposes.

Other studies emphasizing the importance of source text analysis by Troqe and Marchan [24], Rossetti and Gaspari [24], Laviosa, et al. [23], Hansen-Schirra, Czulo, and Hofmann [35], Dicerto [26], [27], [28], and Boukhaffa [29] prove that source text analysis which is an integral part of translation practices show nothing more than its role as the preliminary assessment of the source text prior to its rendering to the target language.

The implementation of machine translation that has been studied since more than five decades ago has been shown by Nirenburg, et al. [30], Birch, Osborne, and Koehn [31], and Kastberg [32] who sees the use of machine translation tools for a professional translator. Meanwhile, Lennon's [33] use of machine translation from the perspective of different cultures and Rozmyslowicz's [34] study on issues related to the use of machine translation disclose the importance of machine translation even though dependency on human work is still required.

The inevitable use of machine translation tools and language aids in translation undertaking necessitate the translator to trim the source text proceeding the source text analysis and before its machine translation. It is this particular point which has made this study significant to be conducted, assuming that such a supplementary phase triggers sound restructuring texts in the target language.

5. Conclusion

As a crucial phase in the rendering process, source text analysis proposed by Nord [1] is constrained by extra- and intra-textual factors, the latter consisting of *subject matter, content, presuppositions, text composition, non-verbal elements, lexis, sentence structure, and suprasegmental features*. Aimed to consider the importance of each of the eight factors, this study find out that source text analysis alone is not adequate to prepare manuscripts written in Indonesian to be rendered into English employing machine translation tools. This has triggered the idea of implementing a process of trimming the manuscripts considering the eight factors. The trimming process, which refers to a process of making a text tidier for inter-lingual machine rendering, is implemented to the texts considering the issues to be more ready for the translation.

Trimming 50 manuscripts in Indonesian for their rendering to Englsih employing machine translation tools, this study disclose several aspects assumed to improve the quality of the manuscripts before their machine translation. The aspects include, among other things, improvement of the eight factors, making the text sound plausible by minimizing redundancies and maximizing clarity as well dependency on the effective uses of macro- and micro-structures of the texts. The trimming process also takes into account the clarity of relations between verbal texts and non-verbal images, the expliciteness in meaning making despite the use of methaphors, parts of speech, connotations, and collocations.

The effectiveness of the trimming process is assessed by comparing the results of rendering the original texts and the trimmed ones in terms of their levels of readaility. The comparison employing the readability indices results in evidence that the trimmed text is a bit more readable that the source or original text and there is a reduction of word count. Given that the intra-textual factors are influential in source text analysis for trimming purposes, it is suggestible to consider them as a supplementary procedure for mechanical rendering of academic manuscripts in Indonesian prior to their translation to English.

Acknowledgments

I would like to express my most heartfelt and sincere gratitude to the Rector and Vice-Rectors of Universitas Negeri Semarang for their encouragement and invaluable support to conduct an academic study as the basis for preparing this article. Their generous and thoughtful assistance meant a great deal to me in completing this article.

References

[1] M. Machado and E. Tao, "Blackboard vs. Moodle: Comparing user experience of learning management systems," *Proc. - Front. Educ. Conf. FIE*, no. December 2006, pp. 7–12, 2007.
[2] O. C. Rodriguez, "MOOCs and the AI-Stanford Like Courses: Two Successful and Distinct Course Formats for Massive Open Online Courses," *Eur. J. Open, Distance E-Learning*, p. 13, 2012.
[3] P. I. Santosa, "Measuring User Experience in an Online Store Using Pulse and Heart Metrics," *J. Ilm. Kursor*, vol. 7, no. 3, pp. 145–152, 2014.
[4] A. Gordillo, E. Barra, S. Aguirre, and J. Quemada, "The usefulness of usability and user experience evaluation methods on an e-Learning platform development from a developer's perspective: A case study," *2014 IEEE Front. Educ. Conf. Proc.*, pp. 1–8, 2014.
[5] Elena Sanchez, "Can UX Design Improve MOOC Completion Rates?" 2013.
[6] C. Colombati, "User Experience (UX) and user engagement : a case study Supervisor : Anna Maria Tammaro Master thesis International Master in Digital Library Learning," 2015.
[7] Elena Sanchez, "Can UX Design Improve MOOC Completion Rates?" 2013.
[8] W. Abeer and B. Miri, "Students' Preferences and Views about Learning in a MOOC," *Procedia - Soc. Behav. Sci.*, vol. 152, pp. 318–323, 2014.
[9] L. Czerniewicz, A. Deacon, M. Fife, J. Small, and S. Walji, "Introduction to MOOCs : definition and characteristics MOOCs in the broader higher education landscape," *Cent. Innov. Learn. Teach.*, 2015.

[10] T. C. Johnston, "What makes a MOOC? Massive Open Online Courses (MOOCs) Compared to Mainstream Online University Courses," *J. Learn. High. Educ.*, vol. 19, no. 1, pp. 17–23, 2014.
[11] J. P. Espada, C. C. Rodriguez, V. Garcia-Diaz, and R. G. Crespo, "Method for analysing the user experience in MOOC platforms," *2014 Int. Symp. Comput. Educ. SIIE 2014*, pp. 157–162, 2014.
[12] H. B. Santoso, R. Y. K. Isal, T. Basaruddin, L. Sadita, and M. Schrepp, "Research-in-progress: User Experience Evaluation of Student Centered E-Learning Environment for Computer Science Program," *2014 3rd Int. Conf. User Sci. Eng.*, pp. 52–55, 2014.
[13] N. Putu, I. Rosita, S. Wibirama, and P. I. Santosa, "Evaluating User Experience of English Learning Interface using User Experience Questionnaire and System Usability Scale," pp. 101–106, 2017.
[14] M. Pušnik, D. Ivanovski, and B. Šumak, "Patterns for improving mobile user experience," *2017 40th Int. Conv. Inf. Commun. Technol. Electron. Microelectron. MIPRO 2017 - Proc.*, pp. 582–587, 2017.
[15] J. Renz, A. Shams, and C. Meinel, "Offline-enabled web-based E-leaming for improved user experience in Africa," *2017 IEEE AFRICON Sci. Technol. Innov. Africa, AFRICON 2017*, pp. 736–742, 2017.
[16] D. S. S. Sahid, P. I. Santosa, R. Ferdiana, and E. N. Lukito, "Evaluation and measurement of Learning Management System based on user experience," *2016 6th Int. Annu. Eng. Semin.*, pp. 72–77, 2016.
[17] A. Sheshasaayee and M. N. Bee, "Evaluating user experience in moodle learning management systems," *2017 Int. Conf. Innov. Mech. Ind. Appl.*, no. Icimia, pp. 735–738, 2017.
[18] K. Rodden, H. Hutchinson, and X. Fu, "Measuring the User Experience on a Large Scale : User-Centered Metrics for Web Applications," *Proc. SIGCHI Conf. Hum. Factors Comput. Syst.*, pp. 2395–2398, 2010.
[19] J. Scholtz, E. Morse, and M. P. Steves, "Evaluation metrics and methodologies for user-centered evaluation of intelligent systems," *Interact. Comput.*, vol. 18, no. 6, pp. 1186–1214, 2006.
[20] P. N. Lestari, P. I. Santosa, and R. Ferdiana, "Pengukuran Pengalaman Pengguna Dalam Menggunakan Sistem Informasi Akademik," *Semin. Nas. Teknol. Inf. dan Komun.*, vol. 2016, no. Sentika, pp. 2089–9815, 2016.
[21] M. L. Khakim, O. O. Sharif, S. Prodi, M. Bisnis, F. Ekonomi, and U. Telkom, "Analisis User Experience Aplikasi Go-Jek Menggunakan Heart Metrics Analysis User Experience Of Go-Jek Applications," vol. 5, no. 1, pp. 189–194, 2018.
[22] S. Informasi, F. Sains, U. I. N. Syarif, and H. Jakarta, "Pratama, Analisis User Experience Aplikasi Academic Information System (AIS) Mobile Untuk User-Centered Metrics Menggunakan Heart Framework 405," vol. 8, no. September, pp. 405–412, 2019.
[23] Consuelo G. Sevilla...et.al, *An intorduction to research methods*. Manila: Rex Book Store, 1988.
[24] Sugiyono, *Metode penelitian kuntitatif kualitatif dan R&D*. 2011.

IOP Conf. Series: Materials Science and Engineering **846** (2020) 012044 doi:10.1088/1757-899X/846/1/012044

On the Periodic Solution of an Oscillator Single Degree of Freedom with a Damping Coefficient that Changes Periodically

S. B. Waluya [1]

[1] Mathematics Department, Universitas Negeri Semarang, Semarang, Indonesia

* s.b.waluya@mail.unnes.ac.id

Abstract. In this paper will be studied a simple one degree of freedom system related to the dynamics of cable-stayed bridges. As a result of wind and rain that hits the bridge cable, it will cause vibrations from the cable to become unstable conditions. The presence of water flowing in the cable changes the cable cross-section as experienced by the wind plane. So, this is what causes the instability of cable vibrations. The idea to model this problem is to consider a horizontal cylinder supported by springs in such a way that only one degree of freedom, i.e. vertical vibration is possible. It will be studied a ridge on the surface of the cylinder parallel to the axis of the cylinder. Additionally, let the cylinder with ridge be able to oscillate, with small amplitude, around the axis such that the oscillations are excited by an external force. In this paper it will be studied the possibilities of the damping of the oscillator that changes periodically by using a perturbation method.

1. Introduction

The mathematical model that describes vibration of the cables in vertical direction because of the galloping phenomenon system of differential equation becomes a. The characteristic of cables of cable-stayed bridges usually smooth polyurethane mantle and have a cross section which is nearly circular. In the rain season the water on the cable may induce aerodynamic instability resulting in vibrations with relatively large amplitudes. The Den Hortog's criteria should be considerate for modeling of vibration from a cable bridge. This criterion considers several assumptions that can cause a galloping phenomenon that is vibrations with small frequencies and large amplitude. (see [[1], [2]. [3], [4]]). The instability mechanism for this type of vibrations is known and can be understood on the basis of quasi-steady modeling and analysis. The experimental validation of a new energy harvesting system based on the wake galloping phenomenon has been studied by Jo Jung and Woo Lee [5].

Piccardo, et.al [1], Van Horssen, et.al, [2], and Waluya [3], [4] have studied oscillation of bridge cable because of wind induce vibration, while mechanism and characteristic of rain-induced vibration on high-voltage transmission line have been studied by Zhou et.al [6]. The dynamic of cable stayed bridges with a time periodic damping coefficient have been studied by Van Horssen, et.al [2] and Waluya [3], [4].

Many types of the oscillators have been studied by many researchers. Afzali, et.al [7] have studied Analysis of the Periodic Damping Coefficient Equation Based on Floquet Theory, Bayat, et.al [8] have studied Non-Linear Oscillation by series solution technique. On the separation of fast and slow motions in mechanical systems has been studied by Blekhman and Sorokin [9]. Analytical solution of strongly nonlinear Duffing oscillators has been studied by El-Naggar and Ismail [10]. Mechanism and characteristic of rain-induced vibration on high-voltage transmission line have been studied by Zhou et.al [6]. Von Wantoch et. Al [11] have studied Adaptive phasor control of a Duffing oscillator with unknown parameters. A novel method for the forced vibrations of nonlinear oscillators have been studied by En Du, et.al [12]. Ismail [13] has studied Analytical Technique for Solving Nonlinear Oscillators of the Motion of a Rigid Rod Rocking Bock and Tapered Beams. Many researcher have studied a various perturbation techniques to analyze nonlinear oscillators, such as a modified homotopy analysis method [14], Floquet Theory [7], Multiple-Scales Lindstedt-Poincaré Method [15], a modified homotopy analysis method [14], multiple time scales perturbation [2]. [3], [4], and many others perturbation technique. In this paper will be studied vibration of an oscillator due to small masses which periodically changing with slow and high frequencies. In [3], [4] used the assumption that external forces were not taken into account in the system, while in this paper will be studied that the force energy give significant impact to the vibration.

This paper is organized as follows. In section 2 of this paper it will be shown the mathematical model. The analysis of the solution will be studied in section 3 of this paper. Finally, in section 4 of this paper some conclusions will be drawn and some remarks will be made.

2. Mathematical Model

In this section will be derived a mathematical model which describe the rain-wind induced vibrations of a simple one degree of freedom system related to the dynamics of cable-stayed bridges. The model is an oscillator in one degree of freedom system consisting of a horizontal rigid cable supported by springs with a solid-state ridge moving with small amplitude oscillations. From the point of view, the type of equation of motion will arrive at a second order differential equation with external forcing. The Model Equation for Rain-Wind Induced Vibrations of a Prototype Oscillator is closely related to the quasi-steady approach as given in [2], [7].

Then the equation of motions can be derived by

$$m\ddot{y} + c_y\dot{y} + k_y y = F_y, \tag{1}$$

Where m is the mass of the cylinder, $c_y > 0$ the damping coefficient of the mechanical structure and $k_y > 0$ the constant linear spring. $\dot{y}$ represents the velocity of the system. Set $v = \dot{y}$, and by using aerodynamic force ($F_y = -(D\sin\phi + L\cos\phi)$) then equation (1) becomes.

$$m\ddot{y} + c_y\dot{y} + k_y y = -\frac{1}{2}\rho dl\sqrt{U^2 + \dot{y}^2}(C_D(\alpha)\dot{y} + C_L(\alpha)U) \tag{2}$$

By using the following new variables and parameters. After linearizing the equation (2) (see also [6], [7] for more details reason), we can obtain :

$$\ddot{Z} + (2\beta + KC_{D0} + KC_{L1})\dot{Z} + Z = -KC_{L0} \tag{3}$$

By using some assumptions an according to the galloping phenomena after using transformation we can obtain

$$Z + Z^{\cdot\cdot} = -Kf(t)[C_{L1} + C_{L3}f(t)^2] + [-2\beta - KC_{D0} - 3KC_{L3}f(t)^2 - KC_{L1}]\dot{Z} \tag{4}$$

Let $f(t) = B\sin(\varphi\tau), \varphi = \frac{\omega}{\omega_y}$, then from equation (4) is obtained :

$$\begin{aligned}\ddot{Z} + Z = &\left[-2\beta - KC_{D0} - \tfrac{3}{2}KC_{L3}B^2 + \tfrac{3}{2}KC_{L3}B^2\cos(2\varphi\tau) - KC_{L1}\right]Z^{\cdot} \\ &+ \left[-KBC_{L1} - \tfrac{3}{4}KB^3C_{L3}\right]\sin(\varphi\tau) + \tfrac{1}{4}KB^3C_{L3}\sin(3\varphi\tau)\end{aligned} \tag{5}$$

Set $\varphi = 1 + \varepsilon\eta$, $\varepsilon \ll 1$ and $(1 + \varepsilon\eta)\tau = \phi$, then equation (10) becomes :

$$\begin{aligned}(1+\varepsilon\eta)\ddot{Z}+Z=(1+\varepsilon\eta)\Big[-2\beta-KC_{D0}-\tfrac{3}{2}KC_{L3}B^2\\+\tfrac{3}{2}KC_{L3}B^2\cos(2\phi)-KC_{L1}\Big]\dot{Z}\\+\Big[-KBC_{L1}-\tfrac{3}{4}KB^3C_{L3}\Big]\sin(\phi)+\tfrac{1}{4}KB^3C_{L3}\sin(3\phi),\end{aligned} \quad (6)$$

with $\dot{Z}=\frac{dZ}{d\phi}$. Let K and β are small, that are $K=\beta=O(\varepsilon)$ or $K=\varepsilon K_1,\beta=\varepsilon\beta_1$, and by neglecting $O(\varepsilon^2)$, then it will be obtained from (6) :

$$\ddot{Z}+Z=\varepsilon\big[-(a_0+a_1\sin(2\phi))\dot{Z}-a_2 sin(\phi)-a_3\sin(3\phi)\big], \quad (7)$$

with

$$a_0=2\beta_1+K_1C_{D0}+\frac{3}{2}K_1C_{L3}B^2+K_1C_{L1}\ a_1=-\frac{3}{2}K_1C_{L3}B^2,$$
$$a_2=K_1BC_{L1}+\frac{3}{4}K_1B^3C_{L3},\ a_3=-\frac{1}{4}K_1B^3C_{L3}$$

3. Solution of The Model

In this section will be studied periodic solution of equation (7). The averaged system of equation (7) can be written as (see how to averaged a system in papaers [6], [7], [8] :

$$\begin{aligned}\dot{\overline{y_1}}&=\varepsilon\Big[-\tfrac{1}{2}a_0y_1+\tfrac{1}{2}a_2-\eta y_2+\tfrac{1}{4}a_1y_1\Big]\\ \dot{\overline{y_2}}&=\tfrac{\varepsilon}{4}[a_1y_1-2a_0y_2+4\eta y_1],\end{aligned} \quad (8)$$

The critical points of equation (8) can be given by :

$$\left(\frac{4a_0a_2}{16\eta^2+4a_0^2-a_1^2},\frac{2a_2(a_1+4\eta)}{16\eta^2+4a_0^2-a_1^2}\right) \quad (9)$$

The existence of the critical point equation (9) is given by $16\eta^2+4a_0^2-a_1^2\neq 0$. If the determinant of the coefficient matrix of equation (8) is not equal to zero (more details see [2] for proof) then system equation (7) has a periodic solution and its stability depends on the eigenvalues of the coefficient matrix. The eigenvalues equation (8) can be given by :

$$\begin{aligned}\lambda_1&=-\tfrac{1}{2}a_0+\tfrac{1}{4}\sqrt{a_1^2-16\eta^2},\\ \lambda_2&=-\tfrac{1}{2}a_0-\tfrac{1}{4}\sqrt{a_1^2-16\eta^2}\end{aligned} \quad (10)$$

The periodic solution can be give by :

$$\begin{aligned}z&=y_1[0]\cos(\phi)+y_2[0]\sin(\phi)\\ \dot{z}&=-y_1[0]\sin(\phi)+y_2[0]\cos(\phi),\end{aligned} \quad (11)$$

with

$$y_1[0]=\frac{4a_0a_2}{16\eta^2+4a_0^2-a_1^2},y_2[0]=\frac{2a_2(a_1+4\eta)}{16\eta^2+4a_0^2-a_1^2}$$

Plot of the solution approximation and numerical result by using Runge-Kutta method for an initial condition and set of parameters can be given in Figure (1)-Figure (3).

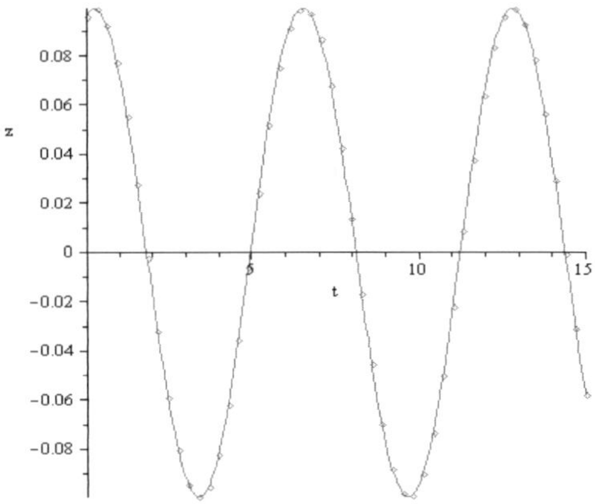

Figure 1. Plot solution approximation versus numeric for $a_0 = 1, a_1 = 0.1, a_2 = 0{,}1. a_3 = 0.1, \varepsilon = 0.01, z(0) = 1, \dot{z}(0) = 0$

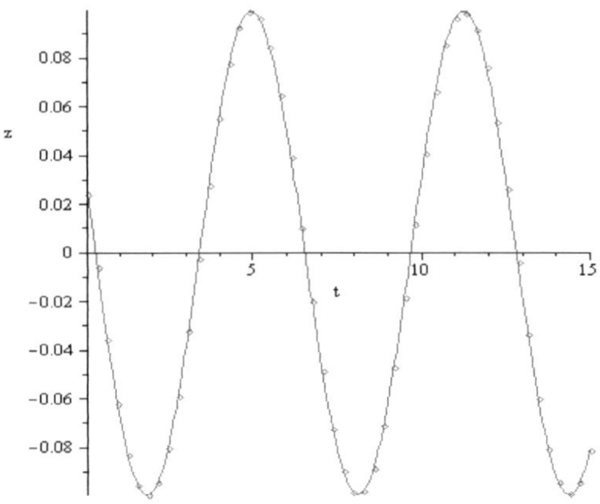

Figure 2. Plot solution approximation versus numeric for $a_0 = 1, a_1 = 0.1, a_2 = 0{,}1. a_3 = 0.1, \varepsilon = 0.01, z(0) = 1, \dot{z}(0) = 0$

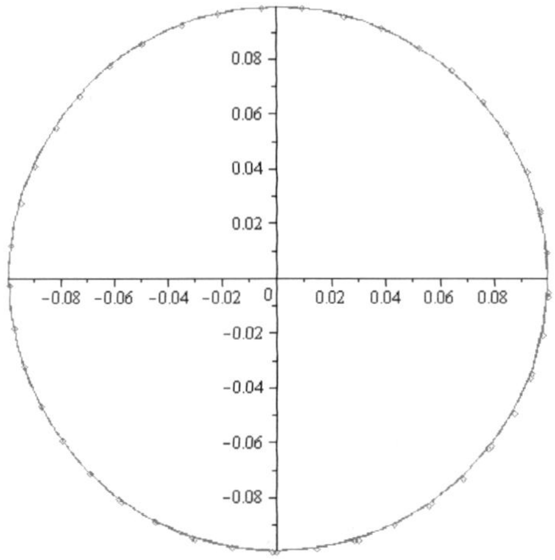

Figure 3. Plot phase portrait approximation versus numeric for $a_0 = 1, a_1 = 0.1, a_2 = 0{,}1. a_3 = 0.1, \varepsilon = 0.01, z(0) = 1, \dot{z}(0) = 0$

From Figure (1) and Figure (3) it can be seen that the approximation of the solution by using perturbation technique also similar with the solution by numerical calculation. From equation (10) can be shown that the stability of critical points (related to periodic solution) if $a_1^2 < \eta^2$, that is the damping coefficient bigger then frequency of the function which respresent of water ridge.

4. Conclusion

In the previous section has been discussed a mathematical model which describe the rain-wind induced vibrations of a simple one degree of freedom system related to the dynamics of cable-stayed bridges. Those vibrations are assumed in the vertical direction due to the galloping phenomena, that is vibrations with small frequencies and large amplitude. The critical points (periodic solutions) can be obtained if the damping coefficient more than the frequency of the function which represent of water ridge. And the periodic solution can be obtained if the damping coefficient bigger then frequency of the function which represent of water ridge.

References

[1] Piccardo, B, Pagnini, L. C, and Tubino. F, "Some research perspectives in galloping phenomena: critical conditions and post-critical behavior," *Continuum Mech. Thermodyn,* vol. 27, p. 261–285, 2015.

[2] Van Horssen, W.T, Pischanskyy, O & Dubbeldam, .J.LA, "On the forced vibrations of an oscillator with a periodically time-varying mass," *Journal of Sound and Vibration,* vol. 329, pp. 721-732, 2010.

[3] Waluya, S. B, "The dynamic of cable stayed bridges with a time periodic damping coefficient," *Int. J, Math, Sci, Engg, Appls, (IJMSEA),* vol. 4, no. 2, pp. 385-394, 2010.

[4] Waluya, S. B, "On the Analysis of an Oscillator with Periodically Time-Varying Mass and Forced Vibration," *Int. J, Math, Sci, Engg, Appls, (IJMSEA,* vol. 6, no. 1, pp. 249-261, 2012.

[5] Jo Jung, H and , Woo Lee, S, "The experimental validation of a new energy harvesting system based on the wake galloping phenomenon," *Smart Mater, Struct,* vol. 20, 2011.

[6] Zhou, C, Liu, Y, and Rui, X, "Mechanism and characteristic of rain-induced vibration on high-voltage transmission line," *J. Mech. Sci. Technol,* vol. 26, no. 8, p. 2505~2510, 2012.

[7] Afzali, F, Acar, G. D, and Feeny, B. F, "Analysis of the Periodic Damping Coefficient Equation Based on Floquet Theory," in *Int. Des. Eng. Tech. Conf. Comput. Inf. Eng. Conf. 8*, Ohio USA, 2017.

[8] Bayat, M, Pakar, I, Barari, A, Nejad, M. G, and Motevalli, M, "On Non-Linear Oscillation: an Analytical Series Solution," *Óbuda Univ. e-Bull,* vol. 4, no. 1, 2014.

[9] Blekhman, I. I., and Sorokin, V. S, "On the separation of fast and slow motions in mechanical systems with high-frequency modulation of the dissipation coefficient," vol. 329, pp. 4936-4949, 2010.

[10] El-Naggar, A. M and Ismail, G. M, "Analytical solution of strongly nonlinear Duffing Oscillators," *Alex. Eng. J,* vol. 55, p. 1581–1585, 2016.

[11] Von Wantoch, T, R¨ock, H, Koschmieder, F, and Benecke, W, "Adaptive phasor control of a Duffing oscillator with unknown parameters," in *19th World Congr. Int. Fed. Autom*, Control Cape Town S. Afr, 2014.

[12] En Du, H, Kang Er, G, and IU, V. P, "A novel method for the forced vibrations of nonlinear oscillators," in *5th Jt. Int. Conf. Multibody Syst. Dyn*, Lisboa, Portugal, 2018.

[13] Ismail, G. M, "Analytical Technique for Solving Nonlinear Oscillators of the Motion of a Rigid Rod Rocking Bock and Tapered Beams," *J. Appl. Comput. Mech,* vol. 1, no. 29-34, p. 2, 2016.

[14] Zurigat, M, Momani, S, and Alawneh, A, "Solving nonlinear oscillators using a modified homotopy analysis method Stud," *Univ, Babe¸s-Bolyai Math,* vol. 57, no. 4, p. 579–588, 2012.

[15] Pakdemirli, M, Karahan, M. M. F. and Boyacı, H, "A new perturbation algorithm with better convergence properties: multiple scales lindstedt poincare method," *Mathematical and Computational Applications,* vol. 14, no. 1, pp. 31-44, 2009.

IOP Conf. Series: Materials Science and Engineering **846** (2020) 012045 doi:10.1088/1757-899X/846/1/012045

Design System Integrated Licensing Service - Based Information Technology

Chris Batara
Department of Electrical Engineering, Universitas Kristen Indonesia Paulus Makassar

* Corresponding author: chrisbatara@gmail.com

Abstract The government management system, which has been a pointed and long hierarchical system of authority and command, must be developed into a network organization management system that can reduce decision making lines and broaden the span of control. Development of E-Government, it needs an in-depth concept to study and determine the workflow of a government process that will be developed and use technical standards and recommendations as a reference. In general, the objectives of this study are: (1) The establishment of quality, fast, and affordable network of public information and services to the wider community, (2) Formation of an interactive relationship with the business world to deal with changes and international trade competition. (3) Establishment of mechanisms and channels of communication with State institutions and the provision of public dialogue facilities. (4) Establishment of management systems and work processes that are transparent and efficient.
In this research specifically, the concept of e-government will be elaborated, which focuses on technological issues and then discusses the modelling of a one-stop licensing service system using the Unified Modeling Language approach which is one of the standard components in the e-government concept language in this study. To achieve these objectives, a method with several stages of activities is carried out as follows, namely: observing the licensing process and observations in several North Toraja District Offices, analyzing the requirements specifications in e-government development, creating a model of the one-stop licensing service mechanism that will be applied , making designs and making applications in the form of prototypes, after that implementation. The benefits obtained from this research are to increase the effectiveness and efficiency of work from the government, the bureaucracy path is more concise, the relationship between the government and the community and business people is increased, has greater opportunities and is open to join the global community.

1. Introduction

1.1 Aspects in E-Government Development

In addition to technological aspects, e-government is related to various non-technological aspects such as cultural aspects, organizational structure, policies and processes that take place in government. To facilitate management, e-government development is reviewed in two aspects or areas:

- Policy and planning aspects
  In this aspect discussed about the vision, mission, and strategy of e-government development, various kinds of challenges that will be faced in the development of e-government such as the

IOP Conf. Series: Materials Science and Engineering **846** (2020) 012045 doi:10.1088/1757-899X/846/1/012045

problem of rules on access to information, privacy, and confidentiality of documents, copyright, and others.

- In terms of application development
  In this aspect, it is discussed about the development of e-government from a technical perspective, including technology choices, determination of required standards, implementation of security, system integration, etc.

Figure 1, visualizing aspects of e-government development above. From another point of view, the two areas are the separation of problems in the field related to technology with non-technological parts, but it needs to be developed to ensure the successful implementation of e-government.

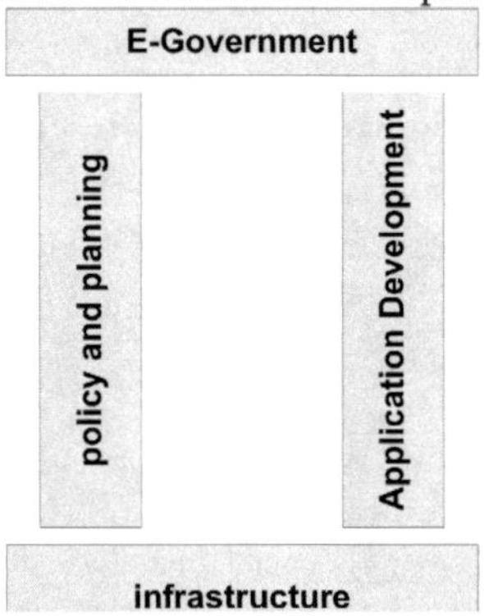

Figure 1. E-government Framework

1.2 E-Government Technology

E-Government was developed using internet-based technology because it is more flexible, more widely used and easier in terms of data interoperability. By using internet-based technology, e-government can be integrated with e-business and various applications that already exist on the internet. A varied and more innovative service can be developed by taking into account data protection and privacy issues. Existing data exchanges use the XML format (eXtensible Markup Language), as shown in Figure 2. XML is the industry standard for determining document formats so that they are easy to manage, interchange and operate.

With a platform like the one above, the government, the community and business people can utilize data together and coherently. Because the internet technology used to develop e-government continues to transform world life in terms of social, economic and cultural aspects, it must be fully understood the potential possessed by the Internet. E-government that is developed must be a system that can be trusted (trusted system) that is able to attract the interest of the community to use it. And security are two fundamental pillars in developing a system that can be trusted.

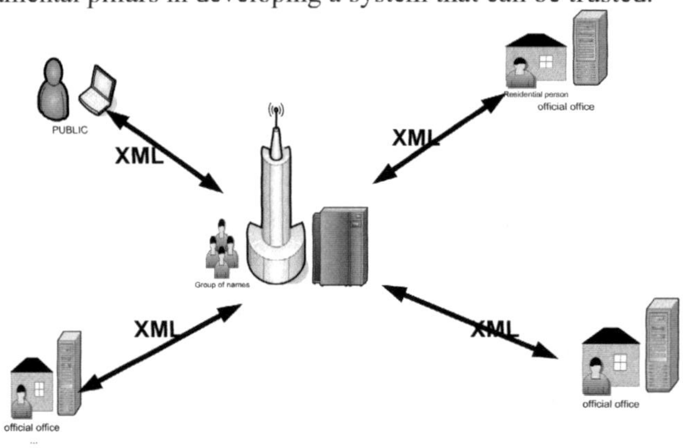

Figure 2. Interoperating e-government networks

1.3 E-Government System.

The e-government system was developed in layers (layered) both in terms of the technology used and the services offered.

Information Technology Framework for E-Government.

The general strategies that can be used in e-government development are as follows:

- Open standard system
- Model of Computing client / server
- Internet / Intranet
- Collaboration for evolutionary integration
- Multimedia services

1.4 E-Government Architecture

The e-government architecture that will be developed consists of three layers, namely transport service, basic service, and cooperative service. The transport service layer consists of the TCP / IP standard as an Internet protocol. The Basic Service Layer consists of services that run on TCP / IP protocols such as e-mail, file transfers, web, and directory access. Middleware at the cooperative service layer serves to facilitate the development and use of applications between departments or sections. Middleware is an architecture that serves to hide the rules or business logic that exists in the server so that the client does not know the process that occurs on the server. The middleware implementation can use COBRA (Comman Object Request Broker Achievement) from OMG.COM (Component Object Model) from Microsoft or EJB (Enterprise Java Beans) from SUN.

Public services provided can be developed directly from the Basic Service Layer or through applications developed that run above the Cooperative Service Layer. Services that run above the Basic Service Layer in the form of information and communication both one-way and two-way. While the services that run above the Cooperative Service Layer are generally in the form of transactions that require collaboration with other parties, such as banking and related departments.

Integrated Services. To provide integrated services, e-government is divided into three layers as follows:

a. Front Office

Functioning as a service center provided to the community. This Front Office has various media (channels) to present services. In addition to electronic services, the public can still use the benefits of non-electronic services using paper-based correspondence. Electronic services are concentrated on Internet based portals.

b. Back Office

Serves as a place to carry out processes and coordination to provide services, both services that are provided or based on community requests. Some services that may be provided in an integrated manner in this back office are, among others, handling certificates, births, deaths and marriages, education, health, taxation, tariffs, licenses, customs, policy rules and others. To provide integrated services, the components that must be met are the common infrastructure, integration enabling mechanism, and specific business processes.

c. Policy Development & Evaluation

Needed to improve the quality of e-government services continuously.

This e-government architecture will cover the complexity of the bureaucracy needed to provide a service. This is very beneficial for the community, because the community does not need to know and be involved in the provision of services. The community only needs to mention the service they need, complete the requirements needed then wait for the process for the service. The community does not need to go to the department / office or office to resolve the issue.

1.5. E-Government Infrastructure

Infrastructure needs to be built to support the e-government system.

- Server with security system

This server is used to provide e-government services. The required server can consist of several servers for several services and departments / services, or one server for all services handled by that server. If each department / service uses its own server, this server can be placed at the ISP or intranet of each department / service. The function of each department / service server is to support e-government servers that are accessed directly by the public.

- Intranet in the department / department involved in e-government.
  This intranet or LAN (Local Area Network) is needed so that local government employees in each department / service can be actively involved in providing e-government services. Because in the implementation of e-government they are dealing directly with the services provided. Intranet is connected to the internet through several alternatives including VSAT, dial up and leased lines. The intranet of each department can be connected to each other using the internet network or independently. If the intranet is interconnected, data exchange can be done between intranets but if it is standalone, the data exchange is done through an e-government server.

1.6 Interpretabilities E-Government

The e-Government system that will be developed is e-government that is able to provide electronic services that are integrated with data and information management carried out in a distributed manner in the departments related to the data and information. So from the community side that uses e-government systems, all data management or information retrieval processes are seen from one application in one server - even though in the process of processing data interoperation occurs between servers that support the e-government system. While from the government side as a service provider, data and information management is carried out on the servers of each relevant department.

This system has advantages for both the community and the government. The public will avoid frustration in using this system, because he only faces one application to take care of various needs. Conversely for the government, information management will be easier, because each department can process information without being interrupted by other departments. To develop an e-government system that has integrated services and data interoperability, two things must be done before the system is implemented, namely:

- Determine metada interoperability standards for the exchange of metadata between e-government servers of each department.
- Determine the exchange mechanism (interchange) and mirroring of e-government server anter metadata.

So the information is actually stored and managed in each e-government server of each department, while what is exchanged is the metadata of that information.

As an illustration, examples of population data are taken. This population data can be in the form of personal data (place of birth date, status, occupation, address, etc.), educational history, employment history, history of crime. It would be difficult if forced to manage this population data in one database server because the data is related to different departments. Personal data is managed in the sub-district office, education history is managed in the education and culture department, work history is managed by the labor department, and a history of crime is managed by the police office. A distributed system can be developed to manage this population data. Each department has its own database server to store and manage the population data that is the responsibility of the department. If complete population data is needed, there will be a mechanism for exchanging data between database servers in each department. The benefits gained from such a system are data integration, data duplication can be overcome , ease in managing and with a distributed system data processing becomes easier, because each department only manages data that is its responsibility.

1.7 E-Government Standards

To guarantee data interoperability, it is necessary to set standards and specifications in the development of e-government. These standards and specifications will be used as a reference for e-government development both at the time of initiation and later when e-government will be developed

IOP Conf. Series: Materials Science and Engineering **846** (2020) 012045 doi:10.1088/1757-899X/846/1/012045

and integrated with other systems. Standards and specifications required include interconnection problems, data integration, and data or information formats. The standards described in this paper are based on standards issued by the Internet Engineering Task Force (IETF) and recommendations made by the W3C (World Wide Web Consortium), as well as several other components that are widely known for their use.

2. Research Method

The research was carried out through several stages as shown in Figure 3 below.

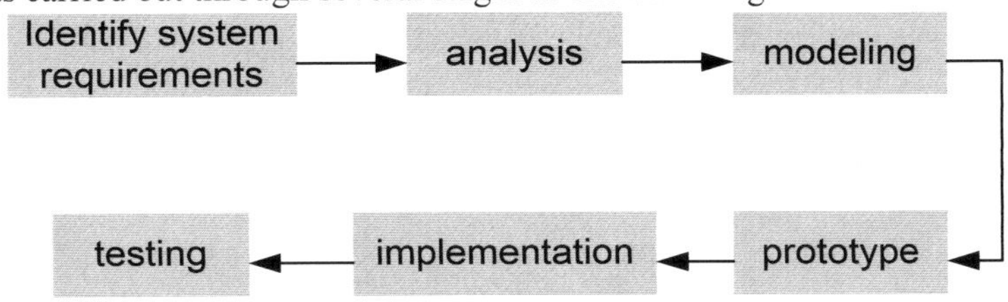

Figure 3. Research Stages

The first phase of this study began with the identification of user needs which was carried out through observation of the licensing service process at the Regional Government Level II North Toraja. The second stage, the results of identification of system requirements are then analyzed to determine the functions or activities that will be applied in the system. The third stage is modeling the functions that will be implemented, doing the process of designing a complete system software using UML (selecting program packages that are appropriately analysis suitable for building systems). The fourth stage is the implementation of modeling in the form of a prototype, the fifth stage is the implementation of the prototype by attempting to register the investment licensing process, the inventory process of meeting results, the process of forming official memorandum and the district head's decision process are the parts that are implemented in the prototype. The sixth stage is testing.

2.1 Research procedure

In conducting this research, the steps are:

A. Determine user needs
Before designing a system, user needs are first determined which includes the capabilities and technical specifications that must be owned by the system. From this first stage will be obtained a need and specifications for the system that will be a reference in the implementation of the next stages.

B. Analysis of System Needs
From the results of the identification of system requirements, an analysis is carried out to determine the functions or activities that will be applied in the system.

C. Making a Model
Model the functions to be applied, then do the system design process in full using UML.

D. Prototipe
At this stage the implementation of the modeling that has been obtained is in the form of a prototype. The model produced at the modeling stage is developed and made more detailed by including the technical aspects of design. From the prototype stage is specifying a design solution that can be easily translated into a programming language.

E. Implementation
The physical architecture of a system describes the structure of hardware and software in a system. The hardware structure includes the hardware needed to run the system and the

relationship between the hardware. Physical architecture also describes the physical structure of program modules and the relationship or dependency for each module with other modules.

Physical architecture is derived from logical architecture that is produced at the detailed design stage. All classes and mechanisms in the logical architecture are mapped to components and nodes in the physical architecture. Physical architecture includes two diagrams namely component diagrams that describe the structure of software and deployment diagrams that illustrate the structure of the hardware.

1. Software Physical Architecture
To implement the system, the Java programming language is used because Java is one of the object-oriented and flexible programming languages, which can run many types of processors and operating systems. By using java, the design and mapping between class classes that are still logical in nature can be transformed and used directly into the component programming code to make prototypes.

2. Hardware Architecture
Web Based Application Architecture

A web-based applications have three main components, namely the web server, the connection to the internet network, and the browser located at the client location. The web server distributes pages containing information in a form that has been formatted to the requesting client. While information on a page is usually stored in a certain format that is fixed in a single file or database system. Requests from clients to access information available on a web server are done through the internet using the HTTP protocol. Data management / storage is carried out by the database server system.

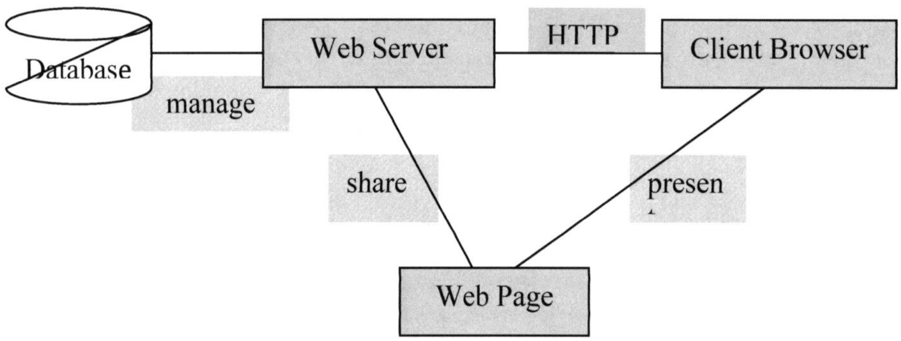

Figure 4. Basic Architecture of Web Based Application

F. Testing

In this stage a trial will be carried out on the design that has been done then make improvements or corrections as needed. To improve services to the community, it is recommended that the North Toraja Regency Government establish an Integrated Licensing Service Unit. Integrated Licensing Services organized by the North Toraja Regency Government, is for licensing administration services in the North Toraja Regency.

3. Results and Discussion

Management of licensing services is carried out as follows:

1. The applicant contacts the information officer to obtain an explanation of the licensing requirements and the registration form,

2. The applicant submits an application to the Office or Section through the Integrated Licensing Services Unit of North Toraja Regency and submitted through the registration counter staff at the Integrated Licensing Services Unit,
3. Documents that have been examined and meet the administrative requirements are submitted to the Secretary of Integrated Licensing, to be informed to the relevant Regional Apparatus Organizations (RAO) / TEAM for technical considerations.
4. The Integrated Licensing Services Unit prepares a joint discussion meeting with certain RAO (Regional Apparatus Organization Unit) and officials appointed for and on behalf of the regent and other parties considered necessary as resource persons.
5. During the discussion meeting of officials appointed for and on behalf of the regent and certain RAO / TEAM and other parties deemed necessary to respond to requests for special permits and permits required,
6. Submission of technical considerations from RAO / TEAM by taking into account academic technical standards, property, strategic planning, RUTR / RDTR, as well as other policies determined by local governments,
7. Secretary, assisted by administrative staff in the Integrated Licensing Services Unit, inventory the technical considerations and recommendations submitted by the relevant RAO, as material for the preparation of minutes and technical suggestions / considerations submitted to the Service or Section,
8. The Service or Section responds to suggestions / considerations submitted by the Secretary of the Integrated Licensing Services Unit, in the form of approval, rejection or certain records that need further discussion.
9. If the Office / Section has provided a response containing approval, the Secretary of the Integrated Licensing Services Unit then drafted the permit decision concept and made and submitted a memorandum of service to the Head of Legal Section of the North Toraja Regency Secretary, for further processing into a Regent Decree.
10. In the decision of the permit contains the determination of permits and includes also the determination of other permits related to the implementation of licensing.
11. If the competent official (Office / Section) has provided a response that contains a rejection, the Secretary of the Integrated Licensing Service Unit drafted a rejection letter signed by the head of the Integrated Licensing Service Unit (regional secretary).
12. The Office / Section decision regarding the signed permit, or the rejection letter that has been signed, is returned to the Integrated Licensing Service Unit, to be given a registration number,
13. Integrated Licensing Services Unit coordinates the calculation of fees from all types of licensing, then submits information on the answer to the decision about the request and the amount of retribution that must be paid by the applicant,
14. The applicant pays retribution through the regional treasury office officer in the North Toraja Regency Integrated Licensing Service Unit,
15. The applicant makes a decision on a permit or rejection letter

Visualization of the licensing service mechanism can be seen in the following figure:

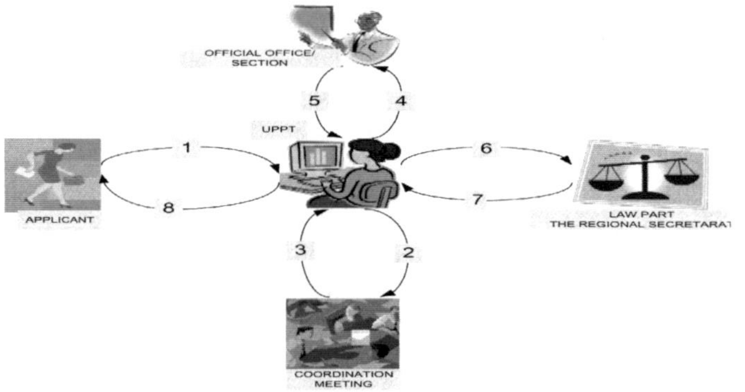

Figure 5. Flow of Licensing Mechanisms

Information:

Submission of Requests; UPPT Meeting; Inventory of Meeting Results; Technical Consideration; Note Memorandum; Service / Section Responses; Request Validation; Ratification of Office / Section Decree; Information on Office / Section Decisions about Permits.

3.1 Modeling the UPPT Licensing System

This section will discuss the licensing modeling of the Integrated Licensing Service System (SPPT). PPT Licensing System Modeling is divided into two layers, namely PPT business process modeling and PPT modeling software.

3.2 PPT Business Process Modeling

The PPT business process explains how the Integrated Licensing Service Unit performs its functions. The business process of PPT is stated in the form of activity diagram. In this study, the business process is explained from the external side, namely the applicant and from the internal side, namely the user

3.3 External Side

Business process from the external side or from the applicant's side is stated in the activity diagram in Figure 7. The applicant first goes to the UPPT office, if the applicant wants to register for a permit application, then the applicant goes to the registration window. At the registration window, the applicant can request information regarding the requirements required to submit an application. If the applicant has completed the requirements, the applicant can submit an application. After completion, the applicant receives a receipt of the application or registration receipt. If the applicant wants to know the outcome of his application, the applicant goes to the information counter and provides a sign of registration. After that the applicant gets information whether the application was accepted or rejected. If the application is accepted, the applicant gets a print out of the levy that must be paid. After paying, the applicant gets proof of payment, which will be exchanged, for a decision letter at the collection counter. Conversely, if the application is rejected, then the applicant gets a receipt to be exchanged for a receipt to get a decision letter of refusal, at the pick up counter.

Figure 6. Applicant Activity Diagram

4. Conclusion

From this study it can be concluded that:

1. At the stage of identifying the system needs can be revealed the weaknesses that occur in the licensing process in the District. North Toraja at this time.
2. With the obstacles that occur at this time in the current licensing process, it can be determined the functions of the system that will be developed in the integrated licensing service. With the model obtained, it can be described an integrated licensing service system in the district. North Toraja.

References

[1] Raumbaught, James, Jacobson, Ivar & Booch Grady, "The Unified Modeling Language Reference Manual, Addison Wesley Longman, Inc., USA, 1999
[2] Heeks, Richard, "Understanding Digital Government Project Failures, Computer, 2001.
[3]] Kominfo, "Pedoman Interoperabilitas Sistem Informasi Instansi Pemerintah, http://www.depkominfo.go.id
[4] Departemen Komunikasi dan Informatika Republik Indonesia, "Blue Print Sistem Aplikasi E-Government", Jakarta 2004.
[5] Departemen Komunikasi dan Informatika Republik Indonesia, "Rancangan Blue Print E-Government", Jakarta 2006.
[6] Dimara Kusuma Hakim, Fitria Ekowati, Implementasi E-Government: Sistem Informasi Aduan Dan Informasi Berbasis Sms Dalam Pelayanan Perijinan,Juni 2009.
[7] R Driana Lusmiarwan, Suhono H. Supangkat, Perancangan Rototype *Single Identity Number* (Sin) Untuk Menunjang E- Government,Prosiding Konferensi Nasional Teknologi Informasi & Komunikasi Untuk Indonesia, Mei 2006,
[8] Dr. Suhono Harso Supangkat, I Made Ari Jaya Negara, Framework Strategi Implementasi E-Government, Prosiding Konferensi Nasional Teknologi Informasi & Komunikasi Untuk Indonesia, Mei 2006,
[9] Rancangan Blue Print E- Government, Direktorat E-Government Direktorat Jenderal Aplikasi Telematika Tahun 2006.

[10] Blueprint Sistem Aplikasi E-Government, Departemen Komunikasi Dan Informatika Republik Indonesia ,Jakarta 2004
[11] Richard Heeks, Prototyping E-Government Applications, Idpm, University Of Manchester, Uk 2001.
[12] Richardus Eko Indrajit, Arsitektur Three-Tier Pada Electronic Government, Seri Artikel E-Government 2001

IOP Conf. Series: Materials Science and Engineering **846** (2020) 012046 doi:10.1088/1757-899X/846/1/012046

Tuning with Ziegler Nichols Method for Design PID Controller At Rotate Speed DC Motor

Nicolaus Allu[1], Apriana Toding[2]

[1,2] Study Program of Electrical Engineering Study Program, Faculty of Engineering, [Universitas Kristen Indonesia Paulus]

* Corresponden Email: nicolaus.allu@ukipaulus.com

Abstract. DC motor is a type of motor that is widely used in industry, electronics and supporting components for some electronic equipment or instrumentation. But now the PID controller is still widely used in various industrial sectors, this is due to its toughness to overcome various problems that exist in the industry. But there is one drawback that is the tuning method, the tuning process must be done by trial and error. More tuning procedures done manually by trial and error method that the result is not necessarily true. To overcome these problems required an alternative approach that can achieve a relatively better value and faster to perform the tuning process. In this case the alternative method used is tuning with Ziegler Nichols. Based on the simulation performed with the controlled torch, the value of the disturbance with the controller Kp = 3,6, Ki = 2,857 and Kd = 1,134 happened to the motor rotation at the lowest interference occurred at 1818 RPM while the highest rotation disturbance reached 2177 RPM from the nominal rotation speed of 2000 RPM for 80 seconds and suffered a friction coefficient of about 20 seconds.

1. Introduction

Direct current or direct current motors are included in the category of most widely used motor types in industrial environments, household appliances to children's toys or as instruments supporting electronic instrument systems. Electric motors are electromagnetic devices that convert electrical energy into mechanical energy. This mechanical energy is used for, for example, rotating the pump impeller, fan or blower, moving the compressor, lifting materials, and others [1]. The physical model of a complete DC motor describing the electrical and mechanical charts can be seen in [2]. PID controller (Proportional - Integral - Derivative) is a combination of three types of controllers. If each of the three types of controllers are independent then the results achieved will be less good because each has its own weaknesses and advantages. To meet the desired system, the three PID parameters must be optimally set. There are several conventional PID tuning or tuning methods that have been developed, such as the cut and try method, the Ziegler-Nichols method, the step response and the analytic method [3]-[5].

From some experiments that have been done with the above methods can be seen that with the use of PID controller in a system has a weakness, namely that the parameters in the controller must always be tuned up (tuned up) when there is a change in the system, the change will cause tuning returns from these PID parameters.

IOP Conf. Series: Materials Science and Engineering **846** (2020) 012046 doi:10.1088/1757-899X/846/1/012046

2. Model of PID CONTROLS

The response of a control system always indicates damped oscillation before it reaches steady-state. Classification of transient response characteristics of a control system to the unit step input [3]. As the name implies, this controller is a combination of three control systems namely, proportional, integral and derivative. If each of the three controllers stands alone, the results are less good because each has its own weaknesses and advantages. Therefore, the combination of these three control systems is expected to eliminate weaknesses respectively and capable of contributing from their respective advantages [3]. In the design of the PID control system that needs to be done is to set parameters P, I or D for the response of the system output signal to certain input as desired. PID tuning can also be based on the Ziegler-Nichols tuning method. This method aims to achieve maximum overshoot (MO): 25% against step input. The shape of the curve of the PID control using the Ziegler-Nichols method [6]-[8].

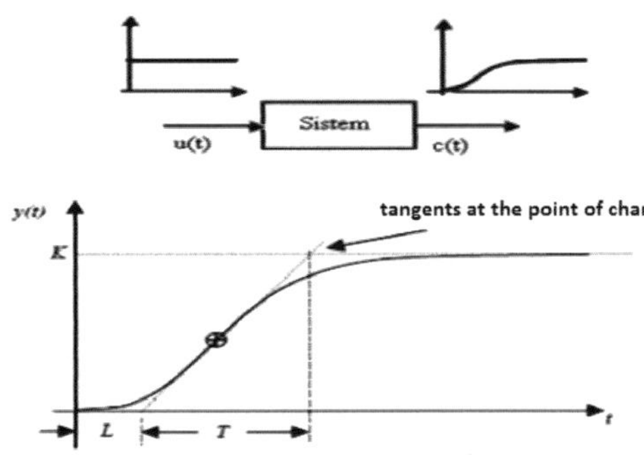

Fig.1 PID Ziegler-Nichols Tuning Curve

Where: L = Delay time
T = Time delay constants

PID value calculation table [6]:

Table 1. ZIEGLER-NICHOLS

Tipe	Kp	Ti	Td
P	T/L	~	0
PI	0.9 T/L	L/0 .3	0
PID	1.2 T/L	2L	0.5 L

3. Direct Current Motors

Direct current motor used is direct current controlled anchor motor type 73186 class 0.3. Direct current motor parameters controlled anchors in the simulation are given in Table 2.

Table 2. Motorcycle Search Parameters Controlled The Media

No	Parameter	Symbol	Value	Unit
1	Power	P	0,3	KWatt
2	Nominal rotational speed	□ nom	2000	RPM
			209,440	rad/sec
3	Anchors Voltage	Ea	220	Volt
5	Anchor current	Ia	1,8	Ampere
6	Anchor resistor	Ra	12	□
7	Moments of Inertia	J	0,177	Nmsec^2/rad
8	Friction constant	B	6,83x10^{-3}	Kg.m^2/rad.sec
9	Motor constants	Km	0,796	N.m/Amp.
10	GGL constants opponents	Kb	0,947	Volt.sec/rad
11	Torque	T	1,432	Kg.m^2/sec^2

Source: Directional Motor Controlled Anchors Type 73186 class 0.3 at the Electrical Engineering Laboratory Faculty of Engineering Department of Electrical Engineering Hasanuddin University.

Direct Current Flow Motor Anchors with Non-Controller Disorder

Under these conditions the motor is in an interference condition as shown in the schematic diagram below:

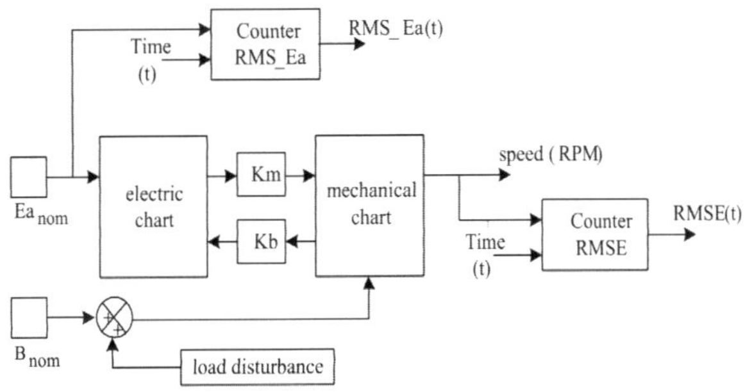

Fig. 2 Chart motion simulation image of direct current uninterrupted disturbance

Changes in rotation speed of a direct current motor occur along with the load changes, meaning that the speed or speed of motor rotation depends on the rise or fall of the motor load. The change of the load itself is then called disturbance. Despite the disturbance that occurs but the rotation speed of the direct current motor wherever possible keeps spinning at its nominal speed. As described above, what is meant by the disturbance condition is how much the load changes are given to the motor that affects the change in the speed of direct current motor. Based on the simulation image of direct current motor without controller (Figure 2) above if executed the graph obtained simulation results as follows

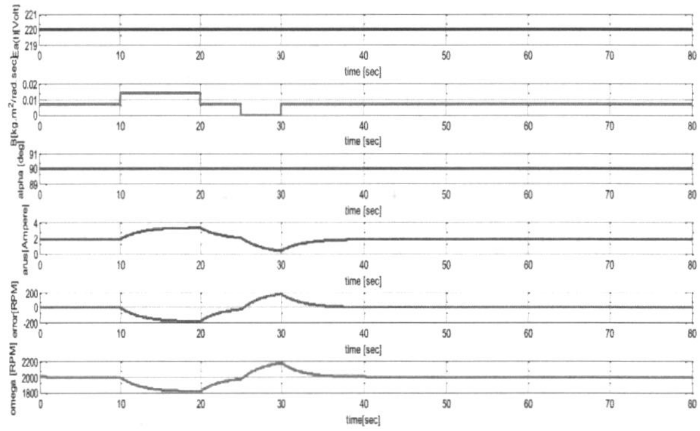

Fig. 3 Graph of simulation results between voltage, load, alpha (angle of ignition), anchor current, error and rotational speed over time without control.

The value of the disturbance that occurs in motor rotation on the graphic image above: The lowest lap disruption occurred in 1872 RPM while the highest lap disruption reached 2214 RPM from the nominal round of 2000 RPM.

Direct current motor with anchor control on the condition of the disorder (Closed loop Control)

The controller used is the Proportional-Integral-Derivative (PID) control. Shown graph pieces simulated direct motor speed simulation results against unmanageable interference conditions such as Fig. 4.

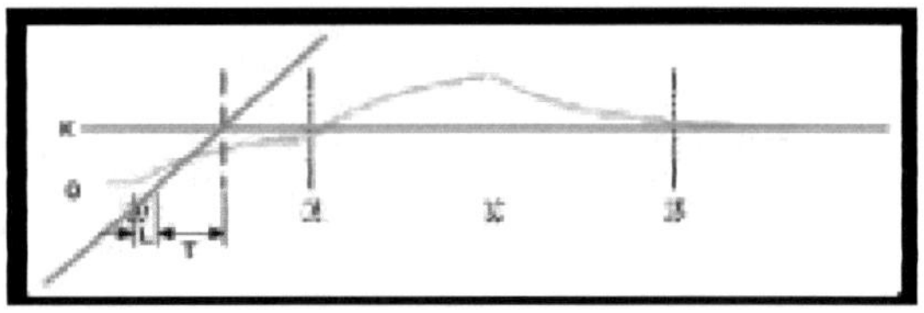

Fig. 4 Graphic pieces of motor speed simulation results on time with uninterrupted interference.

From the picture above graph obtained:
L = 0.63 and T = 1.89.
So obtained:

Kp = 1.2T/L = 1.2 (1.89/0.63) = 3.6
Ti = 2L = 2 x 0.63 = 1.26
Ki = Kp / Ti = 3.6 / 1.26 = 2.857
Td = 0.5L = 0.5 x 0.63 = 0.315
Kd = Kp x Td = 3.6 x 0.315 = 1.134

In making closed-cycle control on direct current motor controlled anchors. More can be seen in the following chart image.

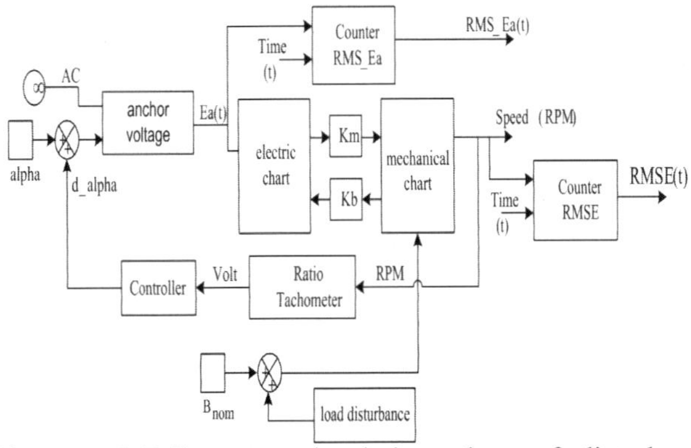

Fig. 5 Picture of DC motor simulation chart of disturbance condition with controller (closed loop control)

When the direct current motor is closed loop control with PID system initialization: Proportional-Integral-Derivative Controller (Kp = 3,6, Ki = 2,857, Kd = 1,134) above is executed the graph of simulation result as follows.

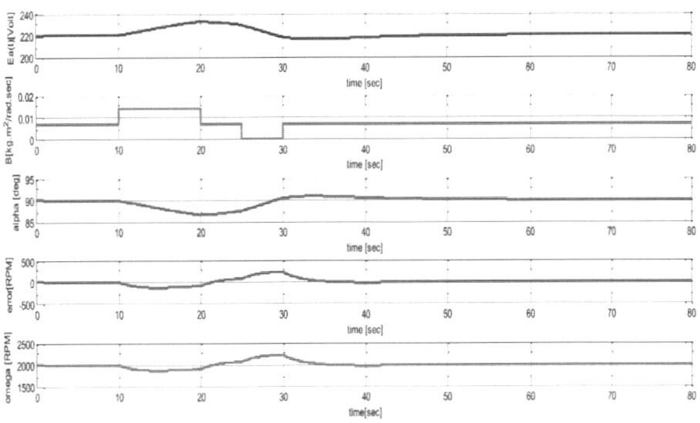

Fig. 6 Graph of DC motor simulation result with controller Kp = 3,6, Ki = 2,857 and Kd = 1,134

The value of the disturbance that occurs in the current and motor rotation in the graphic image above:

The lowest lap disruption occurred at 1818 RPM while the highest lap disruption reached 2177 RPM from the nominal round of 2000 RPM.

4. Conclusion

From the above simulation results can be taken two conclusions:

1. The value of the non-controller disturbance occurring at motor rotation at the lowest spinning disturbance occurred at 1872 RPM while the highest rotation failure reached 2214 RPM from the nominal round of 2000 RPM for 80 s and having a friction coefficient disturbance of about 20 seconds.

2. The value of the disturbance with the controller Kp = 3.6, Ki = 2.857 and Kd = 1.134 which occurred at motor rotation at the lowest rotation disturbance occurred at 1818 RPM while the

highest rotation disturbance reached 2177 RPM from the nominal round of 2000 RPM for 80 seconds and has a friction coefficient of about 20 seconds.

References

[1] A Pinem, "DC Motor Speed Adjustment With Integral Control Cycle (Applications At FT-USU Electrical Energy Conversion Laboratory)", Thesis, 2009.

[2] Harifuddin, "Modeling and Controlling DC Motor Controlled Anchors", Journal of Electrical MEDIA, Volume 3 No. 1, June 2008.

[3] Era Purwanto, Mukti Wibowo, Soebagio, et al. "Development of Self Tuning Method of PID Controller Parameter Using Genetic Algorithm On Induction Motor Setup For Electric Car Drivers", National Seminar on Information Technology Application, Yogyakarta June 20, 2009.

[4] Tatang Sukendar, Berayan Munthe, "Motor Servo AC Adjustment System With PID Self Tuning Controller Based Fuzzy Logic", Journal DiSainTek Vol. 01, No. 01, December 2007.

[5] Ruzita Sumiati, "Analysis of DC Motor Control Using Logic PID With Micro Controller ATMEGA 8535", Journal of Mechanical Engineering Vol. 6, No. 2, December 2009.

[6] Zaidir Jamal, "Implementation of PID Tuning Control of Ziegler Nichols Using Microcontroller", Journal of Informatics, Vol.15, No.1, June 2015.

[7] Erwin Susanto, "Proportional Integral Derivative Control (PID) For DC Motor Using Personal Computer", Department of Electrical Engineering Telkom Bandung Institute of Technology 2009.

[8] Bobal, Vladimir, “Self-Tuning PID Controller”, Department of Automatic Control, Faxculty of Technology Zlin: Proceedings of the 7th Mediterranean Conference on Control and Automation (MED99) Haifa, Israel - June 28-30, 1999.

IOP Conf. Series: Materials Science and Engineering **846** (2020) 012047 doi:10.1088/1757-899X/846/1/012047

Automatic Broadcast Transmission Control With 8 Relay Channels Based On Arduino

Denny Hardiyanto[1*], Sigit Priyambodo[2], Samuel Kristiyana[3], Enrico Colif Ar R[4]
[1,2,3,4] Department of Electrical Engineering, Faculty of Industrial Technology, Institut Sains & Teknologi AKPRIND Yogyakarta, Yogyakarta, Indonesia

[*]Corresponding author: denny.hardiyanto@akprind.ac.id

Abstract. This Automatic broadcast transmission control is a device that functions to replace a manual switch to an automatic switch with a programmed starting timing process. This device consists of an LED VU Display as an indicator of the voice signal input, a monostable multivibrator as a timer, an Arduino Uno as the main control, and an 8 channel relay module as a substitute for a manual switch. This tool works when the voice signal is active, then relays 1 to 8 will be active, but if the voice signal is off, relays 8 to 1 will be deactivated gradually. On off this channel relay is determined by programmed timing. The results of the experiment show that the active and non-active time of the relay can be programmed in stages with a time difference of 1 second for each relay. Automatic Transmission Control with 8 Relays using Arduino Control is able to automate switching from manual switching to programmable automatic switching.

1. Introduction

Nowadays, the development of technology is very rapid to bring humans into the era of modernization. Almost all aspects of human life are very dependent on technology. This is because technology was created to assist humans in completing an activity. A transmitter is a collection of electronic devices that function as an increase in the frequency of the sound signal. The sound signal issued by the mixer has a low frequency then the mixer will be sent to the transmitter unit. In the transmitter unit will be changed from the frequency of a small voice signal to a very high frequency sound signal. Then it will be transmitted in all directions through an antenna with a certain range. In some devices the transmitter still uses the manual method to activate and deactivate the transmitter by manually activating and deactivating the power switch or activating by pressing the manual switch one by one. Aside from being inefficient, it will have a radiation effect on the health of the operator who enters the transmitter room. This is because the power in the transmitter room is very large, causing health problems, ranging from dizziness to the worst, which is infertility. The purpose of this automatic control device is to activate and deactivate the radio transmitter automatically according to the timing that has been programmed using Arduino Uno [1]–[8] so as to minimize the operator for direct contact with the transmitter switch. Arduino is an open-source hardware and software company, project, and user community that designs and manufactures single board microcontrollers and microcontroller kits for building digital devices. Arduino is widely used in research as a hardware controller tool kit that can be programmed as desired. Various researches conducted using Arduino include.

2. Research Method

2.1 System Design

IOP Conf. Series: Materials Science and Engineering **846** (2020) 012047 doi:10.1088/1757-899X/846/1/012047

The design of the tool starts from the design of the system by conducting library study, discussion, and drawing the system to be designed. The ideas that have been obtained are then done using the project board. In this stage the tool design will be designed and tested (trial and error) using the project board. Next design the hardware from the previous design. Then the software design is using Arduino [9][10]–[12] [13][14]–[16] as the mainboard to program the transmitter controller.

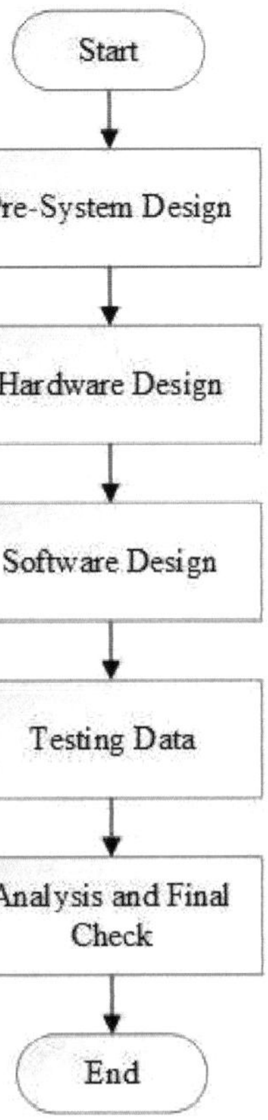

Figure 1. Design Stages

Figure 2. Automatic Transmitter Control with 8 Arduino Based Relay Channels

2.2 Design Result

The results of the design of Automatic Transmitter Control with 8 Arduino Based Relay Channels [11], [16]–[27] show in Figure 2. There is a row of LED VU Display lights that function as an indicator of the sound signal input. At the bottom of the LED VU Display there is a monostable multivibrator circuit with 3 regulator ICs that serves as a voltage-lowering, then Arduino [10], [12], [14], [15], [20], [28]–[33]as the main controller and 12 Volt, 5 Amperes switching power supply. Relay module circuit with 8 channels using relays with specification 12 Volt, 30 Ampere. At the front there are panels, AC input ports, 1 Ampere fuse, AC switches, common ports and normally close (NC) as many as 8 pairs, as well as TS counter batteries.

This tool works when there is an input in the form of a sound signal entering the mixer and leading to the LED VU Display circuit as an input indicator. Then the signal enters the monostable multivibrator circuit, then continues to Arduino. After Arduino gets a high signal (1), Arduino will give commands to the relay module to activate the relay one by one. Likewise, the process of deactivating, when the mixer

IOP Conf. Series: Materials Science and Engineering **846** (2020) 012047 doi:10.1088/1757-899X/846/1/012047

does not have a sound signal, Arduino will get a low signal (0) then give a command to the relay module to deactivate one by one. Figure 3 show the experiment.

Figure 3. Experiment Process of Automatic Transmitter Control

Figure 3 shows the test tool using a mobile phone as a medium to trigger voice signal input. Visible when music is playing, the VU Display light will turn up and down following the voice signal input. Then when the music is playing, relays 1-8 will live one by one.

2.3 Analysis Method

This tool is tested using a mobile phone that functions as a voice signal input. Mobile phones are connected using a 3.5 mm jack to input this tool. After the voice signal from the mobile phone is activated, Arduino will read the voltage and turn on 8 relays. Data is collected by measuring the voltage, observing whether the relay is functioning or not and recording the active and inactive time of 8 relays.

3. Result and Discussion

3.1 Active Relay Time

Testing the active time of the relay is carried out 100 times.

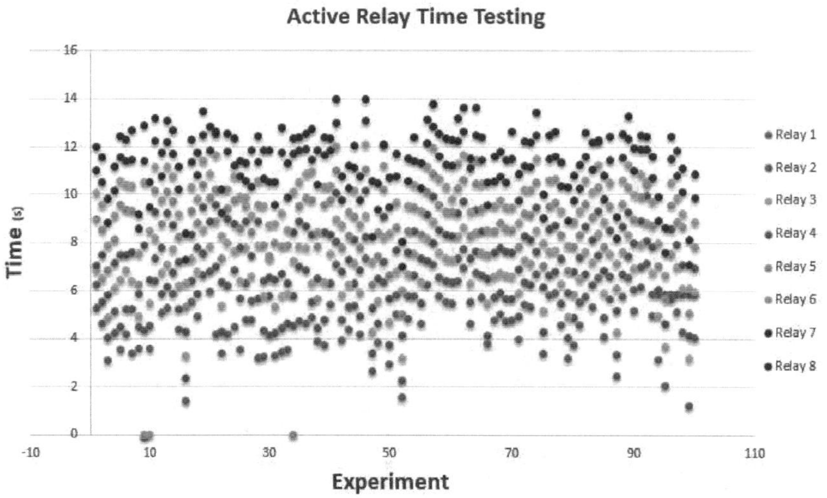

Figure 4. Active Relay Time Testing in 100 Experiments

Figure 4 shows the distribution of Active Time data from relay 1 to relay 8 during 100 experiments. From the distribution of these data, it can be concluded that the average active relay time is shown in Figure 5.

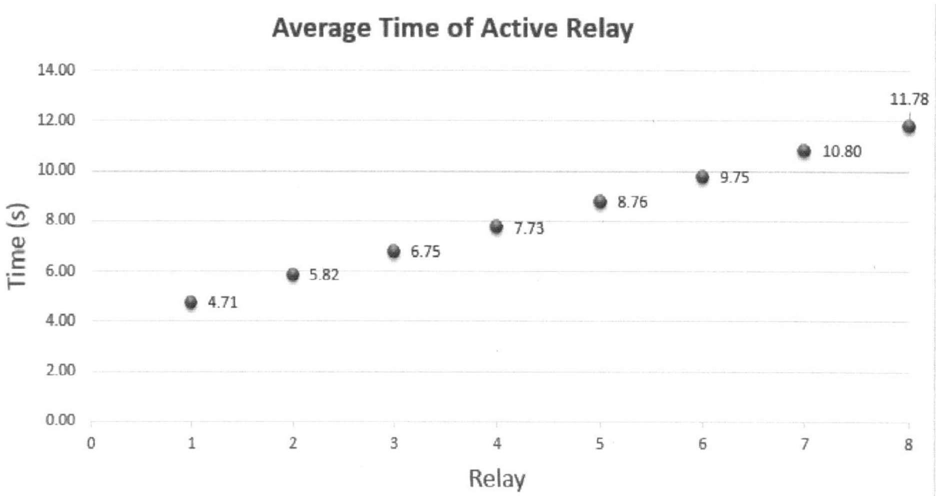

Figure 5. Average Time of Active Relay

Figure 5 shows that the difference in active time of each relay is about 1 second and it is seen that the relay is active in sequence. The active time of the relay is programmed sequentially because the transmitter switch must be active in sequence.

3.2 Non-Active Relay Time

Testing the Non-Active time of the relay is carried out 100 times.

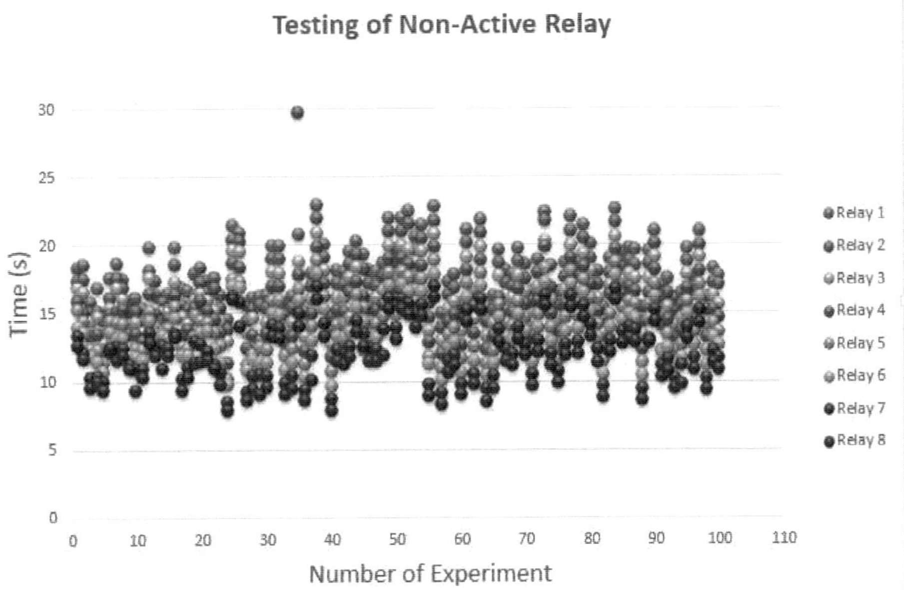

Figure 6. Non-Active Time Testing of Relay in 100 Experiment

Figure 6 shows the distribution of Non-Active Time data from relay 1 to relay 8 during 100 experiments. From the distribution of these data, it can be concluded that the average Non-Active relay time is shown in Figure 7.

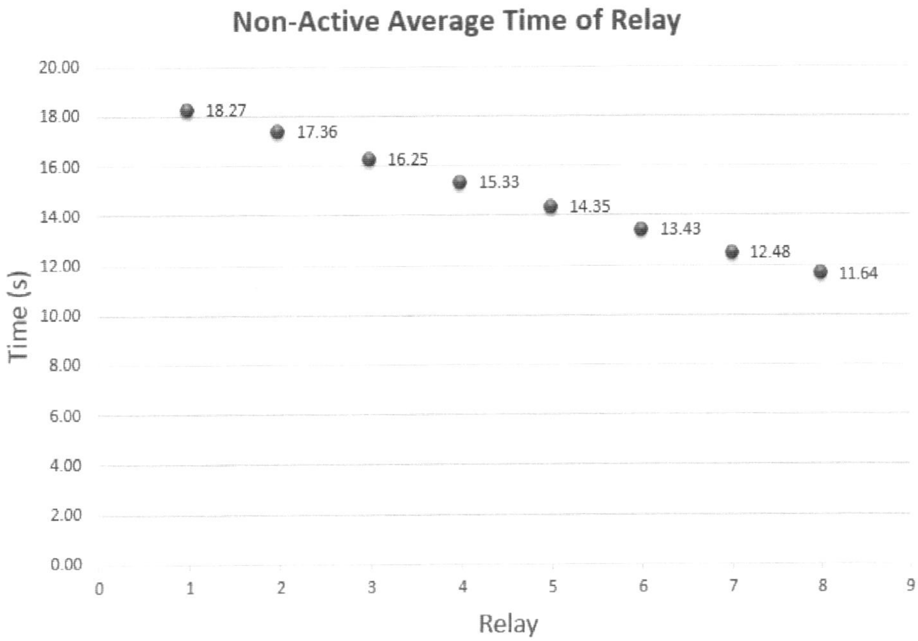

Figure 7. Non-Active Average Time of Relay

In activating or deactivating the transmitter has an SOP (Standard Operating Procedure) that is using time gradually. If the transmitter does not activate or deactivate the transmitter gradually it will suffer damage.

4. Conclusion

The conclusion of this research are:

a. Automatic Transmission Control with 8 Relays using Arduino Control is able to automate switching from manual switching to programmable automatic switching
b. The results of the experiment show that the active and non-active time of the relay can be programmed in stages with a time difference of 1 second for each relay
c. This device can be applied to all radio transmitters.

Acknowledgments

The author would like to thank the electrical engineering study program, IST AKPRIND Yogyakarta and Radio Ista Kalisa for supporting and facilitating this research. In addition, thanks are also addressed to team who helped and participate in this research.

References

[1] P. Megantoro, A. Widjanarko, R. Rahim, K. Kunal, and A. Z. Arfianto, "The Design of Digital Liquid Density Meter Based on Arduino," *J. Robot. Control*, vol. 1, no. 1, pp. 1–6, 2020.

[2] C. Y. Foulis and S. Papadopoulou, "A Portable Low-Cost Arduino-Based Laboratory Kit for Control Education," in *2018 UKACC 12th International Conference on Control (CONTROL)*, 2018, pp. 435–435.

[3] T. Pawlenka and J. Skuta, "Security system based on microcontrollers," in *2018 19th International Carpathian Control Conference (ICCC)*, 2018, pp. 344–347.

[4] V. D. Vaidya and P. Vishwakarma, "A Comparative Analysis on Smart Home System to Control, Monitor and Secure Home, based on technologies like GSM, IOT, Bluetooth and PIC Microcontroller with ZigBee Modulation," in *2018 International Conference on Smart City and Emerging Technology (ICSCET)*, 2018, pp. 1–4.

[5] K. Kunal, A. Z. Arfianto, J. E. Poetro, F. Waseel, and R. A. Atmoko, "Accelerometer Implementation as Feedback on 5 Degree of Freedom Arm Robot," *J. Robot. Control*, vol. 1, no. 1, pp. 31–34, 2020.

[6] M. Pimentel, Y. Alejaldre, A. Avalos, and J. Cerda, "A HIL-based DC Motor Speed Control," in *2018 IEEE International Autumn Meeting on Power, Electronics and Computing (ROPEC)*, 2018, no. Ropec, pp. 1–5.

[7] S. Bipasha Biswas and M. Tariq Iqbal, "Solar Water Pumping System Control Using a Low Cost ESP32 Microcontroller," in *2018 IEEE Canadian Conference on Electrical & Computer Engineering (CCECE)*, 2018, vol. 2018-May, pp. 1–5.

[8] A. Tayab Noman, M. S. Khan, M. Emdadul Islam, and H. Rashid, "A New Design Approach for Gesture Controlled Smart Wheelchair Utilizing Microcontroller," in *2018 International Conference on Innovations in Science, Engineering and Technology (ICISET)*, 2018, no. October, pp. 64–68.

[9] T. P. Tunggal, A. W. Apriandi, J. E. Poetro, E. T.Helmy, and F. Waseel, "Prototype of Hand Dryer with Ultraviolet Light Using ATMega8," *J. Robot. Control*, vol. 1, no. 1, pp. 7–10, 2020.

[10] A. N. N. Chamim, M. E. Fawzi, I. Iswanto, R. O. Wiyagi, and R. Syahputra, "Control of Wheeled Robots with Bluetooth-Based Smartphones," *Int. J. Recent Technol. Eng.*, vol. 8, no. 2, pp. 6244–6247, Jul. 2019.

[11] Iswanto, S. Suripto, F. Mujahid, K. T. Putra, N. P. Apriyanto, and Y. Apriani, "Energy Harvesting on Footsteps Using Piezoelectric based on Circuit LCT3588 and Boost up Converter," *Int. J. Electr. Comput. Eng.*, vol. 8, no. 6, 2018.

[12] Iswanto, J. Syaftriadi, A. Nur, N. Chamim, R. O. Wiyagi, and R. Syahputra, "LED and Servo Motor Control Via Bluetooth Based on Android Applications," *Int. J. Recent Technol. Eng.*, vol.

8, no. 2, pp. 6227–6231, Jul. 2019.

[13] N. H. Wijaya, A. G. Alvian, A. Z. Arfianto, J. E. Poetro, and F. Waseel, "Data Storage Based Heart and Body Temperature Measurement Device," *J. Robot. Control*, vol. 1, no. 1, pp. 11–14, 2020.

[14] I. Iswanto, W. S. Agustiningsih, F. Mujaahid, R. Rohmansyah, and A. Budiman, "Accumulator Charging Control with Piezoelectric Based on Fuzzy Algorithm Scheduling," *TELKOMNIKA (Telecommunication Comput. Electron. Control.*, vol. 16, no. 2, p. 635, Apr. 2018.

[15] I. Iswanto, K. Purwanto, W. Hastuti, A. Prabowo, and M. Y. Mustar, "Smart Smoking Area based on Fuzzy Decision Tree Algorithm," *Int. J. Adv. Comput. Sci. Appl.*, vol. 10, no. 6, pp. 500–504, 2019.

[16] K. Purwanto, Iswanto, T. K. Hariadi, and M. Y. Muhtar, "Microcontroller-based RFID, GSM and GPS for motorcycle security system," *Int. J. Adv. Comput. Sci. Appl.*, vol. 10, no. 3, 2019.

[17] A. Ponniran, A. Hashim, and H. Ali Munir, "A design of single axis sun tracking system," *2011 5th Int. Power Eng. Optim. Conf. PEOCO 2011 - Progr. Abstr.*, no. June, pp. 107–110, 2011.

[18] A. Xhafa, P. Tuset-Peiro, and X. Vilaiosana, "Live demonstration: Wireless PID control of a thermal process using an ultra-low cost LWIR camera," in *2017 IEEE SENSORS*, 2017, vol. 2017-Decem, pp. 1–1.

[19] E. Amareswar, G. S. S. K. Goud, K. R. Maheshwari, E. Akhil, S. Aashraya, and T. Naveen, "Multi purpose military service robot," in *2017 International conference of Electronics, Communication and Aerospace Technology (ICECA)*, 2017, vol. 2017-Janua, pp. 684–686.

[20] A. N. N. Chamim, D. Ahmadi, and Iswanto, "Atmega16 Implementation As Indicators Of Maximum Speed," *Int. J. Appl. Eng. Res.*, vol. 11, no. 15, pp. 8432–8435, Jun. 2016.

[21] N. H. Wijaya, F. A. Fauzi, E. T. Helmy, P. T. Nguyen, and R. A. Atmoko, "The Design of Heart Rate Detector and Body Temperature Measurement Device Using," *J. Robot. Control*, vol. 1, no. 2, pp. 40–43, 2020.

[22] N. H. Wijaya, Z. Oktavihandani, K. Kunal, E. T.Helmy, and P. T. Nguyen, "The Design of Tympani Thermometer Using Passive Infrared Sensor," *J. Robot. Control*, vol. 1, no. 1, pp. 27–30, 2020.

[23] J. Yoon, S. Lee, J. Shin, and S. H. Won, "Performance analysis of wireless relay-based MIMO system using network coding," *Proc. - 7th Int. Conf. Digit. Content, Multimed. Technol. Its Appl. IDCTA 2011*, pp. 188–191, 2011.

[24] M. F. J. Pinkney, D. Hampel, and S. DiPierro, "Unmanned aerial vehicle (UAV) communications relay," pp. 47–51, 2002.

[25] E. A. J. Hansen, M. H. Nielsen, D. E. Serup, R. J. Williams, T. K. Madsen, and R. Abildgren, "On Relay Selection Approaches in Bluetooth Mesh Networks," in *2018 10th International Congress on Ultra Modern Telecommunications and Control Systems and Workshops (ICUMT)*, 2018, vol. 2018-Novem, pp. 1–5.

[26] Z. Dzulfikri, N. Nuryanti, and Y. Erdani, "Design and Implementation of Artificial Neural Networks to Predict Wind Directions on Controlling Yaw of Wind Turbine Prototype," *J. Robot. Control*, vol. 1, no. 1, pp. 20–26, 2020.

[27] I. Prasojo, A. Maseleno, O. Tanane, and N. Shahu, "Design of automatic watering system based on Arduino," *J. Robot. Control*, vol. 1, no. 2, pp. 55–58, 2020.

[28] I. Prasojo, P. T. Nguyen, O. Tanane, and N. Shahu, "Design of Ultrasonic Sensor and Ultraviolet Sensor Implemented on a Fire Fighter Robot Using AT89S52," *J. Robot. Control*, vol. 1, no. 2, pp. 59–63, 2020.

[29] S. Widadi, Z. A. Manzila, I. Ahmad, and O. Tanane, "Aroma Electric Therapy device with Atmega8- Based Heart Rate Monitoring," *J. Robot. Control*, vol. 1, no. 2, pp. 49–54, 2020.

[30] A. N. N. Chamim, M. Heru Gustaman, N. M. Raharja, and I. Iswanto, "Uninterruptable Power Supply based on Switching Regulator and Modified Sine Wave," *Int. J. Electr. Comput. Eng.*,

vol. 7, no. 3, p. 1161, Jun. 2017.

[31] A. Maarif, S. Iskandar, and I. Iswanto, "New Design of Line Maze Solving Robot with Speed Controller and Short Path Finder Algorithm," *Int. Rev. Autom. Control*, vol. 12, no. 3, p. 154, May 2019.

[32] A. Nurjanah, I. Prasojo, F. Anindiyahadi, and N. M. Raharja, "Mosque as a Civilization Center," *Int. J. Innov. Technol. Explor. Eng.*, vol. 8, no. 12, pp. 1072–1074, 2019.

[33] Iswanto, P. Megantoro, and D. V. Senzas, "Calibrator for Temperature Measurement Device with Raspberry Pi-Based Interface," *Int. J. Innov. Technol. Explor. Eng.*, vol. 8, no. 12, pp. 4862–4866, 2019.

Capacitor discharging calculation application based on android

Wisnu Kartika[1*], Erika Loniza[1], Meilia Safitri[1]

[1]Department of Medical Electronics Technology, Vocational Program [Universitas Muhammadiyah Yogyakarta]

* wisnu2007@umy.ac.id

Abstract. At this time, there were many small components to make electronic equipment such as TVs, smartphones and others. This electronic equipment uses passive components and active components. The physical size of the capacitor decreases gradually becoming smaller. Capacitors store energy in the form of charge. If there are two equal charges (for example a positive charge and a positive charge) then both charges will refuse. Conversely, if there are two different charges, they will attract each other. The purpose of this research is to create an Android-based software that can calculate the remaining charge after the discharging process. The method used is to use the calculation method of discharging capacitors. The results of this study found that the calculation can be done well and also the presence of errors in the software design process.

1. Introduction

Capacitors are one of the passive components. Other passive components are inductors and resistors. Each capacitor has a different storage capacity (capacitance). The value of the ability to store charge in a capacitor is printed on the skin of the capacitor wrapper [1]. There are several types of capacitors, namely electrolyte capacitors (often also called ELCO) and ceramic capacitors (mylar or polystyrene) [2]. There are two types of capacitors [3][4], polar capacitors [5] and nonpolar capacitors [6]. Applications using capacitors [7][8][9] can be seen on Personal Computer monitors and smartphone chargers. It can be seen that the capacitor stores the charge a few seconds before the equipment actually dies.

The problem faced at this time is the lack of App [10][11][12] which focuses on the calculation of the capacitor charge value [13][14]. So with this problem, proposed a research scheme based on Android automatic capacitor charge calculation [15]-[21]. This paper will be divided into several sections, namely Introduction, System Design, Results and Discussion and Conclusions.

2. System design

The proposed research scheme is an Android application for calculating the charge remaining on a capacitor based on the initial charge measured on the capacitor, the value of the resistor used, the value of the capacitor used and the charging or discharging time.

This application was developed by MIT App Inventor 2. This application is available online. This application can help design simple block-based applications and interface designs. This research was conducted in stages. First, study the literature review relating to the characteristics of time constant capacitors. After that, the study continued with a simple calculation of the remaining capacitors. The calculations between the proposed applications will then be compared with the comparison application. The selected comparison application is Desmos Scientific Calculator. This Desmos application can be downloaded at Google Playstore.

3. Implementation

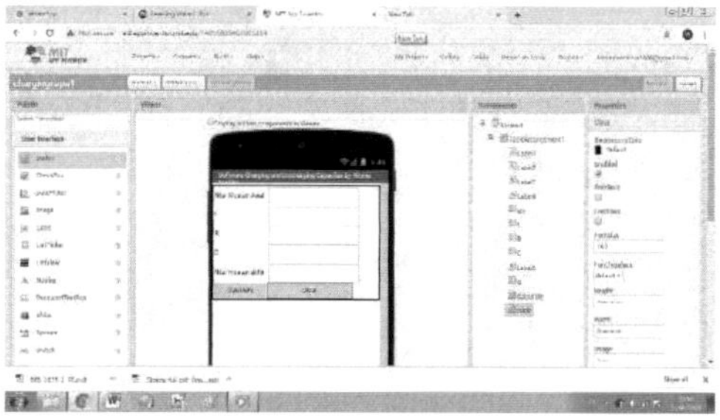

Figure 1. GUI Android Interface

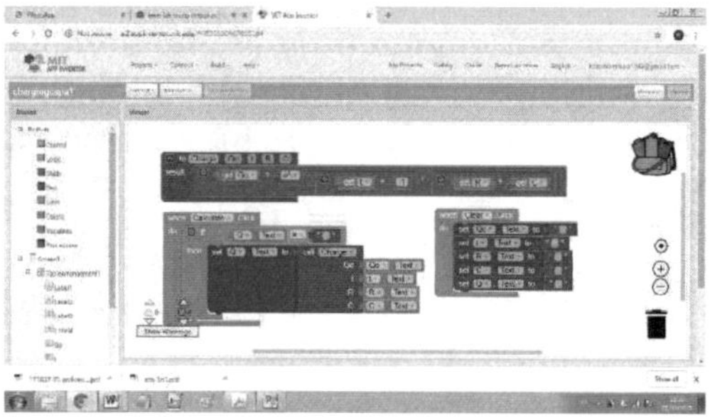

Figure 2. Coding Block App Inventor 2

From the research that has been done, the proposed application through the testing process with 10 times per day with a variety of inputs. This table shows the input and running results of the proposed application.

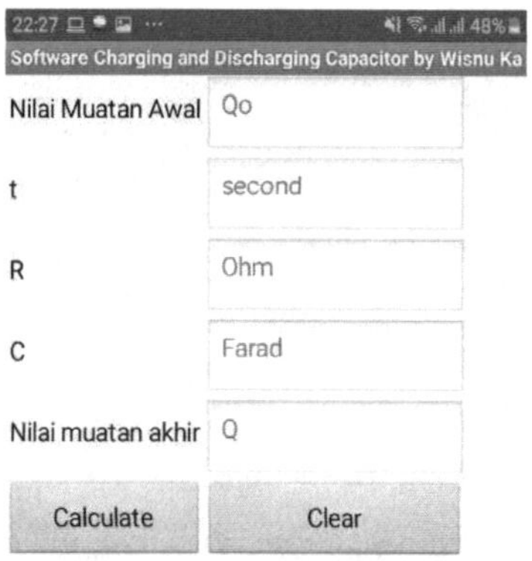

Figure 3. Smartphone Android App Simulation (the initial condition)

IOP Conf. Series: Materials Science and Engineering **846** (2020) 012048 doi:10.1088/1757-899X/846/1/012048

From the Figure 3 can show the new type of creating the Android App. With the MIT App Inventor, using block-shaped coding and connected like a puzzle. This development has been done for a long time. The hardware used to do the MIT App Inventor coding is a set of laptops with the ACER Intel Core i3 brand and an internet connection. While the software used is the Google Chrome browser. This research requires Google Chrome to be able to access the MIT App Inventor application. The MIT App Inventor application is online.

This App is designed one by one part.

Figure 4. Input The First Charge Value on the first row

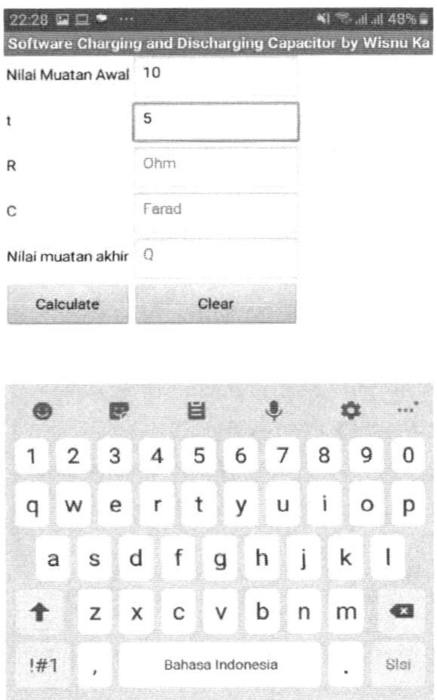

Figure 5. Input the Charging Time on the second row

After entering the initial load value of 10 Coulombs. Then we enter the fill time value of 5 seconds as shown in Figure 5.

IOP Conf. Series: Materials Science and Engineering **846** (2020) 012048 doi:10.1088/1757-899X/846/1/012048

Figure 6. Input the Resistor Value on the third row

After entering the charging time value of 5 seconds, then we enter the value of the installed resistance of 2 Ohms as shown in Figure 6.

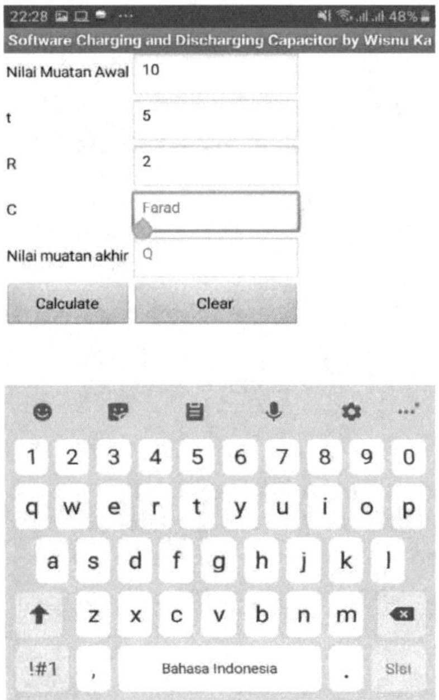

Figure 7. Input The Capacitor Value on the fourth row

After entering the initial load value of 10 Coulombs. Then we enter the fill time value of 5 seconds as shown in Figure 7.

IOP Conf. Series: Materials Science and Engineering **846** (2020) 012048 doi:10.1088/1757-899X/846/1/012048

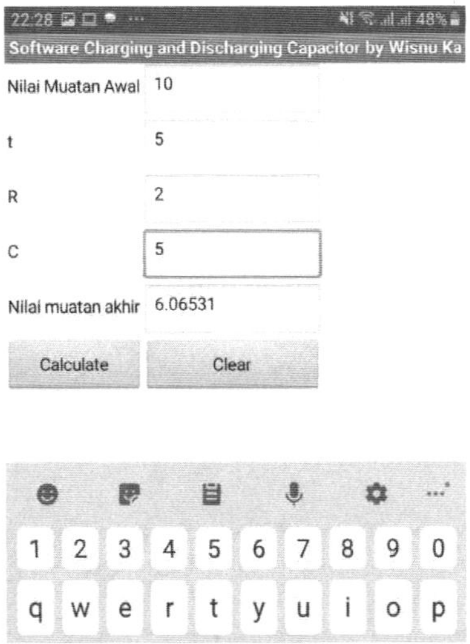

Figure 8. Running The App on the Smartphone by clicking the Calculate button

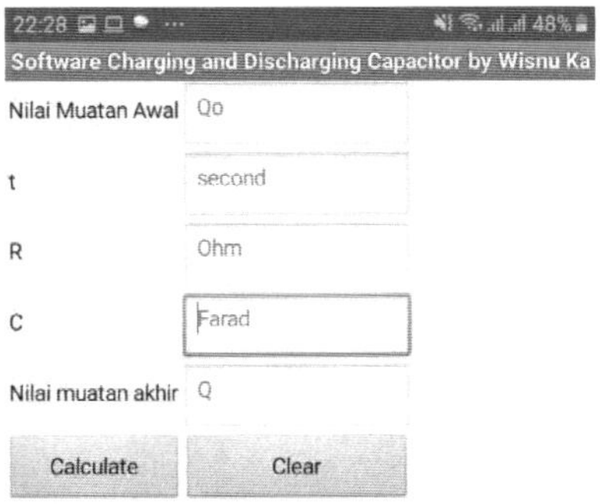

Figure 9. If the Clear Button Clicked

The results of the calculation of this App can be seen on the Figure 8 which the calculate value is 6.06 Coulomb. This app will be compared with the App Desmos Scientific Calculator which has been available on Google Playstore

$$Q = Q_0 e^{-t/RC} \quad (1)$$

By using this equation can show the overall calculation.

The value of the parameter is shown in this table.

The Parameter	The Value	The Notation
Qo	10	Coulomb
t	5	second
R	2	Ohm
C	5	Farad

From the Table can be calculated that

$$Q = Q_o e^{-\frac{t}{RC}}$$

$$Q = 10e^{-\frac{5}{2\times 5}} = 6.06 \text{ C}$$

Figure 10. Calculate on Desmos App (web version)

4. Conclusion

From the research that has been done, it can be concluded that the proposed application can work well.

ACKNOWLEDGEMENT

We say thankful for our research team and D3 Electromedical Technology Study Program and other support staff and my mother and my wife.

References

[1] I. Prasojo, A. Maseleno, O. Tanane, and N. Shahu, "Design of automatic watering system based on Arduino," J. Robot. Control, vol. 1, no. 2, pp. 55–58, 2020.

[2] H. Wang and H. Wang, "A Two-Terminal Active Capacitor," IEEE Trans. Power Electron., vol. 32, no. 8, pp. 5893–5896, Aug. 2017.

[3] A. Latif, K. Shankar, and P. T. Nguyen, "Legged Fire Fighter Robot Movement Using PID," J. Robot. Control, vol. 1, no. 1, pp. 15–19, 2020.

[4] D. S. Wicaksono and F. N. Hakim, "Media Pembelajaran Fisika Interaktif Bahasan Kapasitor Berbasis Flash Dan Xml," J. Speed – Sentra Penelit. Eng. dan Edukasi –, vol. 3, no. 2, pp. 2088–154, 2011.

[5] T. P. Tunggal, A. W. Apriandi, J. E. Poetro, E. T.Helmy, and F. Waseel, "Prototype of Hand Dryer with Ultraviolet Light Using ATMega8," J. Robot. Control, vol. 1, no. 1, pp. 7–10, 2020.

[6] J. Ho, T. R. Jow, and S. Boggs, "Historical introduction to capacitor technology," IEEE Electr. Insul. Mag., vol. 26, no. 1, pp. 20–25, Jan. 2010.

[7] K. Kunal, A. Z. Arfianto, J. E. Poetro, F. Waseel, and R. A. Atmoko, "Accelerometer Implementation as Feedback on 5 Degree of Freedom Arm Robot," J. Robot. Control, vol. 1, no. 1, pp. 31–34, 2020.

[8] H. Nagasaki, P.-Y. Huang, and T. Shimizu, "Characterization of power capacitors under practical current condition using capacitor loss analyzer," in 2016 IEEE Energy Conversion Congress and Exposition (ECCE), 2016, pp. 1–7.

[9] B. A. Saputra, E. P. Hidayat, and A. Z. Arfianto, "Internet of Things-Based Steam Pump Motor Protection Due to Voltage Unbalance," J. Robot. Control, vol. 1, no. 2, pp. 64–69, 2020.

[10] S. Dey, S. Mohon, B. Ayalew, H. Arunachalam, and S. Onori, "A Novel Model-Based Estimation Scheme for Battery-Double-Layer Capacitor Hybrid Energy Storage Systems," IEEE Trans. Control Syst. Technol., vol. 27, no. 2, pp. 689–702, Mar. 2019.

[11] I. Prasojo, P. T. Nguyen, O. Tanane, and N. Shahu, "Design of Ultrasonic Sensor and Ultraviolet Sensor Implemented on a Fire Fighter Robot Using AT89S52," J. Robot. Control, vol. 1, no. 2, pp. 59–63, 2020.

[12] A. Singh, A. V. Agrawal, and A. Kanukotla, "A method to improve application launch performance in Android devices," in 2016 International Conference on Internet of Things and Applications (IOTA), 2016, pp. 112–115.

[13] N. H. Wijaya, A. G. Alvian, A. Z. Arfianto, J. E. Poetro, and F. Waseel, "Data Storage Based Heart and Body Temperature Measurement Device," J. Robot. Control, vol. 1, no. 1, pp. 11–14, 2020.

[14] O. Kaplan and F. Issi, "An android based application and simulation of multiple photovoltaic panels," in 2017 IEEE 6th International Conference on Renewable Energy Research and Applications (ICRERA), 2017, vol. 2017-Janua, pp. 925–930.

[15] P. Megantoro, D. A. Husnan, M. U. Sattar, A. Maseleno, and O. Tanane, "Validation Method for Digital Flow Meter for Fuel Vendors," J. Robot. Control, vol. 1, no. 2, pp. 44–48, 2020.

[16] A. Singhai, R. S. Ramanujam, J. Bose, and V. Kumari, "Implementation and analysis of pluggable Android applications," in 2015 IEEE International Conference on Signal Processing, Informatics, Communication and Energy Systems (SPICES), 2015, pp. 1–5.

[17] Z. Dzulfikri, N. Nuryanti, and Y. Erdani, "Design and Implementation of Artificial Neural Networks to Predict Wind Directions on Controlling Yaw of Wind Turbine Prototype," J. Robot. Control, vol. 1, no. 1, 2020.

[18] C. Zou, N. Zhang, D. Kushner, R. Orchard, C. Mi, and S. Zhang, "High temperature capacitors with high energy density," in 2012 IEEE International Power Modulator and High Voltage Conference (IPMHVC), 2012, pp. 419–421.

[19] N. H. Wijaya, F. A. Fauzi, E. T. Helmy, P. T. Nguyen, and R. A. Atmoko, "The Design of Heart Rate Detector and Body Temperature Measurement Device Using," J. Robot. Control, vol. 1, no. 2, pp. 40–43, 2020.

[20] R. C. Jisha, M. P. Mathews, S. P. Kini, V. Kumar, U. V. Harisankar, and M. Shilpa, "An Android Application for School Bus Tracking and Student Monitoring System," in 2018 IEEE International Conference on Computational Intelligence and Computing Research (ICCIC), 2018, pp. 1–4.

[21] P. Megantoro, A. Widjanarko, R. Rahim, K. Kunal, and A. Z. Arfianto, "The Design of Digital Liquid Density Meter Based on Arduino," J. Robot. Control, vol. 1, no. 1, pp. 1–6, 2020.

Heavy metal toxicity and the influence of water quality in watershed for enhancing fisheries food security

R M D Ujianti[*], A Androva

Faculty of Engineering and Informatics, Universitas PGRI Semarang, Jl. Sidodadi Timur 24. Dr. Cipto Semarang, Indonesia

*rizkymuliani@gmail.com

Abstract. This study investigated the toxicity of heavy metals such as Cu, Fe, Cr, Zn and Mn and the influence of water quality in Garang Watershed Semarang (Banjir Kanal Barat River), Central Java Indonesia for enhancing fisheries food security. The fish biomarker in this research is Nile Tilapia (*Oreochromis niloticus*). The range of Cu, Fe, Cr, Zn and Mn in aquatic environment were 0,15 mg/L; 0,82 mg/L; 0,025 mg/L; 0,562 mg/L and 0,01mg/L. The range of Cu, Fe, Cr, Zn and Mn in Nile Tilapia were 2,547 mg/kg; 306,3 mg/kg; 0,36 mg/kg; 16,69 mg/kg and 35,6 mg/kg. Toxicity in fish is exceeded the limits of food security regulation in WHO and BPOM, and water quality regulation in Government Regulation of Indonesia No 82 /2001 on Water Quality And Water Pollution Management. Management in water quality is and toxicity is important for enhancing fisheries food security. Maintaining food security in the fishery sector is very important for the stability of community nutrition because with good food public health will be guaranteed.

Keywords : food security, toxicity, fish, water quality, watershed

1. Introduction

Fish is one of the main food sources in Indonesia and has high nutritional value through its processed products. The fisheries sector is one of the sectors that support economic and regional development in Central Java [14]. The city of Semarang is the capital of the province of Central Java, which has many industries and directly borders the North Coast of the Laut Jawa and has a river that runs from upstream to downstream. This industrial area produces waste discharged into rivers and spreads to the estuary. As a result of industrial waste, pollution is high enough to result in a polluted water environment and contains heavy metals.

Heavy metal is one type of industrial waste that pollute the waters [11]. Heavy metals in relatively small waters will be very easily absorbed and accumulated biologically by aquatic plants or animals and will be involved in food web systems. The content of heavy metals in aquatic biota will usually increase over time because it is bioaccumulative [12]. Heavy metals are very dangerous for human life, for example when cases of heavy metal poisoning occurred in the Minamata Bay of Japan in the 1950s. Hundreds of lives lost as a result of consuming fish containing mercury [3]. The impact arising from the presence of heavy metals in the waters depends on the presence of metals in water and sediments, their toxicity and concentration in the environment. Heavy metals if pollute into the body of living things will experience bioconcentration, bioaccumulation, and biomagnification [9]. This research should have been conducted to provide information to the public and industry owners regarding the heavy metals contained in fish in river waters, furthermore that risk factors could be minimized.

2. Methodology

2.1. Study Area

The Banjir Kanal Barat river is located on the coast of Semarang City, Central Java Province, Indonesia (Figure 1). The length of the Banjir Kanal Barat river is 5.4 km. On the left and right of the river is a resident's house. The main function of the Banjir Kanal Barat river as the main drainage channel of the city, but the occurrence of sedimentation causes the volume of water to be reduced. Furthermore, there was a flood around it due to the overflow of this river. The Semarang City Government and the Japanese Government through the Japan International Cooperation Agency (JICA) normalized of this river in 2010, furthermore that this normalization was able to restore the main function of the Kanal Banjir river as a flood control channel [51]. Determination of the sampling location using a purposive sampling method [52]. The source of heavy metal input becomes the basis for determining the sampling location. The sampling location is the entry point of heavy metals from the mainland to the sea (estuary) and is the final disposal site of all pollutants from the mainland area [53].

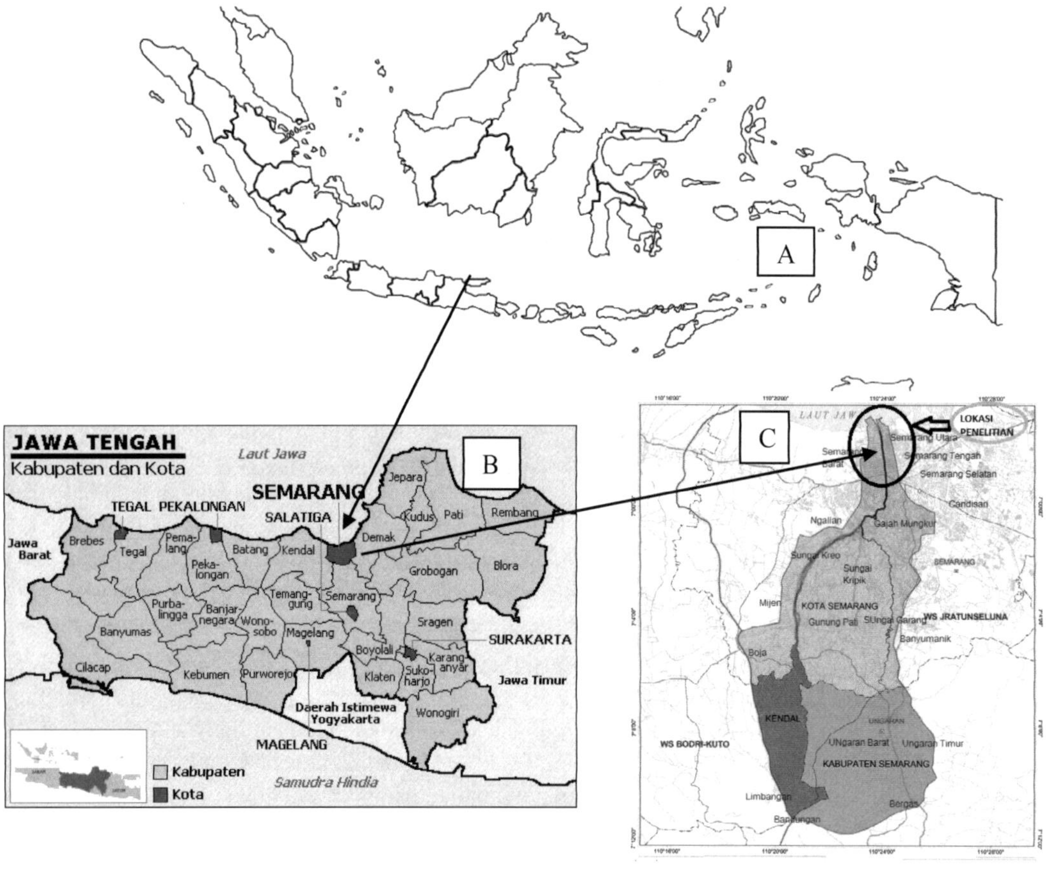

Figure 1. Study area: the map of the Indonesia (A), Central Java Province (B), Location of Banjir Kanal Barat River, Semarang, Central Java Province, Indonesia (C)

2.2. Sampling techniques

Fish collects in Banjir Kanal Barat River is downstream of Garang watershed, Semarang, Central Java, Indonesia. The downstream is used for estuary, settlement, port fishery and aquaculture. In this area, the waste comes from industry, domestic and thus affecting the water quality of the river [41]. This watershed is Garang river, the local name of the river is Banjir Kanal Barat / Kanal Banjir Barat, this river near settlements and downstream area, and this is the estuary zone of Laut Jawa.

2.3. Heavy Metals Analysis

Approximately 25 g (to nearest 0.1 g) of the test portion into a crucible. Dry at a temperature of 135°C to 150°C for 2h. Transfer to the furnace with temperature control. Increase the temperature to 500°C until overnight (16h). Set at room temperature, add 2 mL of HNO_3, and swirl it. Evaporation occurs on a hot plate and a steam bath. Transfer the sample to the furnace, raising the temperature to 500°C, for 1 hour. Afterward, the dish was removed and cooled down. To get clean, C-free ash, HNO_3 graying is repeated. Add 1M HCl, then heat it on the hot plate carefully to dissolve the ash. Transfer the sample to a 25 mL volumetric flask, 5 mL HCl 1M cooled, and diluted with 1M HCl. Dilute Pb 0.02; 0.6; 1,0; 3.0; 5.0 and 10.0 µg Pb / mL each in 1 mL HCl. Set the optimal spectrophotometer conditions at a maximum signal of 283.3 nm. Using the airflow rate - the manufacturer recommends C_2H_2 is a standard condition for Pb. Read the results on a spectrophotometer. Calibrate the concentration mode with solutions containing 0.2 and 10.0 µg Pb / mL. Record the concentration directly after calibrating the instrument. For strip chart readings, adjust the amplification to give an absorption reading of %1% for a working solution of 0.2 µg Pb / mL and prepare a standard curve with µg Pb / mL [42].

3. Result and Discussion

3.1. Heavy Metals inFish Tilapia (Oreochromis niloticus)

3.1.1 Arsenic

Arsenic is a metal that can be found in groundwater. The arsenic content in the study location ranges from <0.029 mg / kg - <0.030 mg / kg, the Maximum Limit of Contamination of Heavy Metals in Processed Food that is equal to 0.25 mg / kg [46]. Arsenic can be dangerous because it accelerates the development of cancer cells, infertility and miscarriage [27]. Source of manganese comes from pesticides, fungicides, metal smelters [16].

3.1.2 Iron (Fe)

Iron (Fe) concentration in Tilapia is 15.86 mg / kg; 256.2 mg / kg; 306.3 mg / kg [45]. The graph of Iron Concentration (Fe) in Tilapia is presented in Figure 1.

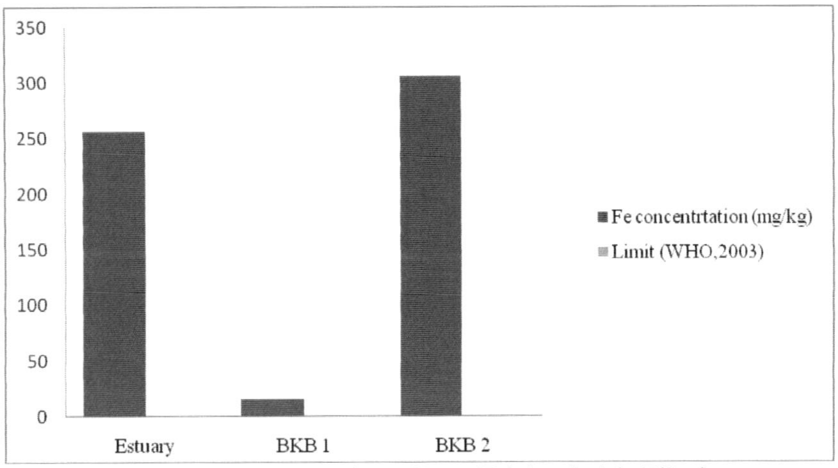

Figure 2. Concentration of Iron (Fe) (mg/kg) in Tilapia

Iron (Fe) concentrations at 3 stations exceed the threshold of heavy metal Fe determined by WHO (2003) [47] is 0.3 mg/kg. Fe is a silvery-white metal, clay and can be formed [24]. Industrial wastes containing Fe metal are toxic to plants, animals, and humans. Excessive amounts of Fe can cause poisoning, death, kidney disease, intestinal wall damage and hepatitis [30]. Tilapia fish (*Oreochromis niloticus*) liver which accumulated by Fe metal will experience pathological changes in the form of fatty cells and necrosis.

3.1.3 Chromium (Cr)

Cr at the study site is 0.29 mg/kg; 0.32 mg/kg and 0.36 mg/kg. Cr in these three locations exceeds the quality standards by BPOM Regulation 5/2018 [46]. The high Cr is thought to originate from the textile industry waste originating from upstream rivers. Another factor that causes high Cr concentrations at this station is the river mouth which is the point of mixing between freshwater and sea, furthermore that this sampling point contains higher of mud [20]. Source of manganese comes from mining, electroplating, textile, tannery industries [16]. The graph of Iron Concentration (Fe) in Tilapia is presented in Figure 2.

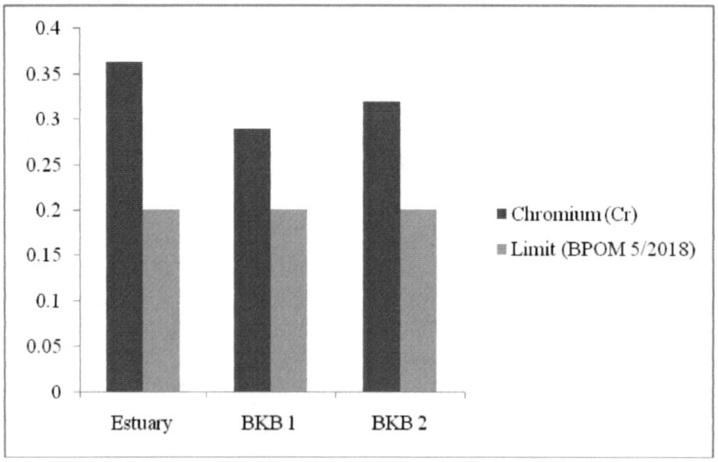

Figure 3. Cr Concentration (mg / kg) in Tilapia

3.1.4 Lead (Pb)

Pb found in tilapia in the study location was <0.387 mg/kg; <0.389 mg/kg and 1.334 mg/kg. The Pb concentration at the BKB2 sampling point exceeds the quality standards set by WHO (2003) [47] is 0.01 mg/kg. Pb can pollute the fish through the body surface, gills, absorption through food, and osmoregulation mechanism. The increased Pb concentration in fish meat occurs due to accumulation. PB that accumulates over time will cause fish death because fish cannot adapt to lead exposure [4]. The accumulation of Pb in the body of the fish will interfere with the synthesis of Hb whose function is to bind oxygen. The inhibition of Hb synthesis causes the ability to bind oxygen to become smaller so that it affects its metabolism [10]. Source of manganese comes from paint, pesticides, batteries, automobile emission, mining, burning of coal [16].

3.1.5 Zinc (Zn)

Zn found in tilapia in the study location was 12.18 mg/kg; 13.09 mg/kg and 16.69 mg / kg. This Zn concentration exceeds the threshold set by WHO (2003) [47] which is 5 mg/kg. The graph of Zn heavy metal concentration is presented in Figure 3

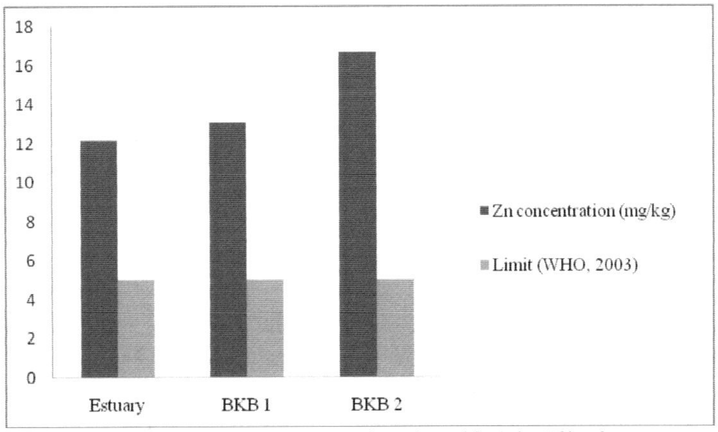

Figure 4. Zn Concentration (mg / kg) in Tilapia

Zn in waters originating from effluent and run-off streams, this pollute the body of aquatic organisms [2]. Detecting heavy metals can be carried out in Zn contaminated fish livers [13]. Muscles are indicators of Zn at low levels [4]. Low Zn concentrations in muscles can be caused by low muscle ability in binding protein (metallothionein) [19]. Source of zinc comes from refineries, brass manufacture, and metal plating [16].

3.1.6 Cadmium (Cd)

The concentration of Cd contained in Tilapia Fish in the study location were: <0.097 mg / kg - <0.099 mg/kg. This concentration is still limit is 0.1 mg / kg (BPOM Regulation 5/2018) [46]. Cd is a dangerous metal sourced from the use of fuel, forest fires, industrial waste and the use of fertilizers and pesticides [1]. The accumulated CDs cause physiological damage to the body of the reproductive glands, bone fragility, sense of smell, kidneys, lungs, blood, and heart [23]. The low Cd content in waters is thought to be due to research time which is the dry season, furthermore that waters are at the lowest discharge so that Cd experiences dilution [17]. Source of cadmium comes from welding, electroplating, pesticides, fertilizer, batteries, nuclear fission plants [16].

3.1.7 Copper (Cu)

Cu concentration contained in Tilapia Fish in the study location was 1.369 mg/kg; 2.118 mg/kg and 2.547 mg/kg. Cu concentration contained in Tilapia Fish originating from BKB 2 sampling point exceeds the threshold of Cu concentration determined by WHO (2003) [47] is 2.25 mg/kg. Graph of Cu Concentration contained in Tilapia in the research location is presented in Figure 4.

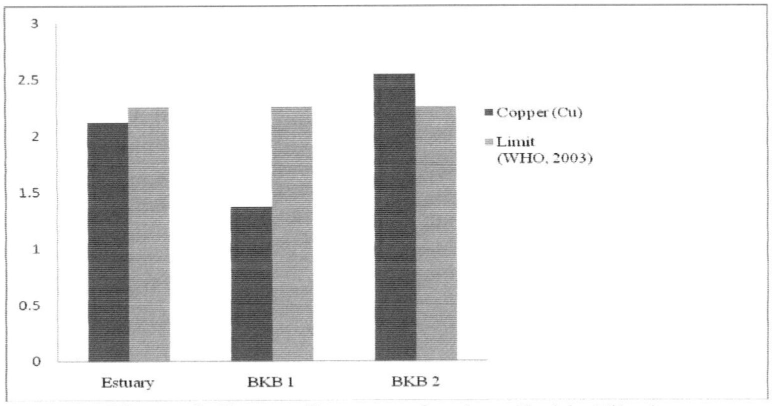

Figure 5. Copper Concentration (mg / kg) in Tilapia

Cu is a metal needed by aquatic organisms for its growth, but if the amount is excessive it will cause toxins [15]. The high Cu in Tilapia Fish found at the study site is thought to be due to the influence of water conditions. Influence factors the accumulation of heavy metals in aquatic organisms depend on the temperature of the aquatic environment, the state of the species, physiological activity and temperature of the water [6], the way the toxins enter the body, the duration and frequency of exposure, the toxicant concentration, shape, and the physical/chemical properties of metals the severity and susceptibility of various species to toxicity [43]. Cu is accumulated continuously in fish, if it will be consumed by humans and experiencing acute poisoning will cause symptoms of gastrointestinal and kidney disease [7]. Source of copper comes from electroplating, pesticides, and mining [16].

3.1.8 Manganese (Mn)

Mn concentrations contained in Tilapia Fish in the study site were 24.79 mg/kg; 34.29 mg/kg and 35.6 mg/kg, this concentration exceeds the threshold set by WHO (2003) [47] is 0.5 mg/kg. Manganese is an important component in metabolism that is the function of flavoprotein and the synthesis of mucopolysaccharides, cholesterol, and hemoglobin. The toxicity of manganese does not depend on the total concentration of manganese, but the concentration of oxidized manganese available. Mn concentrations exceeded the threshold, causing reddish, yellow and black waters, causing an unpleasant taste. The source of cadmium comes from welding, fuel addition, and ferromanganese production [16]. The Mn concentration of Mn in Tilapia Fish at the study site is presented in Figure 5.

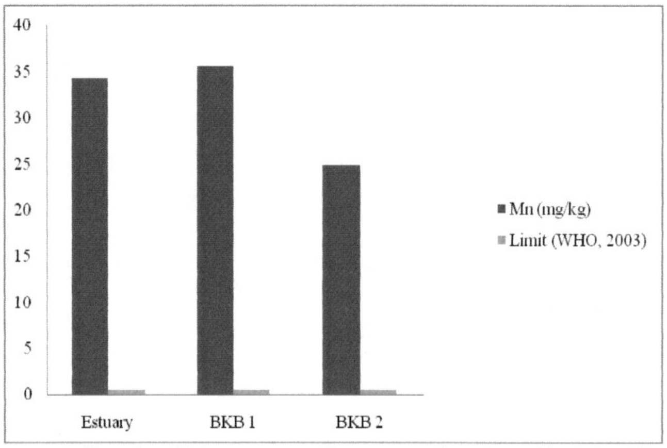

Figure 6. Graph of Manganese Concentration (mg / kg) in Tilapia

3.1.9 Mercury (Hg)

The concentration of Hg contained in Tilapia Fish in the study site was <0.002 mg/kg. The limit is 0.5 mg/kg (BPOM Regulation 5/2018) [46], this concentration is under than the limit. Hg is a metal that is liquid at room temperature. Mercury (Hg) or mercury is a metal that is naturally present, is the only metal that is at room temperature in a liquid form. Hg metal in water and soil is mainly derived from industrial effluents, natural deposits, and volcanic activity. Hg that accumulates continuously can cause nervous system disorders, headaches, chromosomes, allergic damage to brain function, DNA, skin rashes, fatigue, sperm damage, disability in infants and miscarriages [1]. Hg contamination at the research location is suspected to be due to industrial waste contamination at the upstream location. environmentally friendly agricultural systems around rivers. Hg toxicity is influenced by environmental factors of these waters [18]. Source of mercury comes from pesticides, batteries, and paper industries [16].

3.1.10 Nickel (Ni)

The concentration of Ni contained in Tilapia Fish in the study location is <0.097 mg / kg - <0.099 mg / kg. This concentration is below the threshold set by the IAEA (International Atomic Energy Agency)-4017, is 0.6 mg/kg. Nickel can be absorbed by organisms through magnetized food in plants and the waters that become their habitat. Ni accumulates in sediments and aquatic organisms through the process of exposure, bioaccumulation, and biomagnification [39]. Ni is insoluble under anaerobic conditions, and soluble under aerobic conditions [44], the level of Ni toxicity is influenced by the solubility level of Ni compounds [5]. Source of nickel comes from electroplating, zinc base casting and battery industries [16].

3.1.11 Fisheries Food Security
Heavy metals found in fish are very harmful. High heavy metals can reduce the level of food safety in fisheries. Wild tilapia fish have better omega-3 fatty acid content than cultivated tilapia [50], futhrtmore the habitat of wild tilapia (river) is important to maintain its water quality. Freshwater fish contain protein that is important for people and families in developing countries to maintain food security [48]. Characteristics of community who have good food security have adequate amounts of food and good nutrition for public health [49].

4. Conclusion

Heavy metals contained in rivers: As, Fe, Cr, Pb, Zn, Cd, Cu, Mn, Hg, and Ni. Heavy metal concentrations that exceed water quality standards: Fe, Cr, Zn, Cd, Cu, and Mn. The recommendation in this research are: harmonization of upstream and downstream sector policies in the long-term security of water, energy, and food security. Integration of planning and management of water, energy, land, forests, ecosystems, agriculture, and food security for the efficiency of natural resources and reducing environmental impacts. Synchronization between government agencies in terms of water use (for irrigation, energy, navigation, fisheries, and domestic) [54]. Management of disposal of industrial waste into the river, furthermore that the river are not polluted and the fish are not contaminated with heavy metals.

Acknowledgements
Authors would like to thank to LPPM Universitas PGRI Semarang who gave the research funding, with Contract Number: 65/SKK/LPPM/PASCA DOKTOR/VIII/2018.

References

[1] Agustina T 2010 Kontaminasi Logam Berat Pada Makanan dan Dampaknya pada Kesehatan *Teknubuga* **2**(2) pp 53–65
[2] Abdelmoenim M A, Moshelhy K M E, and. Hasan S H 1994 Trace Metals Content in Three Fish Species from Northern Part of the Suez Gulf, Red Sea, Egypt *Journal King Abdul Azis: Marine Science* 7 pp 15-24
[3] Alsuhendra and Ridawati 2013 *Bahan Toksik dalam Makanan* Remaja Rosdakarya Bandung
[4] Connel D W and Miller G J 1995 *Kimia dan Ekotoksikologi Pencemaran* UI Jakarta
[5] Darmono 1995 *Logam dalam Sistim Biologi Makluk Hidup* UI Press Jakarta
[6] Darmono 2006 *Lingkungan Hidup dan Pencemaran* Universitas Indonesia UI Press Jakarta
[7] Fadirubun N A, Daud A and Birawida A B 2015 Kualitas Air dan Sedimen Ditinjau dari Parameter Tembaga (Cu) Studi pada Air Sungai Pangkajene Kabupaten Pangkep
[8] FAO 2014 FIGIS. FishStat(Database) (Latestupdate:31Jan2014) URL (http://data. fao.org /ref /babf3346-ff2d-4e6c-9a40-ef6a50fcd422.html?version=1.0).
[9] Hidayah A H, Purwanto and Soeprobowati T R 2014 Biokonsentrasi Faktor Logam Berat Pb, Cd, Cr dan Cu pada Ikan Nila (*Oreochromis niloticus Linn.*) di Karamba Danau Rawa Pening *Bioma* **16**(1) pp 1-9
[10] Landis G, Wayne R M, Solfield and Hoyu M 2011 Introduction to Environmental Toxicology Molecular Substructures to Ecological Landscapes 4th Edition. CRC Press Taylor & Franciss Group
[11] Maftuch, Marsoedi, Putri V D, Lulloh M H, and Wibisono F K H 2015 Studi Ikan Bandeng (*Chanos Chanos*) yang Dibudidayakan di Tambak Tercemar Limbah Kadmium (Cd) dan Timbal (Pb) di Kalanganyar, Sidoarjo, Jawa Timur Terhadap Histopatologi Hati, Ginjal dan

Insang *Journal of Environmental Engineering dan Sustainable Technology* **02**(02) pp 114-122

[12] Martuti N K T 2012 Kandungan Logam Berat Cu dalam Ikan Bandeng, Studi Kasus di Tambak Wilayah Tapak Semarang *Prosiding Seminar Nasional Pengelolaan Sumberdaya Alam dan Lingkungan*

[13] Moselhy E, Othman K M, El-Azem A I, and Metwally E M E A 2014 Bioaccumulation of heavy metals in some tissues of fish in the Red Sea, Egypt *Egyptian Journal Basic Applied Science* **1**(2) pp 97- 105.

[14] Mudzakir A K 2009 *Peranan dan Kinerja Sektor Perikanan pada Perekonomian Jawa Tengah* Disertasi Institut Pertanian Bogor

[15] Palar H 2004 *Pencemaran dan Toksikologi Logam Berat* Rineka Cipta Jakarta

[16] Paul D 2017 Research on Heavy Metal Pollution of River Ganga: A review *Journal Annals of Agrarian Science* **15** pp 278-286

[17] Priyanto N, Dwiyitno and Ariyani F 2008 Kandungan Logam Berat (Hg, Pb, Cd, dan Cu) pada Ikan, Air, dan Sedimen di Waduk Cirata, Jawa Barat *Jurnal Pascapanen dan Bioteknologi Kelautan dan Perikanan* **3**(1) pp 69-78

[18] Putranto T T 2011 Pencemaran Logam Berat Merkuri (Hg) pada Airtanah *Teknik* **32**(1)pp 62-71

[19] Putri W A E, Bengen D G, Prartono T and Riani E 2016 Accumulation of Heavy Metals (Cu and Pb) in Two Consumed Fishes from Musi River Estuary, South Sumatera *Ilmu Kelautan* **21**(1) pp 45-52

[20] Said I, Jalaluddin M N, Upe A, and Wahab A W 2009 Determination of Heavy Metal Chrome and Lead Concentrations in Estuary Sediment of Matangpondo River Palu *Jurnal Chemica* **10**(2) pp 40-47

[21] Sekarwati N, Murachman B and Sunarto 2015 Dampak Logam Berat Cu (Tembaga) dan Ag (Perak) pada Limbah Cair Industri Perak terhadap Kualitas Air Sumur dan Kesehatan Masyarakat serta Upaya Pengendaliannya di Kota Gede Yogyakarta *Jurnal Ekosains* **7**(1) pp 64-76

[22] Setiadi R, Jawoto S, Sophianingrum M and Rosalia D 2008 Indikator Pembangunan Berkelanjutan Kota Semarang *Jurnal Riptek* **1**(2) pp 1–15

[23] Setiawan H and Subiandono E 2015 Heavy Metals Consentrations in Water and Sediment at Coastal Waters of South Sulawesi Province *Indonesian Forest Rehabilitation Journal* **3**(1) pp 67-79

[24] Slamet J S 2004 *Kesehatan Lingkungan* Gadjah Mada University Press Yogyakarta

[25] Sucipto C D 2015 *Keamanan Pangan Untuk Kesehatan Manusia* Gosyen Publishing Yogjakarta

[26] Sulaeman A and Syarief H 2010 *Tinjauan Ekonomi Penanganan Mutu dan Keamanan Pangan* Seafast Centre. Institut Pertanian Bogor

[27] Sundari D, Hananto M and Suharjo 2016 Heavy Metal In Food Ingredients In Oil Refi nery Industrial Area, Dumai *Buletin Penelitian Sistem Kesehatan* **19**(1) pp 55–61

[28] Superville P J, Ivanovsky A, Bhurtun P, Prygiel J and Billon G 2018 Diel Cycles of Reduced Manganese and their Seasonal Variability in the Marque River (northern France) *Science of the Total* Environment **624** pp 918–925

[29] Suprapti N H 2008 Kandungan Chromium pada Perairan, Sedimen dan Kerang Darah (*Anadara granosa*) di Wilayah Pantai Sekitar Muara Sungai Sayung Desa Morosari Kabupaten Demak, Jawa Tengah *Bioma* **10**(2) pp 36-40

[30] Supriyantini E and Endrawati H 2015 Kandungan Logam Berat Besi (Fe) pada Air, Sedimen, dan Kerang Hijau (Perna viridis) di Perairan Tanjung Emas Semarang *Jurnal Kelautan* **18**(1) pp 38-45

[31] Supriharyono 2002 *Pelestarian dan Pengelolaan Sumber daya Alam di Wilayah Pesisir Tropis* PT Gramedia Pustaka Utama Jakarta

[32] Sultan S N, Abidjulu J and Koleangan H S J 2015 Analisis Kandungan Merkuri dan Sianida di Daerah Aliran Sungai Talawaan, Sulawesi Utara *Jurnal Ilmiah Sains* **15**(1) pp 70-73

[33] Syofyan I, Usman and Nasution P 2011 Studi Kualitas Air untuk Kesehatan Ikan dalam Budidaya Perikanan Pada Aliran Sungai Kampar Kiri *Jurnal Perikanan dan Kelautan* **16**(1) pp

64-70
[34] Taftazani A 2007 Distribusi Konsentrasi Logam Berat Hg dan Cr pada Sampel Lingkungan Perairan Surabaya *Prosiding PPI – PDIPTN. Pustek Akselerator dan Proses Bahan – BATAN*
[35] Valiant R 2014 *Tantangan dalam Pengelolaan Sumberdaya Air untuk Mencapai Lingkungan Lestari Berkelanjutan: Potret Daerah Aliran Sungai (DAS) Brantas* Makalah Seminar Pekan DAS Brantas 2014 Teknik Pengairan Fakultas Teknik Universitas Brawijaya Malang
[36] Widaningrum, Miskiyah and Suismono 2007 Bahaya Kontaminasi Logam Berat dalam Sayuran dan Alternatif Pencegahan Cemarannya *Buletin Teknologi Pascapanen Pertanian* **3** pp 16–27
[37] Yong J 2015 China's Water Security: Current Status, Emerging Challenges and Future Prospects *Journal Environmental Science & Policy* **54** pp 106–125
[38] Youn S J, Taylor W W, Lynch A J, Cowx I G, Beard T D, Bartley D, and Wu F 2014 Inland Capture Fishery Contributions to Global Food Security and Threats to Their Future. *Journal Global Food Security* **3** pp142–148
[39] Yudo S 2006 Kondisi Pencemaran Logam Berat di Perairan Sungai DKI Jakarta *Jurnal Air Indonesia* **2**(1) pp 1-15
[40] Yulaipi S and Aunurohim 2013 Bioakumulasi Logam Berat Timbal (Pb) dan Hubungannya dengan Laju Pertumbuhan Ikan Mujair (*Oreochromis mossambicus*) *Jurnal Sains dan Seni Pomits* **2**(2) pp166-170
[41] Ujianti R M D, Anggoro S, Bambang A N, and Purwanti F 2018 Water quality of the Garang River, Semarang, Central Java, Indonesia based on the government regulation standard. Journal of Physics: Conference Series **1025**
[42] Latimer, G W 2012 *Official Methods of Analysis of AOAC International* Volume 1 book 1 AOAC International
[43] Soemirat J 2003 *Toksikologi Lingkungan* Gadjah Mada University Press Yogyakarta
[44] Effendi H 2003 *Telaah kualitas air: Bagi pengelolaan sumber daya dan lingkungan perairan* Kanisius Yogyakarta.
[45] Ujianti R M D, Anggoro S, Bambang A N, Purwanti F, and Androva A 2019 Environmental Study on Phytoplankton in Garang Watershed, Central Java, Indonesia and Its Water Quality *IOP Conference Series: Earth and Environmental Science* **246**
[46] Peraturan Badan Pengawas Obat dan Makanan Nomor 5 Tahun 2018 Tentang Batas Maksimum Cemaran Logam Berat Dalam Pangan Olahan
[47] WHO 2003 *Malathion in drinking water Background Document for Preparation of WHO Guidelines for Drinking Water Quality World Health Organistaion* (WHO/SDE/WSH/03.04/103) Geneva
[48] Dugan P, Dey M M, and Sugunan V V. 2006. Fisheries and water productivity in tropical river basins: Enhancing food security and livelihoods by managing water for fish *Agricultural Water Management* **80** pp 262–275
[49] Youn S J, Taylor W W, Lynch A J, Cowx I G, Beard T D, Bartley D and Felicia W 2014 Inland capture fishery contributions to global food security and threats to their future *Global Food Security* **3** pp 142–148
[50] Young K 2009 Omega-6 (n-6) and omega-3 (n-3) fatty acids in tilapia and human health: a review *International Journal of Food Sciences and Nutrition* **60** (Suppl. 5) pp 203–211
[51] Puspitasari R A, Setioko B, Pandelaki E E 2015. Persepsi Integrasi Tata Guna Lahan Pada Kawasan Waterfront Development (Studi Kasus: Kanal Banjir Barat Semarang) *Teknik* **36**(1) pp 17-23
[52] Nazir M 2005 *Metode Penelitian* Ghalia Indonesia Jakarta
[53] Chusein A F A and Ibrahim R 2012 Soaking Period of Cooked Blood Cockle (Anadara granosa) Meat in Alginate Solution to Cadmium Content Reduction *Jurnal Saintek Perikanan* **8**(1) pp 20-26
[54] Rasul G 2014 Food, water, and energy security in South Asia: A nexus perspective from the Hindu Kush Himalayan region *Environmental Science and Policy* **39** pp 35-48

Analysis of Electrical Power Quality Disturbances Based on Empirical Mode Decomposition and Statistical Parameters

Faqih Rofii[1], Agus Naba [2], Hari Arief Dharmawan[3], Fachrudin Hunaini[4]

[1,2,3] Departement of Physics, Universitas Brawijaya Malang
[4] Departement of Electrical Engineering, Universitas Widyagama Malang

E-mail: faqihs3fisika@student.ub.ac.id

Abstract. Power quality becomes a severe problem due to the increasing use of nonlinear electrical loads, complex electric power systems on smart grids, inverters in renewable power plants, and electronic control equipment. Power quality problems include variations in voltage or current such as sag, swell, flicker, spike, overvoltage, undervoltage, interruption, transient, harmonics, and frequency fluctuations. Research on power quality disturbances mostly applies signal processing and transformation methods such as Fast Fourier transform, S-transform, and Wavelet. In this paper, we use empirical mode decomposition methods and statistical parameters to analyze power quality disturbance signals. It gives more detailed characteristics of power quality disturbances. We conducted a six-step analysis to get a percentage of each power quality disturbance signal. The developed method provides a preliminary description of the power quality characteristics from the percentage values of RMS, range, and energy levels at the first IMF. The positive percentage indicates the existence of power quality disturbances contains flicker and swell, while negative indicates contain harmonics, sag, and transient.

Keywords : Power quality; EMD; RMS; Energy level; Intrinsic mode functions

1. Introduction

Electric power systems are getting more complex, this can be seen from a number of devices added to the system, such as electronic control equipment, non-linear loads and inverters on renewable energy. [1]. A complex electric power system has the potential to cause power quality problems characterized by variations in voltage, current, and frequency. Maintaining power quality is very important when supplying electrical power to customers. Electric power is good if the electricity supply is constant and meets customer requirements. Power quality is poor when the supply is not constant, the supply voltage exceeds or is less than the acceptable level, the frequency is fluctuating and the sinusoidal waveform is distorted.

According to IEEE 1159-1995 standards [2], power quality disturbance includes various phenomena, namely transients (impulsive and oscillating), short-duration variations (interruptions, sag and swell), frequency variations, long-duration variations (continuous undervoltages and overvoltages) and steady-state variations (harmonics, notches, and blinks) with a time scale ranging from tens of nanoseconds to stable conditions. According to the IEEE 1459-2010 standard it includes measurements of the amount of electrical power in sinusoidal, non-sinusoidal, balanced, and unbalanced conditions [3].

Igual (2018) presented an integrated mathematical model of power quality disturbance with 28 variations of sinusoidal signal distortion [4]. The variations are interruptions, sag, swell, transient/impulse/spike, oscillatory transient, harmonics, flicker, harmonics with sag, harmonics with swell, flicker with sag, flicker with swell, sag with oscillatory transient, swell with oscillatory transient, sag with harmonics, swell with harmonics, notch, harmonics with swell with flicker, sag with harmonics with flicker, swell with harmonics with flicker, sag with harmonics with oscillatory transient, swell with harmonics with oscillatory transient, harmonics with sag with oscillatory transient, harmonics with swell with oscillatory transient, harmonics with sag with flicker with oscillatory transient, harmonics with swell with flicker with oscillatory transient, sag with harmonics with flicker with oscillatory transient and swell with harmonics with oscillatory transient.

Many researchers have attempted to analyze power quality disturbances using a variety of signal processing and transformation methods as well as statistical parameters for detection and classification. These methods are S-transform for electrical power quality recognition [5], classification by wavelet method and support vector machine [6], integration of discrete wavelet transform and fast Fourier transform [7], and statistical estimators for sag and swell characterizations [8]. The method of analyzing non-stationary signals that is gaining attention lately is the empirical mode decomposition (EMD) proposed by Huang [9]. The effective implementation of EMD has been carried out by Rilling [10] to analyze signals that contain three components: two FM sinusoidal components and a gaussian wavepacket. By applying EMD to the analyzed signal, it gets eight IMF and residues. In recent studies, EMD is widely used for signal analysis in various cases, such as wind speed forecasting [11] and electrocardiogram signal analysis [12]. This paper proposes the use of EMD and statistical parameters to analyze power quality disturbance signals in order to obtain a more detailed picture of the characteristics of power quality disturbances.

2. Empirical Mode Decomposition and Statistical Parameters

EMD is a flexible analysis method for extracting nonlinear and non-stationary signals iteratively. EMD breaks down the time series into a number of Intrinsic Mode Functions (IMFs) based on scale separation. The scale is defined as the distance between two consecutive minimum or maximum local extremes. There are two conditions to satisfy an IMF: (1) In the whole data series the number of extrema (sum of the maximum and minimum) and the number of data sets equal to zero must be equal or different at most 1; and (2) The mean of the signal envelope obtained from the local maximum and minimum must be equal to zero at each point [9]. Stopping criteria need to be determined to avoid continuous iteration without stopping because the amplitude threshold of the mean is too small. Excessive iteration causes unnecessary decomposition. Rilling proposes new stopping criteria based on two thresholds $\boldsymbol{\theta_1}$ and $\boldsymbol{\theta_2}$ aimed at ensuring small fluctuations globally while considering large excursions locally.

In the field of signal processing, statistics have an important role in analyzing signal characteristics or interpreting numerical data on acquired signals. Statistics and probabilities can illustrate the deviation of the measured signal from the referral signal [13]. Basic statistical analysis methods including mean, variance, skewness, kurtosis, and confidence limits allow you to calculate the statistical value of a time series. A signal with a sinusoidal waveform will be more easily analyzed using statistical parameters such as mean, root means square, peak to peak, peak to RMS (crest-factor) and even stationarity. [14,15]. For example, to evaluate the density function model on wind power, statistical parameters are needed to determine the characteristics and properties of the wind power density function [16] and also on the formation of features and classifications of power quality disturbance signals [17].

3. Analysis of Electrical Power Quality Disturbance Signals

Any deviations that exceed the standard limits in the form of voltage, current, or frequency that can cause equipment failure or malfunction of electrical/electronic equipment have the potential to be a power quality problem. The main cause of the problem is the increase in non-linear loads, which distort current and voltage waveforms, especially with sensitive modern equipment. Mostly, the waveforms of power quality disturbances are non-stationary and noisy in nature, so an accurate method is needed to analyze and visually identify the components of power quality problems. For an example as shown in Figure 1, two signals are given, the first is pure sinusoid and the second is harmonics with sag with transient oscillation.

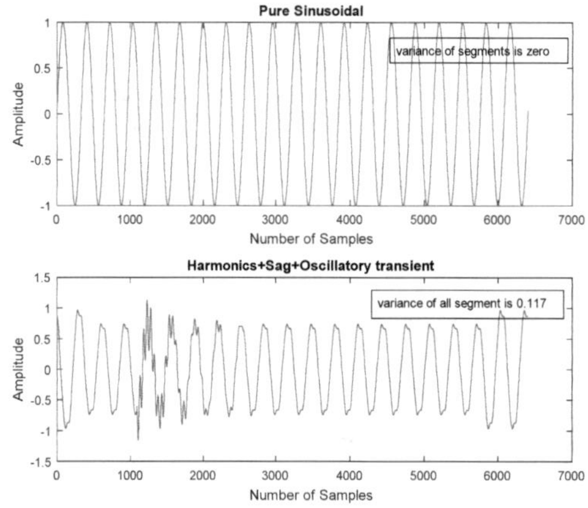

Figure 1. Stationary and non-stationary signals

To find out the stationarity of the two signals, it can be seen from the value of the variance in each segment and all segments. If the variance of all segments converges to zero then the signal is stationer, but if the variance of all segments does not converge to zero, then the signal is non-stationer. After investigation, the variance of all segments in the pure sinusoidal signal is zero, so it is stationer, while the variance of all segment of the second signal is not zero, so it is non-stationer.

In this paper, EMD and statistical parameters are used to analyze the nature of electrical power quality disturbances. EMD is used to decompose power quality signals into a number of IMF, then statistical calculations are applied to each IMF to obtain the signal characteristics. The statistical parameters are mean, standard deviation, root mean square, crest-factor, range, and energy level..

The steps for signal analysis using EMD and statistical parameters are as follows::
1). Generate pure sinusoidal and power quality signals as much as N-data sets

$$X=\left[x_{1,1}(t)\ x_{1,2}(t)\ x_{1,3}(t)\ldots x_{1,N}(t)\quad x_{2,1}(t)\ x_{2,2}(t)\ x_{2,3}(t)\ldots\ldots x_{M,N}(t)\right] \quad (1)$$

2). Decompose data using EMD to produce IMF sets

$$Set_{IMF}=emd(X) \quad (2)$$

3). Apply statistical parameters: mean, standard deviation, root mean square, crest-factor, range, and energy level for each IMF

$$\text{Mean} \rightarrow M=\bar{x}=\frac{1}{N}\sum_{n=1}^{N} x_n \quad (3)$$

$$\text{Root means square} \rightarrow RMS=\left[\frac{1}{N}\sum_{n=1}^{N} x_n^2\right] \quad (4)$$

$$\text{Crest Factor} \rightarrow CF=\frac{x_{n\_max}}{RMS} \quad (5)$$

$$\text{Skewness} \rightarrow SK=\frac{1}{S^3N}\sum_{n=0}^{N-1}(x(n)-\bar{x})^3 \quad (6)$$

$$\text{Range} \rightarrow R=max(x_n)-min(x_n) \quad (7)$$

Energi→ $E=\sum_{n=1}^{N}\left(|x_n|^2\right)$ ………………………………………………… (8)

4). Sum the results of the third step for each IMF in the same signal class, then average it

$$\begin{bmatrix} MM \rightarrow mean\left(sum(M(Set_{IMF_i}C_N))\right),\ i=1,23,4\ \ N=1,2,3,\ldots\ldots 12 \\ MRS \rightarrow mean\left(sum(RMS(Set_{IMF_i}C_N))\right),\ i=1,23,4\ \ N=1,2,3,\ldots\ldots 12 \\ MCF \rightarrow mean\left(sum(CF(Set_{IMF_i}C_N))\right),\ i=1,23,4\ \ N=1,2,3,\ldots\ldots 12 \\ MSK \rightarrow mean\left(sum(SK(Set_{IMF_i}C_N))\right),\ i=1,23,4\ \ N=1,2,3,\ldots\ldots 12 \\ MR \rightarrow mean\left(sum(R(Set_{IMF_i}C_N))\right),\ i=1,23,4\ \ N=1,2,3,\ldots\ldots 12 \\ ME \rightarrow mean\left(sum(E(Set_{IMF_i}C_N))\right),\ i=1,23,4\ \ N=1,2,3,\ldots\ldots 12 \end{bmatrix} \ldots\ldots\ldots\ldots (9)$$

5). Get the results of the subtraction between the average of statistical parameters of the pure sinusoid in step three minus the average of statistical parameters of the power quality disturbance

$$\begin{bmatrix} \Delta MM \rightarrow MM_{C_1}\text{-}MM_{C_N},\ \ N=1,2,3,\ldots 12 \\ \Delta MRS \rightarrow MRS_{C_1}\text{-}MRS_{C_N},\ \ N=1,2,3,\ldots\ldots 12 \\ \Delta MCF \rightarrow MCF_{C_1}\text{-}MCF_{C_N},\ \ N=1,2,3,\ldots\ldots 12 \\ \Delta MSK \rightarrow MSK_{C_1}\text{-}MSK_{C_N},\ \ N=1,2,3,\ldots\ldots 12 \\ \Delta MR \rightarrow MR_{C_1}\text{-}MR_{C_N},\ \ N=1,2,3,\ldots\ldots 12 \\ \Delta ME \rightarrow ME_{C_1}\text{-}ME_{C_N},\ \ N=1,2,3,\ldots\ldots 12 \end{bmatrix} \ldots\ldots\ldots\ldots (10)$$

6). Calculate the percentage in step five for each class of disturbances

4. Results and Discussion

Pure sinusoidal signals and power quality disturbances are generated through computer simulations using MATLAB. We generate 12 power quality signal classes based on mathematical models [4], where each signal class has 100 datasets. The parameters of each generated signal include: the number of samples per class 100 samples / class, the sampling frequency 16 kS / sec, the fundamental frequency 50 Hz with 20 cycles and the amplitude 1 p.u. The signal is generated randomly with a uniform distribution. Figure 2 shows pure sinusoidal and 11 power quality disturbances signal. Data from each signal is then decomposed using EMD to obtain the IMF set. For an example the signal harmonics with sag, after applying EMD to the signal, 6 IMF were obtained as shown in Figure 3.

Tabel 1. The number of IMF of each power quality signal

Class of Signal	Name of Signal	Number of IMF
C1	Pure Sinusoidal	1
C2	Sag	5
C3	Swell	4
C4	Interruption	5
C5	Transient	5
C6	Transient Oscillator	6
C7	Harmonics	5
C8	Harmonics With Sag	7
C9	Harmonics With Swell	6
C10	Flicker	4
C11	Flicker With Sag	5
C12	Flicker with Swell	5

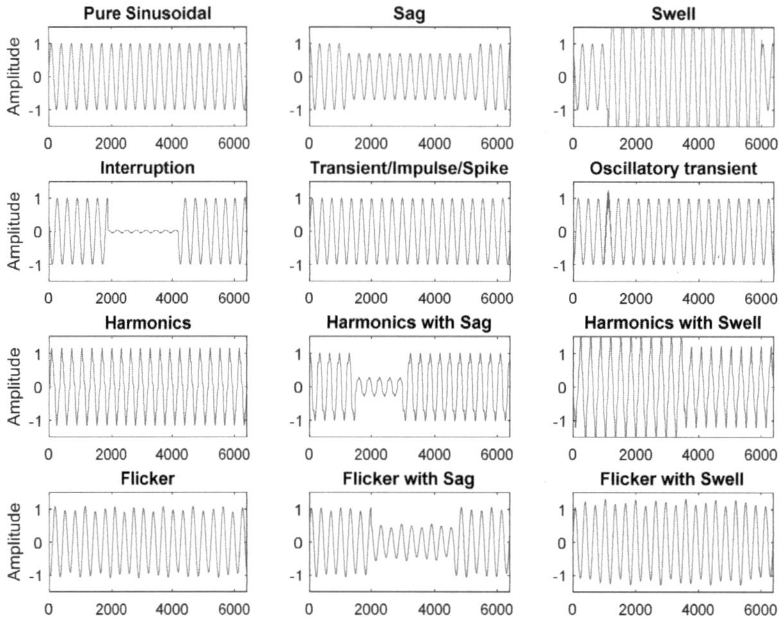

Figure 2. Pure sinusoidal and power quality disturbance signals

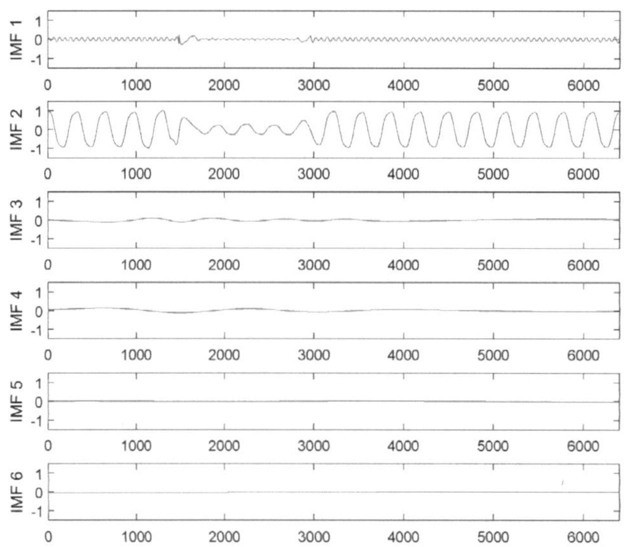

Figure 3. IMF of harmonics with sag signal

Data decomposition using EMD produces a different set of IMF in each class, as summarized in Table 1. Pure sinusoidal signal only produces one IMF while others produce more than three IMFs. The amount of IMF produced is due to the signal components that compose IMF in the form of variations of the amplitude, phase, and frequency of the signal throughout the 20 cycles observed..

In this paper, only the first four IMF's are used for analysis because the fifth and subsequent IMF have a small energy level (i.e. insignificant, as can be seen in Figure 3) so that it is ignored. Signals that only have one IMF then the next three IMF are zero. Figure 4 shows the difference of statistical value between each power quality signals of IMF.

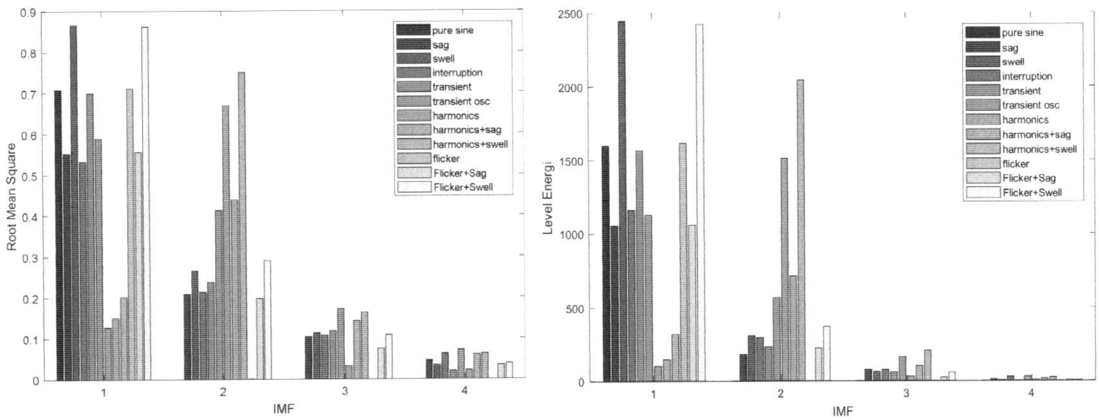

Figure 4. RMS and energy level of each IMF's power quality signal

Different values of statistical parameters are obtained from each signal class, table 2 provides the average value of RMS, range, crest-factor, skewness and energy level at IMF 1. C1, C2 to C12 in table 2 are signal classes as explained in table 1. In this case, C1 or pure sinusoidal signal is determined as a reference signal and will be used as a comparison to other signals.

Table 2. Statistical values of power quality signals at IMF 1

Parameters	C1	C2	C3	C4	C5	C6	C7	C8	C9	C10	C11	C12
RMS	0.707	0.552	0.866	0.533	0.699	0.588	0.127	0.150	0.201	0.711	0.556	0.862
Range	2.000	2.469	3.349	2.364	2.425	2.412	0.526	0.956	0.926	2.138	2.395	3.410
Crest_Factor	1.414	2.407	1.986	2.381	1.794	2.145	2.346	3.779	2.766	1.507	2.272	2.042
Skewness	0.000	-0.003	0.003	-0.006	-0.008	-0.005	0.005	0.009	0.047	0.000	0.001	0.002
Energy Level	1,600	1,055	2,446	1,161	1,566	1,127	103	146	319	1,617	1,059	2,424

To get a clear difference between pure sinusoidal signals and others, we subtract the value of the pure sinusoidal signal (as a reference signal) with other power quality disturbance signals, then get the percentage as presented in Table 3.

Table 3. Percentage of power quality signal statistics to pure sinusoidal signal at IMF 1

Parameters	**Percentage (%)**										
	C1-C2	C1-C3	C1-C4	C1-C5	C1-C6	C1-C7	C1-C8	C1-C9	C1-C10	C1-C11	C1-C12
RMS	-21.96	22.49	-24.61	-1.07	-16.78	-82.00	-78.73	-71.52	0.55	-21.41	21.97
Range	23.46	67.43	18.22	21.27	20.62	-73.71	-52.22	-53.68	6.92	19.76	70.48
Crest-Factor	70.24	40.45	68.35	26.89	51.72	65.89	167.29	95.60	6.61	60.67	44.38
Skewness	-0.26	0.30	-0.60	-0.76	-0.48	0.47	0.89	4.67	0.01	0.07	0.21
Energy Level	-34.04	52.85	-27.41	-2.11	-29.58	-93.59	-90.87	-80.04	1.07	-33.79	51.49

Table 3 illustrates the percentage of power quality disturbance to a pure sinusoid on the parameters of RMS, range, crest-factor, skewness and energy level. A positive percentage value means that the power quality disturbance signal has a higher number than the pure sinusoidal signal, while the negative percentage means the power quality disturbance signal has a lower number than the pure sinusoidal signal. In RMS, signals C10, C12, and C3 are positive, while C7, C8, C9, C4, C2, C11, C6, and C5 are negative. In the range, signal C10, C11.C6, C5, C2, C3, and C12 are positive, while C7, C9, and C8 are negative. In the crest-factor, all signals from C2 to C12 are positive. In skewness, signals C2, C6, C11, C8, C3, C4, and C9 are positive, while C12, C7, C10, and C5 are negative. At the energy level, signals C2, C11, and C8 are positive, while C3, C4, C9, C5, C6, C10, C7, and C12 are negative. Based on the percentage of the RMS value, the range and level of energy obtained, the flicker and swell disturbance signals have a positive percentage, while the interference signals that contain harmonics, sag and transient have a negative percentage. However, on the crest-factor all signals are positive. In skewness, interference signals that contain sag, transient, interruption, harmonic, and swell are positive, while flicker with the swell, harmonics, flicker, and transient are negative.

5. Conclusions

A preliminary analysis of the characteristics of power quality disturbance signals based on basic statistical parameters applied to each IMF resulting from the original signal decomposition has been presented. The EMD method and the statistical parameters were used for the analysis of power quality disturbances. They provide an initial description that the percentage value of RMS and the range and energy level at the first IMF are positive for flicker and swell, while negative for disturbance signals containing harmonics, sag, and transient. The crest-factor parameter did not produce a negative value so it needs to be seen from the resulting numerical value.

References

[1] Adefarati T and Bansal R C 2016 Integration of renewable distributed generators into the distribution system: a review *IET Renewable Power Generation* **10** 873–884

[2] Smith J C, Hensley G and Ray L 1995 IEEE recommended practice for monitoring electric power quality *IEEE Std* 1159–1995

[3] Langella R and Testa A 2010 IEEE standard definitions for the measurement of electric power quantities under sinusoidal, nonsinusoidal, balanced, or unbalanced conditions

[4] Igual R, Medrano C, Arcega F J and Mantescu G 2018 Integral mathematical model of power quality disturbances *Harmonics and Quality of Power (ICHQP), 2018 18th International Conference on* (IEEE) pp 1–6

[5] Kumar R, Singh B, Shahani D T, Chandra A and Al-Haddad K 2014 Recognition of power-quality disturbances using S-transform-based ANN classifier and rule-based decision tree *IEEE Transactions on Industry Applications* **51** 1249–1258

[6] De Yong D, Bhowmik S and Magnago F 2015 An effective power quality classifier using wavelet transform and support vector machines *Expert Systems with Applications* **42** 6075–6081

[7] Deokar S A and Waghmare L M 2014 Integrated DWT–FFT approach for detection and classification of power quality disturbances *International Journal of Electrical Power & Energy Systems* **61** 594–605

[8] Agüera-Pérez A, Carlos Palomares-Salas J, de la Rosa J J G, María Sierra-Fernández J, Ayora-Sedeño D and Moreno-Muñoz A 2011 Characterization of electrical sags and swells using higher-order statistical estimators *Measurement* **44** 1453–60

[9] Huang N E, Shen Z, Long S R, Wu M C, Shih H H, Zheng Q, Yen N-C, Tung C C and Liu H H 1998 The empirical mode decomposition and the Hilbert spectrum for nonlinear and non-stationary time series analysis *Proceedings of the Royal Society of London A: mathematical, physical and engineering sciences* vol 454 (The Royal Society) pp 903–995

[10] Rilling G, Flandrin P and Goncalves P 2003 On empirical mode decomposition and its algorithms *IEEE-EURASIP workshop on nonlinear signal and image processing* vol 3 (NSIP-03, Grado (I)) pp 8–11

[11] Zhang C, Wei H, Zhao J, Liu T, Zhu T and Zhang K 2016 Short-term wind speed forecasting using empirical mode decomposition and feature selection *Renewable Energy* **96** 727–737

[12] Paithane A N and Bormane D S 2015 Electrocardiogram signal analysis using empirical mode decomposition and Hilbert spectrum *Pervasive Computing (ICPC), 2015 International Conference on* (IEEE) pp 1–4

[13] Smith S 2013 *Digital signal processing: a practical guide for engineers and scientists* (Elsevier)

[14] Semmlow J 2018 Chapter 2 - Signal Analysis in the Time Domain *Circuits, Signals and Systems for Bioengineers (Third Edition)* Biomedical Engineering ed J Semmlow (Academic Press) pp 51–110

[15] Semmlow J 2018 Chapter 10 - Stochastic, Nonstationary, and Nonlinear Systems and Signals *Circuits, Signals and Systems for Bioengineers (Third Edition)* Biomedical Engineering ed J Semmlow (Academic Press) pp 449–89

[16] Masseran N 2015 Evaluating wind power density models and their statistical properties *Energy* **84** 533–541

[17] Shen Y, Abubakar M, Liu H and Hussain F 2019 Power Quality Disturbance Monitoring and Classification Based on Improved PCA and Convolution Neural Network for Wind-Grid Distribution Systems *Energies* **12** 1280

IOP Conf. Series: Materials Science and Engineering **846** (2020) 012051 doi:10.1088/1757-899X/846/1/012051

Analysis of Wireless Fire Detector Application to Detect Peat land Fire Based on Temperature Characteristic

N L Marpaung[1], R Amri[2] and E Ervianto[3]

[1,2,3] Electrical Engineering Department, Universitas Riau, Pekanbaru, Indonesia

E-mail: noveri.marpaung@gmail.com

Abstract. Fires of land and forest happened in Riau, give negative impacts to people and its environment. This research discusses about analysis of Wireless Fire Detector as Early Warning System of Peat-land Fire using LM35 Temperature-Sensor, Transmitter-Module(HC-12), Receiver-Module(HC-12), Arduino-ProMini, to produce outputs on LED, LCD, Buzzer. Peat-land fire Detector works when peat land is burning, so its heat spreads through aluminum stem. This temperature is read by Temperature-Sensor1-Temperature-Sensor4. Output of every sensor is sent to each Transmitter-Module(FU1–FU4) in Arduino-ProMini control block. After that, Transmitter-Modules(FU1–FU4) send the readings of sensor to Receiver-Module(FU5). In one time, only one data is received by Receiver-Module from Transmitter-Module. Then, it goes to Arduino-ProMini and processed to get outputs on LCD that display ID of FU1–FU4, Peat-land Status by three activated LEDs, Peat-land temperature. If Green LED is active, means Peat-land not burned. Its temperature is 0°C–21.9°C. LCD displays of SAFE Condition, Buzzer OFF. If Yellow LED is ON, means Peat-land burned underground. Its temperature is 22°C–28.4°C. LCD shows BE-CAREFUL Condition, Buzzer ON. If Red LED is active, means Peat-land burned on the ground. Its temperature is 28.5°C– 50°C. LCD displays DANGER Condition, Buzzer ON. This equipment works as its framework.

1. Introduction

Fires of land and forest in Riau has reached 3,700 hectares in few last years. Based on Riau Forestry, almost areas in districts of Riau Province from 4 hectares up to 2,800 hectares, ever hapeened forest and peat land fires. The negative impacts of forest and peat land fires cause air pollution, smoke misty, people in Riau Province get many diseases, schools and universities are closed, businesslike activities such as restaurants, bank and other trading markets are slack. All of these circumstances ever happened in two weeks in 2015 in Pekanbaru, Riau Province. To overcome this problem, many efforts are needed to avoid any fires of land and forest again in the future. An early warning system can be used to prevent them in advanced. This system is equipped with wireless fire detector to detect any fire in land or forest. Then, the results of the data reading are sent from the burned area to Receiver Modules. So, even every small land fire can be detected and reported to avoid any bigger land or forest fires to the person in charged. This research aims to analyze the wireless Fire Detector as Early Warning System of Peat land Fire using LM35 Temperature-Sensor to produce outputs on LED, LCD, Buzzer.

The scopes of this research are:

1. Maximum Distance of wireless data transmission for each Temperature Sensor can be accepted or not.

2. Range of Wireless Sensor Transmission of Peat land Condition based on Peat land Temperature
3. LM35 temperature sensor is used as temperature detector underground and on the ground of Peat land.
4. Using LCD to display the Temperature, Status and Condition of Peat land,
5. Using three activated LEDs and a Buzzer as indicators of Peat land Conditions.

This problems can be observed by:

1. Determining the Status and Condition of Peat land in three conditions based on the reading of temperatures from the peat land using LM35temperature sensor.
2. Using LCD to display the Temperature, Status and Condition of Peat land.
3. Using three LEDs and a Buzzer to indicate the Peat land Condition.

2. Literature Review

2.1. Previous study

Reference [2] was *Analysis of Controlling Wireless Temperature Sensor for Monitoring Peat-Land Fire* to control wireless Temperature Sensor for monitoring Peat-Land fire by using LM35 Temperature Sensor, Transmitter-Module (HC-12), Receiver-Module (HC-12), Arduino Pro Mini to produce outputs on three different LEDs, Condition of Peat land on LCD, Buzzer if Peat land Condition is DANGER Category. Reference [3] was *Design of Firing Detector System by peat Land With Woody Peat Types Using It's Heat Characteristics* using LM35 Sensor to monitor Peat land fires that has the highest accuracy is ASMC with an accuracy rate of 60%. It took a peat land fire detector design, in order to mitigate catastrophic fires in peat land. The sensor worked by using the heat's characteristics that generated by burnt peat, read from ATMega8 as the microcontroller and LCD 16*2 as the monitor of temperature sensor.

This study uses Temperature Sensors LM35, Arduino Pro Mini, HC-12 Transmitter Module and HC-12 Receiver Module. Sensors can detect Status of Transmitted Data in certain range for different Temperature Sensors, before peat land burned or when peat land burned, for Experiment Tests with obstacles like buildings and trees. This research also test two conditions of Peat land based on their own temperature range. Results of research are displayed on LED, LCD and Buzzer.

2.2. Research Components

Many components are used in this research, such as:

1. Arduino Pro Mini is one type of microcontrollers. This research uses ATMega328P Arduino Pro Mini that has 14 digital pins input/output in one chip of Integrated Circuit (IC). Using this chip, can make connection to FTDI cable to give USB power and can make communication board. This research uses five ATMega328P Arduino Pro Mini. [4].
2. LM35 Temperature Sensor is a temperature sensor module can produce output voltage which is change linearity based on the change of temperature in linear too. LM35 output is in linier scale factor towards the real temperature of 10 mV/^{0}C and can be measured with maximum range from - 55^{0}C up to 150^{0}C. Maximum output voltage of LM35 is 1.5 Volt with 0.5^{0}C accuracy on 25^{0}C temperature. Error level of sensor for 25^{0}C room temperature measurement and no more 2^{0}C, can be calculated by using Equation 1.[5]

   $$Temp\ (^0C) = V_{Out}(in\ mV)\ /\ 10mV \qquad (1)$$
3. HC-12 Module is communication module in serial port using wireless. This module can send and receive serial data through air media with 433 MHz frequency and up to 1800 meter for transmitting distance in open space. HC-12 Module is Data Transmitter/Receiver Module but it cannot transmit and receive data in the same time because it just has link half-duplex between modules. This research uses four HC-12 Modules as Transmitter Modules and one HC-12 Modules as Receiver Module. [6,7]
4. Liquid Crystal Display (LCD) is one type of display media using liquid crystal as main display. This study uses 16x2 LCD.
5. Light Emitting Diode (LED) as indicator lamp with three different colours (Red, Yellow, Green).
6. Buzzer as alarm to indicate in BE CAREFULL Condition or DANGER Condition.

3. Research Methodology

Methods in this are:

1. Literature study.
2. Designing a prototype circuit.
3. Applying designed circuit into a prototype.
4. Testing of Component Activation from each component such as Power Supply, LM35 Temperature-Sensor, Transmitter-Module (HC-12), Receiver-Module (HC-12), Arduino Pro Mini Microcontroller, LED, LCD, and Buzzer.
5. Transmission Distance Testing from Transmitter to Receiver, before peat land burned.
6. Data Transmission Testing from Transmitter to Receiver, when peat land is burned.
7. Measuring temperature characteristic of Peat land.

Flowchart of the research is shown in Figure 1.

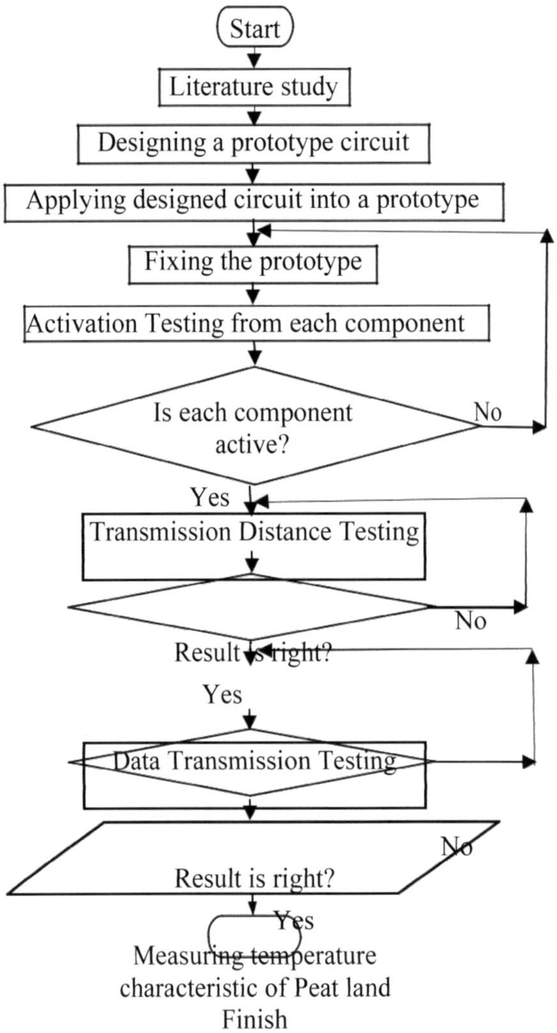

Figure 1. Flowchart of the research

This research is done by injected four aluminum bars into ground up to 60cm depth in different Peat land areas, whereas Temperature Sensor and ATMega328P Arduino Pro Mini are attached to each aluminum bar. All of these Temperature Sensors detect the land temperatures in different area. Each Temperature Sensor sends its temperature data from Transmitter Module (FU1 – FU4) to Receiver Module through Arduino Pro Mini on each aluminum bar. Receiver Module sends each temperature

IOP Conf. Series: Materials Science and Engineering **846** (2020) 012051 doi:10.1088/1757-899X/846/1/012051

data from FU1 – FU4 to other Arduino Pro Mini, to be processed and display it into LCD. Arduino Pro Mini also controls the activated LED and Buzzer depend on the Peat land Status.

4. Results and Discussion

4.1. Testing of transmission distance before peat land burned

Testing of transmission distance before peat land burned is done to know the Peat land condition, whether it is burned or not. Testing is done in the midday in good weather and no windy. In this case, number 1 is set up to state that Peat land is not burned. Number 1 is sent in 8 bits respectively. Data Testing of transmission distance before peat land burned is shown in Table 1.

Table 1. Testing of transmission distance before peat land burned.

Channel Number	Transmitted Data	Environmental Condition	Status of Transmitted Data	Distance (Meter)
FU1	1111 1111		ACCEPTED	50 - 130
			NOT ACCEPTED	140
FU2	1111 1111		ACCEPTED	50 - 270
		Experiment tests :	NOT ACCEPTED	280
		–Buildings		
FU3	1111 1111	–Trees	ACCEPTED	50 - 460
			NOT ACCEPTED	470
FU4	1111 1111		ACCEPTED	50 - 870
			NOT ACCEPTED	880

From Table 1, can be said that testing of transmission distance before peat land burned for Experiment Tests with obstacles like buildings and trees, is performed to get the Status of Transmitted Data for the same data of 1111 1111. Transmitted Data for FU1 is in ACCEPTED Status if transmission distance in 50m – 130m and NOT ACCEPTED when transmission distance in 140m and more. FU2 shows ACCEPTED Status of Transmitted Data if transmission distance is in 50m – 270m, more than that, Transmitted Data is in NOT ACCEPTED Status. For Channel FU3 and FU4, data can be transmitted if transmission distances are 50m – 460m and 50m – 870m, respectively. So, ACCEPTED Status will be seen in Status of Transmitted Data of Channel FU3 and FU4. Data in Channel FU3 and FU4 cannot be sent if transmission distances starts from 470m or more and 870m or more, respectively. In this case, NOT ACCEPTED Status will be shown in Status of Transmitted Data from Channel FU3 and FU4.

4.2. Testing of transmission distance when peat land burned

Testing of transmission distance when peat land burned is done to get information about the Status of Transmitted Data for the same data of 3333 3333, for Experiment Tests with obstacles like buildings and trees. Number 3 is set up to state that Peat land in burned condition. This testing is performed on the midday, good weather and no windy. Testing of transmission distance when peat land burned, can be seen in Table 2.

Table 2. Testing of transmission distance when peat land burned.

Channel Number	Transmitted Data	Environmental Condition	Status of Transmitted Data	Distance (Meter)
FU1	3333 3333		ACCEPTED	50 - 130
			NOT ACCEPTED	140
FU2	3333 3333		ACCEPTED	50 - 270
		Experiment tests :	NOT ACCEPTED	280

Channel Number	Transmitted Data	Environmental Condition	Status of Transmitted Data	Distance (Meter)
		–Buildings		
FU3	3333 3333	–Trees	ACCEPTED	50 - 460
			NOT ACCEPTED	470
FU4	3333 3333		ACCEPTED	50 - 870
			NOT ACCEPTED	880

Based on Table 2, Status of Transmitted Data for FU1 and FU2 are ACCEPTED if transmission distances in 50m – 130m and 50m – 270m, respectively, when peat land burned. NOT ACCEPTED Status of Transmitted Data for FU1 and FU2 when transmission distance in 140m or more and 280m or more, in the same condition. When peat land burned, Channel FU3 and FU4 shows ACCEPTED Status of Transmitted Data if transmission distance are in 50m – 460m and 50m – 870m, sequentially. On the other hand, NOT ACCEPTED Status will be shown in Status of Transmitted Data of Channel FU3 and FU4 if transmission distance in 470m or more and 880m or more, in the same condition. Testing of Peat land Temperature is done to know the temperature reading from land, when Peat land is Not Afire, Burned under Ground and Burned on The Ground. Testing Data of Peat land Temperature is shown in Table 3.

Table 3. Testing Data of Peat land Temperature.

Peat land Temperature (°C)	Status	Temperature Range (°C)	Condition
0 - 20	NOT AFIRE	0 - 21.9	SAFE
22 - 27.7	BURNED*	22 - 28.4	BE CAREFUL
27.8 - 50	BURNED **	28.5 - 50	DANGER

Whereas :

– * : Burned under ground
– ** : Burned on the ground

Based on Table 3, Peat land temperature from 0^0C – 20°C and it is NOT AFIRE, can be said that Peat land in SAFE Condition, because Peat land temperature is in Temperature Range of SAFE Condition (0^0C – 21.9°C). If Peat land is BURNED underground and its temperature in 22^0C – 27.7°C, so Peat land is in BE CAREFULL Condition because its temperature in Temperature Range of BE CAREFULL Condition (0^0C – 28.4°C). If Peat land temperature in 22^0C – 27.7°C and it is BURNED on the ground, so Peat land is in DANGER Condition because its temperature in Temperature Range of DANGER Condition (0^0C – 28.4°C).

5. Conclusion

1. Data can be transmitted and ACCEPTED in distance range of FU1 from 50m-130m, FU2 from 50m-270m, FU3 from 50m-460, FU4 from 50m-870m, before peat land is burned or when peat land is burned with obstacle like buildings and trees. For same condition, if Transmitted Data over the distance range of all Temperature Sensors, so NOT ACCEPTED Status of Transmitted Data will be shown for each Temperature Sensors (FU1 –FU4).
2. If Green LED is Active (ON), means Peat land is not burned with its temperature range is 0°C–21.9°C. In this condition, LCD displays SAFE Condition, Buzzer is OFF. In this case, Temperature-Sensor1 until Temperature-Sensor4 can send temperature value from 50m – 890m.
3. If Yellow LED is ON, means Peat land burned underground with its temperature range is 22°C–28.4°C. In this time, LCD shows BE-CAREFUL Condition, Buzzer is ON. In this case, Temperature-Sensor1 until Temperature-Sensor4 can send temperature value from 50m – 890m.

IOP Conf. Series: Materials Science and Engineering **846** (2020) 012051 doi:10.1088/1757-899X/846/1/012051

4. If Red LED is Active (ON), means Peat land is burned on the ground with its range temperature is 28.4°C–50°C, LCD displays DANGER Condition, Buzzer is ON.

Acknowledgments
Authors would give thanks to Institution of Research and Community Service (Lembaga Penelitian dan Pengabdian kepada Masyarakat) Universitas Riau for funding this research in 2017.

References
[1] Research Centre Riau Province 2014 "Research of Peat Land Fire Preventive System Trough Electromagnetic Reflection Wave Prediction in Riau Province" *Research Centre Pekanbaru*
[2] Noveri, L.M., Rahyul A., Edy E., Nurhalim D.A. 2018 "Analysis of Controlling Wireless Temperature Sensor for Monitoring Peat-Land Fire" *International Journal of Electrical, Energy and Power System Engineering* Vol.1 No.2 e-ISSN: 2654-4644 pp 16-21
[3] Rahyul A., Noveri, L.M., Edy E., Nurhalim 2017 "Design of Firing Detector System by Peat Land with Woody Peat Types Using It's Heat Characteristics" *Published in: Instrumentation, Control, and Automation (ICA) 5th International Conference* Yogyakarta Indonesia DOI: 10.1109/ICA.2017.8068427 IEEE.org IEEE *Xplore* Digital Library | IEEE-SA | IEEE Spectrum
[4] Team Arduino 2014 "Arduino Pro Mini, Spark Fun Electronics" *https://creativecommons.org* Retrieved 12 Juni 2017 at 5 pm
[5] Texas Instruments 2015 "LM35 Precision Centigrade Temperature Sensor" *Texas Instruments Incorporated unpublished* Texas
[6] Dziku 2016 "HC-12 433MHz Wireless Serial Communication Module Configuration" *Electronics Lab, https://quadmeup.com* Retrieved June 12, 2017 at 3pm
[7] Elecrow 2012 "HC-12 Wireless Serial Port Communication Module" *https://www.elecrow.com* Retrieved June 12, 2017 at 2 pm
[8] Arduino Team 2014 "Arduino Pro Mini" *Spark Fun Electronics https://creativecommons.org* Retrieved June 12, 2017 at 5 pm

IOP Conf. Series: Materials Science and Engineering **846** (2020) 012052 doi:10.1088/1757-899X/846/1/012052

Evaluation of Grounding System Values with Grid Type on 150 Kv in Batu Besar Substation, Batam

E Ervianto[1]*, N L Marpaung[1], R Amri[1], and W Zarniadi[1]

[1]Electrical Engineering Department, Universitas Riau, Pekanbaru, Indonesia

*edyervianto@gmail.com

Abstract. Generally, electrical fault occurs in substations, such as phase-to-ground short circuit. The fault currents flowing in Grounding System with Grid-Type can cause voltage gradient different on ground points around substation. The purpose of this study is to evaluate the Grounding System Values with Grid-Type whether meets standard, and protect somebody who is walking around substation or touching equipment stung of 150Kv electric in Batu-Besar Substation, Batam. Grounding System with Grid-Type needs to be evaluated every certain period to get Touch Voltage and Step Voltage that meet safety and security standard of IEEE Std 80-2013. The research method is to collect parameters $\rho, \rho_s, h_s, A, LT, h$, which calculated using formula of Resistance Value, Touch Voltage, Step Voltage of Grounding System with Grid-Type. Then, analyzing and comparing of these data are performed based on results of calculation using the IEEE Std 80-2013. From results of this research, Resistance Value is 0.62Ω, 737.65V and 2,458.43V are Touch Voltage and Step Voltage of Grounding System Values with Grid-Type respectively, based on 0.5second interruption duration with somebody's body weight of 50kg. The Evaluation results show that Grounding System Values with Grid-Type, meet the standard of IEEE Std 80-2013 for somebody who has weight 50kg in Batu-Besar Substation, Batam.

1. Introduction

Substation is the most important part in the electrical power system. It is also a part of the link between the transmission and distribution lines. In addition, the substation contains an electrical power system equipment such as transformers, protection equipment and control equipment.

In substation, there is an electrical fault current of short circuit. It is dangerous. This interference current can damage equipment and also endanger someone in the area of substation because it can cause a voltage gradient on the equipment and the ground surface. The voltage gradient can cause Touch Voltage for someone who touches equipment and Step Voltage to someone who is walking in the area of substation.

To overcome this problem, the substation must have a good and safe grounding system for equipment and especially for someone who is in the area of the substation during normal conditions or in the time of disruption. The grounding system used must be able to prevent danger when the electrical disturbance occurs, so it can not cause a potential different at many points around the disturbance area. There are several parameters that determine whether the grounding system meets safety and security requirements, namely Grounding Resistance, Touch Voltage and Step Voltage. [1] For Grounding System in the substations, there are several standards that used to obtain a good and safe Grounding System. It also can minimize hazards during interruptions. One of the standards used is IEEE Std 80-

2013. In IEEE Std 80-2013, there is a basic concept of the Grounding System for substations until step by step in design and calculation Grounding System for substations are explained clearly and detail. [2] . Based on this standard, the research conduct an evaluation of the 150/20 kV Grounding System to meet the IEEE Std 80-2013 in Batu Besar Substation, Batam. The purpose of this study is to evaluate the Grounding System Values with Grid Type whether meets standard, and can protect somebody who is walking around substation or can protect somebody who is touching equipment stung of 150 Kv electric in Batu Besar Substation, Batam.

2. Theoretical Basic

2.1. Previous study

Reference [3] was Analysis of Grounding System for Substation of Bagan Batu with Grid Construction to plan the grounding system in the form of a grid and to find out the ratio of the amount of touch stress and step voltage in the field with touch voltage and allowable step voltages based on the IEEE 80-2000 Electrical Standards. The results of each Touch Voltage are 154.1 volts from the Touch Voltage permitted 186.7 volts and step voltage of 209.6 volts from permitted step voltage of 214.4 volts.

Reference [4] was Study of Substation Grounding System 150/20 KV Design at Garuda Sakti Substation by taking data in the field and processed to calculate the formulas. Then, compare the results with the IEEE Std 80-2000. Based on the design results, it was obtained 626.38 V mesh voltage, 344.46 V step voltage, 2300 m length of the grid conductor used along with 0.55 Ω resistance and 2/0 AWG (86.59 mm2) conductor with 0.0105 m or 0.414 inch diameter.

Reference [5] was Safety Evaluation on 150 KV Ngawi Substation Grounding System, Surakarta. The results of this study in 150 KV grounding resistance in Ngawi Substation during 2017 period is ≤1. It meets the quality of safety criteria for grounding resistance. The difference of allowable Touch Voltage is 1,430 volts, whereas Touch Voltage of research is 134 volts. The allowable Step Voltage is 4,940 volts, whereas Step Voltage of research is 153 volts and the Transfer Voltage is 3,833.5 volts.

Further Reference [6]was IEEE Guide for Safety in AC Substation Grounding. In this research, it was discussed the importance of the Grounding System at the substation. Disturbances rise in the substation will give hazardous result that can cause accidents to human. So, a proper and safe design of Grounding System that meets the standards for humans is obtained.

2.2. Substation

Substation is a main of electrical power system equipment placed, such as Power Transformers, Control Equipment, Protection Equipment, Current Transformer, Potential Transformer, Conductors and Insulators. Substation functions as a place to increase or decrease high-voltage electric to be low-voltage electric, and viceversa. Substation is used as a place to supervise the equipment operation. Substation in Batu Besar, Batam, can be seen in Figure 1.

Figure 1. Substation in Batu Besar, Batam

Land in the substation has electrical voltage because Grounding System process occurred in the substation. Disturbance current flows to the ground will cause potential differences on the ground surface.

In addition, soil degradation affects the size of resistivity value of the soil type. It also affects the size of the Touch Voltage and Step Voltage. Soil degradation is caused by three factors, namely Physical, chemical and Biological Degradation.

Touch Voltage Value and Step Voltage Value are various. These values depend on location of someone who is standing in the location of the grounding points of an equipment.

2.3. Grounding System with Grid Type

Grounding System on equipment in the substation using conductors that are injected parallel in the ground and connected between one each other to form grid. This grid system is in depth of 0.3m to 0.8m of ground surface. [7] Grid system configuration can be seen in in Figure 2.

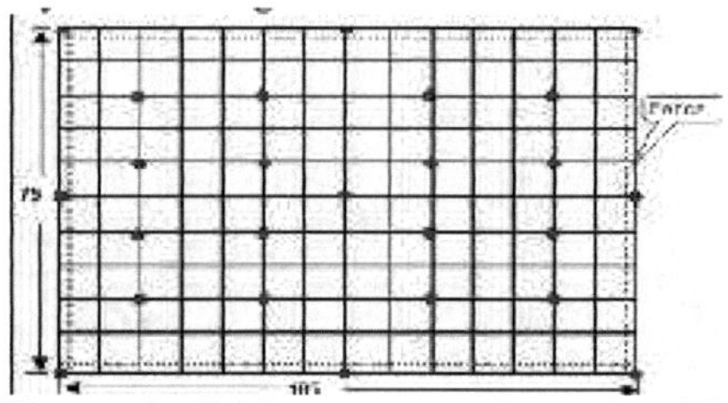

Figure 2. Grid system configuration [7]

2.4. Soil Resistivity

The condition of soil resistivity, structures and soil layers are different in a substation. So. It causes resistance different of soil types. This is caused by geological factor, soil salinity, soil humidity, soil temperature and soil electrolyte. Group of soil types and their resistivity is shown in Table 1.

Table 1. Soil Resistivity [8]

Soil Type	Resistivity (m)
Marshland	30
Organic Soil	100
Wet Sand	200
Wet Gravel	500
Dry Sand and Gravel	1000
Crushed Rock	3000

2.5. Grid Grounding Resistance

For most transmission and other large substation, grid grounding resistance value must be 1Ω or less. Grid Grounding resistance can be determined with simplified calculation, using formula is the Equation 1. [9]

$$R_g = \rho\left[\frac{1}{L_T} + \frac{1}{\sqrt{20A}}\left(1 + \frac{1}{1 + h\sqrt{\frac{20}{A}}}\right)\right] \quad (1)$$

2.6. Touch Voltage

Touch Voltage is voltage occurs between object touched with a point within 1 meter, whereas the object is connected to the grid grounding. The number of disturbance current flowing to someone is limited by the resistance of that person and the contact resistance of that person's feet to the ground. Flowing the Touch Voltage can be seen in Figure 3. The Touch Voltage event is formulated in the equivalent circuit and shown in Figure 4.

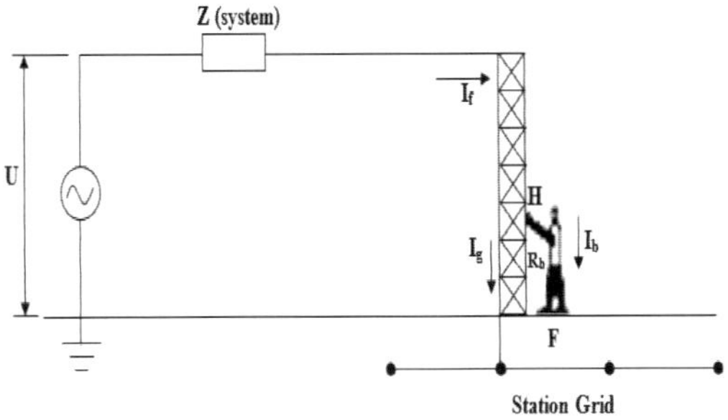

Figure 3. Flowing the Touch Voltage

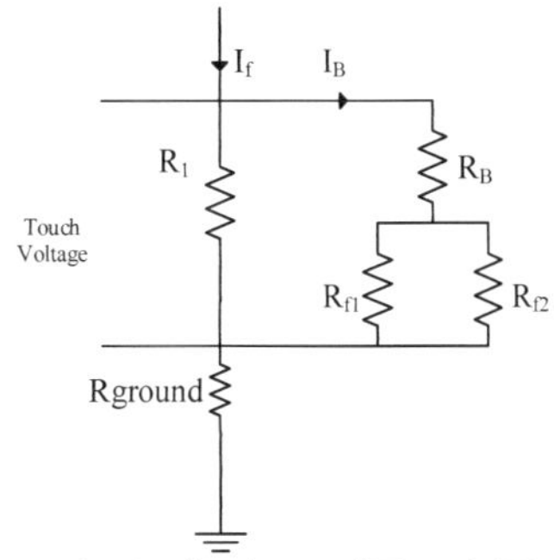

Figure 4. Equivalent of Touch Voltage

From Figure 3, when a short circuit disturbance occurs, it flows to an electrical equipment. Intentionally or accidentally, that equipment is touched by someone.

Touch voltage is stated in Equation 2 and Equation 3 [9]:

$$E_{Touch50} = (850 + 1{,}5C_s \times \rho_s) \, . \, Ib_{50} \tag{2}$$

$$Ib_{50} = \frac{0.116}{\sqrt{t_f}} \tag{3}$$

Resistivity Reduction Factor on the ground surface is stated in Equation 4. [9]:

$$C_S = 1 - \frac{0.09\left(1 - \frac{\rho}{\rho_s}\right)}{2h_s + 0.09} \tag{4}$$

Shock Duration and Tolerable Touch Voltage can be seem in Table 2.

Table 2. Shock Duration and Tolerable of Touch voltage [9]

Shock Duration (s)	Tolerable of Touch Voltage (Volt)
0.1	1,980
0.2	1,400
0.3	1,140
0.4	990
0.5	890
1.0	626
2.0	443
3.0	362

2.7. Step Voltage

Step Voltage is voltage occurs between the two legs of somebody who standing on the ground of substation that flowed of a short circuit fault current to the grid grounding. In this step, assumed voltage distance between two feet of someone is 1 meter and 8 cm of feet diameter. [2] Figure 5 shows the occurrence of Step Voltage caused by the flowing of short-circuit fault current to the grid grounding. At the same time, the same person is standing on the substation area that connected to the grid grounding. Equivalent of Step Voltage is shown in Figure 6.

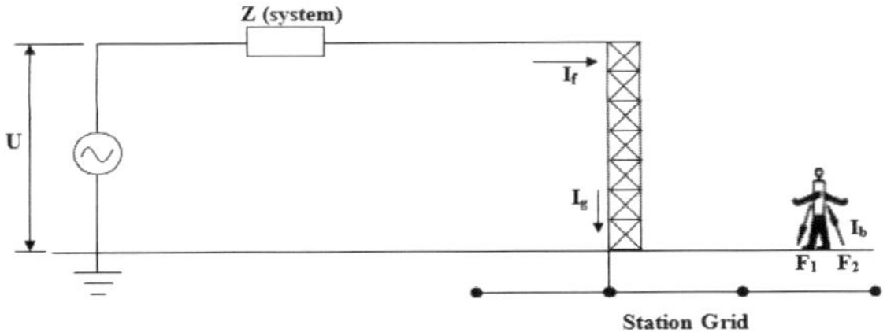

Figure 5. Step Voltage

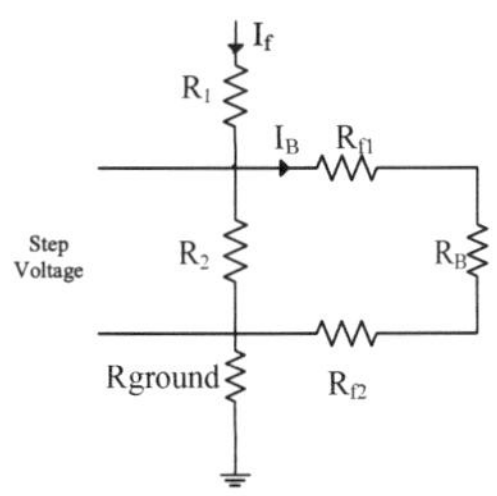

Figure 6. Equivalent of Step Voltage

IOP Conf. Series: Materials Science and Engineering **846** (2020) 012052 doi:10.1088/1757-899X/846/1/012052

Step Voltage is stated in Equation 5 and Equation 6 [2]:

$$E_{Step50} = (1000 + 6C_s \times \rho_s) \, . \, Ib_{50} \quad (5)$$

$$Ib_{50} = \frac{0.116}{\sqrt{t_f}} \quad (6)$$

Table 3 shows the Shock Duration and the Tolerable of Step Voltage

Table 3. Shock Duration and Tolerable of Step Voltage [2]

Shock Duration (s)	Tolerable of Touch Voltage (Volt)
0.1	7,000
0.2	4,950
0.3	4,040
0.4	3,500
0.5	3,140
1.0	2,216
2.0	1,560
3.0	1,280

3. Research Methodology

This research method is to collect parameters of Grounding System with Grid Type, such as ρ, ρs, hs, A, LT, and h. All of these parameters are calculated by using formula of Grid Grounding Resistance, Touch Voltage, and Step Voltage. Then the results of calculations are analyzed and compared with to determine whether all these obtained parameters meet the standard of IEEE Std 80-2013. Procedure of the research can be seen in Figure 7.

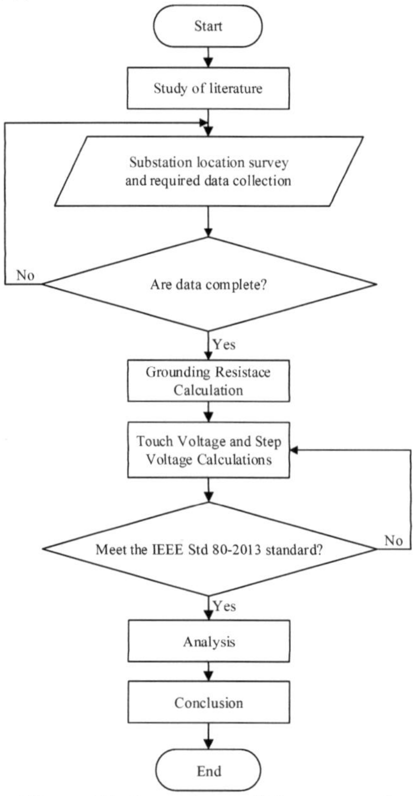

Figure 7. Procedure of the research

4. Results and Discussion

The case study presented in this work is extracted of 150 kV in Batu Besar Substation, Batam Island. Parameters used in the research is shown in Table 4.

Table 4. Parameters of the research

Parameter	Symbol	Value
Soil resistvity		1000 Ω-m
Surface layer resistivity	S	3000 Ω-m
Total length of grid conductor	L_T	1692 m
Total area enclosed by grid	A	6230 m^2
Deep of earth grid conductor	h	0.3 m
Surface layer thickness	hs	0.15 m
Duration time of allowable short circuit	tf	0.5 s

4.1 Ground Resistance Calculation

Calculation of grid ground resistance is carried out with data and Equation 1 as follows:

$$R_g = \rho\left[\frac{1}{L_T}+\frac{1}{\sqrt{20A}}\left(1+\frac{1}{1+h\sqrt{\frac{20}{A}}}\right)\right] = 100\left[\frac{1}{1692}+\frac{1}{\sqrt{20.6230}}\left(1+\frac{1}{1+0.3\sqrt{\frac{20}{6230}}}\right)\right] = 0.62\Omega$$

4.2. Touch Voltage

Calculation of Touch Voltage is carried out with data as follow:

$R_B = 850\ \Omega$, $\rho_s = 3000\ \Omega$-m $t_f = 0{,}5$s

To get *Cs* is used Equation 4 :

$$C_S = 1-\frac{0.09\left(1-\frac{\rho}{\rho_s}\right)}{2h_s+0.09} = 1-\frac{0.09\left(1-\frac{1000}{3000}\right)}{2.0.15+0.09} = 0.777F$$

Touch voltage:

$$E_{Touch50} = (850+1{,}5C_s \text{ x } \rho_s).Ib_{50} = (850+1{,}5 \text{ x } 0.777 \text{ x } 3000).\frac{0.116}{\sqrt{0.5}} = 713 \text{ Volts}$$

Result of Touch Voltage calculation can be seen in Table 5.

Table 5. Result of Touch Voltage calculation

Someone's Weight (kg)	Touch Voltage (E_t) (V)	Tolerable Touch Voltage by IEEE Std 80-2013(V)
50	713	890

Based on Table 5, the Touch Voltage for somebody with 50 kg weight is 713V. This value meets the standard of allowable Touch Voltage based on IEEE Std 80-2013 of 890 V with a duration of 0.5 second interruption. So, with 0.5 seconds, someone is still quite safe to touch electrical equipment without using safety equipment.

4.3. Step Voltage

Calculation of touch voltage is carried out with data as follows:

$R_B = 850\ \Omega$, $\rho_s = 3000\ \Omega$-m $t_f = 0{,}5$s

To get *Cs*, Equation 4 is used:

$$C_S = 1-\frac{0.09\left(1-\frac{\rho}{\rho_s}\right)}{2h_s+0.09} = 1-\frac{0.09\left(1-\frac{1000}{3000}\right)}{2.0.15+0.09} = 0.777F$$

Step Voltage:

$$E_{Touch50} = (1000 + 6C_s \text{ x } \rho_s) \cdot \frac{0.116}{\sqrt{t_f}} = (1000 + 6 \text{ x } 0.777 \text{ x } 3000) \cdot \frac{0.116}{\sqrt{0.5}} = 2{,}458.4 \textit{ Volts}$$

Result of Step Voltage calculation is shown in Table 6.

Table 6. Result of Step Voltage calculation

Someone's Weight (kg)	Touch Voltage (E_t) (V)	Tolerable Touch Voltage by IEEE Std 80-2013(V)
50	2.458,4	3.140

Based on Table 6, Step Voltage for somebody with weight of 50 kg is 2,458.4 V. This value meets the standard of the allowable Step Voltage based on IEEE 80-2013 at 3,140 V with a duration of 0.5 seconds interference. With this result, it is still safe for somebody to stand on both feet without safety equipment in the substation without exceeding 0.5 seconds.

5. Conclusion

From the results, Resistance Value is 0.62 Ω, 737.65 V and 2,458.43 V are Touch Voltage and Step Voltage, respectively, of Grounding System Values with Grid Type based on the duration of 0.5 second interruption with somebody's body weight of 50 kg. The results of Grounding System Values with Grid Type, meet the standard of IEEE Std 80-2013 for somebody who has weight 50 kg in Batu Besar Substation, Batam.

References

[1]. K. Iskak, A. Syamsir and F. Nur 2017 "The Effect of Mesh Size, Number of Rod & Length of Rod Toward Touch Voltage, Step Voltage and Ground Resistance in Grounding System" in *International Conference on High Voltage and Power System.*

[2]. C. Substations 2013 "IEEE Std 80-2013 : IEEE Guide for Safety in AC Substation, United State of America" *IEEE.*

[3]. T. Abrar 2012 " Rod Towards Touch Voltage, Step Voltage and Ground Resistance in Grounding System, Pekanbaru, Indonesia (Analisis Sistem Pentanahan Gardu Induk Bagan Batu Dengan Bentuk Konstruksi Grid (Kisi-kisi), Pekanbaru, Universitas Lancang Kuning)" *University of Lancang Kuning* Pekanbaru Indonesia.

[4]. R. Citra and E. Edy 2017 "Study of 150/20 KV Substation Grounding System Design at Garuda Sakti Substation (Studi Perancangan Sistem Pembumian Gardu Induk 150/20 KV di Gardu Induk Garuda Sakti)," Jurnal Online Mahasiswa *(JOM) FTEKNIK Vol 4 Universitas Riau* Pekanbaru Indonesia.

[5]. A. W. Bimbi 2018 "Safety Evaluation on 150 KV Substation Grounding System at Ngawi, Surakarta (Evaluasi Keamanan Pada Sistem Pentanahan Gardu Induk 150 KV Ngawi, Surakarta" *Universitas Muhammadiyah Surakarta* Surakarta Indonesia.

[6]. S. Simatupang 2011 "Calculation of the Grounding Resistance of Grids on Two Layers of Soil," *Repositori of Universitas Sumatera Utara* Medan Indonesia.

[7]. G. S. Swapnil and R. B. Nitin 2014 "Design of Grounding System For HV/EHV AC Substation (Case Study of 400 kV Substation at Aungarabad)" *International Journal of Advances in Engineering & Technology (IJAET)* India.

[8]. PUIL 2011 "General Electrical Installation Requirements 2011 (Peraturan Umum Instalasi Listrik 2011)" *National Standardization Agency (BSN)* Jakarta.

[9]. T. S. Hutauruk 1999 " Neutral Grounding Power System and Equipment (Pengetanahan Netral Sistem Tenaga dan Pengetanahan Peralatan*)" Erlangga* Jakarta.

Seismic microtremor experiment to determine seismic vulnerability of North Aceh

V H Banyunegoro[1,2], **U Muksin**[1,3], **Y Idris**[3,4]

[1] Department of Physics, Faculty of Natural Sciences, Universitas Syiah Kuala, Banda Aceh, Aceh, Indonesia
[2] Stasiun Geofisika Mata Ie, Indonesian Agency for Meteorological, Climatological and Geophysics (BMKG), Banda Aceh, Aceh, Indonesia
[3] Tsunami Disaster Mitigation Research Center, Banda Aceh, Aceh, Indonesia
[4] Department of Civil Engineering, Faculty of Engineering, Universitas Syah Kuala, Banda Aceh, Aceh, Indonesia

Corresponding Author: vrieslend22@gmail.com

Abstract. North Aceh is an area in Aceh, Indonesia, which is crossed by the active Lhokseumawe Fault. The existence of Lhokseumawe fault makes North Aceh relatively vulnerable to seismic hazard. The seismic vulnerability consists of two main components, natural frequency and amplification factor. Microtremor measurement is necessary in order to obtain natural frequency and amplification factor of certain area. A microtremor experiment has been conducted to understand the seismic vulnerability in north Aceh. The seismic microtremor has been measured at 30 sites which are located relatively evenly in the Lhokseumawe City. The measurement result is processed using Horizontal to Vertical Spectral Ratio (HVSR) method to obtain the main components of seismic vulnerability. The natural frequency and amplification factor are able to be calculated. The seismic vulnerability index calculated form natural frequency and amplification factor is also successfully calculated. The area located in the south of the Lhokseumawe Fault is seismically more vulnerable which is also correlated with the rock's formation of the area. The results are still preliminary that needs more information from the seismic wave velocity representing the characteristics of rocks or soils.

Keywords : Aceh, Microtremor, Seismic Vulnerability

1. Introduction

Aceh is one of the seismically vulnerable areas which is caused by complexity of the tectonic setting of the area [1]. North Aceh in particular, is known to be crossed by Lhokseumawe Fault [2]. Another recently discovered fault, Nisam Fault, near North Aceh adds the source of earthquake that is able to strike this area [3]. The latest major earthquake in North Aceh happened more than one year ago. Badan Meteorologi Klimatologi dan Geofisika (BMKG) reported that the 4.7 magnitude earthquake had been felt in Lhokseumawe with the MMI scale of III [4]. Seismic hazard is closely related to the site effect when earthquakes strike. Site effect has important role in determining how severe the damage caused by earthquake in certain area [5]. Considering that North Aceh has major city, Lhokseumawe, and it is located near the two faults, site effect survey is necessary to be conducted. There are several methods to measure the level of site, one of them is by conducting microtremor survey. Microtremor survey provides two important parameters in assessing site vulnerability which are natural frequency and amplification factor. Natural frequency describes the softness of the rocks composing particular area. The higher the natural frequency, it indicates that the composing rocks are harder. Calculating natural frequency of an area is important for civil engineers to design buildings that are going to be built. When the natural frequency of the rocks is known, then the building's natural frequency should not be similar in order to avoid resonance effect while earthquake waves strike. The other element in assessing site vulnerability is amplification factor. The amplification factor indicates the thickness and softness of the rocks. Thicker and softer the rocks, mean the higher amplification factor will become. The most popular example is Mexico City earthquake that happened in 1985. Mexico City that stands on thick sediments experienced an amplified shake that devastated the city. This phenomenon suggests that amplification factor play an important role in assessing a seismic vulnerability. The Natural frequency and amplification factor produced by the survey are used to generate seismic vulnerability index using Horizontal to Vertical Spectral Ration and shear wave velocity (Vs30) model using inversion method.

2. Microtremor Measurements

The microtremor survey is intended to cover North Aceh and the Lhokseumawe fault which crosses the North Aceh. There are 30 sites that are measured as shown in Figure 1.

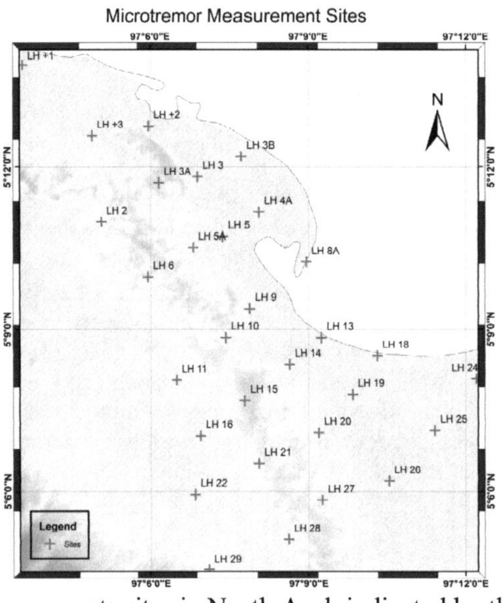

Figure 1. Microtremor measurements sites in North Aceh indicated by the black point covering the Lhokseumawe City and fault.

The sites are originally planned to cover the east and west side of the Lhokseumawe Fault and along the faults. This is intended so that the rock characteristic differences between the west and east side of the fault can be investigated. The spatial grid between sites varies from 1 to 1.5 kilometers. The locations selection are following the guidelines in SESAME [6]. There are some obstacles in the field that made the sites need to be slightly changed. The measurements are carried out for a minimum duration of 30 minutes using short period seismometer. The data resulted from microtremor is processed by using geopsy using Horizontal to Vertical Spectral Ratio (HVSR) method. HVSR is a method to evaluate some characteristics of soft-sedimentary deposits [6]. HVSR works by calculating wave spectral ratio between horizontal and vertical components. The output of HVSR are natural frequency and amplification factor. Those results are the main input in calculating seismic vulnerability index.

3. Preliminary Results

Based on geological map issued by Indonesian Ministry of Energy and Mineral Resources [7] as shown in Figure 2. The Lhokseumawe city is densely populated and is formed of thick sediment consists of mostly sand and clay. This makes the Lhokseumawe city is more vulnerable to seismic hazard based on the geological map. While the less-populated area located on the west of Lhokseumawe Fault consists of slightly higher terrain with the altitude between 10 – 100 m. The sediments are mostly carried by two main rivers in North Aceh, namely Krueng Cunda and Krueng Meuraksa. The rivers are formed along Lhokseumawe fault as direct effect of fault existence. The east side of the Lhokseumawe Fault is mostly a low terrain area with an average altitude between 0 – 10 m where the City of Lhokseumawe is located along the east coast of north Aceh.

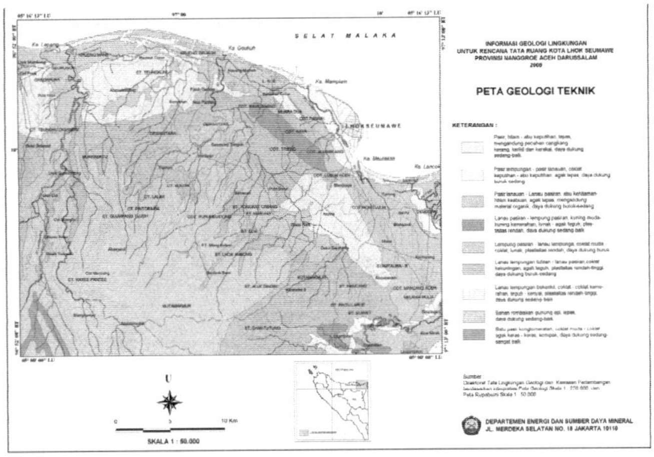

Figure 2. Geological map of North Aceh provided by Departmen Energi dan Sumber Daya Mineral, Indonesia [7].

Five sites have been selected to be analyzed first to provide early illustration of North Aceh ground characteristics. The data collected from microtremor measurements are processed by using Geopsy [8] to obtain natural frequency and amplification factor. The sites selected are LH2, LH3, LH3A, LH25, and LH29. The sites of LH2 and LH29 represent the west side of Lhokseumawe fault, while LH3A and LH25 lied on the opposite side, east area. The LH3 is chosen because it is located close to the Lhokseumawe fault. The result shows that LH29 has the highest natural frequency of 11.5 Hz (Fig. 3a). The lowest natural frequency is located in site LH25 with 1.34 Hz (Fig. 3b). This suggests that area around the LH25 site (in the west) overlaid by softer sediments compared to that of the west area. It is suggested that the higher the natural frequency, the harder the rock that forms that area. This corresponds to geological map issued by Ministry of Energy and Mineral Resources that stated the western part of Lhokseumawe Fault consists of harder rocks formation.

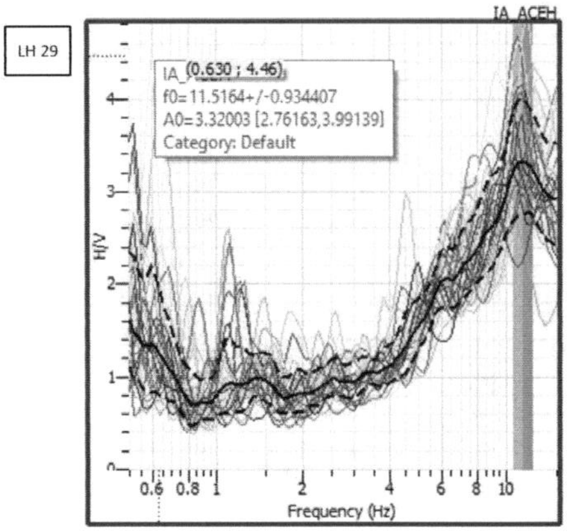

(a)

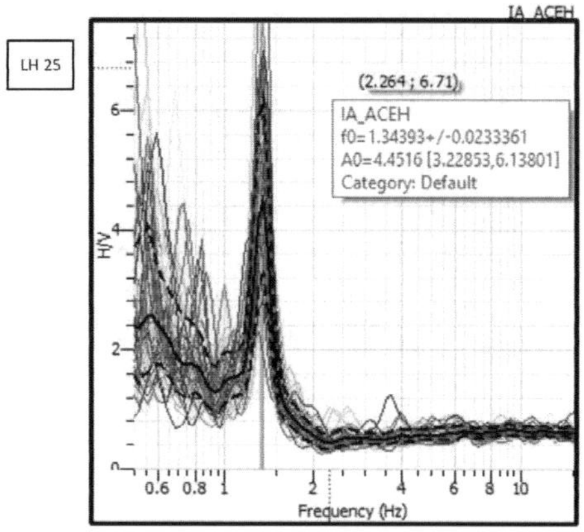

(b)

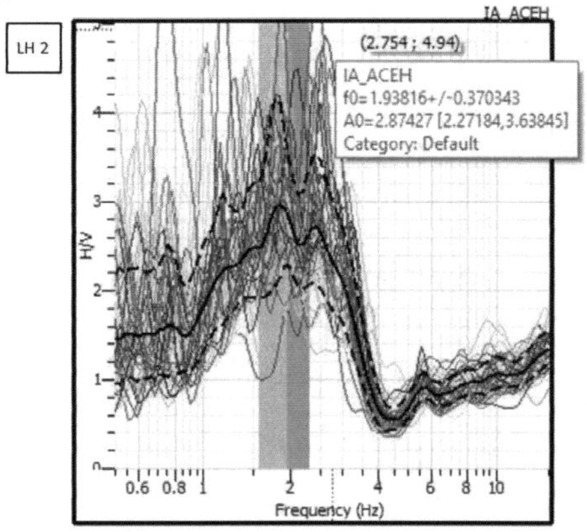

(c)

Figure 3. The results of HVSR processing at sites of LH29 (a), LH25 (b), and LH2 (c).

The second result is site amplification factor. From data processing, the results show that LH25 site has the largest amplification factor with the A0 value of 4.45 (Fig. 3b). The LH2 site has the smallest amplification factor of 2.87 (Fig. 3c). It is suggested that the larger amplification factor, the thicker and softer the sediments of the area. The amplification factor distribution mostly corresponds to geological map and natural frequency distribution. The west side of fault has lower amplification factor and the east side has higher value. Assessing seismic vulnerability based on local site needs the calculation of seismic vulnerability index. The seismic vulnerability index is obtained by amplification factor squared divided by natural frequency [9]. The seismic vulnerability index distribution is plotted in figure 4.

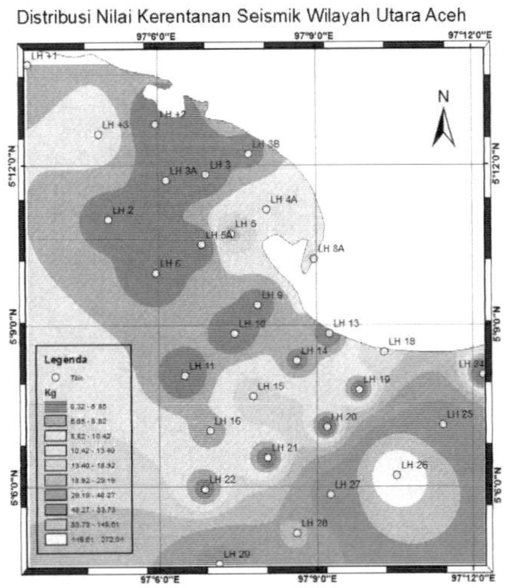

Figure 4. Seismic vulnerability index distribution in North Aceh.

The seismic vulnerability index (Kg) distribution of North Aceh suggests that southern part of study area has higher value than the opposite area. The higher the seismic vulnerability index means that it is more hazardous to seismic hazard based on the natural frequency and amplification factor. The higher seismic vulnerability index area needs extra attention from civil engineers for future development of the area. The natural frequency of the buildings planned to be built needs to differ the natural frequency of the rocks below.

4. Conclusions

Microtremor measurement is capable to determine how vulnerable an area to seismic hazard. Two main outputs of microtremor measurement, natural frequency and amplification factor, are able to describe ground characteristic of certain area. The research project area, North Aceh, is interesting to study as it is crossed by Lhokseumawe fault and located near Nisam fault. The early results show that the project is quite promising to describe vulnerability level of seismic hazard in North Aceh. The results suggest that south side of the Lhokseumawe fault has higher seismic vulnerability index. This means that this area more vulnerable to seismic hazard based on the rock's formation characteristics. \

References

[1] Crow MJ, Barber AJ. Map: Simplified geological map of Sumatra. *Geol Soc London, Mem* 2005; 31: NP.2-NP.
[2] Pustlitbang PUPR. *Buku Peta Gempa 2017*. 2017.
[3] Muksin U, Bauer K, Muzli M, et al. AcehSeis project provides insights into the detailed seismicity distribution and relation to fault structures in Central Aceh, Northern Sumatra. *J Asian Earth Sci* 2018; 171: 20–27.
[4] Badan Meteorologi Klimatologi dan Geofisika. Gempabumi Dirasakan, www.bmkg.go.id (2018).
[5] Nakamura Y. Clear identification of fundamental idea of Nakamura's technique and its applications. *Proc 12th world Conf* ... 2000; Paper no. 2656.
[6] Bard P, Duval A, Koehler A, et al. Guidelines for the Implementation of the H / V Spectral Ratio Technique on Ambient Vibrations Measurements , Processing and Interpretation. *Interpret A J Bible Theol* 2004; 169: 1–62.
[7] Kementerian ESDM. Peta Geologi Teknik.
[8] Geopsy Team. Geopsy Program, http://www.geopsy.org/.
[9] Hafid R, Supriyadi, Kholifatun N, et al. Percepatan Tanah Berdasarkan Data Mikroseismik Wisata Bantir Sumowono Semarang. 2018; 6: 54–59.

IOP Conf. Series: Materials Science and Engineering **846** (2020) 012054 doi:10.1088/1757-899X/846/1/012054

Structural Equation Modelling-Partial Least Square to Determine the Correlation of Factors Affecting Poverty in Indonesian Provinces

Noper Ardi[1], Isnayanti[2]

[1] Electrical and Information Engineering Department, Universitas Gadjah Mada, Yogyakarta,
[2] Mathematics Department, Universitas Gadjah Mada, Yogyakarta

*Email: Noper.ardi@mail.ugm.ac.id

Abstract. Structural Equation Modelling (SEM) is a multivariate statistical technique to analyze the pattern of relationships between latent variables (unobserved) with indicator variables (observed). One aspect that can be solved by SEM is poverty. Poverty is a major problem in economic development in developing countries like Indonesia. The problem of poverty can also be seen from the dimensions of education and health. This research uses SEM model with Partial Least Square (PLS) approach on poverty data in all provinces in Indonesia based on 2015-2017 Badan Pusat Statistik (BPS) data with latent variables of poverty, education, economy, and health. The results show that there is an indicator of latent educational variables that must be excluded from the model because it has a loading factor value > 0.5. The results of the outer model evaluation with Convergent Validity, Composite Reliability (CR), and Average Variance Extracted (AVE) show significant and reliable values that mean the indicators used can explain latent variables well. The SEM-PLS inner model evaluation results can be known from the value of R-Square (R^2) through the bootstrapping stage with 500 sample cases. R-square value for Economics is 0.751 means the model can explain variations of Economy in poverty cases in the Provinces of Indonesia at 75.1%, Education at 0.583 or 58.3% and the poverty model at 0.645 or 64.5%. The structural model for poverty cases in Indonesia in 2015-2017 is obtained Poverty = -0.548 Health - 0.085 Education + 0.128 Economy. The relationship of the poverty model shows that if poverty in Indonesia increases, the health of poor households and Education in Indonesia will decrease by 0.548 and 0.085 units respectively, whereas if poverty in Indonesia increases, the economy of poor households will increase by 0.128 assuming looking at the indicators that form latent economic variables.

1. Introduction

Poverty is a central issue for every country in the world, especially for developing countries, poverty alleviation and creating prosperity for the people is the final goal of a nation. According to Kunarjo [1] a country said to be poor is usually characterized by a low level of income per capita, has a high rate of population growth (more than 2 percent per year), most workers are engaged in the agricultural sector and are shackled in poverty circles. Poverty is also one of the problems faced by the Indonesian government. Poverty is not only related to the issue of a low level of income and consumption but also related to the low level of education, health, powerlessness to participate in the development and various problems on human development. Therefore, alleviating the problem of poverty must be a top priority in economic development, both short and long term.

Indonesia is a developing country consisting of 34 provinces. Based on data from the Central Statistics Agency (BPS), Indonesia's poverty conditions in March 2015, the number of poor people reached 28.59 million people or 11.22% of the total population of Indonesia. Entering March 2016, the sparse population recorded 28.01 million people or 10.86%. In March 2017, the poor community recorded 27.77 million people or 10.64%. Finally, in March 2018 the number of poor people was recorded at 25.95 million people or 9.82%. From the data also mentioned the number of poor people in urban areas in the 2018 period recorded 10.14 million people down 128.2 thousand people compared to the period September 2017 amounted to 10.27 million people.

Meanwhile, in rural areas, it fell by 505 thousand people (from 16.31 million in September 2017 to 15.81 million in March 2018). While in terms of the percentage of poor people in urban areas it was 7.02% lower than the September 2017 period of 7.26%. Meanwhile, the rate of poor people in rural areas in September 2017 was 13.47%, down to 13.20% in March 2018 [2].

The Indonesian government has always launched poverty reduction efforts from year to year. However, the number of poor people in Indonesia has not experienced a significant decrease, although the data in BPS shows a tendency to decrease the number of poor people, qualitatively it has not demonstrated the impact of real changes, but the condition is even more alarming each year [3]. Poor people often suffer from malnutrition, poor health, poor environment, and lack of access to infrastructure and adequate public services. Pockets of poverty spread throughout Indonesia from hamlets in the highlands, forest edge communities, poor small villages, fishing communities, farmers or slums in urban areas.

Factors that influence poverty levels cannot be measured directly because they are concepts (latent variables) such as inability to fulfill basic needs, limited human resource competence, inability to meet supporting needs, and limited access to infrastructure. Noting these factors, an indicator is needed to reflect the concept of poverty. Several statistical methods can be used to determine the relationship of these factors, one of which is Structural Equation Modeling (SEM). SEM analysis requires a strong theoretical foundation and clearly defined, covariance-based SEM method is more appropriate to be applied, but this method requires a large sample, assuming that the data must be a multivariate normal distribution. Real data in the field often show patterns of data that are spread abnormally, so we need a method that is free distribution [4].

A study that collided with small sample size and weak theoretical basis, then the fulfillment of the assumptions becomes difficult, we need an alternative SEM method that can accommodate the constraints of fulfilling the premises. The method is SEM based on variance or component (component-based SEM), which is the Structural Equation Modeling-Partial Least Square (SEM-PLS) method. SEM-PLS allows structural equation modeling with the assumption that the data used do not have to be normally distributed, can use a relatively small sample size, and the indicators used are reflective, formative, or a combination of both. PLS is an analytical method used to confirm theories and can also be used to explain the presence or absence of relationships between latent (unobserved) variables [5].

Until now, research related to poverty using the SEM method has been carried out by various parties. Fitriani and Otok (2013) in their research on developing indicators and determining poor households in East Java Province using Spatial Structural Equation Modeling using the dimensions of health, economy, human resources, in the analysis of poverty dimensions in East Java Province [6]. Afifah and Sunaryo (2013) made a grouping of regencies/cities in Central Java Province in 2011 based on the structure of the poverty model using the Finite Mixture Partial Least Square (FIMIX-PLS) method [7]. Anuraga and Otok (2013) conducted poverty modeling in East Java with Structural Equation Modeling Partial Least Square [8]. In this study, an analysis will be conducted related to the factors that influence poverty in Indonesian Provinces in 2015-2017 using the SEM-PLS method. This research is directed to address the issue of what factors are dominantly affecting poverty in Indonesian provinces. Analysis of the structure of poverty in this study wants to see whether there is a relationship, and how the relationship between the dimensions of poverty, the dimensions of education, economic aspects, and health dimensions. This research can help the government to obtain information about the dominant factors that influence poverty in Indonesia so that in compiling programs to improve the welfare of the people in Indonesia can be done by the potential of each province.

2. Material and Methods

2.1. Data

The data used in this study is secondary data obtained from data from the Badan Pusat Statistik (BPS) in the Provinces of Indonesia in 2015-2017, which consists of 34 Provinces.

2.2. Research Variable

The following are the variables used in the study.

Table 1. Research Variables

Latent Variable	Code	Indicator Variable
Health	X1	Percentage of population that treats in a modern way for a month (Percent)
	X2	Percentage of women aged 15-49 years old and married who are using birth control device (Percent)
	X3	Percentage of women ever married aged 15-49 whose last birth process was assisted by Trained Nurses by Province (Percent)
	X4	Percentage of households that use bottled drinking water (percent)
Poverty	Y1	Percentage of poor population (percent)
	Y2	Poverty depth index (percent)
	Y3	Poverty severity index (percent)
Education	Y4	Percentage of literacy numbers (percent)
	Y5	The average length of school (percent)
	Y6	Percentage of young people (15-24 years) who do not attend school, work or training (percent)
Economy	Y7	Percentage of people in the farming sector (percent)
	Y8	Percentage of Households Using Wood-based as Main Fuel for Cooking (percent)

Based on Table 1. Health latent variables consist of 4 indicator variables namely X1, X2, X3, X4, poverty latent variables comprise 3 indicator variables namely Y1, Y2, Y3, latent educational variables consist of 3 indicator variables namely Y4, Y5, Y6, and economic variables contain 2 indicator variables namely Y7, Y8.

2.3. Poverty

Poverty is a situation or condition experienced by a person or group of people who are unable to carry out their lives to a level that is considered humane [3]. There are four patterns of poverty, persistent poverty, namely poverty that has been chronic or hereditary. The second pattern is cyclical poverty, which is poverty that follows the model of the whole economic cycle. The third pattern is seasonal poverty, which is annual as found in the case of fishermen and food crop farmers. The fourth pattern is accidental poverty, namely poverty, due to natural disasters or the impact of a particular policy that causes a decrease in the welfare level of a community [9].

2.4. Structural Equation Modelling (SEM)

Structural Equation Modeling (SEM) is a statistical technique that can analyze patterns of linear relationships simultaneously between latent variables and indicator variables. Latent variables are variables that cannot be measured directly but can be represented or measured by one or more indicator variables [10]. SEM includes statistical techniques used to build and test statistical models that are usually in the form of causal models. SEM becomes a fairly powerful analytical technique because it considers interaction modeling, nonlinearity, correlated independent variables, measurement

IOP Conf. Series: Materials Science and Engineering **846** (2020) 012054 doi:10.1088/1757-899X/846/1/012054

errors, correlated error terms, multiple latent independent variables in multiple where each is measured by many indicators, and one or two variables depending on latency which is also each measured by several indicators. SEM is based on analysis of covariance so that it gives a more accurate covariance matrix than linear regression analysis.

Data analysis techniques using SEM are used to explain the relationship between variables as a whole that exist in research. The fundamental reasons for using SEM in this research is [11]

a. SEM can estimate relationships between variables that are multiple relationships. This relationship is formed in the structural model (the relationship between the dependent and independent variables).
b. SEM can describe the pattern of relationships between latent variables and indicator variables.

In general, SEM consists of two main parts, namely the structural model and the measurement model [12]. The structural model illustrates the pattern of relationships between endogenous (dependent) latent variables and exogenous (independent) latent variables, as in the following equation:

$$\boldsymbol{\eta} = \boldsymbol{B}\boldsymbol{\eta} + \boldsymbol{\Gamma}\boldsymbol{\xi} + \boldsymbol{\zeta} \quad (1)$$

Where, $\boldsymbol{B}$ is the structural coefficient matrix for the relationship between endogenous latent variables $\boldsymbol{\eta}$ size $m \times m$ with diagonally zero elements ($|\boldsymbol{I} - \boldsymbol{B}| \neq 0$), $\boldsymbol{\Gamma}$ is the structural coefficient matrix for the relationship between exogenous latent variables $\boldsymbol{\xi}$ and endogenous latent variables $\boldsymbol{\eta}$ size $m \times n$, $\boldsymbol{\zeta}$ is an error vector size $m \times 1$ with $E(\boldsymbol{\zeta}) = 0$ and $Cov(\boldsymbol{\zeta}) = \boldsymbol{\Psi}$ (i.e., a diagonal matrix of residual variance for $\boldsymbol{\eta}$ sized $m \times m$), assuming the error term $\boldsymbol{\zeta}$ does not correlate with all other error terms and latent variable $\boldsymbol{\xi}$ [12]. While the measurement model draws a pattern of relationships between latent variables and their indicators, as in the following equation:

$$\boldsymbol{y} = \boldsymbol{\Lambda}_{\boldsymbol{y}}\boldsymbol{\eta} + \boldsymbol{\epsilon} \quad (2)$$

$$\boldsymbol{x} = \boldsymbol{\Lambda}_{\boldsymbol{x}}\boldsymbol{\xi} + \boldsymbol{\delta} \quad (3)$$

Where $\boldsymbol{y}$ is an endogenous latent variable vector size $p \times 1$,, $\boldsymbol{x}$ is an exogenous latent variable vector x size $q \times 1$,, $\boldsymbol{\Lambda}_{\boldsymbol{y}}$ is a loading factor matrix for $\boldsymbol{y}$ size $p \times m$, $\boldsymbol{\Lambda}_{\boldsymbol{x}}$ is a loading factor matrix for $\boldsymbol{x}$ size $q \times n$,, $\boldsymbol{\eta}$ is an endogenous latent variable vector measuring size $m \times 1$,, $\boldsymbol{\xi}$ is an exogenous latent variable vector measuring size $n \times 1$,, $\boldsymbol{\epsilon}$ is a measurement error vector at $\boldsymbol{y}$ scale $p \times 1$, and $\boldsymbol{\delta}$ is a measurement error vector at $\boldsymbol{x}$ size $q \times 1$,, it is assumed that $\boldsymbol{\epsilon}$ and $\boldsymbol{\delta}$ are not correlated with each other [12].

2.5. Structural Equation Modelling-Partial Least Square (SEM-PLS)

Partial Least Square (PLS) is a method that was first introduced by Herman O.A. The world. PLS is an alternative technique in SEM analysis where the data used are generally not distributed multivariate. In SEM with PLS, the value of the latent variable is estimated according to the linear combination of the indicator variables associated with the latent variable and is treated to replace the indicator variable [13]. The advantages of SEM with PLS when compared to SEM-based covariance, SEM with PLS can handle two conditions where:

a. Factor Indeterminacy

Factor indeterminacy is a condition in which the resulting factor score has a different value when calculated from a single factor model. Especially for formative indicators, do not require the existence of a common factor so that it will always obtain a combined latent variable in the form of unity.

b. Inadmissible Solution

Unacceptable solution conditions will not occur in SEM with PLS, because SEM with PLS is based on variance and not covariance so that the singularity matrix problem will never happen. Besides, PLS works on a structural model that is recursive, so that is a problem unidentified, under-identified, or over-identified will never happen.

\According to Monecke and Leisch [14], SEM with PLS consists of three components, namely structural models (inner models), measurement models (outer models), and weight relations. This weight relation section is a unique feature of SEM with PLS and is not present in covariant-based SEM. SEM models with PLS are described as follows:

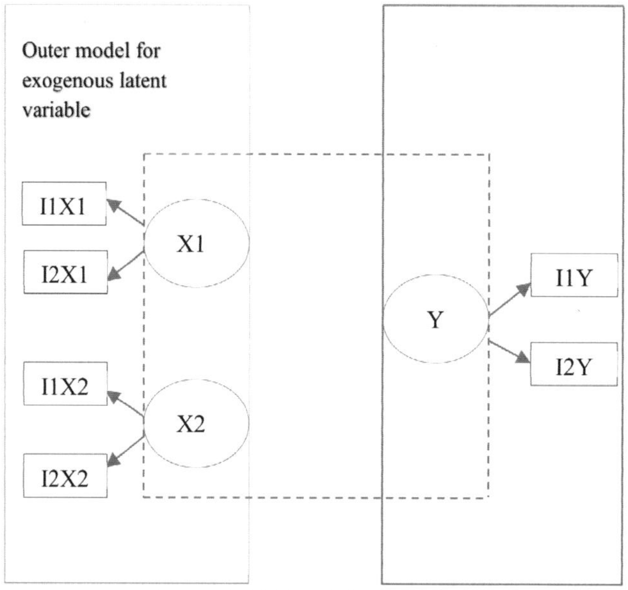

Figure 1. SEM with PLS Model

In Figure 1, the outer model for exogenous latent variables consists of 2 latent variables, namely X1 and X2. The latent variable X1 has 2 indicator variables, namely the I1X1 indicator and the I2X1 indicator. And the latent variable X2 also has 2 indicator variables namely I1X2 indicator and I2X2 indicator. While the outer model for endogenous latent variables consists of 1 latent variable namely Y, which consists of 2 indicator variables, I1Y indicator, and I2Y indicator. Then the inner model draws the relationship between latent variables X1, X2 and Y.

The use of SEM models has an underlying assumption that is multivariate normal distribution data and large sample sizes. The use of small sample size can produce parameter estimates that are not good even not convergent. So that one alternative to SEM that does not require multivariate normally distributed data is the Partial Least Square (PLS) approach. PLS is a powerful analysis method because it does not require the assumption of normality and can use a relatively small sample size [14]. In SEM with PLS only allowed recursive models (cause-effect) and did not allow non-recursive models (reciprocity) as in SEM which is covariant-based and SEM with PLS allows very complex models with many latent variables and indicators [14].

In the path analysis for SEM-PLS, there are three models, namely the inner model, outer model, and weight relation. The inner model shows the relationship between latent variables, the outer model shows the relationship between the latent variable and its indicator variables, and the weight relation shows the estimated value of the latent variable.

2.5.1. Structural Model (Inner Model)
Structural models or inner models describe the relationship model between latent variables that are formed based on the substance of the theory. PLS is designed for a recursive model, so there is a relationship between latent variables called the causal chain system with the following form of an equation:

$$\eta_j = \Sigma\beta_{ji}\eta_i + \Sigma\gamma_{jb}\xi_b + \zeta_j \quad (4)$$

Where $i, \dots, b$ represents the range index along i and b, j represents the number of endogenous latent variables, β_{ji} represents the path coefficient that connects the endogenous latent variable (η) with endogenous (η), γ_{jb} represents the path coefficient linking the endogenous latent variable (η) exogenous (ξ) and ζ represent the level of measurement error (inner residual variable) [14].

2.5.2. Measurement Model (Outer Model)
The measurement model or outer model generally describes the relationship between latent variables and indicators. In the outer model, there are two types of models, namely reflexive and formative indicator models. Reflexive models occur when indicator variables are influenced by latent variables. The equation for the reflexive indicator model is the same as the measurement model in covariance-based SEM. While the formative model assumes that the indicator variables affect latent variables. The direction of causality flows from the indicator variable to the latent variable. The equation for the developmental model is as follows:

$$\xi = \Pi_x \xi X_i + \delta_\xi \quad (5)$$

$$\eta = \Pi_y \eta Y_i + \epsilon_\eta \quad (6)$$

Where, Π_x and Π_y represents the multiple regression coefficients of the latent variable to the indicator, δ_ξ, and ϵ_η represent the level of measurement error (residual error) [14].

2.5.3. Weight Relation
According to Abdillah and Jogiyanto HM [15], the weight relation score shows the relationship of variance values between indicators with their latent variables so that it is assumed to have a mean equal to zero with a variance similar to one to eliminate constants in causality. The equation for weight relation is as follows:

$$\xi_b = \Sigma_{kb} w_{kb} x_{kb} \quad (7)$$

$$\eta_i = \Sigma_{ki} w_{ki} y_{ki} \quad (8)$$

Where w_{kb} and w_{ki} declare the weights *k* used to estimate latent variables ξ_b and η_i.

In the use of PLS, there are several evaluations of structural models (inner models) and measurement models (outer models). In evaluating the measurement model, convergent validity testing is performed, reliability testing uses Composite Reliability (CR) and Average Variance Extracted (AVE) measures. Whereas in the evaluation of structural models the R-squared (R^2) test and the path coefficient estimation test are performed.

2.5.3.1 Convergent Validity
Convergent validity in SEM-PLS is used as an evaluation for the measurement model (outer model). Convergent validity is a type of validity that is related to the principle that measuring a variable must have a high correlation so that it is used to measure the magnitude of the relationship between latent variables and indicator variables in the reflexive measurement model. According to Hair [16], a correlation can be said to meet Convergent validity if it has a standardized loading factor $\lambda \geq 0.5$.

2.5.3.2 Reliability Test
Reliability test in SEM-PLS model is used as one of the evaluations for the measurement model (outer model). To measure the reliability of this research model, Composite Reliability (CR) and Average Variance Extracted (AVE) are used. Latent variables can be said to have excellent reliability if the value of composite reliability (CR) is more significant than 0.7 [16]. CR can be determined using the following formulation:

$$CR = \frac{\left(\Sigma_{i=1}^{n} \lambda_i\right)^2}{\left(\Sigma_{i=1}^{n} \lambda_i\right)^2 + \Sigma_{i=1}^{n} e_i} \quad (9)$$

where, λ_i is the standardized loading factor and e_i is the error value from *i*-th measurement with $e_i = 1 - \lambda_i^2$. Furthermore, latent variables can be said to have excellent reliability if the value of the Average Variance Extracted (AVE) is more significant than 0.5 [16]. AVE can be determined using the following formulation:

$$AVE = \frac{\Sigma_{i=1}^{n} \lambda_i^2}{\Sigma_{i=1}^{n} \lambda_i^2 + \Sigma_{i=1}^{n} e_i} \quad (10)$$

where, λ_i is the standardized loading factor and e_i is the error value from *i*-th measurement with $e_i = 1 - \lambda_i^2$.

2.5.3.3 R-Square (R^2) Test

R-squared (R^2) is a test conducted to measure the level of Goodness of Fit of a structural model. The value of R-squared (R^2) is used to measure how much influence the particular independent latent variable has on the latent dependent variable. According to Chin [17] R^2 Results of 0.67 indicate that the model is well categorized. Results R^2 Between 0.33 and 0.67 show that the model is classified as moderate. While the Results R^2 of 0.33 indicates that the model is categorized as weak.

2.5.3.4 Significance Test

Significance test aims to determine the effect of independent variables on the dependent variable. Significance test in the SEM model with PLS, what is meant by the independent variable is the exogenous latent variable and what is meant by the dependent variable is the endogenous latent variable. Estimated values for path relationships in the structural model are used to determine the significance of the relationships between latent variables. Significant value can be obtained by the bootstrapping procedure developed by Geisser & Stone. The formulation of the hypothesis in the significance test is as follows:
H_0: The independent variable does not significantly influence the dependent variable
H_1: The independent variable significantly influences the dependent variable.
The test statistics used are:

$$T_{statistik} = \frac{b_j}{s(b_j)} \quad (11)$$

Where b_j represents the estimated value for β_j, $s(b_j)$ states the standard error for b_j. The test criteria are the significance level (α). H_0 is rejected if T-statistic > T-table (df, α) or p-value < α [18].

2.6. Bootstrap

Bootstrap is a nonparametric estimation method that can estimate the parameters of a distribution, the variance of a median sample, and can estimate errors [18]. In the bootstrap method, sampling is performed by returning the sample data. Determination of the value of the number of replications (B) is very varied because the size of the value of B can provide different results at each stage in the analysis. A large B value will usually be very good for determining confidence intervals.

2.7. Experimental Design

The analysis used in this study uses the SEM-PLS method with the help of SmartPLS 3.2.8 software. The following is a research method flowchart:

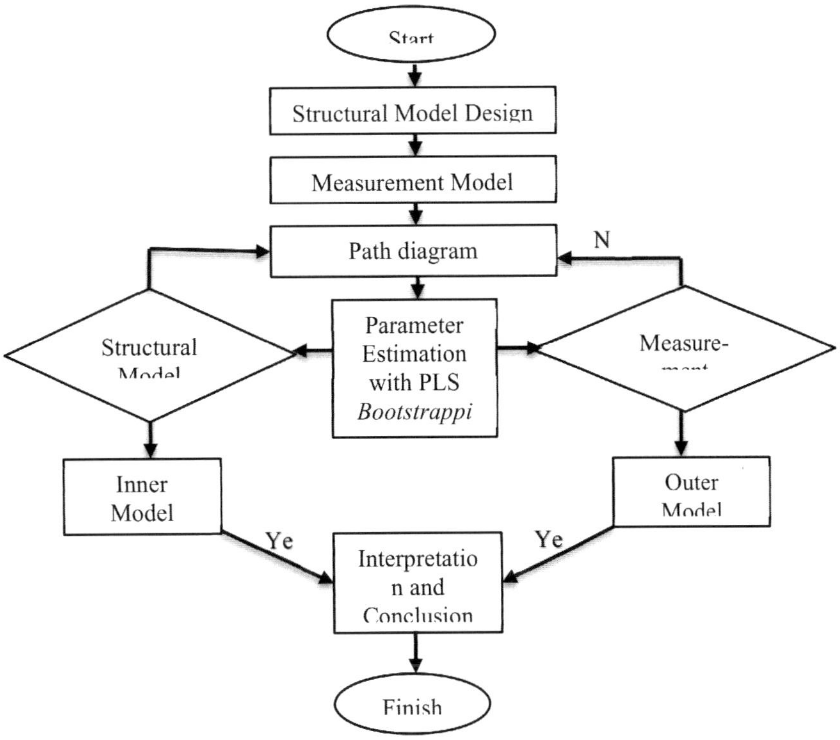

Figure 2. Research Flowchart

The initial stage begins by collecting data from the Badan Pusat Statistik (BPS) in the Provinces of Indonesia in 2015-2017, which consists of 34 Provinces totaling 102 samples. The raw data obtained will be prepared in advance in the SmartPLS 3.2.8 software. To conduct Structural Equation Modeling (SEM) analysis using PLS, it starts from designing structural models and measurement models based on the concepts and theories developed. Next, create a path diagram that explains the pattern of relationships between latent variables and indicator variables. Estimating SEM-PLS parameters through the bootstrapping stage with 500 sample cases, namely evaluating the measurement model (outer model) and evaluating the structural model (inner model). If the evaluation results are significant, then proceed to the interpretation phase of the hypothesis test. From tests conducted on the inner model, a structural model will be obtained for poverty cases in Indonesia.

3. Result and Discussion

3.1 Measurement Model Evaluation (Outer Model)

In this case, evaluation or analysis of the measurement model will be carried out, namely, the relationship between latent variables and their indicators. The purpose of this measurement model analysis is to determine the validity and reliability of indicators of a latent variable.

3.1.1. Convergent Validity

The convergent validity test is used to test whether the indicator variables used are really significant in terms of reflecting latent variables. Convergent validity test consists of validity and reliability tests. The good convergent validity is indicated by the high standardized loading factor (λ) value. Hair [16] suggests a value of $\lambda \geq 0.5$, meaning that a good convergent validity has been achieved.

The results of convergent validity testing in Indonesia can be seen in the outer model loadings, as shown below.

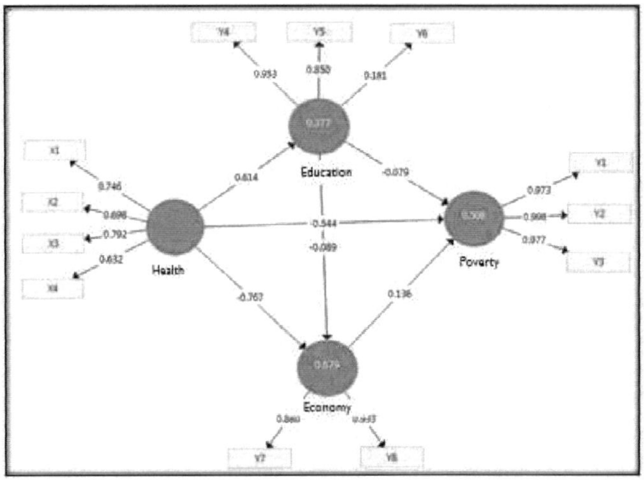

Figure 3. SEM-PLS Poverty Path Diagram in Indonesia

Table 2. Indicator of Health Latent Variables

Indicator of Health Latent Variables	Loading Factor	Evidence
X1	0.746	Valid
X2	0.698	Valid
X3	0.792	Valid
X4	0.632	Valid

Table 2. shows that for the population variable that treats in a modern way (X1), Women who are currently using birth control devices (X2), women whose birth process is assisted by Trained Health Workers (X3), and households that use a Source of Drinking Water in Packaging (X4) can be used to measure latent health variables because the loading factor value is > 0.5.

Table 3. Indicator of Poverty Latent Variables

Indicator of Poverty Latent Variables	Loading Factor	Evidence
Y1	0.973	Valid
Y2	0.998	Valid
Y3	0.977	Valid

Table 3. shows that all indicator variables on the latent variable of poverty are valid because of the loading factor value > 0.5. This means that all indicators can be used to measure the latent variable of Poverty.

Table 4. Indicator of Education Latent Variables

Indicator of Education Latent Variables	Loading Factor	Evidence
Y4	0.953	Valid
Y5	0.850	Valid
Y6	0.181	Not Valid

Table 4. shows that for young age variables (15-24 years) who are not in school, work / attend training (Y6) is invalid because of the loading factor value < 0.5. As for the literacy number indicator (Y4) and the average length of school (Y5) can be used to measure the latent variable of Education.

Table 5. Indicator of Economic Latent Variables

Indicator of Economic Latent Variables	*Loading Factor*	Evidence
Y7	0.880	Valid
Y8	0.933	Valid

Table 5. shows that all variables in the Economic latent variable are valid because the loading factor value> 0.5. This means that all indicators can be used to measure latent economic variables.

3.1.2. Convergence validity after the indicator is removed

Retest after invalid indicators are removed in the provinces of Indonesia, the path diagram is as follows:

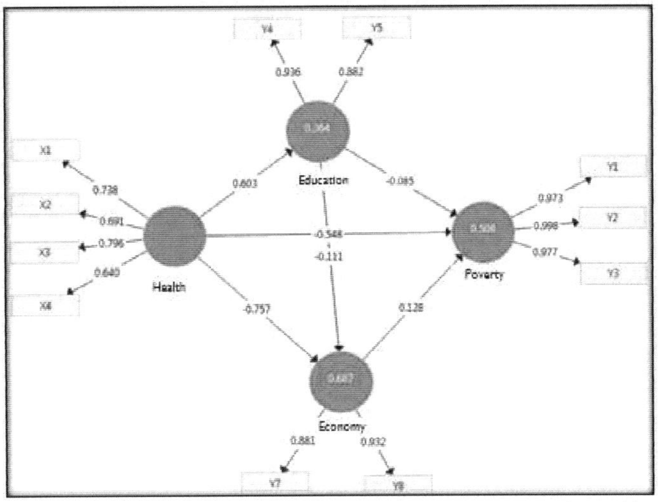

Figure 4. Poverty Line SEM-PLS Diagram in Indonesia after selection of invalid indicators

The feasibility of a model can be shown that the indicator used is valid. This can be seen from the t-statistics value on the results of loading factor measurement models.

Following are the results of retesting after invalid indicators (Y6) are removed from the latent variables of education can be seen in Table 6 below.

Table 6. Bootstrap T-Statistics Significance Test 500

Latent Variable	Indicator	*Loading factor*	Standard Deviation	T-statistic
Health	X1	0.738	0.107	6.919
	X2	0.691	0.137	5.047
	X3	0.796	0.040	19.760
	X4	0.640	0.079	8.094
Poverty	Y1	0.973	0.005	179.426
	Y2	0.998	0.001	738.353
	Y3	0.977	0.004	243.279
Education	Y4	0.936	0.022	43.437
	Y5	0.882	0.030	29.390
Economic	Y7	0.881	0.015	57.917
	Y8	0.932	0.008	121.99

Based on Table 6 shows that the estimated loading factor value for each latent variable > 0.5 is significant, this is indicated by the t-statistic value> t-table 1.98 (2-tailed) at the level of significance (α) = 0.05. This means that all variables are statistically significant and valid in measuring all latent variables.

3.1.3 Reliability Test

The reliability test is used to determine the consistency of measuring indicators of a latent variable. To measure the reliability of this research model, Composite Reliability (CR) and Average Variance Extracted (AVE) are used. The results of reliability testing in Indonesia are as follows:

Table 7. Reliability Test Results

Latent Variable	CR	AVE	Evidence
Health	0.810	0.517	Reliable
Poverty	0.988	0.966	Reliable
Education	0.905	0.827	Reliable
Economy	0.903	0.822	Reliable

Table 7 shows the reliability test values using the CR and AVE values of each latent variable. Based on the CR value on each latent variable shows good results because of> 0.7, while for the value of AVE shows good results because it has a value > 0.5. So that the latent variables of health, poverty, education, and the economy are said to have excellent reliability.

3.2 Structural Model Evaluation (Inner Model)

Structural models or inner models describe the relationship between latent variables. In Partial Least Square, the path parameter coefficient is obtained through the weight of the inner model as seen from the t-statistic value through the bootstrapping stage with 500 sample cases. The hypothesis raised in this study are:

H_1: Health affects Education
H_2: Health affects Economy
H_3: Health affects Poverty
H_4: Education affects Economy
H_5: Education affect Poverty
H_6: Economy affects Poverty

The results of the path parameter coefficients in Indonesia from the structural model can be shown in the following Table 8

Table 8. Path Parameter Coefficients

Variable	Path Parameter Coefficients	*Sample Mean*	Standard Deviation	T-statistic
Health → Education	0.603	0.596	0.094	6.390
Health → Economy	-0.757	-0.765	0.035	21.796
Health → Poverty	-0.548	-0.541	0.125	4.386
Education → Economy	-0.111	-0.111	0.053	2.100
Education → Poverty	-0.085	-0.072	0.131	0.648
Economy → Poverty	0.128	0.135	0.115	1.113

Based on the results obtained, three structural equations are known:

$$\text{Poverty} = -0.548\ \text{Health} - 0.085\ \text{Education} + 0.128\ \text{Economy} \tag{12}$$

$$\text{Education} = 0.603\ \text{Health} \tag{13}$$

$$\text{Economics} = -\ 0.757\ \text{Health} - 0.111\ \text{Education} \tag{14}$$

The poverty model can be seen that health has a negative influence of -0.548 units and is significant, with a T-statistic value of 4.386 (> 1.98) on latent variables of poverty. Education also has a negative effect of -0.085 units but it is not significant with a T-statistic value of 0.648 (<1.98), while the Economic variable has a positive effect of 0.128 units and is not significant with a T-statistic value of 1.113 (<1.98) on the poverty model at 95% confidence interval or t-statistically. For the Education model, it is known that the Health latent variable has a positive influence of 0.603 and is significant, with a T-statistic value of 6.390 (> 1.98) on the latent educational variable. And for the Economic model it can be seen that the latent variables of health and education have a negative and significant correlation to the latent economic variable at 95% confidence intervals, as well as t-statistically with the estimated health variable path coefficient of -0.757 while the education variable path coefficient of -0.111.

The evaluation test value of the structural equation model in SEM-PLS can be known from the value of goodness of fit or R-square (R^2). The results of this research data processing using SmartPLS gives the value of R^2like table 9 below:

Table 9. *R-square* (R^2) Value

Latent Variabl	*R-squar*
Economy	0.751
Education	0.583
Poverty	0.645

The results of the R-Square are for the Economic model of 0.751, for the Education model of 0.583 and for the Poverty model of 0.645, so the model obtained is feasible because the R-Square value > 0.5.

The R-square value (R^2) for the Economy is 0.751 which means that the model can explain the variation of the Economy in the case of poverty in Indonesia's Provinces by 75.1%, Education by 0.583 or 58.3% and the poverty model by 0.645 or 64.5 %.

4. Conclusion

In this study, Structural Equation Modeling (SEM) Partial Least Square is used to determine the relationship between the factors that influence poverty in Indonesian Provinces. The results show that there is a significant negative effect from the health dimension on poverty, a negative influence exerted from the education dimension on poverty and a positive influence given from the economic dimension on poverty in Indonesia but it is not significant.

The structural model for poverty cases in Indonesia in 2015-2017 is as follows:

Poverty = -0.548 Health - 0.085 Education + 0.128 Economy.

For the relationship of the model obtained it can be concluded that if poverty in Indonesia increases, the health of poor households and education in Indonesia will decrease by 0.548 and 0.085 units respectively, whereas if poverty in Indonesia increases, the economy of poor families will increase of 0.128 assuming looking at the indicators that make up the latent economic variable. Thus, health is the variable that has the most dominant influence on poverty in Indonesia.

For further research, modeling the SEM-PLS structural equation can be done with the addition of poverty indicators and new variables to clarify the relationship between latent variables and indicator variables to obtain more significant results and better models.

References

[1] Kunarjo, "Perencanaan dan Pengendalian Program Pembangunan", Jakarta: Universitas Indonesia Press, 2002.

[2] Badan Pusat Statistik, "*Data dan Informasi Kemiskinan* (Indonesia: BPS)", 2018.

[3] A. Suryahadi, G. Hadiwidjaja and S. Sumarto, "Economic Growth and Poverty Reduction in Indonesia Before and After the Asian Financial Crisis", *Bulletin of Indonesian Economic Studies,* pp. 48: 209–227, 2012.

[4] I. Ghozali, "*Structural Equation Modelling Metode Alternatif with Partial Least Square"*(Semarang: Universitas Diponegoro), 2011.

[5] J.F. Hair, C.M. Ringle, S.P. Gudergan, A. Fischer, C. Nitzl and C. Menictas, "Partial Least Squares Structural Equation Modeling-Based Discrete Choice Modeling: An Illustration in Modeling Retailer Choice", *Business Research,* pp. **12:** 115–142, 2019.

[6] E. Saputro and B.W. Otok ,"Pengembangan Indikator dan Penentuan Rumah Tangga Miskin di Provinsi Jawa Timur menggunakan Structural Equation Modelling Bootstrap Aggregating (SEM BAGGING) *Jurnal Sains dan Seni POMITS* pp.**2:** 2337-3520, 2013.

[7] I.N.Afifah and S. Sunaryo, "*Proceeding Seminar Nasional Matematika dan Pendidikan Matematika* (Yogyakarta: FMIPA Universitas Negeri Yogyakarta), 2013.

[8] G. Anuraga and B.W. Otok, "Pemodelan Kemiskinan di Jawa Timur dengan Structural Equation Modelling- Partial Least Square *Jurnal Statistika* pp.**1:** 22- 27, 2013.
[9] S.Djojohadikusumo,"*Perkembangan Pemikiran Ekonomi Dasar Teori Pertumbuhan dan Ekonomi Pembangunan* (Jakarta: LP3ES), 1995.
[10] J. F. Jr. Hair, R.E Anderson, R.L. Tatham and W.C. Black,"*Multivariate Data Analysis (5th ed.)* (USA: Prentice Hall), 1998.
[11] Isnayanti and Abdurakhman,"Metode Diagonally Weighted Least Square (DWLS) Pada Structural Equation Modelling Untuk Data Ordinal: Studi Kasus Dari Pengguna Jasa Kereta Api Majapahit Malang-Pasar Senen", *Jurnal Media Statistika,* pp.**12:** 100-116
[12] K.A. Bollen, "*Structural Equations with latent variables* (New York : Wiley)", 1989.
[13] Yamin and Kurniawan, "*Mengolah Data Penelitian dengan Partial Least Square Path-Modeling",* Jakarta: Salemba Infotek, 2011.
[14] A.Monecke and F. Leisch , " SEMPLS: Structural Equation Modeling Using Partial Least Squares", *Journal of Statistical Software,* pp. **48:** 1-32, 2012.
[15] W. Abdillah and Jogiyanto,"*Partial Least Square (PLS) Alternatif Structural Equation Modelling (SEM) dalam Penelitian Bisnis",* (Yogyakarta: Andi Offset), 2015.
[16] J. F. Jr. Hair, W.C. Black, B.J. Babin , R.E. Anderson ,"*Multivariate Data Analysis, 7th Edition",* London: Pearson Prentice Hall, 2010.
[17] W. Chin, "*The Partial Least Square Approach for Structural Equation Modeling",*London : Lawrence Erlbaum Associates, 1998.
[18] B. Efron and R.J. Tibshirani, "*An Introduction to The Bootstrap",* New York : Chapman & Hall, 1993.

IOP Conf. Series: Materials Science and Engineering **846** (2020) 012055 doi:10.1088/1757-899X/846/1/012055

Android-based energy saving control system integrated with islamic value on lamp switching power

[1]Erina Hertanti, [2]Maswani, [3]Muh. Hifdzil Farasdaq, [4]Fatih Imtiyaz

Universitas Syarif Hidayatullah Jakarta, Indonesia

[1]erina.hertanti@uinjkt.ac.id, [2]maswani@uinjkt.ac.id, [3]hifdzil.farasdaq15@mhs.uinjkt.ac.id, [4]fatih.imtiyaz15@mhs.uinjkt.ac.id

Abstract. Lighting is one of the major consumers of electricity and it is often found to be inefficient in its use. Therefore, this study aimed to provide solutions to these problems through energy-saving behavior that is integrated with Islamic values on lamp switching power. This study use system design build methods. This design results in the form of hardware and software assembly. The results of the performance test of the integration of hardware and software devices provides informations as follow: first, control system of application has not been able to operate on all mobile platforms. Second, the connection range of the control system is able to operate according to its standard frequency range. Third, the connection access of control system is able to pass through the barrier which has a small inhibitory. Fourth, the success of the control system in executing input is high.

Keywords: control system, lamp switch, Android, software, hardware

1. Intoduction

Electrical energy has become a basic needs for human's life, because it is an energy that supports all activities. Various household appliances, offices or public buildings are operated using electricity. Therefore electrical energy has a vital and strategic role in our daily life. However, the problem of electrical energy arises when we start to talk about electricity supply and the ineffectiveness of electricity consumption or usage. Thus far in Indonesia, the source of electricity is still dominated by the use of fossil fuels. A data from Dewan Energi Nasional shows that the use of alternative source for electricity is very low compared to the use of fossil fuel (Edi et al, 2016). Meanwhile, the amount of fossil fuel supply continues to decline while the demand for electricity continues to increase. The increase in demand has even doubled due to Indonesia's economic growth (Otomo and Wildian, 2013). Therefore, the electricity supply in Indonesia itself is now in a state of caution, because the remaining fuel reserves are not widely available (Edi et al, 2016). Another typical problem for electricity is that it's consumption is often found to be inefficient, especially in the lighting sector. For example, the lamp stays on even though it is no longer needed, due to user negligence in terms of switching power from on to off. In fact, lighting is one of the relatively large consumers of electricity. The lamp usage, can reach around 30% of the consumption of electricity used in daily life (Otomo and Wildian, 2013).

The complexity of the above problems eventually led to a crisis of electricity. As a result, alternating power outages can occur in various areas. This problem certainly requires a solution. One important thing in choosing a solution to the problem of electrical energy is that saving electricity is much easier and cheaper than generating it. In some previous studies, energy savings are generally carried out in two ways, first, by increasing the efficiency of technology, and second, by changing the behavior of its users. This study tries to combine the two, namely applying technology that combines electronic engineering, informatics, and architecture to make efficient use of electrical energy based on energy-saving behavior that is integrated with Islamic values. In this case, the concept promoted is smart building and Islamic values. According to Mannan and Muchlis (2012) the principles of smart building have many similarities with the values of Islamic teachings, such as expediency, efficiency, and preventing waste.

In this study, smart building is a design for a lamp switch control system in every room using Android with WiFi, as a control or media that connects the interactions between devices, and the user is a voice command that recites prayers or du'a in accordance with the function of the room. For example, the bathroom lamp will turn on when the user recites du'a upon entering bathroom, and will turn off when recites the du'a upon leaving bathroom. And another example, the bedroom lghts will turn off when the user recites the du'a before going to sleep, and will turn on when recites the du'a upon awakening from sleep. This system has a control distance of approximately around 60 meters, so this system is expected to make it easier for users to control the lamp on and off, without having to interact directly with the power switch. This system can overcome the difficulties in controlling more than one lamp switching power, because all of them will be controlled in one control device. This system can also minimize inefficiencies in the use of electrical energy, so that the available energy can be utilized in the long run. In addition, this system has more value, which is a media liaison between the Creator and His servant.

2. Basic Theory

2.1. Android

Android is a Linux-based operating system on mobile devices that includes operating systems and applications. Android provides an open platform for developers to create their own applications. Features like browser, MMS, SMS, GPS, and many more on Android make it very easy for users to get good information, position, and communication. Some of the advantages of Android, including: (a) Some application can run on multitask (b) There are always notifications for incoming SMS, e-mail, miscall, indicated by a flashing indicator light (c) Easy access to Android applications, and (d) Android phone integrated with Google services. Some disadvantages of Android include: (a) Requires a simultaneous or continuous internet connection (b) Advertisements appeared most of the time when accessing the Android application.

2.2. Smart Building

Smart building is a building that uses the Building Automation System (BAS), also known as the Intelligent Building System (IBS). The concept of smart building combines architectural design, interior design and electrical mechanics. This concept is high in demand, because it uses a system that is integrated with smartphones or other gadgets as a controller for all of their device. In this system, all facility devices allow to be designed and programmed according to needs, desires, and centralized automatic control. The command can be carried out using sound, infrared light, or a remote controller. The concept of smart building is directed to meet 3 criteria: (a) Longevity (b) Energy and efficiency (c) Comfort and satisfaction. Fulfillment of the above criteria uses several approaches, one of which is the interaction between users and buildings. The use of smart building systems is a solution to synergize energy efficiency, comfort, safety and security that transforms buildings into living organisms based on networks, intelligence, sensitivity and adaptability.

2.3. Islamic Value

Islamic values can be interpreted as a concept and belief that is held in high esteem by Muslims regarding some of the main issues relating to Islam. Islamic values can be used as a guide in behavior, both originating from God and the results of human interaction in accordance with the Shari'a. Here are the basic values of Islam: (a) The value of aqeedah, is the value of several cases that must be believed to be true without any doubt. Aqeedah values provide peace of heart and soul, (b) The value of ibadah, is a value that includes all the words and deeds favored and blessed by God, (c) Moral values, are all things related to the nature embedded in the human psyche that appear by themselves without going through thoughts or considerations, and there is no encouragement from outside. The scope of morals, namely morals towards God, morals with humans, and morals with nature.

Islamic values can be instilled in the personal lives of Muslims through a pedagogical civilizing process, because in Islamic life there are Islamic values: (a) Values that can increase the prosperity of life in the world, (b) Values that encourage to achieve a happy life in the afterlife, (c) Value that can combine the interests of worldly life and ukhrawi.

3. Methodology

This study uses system design build methods, the stages of work are as follow:

3.1. Project planning

The activities of this stage include determining the project theme, studying the literature related to the project theme, studying the desired conditions and conditions that exist at the time of implementation, listing the tools and components needed, and planning the budget.

3.2. Preparation of tools and components

The activities at this stage is the selection of tool and component specifications, and then testing the tool and component parts. Part testing is carried out to find out whether the tools and components are functioning properly or not.

3.3. Hardware design

Hardware design consists of designing mechanical systems and designing electrical systems. The main activity carried out at this stage is to make a block diagram for the entire control system in accordance with the specifications of the tools and components. This block diagram serves as a guide in assembly.

3.4. Software design

Software design must prioritize the principle of efficient work. In this stage, the first step is to create a program flowchart, then the program is compiled in a programming language. After compiling, the program is uploaded to the Arduino main component, the microcontroller. Microcontrollers that have been implanted with the program are then assembled together with tools and other electronic components.

3.5. Assembling

Activities at this stage are assembling tools and components based on hardware design and software design. During assembly, tools and components must be ensured to be installed in the correct place and position.

3.6. Testing

At this stage the whole design results are tested. The testing activity is intended to see the accuracy of the design of the control system that has been made, to ensure the performance of the device can function properly. Testing must be done thoroughly and carefully, in order to obtain a reliable control system.

3.7. System optimization

The last working stage in the system design build method is system optimization. This activity is intended to improve the performance and effectiveness of the control system design that is made.

4. Result and Discussion

The results of hardware assembly in the form of mockups and control system circuits consisting of Arduino, Wifi Shield, relays, and external antennas. While the results of the assembly of software in the form of display control system applications on Android-based smartphones and Arduino program listings. Furthermore, when implementing a control system, the results of the software assembly are integrated with the results of the hardware assembly. In this study, the mechanism used to justify the accuracy of the overall system is by testing the performance of hardware and software that has been integrated as a device. Tests carried out directly in each experimental room. Based on the test it can be seen the response of the output (experimental room's lamp) to the input (voice commands or prayers). Linearity between output to input shows the accuracy of the structure of the hardware assembly results, and the functioning of the application and software assembly program results. The following results are the results of testing the integration of hardware and software devices.

4.1. Compatibility test result

Compatibility test is carried out by running the applications that are designed on several Android devices with different mobile platforms and different operating system (OS) versions. This refers to the opinion of Kumar and Cauhan (2014), that the compatibility of application can be measured through testing on various mobile platforms that have different OS versions, different screen sizes, and different resolutions, or testing applications when interacting with other applications that have a different operating system. The results of compatibility test for applications show that applications that are designed can only run on two mobile platforms, namely Oppo with the Oreo operating system and Xiaomi with the Lollipop, Nougat, and Oreo operating systems. While for the other 4 mobile platforms the application can be installed and displayed, but cannot be operated or run. It can be informed here, that Android can be assumed as a software, where the software consists of supporting programs that have different specifications on each operating system. This means that if the features and software architecture of the Android operating system are incompatible and not comprehensive with the application designed, the application's performance and function will be constrained. As a result, applications that are designed cannot be run or operated on several mobile platforms. The more types of mobile platforms and operating systems that an application can run, the better the compatibility level.

4.2. Connection test result

In this study the control system of switching power lamp in each experimental room uses the Wireless Fidelity (WiFi) accessed via WiFi Shield. WiFi Shield is a wireless LAN device that can be used for data communication with a frequency of 2.4 GHz. This type of frequency has a connection range of around 90-100 meters. Therefore, to find out the performance of the WiFi Shield connection range when it is running, a connection range was tested in each experimental room with varying control distances and unobstructed conditions. Connection test results show that WiFi Shield can still be connected at a distance of 20, 40, and 60 meters. The results of this test also show that distance does not affect response time. This is shown from the average response time of the control system, which is not much different for each connection range, therefor it can be said that the transmission process tends to be stable. These results prove that the WiFi Shield signal's range is able to operate according to its standard frequency. In addition, the frequency produced by WiFi Shield (2.4 GHz) is a frequency that is commonly used by almost all wireless devices. This means that there are many supporting devices for signal amplifiers on the market. In this study to overcome interference in the transmission process due to signal deficit, an external antenna is used, because the better the signal quality, the better the connection.

4.3. Penetration test result

In its application, there are several factors that can affect signal transmission, one of which is a barrier. According to Saharuna and Nur (2016), even a human existence can be said to be a temporary barrier, which can affect the signal emission from WiFi devices. Therefore to find out the performance of WiFi Shield when there is a barrier, a penetration test is conducted. The barriers used in this test are mica and plywood. According to Jubilee (2009: 20), the effect of these two barriers on the quality of signal transmission is relatively small. The penetration test results show that the two barriers (plywood and mica) do not hamper the WiFi Shield connection access. The average response time between the two barriers also has a small difference. The average response time of plywood is smaller than the average response time of mica. The difference in response time is due to differences in particle density between the two barriers. The barrier made from mica has higher density than plywood. This means that the denser the particles in a material, the worse the penetration.

4.4. Lamp control test result

Lamp control testing is carried out using voice commands. Voice commands in the form of prayers or du'a are recited with good pronunciation. This test is intended to see whether the application that has been designed can translate voice commands from various sources (male and female) properly. The results of this test can be said to have a very high success rate. Based on the results that have been carried out on this test, the number of failures is very small compared to the number of successes in both the command to turn on and turn off the lights. The main cause of failure in the light control test lies in the execution of voice commands in the form of prayer. In this case the recitation of prayer must be in accordance with the rules of the makhrij al huruf. The aim is to avoid mistakes in reciting letters. Mispronounced letters will affect the meaning of the verse (Alam, 1995: 22-23). In addition, prayer must be said completely, clearly and may not be repeated. The light control test involves the work process of the system from beginning to end. Starting from giving input in the form of voice commands (prayers), then Google Voice will translate voice commands (prayers). An error in pronouncing the voice command will cause a translation failure. The next process involves the work of a smartphone and WiFi Shield for sending web addresses and translating codes according to the web address sent. The final part of the process is the execution of the input (prayer/voice commands) through the output response (lights on/off) according to the control of the electronic switch (relay).

5. Conclucion

Based on the results and discussion, the conclusion is as follows: (1) In this research, an Android-based energy saving control system is integrated with Islamic values on the light switch with the concept of smart building. The resulting control system is very user friendly and has more value on the input, which is in the form of recited du'a that are pronounced with good pronunciation. (2) The control system test results provide the following information: (a) The control system application has not been able to operate on several mobile platforms, (b) The connection range of the control system is able to operate according to its standard frequency, (c) Control system connections are able to penetrate barriers that have relatively small inhibition, (d) The success rate of the control system in executing input is relatively high. Some recommendations that can be used as a follow-up to the research are: (1) The application that created using App Inventor cannot run on several mobile platforms and operating systems. Therefore it is recommended to further develop application makers for Android. Another application maker that can be used to create applications on Android is Android Studio.

References

[1] Alam Sei H. Dt Tombak 1995 Ilmu Tajwid Populer 17 kali Pandai (Jakarta: Bumi Aksara)
[2] Edi et al 2016 Outlook Energi Indonesia 2016 (Jakarta: Dewan Energi Nasional)
[3] Jubilee E 2009 100 Tip dan Trik Wi-Fi (Jakarta: Elex Media Komputindo)
[4] Darwish A S 2015 ISESCO JOURNAL of Science and Technology **12** 35 – 42
[5] Djuric S 2017 Journal of Process Management – New Technologies (JPMNT) **5** 1 – 6
[6] Otomo G and Wildian 2013 Jurnal Fisika Unand **2** 255 – 261
[7] Mannan K A and Muchlis A F 2012 Journal of Islamic Architecture **2** 78 – 81
[8] Shivaji G S and Bhagyashri G J 2016 International Journal of Application or Innovation in Engineering & Management (IJAIEM) **5** 25 – 30
[9] Saharuna Z and Rini N 2016 JURNAL INFOTEL **8** 115 – 123
[10] Sohail A, Ahmad Z, and Ali I International Journal of Advances in Engineering & Technology (IJAET) **6** 678 – 687

IOP Conf. Series: Materials Science and Engineering **846** (2020) 012056 doi:10.1088/1757-899X/846/1/012056

Implementation of Supply Chain Risk Management (SCRM) Using House of Risk (HOR): Case Study on Supply Chain of Craft Bag Industry

Z U Rizqi [1] and A Khairunisa [1]

[1] Departement of Industrial Engineering, Univesitas Islam Indonesia, Yogyakarta, Indonesia

E-mail: ugihzakka@gmail.com

Abstract. The length of supply chain at Ajeng Gallery causes emergence of risks that can harm the company. Moreover, the company has not implemented risk management properly so that the purpose of this study is to implement supply chain risk management properly. By using House of Risk, it is obtained 10 risk events and 9 risk agents. 2 risk agents that have the largest percentage are mitigated. There are 4 priority mitigation strategies to overcome risk agents. The implementation of mitigation strategy can reduce risk levels in risk map based on previous research that discusses the same mitigation plan.

1. Introduction

Supply chain is a very interesting thing to be discussed. Given the many opportunities that can be achieved from the optimal supply chain. In its development, Supply Chain activities involve many business operations and manufacturing activities which are seen as activities that have brought many changes to the environment [1]. The process in the supply chain is quite complex, including product development, procurement, planning & control, production, and distribution [2]. Because the system is so complex, there are many risks that may occur and disrupt business processes in the supply chain.

Risk is uncertainty and can have negative or positive effects. Risk is also a combination of the likelihood and severity of an event. Meanwhile, supply chain risk is the uncertainty of an event that can result in disruption to the smooth supply chain in a company. Risk in the supply chain can be reduced when company implements Supply Chain Risk Management Process [3].

Ajeng Gallery is a company engaged in manufacturing bags-based crafts. Due to the unique and high-quality bag products, this has led to high demand. However, demand is very volatile. The length of the supply chain at Ajeng Gallery and the high dependence on suppliers makes it vulnerable to possible risks. Moreover, the company does not have good risk management. Therefore we need risk management to overcome this problem.

Based on the background description, this study aims to design a supply chain risk management strategy at Ajeng Gallery with the House of Risk (HoR) approach. House of risk is a method that is focused on formulating preventive, reduce and handling risk agents that have the potential to cause more than one risk [4]. In simple terms, the HoR consists of 2 phases, in phase 1 a risk assessment will be carried out where the output is a risk agent priority which is then made an input in phase 2 HoR where the output is a priority mitigation strategy.

2. Research methods

In this study, researchers conducted risk mitigation with a bottom-up approach to the supply chain of Ajeng Gallery which is a handicraft bag company based on cowhide in Yogyakarta.

2.1. Company supply chain mapping

Before conducting risk assessment on the supply chain, knowledge of the existing supply chain system is needed. Ajeng Gallery supply chain network is in Figure 1.

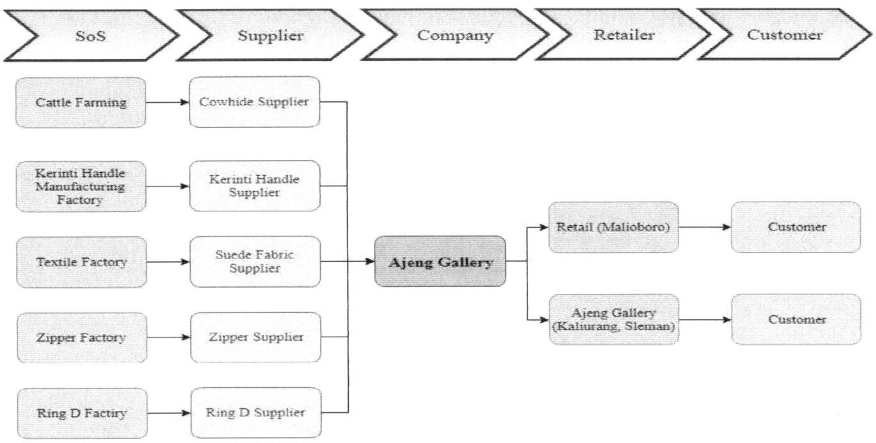

Figure 1. Supply chain network

2.2. Risk assessment

After understanding the supply chain system in the company, the next step is to identify the risk event and severity assessment, then identify the risk agent and occurrence assessment. Risk event identification is done by interviewing the expert. The expert is one of the company's managers who understands the Ajeng Gallery supply chain system. Whereas risk agent identification is carried out using Risk Cause Analysis (RCA). Severity (Si) and occurrence (Oj) assessments are carried out by distributing questionnaires to experts in the form of rating ratings from 1-10 where 1 is the smallest and 10 is the largest. After that, the degree of correlation (Rij) between the risk event and the risk agent is given with the assessment criteria in Table 1. House of Risk (HoR) 1 mapping was then carried out. The output of HoR 1 is the aggregate risk potential (ARPj) which is used to determine the priority of treatment for risk agents that need to be given risk prevention measures first. ARP calculation is obtained using equation 1.

Table 1. Degree of correlation

Weight	Information
0	No relationship
1	Low
3	Moderate
9	High

$$ARP_j = O_j \sum S_i R_{ij} \quad (1)$$

From the calculation of ARP then calculated the percentage of each risk agent. The risk agent chosen has the largest percentage based on the Pareto principle. The Pareto principle is well-known with 80/20, where the failure of 80% is represented by 20%. Therefore, in this study, all risk agents will not be taken to mitigate, but only a few.

2.3. Risk mitigation
After obtaining a dominant risk agent, the next step is to plan risk mitigation to overcome the existing risk agents. At this stage the House of Risk (HoR) phase 2 is also produced. HoR phase 2 is a risk management phase where a selection of treatment strategies that are considered effective to reduce the probability and/or impact caused by the risk agent will be carried out. Handling/mitigation strategies are obtained from Focus Group Discussions with experts. This is because the expert knows more about the possibility of implementing the handling strategy in the company in terms of its financial and technical aspects. The strategies obtained are then determined by the degree of correlation between the existing strategies and all risk agents. The assessment of the degree of correlation is the same as before, which can be seen in Table 1. In addition, the granting of a Degree of Difficulty (DoD) which is the difficulty level of the mitigation design is implemented at the company. DoD was obtained from interviews by the expert with DoD criteria such as Table 2. Then an assessment of the Total Effectiveness (TEk) of each action is based on the correlation of risk agents with a mitigation strategy along with the Effectiveness to Difficulty (ETDk) which is the basis for determining the priority of mitigation handling. The TEk formula uses equation 2 and ETDk in equation 3.

Table 2. Degree of difficulty.

Weight	Information
3	Mitigation actions are easy to implement
4	Mitigation actions are rather difficult to implement
5	Mitigation actions are difficult to implement

$$TE_k = \sum j \; ARP_J \qquad (2)$$

$$ETD_k = TE_k/DoD \qquad (3)$$

3. Results and discussion
The results obtained from the identification of risk events can be seen in Table 3. Risk agents are sought using RCA. The risk agents at Ajeng Gallery can be seen in Table 4.

Table 3. Risk events

Code	Risk Events	Severity
E1	Mistaken of planning for maintenance on production equipment	6
E2	Delayed in design planning	8
E3	Mistaken in production planning	4
E4	Raw materials and supporting materials do not meet the quality desired by the company	8
E5	Lack of raw materials and supporting materials	6
E6	Company's internal communication system is not good	3
E7	Mistaken of budget planning that will be used	6
E8	Uncertainty of orders from customers	8
E9	Delay in planning the selection and use of transportation equipment / logistics providers to customers	6
E10	The production is late	4

Table 4. Risk agents

Code	Risk Agents	Occurence
A1	Lack of involvement and concern of workers in supporting activities in the company	3

Code	Risk Agents	Occurence
A2	Lack of quality control process	7
A3	Suppliers cannot meet company demand in terms of quantity or quality	4
A4	Mistaken in maintenance planning on production equipment	2
A5	The company's internal communication system is not going well	3
A6	The number of orders from customers is uncertain	8
A7	Uncertainty of logistics providers to deliver products to customers	3
A8	Damage to raw materials and or supporting materials during shipping	2
A9	Delayed in supplier delivery	3

3.1. House of risk 1

After scoring the severity and occurance, the researcher analyzes the correlation between the risk event and the risk agent and then mapping the House of Risk Phase 1 and evaluating the ARP (Aggregate Risk Potential) using equation 1. HoR 1 can be seen in Table 5. Based on the ARP values that have been obtained from the calculation of the Table 5, the next is to determine the dominant risk agent with the Pareto approach. Pareto diagrams can be seen in Figure 2.

Table 5. House of risk 1

	A1	A2	A3	A4	A5	A6	A7	A8	A9	Si
E1	1			9	1					6
E2	1				3					8
E3	3		9	1	1			3	3	4
E4		9	9					3		8
E5			9					9	3	6
E6	3				9					3
E7		3	3							6
E8						9				8
E9							9			6
E10	3		3	1	1			3	9	4
Oj	3	7	4	2	3	8	3	2	3	Total
ARP	141	630	768	124	195	576	162	204	198	2998
%	0,047	0,210	0,256	0,041	0,065	0,192	0,054	0,068	0,066	1
Rank	8	2	1	7	4	3	6	5	9	

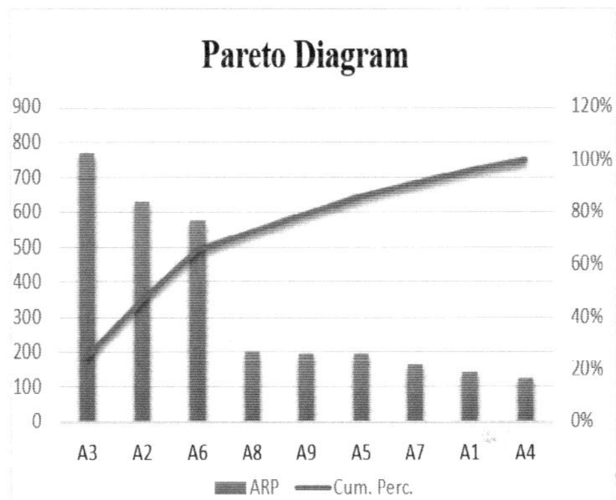

Figure 2. Pareto diagram

From the Pareto diagram, it is known that there is 1 risk agent (A3) out of 9 risk agents which is the dominant cause in the Ajeng Gallery supply chain because it has taken 25.6% of the total risk agents. However, in this study 2 risk agents will be taken so that added risk agents A2 to be mitigated. This is because it provides more comprehensive advice to companies

3.2. House of risk 2

Based on the two risk agents shown in the Pareto diagram, the researcher will then design several mitigation plans or treatment strategies in order to reduce the appearance of these risk agents. The following are mitigations designed by researchers for Ajeng Gallery which can be seen in Table 6. After that the DoD value is given from the mitigation plan, as well as the calculation of Total Effectiveness (TEk) and Effectiveness to Difficulty (ETDk) using equations 2 and 3. The calculation results are mapped into House of Risk 2 in Table 7.

Table 6. Mitigation plan

Risk Agent	Mitigation	Mitigation Code (PAi)
A3	Make a cooperation agreement (MoU) to ensure the availability of raw materials from suppliers	PA1

Risk Agent	Mitigation	Mitigation Code (PAi)
A2	Cooperating with more than 1 supplier	PA2
	Create product standardize quality	PA3
	QC process procurement of raw materials before and during the production process	PA4

Table 7. House of risk 2

Risk Agent	Mitigation				
	PA1	PA2	PA3	PA4	ARP
A2	3	9			768
A3			9	3	630
TEk	2304	6912	5670	1890	1890
DoD	4	4	3	3	
ETDk	576	1728	1890	630	
Rank of Priority (Rk)	3	2	1	4	

Based on HoR 2, it shown the priority order of mitigation strategies that can be applied in advance at Ajeng Gallery. This does not mean that the most recent priority sequence is not done, priority here means the sequence that will be implemented after the next one follows. Implementation The priority order of mitigation plans are PA3, PA2, PA1, and PA4. By carrying out the mitigation strategy the company is expected to be able to reduce and even eliminate the highest risk agent for the company, that is, suppliers cannot meet the company's demand in terms of quantity or quality (A3) and the lack of quality control (A2) processes, which in turn will have an effect on reducing the level of risk events.

4. Conlusion

From the previous discussion, we get the priority order of the mitigation plan. The measurement of mitigation success that is applied at Ajeng Gallery is based on trusted journals related to the mitigation strategies offered that can be seen in Table 8. After implementation, it is estimated that the reduction of risk level can be seen in Figure 3.

Table 8. Implementation of mitigation

Code	After Implementation
PA3	By providing a standardized product quality, customer satisfaction will increase. This is in line with the opinion of Naser et al. in Wibowo (2009) namely in his research produced that customer satisfaction is very dependent on how the level of product quality offered by the company [5].
PA2	According to Govindaraju and Sinulinga (2017) to maintain optimal supply of raw materials, schedule for purchasing raw materials and good cooperation with suppliers. So, collaborating with> 1 supplie is certainly an effective step can help companies achieve the desired production results [6].
PA1	Based on the results of research conducted by Jati (2008), it shows that the indicator of willingness to make cooperation agreements is one of the factors that has a large correlation to long-term relationships with suppliers. That is, if the company cooperates with the supplier, the company will get a guarantee of the availability of raw materials needed by the company for a long period of time [7].
PA4	From the results of research conducted by Elmas (2017), it can be concluded that quality control can minimize failed products. By minimizing failed products, the higher company goal is to provide quality convection products [8].

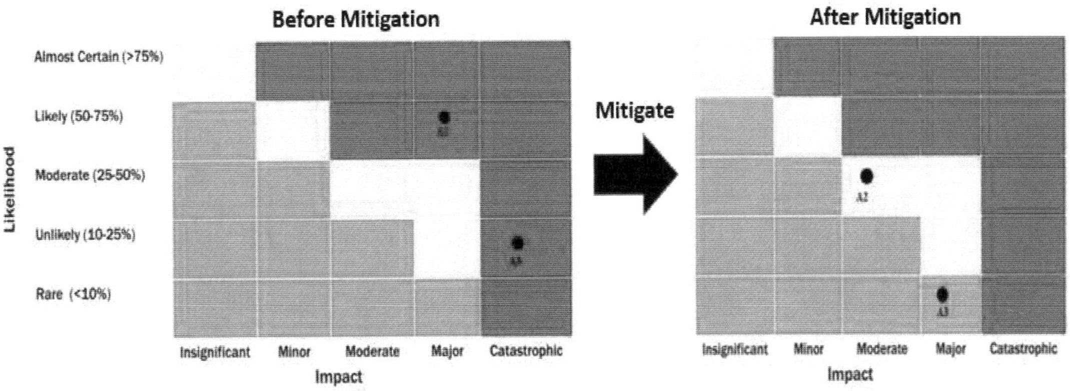

Figure 3. Risk map comparison

So it can be concluded, that there are 9 risk agents that cause risk events. ARP calculation shows that A3 and A2 are the biggest risk agents so the mitigation plan is focused on the two risk agents. There are 4 mitigation plans that can be implemented by companies which based on HoR 2 get priority rankings starting from PA3 PA2 PA1 and PA4. Implementing a mitigation plan can reduce the level of risk agents based on literature studies that discuss the same mitigation efforts.

References

[1] Fortuna I F, SumantrinY and Yuniarti R 2014 Perancangan sistem pengukuran kinerja aktivitas green supply chain management *Jurnal Rekayasa dan Manajemen Sistem Industri* 2 pp 551-562

[2] Astutik W D, Santoso B P and Sumantri Y 2015 Strategi penanganan risiko pada rantai pasok pupuk *organic Jurnal Rekayasa dan Manajemen Sistem Industri* 3 pp 558-567

[3] Tummala R and Schoenherr T 2011 Assessing and managing risk using the supply chain risk management process (SCRMP) *Supply Chain Management: An International Journal* 16 pp 474-483

[4] Pujawan I N and Geraldin I H 2009 House of risk: a model for proactive supply chain risk management *Business Process Management Journal* 15 pp 953-967

[5] Wiwobo A 2009 Analisis pengaruh kualitas produk terhadap kepuasan pelanggan (studi pada perumahan Sembungharjo permai pengembang PT. sindur graha tama) *Jurnal Sains Pemasaran Indonesia* 2 pp 173 -186

[6] Govindaraju R and Sinulingga J P 2017 Pengambilan keputusan pemilihan pemasok di perusahaan manufaktur dengan metode fuzzy ANP *Jurnal Manajemen Teknologi* 1 pp 1-6

[7] Jati B P 2008 Membangun hubungan jangka panjang melalui rasa percaya dan komitmen pengaruhnya pada keunggulan bersaing *Jurnal Bisnis Strategi* 2 pp 160-176

[8] Elmas S M H 2017 Pengendalian kualitas dengan menggunakan metode statistical quality control (SQC) untuk meminimumkan produk gagal pada toko roti barokah bakery *Jurnal Penelitian Ekonomi WIGA* 7 pp 15-22

Pricing and Contract Design for Software Products in the Presence of Requirements Change

Pandu Dwi Luhur Pambudi [1,2*], Cheng-Han Wu [1], Mahendrawathi ER [2]

[1] National Cheng Kung University, Tainan, Taiwan
[2] Institut Teknologi Sepuluh Nopember, Surabaya, Indonesia

* luhur.pandu@gmail.com, wuchan@mail.ncku.edu.tw, mahendra_w@is.its.ac.id

Abstract. The presence of software requirements change (RC) during project development is a critical challenge for the developer to offer software contract designs. Because under the presence of RC, the decisions toward the contract offer will impact to project's price spent by the developer. Managers of software companies must decide what contract designs to offer to clients in the development of software. Abstracting from an example drawn from the software industry, we exhibit three designs of software contracts incorporating fixed price (FP) and time-and-materials (T&M) policies. Specifically, a software company offers a fixed-price and agree to RC with adjustment cost (SW contract), or initially provides a fixed price and then charges an additional fee based on the time-and-material in response to RC (MP contract).We examine the strategic choices of two contract designs in a two-stage game. We carry out a full analysis of distinct settings of monopoly model. We characterize the conditions under which the contracts can be the best decision for developers. As the result, we found that the MP contract performed better when accommodating the presence of RC instead of the SW contract.

1. Introduction

In every software development project, we use contracts to regulate project work and results. Throughout substantial evidence and proven theories over decades, some experts introduce and promote success variety of contract types, but two contract types have dominated and still dominate most work of software projects: fixed price (FP) and time and materials (T&M) types of contracts. The two contracts are frequently used in the software industry, as discussed in [1]–[3].

The FP type of contract is when the client agrees to pay the developer a specific price for a delivered software product. Meanwhile, A T&M type of contract is when the client agrees to pay for the effort spent by the developer, usually based on the price agreed by both sides per hour of work for different types of competence and other necessary expenditures [3]. In the FP contract, the price for completing the project is predetermined in advance. Conversely, T&M contract does not specify a price, but rather reimburse the vendor for its costs plus a predetermined profit [1]. Despite the extensive study in software contract, the issues related to contract choice and performance for software projects remain unclear. Those conditions explained by [4]; because of the high complexity in software engineering processes, software contracts pose many unique challenges that are usually not seen in contracts in other industries. We capture two main problems that challenge a developer to offer a software contract due to the presence of requirement change (RC).

Incomplete initial requirements and difficulty in accommodating the RC are two of these challenges. These problems could make software contract challenging to manage by often incurring significant cost overruns from RC. However, there is little research that provides a clear understanding and guidance on the choice and performance of contracts in terms of price perspective. Based on the problems, we are motivated to study the contract choice and design. We examine the adoption of the two contracts in the software industry: fixed-price (FP) contract and time-and-materials contract (T&M). Our research contributions are threefold, as explained below:

a. Formulation of the determinants of contract choices. This study is one of the papers to analytically study the determinants of contract choice in the software industry. Our paper is one of the papers to analytically study the determinants of contract choice in the software industry including [4]–[6] but of course we design our model differently with the two-stages game. Although there is some discussion of the contract types in the [1]–[3], [7], [8], they discussed their study in an empirical way.
b. Discussion on the impacts of competition. We discuss this study into one setting, which is monopoly model.
c. Investigation of the equilibrium pricing decisions and contract choices. We address the linkage between contract designs and project profits in the presence of RC. We aim to study this problem at a project level by modeling the characteristics of a software contract based on established theories in software development.

2. Literature Review

This study relates to three streams of literature: one examines the RC present analytically in the software industry, second provides a survey on software pricing, and the third are types of software contract design. We present the primary previous studies considerations, specifically the analytical and theoretical of the observed studies.

2.1. RC in The Software Industry

In any software development activity, RCs are inevitable and can occur due to changes in user requirements, increased understanding of the stakeholders' needs, customer organizational re-structure, and availability of new technologies [9], [10]. With every change in requirements, a developer can be affected by the change of the overall cost, quality, and schedule of the software, which is why the RCs as one of the significant causes of software failure [11]. However, dealing with RCs not only poses a risk to the successful delivery of software but also provide an opportunity for improved usability, value, and enhance software development process [12], [13]. Topics related to the first stream have been examined in the field of computer science and software development projects, e.g., [14]–[16]. These topics mainly discussed how RC can affect software development in theoretical and analytical ways.

In this study, we use the term of Requirements Change (RC) as a description from [15]. RC is the number of changes (addition, deletion, and modification) in each period of the development life cycle. It means that every addition, deletion, and modification of software requirements are referred to as RC.

2.2. Pricing in The Software Industry

Software companies often struggle with communicating the value of their solution to their customers and, as a result, lack of profits at the end. Many researchers tried to optimize and analyze the software pricing with some scenarios of strategy. They develop a pricing strategy that is tailored to customer segments that optimizes a company's financial goals. Here we summarize the second stream surveys about pricing in the software industry. Studies related to this topic are broadly discussed in the field of information systems, operation management, and industrial management. e.g., [17]–[22].

IOP Conf. Series: Materials Science and Engineering **846** (2020) 012057 doi:10.1088/1757-899X/846/1/012057

2.3. Software Contract Policy

The third stream examines software contract policy. A software contract is a binding agreement between the software developer and the client in the outsourcing mechanism. The contract provides the development policy of the software to the client. A typical software development contract must deal with a variety of closely related issues such as the quality of the developed system, the timeliness of delivery, the effort and cost associated with the project, and the support [4]. The economists have studied contract design for at least 40 years (see McCall, 1970) in [23]. Meanwhile, [8] examined the use of fixed-fee and time-and-materials (T&M, or cost-plus) contracts and a hybrid contract that consisted of a T&M contract with a cap.

We resume our survey related to the previous studies into the table as seen in Table 1. We survey those studies based on the criteria set in the beginning through keywords searching from reputable journal databases. As we can see in Table 1, we highlight our study with a thick font that describes if this study has a novelty aspect compared to the previous studies. We consider the presence of RCs and duopoly market as the external competition which never discussed before in the contract of software development.

Table 1. Survey of the previous study

Author(s) (Year)	RC Presence	External Competition	Software Pricing	Software Contract Type FP	Software Contract Type T&M	Decision Variable
This study	√	√	√	√	√	Price
[17]	-	√	√	-	-	Price
[16]	√	-	-	-	-	RC
[5]	√	-	√	√	√	Price
[20]	-	-	√	√	-	Price
[19]	-	√	√	√	-	Price
[24]	-	√	√	-	-	Price, quality
[6]	√	-	-	√	√	Work rate
[21]	-	√	√	-	-	Price
[25]	-	√	√	-	-	Price
[4]	√	-	-	√	√	Effort, release time
[26]	-	√	√	-	-	Price, time

3. The Model

3.1. Model Setup

In this section, we analyze the two most common types of contracts, namely fixed-price (FP) contract and time and materials (T&M) contract. We extend the FP contact into one contract design. That is single price with second stage adjustment (SW) and T&M as the manual price (MP) contract design.

We build our model in a monopoly model between developer and client (see Figure 1). The developer will offer two different software contracts. As the model is built in a monopoly market, we do not consider the competition with other developers, because there is a case when the client chooses directly one software developer to execute their project. This condition usually happens when the developer is a loyal vendor for that client. We want to highlight under which software contract the developer as monopolist can obtain the best profit.

3.2. Two-stages model

We set this model into two stages of contract offers. At the first stage of the contract, the developer offers a price portion model to accommodate the main wage cost coming from the initial requirement of software in stage 1. Meanwhile, at the second stage of the contract, the developer offers a price portion model to accommodate the RC cost after stage 1 is completed.

3.3. Monopoly Model

In this section, we study contract designs decision for a monopoly selling a software product based on three contract designs. We derive the conditions under which the developer as a monopolist will offer SN, SW, or MP to the client. In this monopoly model, we model the contract designs to complete the initial requirements with the price portion of α in stage 1 and accommodate the RC with the price portion of $1-\alpha$ in stage 2. These price portion models are developed for SN and SW models due to the explanation in Chapter 3.1. Otherwise, The MP contract does not consider the price portion, so we remove the portion of α in stage 1 and $1-\alpha$ in stage 2. Following prior literature [2], [4]. We also use contract design profit or surplus to measure the performance of the contract by maximizing the price and effort. But for this proposal, we firstly execute the price.

The client's willingness to pay (WTP) as S increases with the quality of the software q and consumer valuation q. Thus, the client' WTP can be expressed $S = \theta\, q$. Given the software quality q as a function of effort as e and time to complete effort as τ. We describe the model as $q = e\,\tau$. We assume that the effort e and the software quality q are private information of the developer and are not directly observable by the client. However, in all settings, software quality may be either higher or lower in WTP. So, in our model, all types of clients coexist (1) low quality, low WTP, and (2) high quality, high WTP. But in our model, we assume that the consumer valuation θ is always positive. Let θ denote the consumer valuation to quality that closely related to the features of the software development product. Even though some clients with greater knowledge and experience in software development usually understand the contract designs in more detailed terms and conditions. That is why we assume both the client and the developer are risks neutral.

We consider the benchmark case where the developer offers three contract designs. We build the model step by step by solving the utility function, demand function, and developer's problem by solving the profit, after that, we derive the optimal price for the developer's problem. We explain the steps as follows:

3.3.1. Utility and Demand Function

Let N be the number of clients with a positive valuation for the software contract designs. We normalize N to 1 for the sake of simplicity and to capture the real event. Client's valuation for the software is denoted by θ and uniformly distributed over [0,1]. In our model, we assume clients always in positive valuation. Figure 1 shows the demand for the software contract designs.

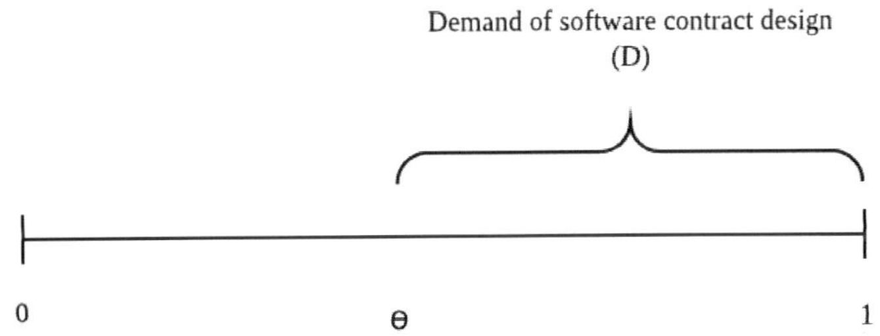

Figure 1. Demand for software contract designs

We assume the demand D_A as a possible demand for the client to choose the contract designs. The client's valuation for the software is denoted by θ and uniformly distributed over [0,1]. Therefore, it corresponds to the demand D_A for all of the software contract, a possible demand can be expressed as $D_A = 1 - \theta$.

The monopoly offers one software development product under SN contract with software utility $U = S - \alpha\, P_A$ and SW contracts with software utility $U = (1+\delta)S - P_A$ and under MP contract with software utility $U = (1+\delta)S - P_{A1} - P_{A2}$. We model WTP as $S = \theta\, q$, let θ as client's positive valuation for the software and q as quality at a price P_A this function applied in the stage 1. Meanwhile, in the stage 2 when the RC is acted, the δ added as primary value addition due to RC. In this phase, we assume that the project is always delivered on time to the client, so there is no variable to decrease the utility of the software due to the project delay. Thus, the expected software utility of the client for SN contracts who have zero net utility can be expressed as $S - \alpha\, P_A = 0$. The software utility of the client for SW contracts who have zero net utility can be expressed as $(1+\delta)S - P_A = 0$ while software utility of the client for MP contract who has zero net utility can be expressed as $(1+\delta)S - P_{A1} - P_{A2} = 0$. After that, we solve θ and get $\theta = \alpha\, P_A/e\,\tau$ for SN, $\theta = P_A/e(1+\delta)\tau$ for SW and $\theta = P_{A1} + P_{A2}/e(1+\delta)\tau$ for MP contract. A client with a higher primary valuation of quality q than the consumer valuation θ has higher WTP S, that leads to positive utility and will buy the software. Otherwise, a client with a lower primary valuation of quality q than the consumer valuation θ will not buy the software. In SW and MP contract we have δ as the addition of primary valuation due to RC. We solve the demand D_A^{SN} for SN contract $D_A^{SN} = 1 - \alpha\, P_A/(e\,\tau)$, demand D_A^{SW} for SW contract $D_A^{SW} = 1 - P_A/e(1+\delta)\tau$ and demand D_A^{MP} for MP contract $D_A^{MP} = 1 - P_{A1} + P_{A2}/e(1+\delta)\tau$.

3.3.2. SW Contract Design

We model the developer's problem under SW contract, as follows:

$$\underset{P_A}{\text{Max}}\, \pi_A^{SW} = \left[(\alpha\, P_A - C_1)D_A^{SW} + \left((1-\alpha)P_A - C_2\right)\right]D_A^{SW} \tag{1}$$

The direct development cost incurred by the developer denoted as $C_1 = c_1\, e^2$. The support cost incurred by the developer denoted as $C_2 = c_2\, \Delta^2$. c_2 is the adjustment cost and Δ is the level of RC, where $\Delta \leq 1$ is always less than equal with 1. The support cost captures the fact that higher quality software will incur a higher support cost and vice versa. This condition also related to the level of RC, the higher level of RC Δ the higher support cost C_2 and the higher software quality obtained. First, we simply set the two different costs without discounting them even though they are typically incurred at different time points. This is because we assume that the cost parameters (c_1, and c_2) in the model already account for the necessary discounting.

Proposition 1. *The optimal price for the developer's problem under SW is*

$$P_A^{SW^*} = \frac{1}{2}\left(c_2\Delta^2 + e(c_1 e + \tau + \delta\tau)\right) \tag{2}$$

Proof of Proposition 1. The first-order condition for the developer's problem in SW contract is

$$\frac{\partial \pi_A^{SW}}{\partial P_A^{SW}} = [(\alpha\, P_A - C_1)D_A^{SW} + \left((1-\alpha)P_A - C_2\right)]D_A^{SW} \tag{3}$$

and from Eq. (3), we get optimal price $P_A^{SW^*}$

$$P_A^{SW^*} = \frac{1}{2}\left(c_2\Delta^2 + e(c_1 e + \tau + \delta\tau)\right) \tag{4}$$

Because we only consider price as our decision variable, we derive second profit to see whether the objective function in $P_A^{SW^*}$ is concave or convex, so we get the result of second derivation, as follows:

$$\frac{\partial^2 \pi_A^{SW}}{\partial P_A^{SW^2}} = -\frac{2(1-\alpha)}{e(1+\delta)\tau} - \frac{2\alpha}{e(1+\delta)\tau} \tag{5}$$

Therefore, the objective function is concave because $\frac{\partial^2 \pi_A^{SW}}{\partial P_A^{SW^2}} < 0$ and is the optimal solution.

3.3.3. MP Contract Design

We model the developer's problem under the MP contract by solving it using a backward induction. We firstly solve the price 2 as P_{A2}^{MP} in stage 2 by solving the Eq. (6). After that, we calculate both the profit 1 and profit 2 as $\pi_A^{MP} = \pi_{A1}^{MP} + \pi_{A2}^{MP}$ and we derive the price 1 as P_{A1}^{MP} in stage 1 by solving Eq. (7). Then, we backward the P_{A2}^{MP} by solving $\pi_A^{MP} = \pi_{A1}^{MP} + \pi_{A2}^{MP}|_{P_{A2}^{MP}}$ in Eq. (8). We express our model, as follows:

and
$$\underset{P_{A2}}{\text{Max}}\, \pi_{A2}^{MP} = [(P_{A2} - C_2)]D_A^{MP} \tag{6}$$

and
$$\pi_A^{MP} = \pi_{A1}^{MP} + \pi_{A2}^{MP}|_{P_{A2}^{MP}} \tag{7}$$

$$\underset{P_{A1}}{\text{Max}}\, \pi_{A1}^{MP} = [(P_{A1} - C_1)]D_A^{MP} \tag{8}$$

In this MP contract design, we don't include the price portion α both in stages 1 and 2, but we include δ as addition towards primary valuation due to the level of RC.

Proposition 2. *The optimal prices for the developer's problems under MP are*

$$P_{A2}^{MP^*} = \frac{1}{2}(\Delta^2 c_2 + e(c_1 e + \tau + \delta\tau) \tag{9}$$

and

$$P_{A1}^{MP^*} = e^2 c_1 \tag{10}$$

Proof of Proposition 2. The first-order condition for the developer's problem in MP contract is using standard backward induction. We first examine the first-order condition for the developer's problem in Eq. (7,8, and 9). The first-order condition is

$$\frac{\partial \pi_{A2}^{MP}}{\partial P_{A2}^{MP}} = [(P_{A2} - C_2)]D_A^{MP} = 0 \tag{5}$$

and from Eq. (11), we get

$$P_{A2}^{MP} = \frac{1}{2}(c_2\Delta^2 + e(1+\delta)\tau - P_{C1}) \tag{6}$$

After that, we calculate the profit 1 and profit 2 as $\pi_A^{MP} = \pi_{A1}^{MP} + \pi_{A2}^{MP}$ and get the result, as follows:

$$\pi_A^{MP} = (-c_1 e^2 + P_{A1})\left(1 - \frac{P_{A1} + P_{A2}}{e(1+\delta)\tau}\right) + (-c_2\Delta^2 + P_{A2})\left(1 - \frac{P_{A1} + P_{A2}}{e(1+\delta)\tau}\right) \tag{7}$$

and derive the price 1 as P_{A1}^{MP} by substituting Eq. (13) into Eq. (12). We get the optimal price 1 $P_{A1}^{MP^*}$, as follows:

$$P_{A1}^{MP^*} = e^2 c_1 \tag{8}$$

After that, we backward the price 2 P_{A2}^{MP} by solving $\pi_A^{MP} = \pi_{A1}^{MP} + \pi_{A2}^{MP}|_{P_{A2}^{MP}}$. We get the optimal price 2 $P_{A2}^{MP^*}$, as follows:

$$P_{A2}^{MP^*} = \frac{1}{2}(\Delta^2 c_2 + e(c_1 e + \tau + \delta\tau) \tag{9}$$

Now, we look at the Hessian matrix, which is given by

$$\det(\mathbf{H^{MP}}) = \begin{bmatrix} \frac{\partial^2 \pi_A^{MP}}{\partial {P_{A2}}^2} & \frac{\partial \pi_A^{MP}}{\partial P_{A2}\,\partial P_{A1}} \\ \frac{\partial \pi_A^{MP}}{\partial P_{A1}\,\partial P_{A2}} & \frac{\partial^2 \pi_A^{MP}}{\partial {P_{A1}}^2} \end{bmatrix} = 0 \geq 0.$$

It is easy to see that $\mathbf{H}^{MP}$ is negative semidefinite for any value of P_{A2}^{MP} and P_{A1}^{MP}. This is because $\det(\mathbf{H}^{MP}) \geq 0$. We derive second profit to see whether the objective function in P_{A2}^{MP} and P_{A1}^{MP} are concave or convex, so we get the result of second derivation, as follows:

$$\frac{\partial^2 \pi_A^{MP}}{\partial P_{A2}^2} = -\frac{2}{e(1+\delta)\tau} < 0 \tag{16}$$

and

$$\frac{\partial^2 \pi_A^{MP}}{\partial P_{A1}^2} = -\frac{2}{e(1+\delta)\tau} < 0 \tag{17}$$

Therefore, the objective functions in P_{A2}^{MP} and P_{A1}^{MP} are concave, and the equilibrium results are the optimal solutions.

4. Model Analysis and Result

Proposition 1 suggests that under SW, the developer does accommodate development cost C_1 in stage 1 and support cost C_2 in stage 2. This SW contract is designed to solve the problem found in the SN contract. This means after completing the initial requirement in stage 1, the developer will continue to perform stage 2 for executing the RC. The cost of mainly wages c_1 will increase when the effort e spent to do the initial requirement in stage 1 is high. Meanwhile, when the level of RC increases Δ, the adjustment cost c_2 will increase. This is because the developer requires to do the additional RC proposed by the client.

We can also see the quality of software q. When the effort e increases, the time to complete the effort τ will also increase. At the end it will leads to the higher mainly wages c_1 and adjustment cost c_2. This indicates when the initial requirement proposed by the client is high, the developer will set up the higher effort e and the higher mainly wages c_1 in advance. Meanwhile, the higher level of RC Δ, means the higher adjustment cost c_2. This indicates when the additional RC proposed by the client is high, the developer will set up a higher level of RC Δ by working on that as well as the primary value will increase due to the addition of RC δ. That is why it reflects the higher adjustment cost c_2. We suggest these contracts for a client who wants to improve their software quality by adding the RC.

Meanwhile, Proposition 2 suggests that under MP, the developer accommodates the support cost C_2 in stage 2 first and continue with the development cost C_2 in stage 1. The cost of mainly wages c_1 will increase when the effort e spent to do the initial requirement in stage 1 is high. Meanwhile, when the level of RC increases Δ, the adjustment cost c_2 will increase. This is because the developer requires to do the additional RC proposed by the client. We can also see the quality of software q. When the effort e increases, the time to complete the effort τ will also increase. At the end it will leads to the higher mainly wages c_1 and adjustment cost c_2. This indicates when the initial requirement proposed by the client is high, the developer will set up the higher effort e and the higher mainly wages c_1 in advance. Meanwhile, the higher level of RC Δ, means the higher adjustment cost c_2. This indicates when the additional RC proposed by the client is high, the developer will set up a higher level of RC Δ by working on that as well as the primary value will increase due to the addition of RC δ.

We design this contract under backward induction by solving stage 2 first and obtain the price 2 and following stage 1 and obtain the price 1. We intend if the client will propose the additional RC during the project that will influence the initial requirement in stage 1. In this MP contract, we do not require the client to complete stage 1 first and continue to stage 2, but we want to approach the condition when the client can flexibly adding RC as much as they want and capture the initial requirement after it. That is why we will charge the adjustment cost c_2 in MP contract always be higher than the adjustment cost c_2 in SW contract $c_2^{MP} \geq c_2^{SW}$ as a support cost by the developer to deal with the increasing level of RC Δ. We also do not include the price portion α for both stages since this MP contract is designed for a client who wants a flexible setting to accommodate the RC. We illustrate the result of Proposition 1 and Proposition 2 using parameter values $e = 0.4, c_1 = 0.2, c_2 = 0.2, \tau = 0.2, \alpha = 0.3, \delta = 0.3$. Meanwhile, for MW contract we set $c_2 = 0.4$ or higher than the SW contract due to the developer's additional consultant for RC.

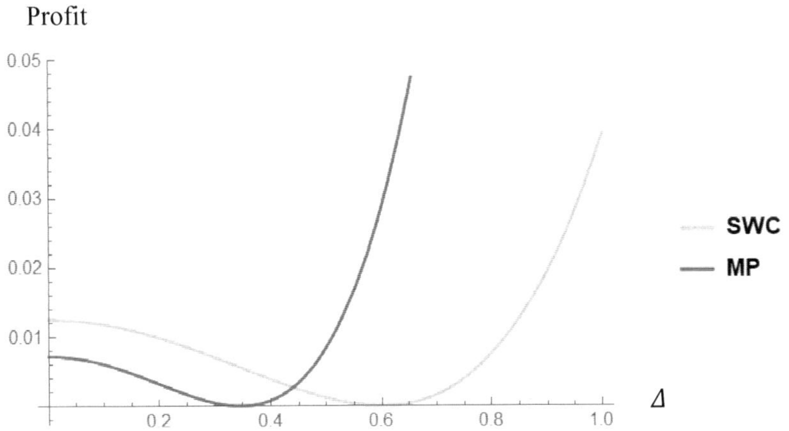

Figure 2. Developer's profit with the presence of RC

We can see in the figure 2. When level of RC Δ increased, the profit received by developer will also increase. We see that in MP contract the developer will get more profit with the same level of RC Δ in SWC contract.

5. Conclusion

In this study we model a monopoly model to know which contract would perform better under the presence of RC. From the result we can see, if the MP contract performed better when accommodating the RC instead of the SW contract. The higher level of RC will lead to the higher profit that developer get in the project.

Our future study will focus on the analysis of model and explore another potential of decision variable. We also want to see this model in a competitive market when dealing with another developers.

References

[1] L. Fink, Y. Lichtenstein, and S. Wyss, "Ex post adaptations and hybrid contracts in software development services," *Appl. Econ.*, vol. 45, no. 32, pp. 4533–4544, 2013.

[2] A. Gopal, K. Sivaramakrishnan, M. S. Krishnan, and T. Mukhopadhyay, "Contracts in Offshore Software Development: An Empirical Analysis," *Manage. Sci.*, vol. 49, no. 12, pp. 1671–1683, 2003.

[3] M. Jørgensen, P. Mohagheghi, and S. Grimstad, "Direct and indirect connections between type of contract and software project outcome," *Int. J. Proj. Manag.*, vol. 35, no. 8, pp. 1573–1586, 2017.

[4] D. Dey, M. Fan, and C. Zhang, "Design and analysis of contracts for software outsourcing," *Inf. Syst. Res.*, vol. 21, no. 1, pp. 93–114, 2010.

[5] S. Oh, J. Rhodes, and R. Strong, "Impact of cost uncertainty on pricing decisions under risk aversion," *Eur. J. Oper. Res.*, vol. 253, no. 1, pp. 144–153, 2016.

[6] H. Dharma Kwon, S. A. Lippman, and C. S. Tang, "Optimal time-based and cost-based coordinated project contracts with unobservable work rates," *Int. J. Prod. Econ.*, vol. 126, no.

2, pp. 247–254, 2010.

[7] M. Suprapto, H. L. M. Bakker, H. G. Mooi, and M. J. C. M. Hertogh, "How do contract types and incentives matter to project performance?," *Int. J. Proj. Manag.*, vol. 34, no. 6, pp. 1071–1087, 2016.

[8] A. Kalnins and K. J. Mayer, "Relationships and hybrid contracts: An analysis of contract choice in information technology," *J. Law, Econ. Organ.*, vol. 20, no. 1, pp. 207–229, 2004.

[9] N. Ali and R. Lai, "A method of requirements change management for Global software development," *Inf. Softw. Technol.*, vol. 70, pp. 49–67, 2016.

[10] S. Basri, N. Kama, F. Haneem, and S. A. Ismail, "Predicting effort for requirement changes during software development," pp. 380–387, 2016.

[11] N. Nurmuliani, D. Zowghi, and S. P. Williams, "Requirements Volatility and Its Impact on Change Effort: Evidence Based Research in Software Development Projects."," *Aust. Work. Requir. Eng.*, 2006.

[12] A. AlSanad and A. Chikh, "The Impact of Software Requirement Change – A Review," in *Á. Rocha et al. (eds.), New Contributions in Information Systems and Technologies, Advances in Intelligent Systems and Computing 353*, 2015, vol. 353.

[13] S. McGee and D. Greer, "Towards an understanding of the causes and effects of software requirements change: Two case studies," *Requir. Eng.*, vol. 17, no. 2, pp. 133–155, 2012.

[14] C. Gaebert, "Dilemma structures between contracting parties in software development projects," *ICSOFT-EA 2014 - Proc. 9th Int. Conf. Softw. Eng. Appl.*, pp. 539–548, 2014.

[15] C. Mao, Y. Lu, and X. Wang, "A study on the distribution and cost prediction of requirements changes in the software life-cycle," *Lect. Notes Comput. Sci. (including Subser. Lect. Notes Artif. Intell. Lect. Notes Bioinformatics)*, vol. 3840 LNCS, pp. 136–150, 2006.

[16] Z. Tong, X. Su, L. Cen, and T. Wang, "Greening Software Requirements Change Management Strategy Based on Nash Equilibrium," *Wirel. Commun. Mob. Comput.*, vol. 2017, pp. 1–10, 2017.

[17] R. Bala and S. Carr, "Usage-based pricing of software services under competition," *J. Revenue Pricing Manag.*, vol. 9, no. 3, pp. 204–216, 2010.

[18] H. K. Cheng, S. Li, and Y. Liu, "Optimal software free trial strategy: Limited version, time-locked, or hybrid?," *Prod. Oper. Manag.*, vol. 24, no. 3, pp. 504–517, 2015.

[19] V. Choudhary, A. Ghose, T. Mukhopadhyay, and U. Rajan, "Personalized pricing and quality differentiation," *Manage. Sci.*, vol. 51, no. 7, pp. 1120–1130, 2005.

[20] Y. Liu, H. K. Cheng, Q. C. Tang, and E. Eryarsoy, "Optimal software pricing in the presence of piracy and word-of-mouth effect," *Decis. Support Syst.*, vol. 51, no. 1, pp. 99–107, 2011.

[21] A. Mehra, R. Bala, and R. Sankaranarayanan, "Competitive behavior-based price discrimination for software upgrades," *Inf. Syst. Res.*, vol. 23, no. 1, pp. 60–74, 2012.

[22] G. Nan, F. Shi, R. Dou, and M. Li, "Duopoly pricing of software products under free strategy: Limited-feature vs. seeding," *Comput. Ind. Eng.*, vol. 100, pp. 13–23, 2016.

[23] S. A. Lippman, K. F. McCardle, and C. S. Tang, "Using Nash bargaining to design project management contracts under cost uncertainty," *Int. J. Prod. Econ.*, vol. 145, no. 1, pp. 199–207, 2013.

[24] H. K. Cheng and Q. C. Tang, "Free trial or no free trial: Optimal software product design with network effects," *Eur. J. Oper. Res.*, vol. 205, no. 2, pp. 437–447, 2010.

[25] G. Nan, F. Shi, R. Dou, and M. Li, "Duopoly pricing of software products under free strategy: Limited-feature vs. seeding," *Comput. Ind. Eng.*, vol. 100, pp. 13–23, 2016.

[26] H. K. Cheng, S. Li, and Y. Liu, "Optimal software free trial strategy: Limited version, time-locked, or hybrid?," *Prod. Oper. Manag.*, vol. 24, no. 3, pp. 504–517, 2015.

IOP Conf. Series: Materials Science and Engineering **846** (2020) 012058 doi:10.1088/1757-899X/846/1/012058

Modeling and the Validation Model of PM10 Concentration due to the Changes in the Dominant Wind Direction to the Road in the Roadside Area

Vera Surtia Bachtiar[1], Purnawan[2], Reri Afrianita[1], Iqbal Mustofa[1], Randa Anugerah[1]

[1]Environmental Engineering Department, Andalas University, Indonesia
[2]Civil Engineering Department, Andalas University, Indonesia

*verasurtia@eng.unand.ac.id

Abstract. This study aims to model and validate the PM10 concentration from the transport sector and its relationship to the characteristics of traffic, wind speed and wind direction. PM10 was analyzed using the gravimetric method. Sampling were taken twice, with the first sampling carried out at 3 monitoring points representing the different angles of arrival of the wind direction towards the road (α), i.e. 0°, 30° and 60°. For validation, a second sampling was carried out at 0°, 45° and 90° angles. The statistical analysis used correlation and regression tests. Validation tests were carried out using the Pearson Product Moment formula and the Two Variance Test. The sampling results showed that the highest concentration was on a road with an angle of 60° with a concentration of 107.748 µg/Nm3 (number of vehicles 3,219 units and wind speed 0.3 m/s). The lowest concentration was at an angle of 0° with a concentration of 19,298 µg/Nm3 (number of vehicles 680 units and wind speed of 0.98 m/s). Increased traffic volume and density was found to be proportional to the increase in PM10 concentration. Vehicle speed and wind speed have an inversely related relationship with increasing PM10 concentration. There was a decrease in PM10 concentration in traffic volume from α 90° to 60° by 26%, α 90° to 30° by 29% and α 90° to 0° by 61%. There was a decrease in PM10 concentration at the traffic density from α 90° to 60° by 17%, α 90° to 30° by 33% and α 90° to 0° by 51%. The results of the Two Variance Test showed there to be no significant difference in the yield variability between the two PM10 parameter concentrations. This is indicated by the value of the Test Ratio (RUf) being lower than the Critical Point. The validation test conducted using the Pearson Product Moment formula shows that the two equations tested can be declared valid and used to determine the concentration of PM10. This is indicated by the value of R (correlation coefficient) having been calculated as greater than the value of the R table for all parameters.

1. Introduction

Transport activities are the main source of air pollution in urban areas. The growth in the number of ownerships of vehicle use in cities has had a severe impact on the environment. The data shown by [1] confirms that motor vehicles are a major source of pollution in urban areas. Motorized vehicles, inclusive of trucks, buses, cars and motorbikes, produce air pollutants when the vehicle is used, refueling, in the manufacturing processes and post-use. The development of the number of motorized vehicles in big cities increasingly makes the transportation sector the dominant pollutant source [2].

One type of air pollutant from the vehicle sector that has a major impact on human health is particulate matter (PM10) because it is respirable. This triggers respiratory problems. Particulate matter is a mixture of particles in the form of solids and liquids suspended in the air. These particles can be produced from motorized vehicles. Small particles of 10 microns (PM10) can enter the respiratory tract and lungs which can cause asthma, respiratory diseases, cardiovascular disease and death. Meteorological factors include wind speed and direction, air humidity, air temperature and air pressure. These factors have an influence on the PM10 concentration. According to [3], a change in the PM10 concentration was caused by changes in the meteorological conditions. The meteorological factor does not necessarily cause an increase in PM10 concentration but it can decrease the PM10 concentration. One example of meteorological factors is that wind speed affects the level of particulate concentration. When the wind blows hard, the particulates will spread more widely and thus reduce their concentration. Conversely, when the wind blows slowly, the particulates do not spread and instead accumulate in one area, and thus the concentration increases.

Padang city does not only having roads that are perpendicular to the direction of the dominant wind. There are many variations of the road angle towards the dominant wind direction. Thus there is a need for research on the effect of α (angle of wind direction on the road) on the dispersion of the concentration of PM10 on the roadside air in the city of Padang. In this study, the measurements were focused on the changes in α (the angle of arrival of the wind towards the road) using roads with angles of 0°, 30° and 60° towards the dominant wind direction in the city of Padang.

2. Methodology

The research consisted of field sampling and laboratory analysis. The sampling of PM10 was undertaken twice, first for the modeling and the second for validation.

Sampling for PM10 Modelling

For the PM10 modeling, the first sampling of PM10 concentration was accomplished in 9 days over 3 monitoring points regarding the meteorological factors and traffic characteristics. Sampling was conducted using a Low Volume Sampler with a digital pocket weatherman, compass and anemometer. At each monitoring point, the sampling was carried out for 3 days with measurements every 1 hour for 10 hours/day starting at 07.00-17.00 WIB. The total number of samples collected for the 3 points was 90 samples. The sampling location chosen for α 0^0 was on A. Yani Road, 30^0 was located on Andalas Road, and 60^0 was on Prof. Dr. Hamka Road. The sampling locations were chosen based on the consideration of the dominant wind direction with reference to the Padang City windrose in 2012 - 2016. For more details, the sampling locations can be seen in Figures 1 through to 3. Meanwhile, the windrose on Padang City between 2012 and 2016 can be seen in Figure 4.

Figure 1. Sampling location on Jend A. Yani Road, $\alpha = 0^0$

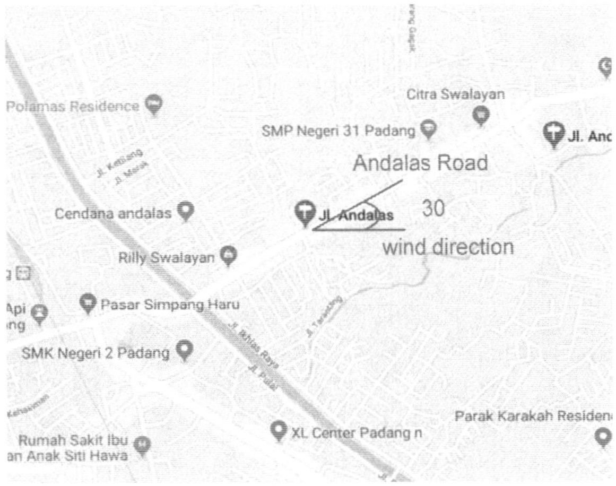

Figure 2. Sampling location on Andalas Road, $\alpha = 30^0$

Figure 3. Sampling location on Prof. Dr. Hamka Road, $\alpha = 60^0$

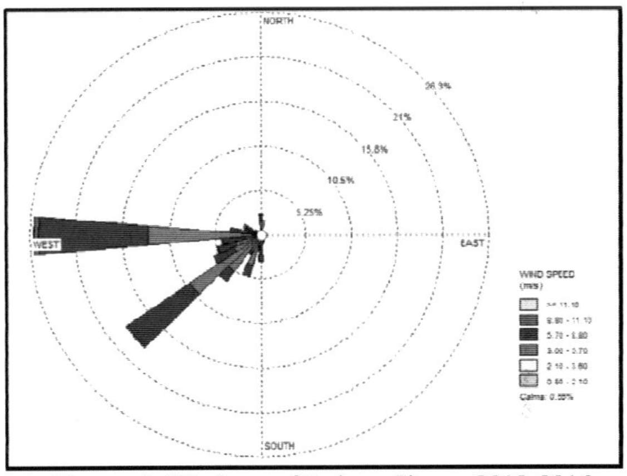

Figure 4. Windrose of Padang City on 2012-2016

Sampling for PM10 Model Validation

The second sampling for validation was done over 3 days with measurements every 1 hour for 10 hours/day for each point starting at 07.00 - 17.00. The total number of samples collected for the 3 validation points was 30 samples. The sampling was performed using a Low Volume Sampler with a digital pocket weatherman, compass and anemometer. The sampling location chosen for α 0^0 was on A. Yani Road, 45^0 was located on By-Pass Road, and 90^0 was on Prof. Dr. Wahidin Road. The α was chosen from 0°, 45° and 90° to represent an α value that ranged between 0° and 90° [4]. The sampling locations can be seen in Figures 5 through to 7.

Figure 5. Sampling location on Jend A. Yani Road, $\alpha = 0^0$

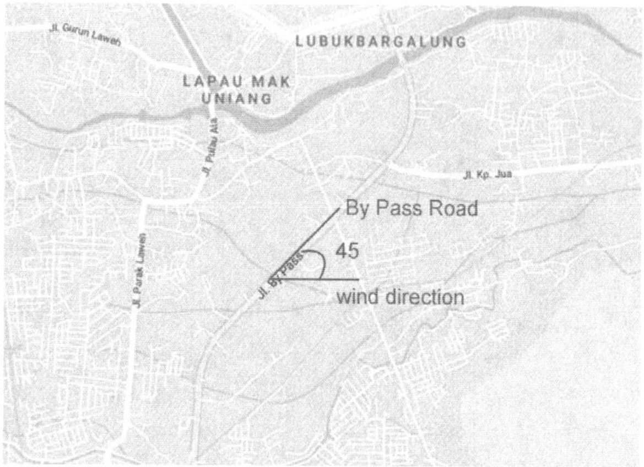

Figure 6. Sampling location on By-Pass Road, $\alpha = 45^0$

Figure 7. Sampling location on Prof. Dr. Wahidin Road, $\alpha = 90^0$

Modelling of PM10 Concentration

The PM10 measurement data from the first sampling at 3 monitoring locations was used to create multiple linear regression models. The traffic characteristics measured were traffic volume, traffic speed and traffic density. The next step was the calculation and processing of the data to analyze the relationship of PM10 concentration with the characteristics of the traffic, the meteorological factors and the magnitude of changes in concentration due to changes in α (the wind angle against the road).

Validation of PM10 Model

Model validation is a simulation process that is used to ascertain whether a model can be used to model a given parameter. The data from the validation sampling was used to test the validity of the PM10 pollutant concentration equation due to the influence of wind direction on roadside ambient air using the Two Variance Test formula [5] and the Pearson Product Moment formula [6].

3. Results and Discussion

Relationship between PM10 and Meteorological Conditions

Figure 8 shows the relationship between PM10 concentration and wind speed in (A) α = 0°, (B) α = 30°, (C) α = 60°.

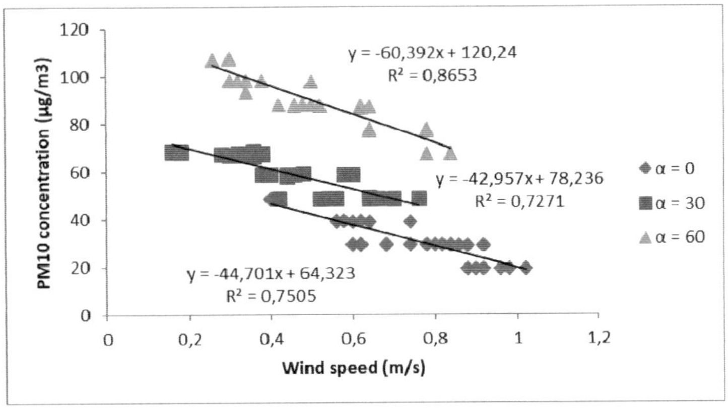

Figure 8. Relationship between PM10 Concentration and wind speed with α = 0°, 30°, and 60°.

Based on Figure 8, the PM10 concentration shows a strong correlation with wind speed, with the coefficient of the correlation(R) values at α = 0°, 30° and 60° being 0.866, 0.852 and 0.930 respectively. The coefficient of determination (R^2) at α = 0°, 30°, and 60° is 0.8653, 0.7271 and 0.7505 respectively. This means that 86.5%, 72.7% and 75.0% of the PM10 concentration is influenced by wind speed. Figure 8 shows that the PM10 concentration will decrease with the increasing wind speed [7]. The higher the wind speed, the wider the area affected by PM10 so therefore the PM10 concentration becomes lower.

Relationship between PM10 Concentrations and Traffic Characteristics

The relationship between PM10 concentration and traffic density has been shown in Figure 9.

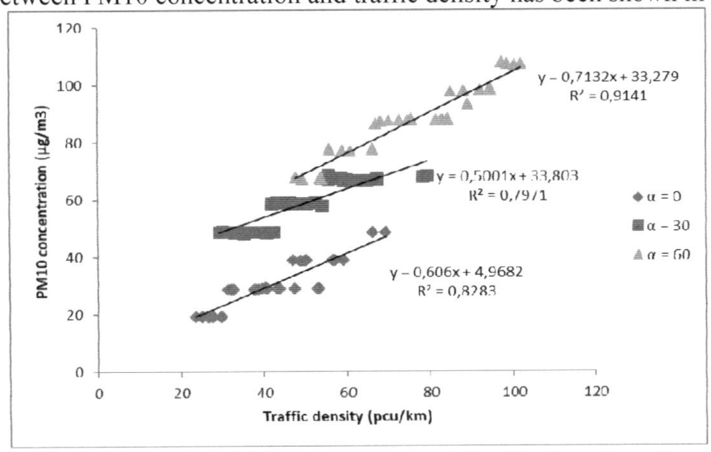

Figure 9. Relationship between PM10 Concentration and Traffic Density with α = 0°, 30°, and 60°

Based on Figure 9, the R value on α = 0°, 30° and 60° was 0.910, 0.892 and 0.956 respectively. The correlation between the concentration of PM10 and the traffic density is very strong on α = 0° and 30°. Meanwhile, the correlation is strong on α = 60°. The coefficient of determination (R^2) on α = 0°, 30° and 60° was 0.8283, 0.7971 and 0.9141 respectively, which means that 82.8%, 79.7% and 79.71% of the PM10 concentration was influenced by traffic density. The above results show that the higher the density of traffic, the higher the PM10 concentration.

The relationship between PM10 concentration and traffic volume has been shown in Figure 10.

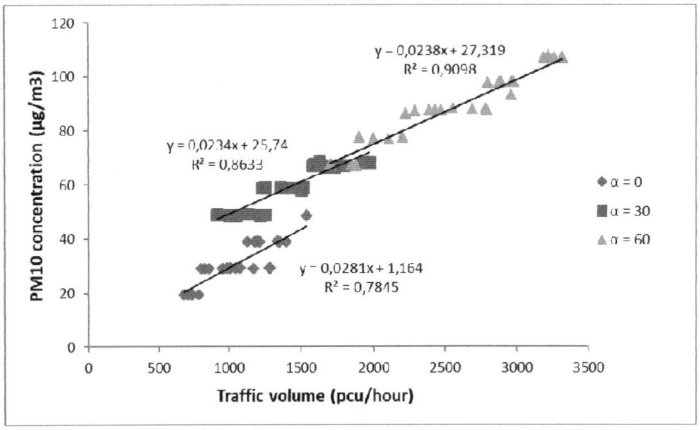

Figure 10. Relationship between PM10 Concentration and Traffic Volume with $\alpha = 0^o$, 30°, and 60°

Based on Figure 10, the R value for $\alpha = 0^o$, 30° and 60° is 0.785, 0.929 and 0.954 respectively. The correlation between the concentration of PM10 and the traffic density is very strong at $\alpha = 0^o$ and 30°. Meanwhile, the correlation is strong at $\alpha = 60^o$. The coefficient of determination (R^2) on $\alpha = 0^o$ (A. Yani Road), $\alpha = 30^o$ (Andalas Road) and $\alpha = 60^o$ (Prof. Dr. Hamka Road) is 0.7845, 0.8633 and 0.9098 respectively. This means that the 78.45%, 86.33% and 90.98% PM10 concentrations were influenced by the traffic volume. The above results show that traffic volume caused the increase in pollutant concentration [8].

Dispersion of PM10 Concentration based on the Angle of Dominant Wind Direction
The relationship between PM10 concentration and the angle of dominant wind direction is as shown in Figure 11.

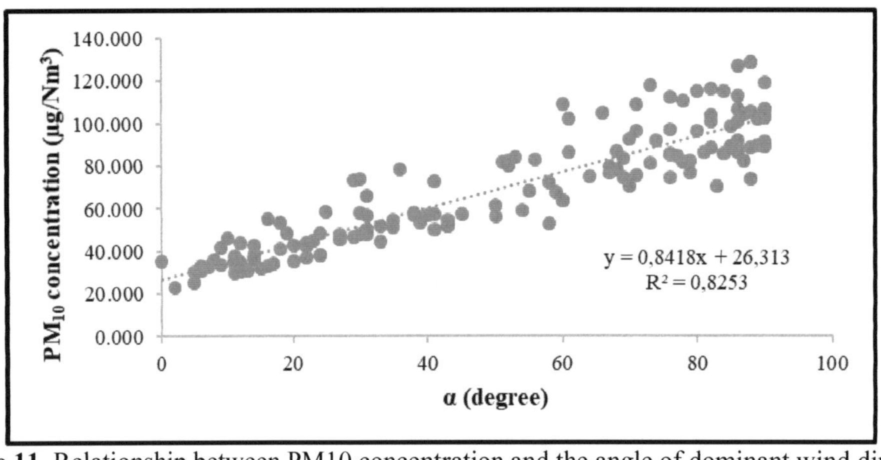

Figure 11. Relationship between PM10 concentration and the angle of dominant wind direction

Based on Figure 11, the determination value (R^2) obtained was 82.5% with a coefficient of correlation (R) of 0.908. This shows that the correlation is very strong. The positive variable x in the equation on the chart indicates that the relationship of PM10 concentration with the angle of dominant wind direction to the road is directly proportional. If there is an increase of α every 10 degrees, the PM10 concentration will increase for 8,418 μg/Nm3. The higher the angle of arrival of the wind on the road, the higher the concentration of PM10. This means that the pollutant concentration can be affected by the wind angle direction to the road [9].

PM10 Concentrations Models

The PM10 concentration model was created using variations in the concentration related to traffic characteristics, wind speed and the dominant wind angle direction to the road. The model produced:

- *Model A*
  Model A shows the PM10 concentration with the influence of traffic volume, vehicle speed, wind speed and the dominant wind direction angle on the road.
- *Model B*
  Model B shows the PM10 concentration with the influence of traffic density, wind speed and the dominant wind direction angle on the road.

The regression models produced can be seen in Table 1.

Table 1. PM10 Concentration Models

Model	Formula	R
A	$C_{PM} = 0{,}011V_T - 0{,}579S_T - 4{,}370S_W + 0{,}695\alpha + 35{,}286$	0,971
B	$C_{PM} = 0{,}368D_T - 2{,}52S_W + 0{,}712\alpha + 16{,}029$	0,970

Where:

C_{PM} = PM10 concentration (μg/Nm^3)
S_T = traffic speed (km/hour)
V_T = traffic volume (pcu/hour)
D_T = traffic density (pcu/km)
S_W = wind speed (m/s)
α = wind angle direction to the road

Table 1 shows Model A and Model B with a coefficient correlation of (R) 0.971 and 0.970. This correlation shows that there is a very strong relationship between the PM10 concentration and traffic volume, traffic density, traffic speed, wind speed and the wind angle direction to the road.

Model Validations

The validation of the PM10 model using the Two-Varians Test and the Pearson Moment Formula can be seen in Tables 2 and 3.

Table 2. Validation using the Two-Varians Test

Parameter	Critical Value	RU_f Density	RU_f Volume
PM_{10}	2,102	1,449	1,510

Table 2 shows the two-variance test between the PM10 sampling model and the PM10 sampling for validation. The value of the test ratio between the sampling concentration and the validation concentration is 1.449. This value indicates that RUf is smaller than the value of the critical point (RUf<2.102). This means that there is no significant difference in the yield variability of the two concentrations. The results of the two-variance test between the sampling concentration and the concentration of the results of the volume equation show that the value of the test ratio (RUf = 1.510) is smaller than the value of the critical point (RUf<2.102). This also shows that there is no significant difference between concentrations of PM10 sampling model and concentrations of PM10 for validation.

Table 3. Validation using the Pearson Moment Formula

Parameter		R Value	R Tabel Value	
PM10	Model based on density	0,965	0,361	valid
	Model based on volume	0,961	0,361	Valid

Table 3 shows the validation uses of the Pearson product moment formula. The PM10 parameter shows that the calculated R value uses the Pearson product moment formula of 0.965 for the concentration results of the density equation and 0.961 for the concentration resulting from the volume equation. Both R values indicate that the value is greater than the table R value which is equal to 0.361. The 2 tested equations are valid and can be used to predict the PM10 concentration.

PM10 Concentration Map Based on the PM10 Model

Figure 12 is a mapping of the PM10 concentrations obtained from the validated PM10 model. The lowest to highest concentrations in a row are denoted by navy blue, pink, hard blue, purple and maroon colors.

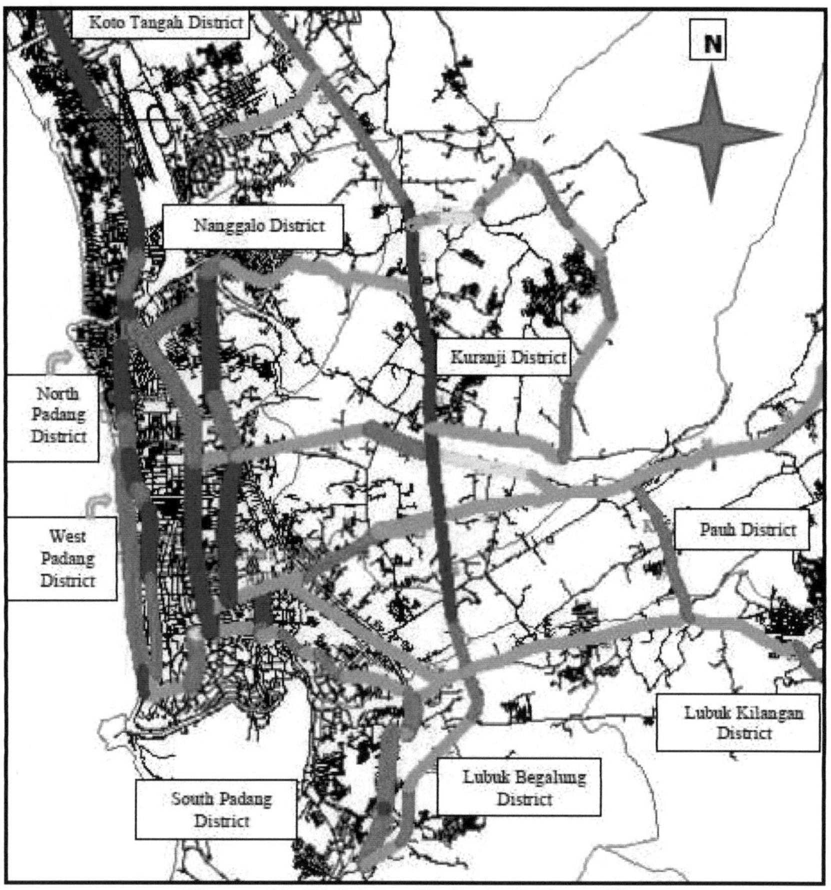

Figure 12. PM10 Predictionin Padang City

Note :

Navy Blue	: 0 – 30 µg/Nm^3	
Pink	: 31 – 60µg/Nm^3	
Hard Blue	: 61 – 90 µg/Nm^3	
Purple	: 91 – 120 µg/Nm^3	
Marron	: 121 – 150 µg/Nm^3	

The highest PM10 concentration was 128,395 µg/Nm^{3}with traffic volume 6.832 smp/hour, traffic speed 28,87 km/hour, wind speed 0,36 m/s and α 88^0. The lowest PM10 was 23,074 µg/Nm^{3}with traffic volume 677 smp/hour, traffic speed 31,06 km/hour, wind speed 0,84 m/s and α 2^0.

4. Conclusion

The validation test conducted using the Two Variance Test obtained a critical point value of 2.102. The Test Ratio Value (RUf) had a value that was smaller than the critical point value. This shows that there is no significant difference between the concentration of the model and the concentration of the validation data. The small value RUf value from the critical point was declared to be valid and this can be used to find the PM10 pollutant concentrations. The validation test conducted using the Pearson Moment Formula produced an R table value of 0.361. The validation of the PM10 model shows that the calculated R value is greater than that of the R table. This shows that all of the equations are valid to use.

This research is supported by the Engineering Faculty of Andalas University under contract No. 026/UN.16.09.D/PL/2019.

References

[1] A. Suleiman, M. R. Tight, And A. D. Quinn, "Assessment And Prediction Of The Impact Of Road Transport On Ambient Concentrations Of Particulate Matter Pm10," *Transp. Res. Part D Transp. Environ.*, Vol. 49, Pp. 301–312, 2016.

[2] V. S. Bachtiar And M. Ammar, "Diurnal Variation Of Tropospheric Ozone (O3) Anditsprecursors (Co And No 2) Due To Transportationactivity In The Roadside Areas In Padang City , Indonesia," Vol. 12, No. 24, Pp. 7012–7023, 2017.

[3] F. Al Jallad, E. Al Katheeri, And M. Al Omar, "Concentrations Of Particulate Matter And Their Relationships With Meteorological Variables," *Sustain. Environ. Res.*, 2013.

[4] V. S. Bachtiar, Purnawan, R. Afrianita, And R. Anugerah, "Validation Of Co Dispersion Model Due To The Road Position On The Dominant Wind Direction On Transport Sector," In *Matec Web Of Conferences*, 2018.

[5] M. Neuhäuser, "Two-Sample Tests When Variances Are Unequal," *Animal Behaviour*. 2002.

[6] J. W. Gooch, "Pearson's Product-Moment Correlation Coefficient," In *Encyclopedic Dictionary Of Polymers*, 2011.

[7] B. Sharratt And H. Pi, "Field And Laboratory Comparison Of Pm10 Instruments In High Winds," *Aeolian Res.*, 2018.

[8] V. S. Bachtiar, Purnawan, R. Afrianita, And S. H. Ritonga, "Modelling Of No2 Dispersion Based On Receptor Position Due To Transport Sector In Padang City, Indonesia," *J. Environ. Sci. Technol.*, 2017.

[9] V. S. Bachtiar, P. Purnawan, R. Afrianita, And N. Dahlia, "Effect Of Wind Angle Direction On Carbon Monoxide (Co) Concentration Dispersion On Traffic Flow In Padang City," In *Iop Conference Series: Materials Science And Engineering*, 2018.

IOP Conf. Series: Materials Science and Engineering **846** (2020) 012059 doi:10.1088/1757-899X/846/1/012059

Removal of nitrate from groundwater by column using pumice as adsorbent as an effort for water resources conservation

Denny Helard*, Shinta Indah, Dhea Pratiwi Kiflia

Department of Environmental Engineering, Faculty of Engineering, Universitas Andalas, Kampus Unand Limau Manis, Padang 25163

*Corresponden Email: dennyhelard@eng.unand.ac.id

Abstract. As an effort for water resources conservation, a fixed-bed column was applied to remove nitrate from groundwater using pumice as adsorbent. The column performances were evaluated by varying the influent concentration (50 mg/L dan 80 mg/L) and influent flow rate (2 - 3 gpm/ft^2 equal to 43–65 mL/min) with 85 cm of adsorbent bed depth. The results indicated that the increase in influent concentration increased the amount of nitrate and cause the earlier breakthrough time. Furthermore, the increase in influent flow rate caused the column exhaustion time to occur earlier resulted in the shortened lifespan of the column. The column system with a bed depth of 85 cm and flow rate of 2 gpm/ft^2 (43 mL/min) for 80 mg/L of influent concentration showed the best nitrate uptake performance in this study with a total removal of 31.42% and adsorption capacity of 1.394 mg/g. The results revealed that the pumice in column has potential for nitrate removal from groundwater.

1. Introduction

In soil, fertilizers containing inorganic nitrogen is first decomposed to give ammonia, which is then oxidized to nitrite and nitrate. The nitrate is taken up by plants during their growth and used in the synthesis of organic nitrogenous compounds. Excessive application of nitrogenous fertilizer not only pollutes surface water, it also pollutes groundwater [1]–[3]. Beside as a consequence of agricultural activity, nitrate also can reach groundwater from wastewater treatment and oxidation of nitrogenous waste products in human and animal excreta, including septic tanks. The natural nitrate concentration in groundwater under aerobic conditions is a few milligrams per liter, however, as a result of agricultural activities, the nitrate concentration can easily reach several hundred milligrams per liter [4]. For example, in an agricultural area of India, concentrations of up to 1500 mg/L were found in groundwater [5].

The maximum permissible level 50 mg/L of nitrate concentration in drinking water was recommended by WHO [4]. When drinking water loaded with higher level of nitrogen in the form of nitrate, it causes potential health problems among people including blue-baby syndrome or methemoglobinemia in infants, cyanosis, shortening of breath, enlargement of the thyroid gland, hyper tension, very fast heartbeat, frequent urination etc [6].

For removal of nitrate from water, various purification technology were applied [7] such as denitrification, electro-dialysis, reverse osmosis, ion exchange, adsorption etc. [8], [9]. Among them, adsorption technology is the most effective and easy to perform and regenerate the exhausted adsorbent

[10], [11]. In the previous literature, many adsorbents are reported for removal of nitrate from groundwater such as clay [12], wheat straw [13], amino clay [14] and chitosan modified zeolite [15]. Many studies also have been conducted to minimize the operational costs of adsorption process by using natural materials which are abundantly available for alternative adsorbents. Pumice is one of the natural materials which has a low weight, porous structure (up to 85%) and a high specific surface area. Recently, pumice also has been utilized as adsorbent for pollutant removal from water and wastewater [11].

In the present study, the performance of pumice through a fixed-bed column for nitrate removal from aqueous solution as simulated groundwater was evaluated. The natural pumice was collected from Sungai Pasak, West Sumatera, Indonesia as byproduct of sand mining process in that location and has potential for removal of iron and manganese from water, as previous investigations [11], [16]. The application of column adsorption is practical and economic as the operation is performed continuously and the process is controllable [17], [18]. The effect of parameters including influent concentration and flow rate on the shape of the breakthrough curve was studied. The total removal percentage of nitrate ions and the adsorption capacity of adsorbent was calculated to examine the column performance.

2. Methodology

Preparation of adsobent and aqueous solution

Pumice were washed with distilled water several times, dried out at room temperature, crushed and sieved to obtain the desired particle size fractions. To observe the surface morphology and the oxide content of pumice, a scanning electron microscopy (SEM, model S-3400N, Hitachi, Japan) and eEnergy dispersive x-ray (EDX) spectroscopy were employed. Nitrate aqueous solutions were prepared by dissolving potassium nitrate (KNO_3) in tap water.

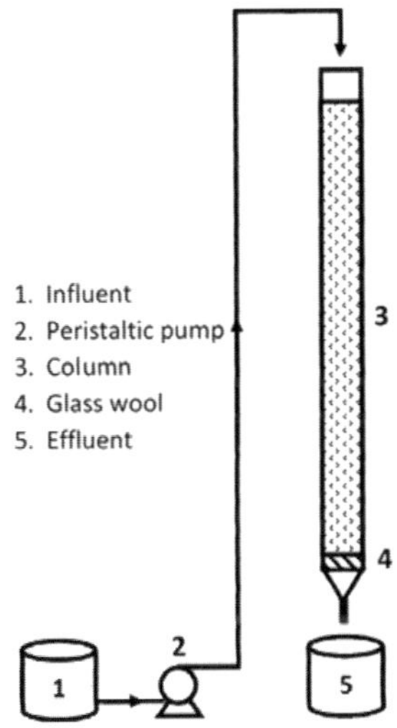

Figure 1. A schematic of fixed-bed column of pumice for nitrate.

Fixed-bed column

A glass column with a length of 130 cm and an inner diameter of 2.6 cm was used for the column experiments. A glass wool were placed at the bottom of the column to avoid loss of adsorbent during the adsorption process. A known weight of adsorbent was packed into the column to obtain 85 cm of the bed depth. The particle size of pumice was 0.5 – 1 mm. To withdraw the trapped air between the particles, before the experiment started, the adsorbent packed in the column was wetted with deionised water in downward flow direction. The solution sample were continuously fed downward into the column by a peristaltic pump (Kamoer, China). The experiment was conducted at room temperature 27

± 2°C. The effects of influent concentration (50 mg/L and 80 mg/L) and flow rates (2 gpm/ft^2 and 3 gpm/ft^2 equal to 43 and 65 mL/min) were examined. Every 60 minutes samples were collected from the bottom of the column and were measured for the remaining nitrate by Spectrometer UV-Vis (Shimadzu UV-2600). The column performance was investigated by calculating the breakthrough time and adsorption capacity. The schematic of fixed-bed column system is shown in **Figure 1**.

Mathematical description of fixed bed column studies

The performances of column were evaluated through the breakthrough curve of the continuous fixed bed system. The breakthrough curve was expressed by C_t/C_0, in which C_t and C_0 represent the effluent and influent concentrations, respectively.The curve was described as C_t/C_0 against the contact time. The adsorbed ion concentrations in the column were confirmed by a plot of the adsorbed concentration (C_{ad}= inlet concentration (C_0) - outlet concentration (C_t)) or normalized concentration assigned as the ratio of effluent concentration to influent concentration (C_t/C_0) as a function of time or volume of effluent (V_{eff}), as presented in equation (1) [19].

$$V_{eff}(mL) = Qt_{total} \qquad (1)$$

The Q and t_{total} describe the volumetric flowrate (mL/min) and total flow time (min), respectively. The total adsorbed ion (q_{total}) by the column can be calculated by integrating the plot of adsorbed concentration (C_{ad}) versus the flow time (t). The area (A) under this integrated plot is substituted in equation (2) to determine q_{total}.

$$q_{total}(mg) = \frac{QA}{1000} = \frac{Q}{1000}\int_{t=0}^{t=t_{total}} C_{ad}\, dt \qquad (2)$$

The total amount of ions delivered to the column system (m_{total}) is gained from equation (3).

$$m_{total}(mg) = \frac{C_0 Q t_{total}}{1000} \qquad (3)$$

The column performance can be examined by the total removal percentage of ions from the ratio of total adsorbed ions in the column to the total amount of ions delivered to the column, as shown in equation (4).

$$Total\ removal\ of\ metal\ ions\ (\%) = \frac{q_{total}}{m_{total}} \times 100 \qquad (4)$$

The equilibrium adsorption was also calculated from the column data to obtain the capacity of adsorbent required for ions removal. Equation (5) declares the equilibrium ion uptake (q_{eq}), also known as the column maximum capacity.

$$q_{eq}(mg/g) = \frac{q_{total}}{X} \qquad (5)$$

where, X is the unit mass of adsorbent packed in the column..

3. Result and Discussion

3.1 Physical characteristics of natural pumice

As determined by EDX, Si, Al and Fe are the major elements in natural pumice from Sungai Pasak, as shown in Table 1. Other elements, except K, Ca, Na and Mg were present in relatively smaller amounts (less than 3%). The elemental compositions of the pumice also indicate the absence of hazardous or carcinogenic substances; thus the pumice is considered appropriate as adsorbent to treat polluted water.

The SEM image showed the surface morphology of natural pumice from Sungai Pasak, West Sumatra was displayed in **Figure 2**. The image indicated that the pumice had a smooth surface, cellular, irregular texture and highly porous with larger cavities, which serves suitable sites for adsorption.

Table 1 Chemical components of pumice from Sungai Pasak, West Sumatra, Indonesia

Oxide	**% weight**
MgO	0,876
Al_2O_3	13,913
SiO_2	76,586
P_2O_5	0,822
K_2O	3,604
CaO	2,11
TiO_2	0,197
MnO	0,044
Fe_2O_3	1,485
ZnO	0,006

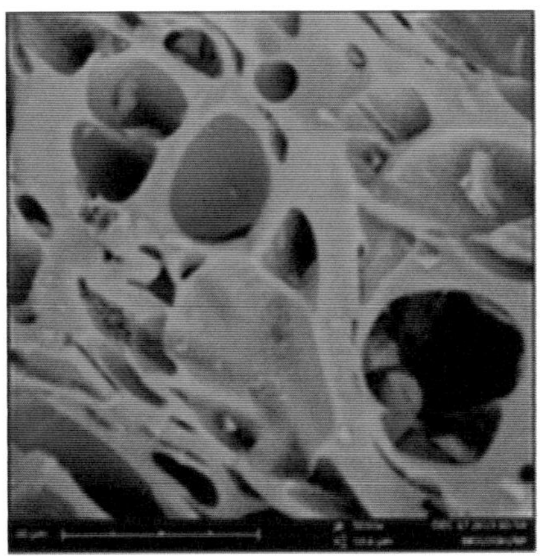

Figure 2. SEM image of pumice from Sungai Pasak, West Sumatra, Indonesia

3.2 Column studies

3.2.1 Effect of influent concentration

In order to determine the effect of the influent concentration on the breakthrough curves, samples of nitrate solution with initial concentrations of 50 mg/L and 80 mg/L were passed through the column of pumice at a bed depth of 85 cm and flow rates of 2 gpm/ft^2 and 3 gpm/ft^2. The effect of influent concentration on the breakthrough curve was evaluated plotting the effluent/influent concentration ratio versus elution time (**Figure 3**). From the **Figure 3**, at all of various flow rates, it was observed that the higher nitrate concentration provided the earlier breakthrough time. This earlier breakthrough time at higher concentration could be explained based on the improved diffusion of nitrate compounds into the pores of pumice. Abdolali et al. [20] found that at higher influent concentration due to existence of greater concentration gradient lesser resistance is encountered during mass transfer of solute particles.. **Figure 1** shows that steepness of the breakthrough curve is dependent upon influent concentration i.e. greater the concentration steeper is the curve.

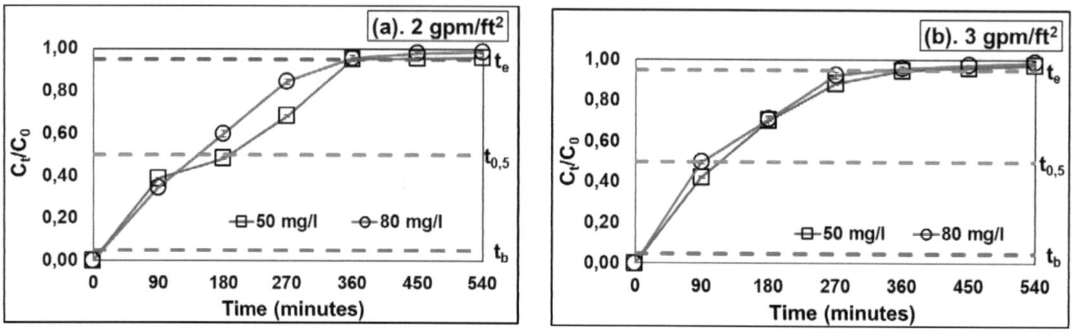

Figure 3. Effect of various influent concentrations on the breakthrough curve of nitrate adsorption onto pumice at various flow rates: (a). 2 gpm/ft^2 dan (b) 3 gpm/ft^2 (bed depth = 85 cm).

3.2.2 Effect of flow rate

The effect of flow rate in the fixed bed of pumice was investigated from 2 gpm/ft^{2}and 3 gpm/ft^2 with 85 cm of bed depth. The breakthrough curve, C_t/C_0 versus time with various flow rates at various influent concentration is shown in **Figure 4**. The results indicate that a decrease in flow rate at all various influent concentration increased the breakthrough time (t_b). Form **Figure 4**, it was shown that the shape of the breakthrough curve is saturated earlier at higher flow rates because the front of the adsorption zone quickly reached the top of column. In addition, the increased flow rate resulted the shorter contact time between the adsorbate and adsorbent. It may cause uncompleted adsorption and led to steep breakthrough results at the beginning of the operation [6], [16]. The results indicated that lower flow rate or longer contact time would be required for nitrate adsorption in the column of pumice.

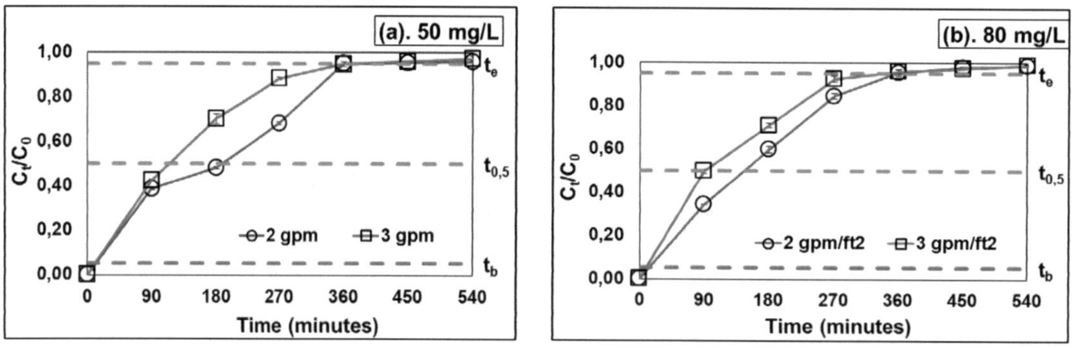

Figure 4. Effect of various flow rates on the breakthrough curve of nitrate adsorption onto pumice at various influents: (a). 50 mg/L dan (b) 80 mg/L (bed depth = 85 cm).

3.3 Total removal and adsorption capacities

To obtain information on the influence parameters, the column data were calculated into the mathematical theories of the column system. The equilibrium uptake of nitrate ions were observed to increase with the increase in influent concentrations (**Table 2**). The unadsorbed nitrate concentrations at the equilibrium also showed increased along with the higher influent concentration. This also indicated that the bed saturates faster when higher amounts of nitrate ions were introduced to the adsorbent column. **Table 2** shows that exhaustion time of 80 mg/L of influent concentration is faster than 50 mg/L. Direct proportionality of adsorption capacity with concentration gradient of sorbate is also confirmed by experimental results of Dong and Lin [21].

Table 2 Removal percentages and adsorption capacities of nitrtae by column of pumice at various concentrations of influent and flow rates

Bed Depth (cm)	Concentration of influent (mg/L)	Flow rate (gpm/ft^2)	Removal percentages (%)	Adsorption capacities (mg/g)	Exhaustion time (t_e) (minutes)
85	50	2	30.92	1.073	450
		3	21.75	1.178	450
	80	2	31.42	1.394	360
		3	22.70	1.490	360

4. Conclusion

For the conservation of groundwater as one of water resources, a fixed-bed column using pumice as adsorbent was investigated for removal of nitrate from aqueous solution as simulated groundwater. The effect of influent concentrations and flow rates on the adsorption process were examined. The results indicate that the nitrate adsorption through fixed-bed columns is dependent on the influent concentration and flow rate. The higher nitrate concentration provided the earlier breakthrough time and the increase in flow rate led to accelerate the exhaustion of the column. The total removal percentages of nitrate were increased with the increase in the influent concentration and decrease of the flow rate. For the optimum performance of the column operation system, suitable parameters are necessary to be investigated. The obtained results suggested that pumicein the column system has potential for the removal of nitrate from groundwater or other polluted waters.

5. Acknowledgement

The authors would like to thank Universitas Andalas, Indonesia (Grand No. T/12/UN.16.17/PP.IS-KRP2GB/LPPM/2019) for supporting this work

References

[1] EPA, "Criteria and Standards Division Office of Drinking Water Environmental Protection," no. 2, 1990.

[2] S. Ferrario-Méry, C. Meyer, and M. Hodges, "Chloroplast nitrite uptake is enhanced in Arabidopsis PII mutants," *FEBS Lett.*, vol. 582, no. 7, pp. 1061–1066, 2008.

[3] Y. Gao, G. Yu, C. Luo, and P. Zhou, "Groundwater nitrogen pollution and assessment of its health risks: A case study of a typical village in rural-urban continuum, china," *PLoS One*, vol. 7, no. 4, pp. 1–8, 2012.

[4] N. Graham, "Guidelines for Drinking-Water Quality, 4th Ed," *Who*, vol. 1, no. 2, p. 183, 1999.

[5] G. Jacks and V. P. Sharma, "Nitrogen circulation and nitrate in groundwater in an agricultural catchment in Southern India," *Environ. Geol.*, vol. 5, no. 2, pp. 61–64, 1983.

[6] L. Fewtrell, "Drinking-water nitrate, methemoglobinemia, and global burden of disease: A discussion," *Environ. Health Perspect.*, vol. 112, no. 14, pp. 1371–1374, 2004.

[7] M. Islam, P. Chandra Mishra, and R. Patel, "Physicochemical characterization of hydroxyapatite and its application towards removal of nitrate from water," *J. Environ. Manage.*, vol. 91, no. 9, pp. 1883–1891, 2010.

[8] A. Abou-Shady, C. Peng, J. Bi, H. Xu, and J. Almeria O, "Recovery of Pb (II) and removal of NO 3- from aqueous solutions using integrated electrodialysis, electrolysis, and adsorption process," *Desalination*, vol. 286, pp. 304–315, 2012.

[9] A. Bhatnagar and M. Sillanpää, "A review of emerging adsorbents for nitrate removal from water," *Chem. Eng. J.*, vol. 168, no. 2, pp. 493–504, 2011.

[10] P. Loganathan, S. Vigneswaran, and J. Kandasamy, "Enhanced removal of nitrate from water using surface modification of adsorbents - A review," *J. Environ. Manage.*, vol.

131, pp. 363–374, 2013.

[11] S. Indah, D. Helard, and A. Binuwara, "Studies on desorption and regeneration of natural pumice for iron removal from aqueous solution," *Water Sci. Technol.*, vol. 2017, no. 2, 2017.

[12] M. El Ouardi, S. Qourzal, S. Alahiane, A. Assabbane, and J. Douch, "Effective Removal of Nitrates Ions from Aqueous Solution Using New Clay as Potential Low-Cost Adsorbent," *J. Encapsulation Adsorpt. Sci.*, vol. 05, no. 04, pp. 178–190, 2015.

[13] X. Xu *et al.*, "Nitrate removal from aqueous solution by Arundo donax L. reed based anion exchange resin," *J. Hazard. Mater.*, vol. 203–204, no. 3, pp. 86–92, 2012.

[14] Y. C. Lee, E. J. Kim, H. J. Shin, M. Choi, and J. W. Yang, "Removal of F -, NO 3-, and PO 43- ions from aqueous solution by aminoclays," *J. Ind. Eng. Chem.*, vol. 18, no. 3, pp. 871–875, 2012.

[15] S. Chatterjee, D. S. Lee, M. W. Lee, and S. H. Woo, "Nitrate removal from aqueous solutions by cross-linked chitosan beads conditioned with sodium bisulfate," *J. Hazard. Mater.*, vol. 166, no. 1, pp. 508–513, 2009.

[16] S. Indah, D. Helard, T. Edwin, and R. Pratiwi, "Utilization of pumice from Sungai Pasak, West Sumatera, Indonesia as low-cost adsorbent in removal of manganese from aqueous solution," in *AIP Conference Proceedings*, 2017, vol. 1823.

[17] V. Vinodhini and N. Das, "Packed bed column studies on Cr (VI) removal from tannery wastewater by neem sawdust," *Desalination*, vol. 264, no. 1–2, pp. 9–14, 2010.

[18] A. D. Dorado, X. Gamisans, C. Valderrama, M. Solé, and C. Lao, "Cr(III) removal from aqueous solutions: A straightforward model approaching of the adsorption in a fixed-bed column," *J. Environ. Sci. Heal. - Part A Toxic/Hazardous Subst. Environ. Eng.*, vol. 49, no. 2, pp. 179–186, 2014.

[19] Z. Aksu, F. Gönen, and Z. Demircan, "Biosorption of chromium(VI) ions by Mowital®B30H resin immobilized activated sludge in a packed bed: Comparison with granular activated carbon," *Process Biochem.*, vol. 38, no. 2, pp. 175–186, 2002.

[20] A. Abdolali *et al.*, "Application of a breakthrough biosorbent for removing heavy metals from synthetic and real wastewaters in a lab-scale continuous fixed-bed column," *Bioresour. Technol.*, vol. 229, pp. 78–87, 2017.

[21] Y. Dong and H. Lin, "Competitive adsorption of Pb(II) and Zn(II) from aqueous solution by modified beer lees in a fixed bed column," *Process Saf. Environ. Prot.*, vol. 111, no. Ii, pp. 263–269, 2017.

IOP Conf. Series: Materials Science and Engineering **846** (2020) 012060 doi:10.1088/1757-899X/846/1/012060

The Performance between Two Long-span Roof Trusses of Cold Formed Steel with Back to Back Channel Section

Ridho Aidil Fitrah[1], **Nofriyandi R**[2]

[1]Civil Engineering Department, [Universitas Dharma Andalas]*
[2]Mechanical Engineering Department, [Politeknik Negeri Padang]

*ridho.af@unidha.ac.id

Abstract. Cold formed steel (CFS) is mostly used for short span structures in Indonesia because its advantages in cost and time efficiency. However, there is still limited number of long-span construction using cold formed steel because its buckling issue. Throughout the time, there are many researches that revealed the buckling issues can be prevented; one of them is combining their section. Then, CFS structure should give full structural performance. This research aims to analyze the performance of long-span roof truss, 24 m, using CFS back to back channel section. Two types of roof trusses, model 1 and model 2, are compared to get better configuration for long-span roof truss. Analysis is conducted by 2D model and analytically calculated based on SNI 7971:2013. The results are the maximum axial tensile force and axial compressive force, and also deflection. They will be compared to CFS back to back channel section capacity. This analysis obtained the stress ratio and deflection of both configurations. Based on analysis results, CFS back to back channel section safely withstands resisting axial forces by giving compressive ratio number around 0,54-0,72. Then, model 2 has given better structural performance, stress ratio and deflection, than model 1.

1. Introduction

In this recent time, demanding of cold-formed steel (CFS) increases for roof structure construction, whether for simple residential houses or medium-rise buildings. There are many advantages can be obtained using cold-formed steel structure, such as time and cost efficiency as well as getting strong structure through its properties [1,9]. Compared to hot rolled steel, the construction of cold-formed steel causes the mass of the building can be lighter. It means the effect of the earthquake also can be minimized [5,6]. However, cold-formed steel (CFS) structure still has the stability issue where buckling occurs due to the ratio of width-thickness [9,12]. Local buckling, one of major buckling, commonly occurs in the elastic zone when the axial-compression subjected to compression members [2,3].

Many kinds of research have been conducted to find the proper methods to prevent buckling in cold-formed steel structure; one of them is combining a single section into double sections such as the back-to-back section [7, 8,10]. This method has been studied analytically and experimentally to prove that CFS stiffness and capacity can be increased to 50%. Furthermore, the ductility of this section also increased to 70%. According to these researches, it can be determined that the cold-formed steel structure can be enhanced for not the only short-span structure but also the long-span structure, specifically for roof structure. This type of structure should be applied for plant, factory storage, sports hall, market, and hangar [7]. However, the study for long-span truss of cold-formed steel is slightly conducted to obtain appropriate analysis and design.

This paper aims to analyze the optimum configuration for long-span roof truss using cold-formed steel back-to-back section. The main configuration that is analytically studied in this paper is Howe Slope Parallel Chord where it can be modified based on force ratio concept and structure deflection which are obtained from the two-dimensional analysis. Fitrah [7] has concluded that howe slope parallel chord must be continuously developed to get better configuration. The comparison of these results will determine which one of the configuration that might be good for long-span roof truss.

2. Methodology

The two types of howe slope parallel chord roof truss 24 m long-span were analyzed to obtain comparison of each structural performance such as deflection and stress ratio. This span was taken as common span for long-span constructions such as sport hall, plant, hangar, etc. They were constructed using the same CFS section of back to back channel. Both trusses, named model 1 and model 2, have the same span and height, as shown in Figure 1(a) and 1(b). Model 2 has modification that 6 m span of bottom chord is straight.

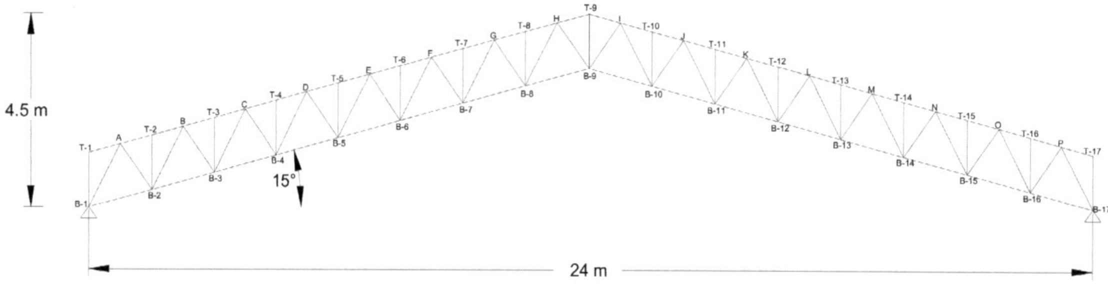

(a)

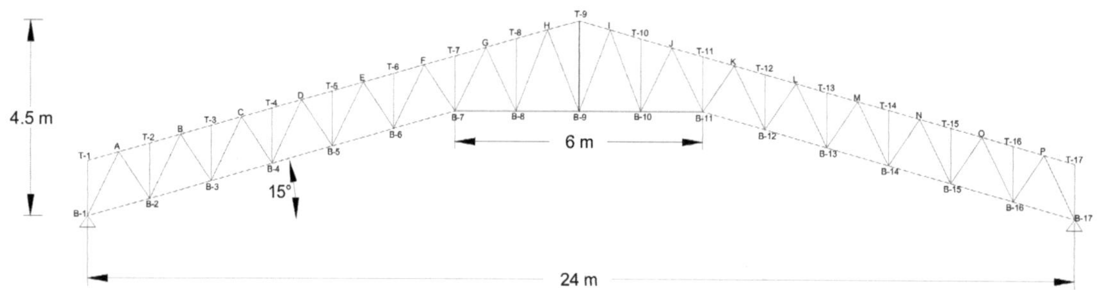

(b)

Figure 1. (a) Model 1 (b) Model 2

Model 1 and Model 2 were constructed by using CFS back to back channel with double section of 150.75.0.8 for top and bottom chords and single section of 75.35.0.75 for web members, as shown in Figure 2. The length of each chord is specifically illustrated in Table 1.

Analysis of both trusses was assumed that they are connected to the supporting column using a few bolts and it does not permit to any lateral movement. It means that both trusses used pin supports and most of truss members are subjected to compressive load. However, the connection designs for support and truss members are not presented in this paper.

Material properties of CFS were obtained from previous research conducted by Fitrah [7]. The average yield stress and average elastic modulus of CFS are 495 MPa and 200.000 MPa respectively.

Table 1. The length of tensile and compressive member

Model	Member Code	Length (m)
1	T-8 – T-9 ; B-8– B-9	1.6
	B-8– H	1.69
2	T-8 – T-9	1.6
	B-1– A	1.69
	B-7– G	1.69
	B-8– H	2.07
	B-8 – B-9	1.54

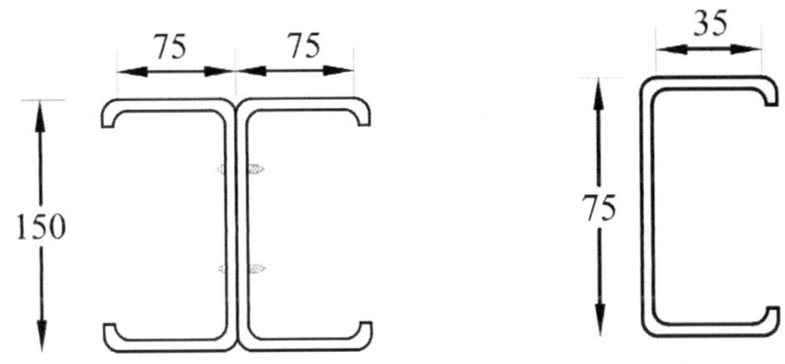

Figure 2. (a) The cross section for top chord and bottom chord (b) The cross section for web

The analysis was conducted by modeled both trusses using SAP2000. Truss connection in each node was assumed as the rigid connection. According to the Indonesian Code, 4 type loads were determined such as the dead load (D), live load (L), wind load (W), and rain load (R). However, the earthquake load (E) was simply calculated by distributing 10% of the dead load for each node. These loads were calculated to obtain the load distribution between two adjacent trusses where 1.5 m distance was used in this design. Furthermore, the dead loads (D) were generally determined based on material type and CFS section used in truss design. CFS purlin (0.01 kN/m) and metal roof (0.18 kN/m2) were also defined as point load in each node that is subjected to truss.

Load combinations subjected to each model were determined based on SNI 1727-2013 verse 2.3.2 [11] below:

1,4 D (1)
1,2 D+ 1,6 L + 0,5 W (2)
1,2 D + 1,6 R + 0,5 W (3)
1,2 D + 1,6 R + 1,0 L (4)
1,2 D + 1,0 W + 1,0 L +0,5 R (5)
0,9 D + 1,0 W (6)
0,9 D + 1,0 E (7)

3. Result and Discussion

The axial tensile capacity of back to back channel is calculated based on SNI 7971:2013 verse 3.2.1 (Equation 8,9) where N_t is the nominal tensile capacity, $\emptyset_t$ is the reduction factor for tension element where 0,9 is taken for the calculation. k_t is the correction factor for the force distribution where 0,85 and 1,0 are taken for the calculation because the section is single and double channel [4]. A_g is the gross cross-sectional area, A_n is the net cross-sectional area, and f_u is the tensile strength of CFS based on aforementioned data.

$$N_t = \emptyset_t A_g f_y \tag{8}$$

$$N_t = \emptyset_t 0.85 k_t A_n f_u \tag{9}$$

The axial compressive capacity calculation is also based on SNI 7971:2013 verse 3.4.1 (Equation 10,11) where Nc is the nominal compressive capacity, $\emptyset_c$ is the reduction factor for compressive element where 0,85 is taken for the calculation. A_e is the effective cross-sectional area when f_n occurred. f_n is critical strength based on non-dimensional buckling strength [4].

$$N_c = \emptyset_c A_e f_y \tag{10}$$

$$N_c = \emptyset_c A_e f_n \tag{11}$$

Based on the calculation results, the axial tensile and compressive for section double back to back channel and single channel can be seen in Table 2. Based on analysis, all top and bottom chord of both model occurred the compressive force. Then, the capacity is presented for top and bottom chord are compressive capacity, N_c, only. Meanwhile, the tensile and compressive force occurred in web.

Top chord (T-8 – T-9) of model 1 and model 2 has the same section and length. Then, the compressive capacity of both also has same result. It can be seen that for straight chord in bottom chord (B-8– B-9) of model 2, the compressive capacity is slightly higher than model 1.

Table 2. The axial tensile and compressive capacity for member

Model	Member	N_c (kN)	Member	N_t (kN)	N_c (kN)
1	T-8 – T-9	39,76	B-8 - H	20,7	17,75
	B-8– B-9	39,76			
2	T-8 – T-9	39,76	B-8 - H	22,3	16,51
	B-8– B-9	41,33	B-7– G	20,7	17,75

According to structural analysis result, the maximum axial compressive for model 1 is 28,7 kN and for model 2 is 25,43 kN. It is occurred at each bottom chord (T-8 – T-9). Axial tensile and compressive are occurred in their webs. Table 3 showed the comparison between axial tensile and compressive force that is occurred for each truss and axial tensile and compressive capacity of CFS section. Truss performance is determined by compressive and tensile ratio, and structure deflection. The compressive ratio is obtained by dividing the required capacity (N*) and axial compressive capacity (N_c). Then, the tensile ratio is obtained by dividing the required capacity (N*) and axial tensile capacity (N_t).

Table 3. Ratio and deflection results

Model	Member	Required Capacity, N* (kN)	Axial Compressive Capacity,N_c (kN)	Axial Tensile Capacity,N_c (kN)	Compressive Ratio	Tensile Ratio	Deflection (mm)
1	T-8 - T-9	27.23	39,76		0,68		
	B-8 - B-9	28.7	39,76		0,72		12.7
	B-8 - H	13.4		20,7		0,65	
2	T-8 - T-9	21.66	39,76		0,54		
	B-8 - B-9	25.43	41,33		0,62		
	B-8 - H	12.8		22,3		0,57	6.2
	B-7– G	10,1		20,7		0,49	

Based on calculation, Compressive ratios for top and bottom chord in model 1 are 0,68 and 0,72 respectively. In the other hand, model 2 has smaller compressive ratios than model 1, they are 0,54 and 0,62. It means both trusses have well performance and safely withstand for maximum tensile and compressive force by using CFS back to back channel section in the chord. Deflection that occurred in model 1 and model 2 is 12,7 mm and 6.2 mm respectively. It also showed that maximum deflection of both trusses is lower than the allowable maximum deflection for 24 m span, 200 mm. Besides that, CFS back to back channel already proved can increase the stiffness of trusses as well.

4. Conclusions

This research aimed to analytically conduct the comparison of two types of CFS roof trusses for long-span 24 m. conclusions on this research can be listed as follows:

1. Analysis showed that CFS back to back channel section can safely construct for long-span roof truss because it can withstand the maximum forces in the truss whether axial tensile and axial compressive. It also increases the stiffness of trusses.
2. Based on the performance of roof trusses, model 2 has given well performance compared to model 1. It clearly seen that both stress ratio and deflection in model 2 is smaller than model 1.

References

[1] I. S. Irawati, A. Awaludin, and N. P. Sebastian, "The performance of cold-formed steel long-span roof structure combined with laminated timber : cold-formed steel-laminated timber composite," *Proceedia Engineering,* vol. 171, pp. 1242–1249, 2017.

[2] M. Reda, T. Sharaf, A. ElSabbagh, and M. Elghandour, "Behaviour and design for component and system of cold-formed steel roof trusses" *J. Thin-Walled Structures.*, vol. 135, pp. 21–32, 2019.

[3] R.A. Fitrah and H. Herman, "Studi Eksperimental Perilaku Tekan Baja Ringan Dengan Variasi Profil Penampang" *Rang Teknik Journal UMSB.*, vol. 2, no. 1, pp. 127–131, 2019.

[4] Badan Standarisasi Nasional, "Struktur Baja Canai Dingin SNI 7971:2013" *Badan Standarisasi Nasional,* Jakarta, Indonesia, 2013.

[5] S. Selvaraj and M. Madhavan, "Structural design of cold-formed steel face-to-face connected built-up beams using direct strength method" *J. Constructioal Steel Research.*, vol. 160, pp. 613–628, 2017.

[6] J. Ye, S.M Mojtabei, I. Hajirasouliha, "Seismic performance of cold-formed steel bolted moment connections with bolting friction-slip mechanism" *J. Constructioal Steel Research.*, vol. 156, pp. 122–136, 2019.

[7] R.A. Fitrah and H. Herman, "Analisis Kekuatan Elemen Baja Ringan Penampang Kanal Ganda Pada Konstruksi Kuda-Kuda Bentang Panjang dengan Konfigurasi Howe Slope Parallel Chord" *Prosiding 4th Andalas Civil Engineering (ACE) Conference,* pp 637-644, 2017.

[8] S. Haris, A. Prasetyo, R. Thamrin, and H. Herman, "An Experimental Study of Bending Behaviour of Double Channel and Hollow Section of Light Steel Gauge " *I. J. Advanced Science, Engineering, and Information Technology.*, vol. 8, no. 3, pp 882-888, 2018.

[9] W. Wuryanti, "Kapasitas Rangka Atap Baja Canai Dingin Dengan Berbagai Model Geometris Dalam Memikul Beban Statis" *J. Permukiman,* vol. 13, no. 2, pp 90-103, 2018.

[10] J. Ye, S.M Mojtabei, I. Hajirasouliha, P. Shepherd, and K. Pilakoutas "Strength and deflection behaviour of cold-formed steel back-to-back channels " *J. Engineering Structures.*, vol. 177, pp 641-654, 2018.

[11] Badan Standarisasi Nasional, "Beban minimum untuk perancangan bangunan gedung dan struktur lainnya SNI 1727:2013" *Badan Standarisasi Nasional,* Jakarta, Indonesia, 2013.

[12] J. Dawe, J. Wood, "Small-scale test behaviour of cold-formed steel roof trusse" *J. Struct. Eng. ASCE,* vol. 132, pp. 608-615, 2006.

IOP Conf. Series: Materials Science and Engineering **846** (2020) 012061 doi:10.1088/1757-899X/846/1/012061

THE ROLE OF THE CHANGE ORDER ON TIME PERFORMANCE IN CONSTRUCTION PROJECT

Fajar Susilowati[1], Temi Ariati Budi[2]

[1]Civil Engineering, Universitas Tidar, Jl. Kapten Suparman 39 Magelang
[2]Civil Engineering, Politeknik Negeri Jakarta, Jl. G. A. Siwabessy, Campus of UI Depok

Email: fajar.susilowati@untidar.ac.id

Abstract. This research was conducted to determine how the conditions change order and time performance on a construction project. The population in this study were employees of one project in Jakarta with a sample of 30 employees. Analysis of the data used in this research is descriptive analysis. The results obtained showed that the change order is including in the not good category with a percentage value of 49.79%. While the time performance included in the category is also not good with a percentage value of 42.67%. Based on the results of calculations in the regression analysis, the results show that change order has a positive effect on increasing the time of implementation of construction project work (time performance). The most dominant change order indicator influencing time performance is the lack of information when planning a project. Based on the calculation of the Pearson Product-Moment Correlation, the change order gives a strong relationship to time performance equal to 0.556. So we can be explaining that the change order affects the time performance equal to 30.9% and the rest influenced by other variables.

1. Introduction

Construction is one of the most dangerous industries because it is unique, dynamic, and temporary [1]. The on-time construction works (at scheduled times) are very important for investors and contractors. In the construction projects, even those that are excellently planning and organizing, run the risk of being delayed. It is due to various changes that exist in the construction project [2]. Change is a reality in industrial processes, and the danger of significant accidents must be well managing. The change order is part of safety management. However, changes are often complex and usually involve technical and organizational aspects [3]. Change orders also play an essential role in ensuring the quality of the final product and moving the organization towards the desired goals. The changes that occur are usually very dependent on humans as the main driver in the success of change order [4]. In addition to human factors, the factor of management strategy and technology is no less critical. In previous studies explained that there is a significant relationship between change order and management strategies as well as technological innovation [5].

There are so many factors that cause change orders, one of which is the change order requested by the owner during the design and construction process. It can lead to budget and schedule overrun and increased uncertainty for shipping goods to the project. Practical strategies for managing change can be used by the project team to reduce the number of change orders during construction [6]. In other studies also explained related to the development of the concept of successful construction management is accompanied by the management of change orders in detail and well planned [7].

Change orders also significantly affect the effectiveness of project work where its sustainability depends on three interrelated components, namely quality, time, and cost. Not a few of the projects in Indonesia are experiencing delays caused by change orders. It is because the change order process must go through several stages and involve several parties who do have different authority. So this change order often results in additional implementation time and project costs [8].

Change orders often occur in the implementation of construction projects start at the beginning until the end of project implementation. The change order is an effort to improve the performance of the construction process to reach the perfect results of the work by the owner's needs. Although it objective to improve work, this change order can result in additional implementation time and project costs during the construction process [9].

The implementation of work on construction projects also often changes planning, so it is interfering with the time of implementation, the addition or reduction of costs on works, or changes in the scope of work. Changes in the scope of work can result in new requests that direct or indirect effect planned work items. These changes will automatically affect the volume, implementation time, and costs required [10].

The importance of managing change orders in construction projects is growing to make an integrated part of project management. Although change orders seem systematic handling of change orders has not been discussed a lot until the late 1990s. However, there has been progressing in managing the change order since the early 2000s [11].

One of the problems at the time of implementation of the Building Construction Projects in the Jakarta region, which was the object of this study, showed that the change of work methods on pile foundations became bored pile foundations. Given the much of work volume in the construction, it needs to conduct various studies related to change orders about project performance, to be able to manage change orders systematically. Therefore, the study is required to explore more related to the main factors that cause change requests and their effects on project performance, specifically related to the time performance of project implementation.

a. Change Order in Construction Project

Change orders are changes in writing between the owner and the contractor to change the conditions of the initial contract documents, by adding or subtracting work. These changes can change contract costs and project implementation schedules. In short, a change order can define as a modification of the original contract. The change order is an agreement in the form of a written agreement signed between owner and contractor after agreeing to confirm changes and the number of compensation costs and implementation time to the contractor that occurs at the project implementation stage, after the signing of a work contract between the owner and the contractor [6].

Types of change orders can be grouping into Direct Change (Formal Change) and Construction Change (Informal Change). Direct Change is a change submitted in written form, proposed by the contractor to the owner to change the scope of work, implementation time, costs, or other things that are different from those specified in the contract document. These provisions usually give the owner unilateral freedom to change the scope of work and require the contractor to keep up with these changes. Formal changes know before doing the work. Constructive Change is an informal action to order a modification of a contract in the field that occurs due to the request of the owner, planner, or contractor [12].

b. The Factors Affect of Change Orders

Change orders also often occur in contracts. Based on the previous study can be known several factors that cause Change orders on contracts lead to three variables, namely changes in contract value, changes in contract completion time, changes in contract administration. Of the three variables, changes in contract administration are the main of all changes. The most frequent amendments are changes in contract value caused by escalation (price adjustment), lack of work added based on calculations, and design changes [13].

In the other studies also found 5 (five) dominant factors causing change orders. There is the owner's request to optimize building functions, mismatches between drawings and field conditions, design errors or drawings from planning consultants, significant differences in volume between drawings and field conditions (bill of quantity), and the state of the change order is not apparent in the construction contract. The dominant influence of change orders on the performance of contractors is the availability of material, disruption of cash flow, availability of labor, funding or capital that must be spent by service providers or contractors, and the availability of work equipment [14].

In a study related to other construction, also found 5 (five) the most dominant factors causing Change Order is the contractor's problems, the physical condition of the field, Change of scope, the problem of the consultant superintendent or internal supervisor, and errors or omissions in the design. Besides the problems at the project site related to the policy of the project owner, security or safety constraints, and project financing problems are some of the elements that cause Change Orders, which are mostly caused by service providers [15].

Whereas 6 (six) Change Order factors that contribute to minimizing resistance to change, namely project scope, project size, project duration, organizational expectations in implementing faster changes, the establishment of official units to handle of change, and unit involvement which handles changes with the implementation of activities [16].

c. Impact of Change Orders

The impact of change orders occurs on costs because the cost increases have implications for additional work volumes. The impact on time, every time there is an increase in the scope of work or the volume, may not have an impact on the adding of implementation time. The most dominant effect with change orders is project implementation delays, disputes, and cost overruns [14].

In another study also explained that due to frequent change orders, where the administrative process was not carried out according to procedures, there were often disputes between owners and contractors. Change orders also result in projects being delayed and costs soaring [17].

Change orders are often carried out on construction projects, resulting in increased costs on some works and an impact on completion time (as well as) quality of work. These results occur in additional work from the initial contract [10].

There are five most common Change Order impacts in the construction, namely increasing project costs, increasing the duration of activities of each work, delaying completion of work schedule, adding a budget for the contractor, and delay in payment. So companies or organizations in construction projects must collect data and information about their projects to support their decision making to achieve cost efficiency, and this is usually the reason for delays in the completion of work schedules [15].

d. Project Time Performance

Project scheduling is one of the elements of planning results, which can provide information about the planned schedule and project progress of resource performance in the form of costs, labor, equipment, and materials as well as the project duration plan and progress time to complete the project. Schedule or time control is the main factor in achieving goals, providing resources such as materials, equipment, and labor [18].

Performance interpreted as the ratio of input to output, where the input of project performance is a series of activities measured from its output. Performance in the construction industry interpreted as efficiency in organizations or individuals. The economic value measured from the output produced by the material, equipment, and effort of labor in creating a product [9].

Time performance is a comparison between the time planned, and the agreed time between the owner and the contractor with the actual work, the differences of smaller percentage is the better the performance. Delay in completion of works can occur due to the late start of the activity or an extension of the duration of the activity. Delay of an activity that is on the critical path of the project will be able to make the overall project delay [19].

Based on the study from several countries can be explained about several factors that influence project time performance. Delays in infrastructure projects are caused by 20 factors, namely terrain conditions, weather conditions, variation of orders, availability of labor, design errors, planned costs for project development, market conditions (availability of resources), fluctuations in material prices, time planned for project construction, Emergency work, quality of equipment and raw materials, Late payment, Variation Order Costs, Government requirements, Poor processing of material quality, Delay in decision making, Management of work relations, Sequence of work on schedule, Poor project management, and Delays in decision making [20]. In another study, it explained that the most significant factor influencing delays in construction work was the error factor in design documentation [2]. Project delays are also affected by several factors, which include: inadequate contractor experience, inadequate planning, inflation, an incessant sequence of variations, and changes in project design [21]. Construction materials that are not available for installation are also a cause of delays in the construction project supply chain and pose a risk to overall project delays [22].

2. Methodology

The method used to explain the contribution of a change order to time performance is Descriptive Analysis, Simple Linear Regression Analysis, Multiple Linear Regression Analysis, Hypothesis Test, Correlation Analysis, and Determination Coefficient Test. Then the results of research calculations are compared with data in the field which objective to strengthen the answers to the research problem. Based on the comparison between the calculated data with the interview data and the observation data in the field, it can be concluding that this is a match the calculated data with the project data.

a. Independent Variable

Based on various references, the independent variable used in this study is a change order which consists of seven indicators, namely:

1. Design
2. Natural
3. Work added or less
4. Information
5. Schedule
6. Method
7. Contract

b. Dependent Variable

Based on various references, the dependent variable also used in this study is time performance which consists of five indicators, namely:

1. Equipment and materials
2. SHE
3. Mobilization
4. Damage
5. Complexity

Descriptive analysis was conducted to find out the respondents' answers about the variables that exist so that an overview of change orders and time performance in the research object can be known. The responses of respondents regarding existing variables measured from respondents' answers to the questionnaire, accompanied by five possible answers to be chosen and considered according to the respondent. From these answers, the assessment criteria for each item based on percentage are then prepared using the following steps:

1. Cumulative value is the sum of the values of each statement item, which is the answer of 30 respondents. The percentage of cumulative value divided by the frequency value multiplied by 100%.

2. The number of respondents is 30 people with the highest measurement scale is five, and the lowest measurement scale is one, so the highest cumulative number is obtained 30 x 5 = 150 with the lowest cumulative number 30 x 1 = 30.

The highest percentage value is 100%. While the lowest percentage value is 20%. The range value is 100% - 20% = 80%. If the range value is divided by 5 measurement scales, a percentage interval value of 16% will be obtained. So that obtained percentage rating classification that can be seen in table 1

Table 1 Interpretation Value Category

Value	**Percentage**	**Value Category**
1	20% - 36%	Bad
2	>36% - 52%	Not Good
3	>52% - 68%	Good Enough
4	>68% - 84%	Good
5	>84% – 100%	Very Good

The results of the total value of each variable can see in Figure 1.

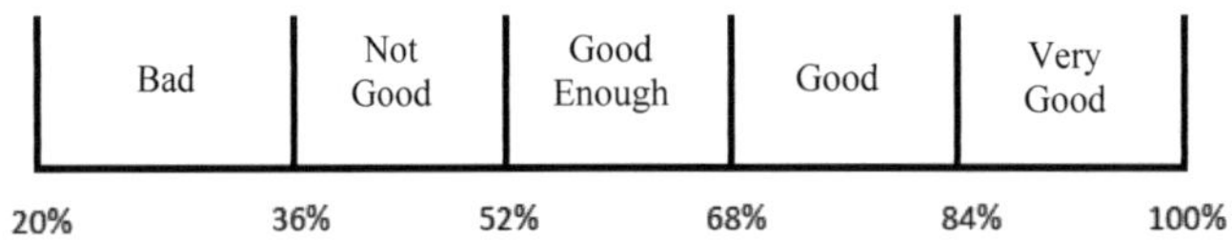

Figure 1 Category Interpretation of Values in the Continum Line

3. Results and Discussion

a. Descriptive Change Order Analysis (X)

Based on the analysis of respondents' responses regarding Change Order, it shows that the total value obtained is 1195, then after being entered into the continuum line (figure 1) the following values are obtained:

1. Maximum Index Value = 5 x 16 x 30 = 2400
2. Minimum Index Value = 1 x 16 x 30 = 480
3. Interval Distance = [maximum value - minimum value]: 5
   = [2400 - 480]: 5
   = 384
4. Value Percentage = [(total value): maximum value] x 100%
   = [1195: 2400] x 100%
   = 49.79%

Based on Figure 1, percentage of the Change Order Continuous Line can see the ideal value obtained for respondents 'answers to 16 statements is 2400. From the recapitulation of respondents' data shows the value obtained is 1195 or 49.79% of the maximum value of 150. It shows that the Change Order on the projects, which are the objects of this research in the not good category.

b. Descriptive Analysis of Time Performance (Y)

Based on the analysis of respondents' responses regarding Time Performance, it shows that the total value is 764, then after being entered into the continuum line (figure 1) the following value are obtained:

1. Maximum Index Value = 5 x 12 x 30 = 1800
2. Minimum Index Value = 1 x 12 x 30 = 360
3. Interval Distance = [maximum value - minimum value]: 5
   = [1800 - 360]: 5
   = 288
4. Value Percentage = [(total value): maximum value] x 100%
   = [764: 1800] x 100%
   = 42.67%

Based on Figure 1, Percentage of Time Performance Continuous Line can be seen the ideal value obtained for respondents' answers to 12 statements is 1800. From the calculations in the table shows the value obtained 764 or 42.67% of the maximum value of 150. This shows that the Time Performance on the projects, which are the objects of this research also in the not good category.

c. Regression Analysis

Simple Linear Regression

The effect of change order (X) to time performance (Y) simple linear regression analysis with the following equation used:
Y = a + bX

Where :
Y = Time Performance
a = constant
b = direction of the regression coefficient
X = Change Order

The SPSS software processing results for simple linear regression analysis is presenting in table 2.

Table 2 Simple Linear Regression

Model	Unstandardized Coefficients		Standardized Coefficients	t	Sig.
	B	Std. Error	Beta		
1 (Constant)	.389	.229		1.698	.101
Change Order	.792	.016	.994	48.749	.000

Dependent Variabel: Time Performance

Based on the calculation results in table 2, the simple linear regression equation obtained as follows:

$$Y = 0.389 + 0.792X$$

The simple linear line equation above, a constant of 0.389, means that if there is no increase in the value of the change order variable, it is worth 0.389. The regression coefficient of 0.792 states that each addition of one value to the change order variable, the addition of time will increase by 0.792. A positive coefficient means that there is a positive influence between change order to time performance. Change orders affect the increase in the time of implementation of construction project work (performance time).

The calculation results of Simple Linear Regression Analysis, to determine the normal distribution or not is to use the normal probability plots. The graph of normal probability plots can see in Figure 2.

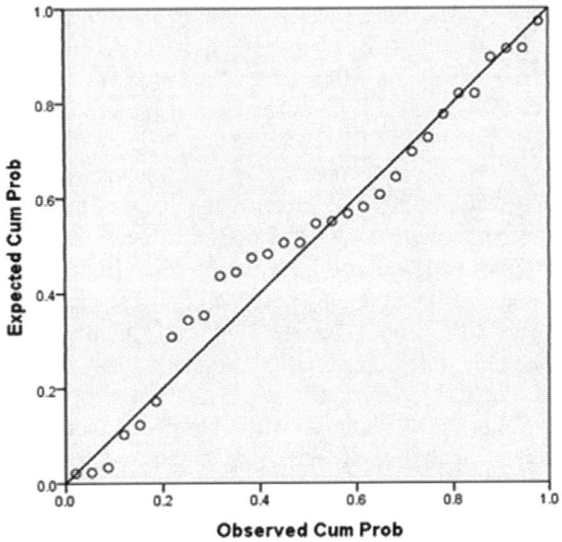

Figure 2 Normal Probability Plots

Based on Figure 2, Normal Probability Plots show that the data spread around the diagonal line and follow the direction of the diagonal line. The data processed is normally distributed data, so the linearity test is accepted.

Multiple Linear Regression

To see the contribution of change order (X) to time performance (Y) used multiple linear regression analysis with the following equation:
$Y = \alpha + \beta_1 X_1 + \beta_2 X_2 + ... + \beta k X k + e$

Where :
Y = Time Performance
α = Constant
β = direction of the regression coefficient
X = Change Order

The results of SPSS software processing for multiple linear regression analysis are presented in table 3.

Table 3 Multiple Linier Regression

Model	Unstandardized Coefficients		Standardized Coefficients	t	Sig.	Collinearity Statistics	
	B	Std. Error	Beta			Tolerance	VIF
1 (Constant)	9.069	5.554		1.633	.117		
X1	.094	.637	.036	.148	.884	.319	3.139
X2	1.251	1.849	.144	.677	.506	.410	2.438
X3	.345	.549	.178	.628	.536	.230	4.357
X4	2.177	1.820	.246	1.196	.245	.436	2.293
X5	-.702	1.819	-.073	-.386	.703	.514	1.945
X6	-.319	1.573	-.034	-.203	.841	.644	1.554
X7	2.033	.775	.443	2.623	.016	.647	1.545

Dependent Variable: Time Performance

Based on the results of calculations in table 3, Multiple Linear Regression Analysis, obtained the form of multiple linear regression equations as follows:

$$Y = 9.069 + 0.094X1 + 1.251X2 + 0.345X3 + 2.177X4 - 0.702X5 - 0.319X6 + 2.033X7$$

The equation of the multiple linear lines above, a constant of 9.069, means that if there is no increase in the value of the change order variable, it is worth 9.069.
a) Every increase of 1 unit of variable X1 will increase the value of the variable Y by 0.094 units.
b) Every increase of 1 unit of variable X2 will increase the value of the variable Y by 1,251 units.
c) Every increase of 1 unit of variable X3 will increase the value of the variable Y by 0.345 units.
d) Every increase of 1 unit of variable X4 will increase the value of the variable Y by 2,177 units.
e) Every increase of 1 unit of variable X5 will decrease the value of the variable Y by -0.702 units.
f) Every increase of 1 unit of variable X6 will decrease the value of the variable Y by -0.319 units.
g) Every increase of 1 unit of variable X7 will increase the value of the variable Y by 2,033 units.

A positive coefficient means that there is a positive contribution between change order and time performance. More of the change order (changes in construction work) will increase the time performance (increase in the time of the construction project work).

To find out the most dominant variable contributing can be determined by the largest regression coefficient. Based on the results of calculations on the Multiple Linear Regression Analysis table, the largest regression coefficient is 2.177. It shows that the most dominant independent variable contributing to the dependent variable is the X4 variable, namely the lack of information when planning the project.

The related contribution of Change Orders to Project Time Performance can be seen in Figure 3 as follow.

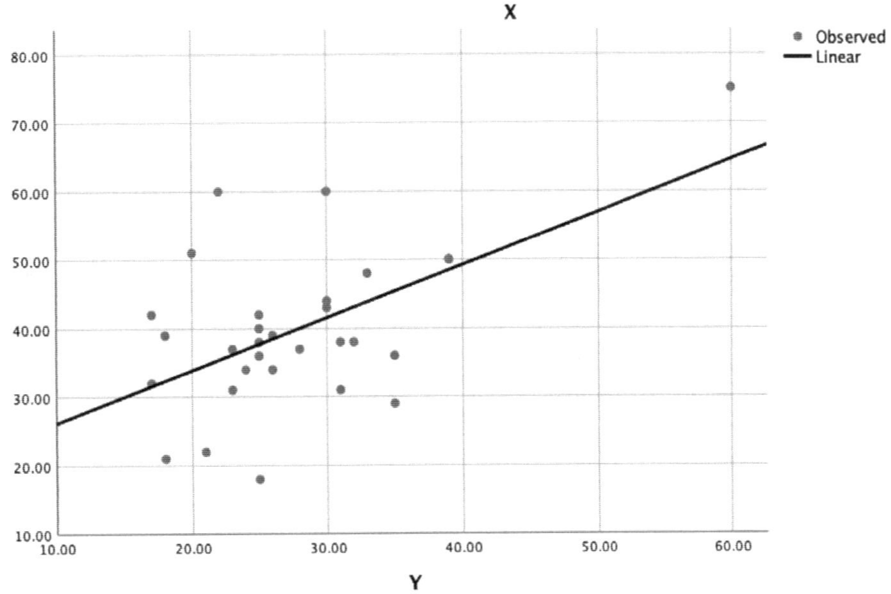

Figure 3 Contribution of Change Orders to Project Time Performance

Figure 3 shows that change order has a relation to project time performance in linear regression. If the change order increased will cause additional time in the project.

d. Hypothesis Testing

The t statistic test shows how far the partial effect of the independent variable on the dependent. The t-test is done with comparing count with t-table and significance level. The hypothesis as follow :

1. H_0: $\rho\_ = 0$ Change orders do not have a significant contribution to time performance.
2. H_a: $\rho\_ \neq 0$ Change order has a significant contribution to time performance.

This research was conducted by looking at the significance value of each variable in the output of the regression results using SPSS. If the profitability value of t-count <0.05, then there is a significant influence between the independent variable and the dependent variable, and vice versa, if the value of profitability t-count> 0.05, then there is no significant effect between the independent variable and the dependent variable. The table of t-test partial result also can be seen in table 2.

Based on table 2 shows that the t-count for change orders is 48.749, so t-count> t-table (48.749> 2.042) and a significant level of 0.000 <0.05. Based on the results of the analysis, it knows that H_0 is not accepted, meaning that change orders have a significant contribution to time performance.

e. Correlation Coefficient (R)

Pearson Product Moment Correlation Analysis (R) is a value used to measure the magnitude of the literature relationship between two or more variables. Pearson Product Moment Correlation obtained from the correlation coefficient. Correlation value (R) obtained is 0.556. The results of the calculation of the coefficient of determination can see in table 4.

Table 4 Correlation Coefficient (r)

Model	R	R Square	Adjusted R Square	Std. Error of the Estimate
1	.556	.309	.284	7.27262

Predictors: (Constant), Change Order
Dependent Variabel: Time Performance

Based on table 4, Pearson Product Moment Correlation shows that the correlation obtained from the value of R is 0.556. The results of the correlation value obtained to compare with the value of the R Correlation Coefficient Interpretation Table R. This shows that the change order provides a strong enough relationship of 0.556 to time performance.

f. Coefficient of Determination (R^2)

The coefficient of determination (R^2) is the value used to determine whether there is an influence relationship between two variables. The coefficient of determination obtained from the results of the square of the correlation coefficient. The R Square coefficient value obtained a figure of 0.309. The results of the calculation of the coefficient of determination can see in table 4.

Based on table 4, the coefficient of determination shows that the coefficient of determination obtained from the R square value of 30.9%. It shows that change order has an effect of 30.9% for time performance, and the rest influenced by other variables not explained in this study.

g. Overview of Change Orders and Project Time Performance

The change order conditions in the project that was the object of this study included in the not good category with the acquisition value of 49.79%. Change orders that occur in this project caused by changes in plan drawings, changes in work methods, and changes in work. To find out the condition of time performance on this project is also included in the category that is not good, with a value of 42.67%. The time performance that occurred in this project was due to a lack of good communication, lack of employee performance, material delays, and delays in carrying out the work.

Based on the results of the analysis, compared with data in the field, it was done to strengthen the results of the analysis. Based on the comparison between the data analysis results with interview data and observation data in the field, it concluded that there is a match between the results of the analysis with the conditions in the field. The situation on the ground also shows a time delay due to several change orders that occur during implementation.

h. Contribution of Change Orders to Project Time Performance

Simple Linear Regression Analysis Results obtained equation $Y = 0.389 + 0.792X$. In the simple linear regression equation, there is a constant of 0.389, meaning that if there is no increase in the value of the change order variable, the time performance is 0.389. The regression coefficient of 0.792 states that each addition of one value to the change order variable, time performance increases by 0.792. A positive coefficient means that there is a positive contribution between change order and time performance. The more change order (changes in construction work), the more time the construction project implementation time (performance time) will also increase.

Results of Multiple Regression Analysis obtained by the equation $Y = 9.069 + 0.094X1 + 1.251X2 + 0.345X3 + 2.177X4 - 0.702X5 - 0.319X6 + 2.033X7$. Based on these equations, it can see that the biggest regression coefficient value is X4 of 2.177. It shows that the most dominant change order variable contributing to time performance is the X4 variable, namely the lack of information when planning a project.

The Pearson Product Moment Correlation Analysis results show that the correlation obtained from the R-value of 0.556. The results of the correlation value obtained show that change order provides a strong enough relationship of 0.556 to time performance. The Determination Coefficient Test results show that the coefficient of determination obtained from the R square value of 30.9%. It shows that change order contributed 30.9% to the time performance, and the rest was influenced by other variables not explained in this study such as costs, lack of communication between owner and contractor, employee performance, etc.

The results of the Hypothesis Test using the t-test obtained t value> t table (48.749> 2.042) and a significance level of 0.000 <0.05 means that there is a positive contribution between change order to time performance, meaning that change order affects increasing the time of construction project work (time performance).

4. Conclusion

Based on the results of the analysis, can concluded that are,

a. Change orders that occur in the Building Construction Project in the Jakarta region, which is the object of this study, are included in the not good category with the acquisition of a value of 49.79%. Change orders in this project caused by changes in the plan drawing, the work methods, and the addition of works. Time Performance on this project also included in the not good category with a value of 42.67%. The time performance that occurred in this project caused by a lack of communication, lack of employee performance, materials delays, and works delays.

b. The results of the Regression Analysis shows that the most dominant independent variable contributes to the dependent variable is the lack of information project planning, so have affected changes in the plan drawing. In the regression equation, it can be explaining that there is a positive influence between change orders to time performance. The change order had a substantial impact on the on-time performance of 0.556. The results of the correlation analysis obtained the results that the change order contributed to the time performance of 30.9%.

References

[1] A. Mohammadi, M. Tavakolan, and Y. Khosravi, "Factors influencing safety performance on construction projects: A review," *Saf. Sci.*, vol. 109, no. June, pp. 382–397, 2018.

[2] M. Głuszak and A. Lešniak, "Construction Delays in Clients Opinion - Multivariate Statistical Analysis," *Procedia Eng.*, vol. 123, pp. 182–189, 2015.

[3] M. Gerbec, "Safety change management – A new method for integrated management of organizational and technical changes," *Saf. Sci.*, vol. 100, pp. 225–234, 2017.

[4] N. H. Al-Moosa and N. Sharts-Hopko, "Using Change Management to Redesign Oman's Health Professions Education Sector," *Heal. Prof. Educ.*, vol. 3, no. 2, pp. 108–112, 2017.

[5] S. Wadood, B. Gharleghi, and B. Samadi, "Influence of Change in Management in Technological Enterprises," *Procedia Econ. Financ.*, vol. 37, no. 16, pp. 129–136, 2016.

[6] R. C. Okada, A. E. Simons, and A. Sattineni, "Owner-requested Changes in the Design and Construction of Government Healthcare Facilities," *Procedia Eng.*, vol. 196, no. June, pp. 592–606, 2017.

[7] M. Almanei, K. Salonitis, and C. Tsinopoulos, "A conceptual lean implementation framework based on change management theory," *Procedia CIRP*, vol. 72, pp. 1160–1165, 2018.

[8] Dzulqarnain, "Analisa Faktor Penyebab Dan Akibat Contract Change Order Terhadap Biaya Dan Waktu Pada Proyek Konstruksi Jalan Di Sulawesi Selatan," *J. Tugas Akhir, Univ. Hasanuddin*, 2017.

[9] M. R. A. S. dan A. S. Pamungkas, "Kajian Awal Peran Change Order Untuk Dapat Meningkatkan Kinerja Proses Konstruksi Pada Proyek Gedung PT. X Di Jakarta," in *Konferensi Nasional Teknik Sipil 12*, 2018, no. September, pp. 18–19.

[10] E. Rita and D. N. Carlo, "Studi Lingkup Pekerjaan yang Dominan Mengalami Perubahan Pada Proyek Jalan di Provinsi Sumatera Barat," *J. REKAYASA*, 2017.

[11] E. Eshtehardian and S. Khodaverdi, "Analytical Survey of Construction Change Systems: Gaps & Opportunities," in *Procedia Engineering*, 2016.

[12] H. Sulistio and M. A. Wibowo, "Pengaruh Model Change Order pada Kehilangan Produktivitas dan Performa Proyek KOnstruksi Jalan dari Perspektif System Teori," *MEDIA Komun. Tek. SIPIL*, 2014.

[13] A. Maulana, "Faktor Penyebab Terjadinya Contract Change Order (CCO) Dan Pengaruhnya Terhadap Pelaksanaan Proyek Kontruksi Pembangunan Bendung," *Univ. Katolik Parahyangan*, 2017.

[14] A. Y. Y. Martanti, "Analisis Faktor Penyebab Contract Change Order dan Pengaruhnya Terhadap Kinerja Kontraktor Pada Proyek Konstruksi Pemerintah," *Rekayasa Sipil*, 2019.

[15] Dikdik Moh. NS1, "Analisis Faktor Change Order Pada Proyek Pekerjaan Bangunan Air,"

Rekayasa Sipil Mercu Buana, vol. 6, no. 1, pp. 1–8, 2017.

[16] B. C. Lines, K. T. Sullivan, J. B. Smithwick, and J. Mischung, "Overcoming resistance to change in engineering and construction: Change management factors for owner organizations," *Int. J. Proj. Manag.*, vol. 33, no. 5, pp. 1170–1179, 2015.

[17] Ningsih and Syahrudin, "Identifikasi Dan Analisis Penyebab Dan Akibat Contract Change Order Terhadap Biaya Dan Waktu Pada Proyek Konstruksi," *Tek. Sipil FT.Untan*, 2015.

[18] A. Hidayat, "Analisa Kinerja Biaya Dan Waktu Pada Pelaksanaan Proyek Konstruksi Dengan Metode Earned Value (Studi Kasus Proyek Konstruksi Mall Dan Hotel X Di Pekanbaru)," *J. Ilmu Tek. dan Komput.*, vol. 1, no. 1, 2017.

[19] M. Marlee, "Analisis Korelasi Faktor Penerapan K3 Terhadap Kinerja Waktu," *J. Mitra Tek. Sipil*, vol. 1, no. 1, pp. 220–229, 2018.

[20] N. Al-Hazim, Z. A. Salem, and H. Ahmad, "Delay and Cost Overrun in Infrastructure Projects in Jordan," *Procedia Eng.*, vol. 182, pp. 18–24, 2017.

[21] L. M. Amusan, A. Afolabi, R. Ojelabi, I. Omuh, and H. I. Okagbue, "Data exploration on factors that influences construction cost and time performance on construction project sites," *Data Br.*, 2018.

[22] H. Wang and B. Hubbard, "A Survey Study on Industrial Construction Project Supply Chain: On Time Performance and Practices of Structural Steel and Pipe Spools," *Procedia Eng.*, vol. 196, no. June, pp. 653–659, 2017.

IOP Conf. Series: Materials Science and Engineering **846** (2020) 012062 doi:10.1088/1757-899X/846/1/012062

Analysis of Time Lapse Microgravity Anomaly Data Case Study in Kota Lama Tourism Area Semarang Indonesia

S Supriyadi [1], K Khumaedi [1], S Sugiyanto[1], Fajar Setiawan[1], M. Yani[1], T Teguh[1]

[1] Physics Department, Universitas Negeri Semarang, Semarang - Indonesia

* supriyadi@mail.unnes.ac.id

Abstract. Time lapse microgravity survey has been carried out in Kota Lama Semarang using the Gravimeter Scintrex CG5 on January 2019 and July 2019 with a total of 75 points spread evenly at research field. There are three possible values of gravity anomaly between time, positive, negative, and zero does not change. The results which is an anomaly which is known as time lapse microgravity anomaly show that maximum positive 0,29 mGal in the north and south, and minimum negative -0,11 mGal in northeast and southwest. For the time lapse microgravity vertical gradient anomaly show that maximum positive 0,0013 mGal in west, center, east, and minimum negative -0,0082 mGal in north and south. Conclusion are rainfall data throughout the study range show a downward trend and calculation of groundwater reduction using this data obtained a groundwater level decrease an average of -3.63 mm.

1. Introducing

Gravity method is one of the oldest methods in geophysics, but its application to sources of anomalies near the surface and those related to the environment is not as intensive as applications for geodynamic studies or exploration in the estimation of relatively large geological structures. This is due to the accuracy of the anomaly still in the order of mGal or 10^3 μGal. Whereas in terms of gravitimeter devices, an accuracy of 10 μGal was achieved in the early 1970s, but the accuracy of the readings is entirely dependent on the accuracy of the operator because the gravimeter still uses a mechanical reading device.

The time lapse microgravity method is a development of the gravity method with its fourth dimension being time. The principle of this method is measuring gravity repeatedly both daily, weekly, monthly and yearly using a careful gravimeter in the order μGal and meticulous elevation measurements.1 Any changes or differences in the results of observational gravity in the first period with the next period are called microgravity anomalies. Changes in observation gravity can be caused by dynamics around very points, such as changes in the depth of the ground water level and subsidence.

In line with the rapid increase in digital technology in the late 1980s, the problem of reading accuracy can be increased by the use of digital reading systems. At the end of 2000, LaCoste & Romberg released a full digital gravimeter called graviton with an accuracy of 1 μGal and semi digital gravimeter which is the development of LaCoste & Romberg type G which is equipped with a digital reading system with an accuracy of 1-5 μGal. Early in 2002 Scintrex also issued a full digital gravimeter called Scintrex Autograv CG5 with an accuracy of 1 μGal. With the availability of digital gravitmeter systems in full or semi-digital, the obstacles associated with tools for observing changes in the gravity field in the order μGal can be removed.

An increase in gravimeter accuracy and development of digital systems, the application of gravity methods for sources of anomalies near the surface and those related to the environment and for monitoring purposes are increasingly being used, including for monitoring geothermal reservoirs, oil and gas. The process of steam production and water injection in geothermal reservoirs must be monitored properly with the aim that the geothermal reservoir remains stable, steam production is stable so that the geothermal reservoir can last long [2]-[6].

Utilization of microgravity for monitoring, including the following: survey of geothermal potential in Mount Ungaran [7], monitoring of volcanic activity [8]. Utilization for monitoring environmental quality, for example the potential for subsidence zones in Jakarta [9]. Kennedy conducted a survey of changes in subsurface mass [10]. Monitoring changes in surface water depth [11]

Based on the description above, this study aims to analyze the time lapse microgravity data in the tourism area Kota Lama Semarang. From the analysis of the data it is expected that microgravity anomalies will be obtained time lapse anomaly. This anomaly data is related to the source of the anomaly that is the target of this study.

2. Method

In this study using the time lapse microgravity gravity method between time, which is measuring the gravity at the same point with a certain time interval, namely January and July 2019 with the number of 75 points that are spread evenly in the study location (Kota Lama area). The tool used is autograv CG-5 gravimeter. Stages of methode as follows:

2.1. First

Before measuring gravity at the research location, modeling is done first. Modeling aims to determine the various values of micro-gravity anomalies over time produced by different sources of anomalies. For modeling the source of the anomaly and the magnitude of the anomaly caused by using some Equations according to needs.

a. Equation which states the water level depth housing with an intermittent microgravity anomaly [12] which is stated by Eq.1.

$$\Delta g_w = 41{,}93\rho_w\phi\Delta d \tag{1}$$

where: Δg_w, G, ρ_w, ϕ, Δd each is the change in gravity value due to changes in groundwater level depth, general gravity constant, water mass density (gr/cm3), porosity (%), and change in depth groundwater level (meters).

The amount of precipitation will affect changes in groundwater level. In the rainy season the ground water level will be at its maximum value and then gradually decrease with the change of seasons. Changes in groundwater level can be calculated based on the Eq.2.

$$H(t)= H_1+\alpha\Sigma_n R_n \exp\{-c(t - t_n)\} \tag{2}$$

where H_1 is groundwater level at first, t is time, t_n is nth time, α is absorption constant (0.00932), c is evaporation constant (0.00985), and R_n is precipitation in the month.

b. Modeling time lapse microgravity anomaly using Eq. 3 is used for this purpose [13]

$$(g_{obs(2)} - g_{obs(1)}) = \left(G\int_0^{\infty}\int_{-\infty}^{\infty}\int_{-\infty}^{\infty} \frac{\Delta\rho(\alpha,\beta,\gamma,\Delta t)(z-\gamma)}{\left[(x-\alpha)^2+(y-\beta)^2+(z-\gamma)^2\right]^{3/2}}\, d\alpha d\beta d\gamma \right) + c_1(h_2 - h_1) \tag{3}$$

where g_{obs2} is the observational gravity of measurements at t_2, g_{obs1} is the observational gravity of mrasuremnet at t_1, G is universal gravity constant (6,67x10^{11} m^3/kg.sec^2), $\Delta\rho$ is dnesity change, α, β, γ is density coordinate, x, y, z is the measuring point coordinate.

Eq.3 shows that the difference in the gravity value of the measurement results due to changes in subsurface mass density associated with changes in the depth of the ground water level and subsidence

2.2. Second
Perform gravimetric calibration to ensure that the tool can be used for research. The gravimeter used has accuracy in μGal with the intention of being able to measure changes or small anomalies.

2.3. Third
Determine the location and position of the measuring point. For the necessity of measuring gravity over time, the location of the measuring point is sought to be stable and predicted not to experience physical damage during the study.

2.4. Fourth
The next step is to measure gravity and a measurement period has been planned in the next period. The next interval usually uses the season guidelines, which are the rainy season and the dry season. Note The sequence of points to measure the gravity is recorded, and in the measurement of the next period the order of points in a loop must be the same as the previous period.

2.5. Fifth
In the same way at each point, the measurement is repeated by placing the gravimeter at a height of 25 cm.

2.6. Sixth
Make an initial correction ie tide correction using Eq. 4 and drift correction with Eq. 5 every day after measurement in the field [14]-[17].

$$tide=G(r)[(c/R)^3(sin2\theta_m+1/3)+1/6r/c\ (c/R)^4\ (5cos3\theta_m+\ 3cos\theta_m\ )] \quad (4)$$

where *tide* is tide correction, *G(r)* is Universal gravity constant, *c* is average distance to the month, *R* is earth's diameter, and θ_m is zenith angle of moon.

$$drift_n=[(t_n-\ t_1)/(t_N-t_1\ )](g_N-g_1\ ) \quad (5)$$

where $drift_n$ is drift correction, t_n is reading time at the nth station, t_N is reading time at base station (end looping), t_1 is reading time at base station (initial looping), g_N is tidal-corrected gravimeter reading at base station (end of loop), and g_1 is tidal-corrected gravimeter reading at base station (initial looping).

Tide correction is done to eliminate the gravitational influence of objects outside the earth like the moon and the sun, which changes with latitude and time. In practice, tidal correction is done by measuring the gravity value at the same station (base) at a certain time interval. Then the gravimeter reading is plotted against time to produce an Equation used to calculate tidal correction. This tidal correction value is always added to the gravity reading.

Drift correction is due to differences in gravity readings from the same station at different times, which is caused by the shock spring of the gravimeter during the transportation process from one station to another. To eliminate this effect, gravity data acquisition is designed in a closed circuit (loop), so that the magnitude of the deviation can be identified and assumed to be linear at a certain time.

2.7. Seventh
Finally, make further corrections, which include elevation correction, correction of changes in ground water level depth. At this stage the aim is to obtain anomalous sources (research targets), and proceed with analysis of gravity anomaly data between times.

3. Results and Discussion
General research results include (1) modeling, (2) measurement results, and (3) analysis of measurement results. Explanation of each result as follows :

3.1. Response anomaly modeling

Using Eq. 1 and assuming rock porosity (groundwater reservoir) of 30%, any change in groundwater level of 1 meter during an interval of time causes time lapse microgravity anomaly 12,579 μGal as shown in Figure 1.

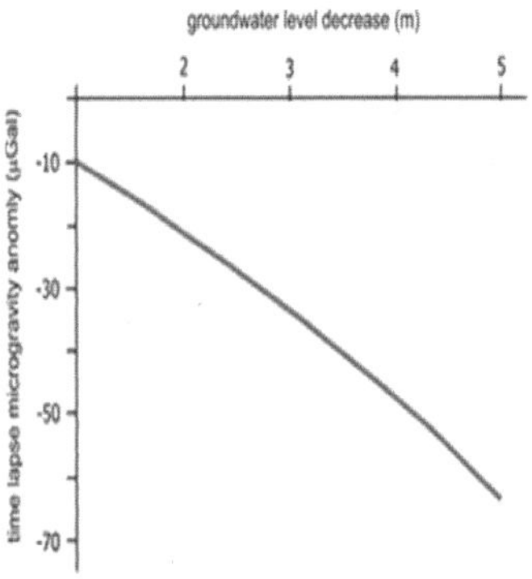

Figure 1. Time lapse microgravity anomaly (μGal) versus changes in groundwater level (m)

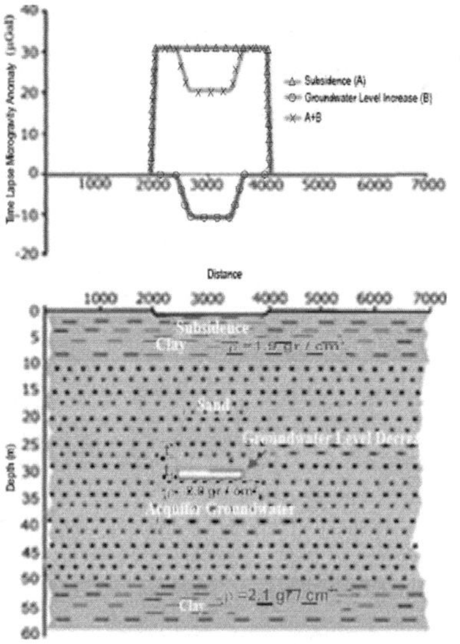

Figure 2. Models response time lapse microgravity anomaly caused by subsidence and groundwater level decrease

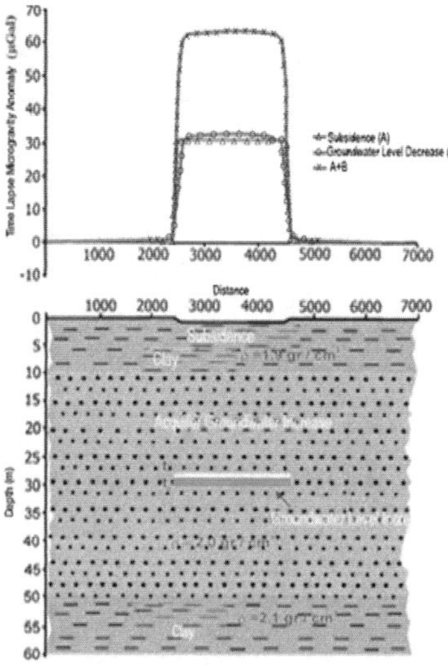

Figure 3. Models response time lapse microgravity anomaly caused by subsidence and groundwater level increase

Using Eq. 2 and the model parameters used in the form of a three-layer earth model extending horizontally with the physical properties as follows:

a. Layer 1 in the form of clay has a thickness of 10 m and $\rho = 1.9$ gr/cm^3
b. Layer 2 is sand with $\rho = 2.0$ gr/cm^3. Aquifer porosity is 30%, changes in mass density due to a decrease in groundwater level $\Delta\rho = -0.3$ gr/cm^3
c. Layer 3 is clay with a thickness of 10 m and $\rho = 2.1$ gr/cm^3

It is assumed (a) that subsidence occurs at coordinates 2000 until 4000 m, the magnitude of which is at $t_0 = 0$ and $t_1 = 10$ cm. For the decrease in ground water level occurs at coordinates 2500 until 3500 m a ground water by 1 m with a ground water level position at $t_0 = 30$ m and $t_1 = 31$ m. The maximum response time lapse microgravity anomaly caused subsidence and groundwater level decrease is 30,856 μGal (Figure2).

It is assumed (b) that subsidence occurs at coordinates 2500 - 4500 m, the magnitude of $t_0 = 0$ and $t_1 = 10$ cm, and the increase is assumed to occur at coordinates 2500 - 4500 m at 2.5 m with ground water level at $t_0 = 30$ and $t_1 = 27.5$ m. The maximum response of time lapse microgravity anomaly ssubsidence and groundwater level increase 63,546 μGal (Figure3).

3.2. The result of observation gravity

The gravity measurement data was carried out in January 2019 which is shown in Figure4. The gravity measurement value in January with a minimum gravity of 978118.44 mGal and maximum gravity of 978118.88 mGal, while the value of the gravity measurement in July 2019 is shown in Figure5 with a minimum gravity of 978118.62 mGal and a maximum of 978118.88 μGal.

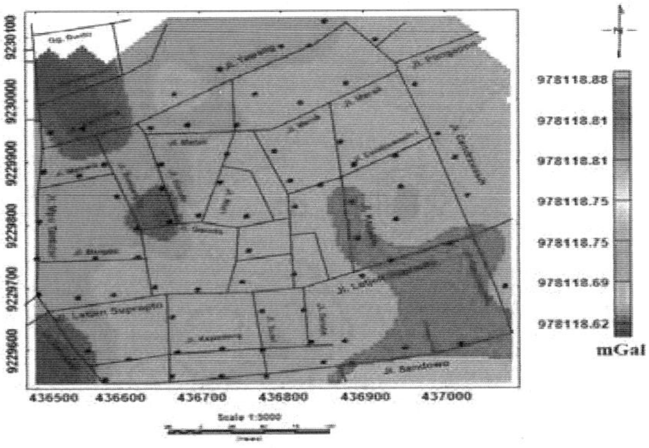

Figure 4. Observation gravity value on January 2019

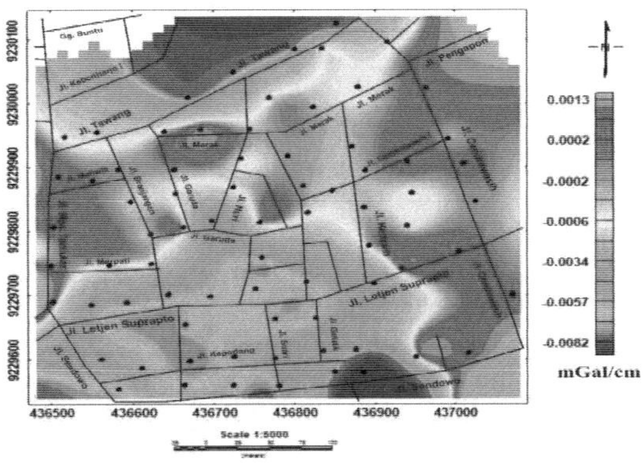

Figure 5. Observation gravity value on July 2019

In Figure 4 and Figure 5 shows the areas that have low observed gravity values have a higher topography such as the western part of the city on Mpu Tantular street and the north side of Tawang Station which is shown by blue to dark blue images. on the other hand, regions that have high observational gravity values have low topography as shown almost evenly distributed in the Kota Lama which is shown by green to pink imagery. In the January 2019 measurement shown in Figure 4. and the July 2019 measurement shown in Figure 5, there is a significant difference in the contour pattern which is caused by several factors, one of which is due to the increase in the rock density which causes the elevation at a measurement point bias can increase or reduced. Changes in gravity values from the period January 2019 to July 2019 can be seen in Figure6 below.

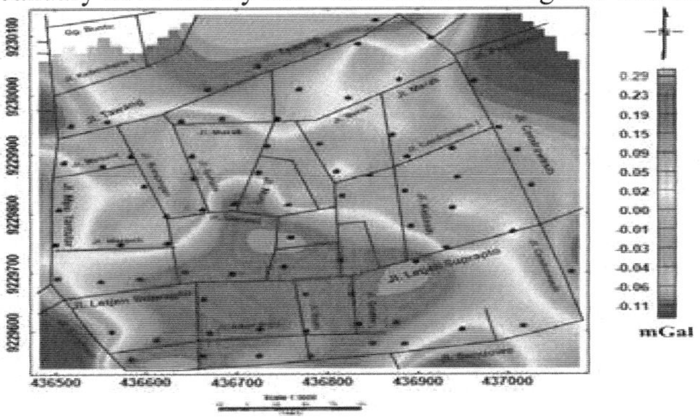

Figure 6. Time lapse microgravity anomaly value on January until July 2019

Based on Figure 6. can be seen measurement points or areas that have increased or decreased the value of the observed gravity. The results of observational gravity measurements from the period January 2019 to the period July 2019 showed a value that was almost the same as the average gravity value of the earth, namely 980000 mGal or 9.8 m/s^2. Furthermore, the gravity measurement data for each period is corrected to obtain a simple bouguer anomaly contour map in January 2019 and July 2019. The corrections used to reduce the factors that affect the gravity value are float correction, tidal correction, and Bouguer correction on the observed gravity value. The next stage was to determine the value of gravity anomalies in the study area with Oasis Montaj software to describe the contours of the study area.

3.3. Microgravity vertical gradient

The value of the vertical gradient microgravity anomaly in the January 2019 period (Figure7) shows the value -0.0039 to -0.0063 mGal/cm. The southern part of Kota Lama such as on the Sendowo street and the northern part around the Pengapon street have an anomaly range of vertical gravity gradient values of -0.0039 mGal/cm until -0.0030 mGal/cm indicated by gradations of blue. Most of the Kota Lama Region has a vertical gradient gravity anomaly value of -0.0030 mGal/cm to -0.0063 mGal/cm.

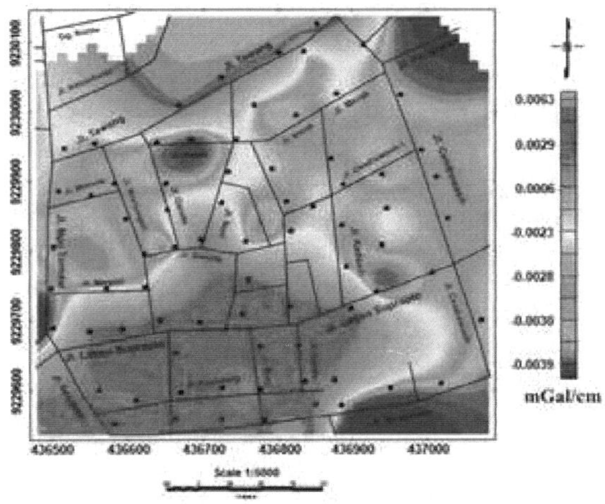

Figure 7. Microgravity vertical gradient value on January 2019

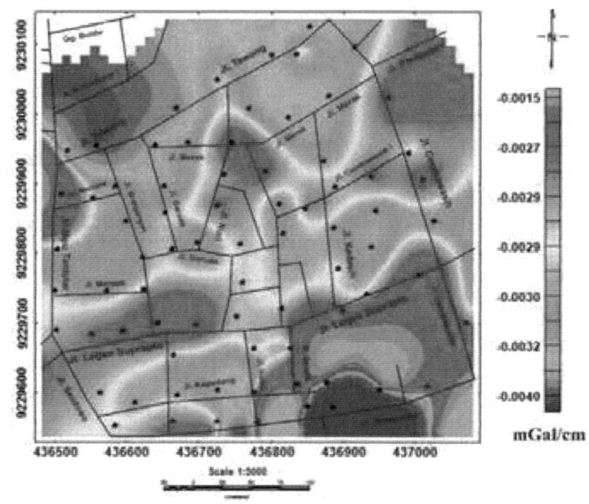

Figure 8. Microgravity vertical gradient value on July 2019

The value of the microgravity vertical gradient on July 2019 (Figure8) shows the value of -0.004 until -0.0015 mGal/cm. The southern part of Kota Lama such as on Sendowo street and the northern part around Tawang street have an anomaly range of -0.004 mGal/cm until -0.0032 mGal/cm indicated by gradations. Most of the Kota Lama region has a vertical gradient gravity anomaly value of -0.003 mGal /cm until -0.0015 mGal /cm.

Difference in the gravity gradient microgravity between the two periods will be obtained time lapse microgravity vertical gradient anomaly values (Figure 9) shows the change in the value of the vertical gravity gradient anomaly. The southeast area of the Kota Lama area around Sendowo and Cendrawasih street and points that show gradations of pink and green experience anomalous vertical gradient gravity values of positive values of 0,0002 mGal/cm to 0.0013 mGal/ cm, while the points points that point to gradations of green to blue such as on Tawang Street have decreased the vertical gravity anomaly values of -0,0006 mGal/cm to -0.0082 mGal/cm.

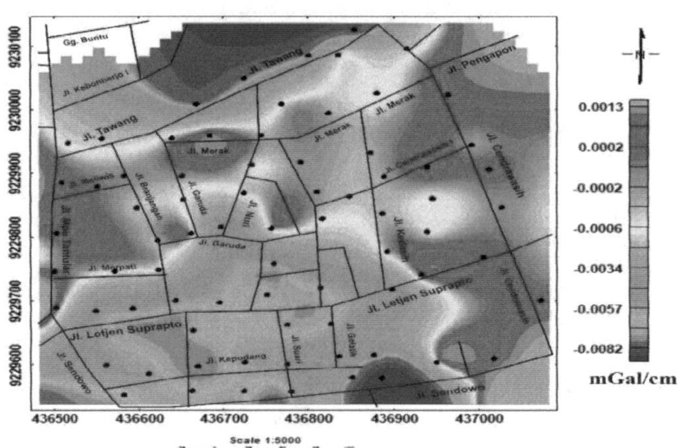

Figure 9. Time lapse microgravity vertical gradient anomaly on January until July 2019

In general, the results of the measurement and analysis above are simplified as in Table 1 below[16]. This table will make it easier to help during research on time lapse microgravity, repetition of gravity shifts at the same point at certain intervals, usually in different seasons, namely the rainy season and the dry season.

Table 1. Relationship between time lapse microgravity – time lapse microgravity vertical gradient and groundwater level

No	Time lapse microgravity (TLM)	Time lapse microgravity vertical gradient (TLMVG)	Explanation
1	+	+	Groundwater recharging, subsidence
2	+	-	groundwater decreasing, Subsidence
3	-	-	Groundwater decreasing
4	0	0	Stable, no change

Furthermore, using gravity data and the vertical gradient of gravity for each period, between time and Table 1, it can be explained that in some research zones in the Kota Lama area there are potential subsidence and subsidence. Remembering this latest condition that the Kota Lama area has been revitalized to become a tourist visiting area and paving is done to prevent flooding (Figure10). This condition directly affects the results of the study. Based on Eq.3, it can be explained that the time lapse microgravity anomaly to monitor subsidence occur less effectively, given that paving causes elevation of surface at Kota Lama and its surroundings to change. This surface's elevation change affects the elevation of each measuring point. Changes referred to the initial height used as a reference increasingly high.

Figure 10. Paving in Kota Lama Semarang to prevent flood

In addition to subsidence other information obtained based on Eq. 2 is a groundwater level decrease that occurred in the Kota Lama as indicated by microgravity anomaly and vertical gradient microgravity anomaly is negative value. In addition, rainfall (Figure 11) that occurred during the January-July 2019 in Kota Lama showed a declining trend. Based on the calculation of rainfall data from January to June 2019 using Eq.2, the results of groundwater level decrease of -3.63 mm.

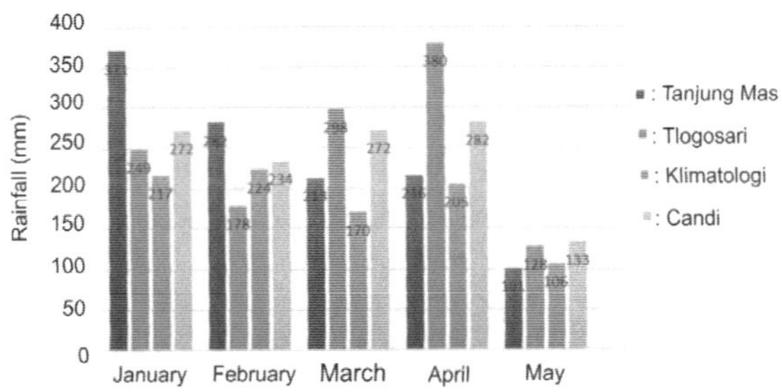

Figure 11. Rainfaal data from record station located around Kota Lama

4. Conclusion

Based on the results of the study it can be concluded, that the measurement of repetitive gravity at the same point in the Kota Lama provides valuable anomaly information (+) caused by subsidence, (-) caused by a groundwater level decrease, and (0) no change occurred. For the January to July 2019 research period in Kota Lama based on microgravity and vertical gradient data, there was no subsidence, this is because when the research was conducted revitalizing the Kota Lama made it difficult to monitor the subsidence, the gravity measurement point experienced a change in elevation. However, using rainfall data can be attributed to a groundwater level decrease which gives an average of -3.63 mm.

Acknowledgments

Research tim thanks the DRPM (Directorate of Research and Community Service) Kemenristekdikti for funding this research through the National Competitive Basic Research scheme with the Research Contract Agreement Letter no. 192 / SP2H / LT / DRPM / 2019, date March 11, 2019.

References

[1] L. Hasanah , A. Aminudin, N. D. Ardi, A. S. Utomo, H. Yuwono, Kamtono, D. D. Wardhana, K. L. Gaol, and M. Iryanti, Graben Structure Identification Using Gravity Method, *IOP Conf. Series: Earth and Env. Sci. 29 (2016),* International Symposium on Geophysical.

[2] M.Y. Kamah, C. Negara, I. Pulungan,, and Budiardjo, Application of Microgravity Method on Monitoring Geothermal Reservoir Changes during Production of Steam in The Kamojang Geothermal Field, West Java Indonesia, *in* 5th *SEGJ Int. Sym. – Imag. Tech.*, (Tokyo, Japan, 24-26 January 2001).

[3] O. Eiken, M. Zumberge, and G. Sasagawa, Gravity Monitoring of Offshore Gas Reservoir, *SEG Expanded Abst*ract, **19** (2000), 431.

[4] C. Akasaka, and S. Nakanishi, An Evaluation of The Background Noise for Microgravity Monitoring in The Oguni Field, Japan, in *Proc. of 25th Stanford Geoth. Workshop,* (24 - 26 January 2000).

[5] N.O. Mariita, Application of Precision Gravity Measueremnt to Reservoir Monitoring of Olkaria Geothermal Field, Kenya, in *Proc. World Geoth. Congr.* (Kyushu – Tohoku, Japan, May 28 – Jun 10, 2000).

[6] J. Nishijima, Y. Fujimitsu, S. Ehara, and M. Yamauchi, Microgravity monitoring and repeated GPS survey at Hatchobaru geothermal field, Central Kyushu, Japan., *in Proc. World Geoth. Congr. 2005.*

[7] H. Saibi & E. Aboud, A. Setyawan, S. Ehara & J. Nishijima, Gravity data analysis of Ungaran Volcano, Indonesia. *Arab J Geosci* 5 (2012), pp.1047–1054.

[8] M. Bagnardi, M.P. Poland, D. Carbone, S. Baker, M. Battaglia, and F. Amelung, Gravity changes and deformation at Kilauea Volcano, Hawaii, associated with summit eruptive activity, 2009 2012. *Eophys Res-SOL EA*, 10.1002/2014JB011506, (2014). pp.7288-7305.

[9] A. Setyawana, Y. Fukudab, J. Nishijimac and T. Kazamab, Detecting Land Subsidence Using Gravity Method in Jakarta and Bandung Area, Indonesia. in *Procedia Env. Sci.* 23 (2015) 17 – 26. International Conference on Tropical and Coastal Region Eco-Development 2014(ICTCRED 2014).

[10] J. Kennedy, T. P. A. Ferré, A. Güntner, M. Abe, and B. Creutzfeldt, Direct measurement of subsurface mass change using the variable baseline gravity gradient method. ***Geophys Res Lett***, 10.1002/2014GL059673, (2014) pp. 2827-2834.

[11] L. Handayani, D. D. Wardhana, P. Hartanto, R Delinom, Sudaryanto, RF Lubis, and H Bakti. *Gravity survey of groundwater characterization at Labuan Basin,* in *IOP Conf. Series: Earth and Env. Sci.* ***118*** (2018) 012015, Global Colloquium on GeoSciences and Engineering 2017.

[12] S. N. E. Wibowo, G. E. Mamuaya, and R. Djamaluddin . Land Subsidence Analysis of Reclaimed Land using Time-Lapse Microgravity Anomaly in Manado, Indonesia, *Indonesia Journal of Spatial and Regional Analysis. Forum Geografi*, 32(1), pp-53-63. DOI: 10.12917/forgeo.v.31i2.5882.

[13] D. Santoso, W.G.A. Kadir, M. Sarkowi, Ardiansyah, dan Waluyo, Time Lapse Microgravity Study for Rejection Water Monitoring of Talang Jimar field, in *Proc. of the 7th SEGJ Int. Sym.* (Sendai-Japan 24-26 November 2004), pp.497-502.

[14] W.G.A Kadir, D. Santoso, dan M. Sarkowi, Time Lapse Vertical Gradient Microgravity Measurement for Subsurface Mass Change and Vertical Ground Movement (Subsidence) Identification, Case Study : Semarang alluvial plain, central Java, Indonesia, *in Proc. of the 7th SEGJ Int. Sym.* (Sendai – Japan 24 – 26 November 2004), pp. 421-426.

[15] R. R. Amarante and J. L. A. Trabanco. Calculation of The Tide Correction Used in Gravimetry, *Revista Brasileira de Geofisica* 34(2) (2016), pp.193-206.

[16] Y. Wang, J. Schnitzbauer, Z. Hu,, X. Li, Y. Cheng, H. Zhen-Li, and H. Bo, Localization events-based sample drift correction for localization microscopy with redundant cross-correlation algorithm. *OPT. EXPRESS* 15982. 22(13)| (2014), pp. 15982-15991

[17] Salit and G. C. Turk, A Drift Correction Procedure. *Anal. Chem.*, 70, (1998), pp. 3184-3190

IOP Conf. Series: Materials Science and Engineering **846** (2020) 012063 doi:10.1088/1757-899X/846/1/012063

First-Order Fuzzy Time Series based on Frequency Density Partitioning for Forecasting Production of Petroleum

R. Wulandari* **Farikhin, B Surarso, B Irawanto**

Universitas Diponegoro, Semarang, Indonesia

*Dariwulan711@gmail.com

Abstract. Forecasting method based on fuzzy time series has been widely developed in recent years. In this paper, we propose a new improvement at determining universe of discourse, variation historical data and partitioning stage. At early stage, we define the universe of discourse then calculate the basis value to find out how much interval should be used with variatin historical data. Secondly, we are partitioning the main intervals into several numbers of sub-intervals. The empirical analysis shows that sub-interval caused the fuzzy number getting closer to crisp value. It causes the better forecasting value. We use the data of yearly production petrolium Indonesia for simulation. We compare the forecasting results and error value of the method with previous existing methods The modifications give better forecasting results than previous methods indicated with smaller The Means Squared Error (MSE) and Average Forecasting Error (AFER).

Keywords: First-Order Fuzzy time series, Frequency density, Partitioning, Sub-intervals.

1. Introduction

Mathematics as a fundamental science in the knowledge of mathematical causes can be applied in various disciplines in the world both science, technology, Economics and other sciences (Masykur 2008:43). The set of fuzzy was introduced by Zadeh (1965:338-353) and is expanding in various applications in the field of science one of which is in the field of forecasting, the forecasting using the fuzzy is commonly referred to as fuzzy forecasting, in different Forecasting using statistical methods such as ARMA, regression and others that can directly be in the process with existing values, in fuzzy forecasting in large lines there are 3 steps that is step Fuzzyfikasi, must change the existing values or Commonly called a Crips value to a linguistic value first, then fuzzy relation seeks the right relation to the value of linguistic values, subsequent defuzzyfication then by using a method to return the linguistic value to the Crips value , the advantage of using linguistic values is that it can be applied even if some data is lost (Klir, 1998). The Fuzzy time series was first introduced by song and Chissom in 1993 and grew rapidly in recent years, later (Chen,1996), (Ismail,2011), (Chissom,1993), (Chissom,1994) introducing several forecasting methods To foresee one object i.e. enrollment at the University of Alabama, Chen brought about an easier and simpler method. (Malike and Konstantin,2005) developed the first-order fuzzy time series to complete the university's forecasting of Alabama, then (Jilani TA,2007) writes that forecasting based on Frequenscy density has a smaller error value Compared to the previous metoe. From the above, we modified the method belonging to Jilani with Malike and Constantine's methods became a new method which was later used to find forecasting of data Petroleum Indonesia in 1996-2017. The result of each forecast method will be Evaluated by comparing the value of MSE and AFER.

2. Methodology

2.1 Fuzzy Time Series

The main concepts of fuzzy time series is a fuzzy set A of the universe of discourse U, $U = \{u_1, u_2, \dots, u_n\}$, is defined as follows : $A = \{f_A(u_i)\}$, $i = 1,2,\dots,n$ where f_A is the membership function of the fuzzy set A, $f_A\colon U \to [0,1], f_A(u_i)$ denotes the grade of membership of u_i in the fuzzy set A, and $1 \le i \le n$.

Definition 1

Imprecise data at equally spaced discrete time points are modeled as fuzzy variables. The set of this discrete fuzzy data forms a fuzzy time series.

Definition 2

Chronological sequences of imprecise data are considered as time series with fuzzy data. A time series with fuzzy data is referred to as fuzzy time series.

Definition 3

Let $Y(t), (t = \cdots, 0,1,2, \dots)$ be universe of discourse and $Y(t) \subseteq R$. Assume that $f_i(t), i = 1,2, \dots$ is defined in the universe of discourse $Y(t)$ and $F(t)$ is collection of $f(t_i), (i = \cdots, 0,1,2, \dots)$. Then $F(t)$ is called a fuzzy time series of $(t), i = 1,2, \dots$. Using fuzzy relation, we define $F(t) = F(t-1) \circ R(t, t-1)$ where $R(t, t-1)$ is a fuzzy relation and " $\circ$ " is a max-min composition operator, then $F(t)$ is caused by $F(t-1)$ where $F(t)$ and $F(t-1)$ are fuzzy set.

Definition 4

The membership function of the triangle is marked by the presence of three (3) parameters fuzzy $\tilde{a} = (\alpha_1, \alpha_2, \alpha_3)$, with the$(\alpha_1, \alpha_2, \alpha_3) \in \mathbb{R}$R that will determine x coordinate from three angles. This curve is essentially a combination of the two lines. As for the triangular form of the equation for this is:

$$\mu_{\tilde{a}}(x) = \begin{cases} \dfrac{x - \alpha_1}{\alpha_3 - \alpha_1} & , \quad \alpha_1 \le x \le \alpha_2 \\ \dfrac{\alpha_3 - x}{\alpha_3 - \alpha_2} & , \quad \alpha_2 \le x \le \alpha_3 \\ 0 & , \quad other \end{cases}$$

2.2 First order method

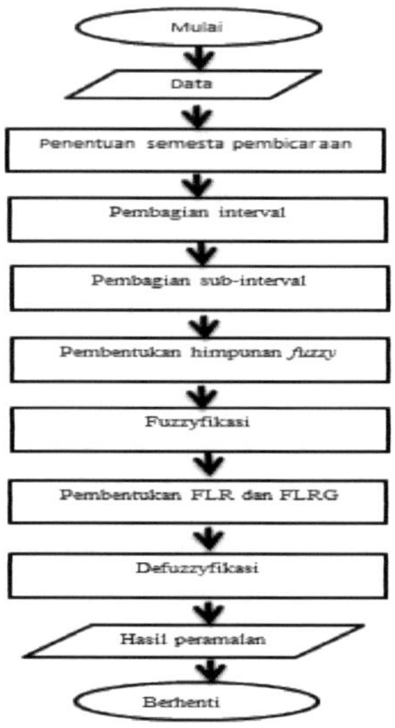

Figure 1. Flowchart of first order forecasting

2.3 Simulation

Step 1 Determine the universe of discourse

Determine the universe of discourse[2]. From the table, we knew that the value of
$X_{min} = 548648.30$ dan $X_{max} = 286814.20$

$$U = [X_{min} - D_1, X_{max} + D_2]$$

$$U = [250000\,; 550000]$$

Step 2 Find the number of intervals

Find the number of intervals with the mean[3].

$|\bar{X}| = 16536.045$

$J = X_{max} - X_{min} = 81719.6$

$\frac{|\bar{X}|}{2} = 8268.0225$

$$I = \frac{J}{basis} = \frac{81719.6}{8000} = 10.21495$$

Since the data is single data and single median data, the number of intervals is an odd number. 10.21495 are rounded to the odd 11 number of the nearest number.

Step 3 Divide into sub-interval

Divide into sub-interval with partitioning based on mean[4]. The division returned the average base intervals according to their respective frequency, if there is no data that is distributed in the interval (frequency is zero) then the interval will be discarded.

Table 1. Fuzzy sub-interval using mean based partitioning

	Sub-Interval
U1	[250000 ; 277272.73]
U2	[277272.73 ; 304545.45]
U3	[304545.45 ; 331818.18]
⋮	⋮
U11	[522727.27 ; 550000]

We have assumed 11 partitions of the universe of discourse of the main factor fuzzy time series.

Step 4 define a fuzzy triangle based on table

Let $A_1, A_2, \dots, A_k$ be the fuzzy sets which are linguistic values of the linguistic values of the linguistic varible IHSG data. In this paper, there are 11 linguistic values, as formula .

From u_1 triangular membership functions can be made from the interval [250000 ; 277272.73]

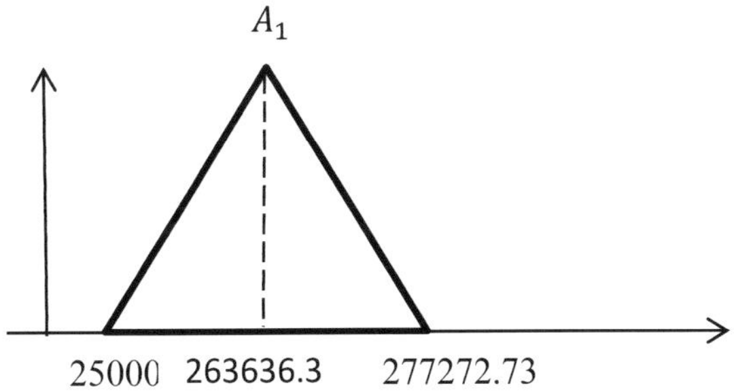

Figure 2. Membership Function Of Triangular Fuzzy Numbers A_1

$$\mu_{\tilde{a}}(x) = \begin{cases} \dfrac{x - 25000}{277272.73 - 263636.3}, & 4194 \le x \le 4242{,}5 \\ \dfrac{277272.73 - x}{277272.73 - 263636.3}, & 4242{,}5 \le x \le 4291 \\ 0, & other \end{cases}$$

And further by replacing the as linguistic . Based on the fuzzify historical enrollments obtained in step 4, we can get the fuzzy logical relationship group (FLRG).

Table 2. Fuzzified historical data

Year	Actual Data	Fuzzification
1996	548648.30	A1
1997	548648.30	A1
⋮	⋮	⋮
2015	286814.20	A2
2017	292373.80	A2

Step 5 Determine first-order relation (FLR &FLRG)

Logic Relationship between two consequent variationsis as show **Table 2** we assume that the fuzzy logical relationship "$A_p \to A_q$" denotes that " if the fuzzified enrollments of year q are A_q respectively, then the fuzzified enrollments of year (p) is A_p".

FLRG for A_1 is $A_1 \to A_1, A_1 A_9$
Then FR left group strarting from identical left –hand sides
The same rules for FLRGA_2 , For A_2 FLRGnya $A_2 \to A_2, A_2 A_2$

Step 6 Defuzzyfikasi

For the determination of the value of forecasting by determining the value prediction by defuzzification of linguistic forms through the Song and Chissom, its used principles that following :

$$if \begin{cases} 0 \quad \forall \mu = 0 \\ mid\ point \quad \forall\ \mu\ has\ one\ maximum \\ mid\ point\ of\ corresponding\ conjunct\ interval \quad \forall\ \mu\ has\ more\ than\ one\ maximum \end{cases}$$

Step 7 fault finding forecasting

Jilani, Burney, and Ardil (2007) using the AFER (Average Forecast Error Rate) and MSE (Mean Squared Error) methods to find the magnitude of the irregularities which occurred in the data results of the forecasting actual data. As for the calculation of AFER:

$$AFER = \frac{\frac{|X_i - F_i|}{X_i}}{n} \times 100\%$$

$$MSE = \frac{\sum_{i=1}^{n}(X_i - F_i)^2}{n}$$

We use MSE and AFER to compare the forecasting result of different forecasting methods, where A_i denotes the actual enrollment and F_i denotes the forecasting enrollment of year i.

Table 3. Criteria of AFER

Criteria AFER	
AFER	Criteria
<10%	Very good
10%-20%	Good
20%-50%	Good enough
>50%	Bad

3 Results and Discussion

In this paper, we apply a new algoritm method to forecast the petrolium data from the year 1996-2017 which obtained from bps.co.id. Using the first-order fuzzy time series based on frequency density partitioning method, we compare the result of first-order fuzzy time series based on frequency density partitioning with error 0,033 % and Error of the first-order at the (Melike Sah and Konstantin 2005) give about 2% until 3 %.

Table 4. Forecasting result, AFER and MSE

Year	**Actual Data**	**Proposed method**
1996	548648.30	336363.6364
1997	543752.60	336363.6364
⋮ 2010	⋮ 344888.00	340000
2017	292373.80	290909.0909
MSE		6382409916
AFER		0.0330083%

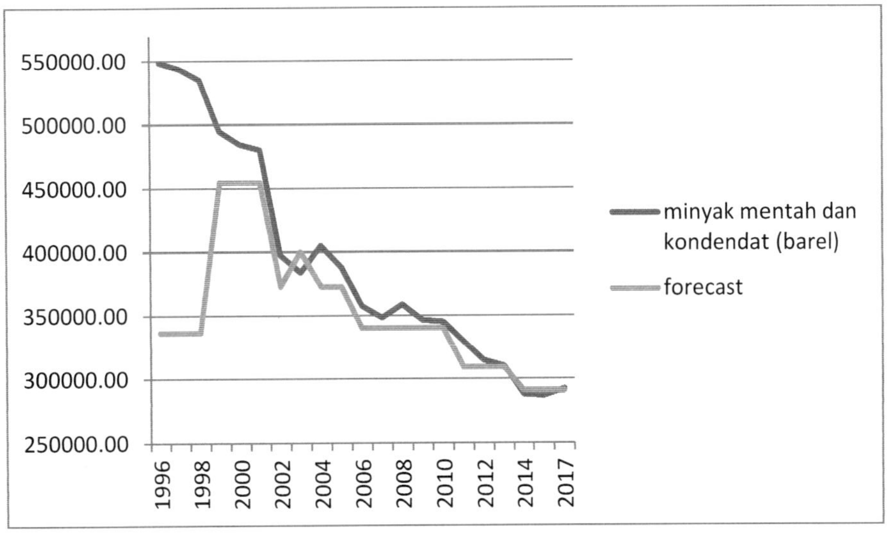

Figure 3. Forecasting result and actual data

4 Conclusion

Using the first-order fuzzy time series based on frequency density partitioning method, we compare the result of first-order fuzzy time series based on frequency density partitioning with error 0,033 % and Error of the first-order at the (Melike Sah and Konstantin 2005) give about 2% until 3 %.

References

[1] M. Sah and Konstantin, "Frecasting Enrollment Model Based and First-Order Fuzzy Time Series," *Proceedings of World Academy of Science Engineering and Technology.*, vol. 1, pp. 1307-6884, 2003.

[2] T.A. Jilani, S.M.A. Burney and C. Ardil "Fuzzy Metric Approach for Fuzzy time series Forecasting based on Frequency Density based Partitioning," *International Journal of Computer and Information Engineering*, vol. 4, pp. 1194 – 1199, 2007.

[3] L. Abdullah, "Performance of exchange rate forecast using distance-based fuzzy time series,"*International Journal of Engineering and Technology (IJET)*, vol.5. pp. 452 – 459, 2007.

[4] W.Wu, "Fuzzy reasoning and fuzzy relational equations,", *Fuzzy Sets and Systems*, vol. 20, pp. 67 – 78.

[5] S.M. Chen, "Forecasting Enrollments Based on Fuzzy Time Series,", *Fuzzy Sets and Systems*, vol. 81, pp. 311–319, 1986.

[6] S.M. Chen and K. Tanuwijaya,"Fuzzy Forecasting Based on High-order Fuzzy Logical Relationships and Automatic Clustering Techniques," *Expert Systems with Applications,* vol.3, pp. 15425–15437, 2011.

[7] R.C. Tsaur, "A fuzzy time series-markov chain model with an application to forecast the exchange rate between the Taiwan and US Dollar,"*International Journal of Innovative Computing. Information and Control*, vol.8, pp. 4931 – 492, 2011.

[8] K. Huarng, "Heuristic models of fuzzy time series for forecasting," *Fuzzy Sets and Systems*, vol.123, pp. 369 – 386, 2001.

[9] S.M. Chen, "Forecasting Enrollment on Fuzzy Time Series,"*Fuzzy Sets and System,* vol.81, pp. 311-319,1996.

[10] Q. Song and B,S. Chissom, "Forecasting Enrollment with Fuzzy Time Series Part-1,"*Fuzy Sets and System,* vol.pp.54 1-9,1993.

[11] Q. Song and B,S. Chissom, "Forecasting Enrollment with Fuzzy Time Series Part-2", *Fuzy Sets and System*, vol.62.pp. 1-8 1994.

[12] A.M. Masykur dan F.A. Halim, " Mathematical Intelegence," *AR-Ruzz Media* 2008.

[13] L. A. Zadeh, "Fuzzy Sets", *Information and Control*, vol.8.pp.338-353,1965.

IOP Conf. Series: Materials Science and Engineering **846** (2020) 012064 doi:10.1088/1757-899X/846/1/012064

Double Seasonal ARIMA for Forecasting Electricity Demand of Kuaro Main Gate in East Kalimantan

M Azka[1*], SA Wiradinata[1], M Faisal[1], and Suhartono[2]

[1] Department of Mathematics, Institut Teknologi Kalimantan, Indonesia.
[2] Department of Statistics, Institut Teknologi Sepuluh November, Indonesia.

*Email: muhammad.azka@lecturer.itk.ac.id

Abstract. This study uses intraday electricity load demand data from Kuaro Main Gate data in East Kalimantan as the basis of an empirical comparison of Double Seasonal ARIMA models for prediction up to a day ahead. For the purpose of this study, a one-year hourly Kuaro Main Gate data load demand from 1 January 2018 to December 2018 measured in Megawatt (MW) is used. In multiple times of load demand data, in addition to intraday and intra week cycles, and intra year seasonal cycle is also apparent. We extend the Double Seasonal ARIMA methods in order to accommodate the Intra year seasonal cycle. The mean absolute percentage error (MAPE) is used as the measure of forecasting accuracy. A notable feature of the time series is the presence of both an intraweek and an intraday seasonal cycle. We also propose that a Double Seasonal ARIMA model with the one-step-ahead forecast as the most appropriate model for forecasting the two-seasonal cycles Kuaro Main Gate data load demand time series. We use the Statistical Analysis System package to analyze the data. Using the least-squares method to estimate the coefficients in a Double Seasonal ARIMA model, followed by model validation and model selection criteria, we propose the ARIMA $(1,1,1)(0,1,1)^{24}(0,1,1)^{168}$ within-sample MAPE of 0.000992 as the best model for this study. Comparing the forecasting performances by using k-step ahead forecasts and one-step-ahead forecasts, we found that the MAPE for the one-step ahead out-sample forecasts from any horizon ranging from one week lead time to one month one week lead time are all less than 5%. Therefore we propose that a double seasonal ARIMA model with a one-step-ahead forecast must be considered in forecasting time series data with two seasonal cycles.

1. Introduction

East Kalimantan-Indonesia's geographical status as an archipelago makes domestic electricity coverage and power generation a particularly daunting challenge. The issue of electricity pricing is now dominating public discourse on the sector as the government seeks to reduce the burden of electricity subsidies on the state budget. Electricity demand forecasting is of great potential for the management of power systems. Long-term forecasts of the peak electricity demand are needed for capacity planning and maintenance scheduling [1]. Medium-term demand forecasts are required for power system operation and planning. Short-term load forecasts are required for the control and scheduling of power systems. Short-term forecasts are also required by transmission companies when a self-dispatching market is in operation, power generation, and has exclusive powers over the transmission, distribution, and supply of electricity to the public. There are several such markets in Kalimantan and Indonesia. For example, in Indonesia, one hour-ahead forecasts are a key input to the balancing market, which operates on a rolling one hour-ahead basis to balance supply and demand after the closure of bi-lateral trading between generators and suppliers [2][3].

More generally, error in predicting electricity load has significant cost implications for companies operating in competitive power markets [4]. It is well recognized that meteorological variables have a very significant influence on electricity demand (see, for example, [5]). However, in short-term forecasting systems, multivariate modeling is usually considered impractical [6]. In such systems, the lead times considered are less than a day ahead, and univariate methods can be sufficient because the meteorological variables tend to change in a smooth fashion, which will be captured in the demand series itself. Indeed, univariate models are often used for prediction up to about four to six hours ahead, and, due to the expense or unavailability of weather forecasts, univariate methods are sometimes used for longer lead times. In a recent study [7], methods for short-term load forecasting are reviewed, and two intraday load time series are used to compare a variety of univariate methods. One of the aims of this paper is to validate the results of that study. It concluded that a double seasonal ARIMA was the most accurate method, with a new approach based on Box-Jenkins also performing well.

Due to academic interests and industrial needs, short term load forecasting has gained great attention compared to the others. Short term load forecasting plays an important role in a utility company because the accuracy of prediction will affect the power system operations. Forecast errors result in an unbalance between power supply and demand hence increase operations cost. Double seasonal ARIMA modeling has been used by many researchers, in a univariate modeling framework, as a comparison for evaluating alternative models (see [8]). One of the main problems that arise in modeling the use of short-term energy electricity with double seasonal ARIMA is the determination of the optimal order of AR and/or MA models. This is due to the complexity of the variable lag that can be used as the best candidate in the double seasonal ARIMA model. Until now, there is no single time series analysis literature that discusses double seasonal ARIMA models, especially the identification stages for determining the order of models that are in accordance with the data pattern. These orders tend to involve orders ranging from short lags too long lags, especially in the second seasonal order.

This paper will be structured as follows. We start by presenting the Box-Jenkins, double seasonal ARIMA model, specifically formulated to deal with the double seasonality that typically arises in load demand data. This seasonality involves intraday and intraweek seasonal cycles, and describe the details results of a double seasonal ARIMA model. We would hope that the best performing method in our study can serve as benchmarks in future studies with double seasonal ARIMA. Finally, we give our conclusions based on the forecasting evaluation method presented in this study.

The General ARIMA Models More generally, we can write a general ARIMA model as follows:

$$\prod_{j=1}^{M}\phi_j(B)\prod_{i=1}^{K}\left(1-B^{s_i}\right)^{d_i} Z_t = \theta_0 + \prod_{k=1}^{N}\theta_k(B)a_t \tag{1}$$

Thus, the model may contain *K* differencing factors, *M* autoregressive factors, and *N* moving average factors. This extension is useful in describing many nonstandard time series that, for example, may contain a mixture of seasonal phenomena of different periods. Because it is this general form that most time series software use, we now explain this general model in more detail.

Generally, a nonseasonal time series can be modeled as a combination of past values and past error, denoted as ARIMA*(p,d,q)* can be written as follows:

$$\phi_j(B)\left(1-B^{s_i}\right)^{d} Z_t = \theta_q(B)a_t \tag{2}$$

$$\phi_p(B) = 1-\phi_1 B - \phi_1 B^2 - \cdots - \phi_p B^p$$
$$\phi_q(B) = 1-\phi_1 B - \phi_2 B^2 - \cdots - \phi_q B^q$$

Where Z_t is appropriately transformed load demand in period t; $(1-B)^d$ is the nonseasonal differencing operator; B is the backward shift operator; and a_t is the purely random process. As an example let hence the model can be expressed as ARIMA (0,0,1). Consider the model can be written as follows:

$$\begin{aligned} Z_t &= (1+0,8395B)a_t \\ &= a_t + 0,8395a_{t-1} \end{aligned} \quad (3)$$

2. Methodology

A Modelling Strategy For The Double Seasonal ARIMA Model: In 1972 George E.P Box and Gwilym M. Jenkin developed a method for analyzing stationary univariate time series data. In this section, the importance and general nature of the ARIMA approach to time series analysis are discussed. The novel contributions of this method and limitations are explained. Prerequisites of Box-Jenkins models are defined and explored. Different types of nonstationarity are elaborated. We also discuss tests for detecting these forms of nonstationarity and expound on transformations to stationarity. We then review problems following from the failure to fulfill these prerequisites, as well as common means by which these problems may be resolved. Programming examples with both SAS and MINITAB are included. The modeling procedure of Box-Jenkins ARIMA model involves an iterative five-stage process as follows:
Step 1: Preparation of data including transformation and differencing.
Step 2: Identification of the potential models by looking at the sample autocorrelations and the partial autocorrelations.
Step 3: Estimation of the unknown parameters by some optimization methods.
Step 4: Checking the adequacy of the fitted model by performing a normal probability plot, ACF and PACF on model residuals.
Step 5: Forecast future outcomes based on the known data.

Measuring Forecast Accuracy: The Principal question is how can one compare and contrast several competing models to determine which is the better model. The better model will usually fit the data well. The general model goodness of fit needs to be evaluated. Commonly used measures of goodness/lack of fit include the mean error, the mean percent error, the mean absolute error, and the mean absolute percentage error. Applied to forecasts, these measures are:

Mean absolute error $= \frac{1}{T}\sum_{t=0}^{T}|y_t - \hat{y}_t|$

Mean absolute percentage error $= \frac{100}{T}\sum_{t=0}^{T}\left|\frac{(y_t - \hat{y}_t)}{y_t}\right|$

Where T is the total number of time periods (number of observations), is the actual value and is the forecast value at time t. These are average measures of percent and absolute error that can be used as indicators of forecast accuracy. Although there is no single absolute level above which the model is unacceptable. Data Set: The data used is half year hourly load demand measured in Megawatt (MW) from August 01, 2017 to December 31, 2017. They are gathered from PLN AP2B SISTEM KALTIM-Balikpapan, East Kalimantan electricity utility company, Balikpapan, Indonesia. PLN (The State-owned electricity company) is one of the biggest and most well-managed power companies in Indonesia in which this utility company has powered for decades through the generation, transmission, and distribution of electricity.

3. Result and Discussion

Plotting the load demand series after three-time differencing which are non-seasonal differencing (d = 1), daily seasonal differencing (D1 = 1, s1 = 24) and weekly seasonal differencing (D2 = 1, s2 = 168) in Figure 1 indicates that load series is stationary series. In order words, this identification step shows that the load data have two seasonal periods, which are daily and weekly seasonality with length 24 and 168 respectively. We then present the ACF and the PACF of the stationary load demand series after three-time differencing in Figure 2.

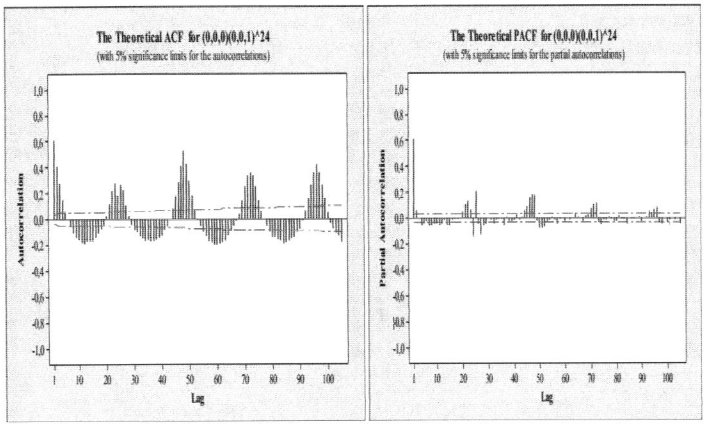

Figure 1. Estimated autocorrelation function of hourly electricity demand in Kuaro Main Gate in East Kalimantan

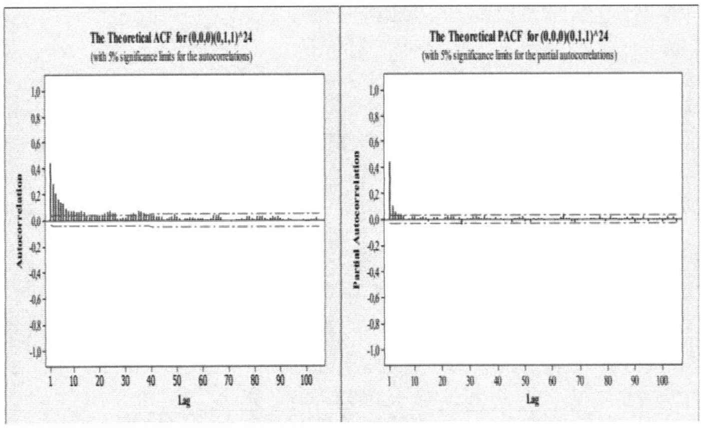

Figure 2. Estimated autocorrelation function of hourly electricity demand in Kuaro Main Gate-East Kalimantan

A Modeling Strategy For The Double Seasonal Arima Model

1. Model 1 : Double Seasonal ARIMA(0,1, 1) $(0,1,1)^{24}$ $(0,1,1)^{168}$

All the parameters of this model are significant at alpha 5% significance level with white noise residuals based on Ljung-Box statistic Q* until lags 24. This model gives 5 extreme residual values. In terms of the magnitude of the residuals, there are at Observations. The model's residuals, however, do not satisfy the Normal Distribution because of the presence of outliers in the data. We considered lag polynomials up to order three. This choice was made arbitrarily, but it is consistent with other load forecasting studies, and it was supported by experimentation with several of the series. We based model selection on the Schwarz Bayesian Criterion, with the requirement that all parameters were significant (at the 5% level). We compared the Schwartz Bayesian information criterion (SBC) for an extensive range of different ARIMA models. The AIC and SBC of this model are 62680.53 and 62800.14 respectively.

2. Model 2 : Double Seasonal ARIMA([2,4,5,7,9,11,12,20,21,22,23,24,48],1,1)$(0,1,1)^{24}(0,1,1)^{168}$
All the parameters of this model are significant at alpha 5% significance level with white noise residuals based on Ljung-Box statistic Q* until lags 24. This model gives 5 extreme residual values. In terms of the magnitude of the residuals, there are at Observations. The model's residuals, however, do not satisfy the Normal Distribution because of the presence of outliers in the data. We compared the Schwartz Bayesian information criterion (SBC) for an extensive range of different ARIMA models. The AIC and SBC of this model are 12914.73 and 13013.2 respectively.

3. Model 3 : Double Seasonal ARIMA (1,1,[1,2,4,5,7,9,11,12,20,21,22,23,24,48]) $(0,1,1)^{24}$ $(0,1,1)^{168}$
All the parameters of this model are significant at alpha 5% significance level with white noise residuals based on Ljung-Box statistic Q* until lags 24. This model gives 5 extreme residual values. In terms of the magnitude of the residuals, there are at Observations. The model's residuals, however, do not satisfy the Normal Distribution because of the presence of outliers in the data. We compared the Schwartz Bayesian information criterion (SBC) for an extensive range of different ARIMA models. The AIC and SBC of this model are 12752.78 and 12857.4 respectively.

4. Conclusion
We investigated the double seasonal ARIMA model for forecasting the double seasonal (daily and weekly Kuaro Main Gate-East Kalimantan data load demand time series. Comparing the forecasting performances by using k-step ahead forecasts and one-step-ahead forecasts, we found that the MAPE for the one-step ahead out-sample forecasts for any horizon ranging from one week lead time to one month lead time as tabulated in Table 3 are all less than 5%. The MAPE of the k-step ahead forecasts, on the other hand, increases as the time scale increases. It can be concluded that the one-step-ahead forecasts are not greatly influenced by the lead times and are more accurate than k-step ahead forecasts in forecasting Kuaro Main Gate-East Kalimantan load demand.

Acknowledgment
This research was supported by Institut Teknologi Kalimantan under its LPPM with contract number: 1996/IT10.III/PPM.01/2019. The researchers are grateful for this support. The researchers would also like to thank AP2B SITEM KALTIM PLN Balikpapan, East Kalimantan for providing the load data. We also wish to express our appreciation to all friends who help us directly and indirectly.

References
[1] P. E. Mcsharry, S. Bouwman, and G. Bloemhof, "Probabilistic Forecasts of the Magnitude and Timing of Peak Electricity Demand," vol. 20, no. 2, pp. 1166–1172, 2005.
[2] E. González-romera, M. Á. Jaramillo-morán, and D. Carmona-fernández, "Monthly Electric Energy Demand Forecasting Based on Trend Extraction," vol. 21, no. 4, pp. 1946–1953, 2006.
[3] J. W. Taylor, "Density forecasting for the efficient balancing of the generation and consumption of electricity," vol. 22, pp. 707–724, 2006.
[4] D. W. Bunn, "Forecasting Loads and Prices in Competitive Power Markets," vol. 88, no. 2, pp. 163–169, 2000.
[5] J. W. Taylor and R. Buizza, "Neural Network Load Forecasting with Weather Ensemble Predictions," no. September, p. 2002, 2002.
[6] D. W. Bunn, "Short-Term Forecasting : A Review of Procedures in the Electricity Supply Industry," vol. 33, pp. 533–545, 1982.
[7] J. W. Taylor, L. M. De Menezes, and P. E. Mcsharry, "A comparison of univariate methods for forecasting electricity demand up to a day ahead," vol. 22, pp. 1–16, 2006.
[8] G. A. Darbellay and M. Slama, "Forecasting the short-term demand for electricity Do neural networks stand a better chance ?," vol. 16, pp. 71–83, 2000.

IOP Conf. Series: Materials Science and Engineering **846** (2020) 012065 doi:10.1088/1757-899X/846/1/012065

Comparative Analysis of Mean Variance Efficient Frontier and Resampled Efficient Frontier For Optimal Stock Portfolio Formation

Emy Siswanah[1]

[1]Matemathics Departement UIN Walisongo Semarang

emysiswanah@walisongo.ac.id

Abstract. One way to form an optimal stock portfolio is to use the mean-variance efficient frontier (MVEF) method. However, the MVEF method is susceptible to changes in input. An optimal stock portfolio formation method has been developed to overcome this problem, known as the Resampled Efficient Frontier (REF) method. The REF method is a refinement of the MVEF method. REF is done by generating data repeatedly using the Montecarlo simulation. The parameters needed to form a stock portfolio using REF are the average input return μ and the covariance variance matrix Σ. The calculation results show that the weight of the REF portfolio is not negative (there is no short selling strategy). The Sharpe ratio of the REF portfolio is also higher than the MVEF. A larger Sharpe ratio indicates that the REF portfolio has a better performance compared to MVEF. However, empirically based on a case study simulation, the MVEF method experienced a smaller loss compared to the REF method.

1. Introduction

Modern portfolio optimization theory was put forward by Markowitz [1], known as the Mean-Variance Efficient Frontier (MVEF). In mean-variance optimization, it needs to mean parameter and covariance matrix estimated from the return data sample. MVEF itself also developed by many researchers such as Brito and Vicente [3], Guo et.el. [4], Ismail and Pham [5], Halliwell [6], Suzuki and Tanaka [7], Steinbach [8], Gasser et al. [9] for perfecting the portfolio optimization theory.

One of the purposes of MVEF development is to solving sensitivity towards input parameter changes. The mean parameter estimation and covariance matrix on the MVEF theory creates a fluctuating portfolio weight over time. Small input change can cause a significant change in portfolio composition. This change makes bad portfolio results for an extended period. To solving the mean-variance instability in the Markowitz model is to use the Resample Efficient Frontier (REF) method. REF is a concept in asset management introduced by Michaud [2]. A comprehensive Resampled Efficient Frontier (REF) technique is used to perfect Markowitz's Mean-Variance input in determining an efficient portfolio [10]. Maginn et al. [11] state that the efficient portfolio composition shown by Mean-Variance optimization is susceptible to changes in input. This was also explained by Fong [12], who stated that the limitations on the efficiency of Mean-Variance, especially on input information that was too specific. Mean-Variance Efficiency assumes a 100% confidence level for optimization inputs, a condition that never exists so that asset managers do not have perfect forecast information and find it challenging to implement optimization procedures.

According to Maginn et al. [11] and Fong [12], the results of a single Mean-Variance Optimization (Single Period) are less convincing. Sometimes professional investors carry out repeated optimization using the input range around the estimation point to measure the sensitivity of the optimization results on various inputs. Some researchers have tried to overcome the problem of input sensitivity optimization of Mean-Variance, among others Jobson and Korkie [13-14], which showed a bias in optimizing portfolio weights and then developed by Michaud [2, 15] and Jorion [16]. The asset allocation approach developed by Michaud uses a Mean-Variance optimization simulation and a set of historical data [11]. The estimated effect of an error on a portfolio can be summarized using a Monte Carlo procedure known as resampling [17]. Jiao [18] states that the resampled portfolios method shows a smooth transition of asset proportions for each desired change in return. In the REF method, the data generated by the Monte Carlo simulation will produce a different efficient frontier. Then the average is taken to get a resampled efficient frontier. The REF method guarantees that the total portfolio weight will remain equal to one.

2. Methodology

This study is case study research by using Markowitz's portfolio selection [1] and optimizations [2] to obtain the formula of the mean-variance efficient frontier and resampled efficient frontier. After obtaining formulas from the two methods, those formulas were applied to several shares to obtain a percentage of the weight of each share. Then the share percentage results are compared based on the Sharpe ratio. Furthermore, the percentage weight of shares is compared empirically through a stock purchase simulation. The results of the purchase of shares are compared to find out whether the portfolio gains or loses. Sharpe ratio comparison and stock purchase simulation are then analyzed and concluded.

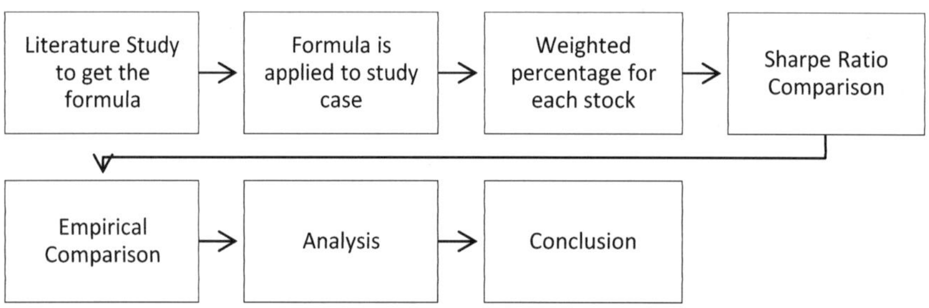

Figure 1. Research Flow

3. Result and Discussion

In portfolio formation, each stock has a contribution to total capital with weight $\mathbf{w}$ where $\mathbf{w} = (w_1, ..., w_k)^T$. Suppose the investor has a portfolio that consists of k stock. Defined p_{ij} is asset price j on time i and *return* from an asset j for one period of time is

$$r_{ij} = \frac{p_{i,j} - p_{i-1,j}}{p_{i-1,j}}$$

Data collected from daily return can be shown in the matrix $\mathbf{R} = (r_{ij})$. Lorentz [19] said that Portfolio *Return* is weighted addition from *k stock return* which is

$$R_p = w_1 r_1 + \cdots + w_k r_k = \sum_{i=1}^{k} w_i r_i = \mathbf{w}^T \mathbf{r} \quad (1)$$

where $\mathbf{w} = (w_1, w_2, \dots, w_k)^T$ and $\mathbf{r} = (r_1, r_2, \dots, r_k)^T$ is a weight vector and average stock return vector from each stock. Portfolio expected return is a weighted addition from the expected return of each stock

$$E(R_p) = w_1 \mu_1 + \cdots + w_k \mu_k = \sum_{i=1}^{k} w_i \mu_i = \mathbf{w}^T \boldsymbol{\mu} \quad (2)$$

where $\boldsymbol{\mu} = (\mu_1, \mu_2, \dots, \mu_k)^T$ is an expected return vector from each stock. Portfolio risk is a value of expected profit deviates from its expectation. Portfolio risk is calculated from the volatility of the portfolio, which is the variance from R_p.

$$Var(R_p) = \mathbf{w}^T \Sigma \mathbf{w} \quad (3)$$

where $\Sigma = (\sigma_{ij})_{ij}$ is the variance-covariance matrix of stock return.

MVEF Markowitz's portfolio formation, according to Nyholm [20], is conducted with minimizing the risk on a particular return. Portfolio weight is obtained with the specification of minimizing $\frac{1}{2}\mathbf{w}^T \Sigma \mathbf{w}$ with rule $\mathbf{w}^T \boldsymbol{\mu} = \check{r}$ dan $\mathbf{w}^T \mathbf{1} = 1$. $\mathbf{w}$ notate weight of each stock from the portfolio, Σ notate covariance matrix, $\boldsymbol{\mu}$ shows vector return of each stock, and $\check{r}$ shows expected return According to Markowitz [21] optimation problem $\frac{1}{2}\mathbf{w}^T \Sigma \mathbf{w}$ with rule $\mathbf{w}^T \boldsymbol{\mu} = \check{r}$ and $\mathbf{w}^T \mathbf{1} = 1$ can be solved with the Lagrange method :

$$L(\mathbf{w}, \lambda_1, \lambda_2) = \frac{1}{2}\mathbf{w}^T \Sigma \mathbf{w} - \lambda_1(\mathbf{w}^T \boldsymbol{\mu} - \check{r}) - \lambda_2(\mathbf{w}^T \mathbf{1} - 1) \quad (4)$$

There are two constraints on the model (1) which $\mathbf{w}^T \boldsymbol{\mu} = \check{r}$ dan $\mathbf{w}^T \mathbf{1} = 1$. Optimation result of the model (4) with the Lagrange method is

$$\mathbf{w} = \lambda_1 \Sigma^{-1} \boldsymbol{\mu} + \lambda_2 \Sigma^{-1} \mathbf{1} \quad (5)$$

Model (5) can be explained below

$$\begin{bmatrix} \check{r} \\ 1 \end{bmatrix} = \begin{bmatrix} \boldsymbol{\mu}^T \Sigma^{-1} \boldsymbol{\mu} & \mathbf{1}^T \Sigma^{-1} \boldsymbol{\mu} \\ \mathbf{1}^T \Sigma^{-1} \boldsymbol{\mu} & \mathbf{1}^T \Sigma^{-1} \mathbf{1} \end{bmatrix} \begin{bmatrix} \lambda_1 \\ \lambda_2 \end{bmatrix} \quad (6)$$

If supposed $P = \boldsymbol{\mu}^T \Sigma^{-1} \boldsymbol{\mu}$; $Q = \mathbf{1}^T \Sigma^{-1} \boldsymbol{\mu}$; and $R = \mathbf{1}^T \Sigma^{-1} \mathbf{1}$ the model (3) can be simplified into

$$\begin{bmatrix} \check{r} \\ 1 \end{bmatrix} = \begin{bmatrix} P & Q \\ Q & R \end{bmatrix} \begin{bmatrix} \lambda_1 \\ \lambda_2 \end{bmatrix} \quad (7)$$

Model (4) system can be solved with using Cramer rule, and the value is

$\lambda_1 = \frac{R\check{r} - Q}{S}$ dan $\lambda_2 = \frac{P - Q\check{r}}{S}$ dengan $S = PR - Q^2$.

Portfolio weight to obtain minimum risk portfolio on a particular return level based on the mean-variance efficient frontier method is

$$\mathbf{w}_p = \frac{1}{S}\left[(P\Sigma^{-1}\mathbf{1} - Q\Sigma^{-1}\boldsymbol{\mu}) + (R\Sigma^{-1}\boldsymbol{\mu} - Q\Sigma^{-1}\mathbf{1})\check{r}\right] \quad (8)$$

$$\boldsymbol{w}_{spec} = \frac{1}{S}(R\Sigma^{-1}\boldsymbol{\mu} - Q\Sigma^{-1}\bar{\mathbf{1}})\check{r} \quad (9)$$

$$\boldsymbol{w}_{min} = \frac{1}{S}(P\Sigma^{-1}\bar{\mathbf{1}} - Q\Sigma^{-1}\boldsymbol{\mu}) \quad (10)$$

Scherer [17] states that optimal portfolio weights can be split into minimum variance weights (equation (10)) and speculative weights (equation (9)), known as two-fund separation (two-fund separation). Calculation of optimal weights for assets in the Global Minimum Variance Portfolio (GMVP) can also use equation (11).

$$\boldsymbol{w} = \frac{\Sigma^{-1}\bar{\mathbf{1}}}{\bar{\mathbf{1}}^T\Sigma^{-1}\bar{\mathbf{1}}} \tag{11}$$

Meucci [10], Scherer [17], and Kohli [22] state that the Mean-Variance approach is a two-step simplification of the problem of asset allocation, namely:

1. Determine assets and calculate returns and standard deviations or volatility for each asset based on historical data
2. Determine Frontier Efficient

Portfolio weights are determined using equation (8) with a certain number of returns ($\check{r}$). Portfolios formed are Efficient Frontier. Efficient Frontier is a collection of portfolios that have the smallest risk at a specific rate of return. A portfolio can be selected from efficient portfolios on the Efficient Frontier based on a certain level of risk or maximum Sharpe ratio.

The Mean-Variance optimization problem with I shows the number of efficient frontiers can be formulated as follows:

$$\mathbf{w}^{(i)} = \min_{\substack{\boldsymbol{w}^T\bar{\mathbf{1}}=1 \\ \boldsymbol{w}^T\boldsymbol{\mu}}} \mathbf{w}^T\Sigma\mathbf{w}, \qquad i = 1,2,..,I \tag{12}$$

The μ and Σ symbols in equation (12) indicate the expectation and covariance values. Whereas the target return value for each point on the efficient frontier is $\check{r} = \{\check{r}^{(1)}, \check{r}^{(2)}, \dots, \check{r}^{(I)}\}$.

Procedure for establishing MVEF

1. The formation of efficient frontier using the Mean-Variance method begins by calculating the mean and return covariance as input to the optimization process
2. Determine the upper and lower limits of return expectations
   The lower limit (LB) in Efficient Frontier is obtained by calculating the return from the global minimum variance portfolio (GMVP) while the upper limit (UB) efficient frontier is the biggest mean return among all stocks.
3. Determine the increment
   The efficient frontier increment is determined by counting
   $$\delta = \frac{(UB - LB)}{I - 1}$$
   so get I different expected return value, i.e.
   $$\check{r} = \{LB, LB + \delta, \dots, UB - \delta, UB\}$$
4. Optimization
   Determination of portfolio weights uses equation (8) with a number of I different expected return values ($\check{r}$) to obtain I portfolio. The optimization process is carried out at each point on the efficient frontier to obtain the optimum portfolio.

The determination of the optimal portfolio using the REF method is not much different from the MVEF method. The difference is that after obtaining input data in the form of $\boldsymbol{\mu}$ and Σ, a Monte Carlo simulation is performed to generate a number of data. The Monte Carlo simulation is done by generating random return data based on the values $\boldsymbol{\mu}$ and $\boldsymbol{\Sigma}$ obtained. Data are generated randomly from the normal distribution with an average $\boldsymbol{\mu}$ and standard deviation $\boldsymbol{\sigma}$, as many observations made. The process of resampling to determine the weight of an efficient portfolio is:

1. Estimate input value $\hat{\mu}_0$ dan $\hat{\Sigma}_0$ as the working structure of Mean-Variance from observation for T period and asset of k on historical data.

$$\hat{\mu}_0 = \begin{bmatrix} \hat{\mu}_1 \\ \hat{\mu}_2 \\ \vdots \\ \hat{\mu}_k \end{bmatrix} \quad dan \quad \hat{\Sigma}_0 = \begin{bmatrix} \hat{\sigma}_1^2 & \hat{\sigma}_{12} & \cdots & \hat{\sigma}_{1k} \\ \hat{\sigma}_{21} & \hat{\sigma}_2^2 & \cdots & \hat{\sigma}_{2k} \\ \vdots & \vdots & \ddots & \vdots \\ \hat{\sigma}_{k1} & \hat{\sigma}_{k2} & \cdots & \hat{\sigma}_k^2 \end{bmatrix}$$

with

$$\hat{\mu}_a = \frac{1}{T}\sum_{t=1}^{T} R_t$$

And

$$\hat{\sigma}_k^2 = \frac{1}{T}\sum_{t=1}^{T}(R_t - \hat{\mu}_a)^2, \quad a = 1,2,\dots,k$$

Observed return realization :

$$i_k = \{r_1, r_2, \dots, r_k\} \quad or \quad r_1 \sim N(\hat{\mu}_1, \hat{\sigma}_1^2) \quad \vdots \quad r_k \sim N(\hat{\mu}_k, \hat{\sigma}_k^2) \tag{13}$$

2. Based on return i_A, distribution of return for every asset :

$$I_k = \{R_1, R_2, \dots, R_k\} \tag{14}$$

Make assumptions about the results of the return distribution (13), for example, with the assumption of normality and set the estimated parameters as the actual parameters to determine the return distribution.

$$R_a \sim N(\hat{\mu}_0, \hat{\Sigma}_0) \tag{15}$$

Resample the realization of Q as many as the Monte Carlo scenario in equation (13) based on the parameters in the distribution (14)

$$i_{qA} = \{r_{q1}, r_{q2}, \dots, r_{qA}\} \qquad q = 1,2,..,Q \tag{16}$$

By performing the sampling procedure repeatedly Q times, obtained as many as Q optimization input sets $(\hat{\mu}_1, \hat{\Sigma}_1)$ until $(\hat{\mu}_Q, \hat{\Sigma}_Q)$

3. Estimating inputs $\hat{\mu}_q$ and $\hat{\Sigma}_q$ based on the Mean-Variance framework from the results of the resample in step 2 in a manner like step 1

$$\hat{\mu}_1 = \begin{bmatrix} \hat{\mu}_1 \\ \hat{\mu}_2 \\ \vdots \\ \hat{\mu}_k \end{bmatrix} \quad and \quad \hat{\Sigma}_1 = \begin{bmatrix} \sigma_1^2 & \sigma_{12} & \cdots & \sigma_{1k} \\ \sigma_{21} & \sigma_2^2 & \cdots & \sigma_{2k} \\ \vdots & \vdots & \ddots & \vdots \\ \sigma_{k1} & \sigma_{k2} & \cdots & \sigma_k^2 \end{bmatrix}$$

$$\vdots$$

$$\hat{\mu}_Q = \begin{bmatrix} \hat{\mu}_1 \\ \hat{\mu}_2 \\ \vdots \\ \hat{\mu}_k \end{bmatrix} \quad and \quad \hat{\Sigma}_Q = \begin{bmatrix} \sigma_1^2 & \sigma_{12} & \cdots & \sigma_{1k} \\ \sigma_{21} & \sigma_1^2 & \cdots & \sigma_{2k} \\ \vdots & \vdots & \ddots & \vdots \\ \sigma_{k1} & \sigma_{k2} & \cdots & \sigma_k^2 \end{bmatrix}$$

4. Determining the point of Efficient Frontier
   a. Calculating global minimum variance portfolio (GMVP) for every optimation input

$$\mathbf{w}_{qMV} = \min_{w^T\mathbf{1}=1} \mathbf{w}^T\hat{\Sigma}_q\mathbf{w} \qquad q = 1,2,..,Q \tag{17}$$

   b. Calculating estimation of every expected value

$$\check{\underline{r}}_q = (\mathbf{w}_{qMV})^T \hat{\mu}_q \qquad q = 1,2,..,Q \tag{18}$$

   c. Calculating estimation of maximum expected value

$$\check{\bar{r}}_q = \max\{\hat{\mu}_q^T\sigma^{(1)}, \hat{\mu}_q^T\sigma^{(2)}, \dots, \hat{\mu}_q^T\sigma^{(N)}\} \tag{19}$$

A basis in vector space is a set of linear and independent elements in that space which is $e^{(n)}$ with $n = 1, 2, \ldots, N$, and then δ is canonical basis vector itself.

$$\delta^{(1)} = (1, 0, \ldots, 0)^T$$
$$\vdots$$
$$\delta^{(N)} = (1, 0, \ldots, 0)^T$$

Determine the target return value $\left\{\check{r}_q^{(1)}, \check{r}_q^{(2)}, \ldots, \check{r}_q^{(I)}\right\}$ for each q :

$$\check{r}_q^{(1)} = \underline{\check{r}}_q$$
$$\vdots$$
$$\check{r}_q^{(i)} = \underline{\check{r}}_q + \frac{\overline{\check{r}}_q - \underline{\check{r}}_q}{I - 1}(i - 1)$$
$$\vdots$$
$$\check{r}_q^{(I)} = \overline{\check{r}}_q \tag{20}$$

Solving Mean-Variance problem for every Monte Carlo q scenario = 1, 2, …, Q and every target return value $i = 1, 2, \ldots, I$:

$$\mathbf{w}_q^{(i)} = \min_{\substack{w^T\mathbf{1}=1 \\ w^T\hat{\mu}_q}} \mathbf{w}^T \hat{\Sigma}_q \mathbf{w} \tag{21}$$

5. Define *Resampled Efficient Frontier*

$$\mathbf{w}_{resampled}^{(i)} = \frac{1}{Q}\sum_{q=1}^{Q} \mathbf{w}_q^{(i)}, \qquad i = 1, 2, \ldots, I \tag{22}$$

In summary the steps for Resampled Efficient Frontier (REF)

1. The process of forming an efficient frontier begins by determining the estimated value of the parameters (μ and ∑) used to generate return data as many as T observations according to the number of observations on historical data
2. Based on the estimated parameter values, data returns are generated repeatedly randomly from the normal distribution according to the number of simulations
3. Determine the lower limit (LB) which is calculated based on the return on the global minimum variance portfolio (GMVP) and the upper limit (UB) expectation of return which is the largest mean return for each data set
4. Increment in the efficient frontier is determined by counting
$$\delta = \frac{(UB - LB)}{I - 1}$$
so get I different expected return value, i.e.
$$\check{r} = \{LB, LB + \delta, \ldots, UB - \delta, UB\}$$
5. Calculate portfolio weights using equation (8) for each particular expected return value ($\check{r}$) for each data set q
6. Calculate the average weight of each stock at each efficient frontier point to get the resampled weight

The application of MVEF and REF was carried out on the formation of 2 stock portfolios, each consisting of five shares. The shares are used, as shown in tables 3.1 and 3.2. The data used are daily stock price closing data for three years, 23 September 2016 - 20 September 2019. The data was taken on Yahoo Finance [26].

Table 1 Stock Data on Portfolio 1

No	Stock Code	Company Name
1	ASII	Astra International Tbk
2	BBRI	Bank Rakyat Indonesia (Persero) Tbk
3	BUMI	Bumi Resources Tbk
4	INTP	Indocement Tunggal Prakarsa Tbk
5	PGAS	Perusahaan Gas Negara (Persero) Tbk

Table 2 Stock Data on Portfolio 2

No	Stock Code	Company Name
1	BBNI	Bank Negara Indonesia Tbk
2	BMRI	Bank Mandiri (Persero) Tbk
3	INCO	Vale Indonesia Tbk
4	PANS	Panin Sekuritas Tbk
5	SMGR	Semen Indonesia (Persero) Tbk

Based on stock return data on table 1 and 2 weighted percentage portfolio calculation is conducted from each stock. The observation data that used from each stock is 771 with 51 efficient points and 700 simulations. Results of the weighted portfolio with MVEF and REF are written in this table.

Table 3. Weighted stock percentage with MVEF and REF in Portfolio 1

No	Stock Code	MVEF (%)	REF (%)
1	ASII	41.45	34.74
2	BBRI	38.95	33.82
3	BUMI	3.00	4.09
4	INTP	10.28	15.93
5	PGAS	6.32	11.42

Table 4 Weighted stock percentage with MVEF and REF in Portfolio 2

No	Stock Code	MVEF (%)	REF (%)
1	BBNI	21.06	25.89
2	BMRI	27.73	29.22
3	INCO	7.91	10.58
4	PANS	29.29	18.48
5	SMGR	14.01	15.83

Portfolio 1 and Portfolio 2 are compared based on their Sharpe Ratio [23] using stock return data. Sharpe Ratio is shown in table 5.

Table 1 Sharpe Ratio (SR)

No	Portfolio	SR MVEF	SR REF
1	Portfolio 1	0.026	0.027
2	Portfolio 2	0.010	0.023

Table 5 shows that all portfolios show SR REF > SR MVEF. Based on Sharpe Ratio, the REF portfolio is considered more efficient in determining portfolio weight. SR value shows the performance of the portfolio. The higher the SR value, the better the performance of the portfolio [23]. Return and risk input that is used in the portfolio optimization process is sure to have uncertainty. This uncertain input cause portfolio performance becomes terrible. Portfolio performance is more better on REF because REF combines the input uncertainty with optimization theory using Monte Carlo simulation. On the other side, MVEF only depends on the input parameter. If the input parameter change slightly, it can cause significant change to the MVEF portfolio weight composition. Fletcher and Hillier [24] stated that the REF performance based on Sharpe Ratio was generally better than MVEF. REF provides a better advantage compared to MVEF when using input parameters in the form of mean and covariance matrix. According to Frahm [25], REF gives better performance because REF applies the random-walk hypothesis on Monte Carlo simulation under iid (identic and independent distribution) assumption. In Monte Carlo simulation, some data are generated so it can create a hundred even thousand portfolio possibilities. REF can predict portfolio variety with a particular return, so the best portfolio weight mean can be obtained based on the simulation. Fletcher

From the Sharpe ratio comparison, it is recommended that the REF portfolio be recommended. In addition to the Sharpe ratio, the MVEF and REF comparison can also be seen from the performance of the method in the application of stock prices in the market. Suppose an investor has a fund of Rp 1 billion. The fund of Rp 1 Billion is invested in 2 portfolios with weights as in Table 3.3 and Table 3.4. The shares were bought on September 20, 2019, and sold on October 2, 2019. Which method will provide a higher return. The price of the purchase of shares and the sale price uses the opening price. The stock price can be seen in the following table.

Table 6 Table of Buying and Selling Prices of Portfolio in Portfolio 1

No	Stock Code	Buy Price	Sell Price
1	ASII	6575	6500
2	BBRI	4200	4050
3	BUMI	95	90
4	INTP	21000	18425
5	PGAS	2170	2020

Table 7 Table of Buying and Selling Prices of Portfolio in Portfolio 2

No	Stock Code	Buy Price	Sell Price
1	BBNI	7725	7225
2	BMRI	7100	6875
3	INCO	3720	3420
4	PANS	1600	1555
5	SMGR	12500	11200

Simulations of the results of stock purchases, stock sales, and profits using the MVEF and REF methods are illustrated in the following table.

Table 8 Table of MVEF and REF Method Performance in Portfolio 1

NO	STOCK	MVEF Portfolio Value	REF Portfolio Value	MVEF Profit	REF Profit
1	ASII	Rp 409,771,863	Rp 343,437,262	Rp (4,728,137)	Rp (3,962,738)
2	BBRI	Rp 375,589,286	Rp 326,121,429	Rp (13,910,714)	Rp (12,078,571)
3	BUMI	Rp 28,421,053	Rp 38,747,368	Rp (1,578,947)	Rp (2,152,632)
4	INTP	Rp 90,194,762	Rp 139,766,786	Rp (12,605,238)	Rp (19,533,214)
5	PGAS	Rp 58,831,336	Rp 106,305,991	Rp (4,368,664)	Rp (7,894,009)
	Total	Rp 962,808,299.77	Rp 954,378,835.85	Rp (37,191,700)	Rp (45,621,164)

Table 9. Table of MVEF and REF Method Performance in Portfolio 2

NO	STOCK	MVEF Portfolio Value	REF Portfolio Value	MVEF Profit	REF Profit
1	BBNI	Rp 196,968,932	Rp 242,142,718	Rp (13,631,068)	Rp (16,757,282)
2	BMRI	Rp 268,512,324	Rp 282,940,141	Rp (8,787,676)	Rp (9,259,859)
3	INCO	Rp 72,720,968	Rp 97,267,742	Rp (6,379,032)	Rp (8,532,258)
4	PANS	Rp 284,662,188	Rp 179,602,500	Rp (8,237,813)	Rp (5,197,500)
5	SMGR	Rp 125,529,600	Rp 141,836,800	Rp (14,570,400)	Rp (16,463,200)
	Total	Rp 948,394,011.22	Rp 943,789,901.23	Rp (51,605,989)	Rp (56,210,099)

From tables 3.8 and 3.9, it can be seen that all share prices have decreased, resulting in the formed stock portfolio suffered losses. The loss of a portfolio of 1 share with the MVEF method was Rp. 37,191,700, and the REF method was Rp. 45,621,164. The portfolio loss of 2 shares using the MVEF method was Rp. 51,605,989, and the REF method was Rp. 56,210,099. It can be seen that the portfolio using the MVEF method provides fewer losses compared to the REF method because the MVEF method has the characteristics of minimizing risk (avoiding risk).

Based on the calculation results, it is found that the REF portfolio is more recommended than MVEF because the Sharpe ratio of the REF portfolio is higher than that of MVEF. However, empirically (in the case of the share purchase on September 20 and the sale of shares on October 2) both methods, both MVEF and REF both suffered losses because in that period, the stock price had a declining trend. In this case, the MVEF method provides less loss than the REF. Frahm [25] argues that REF could have excellent performance based on Sharpe Ratio, but it is not profitable when it applies in real life, such as empirical study case. This matter links to Monte Carlo simulation. Monte Carlo simulation on REF has the purpose of eliminating variance from investment strategy, so it is not appropriate if applied to the stock which high variance.

Besides, the ineffectiveness of the REF is due to the short investment period taken, only nine trading days. This short time makes it difficult for investors to avoid selling shares when the stock price falls. Because in the simulation performance of the MVEF and REF methods to see whether investors gain profits or experience losses, it involves buying and selling shares. Besides, the declining stock trend also contributed to the ineffectiveness of the REF portfolio. All shares used in forming a portfolio experience a closing price that is declining. From 9 trading days, almost every day, the closing price of the stock decreased from the previous day's closing. Declining stock prices will have a large loss impact on the weight of a large portfolio as well. Portfolio accomplishment still depends on the investor's capability in processing the input parameter information. The investor must be able to use the collected input to evaluate portfolio formation so it can increase the portfolio performance.

4. Conclusion

The REF method can be used as an alternative to determining the weight of a stock portfolio. The REF method can overcome the problem of sensitivity to input changes that exist in the MVEF method. The problem is solved by generating data using Monte Carlo simulation. The results of the weight of the stock portfolio with the REF method have a higher Sharpe ratio. The greater the Sharpe ratio, the better the portfolio performance, which means that the REF portfolio has better performance than the MVEF portfolio. However, it turns out that the REF method is not suitable to be applied to portfolios whose share prices have a downward trend from the buying period to the sale of shares. This can be seen from the performance of the REF and MVEF methods on stock prices in the market. Both the REF and MVEF methods suffer losses but the MVEF method losses are smaller than the REF method.

References

[1] H. Markowitz, "Portfolio Selection," *Journal of Finance*, vol. 7, no. 1, pp. 77–91, 1952.
[2] R. O. Michaud, "The Markowitz Optimization Enigma: Is Optimization Optimal?," *Financial Analysts Journal*, vol. 45, no. 1, pp. 31-42, 1989.
[3] R. P. Brito and L. N. Vicente, "Efficient Cardinality/ Mean-Variance Portfolios," *in Proc. CSMO: IFIP Conference on System Modeling and Optimization*, Klagenfurt: Springer, pp. 52 – 73, 2014.
[4] X. Guo, R. H. Chan, and W-K. Wong, "Mean-Variance, Mean-VaR, and Mean-CVaR Models for Portfolio Selection with Background Risk," *Risk Management*, vol. 21, no. 2, pp. 73–98, 2018.
[5] A. Ismail and H. Pham, "Robust Markowitz Mean-Variance Portfolio Selection Under Ambiguous Covariance Matrix," *Mathematical Finance*, vol. 29, no. 1, pp. 174–207, 2019.
[6] L. H. Halliwell, "Mean-Variance Analysis and the Diversification of Risk," *Incorporating Risk Factors in Dynamic Financial Analysis*, 1995
[7] T. Suzuki and K. Tanaka, "Mean-Variance Portfolio Model Modified by Nonlinear Bagging Predictors," *Jurnal of Signal Processing*, vol. 18, no. 6, pp. 238–290, 2014.
[8] M. C. Steinbach, "Markowitz Revisited: Mean-Variance Model in Financial Portfolio Analysis," *SIAM Review*, vol. 43, no. 1, pp. 31-85, 2001.
[9] S. M. Gasser, T. Kremser, M. Rammerstorfer, and K. Weinmayer, "Markowitz Revisited: Social Portfolio Engineering", *European Journal of Operational Research*, vol. 258, no. 3, pp. 1181-1190, 2017.
[10] A. Meucci, *Risk and Asset Allocation*, New York: Springer Finance, 2005.

[11] J. L. Maginn, D. L. Tuttle, D. W. McLeavey, J. E. Pinto, *Managing Investment Portfolios: A Dynamic Process*, New Jersey: John Wiley and Sons, 2007.
[12] H. G. Fong, *The World of Risk Management*, London: World Scientific Publishing, 2006.
[13] J. D. Jobson and B. Korkie, "Estimation for Markowitz Efficient Portfolios," *Journal of the American Statistical Association*, vol. 75, no. 371, pp. 544–554, 1980.
[14] J. D. Jobson and B. Korkie, "Putting Markowitz Theory to Work," *Journal of Portfolio Management*, vol. 7, no. 4, pp. 70-74, 1981.
[15] R. O Michaud and R. O. Michaud. *Efficient Asset Management: A Practical Guide to Stock Portfolio Optimization and Asset Allocation*, Second edition, New York: Oxford University Press, 2008.
[16] P. Jorion, "Portfolio Optimization in Practice," *Financial Analysts Journal*, vol. 48, no. 1, pp. 68-74, 1992.
[17] B. Scherer, "Portfolio Resampling: Review and Critique," *Financial Analysis Journal*, vol. 58, no. 6, pp. 98 – 109, 2002.
[18] W. Jiao, "Portfolio Resampling and Efficiency Issues," Master of Science Thesis, Department of Mathematics, Humboldt Universit¨at zu Berlin, 2003.
[19] P. Lorentz, "A Modified Sharpe Ratio Based Portfolio Optimization," Master of Science Thesis, Department of Mathematics, KTH Royal Institute of Technology, Stockholm, 2012.
[20] K. Nyholm, *Strategic asset allocation in fixed-income markets: a MATLAB-based user's guide*, New Jersey: John Wiley and Sons, 2008.
[21] H. Markowitz, *Portfolio selection: Efficient diversification of investments*. New York: Yale University Press, 1959.
[22] J. Kohli, "An Empirical Analysis of Resampled Efficiency," Master of Science Thesis, Mathematical Sciences Department, Worcester Polytechnic Institute, 2005.
[23] W. F. Sharpe, 1964. "Capital Asset Prices: A Theory of Market Equilibrium under Conditions of Risk." *Journal of Finance*, vol. 40, no. 3, pp. 425–442, 1964.
[24] J. Fletcher and J. Hillier, "An Examination of Resampled Portfolio Efficiency", *Financial Analysts Journal*, vol. 57, no. 5, pp. 66-74, 2001.
[25] G. Frahm, "A theoretical foundation of portfolio resampling," *Theory and Decision*, vol. 79, no. 1, pp. 107–132, 2015.
[26] Yahoo!finance. *Business Finance, Stock Market, Quotes, News*. 2019. [Online]. Available: https://finance.yahoo.com/ [Accessed: 20 September 2019]

IOP Conf. Series: Materials Science and Engineering **846** (2020) 012066 doi:10.1088/1757-899X/846/1/012066

THE EFFECT OF AGING TREATMENT ON MECHANICAL PROPERTIES AND MICROSTRUCTURES OF Ti-12Cr IN ORTODONTIC APPLICATIONS

Sir Anderson[1,3], Annisa [2a], Jon Affi[2b], Yuli Yetri[3*], Gunawarman[2c*]

[1]Mechanical Engineering, [Doctoral of Student Andalas University, Padang, Indonesia]
[2] Mechanical Engineering, [Faculty of Engineering, Andalas University, Padang, Indonesia]
[3] Mechanical Engineering, [Politeknik Negeri Padang, Padang, Indonesia]

*Corresponding author (E-mail): gunawarman@eng.unand.ac.id , yuliyetri@pnp.ac.id

Abstract. Many adults experience tooth loss due to accidents and lack of care. This causes various problems, one of which is a problem in chewing food. One solution that can be used is the use of dental implants. Dental implants that are currently made of titanium Ti-6Al-4V at a very expensive price. One of the newest types of titanium alloy material that can be used with relatively cheap prices is Ti-12Cr. As a new material, the effect of heat treatment processes on mechanical properties has not been investigated in more detail. So that research is needed on the effect of heat treatment with the aging treatment process on the mechanical properties and microstructure of the Ti-12Cr material as a substitute for cheaper titanium implants. In this research, a solution treatment (ST) heat treatment was carried out at 885oC and an aging treatment (AT) at 540^{0}C. Aging treatment is carried out with variations in holding time of 30 ks and 60 ks. Then the microstructure inspection and mechanical properties testing are carried out through hardness and tensile tests. With the aging treatment of the microstructure of the material, Ti-12Cr experienced grain refinement and the hardness obtained was higher than that of Ti-6Al-4V. The hardness values of Ti-12Cr ST, Ti-12Cr AT 30 ks, Ti-12Cr AT 60 ks and Ti-6Al-4V are 457 VHN, 605 VHN, 772 VHN, and 349 VHN. However, the tensile strength of Ti-12Cr is lower than the tensile strength value of Ti-6Al-4V.

1. Introduction

Many people lose teeth (toothless teeth). This is caused by various factors, such as accidental damage and lack of maintenance. Statistics show that 70% of adults expelled from 35-44 years of age increase. One in four adults over 74 years has lost all their teeth.

Tooth loss cannot only reduced appearance but can cause various problems such as difficulties when chewing food, and difficulty speaking. Currently, various solutions are available to replace missing teeth, one of which is dental implants. Dental implants are part of orthodontics. The main components of orthodontics are dental implants, brackets, auxiliary, and orthodontic wire (archwire) [1]. Some of the options available are complete dentures made of acrylic which are usually used if all the teeth have been lost, partial dentures that can be used when only lose a few teeth. Besides, there is a dental bridge that is a denture that is permanently placed and finally a dental implant whose use involves placing a small metal rod into the jawbone. This stem functions like a tooth root, when the implant is attached to the jawbone, the dental crown can be attached to it.

Generally, metal implants that are often used and are quite popular are stainless steel and titanium. Both have high strength and good tenacity [2]. The stainless steel used is the type of SUS 316L (ASTM F138)[3]. Stainless steel is used because it has corrosion-resistant properties. This type of material still causes problems that are allergens to some consumers due to hypersensitivity reactions[4]. The main factor causing allergies is the element content of Nickel (Ni). Especially titanium alloy is a safe material for the body, because it has the best biocompatibility among metals[5], and has high strength, more corrosion resistance, and lower elastic modulus [6] [7]. Titanium used for dental implants today is a type alloy ($\alpha+\beta$) Ti-6Al-4V-ELI alloy (ASTM F136) [8] with aluminum and vanadium alloys. The mechanical properties of these metal biomaterials can be seen in Table 1. From the Table, it can be seen that compared to other types of biomaterials, Ti-6Al-4V alloys have the best tensile properties for biomaterials. Ti-6Al-4V has the highest yield and tensile strength even though the elasticity is slightly higher than Cp-Ti. Titanium has a very expensive price of around 15 million / tooth shaft [7]. To overcome the problem of relatively expensive prices, alternative materials are needed that are not too expensive but have high tensile strength and yield strength. The titanium that is being developed at the moment β-type titanium including TNTZ (Ti-29Nb-13Ta-4,6Zr) [5][9] and the latest Ti-12Cr with a few and inexpensive alloy elements, such as Cr [10]. Research on Ti-12Cr alloys was only carried out at the mechanical properties testing stage, namely bending test to determine the elastic modulus (Young's modulus) of the alloy [11]. The modulus of elasticity and strength of Ti-12Cr alloys are suitable for applications in backbone support [11]. It has been concluded that Ti-12Cr is more recommended for backbone applications [12]. From the description above, it can be seen that previous studies on the Ti-12Cr material were limited to bone implants. However, research on the effect of the heat treatment process on the mechanical properties of Ti-12Cr for dental implants is not yet clearly known. Therefore, this research will examine how the effect of aging treatment on Ti-12Cr material by looking at mechanical properties, especially tensile properties and microstructure in the hope that it can approach the properties of Ti-6Al-4V.

Table 1. Comparison of some mechanical properties of several metals commonly used as biomaterials [13].

Alloy Type	Modulus of Easticity		Yield Stregth		Maximum Tensile Strength	
	GPa	10^{-6}psi	MPa	10^{-6}psi	MPa	10^{-6}psi
Stainless steel	200	29	150-170	25-110	465-950	65-140
Co-Cr-Mo	200-230	29-33	275-1583	40-230	600-1795	90-260
Commercial Pure Ti	105	15	692	100	785	115
Ti-6Al-4V	110	16	850-900	120-130	960-970	140-141

The expected goal of this study is to determine the effect of aging treatment on microstructure and mechanical properties, namely the value of hardness and tensile strength of Ti-12Cr as material for dental implants.

2. Methodology

a. Sample Preparation

The samples tested were β-type Ti-12Cr titanium in the form of plates with a thickness of 3 mm and each test of 3 samples. Then the cutting process is done using wire cutting with a size of 12X10 mm. Next, level the surface using sandpaper starting from the smallest 400, 800, 1200, 1500 mesh, the last mesh 2000 and polished to produce a shiny surface free from scratches.

b. *Aging Treatment Process*

The aging treatment process begins with heating using a Ney Ceramfires vacuum furnace. Before the aging treatment process, the solution treatment process is carried out by heating the specimen to a temperature of 885°C degree of temperature increase of 10°C / s, holding time for 1 hour. Next, do a quick cooling by dipping the specimen in water (water quenching). After that, the heating process is re-heated at 540˚C withholding for 30 ks (about ± 8 hours) and variations in the heating process to 60 ks (about ± 16 hours) are also carried out. Then for each per variation carried out rapid cooling using water. Seen in Figure 1, the aging treatment process occurs.

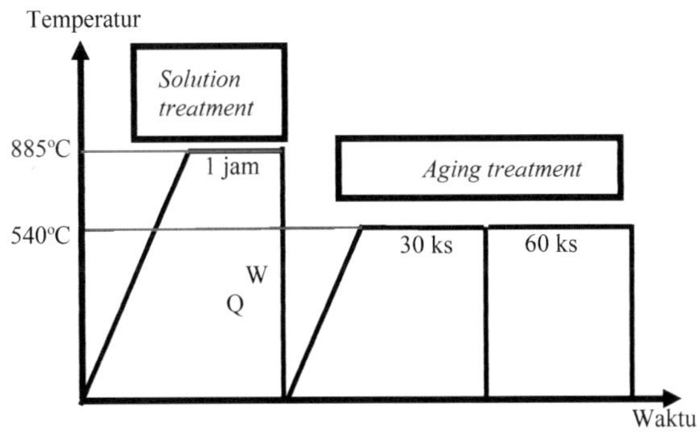

Figure 1. Scheme of the aging treatment process

c. Inspection of Micro Structure

Testing using an Olympus optical microscope to see the microstructure of titanium Ti-12Cr material that has been carried out the heat treatment process with three treatments namely solution treatment, aging treatment withholding time 30 ks, and aging treatment withholding time 60 ks. The magnification variation used starts from 20 times, 50 times and 100 times.

d. Mechanical Testing

Mechanical testing is done by a hardness test and tensile test. A hardness measurement of Ti-12Cr refers to ASTM 384; Standard Test Methods for Knoop and Vickers Hardness Materials [14]. The indenter used in this test is a rectangular pyramid with a 136° angle and a loading value of 9.8 N with an indentation time of 10-15 seconds. Tensile testing with Universal Testing Machine (UTM).

3. Result and Discussion

A. Ti-12Cr Micro Structure following Aging Treatment Process.

The process carried out to see the results of the Ti-12Cr microstructure description is a metallographic process.

1. Solution Treatment

This water quenching treatment is carried out after giving a treatment solution that changes the β phase to the α phase on heating at 890°C. The results of observing microstructure with solution treatment can be seen in Figure 2.

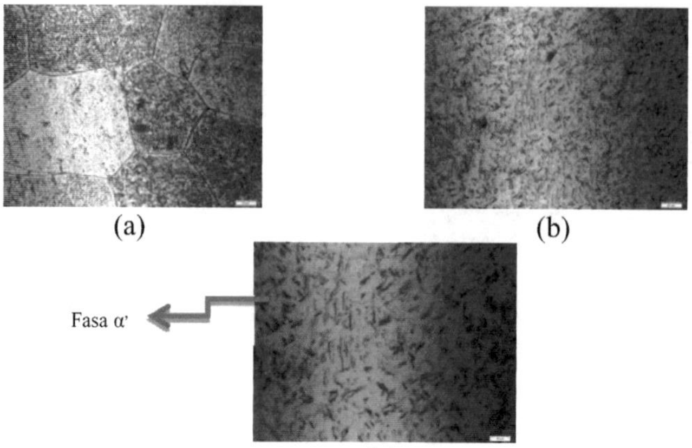

(c)

Figure 2. The results of the micro-structure Ti-12Cr treatment solution treatment (a) magnification 20 times, (b) magnification 50 times, and (c) magnification 100 times.

In Figure 2, it is seen that the results of the solution treatment process have a martensite phase (α phase) with a needle-shaped microstructure. At a magnification of 20 times, a phase that resembles very tight needles is formed. In Figure 2 (b) with a magnification of 50 times, the needling phase is not too tight. Whereas in Figure 2 (c) with a magnification of 100 times the shape of the needle is seen more clearly.

2. *Aging Treatment* 30 ks

This aging treatment occurs after the existence of solution treatment (water quenching). The aging treatment is carried out in the form of heating withholding for 30 ks (± 8 hours), then cooling again using water. The results of the 30 ks aging treatment microstructure can be seen in Figure 3. The phases formed in this treatment are precipitated α and β phase. Precipitation α is formed after the aging treatment. The longer the aging treatment time, the more smooth the precipitate phase occurs.

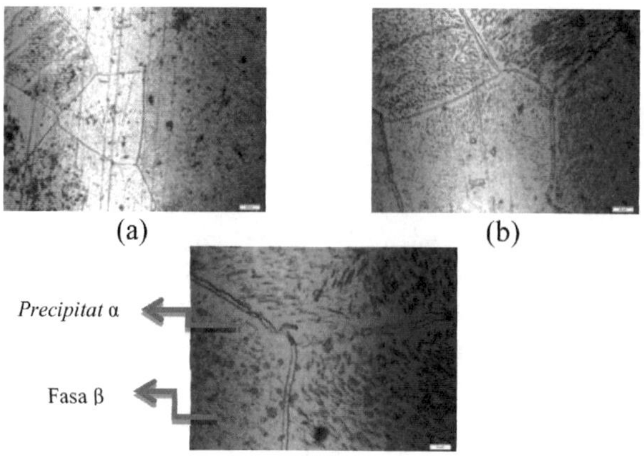

(c)

Figure 3. Ti-12Cr microstructure results of aging treatment 30 ks, (a) magnification 20 times, (b) 50 times magnification, and (c) magnification 100 times.

In Figure 3 you can see the dark-colored precipitate α and the light-colored β phase. In Figure 3 (a) it is seen that the α-shaped precipitate specimen is small in magnification 20 times. In Figure 3 (b) the precipitate α appears to be greater than 20 times magnification, whereas in Figure 3 (c) the specimen has a greater precipitate α compared to 50 times magnification. Precipitate α is formed after aging treatment. The microstructure is smooth compared to water quenching treatment.

3. *Aging Treatment* 60 ks

The 60 ks aging treatment is the same as the 30 ks aging treatment. The difference occupies in the length of detention, which is at 60 ks (± 16 hours). The results of observing the microstructure of the 60 ks aging treatment can be seen in Figure 4. The phases formed in this treatment are precipitated α and β phase. The precipitate α that is formed in the 60 ks aging treatment is finer than the 30 ks aging treatment. The finer the precipitate formed, the harder the material [15].

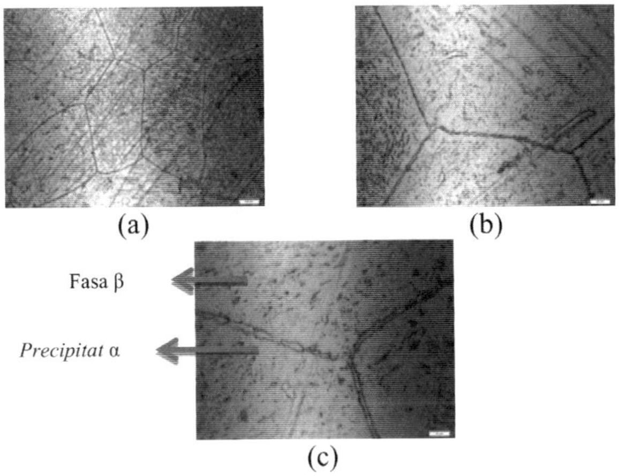

Figure 4. The results of the Ti-12Cr microstructure aging treatment treatment 60 ks (a) magnification 20 times, (b) magnification 50 times, and (c) magnification 100 times.

In Figure 4 there is a dark-colored precipitation α and a light-colored β phase. In Figure 4 (a) the specimen looks precipitate α, small in size with a magnification of 20 timesIn Figure 4 (b) precipitation α looks greater than a magnification of 20 times. In Figure 4 (c) the specimen has a precipitate α greater than 50 times magnification. The precipitation α formed in Figure 4 is finer than the precipitation α found in the results of the 30 ks aging treatment test.

B. Hardness Ti-12Cr

Hardness value was obtained from 3 tests, namely testing the average hardness value of the Ti-12Cr titanium alloy was carried out at 5 points for each test. As shown in Table 2 the value of hardness shown is the average price of the hardness of alloys that have been treated with solution treatment and aging treatment. Based on the data summarized in Table 2, a comparison graph is obtained from the average hardness of the aging treatment (Figure 5).

Table 2. Hardness value of Ti-12Cr results of testing of treatment and aging treatment solutions.

No	Treatment	Hardness value	Mean
1		404	
		463	
	Solution treatment	488	457
		499	
		430	
		553	
		631	
2	Aging treatment 30 ks	625	605
		688	
		528	
		665	
		708	
3	Aging treatment 60 ks	745	772
		898	
		843	

In Table 2 and Figure 5, it can be seen that the Ti-12Cr hardness value obtained is quite significant after undergoing the aging treatment process. The average value of violence increased from 457 HVN in the treatment solution to 605 HVN in the 30 ks aging treatment as well as for aging treatment 60 ks in the average hardness to 772 HVN. The highest hardness value obtained from aging treatment 60 ks. This is in line with previous research as shown in Table 3, the value of violence increased after aging treatment [16]. Based on Table 2 and Table 3, the hardness value of Ti-12Cr after experiencing aging treatment is higher than Ti-6Al-4V.

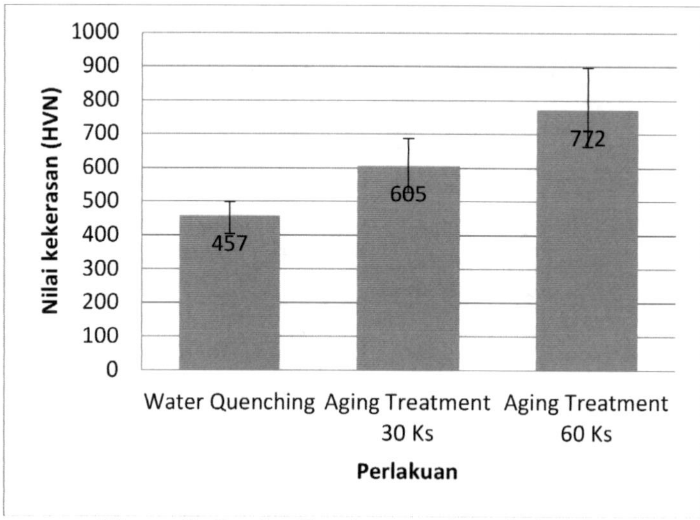

Figure 5. Graph of hardness testing of Ti-12Cr

Table 3. Hardness of Ti-6Al-4V with solution treatment and aging treatment [14].

Titanium alloy Ti6Al4V Hardness In as-cast condition 312 HV 10	**Solution treatment**	1050ºC/ 1h/water	1050ºC/ 1h/air	950ºC/1h/ water	950ºC/1h/air	800ºC/1h/ water	800ºC/1h/air
	Hardness HV 10	405	320	395	311	344	319
	Solution treatment + aging	1050ºC/ 1h/water + 550ºC/4h	1050ºC/ 1h/air + 550ºC/4h	950ºC/1h/ water + 550ºC/4h	950ºC/1h/ water + 550ºC/4h	800ºC/1h/ water + 550ºC/4h	800ºC/1h/ water + 550ºC/4h
	Hardness HV 10	428	338	414	352	368	346

C. Ti-12Cr Tensile Strength Test

The test was carried out using a UTM machine with a scale A load and a total load of 3 tons. This test uses support for supporting the specimen because of its small size. The following is the attachment of the tensile strength test data using the UTM machine as shown in Table 4.

Table 4. Tensile strength value Ti-12Cr

No.	Treatment	Test	Tensile strength (MPa)
1	*Solution treatment*	1	925
		2	905
		3	900
2	*Aging treatment* 30 ks	1	310
		2	450
		3	435
3	*Aging treatment* 60 ks	1	120
		2	110
		3	105

Based on Table 4, it can be seen that the strength value of the specimens that experienced higher treatment treatments is around 900 MPa, followed by 30 ks aging treatment test and the lowest value is 60 aging ks treatment. Noticeable differences occur for the results obtained between strength testing and hardness testing. According to his theory violence and strength are directly proportional [13]. Whereas based on the results of the hardness test, the highest hardness was obtained in the 60 ks aging treatment test, followed by the 30 ks aging treatment, and the lowest hardness in the solution treatment test.

From the results of this study, it is known that violence increases while strength decreases. This shows that there was an error in data retrieval that allegedly occurred during the tensile testing. In the tensile test, the speed used is too large while the specimen is small so that the specimen experiences a decrease in strength and causes agility. Then the data obtained is not a tensile test data but impact test data because the shape of the fracture that occurs in the form of brittle and there is no cross-section at all (can be seen in Figure 4.10, Figure 4.11, and Figure 4.12). In comparison to the results obtained from the study of Zhao, et al. (2012) [9] can be seen in Figure 6.

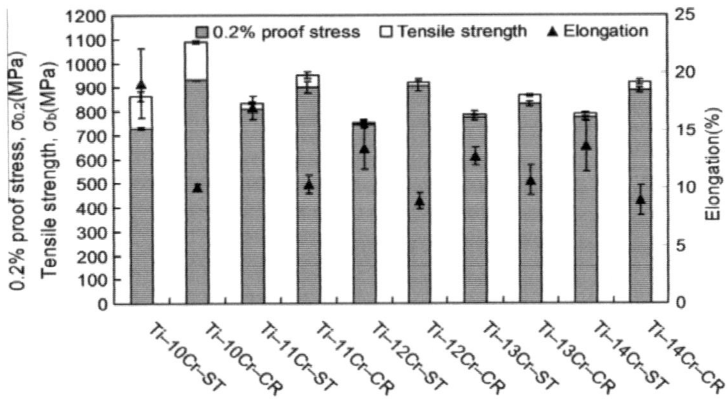

Figure 6. Tensile strength of Ti-xCr alloy undergoing treatment and cold rolling [9].

In Figure 6, it can be seen that the strength of the treatment solution ranges from 700-800 MPa. The value of tensile strength obtained in this study is close to the value of tensile strength carried out by Zhao, et al. (2012) [9].

The tensile test carried out obtained a curve as shown in Figure 7. The highest strength value is obtained in the treatment solution, while the lowest is in the aging treatment test.

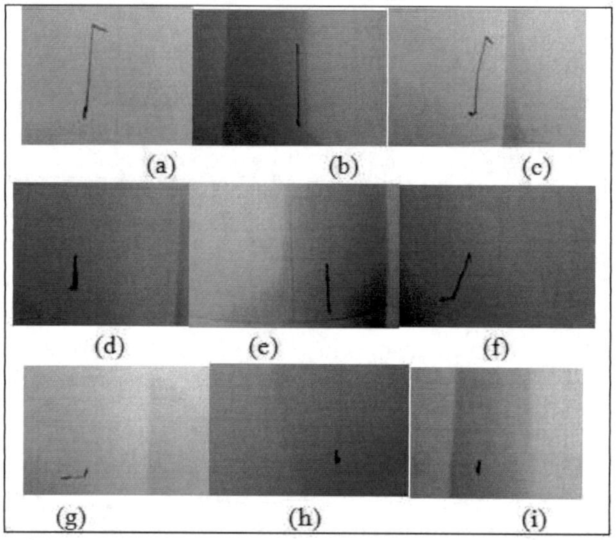

Figure 7. Graph of the tensile test results of Ti-12Cr (a) *solution treatment* 1, (b) *solution treatment* 2, (3) *solution treatment* 3, (d) *aging treatment* 30 ks 1, (e) *aging treatment* 30 ks 2, (f) *aging treatment* 30 ks 3, (g) *aging treatment* 60 ks 1, (h) *aging treatment* 60 ks 2, dan (i) *aging treatment* 60 ks 3

The location of the fracture in the treatment solution can be seen in Figure 8.

specimen (1) specimen (2) specimen (3)

Figure 8. The area of fracture in the treatment solution

Figure 8 shows that the fault occurred in the reduced cross-section. The area is the test area to be tested tensile. In this test, it can also be seen that the cross-section has been reduced before experiencing a break. The location of the fracture in the 30 ks aging treatment can be seen in Figure 9.

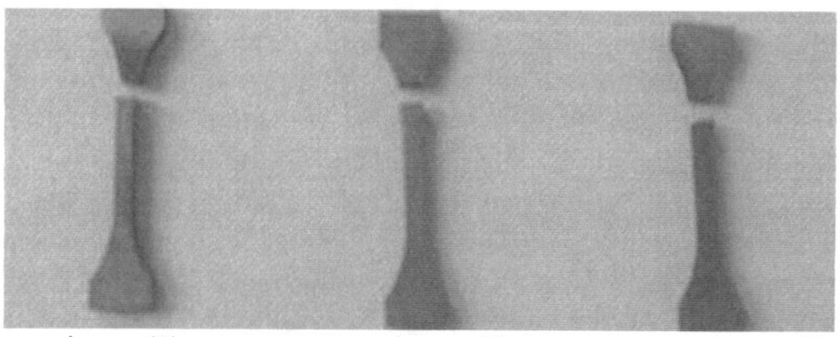

specimen (1) specimen (2) specimen (3)

Figure 9. Areas of fracture treatment aging treatment 30 ks.

In Figure 9 there is a fault that occurred in the test area near the cross-section of the exemplar. In this test, it can also be seen that the specimen did not experience the slightest reduction in cross-section before the occurrence of fracture. The area of fracture with 60 ks aging treatment can be seen in Figure 10.

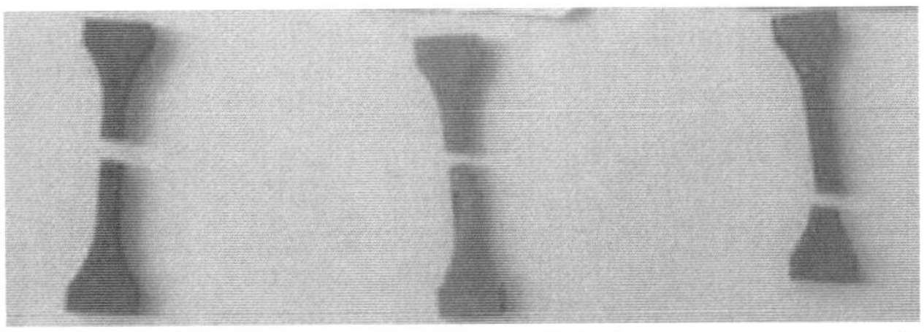

specimen (1) specimen (2) specimen (3)

Figure 10. Fractional area of aging treatment 60 ks

In Figure 10 it appears that the fault happened in the area that should be, but for exemplar 3 the difference in fracture occurred in the area near the cross-section of the specimen. In the tests carried out did not experience the slightest reduction in cross-section before the occurrence of a fracture.

In testing the strength of the Ti-12Cr, the results obtained for the treatment of fracture treatment as shown in Figure 11.

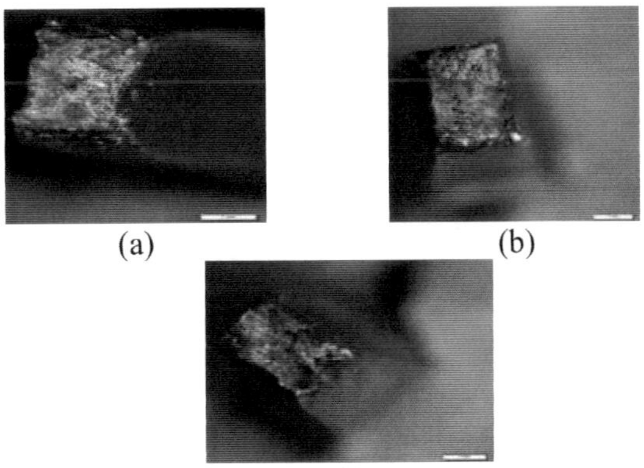

(a) (b) (c)

Figure 11. The shape of the fracture in the treatment solution (a) specimen 1, (b) specimen 2, and (c) specimen 3

In Figure 11, it can be seen that the fracture that occurs in the solution treatment is ductile, the texture is fibrous and gray. Based on these results it can be seen that the material resulting from this treatment has a ductile fracture. The results of fractures that occur in the 30 ks aging treatment can be seen in Figure 12.

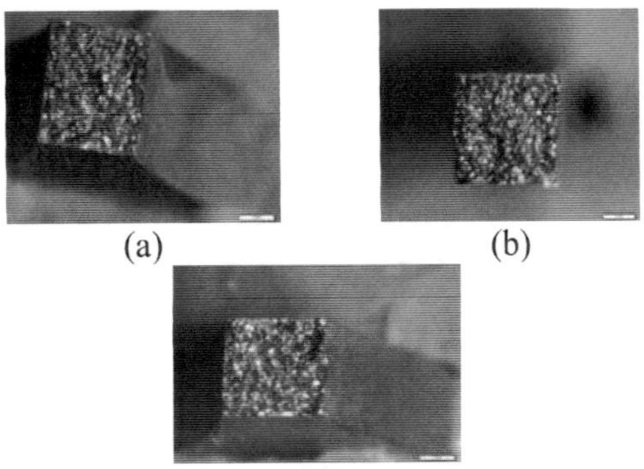

(a) (b) (c)

Figure 12. Form of fractures in the treatment of 30 ks, (a) specimen 1, (b) specimen 2, and (c) specimen 3

In Figure 12 it can be seen that the fracture has shiny features with a relatively flat shape. Based on the features seen, it can be said that the fracture in the 30 ks aging treatment has a brittle fracture. The fracture that occurred did not experience the slightest cross-section reduction. For results of fractures in the 60 ks aging treatment can be seen in Figure 13.

(a)

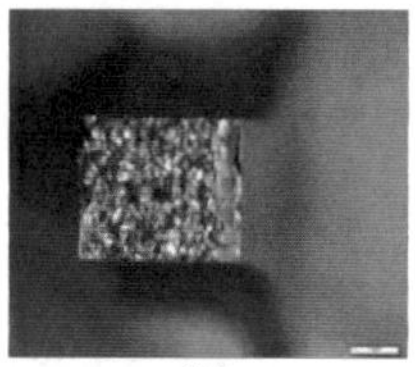
(b)

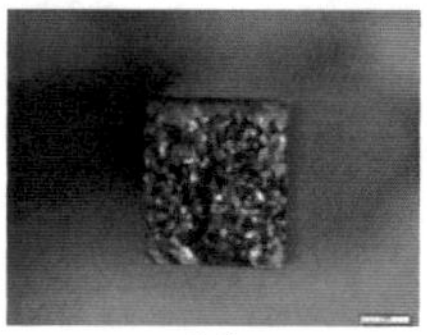
(c)

Figure 13 Form of fracture in aging treatment 60 ks, (a) specimen 1, (b) specimen 2, and (c) specimen 3

Figure 13 shows that the shape of the fault is almost the same as the 30 ks aging treatment because the shape of the fracture is flat and also looks shiny.

4. Conclusion

Aging treatment causes changes in the microstructure of Ti-12Cr, where the grain obtained is getting smoother.

Aging treatment increases the hardness value of the Ti-12Cr material. A significant increase from 457 VHN in the treatment solution to 605 VHN in the 30 ks aging treatment test and finally 772 VHN in the 60 ks aging treatment.

Aging treatment causes the tensile strength of titanium to decrease due to high loading speeds so it does not qualify as static loading.

References

[1] P. A. Machfudzoh, M. N. Amin, L. Sandra, D. Putri, F. K. Gigi, and U. Jember, "Efektivitas Ekstrak Daun Belimbing Wuluh sebagai Bahan Inhibitor Korosi pada Kawat Ortodonsi Berbahan Dasar Nikel-Titanium (Effectiveness of Bilimbi Leaves Extract as Corrosion Inhibitor on Nickel-Titanium Orthodontic Wire)," no. kelompok 3, 2014.

[2] S. M. Perren, R. Mathys, and O. Pohler, "AO Principles of Fracture Management: Implants and Materials in Fracture Fixation." New York: AO Publishing, 2000.

[3] *ASTM Handbook. Standard Specification for Wrought 18Chromium-14Nickel-2.5Molybdenum Stainless Steel Bar and Wire for Surgical Implants (UNS S31673). Ohio: ASTM International Vol. 3, (2011): 103-115.* 2011.

[4] S. A. S. Goenharto, "Breket Titanium.," *J. Fak. Kedokt. Gigi Univ. Airlangga, Surabaya.2005, Vol 38, No. 3.*, vol. 38, pp. 120–123, 2005.

[5] M. Niinomi, "Biologically and Mechanically Biocompatible Titanium Alloys Ti-6Al-4V ELI," vol. 49, no. 10, pp. 2170–2178, 2008.

[6] A. Ajiz and J. Affi, "The Effects of Short-Time Solution Treatment and Short-Time Aging on Mechanical Properties of Ti-6Al-4V for Orthopaedic Applications," vol. 5, no. 4, pp. 329–334,

2015.
[7] M. Gunawarman, Niinomi, D. Eylon, S. Fujishiro, and C. Ouchi, "Effect of b Phase Stability at Room Temperature on Mechanical Properties in b -Rich a ⁄ b Type Ti – 4 . 5Al – 3V – 2Mo – 2Fe Alloy," vol. 42, no. 2, pp. 191–199, 2002.
[8] *ASTM Handbook. Standard Specification for Wrought Titanium-6Aluminum-4Vanadium ELI (Extra Low Interstitial) Alloy for Surgical Implant Applications (UNS R56401) Ohio: ASTM International Vol. 2, : 67-89.* 2011.
[9] M. T. Mohammed, Z. A. Khan, and A. N. Siddiquee, "Beta Titanium Alloys : The Lowest Elastic Modulus for Biomedical Applications : A Review," *World Acad. Sci. Eng. Technol. Int. J. Chem. Nucl. Metall. Mater.*, no. April 2016, 2014.
[10] M. Abdel-Hady Gepreel and M. Niinomi, "Biocompatibility of Ti-alloys for long-term implantation," *."Journal Mech. Behav. Biomed. Mater. 1-9.*, vol. 20, pp. 407–415, 2013.
[11] X. Zhao *et al.*, "Acta Biomaterialia Optimization of Cr content of metastable b -type Ti – Cr alloys with changeable Young ' s modulus for spinal fixation applications," *Acta Biomater.*, vol. 8, no. 6, pp. 2392–2400, 2012.
[12] H. Liu, M. Niinomi, M. Nakai, J. Hieda, and K. Cho, "Bending springback behavior related to deformation-induced phase transformations in Ti – 12Cr and Ti – 29Nb – 13Ta – 4 . 6Zr alloys for spinal fi xation applications," *J. Mech. Behav. Biomed. Mater.*, vol. 34, pp. 66–74, 2014.
[13] *ASM Handbook. Fundamentals of Medical Implant Materials: Materials for Medical Devices. Ohio: ASTM International Vol. 23: 303-325.* 2012.
[14] *ASTM 384. Standard Test Methods for Knoop and Vickers Hardness Materials.ASTM International, 100 Barr Harbour Dr., PO Box C700 West Conshohocken, PA.19428-2959, United States. .*
[15] W. D. Calister, *Material Science and Engineering: An Introduction. John Willey & Sons: Singapore, 2007. .*
[16] P. Pinke, M. Réger, and T. Kovacs, "Heat Treatment of the casted Ti6Al4V titanium alloy," *Mater. Sci. Technol.*, vol. 5, no. 1, pp. 1–6, 2005.

Reducing the High Temperature of Water Radiator Dump Truck (HD 785-7 Type) by Making Special Tool

Vuko A.T Manurung[1], Yohanes Tri Joko[2], Agus Santosa[3]

[1,3]Program Studi Mesin Otomotif, Politeknik Manufaktur astra Jl. Gaya Motor Raya no 8 Sunter II, Jakarta 14330
[2]Program Studi TP3, Politeknik Manufaktur astra Jl. Gaya Motor Raya no 8 Sunter II, Jakarta 14330

* Email: Vuko.manurung@polman.astra.ac.id

Abstract. The high temperature of the engine in the dump truck vehicle will cause a decrease in the performance. Therefore, it is necessary to monitor and analysis the causes high temperature by creating a program called Intelligence Availability. This program is monitor and control the critical engine conditions and to make sure the engine coolant temperature can't exceed 85 °C. In March 2018, 50 units HD 785-7 type have temperature above 85 °C. The main cause is the dirty core of the radiator and the absence of tools to clean the cores, so it is necessary to make a special tool to clean the radiator cores by flushing method. This flushing method can reduce the high temperature to be normal.

1. Introduction

Dump trucks are generally used for materials from medium to long distances, where the material transported can be filled by excavators, wheel loaders, or shovels [1] [2]. In this research, dump trucks manufactured by KOMATSU are used to transport solid objects in the form of top soil and overburden and liquids (water tanks or oil tanks). To support the coal production process, coal mining equipment is needed in excellent performance. Intelligent Availability (IA) is a program created to measure capability and maintain equipment availability by using knowledge and technical expertise (Skill). Thus IA becomes a parameter in the productivity of the unit [3].

In this mining area the achievement of program maintenance has not been maximally and still focused on short-term corrective & preventive activities. The use of predictive tools (VHMS-Vehicle Healthy Monitoring System- & LAP-Lubricant Analysis Program-) has not been used optimally in maintaining equipment availability. Exploration from the predictive side of maintenance still needs improvement, and data-based decision making. [4] [5]

IA (Intelligence Availability) has several assessment variables that will continue to be monitored according to that units. Variable in HD785-7 units are engine oil temperature, coolant temperature, blow by pressure, EGT sensor, boost pressure, and oil temperature transmission. IA data retrieval process can be taken from downloading VHMS, LAP Engine, LAP Final drive. This research is intended to increase IA units to be able to produce maximally. [6] [7] [8] [9] [10]

2. Methodology

Observation in the field has been a very significant decrease in IA so that it needs to be improved to its original target event though if it is possible to exceed the target set. The method used is the corporate culture approach where this research is conducted, which has known continuous improvement. The process starts from the data that is available then analyzed and compared with the targets that have been set. If based on that results of the data below the target set, it is necessary to analyze why until it is below the target, then find the root of the problem using the tool management fishbone analysis. After finding the root of the problem, the improvement process is carried out by involving all stakeholders so that the same mistakes will not occur in the future. [11] [12] [13] [14] [15] [16]

3. Background Data

Based on data in Heavy Equipment Plant 1, the HD785-7 is the most populated units with a total 323 units (March 2018), and the 323 units are divided into several parts, 243 units as Hauler dump trucks, 56 units dump trucks ex remanufacturing, 17 units of tailgate dump truck and 17 units as water tank.

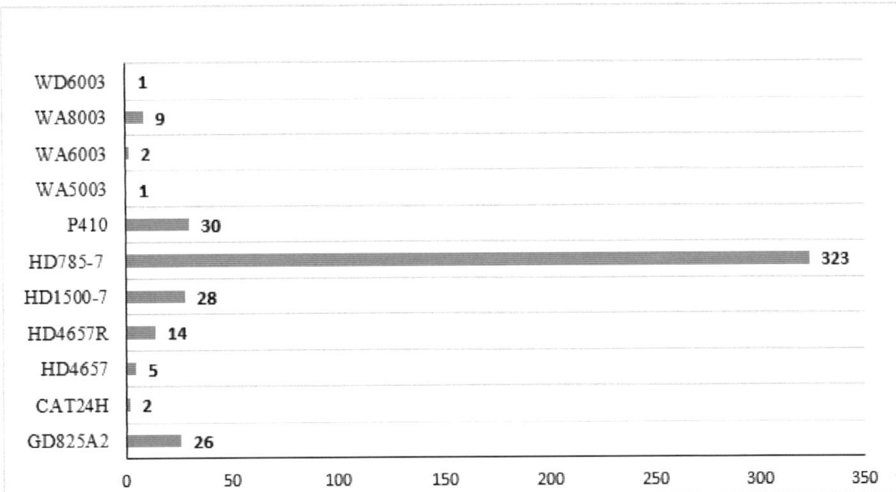

Figure 1. Unit's population (March 2018)

Site observations indicate that the achievement of IA for HD 785-7 units from January to March 2018 is as follows:

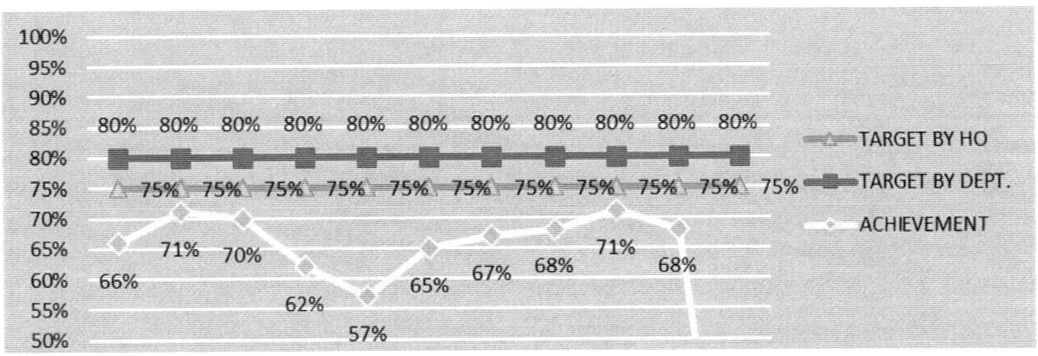

Figure 2. Trend Data IA HD 785-7 from January - March 2018. The vertical axis is achievement (%) and horizontal axis is months (January – March).

Data at figure 2, that HO target (head office) for IA is 75% while the target of the site is 80%. The actual achievement is still below the target. Observations at the workshop showed that there were 50 units of HD 785-7 that had critical coolant high temperature problems and having small lifetime engines while compared to the engine lifetime overhaul limit at 36,000 HM (Hour Meter).

4. Result and Discussion

Based on table 1, then the improvement was chosen based on consideration of effectiveness, processing time and costs for making improvements so that the chosen solution could solve the existing problems. For this reason, the cheapest and most effective solution is to make radiator core flushing tools

Table 1. Alternative Solution Ideas

	NO	ROOT CAUSES	SOLUTION ALTERNATIVE	EFFEC TIVENESS	TIME	COST	REMARK
MAN	1	**The numbers of New mechanics at PHE 1**	Training to all new mechanics equally to increase the competency of maintenance regarding the HD 785-7	Effective	Too long	Expensive	Training to new mechanics, it is expected that the mechanic's individual competence will get same level
MATERIAL	2	**There is not program for Rehose Cooling System**	A hose cooling system program is held for HD 785-7	Effective	Too long	Expensive	In terms of cost, it is expensive, but by doing rehose can improve the quality of the hose
MACHINE	3	**There is no tool to clean the core radiator**	Making a Flushing Core Radiator's Tool	Effective	Too long	Cheap	With the flushing core raditaor's tool it can clean the dirt inside the core, without having to split the cores
METHODE	4	**The radiator cleaning program is not standardized**	Maintenance of radiator every PS 2000 is done by spraying using water	Effective	Faster	Cheap	With fast time and low costs, radiator maintenance can be done at any Periodic Service dilakukan setiap Periodic Service

After number of feeding trials, a tool was obtained as shown in figure 3.

IOP Conf. Series: Materials Science and Engineering **846** (2020) 012067 doi:10.1088/1757-899X/846/1/012067

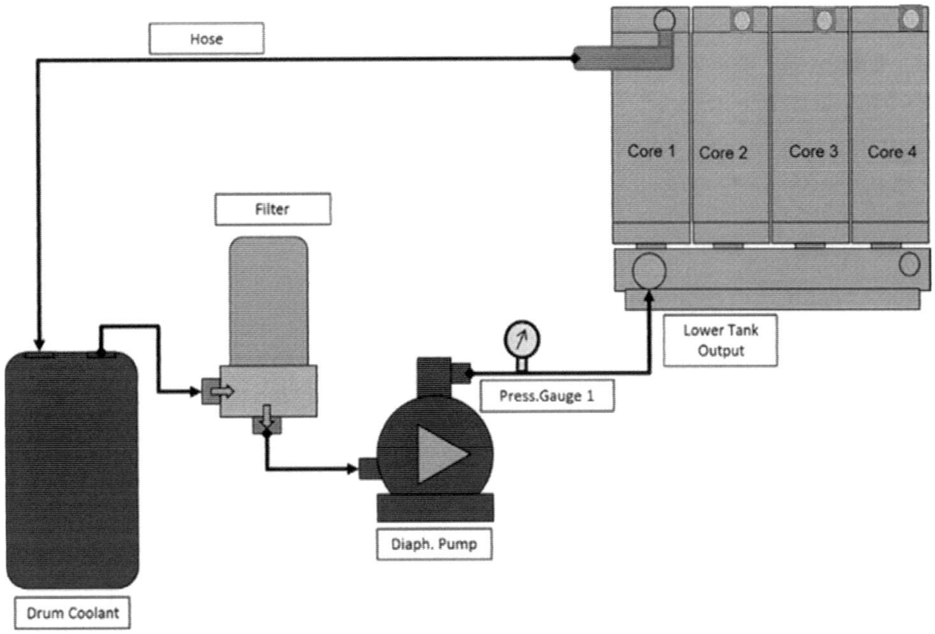

Figure 3. Flushing Tool Core Radiator's scheme.

This tool can remove dirt particles from the radiator core so that there is no blockage in the radiator tube core, which can cause engine coolant heat exchanger processes to be hampered. The principle of this tool is to pump coolant into the radiator core then filter it using a filter with a capability of 10 micrometers. Pump discharge is 350 liters/minute. The testing process was carried out on 28 units of HD 785-7 that overheating. The results are in the following table 2.

Table 2. Result of this Tool Applied to HD 785-7 Units

NO	CN UNIT	TEMPERATURE BEFORE	TEMPERATURE AFTER
1	DT3606	87	85
2	DT3584	87	82
3	DT3888	86	85
4	DT3837	88	84
5	DT3746	86	76
6	DT3914	90	82
7	DT3586	88	84
8	DT3276	86	77
9	DT3506	87	78
10	DT3619	86	83
11	DT3823	89	85
12	DT3837	86	81
13	DT3666	87	85
14	DT4118	88	80
15	DT3531	87	76

16	DT3645	87	80
17	DT4261	90	82
18	DT3649	88	81
19	DT3896	87	85
20	DT3575	87	82
21	DT3929	86	82
22	DT3839	88	78
23	DT3880	87	85
24	DT3473	88	84
25	DT3804	87	85
26	DT4116	86	82
27	DT4103	87	82
28	DT3574	90	84

IA's Intelligence Availability period January – June 2018 has increased, in March IA's achievement was only 68%, after this improvement there was an increase of 83.9% and had reached the target from Head Office and Plant Heavy Equipment 1 Department (figure 4)

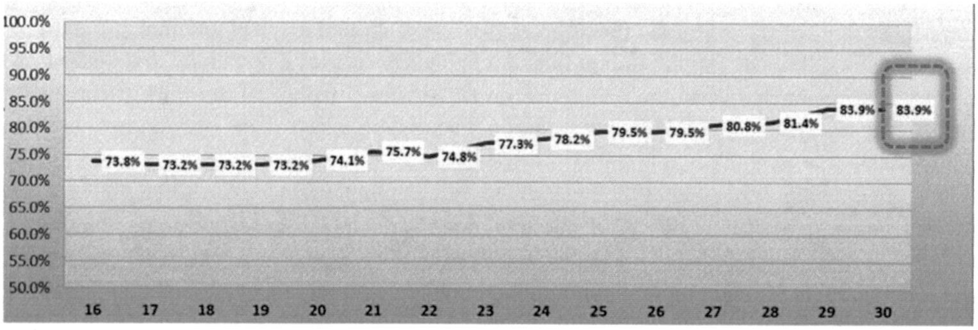

Figure 4. Flushing Tool Core Radiator's scheme[3]. The vertical axis is achievement (%) and horizontal axis is June16-30.

The High Temperature Coolant's trend in the period January – June 2018 has decreased in March 2018 there are 50 units with coolant abnormal temperature and in June 2018 it has decrease to 27 units. This is an effect on IA achievement because the percentage coolant temperature to IA achievement is 17% (figure 5).

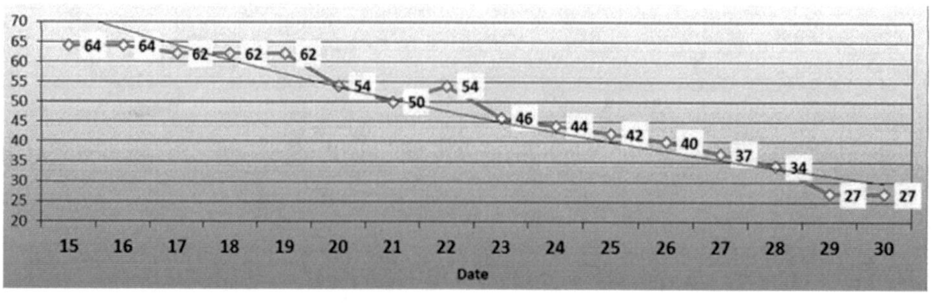

Figure 5. Achievement of IA (Intelligence Availability)[3]. The vertical axis is quantity.

5. Benefit Using this Tool

If using this tool, the potential losses that can be saved are:

a. Radiator Unit.

If there are 50 radiator's to be replaced cause Coolant High temperature
Core Radiator No.1 (Part Number 561-30-81640) = 50 pcs x @ 70,877,408 IDR = 3,543,870,400 IDR
Core Radiator No.2 (Part Number 561-30-81650) = 50 pcs x @ 54,970,648 IDR = 2,748,532,416 IDR
Core Radiator No.3 (Part Number 561-30-81630) = 50 pcs x @ 70,877,408 IDR = 3,543,870,400 IDR
Core Radiator No.4 (Part Number 561-30-81630) = 50 pcs x @ 70,877,408 IDR = 3,543,870,400 IDR
Total = 13,380,143,000 IDR

b. Engine Unit.

In early 2017, engine lifetime's target to be remanufacturing is 33,000 HM (Hour Meter). Whereas in 2018 the target changed to 38,000 HM. With this flushing program, the target can be achieved. The difference of lifetime's engine that can be saved is 5000 HM for each unit. If cost per hour = 25,999 IDR, for 50 units: 50 x 5000 HM x 25,999 IDR = 6,499,767,000 IDR.

The total cost that can be saved from the components are: radiator component cost + engine component cost = 13,380,143, 616 IDR + 6,499,766,737 IDR = 19,879,910,353 IDR

c. Cost of Loss Due to Lead Time

If 50 units having coolant high temperature problem are replaced, all of them will become Unscheduled Breakdown. The lead time processing is 18 hours. 1 hour HD 785-7 can transport coal as much as 2 rotations. If 1 rotation can transport 45 bcm (bank cubic meter) then in 18 hours it transports 1620 bcm. The price of OB is 1 bcm, 2 USD. (1 USD = 13,888 IDR at that time), the cost wasted is 44,997,120 IDR. If there are 50 units, the total lost is 2,249,856,000 IDR

d. Man Hour Cost.

The number of mechanics for the repair process per unit is 3 peoples and the lead time is 18 hours. If a man hour is paid USD 25, then the mechanical cost for repairs per unit is 18 x 3 x (25 USD) x 13,888 = 18,748,800 IDR. For 50 units, the total man hour cost is 937,440,000 IDR.

The total potential loss if it is not use this tool for 50 units are: Cost due to component damage + Costs due to loss of lead time + cost of loss to pay man-hour = 23,067,206,000 IDR.

Total cost of producing tool is 53,821,030 IDR and the lead time of flushing per unit 9 hours with number of mechanics of 2. Thus for 50 units of flushing costs are 277,760,000 IDR. The total cost that can be saved for 50 units are 23,067,206,000 IDR - 331,581,030 IDR = 22,735,625,000 IDR

6. Conclusion

By using this tool, the coolant temperature HD 785-7 can be normal according to the standard of the manufacturer. The flushing process can save component damage to 22,735,625,000 IDR.

Acknowledgment

We are very grateful for the financial assistance and facilities from the PT PAMA ADARO site so that this research can take place according to plan.

References

[1] komatsu Ltd, Shop Manual Komatsu HD 785-7, Japan: Komatsu Ltd, 2011.

[2] k. Ltd, Operation and Maintenance Manual Komatsu HD785-7, Japan: Komatsu Ltd, 2011.

[3] PAMA, Pamawebproud/apps/Mamadata/Default/785-7/IA (Inteligence Availability), Jakarta: PAMA, 2018.

[4] Vuko A T Manurung, Yohanes Wibowo, Muhammad Ramadlan, "Oil Flushing Method to Avoid Unschadule Breakdown Final Drive Components at Komatsu Dump HD 465-7R in PT ATP," in ICETASIA, Solo, 2018.

[5] vuko A T Manurung, Yohanes Wibowo, Muhammad Ramadlan, "Oil Flushing to Avoid Unschedule Breakdown Final Drive Components at Komatsu Dump HD465-7R in PT ATP," in ICETASIA, SOLO, 2018.

[6] Michel J Maron, Haword N Saphiro, fundamental of Engineering Thermodynamics, England: John Wiley & Son Inc., 2006.

[7] F. Justin Dhiraviam1, V. NaveenPrabhu, T. Suresh,, "Improved Efficiency in Engine Cooling System by Repositioning of Turbo Intercooler," Applied Mechanics and Materials , vol. 787, pp. 792-796, 2015.

[8] S. Bennett, Heavy Duty Truck System, vol. 3, USA: Beaverton: Ringgold Inc., 2016.

[9] Mavropoulos, Georgios C, Katsanos, Christos, Knecht, Walter, "Mavropoulos, Georgios C, Katsanos, ChristoImprovement of bottoming cycle efficiency and heat rejection for HD truck applications by utilization of EGR and CAC heat," vol. 53, no. 1, pp. 19-32, 2012.

[10] Gilles, Tim, Automotive Service Inspection, Maintenance, Repair, vol. 86, USA: Delmar, 2012, pp. 2632-2643.

[11] M. Thomas, Truck and Trailer Systems, USA: MacGrow-Hill, 2014.

[12] Liao, Yidai, Zhuo, et al, Liao, Yidai, Zhuo, et al, Heat Flow Characteristics of a Newly-Designed Cooling System with Multi-Fans and ThermLiao, Yidai,Heat Flow Characteristics of a Newly-Designed Cooling System with Multi-Fans and Thermal Baffle in the Wheel Loader,, Applied Science, 2017.

[13] Mavropoulos, Georgios C, Katsanos, et al, "Mavropoulos,Improvement of bottoming cycle efficiency and heat rejection for HD truck applications by utilization of EGR and CAC heat," Mavropoulos, Georgios C, Katsanos, et al, Improvement of bottoming cycle efficiency and heat rejection for HD truck Energy Conversion & Management, vol. 53, pp. 19-32, 2012.

[14] Chong, Kok-Keong, Tan, Woei-Chong, "Study of automotive radiator cooling system for dense-array concentration photovoltaic system," Solar Energy, vol. 86, no. 9, pp. 2632-2643, 2012.

[15] M. Thomas, Truck and Trailer Systems, USA: McGrow-Hill, 2014.

[16] Hountalas, Dimitrios T, Mavropoulos, Georgios C, Katsanos, Christos, Knecht, Walter, "Hountalas, Dimitrios T, Mavropoulos, Georgios Improvement of bottoming cycle efficiency and heat rejection for HD truck applications by utilization of EGR and CAC heat," Energy Conversion & Management., vol. 53, no. 1, pp. 19-32, 2012.

[17] Gilles, Tim, Automotive Service Inspection, Maintenance, Repair, USA: Delmar, 2012.

Estimation of Shear Wave Velocity of Darul Imarah District, Aceh Besar, Indonesia by Using 1D HVSR Inversion

Yusran Y[1,2,*], Aulia Khalqillah[3], Umar Muksin[3,4], Muhammad Syukri[3], Syamsul Rizal[5], Nazli Ismail[3]

[1] Doctoral Program of Mathematics and Applied Sciences, Universitas Syiah Kuala, Banda Aceh, Indonesia
[2] Education of Physics, Universitas Islam Negeri Ar-Raniry, Banda Aceh, Indonesia
[3] Department of Physics, Universitas Syiah Kuala, Banda Aceh, Indonesia
[4] Tsunami and Disaster Mitigation Research Center (TDMRC), Universitas Syiah Kuala, Banda Aceh, Indonesia
[5] Department of Marine Sciences, Universitas Syiah Kuala, Banda Aceh, Indonesia

*Corresponding author: yusran775@gmail.com

Abstract. A total of 19 locations of Horizontal to Vertical Spectral Ratio (HVSR) measurements was conducted in Darul Imarah District, Aceh Besar, Indonesia. The objective of the research is to investigate the shear wave velocity (V_s) structure. This survey is important to analyze a structural characteristic through a V_s, considering the measurement points intersect to Aceh Segment Fault. One of the methods to obtain V_s is 1D HVSR inversion. In this paper, we analyzed only four of 19 measurement points of HVSR data to estimate V_s, they are AB03, AB08, AB09 and AB19 points. The inversion results show that the four points have four layers with different V_s. The average of V_s30 of AB03, AB08, AB09 and AB19 were calculated, they are 150.90 m/s, 197.16 m/s, 150.45 m/s and 316.80 m/s, respectively. The AB09 has the lowest average of V_s30 (highest amplification) than the other points and has soft soil composition. Furthermore, based on the average of V_s30 and amplification, the subsurface of AB03, AB08 and AB19 may have combination soft soil and stiff soil in 30 m depth. The complete analysis of average of V_s30 structure will be conducted at all sites including a 3D velocity structure will be derived from the interpolation of 1D velocity.

Keywords: V_s structure, HVSR inversion, Microtremor

1. Introduction

Since 1999 to 2019, as many as 697 earthquakes with magnitude in range M3.6-M6.5 were recorded by United States Geological Survey (USGS) around Aceh Besar, as shown Figure 1a. Aceh Besar is one of the most earthquake prone areas in the world where earthquakes could occur along the Sumatran Fault (Aceh segment), and Seulimeum Fault, as well as the Indo-Australian and Eurasian plate convergence. Therefore, the impact of the earthquake hazards of the area should be studied in detail. The study of the impact of earthquake hazard can be derived from the microtremor data. Microtremor data make use of natural vibration (non-earthquake) of earth surface that always in motion at seismic frequency to analyze a structure characteristics of a site [1].

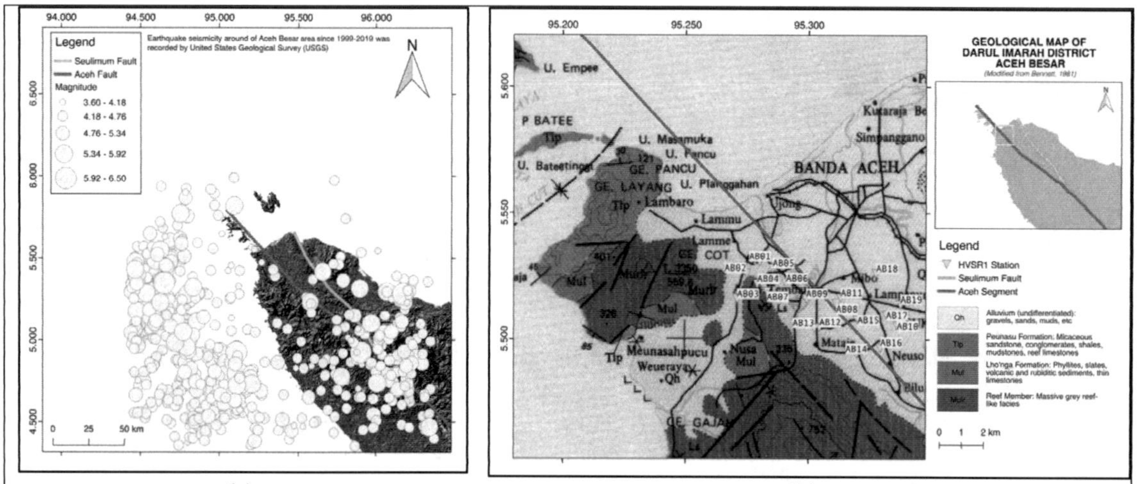

(a) (b)

Figure 1. (a) Since 1999-2019, as many as 697 earthquakes in range M3.6-M6.5 around Aceh Besar were recorded by United States Geological Survey (USGS). (b) Geological map around Darul Imarah District, Aceh Besar Indonesia (modified from Bennett, 1981)

One of the parameters that can be used to explain a structure characteristic is shear wave velocity. The shear wave velocity is very important to seismic hazard evaluation of an area [2], [3]. A low shear wave velocity (less than 200 m/s) indicates high amplification, in other words, a ground is vulnerable to shaking [4]–[6]. The shear wave velocity can be obtained from 1D Horizontal to Vertical Spectral Ratio (HVSR) inversion method of microtremor data. Therefore, the objectives of the research are to estimate of shear wave velocity (V_s), in particularly to estimate an average of shear wave velocity in 30 m depth (V_s30) of Darul Imarah District, Aceh Besar, Indonesia, where only four points microtremor data are used to analyze in this research.

2. Geological Settings

Geological settings around Darul Imarah District, Aceh Besar has been well explained and mapped by [7], as shown in Figure 1b. The lithology around these areas is composed by Qh (alluvium), Tlp (conglomerates, shales, mudstones, reef limestones), Mul (phyllites, volcanic and turbiditic sediments, thin limestones), and Mulr (reef-like facies). Specifically, the lithology around of the HVSR survey points (inverted cyan triangle in Figure 1b) is dominated by alluvium.

3. Methodology

3.1 Data

A total of 19 HVSR measurements points (Figure 2) was conducted in Darul Imarah District, Aceh Besar, Indonesia, considering the survey area is near to Aceh Segment Fault (red line in Figure 2). In this research, we only use four measurement points to analyze of shear wave velocity, there are AB03, AB08, AB09, and AB19 points. The locations of AB03, AB08, AB09 and AB19 intersect to the Aceh Segment, where the AB03 and AB09 are in west-side of Aceh Segment (both points are separated by hill), the AB08 and AB19 are in east-side of Aceh Segment. The HVSR was recorded for 30 minutes by using short-period seismometer Model L-4-3D and data logger of DATA-CUBE^{3}ext. Furthermore, the data processing of HVSR is conducted by setting several parameters, such as bandpass filter 0.5-10

Hz, window length of 25 s, STA window of 1 s, LTA window of 30 s, minimum threshold of 0.2, maximum threshold of 2.0, and Konno-Ohmachi smoothing bandwidth of 40.

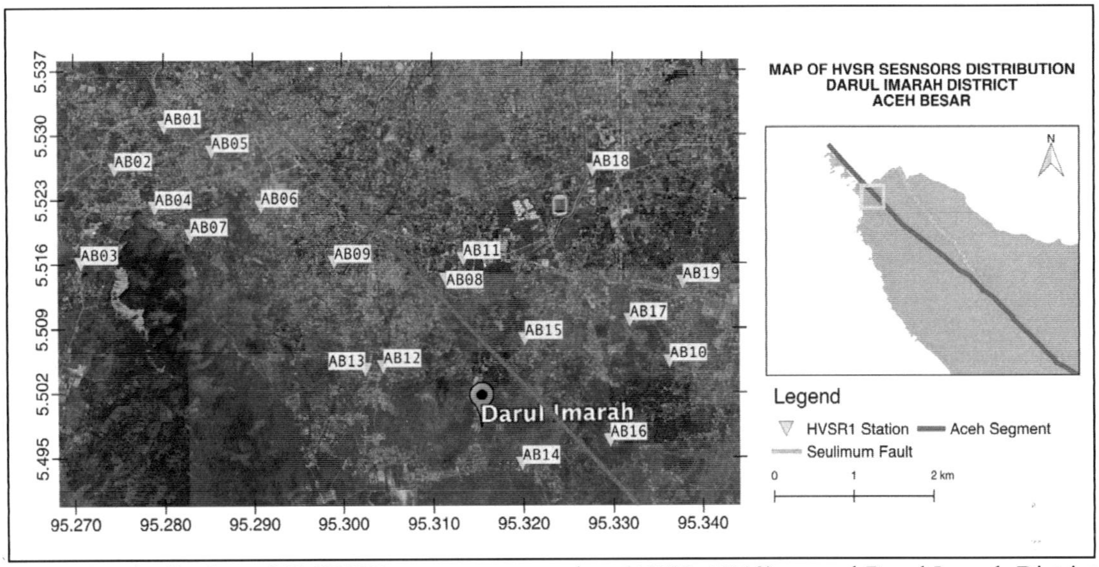

Figure 2. Distribution of 19 HVSR measurement points (AB01-AB19) around Darul Imarah District, Aceh Besar, Indonesia.

3.2. Horizontal to Vertical Spectral Ratio (HVSR) Method

The HVSR is a method to estimate the dominant frequency (f_0) and high amplification from three components (two horizontal components and one vertical component) of microtremor data. The method was introduced by Nogoshi and Igarashi in 1971 to identify the dominant frequency and amplitude characteristic of microtremor data through the Rayleigh wave [8]. However, HVSR method was re-popularized by Nakamura (1989) who mentioned that the microtremor data can be explained by vertical incident SH wave [9]. The Nakamura statement was supported by explanation of Sungkono and Santosa through numerical modelling of HVSR [10]. The HVSR is calculated by using equation 1.

$$\frac{H}{V} = \frac{\sqrt{NS^2+EW^2}}{2V} \quad (1)$$

NS = North-South component (horizontal 1)
EW = East-West component (horizontal 2)
V = Vertical Component

By using the HVSR data (dominant frequency and amplification), we can estimate the structure characteristic beneath a surface, e.g. shear wave velocity (V_s) by using 1D HVSR inversion method.

3.3 1D HVSR Inversion Method

A Neighbourhood Algorithm (NA) method is the inversion method that used to obtain the 1D ground profile of shear wave velocity (V_s) in this research by using the HVSR data. The NA method was introduced by [11] and improved by [12] and implemented in Geopsy software. The NA method is a direct search method that has been widely used to solves inversion problems. The basic idea of those algorithm is to find an optimal models of acceptable data fit in a multidimensional parameters space. This algorithm makes use of the geometrical constructs called Voronoi cells to derive the optimal model in parameter space from each iteration. Therefore, for each new iteration of the models are concentrated around the lowest misfit that is the optimal model in the neighbourhood algorithm [11].

IOP Conf. Series: Materials Science and Engineering **846** (2020) 012068 doi:10.1088/1757-899X/846/1/012068

4. Results

The HVSR data of AB03, AB08, AB09 and AB19 points had been analyzed to obtain the dominant frequency (f_0), amplification (A_0) and seismic vulnerability (Kg), as shown in Table 1. Furthermore, the HVSR data (dominant frequency and amplification) are inverted by using neighbourhood algorithm in Geopsy to obtain the shear wave velocity. As many as 10000 a priory velocity models were imposed to obtain reliable 1D ground profile of shear velocity model for each data from inversion. After the inversion process is complete, the solutions of the 1D shear velocity model of the AB03, AB08, AB09 and AB19 points are shown in Figure 4, respectively. The ground profile of AB03, AB08, AB09 and AB19 consists of four different layers indicated by different shear wave velocity (V_s) as shown in Table 2.

Table 1. Results of data processing of HVSR observation

Point	F_0 (Hz)	A_0	Kg	Vs30 (m/s)
AB03	5.524	0.24	0.01043	150.902
AB08	2.076	0.59	0.16765	197.160
AB09	2.739	1.02	0.38167	150.450
AB19	2.809	0.73	0.19389	316.804

Table 2. The depth of shear velocity (V_s) for each layer

Layer	Depth (m)			
	AB03	AB08	AB09	AB19
Layer 1	0-30.96	0-7.92	0-56.81	0-7.17
Layer 2	30.96-53	7.92-12.77	56.81-60.92	7.17-19.01
Layer 3	53-98.25	12.77-98.25	60.92-99.23	19.01-92.55
Layer 4	98.25-103.17	98.25-103.17	99.23-104.2	92.55-97.19

By using shear wave velocity (V_s) data, an average of shear wave velocity in 30 m depth (V_s30) had been calculated to identify a structure characteristic by using equation 2. The AB03 has V_s30 of 150.902 m/s, AB08 has V_s30 of 197.160 m/s, AB09 has V_s30 of 150.450 m/s and AB19 has V_s30 of 316.804 m/s. The V_s30 of AB03, AB08, and AB09 are categorized to low shear velocity (less than 200 m/s), while the V_s30 of AB19 is categorized to moderate shear velocity. However, according to site classification of V_s30 by SNI 1726-2012 [13], the AB03 and AB09 are in west-side of Aceh Segment categorized to SE class or soft soil with low shear velocity. Meanwhile, the AB08 and AB19 are in east-side of Aceh Segment categorized to SD class or medium soil with moderate shear velocity.

$$\bar{v}_s = \frac{30}{\sum_{i=1}^{n} \frac{d_i}{v_{si}}} \qquad (2)$$

d_i = thickness for each layer between 0-30 m depth
v_{si} = shear velocity for each layer between 0-30 m depth
$\bar{v}_s$ = average of shear wave velocity (m/s)
n = a total of layer in 30 m depth

The low velocity indicates high amplification and high seismic vulnerability if a natural vibration (e.g. earthquake) occurs in an area. According to the results of V_s30 in Table 1, the AB09 has lowest of V_s30 (highest amplification) than the other points. Furthermore, the AB03 has second lowest V_s30 after AB09 and lower than V_s30 of AB08 and AB19. However, the amplification of AB03 is lower than amplification of AB08 and AB19. The same problems apply to AB08 compared with AB19.

The low V_s30 and low amplification of AB03 may be affected by natural vibration that coming from the hill (east-side of AB03 point) that has high frequency, because the observation point of AB03 is close to the hill. The AB03 may be due to combination of soft soil and stiff soil. Furthermore, the V_s30 of AB08 is higher than AB03, that means the subsurface of AB08 may has a composition

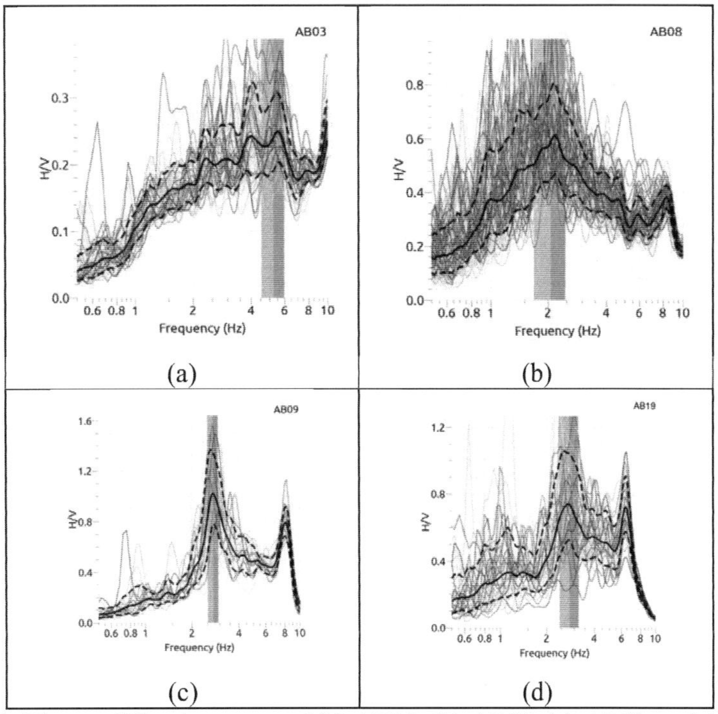

Figure 3. The amplification curve of (a) AB03, (b) AB08, (c) AB09, and (d) AB19

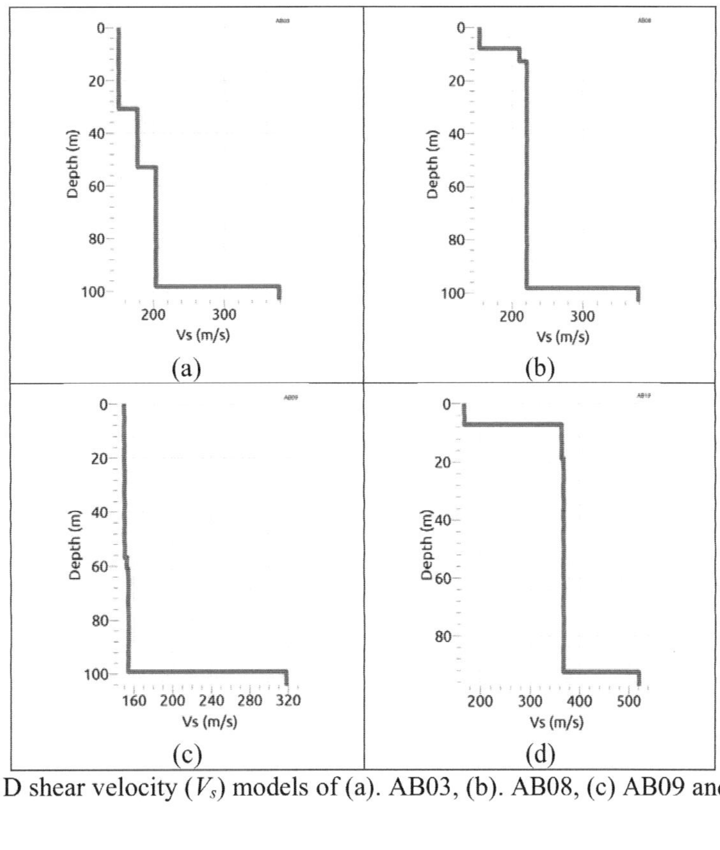

Figure 4. 1D shear velocity (V_s) models of (a). AB03, (b). AB08, (c) AB09 and (d) AB19.

between soft soil and stiff soil with high amplification than the AB03. The AB19 has the highest V_s30 than other points. In this condition, the AB19 has more stiff soil than the other points with the high amplification after AB09. However, the amplification of AB19 is higher than amplification of AB03 and AB08. It means, the AB19 may has a composition between soft soil and stiff soil. To ensure these results, boreholes survey have to be carried out. The complete analysis all of HVSR data of this research will be reviewed in the next article.

5. Conclusion

By using the HVSR inversion of four points (AB03, AB08, AB09 and AB19) the velocity structure, V_s, consists of four layers by different shear wave velocity. The average of V_s30 of AB03, AB08, AB09 and AB19 are 150.90 m/s, 197.16 m/s, 150.45 m/s, and 316.80 m/s, respectively. According to site classification of V_s30 by SNI 1726-2012, the AB03 and AB09 are in west-side of Aceh Segment categorized to SE class or soft soil and the AB08 and AB19 are in east-side of Aceh Segment categorized to SD class or medium soil. The AB09 has the lowest V_s30 (highest amplification) than the other points and has soft soil composition. Furthermore, based on the V_s30 and amplification, the subsurface of AB03, AB08 and AB19 may have combination soft soil and stiff soil in 30 m depth. To ensure these results, boreholes survey have to be carried out.

References

[1] H. Okada and K. Suto, *The Microtremor Survey Method*. 2010.
[2] A. Cipta *et al.*, "Seismic velocity structure of the Jakarta Basin, Indonesia, using trans-dimensional Bayesian inversion of horizontal-to-vertical spectral ratios," *Geophys. J. Int.*, vol. 215, no. 1, pp. 431–449, 2018.
[3] D. Fäh, F. Kind, and D. Giardini, "Inversion of local S-wave velocity structures from average H/V ratios, and their use for the estimation of site-effects," *J. Seismol.*, vol. 7, no. 4, pp. 449–467, 2003.
[4] N. Sundararajan and T. Seshunarayana, "Surface wave studies for shear wave velocity and bedrock depth estimation over basalts," *Arab. J. Geosci.*, vol. 7, no. 9, pp. 3791–3799, 2014.
[5] N. Sundararajan *et al.*, "Shear Wave Velocity Characteristics in Parts of Muscat, Sultanate of Oman — A Measure of Earthquake Hazard Assesment," *J. Geol. Soc. India*, vol. 93, no. 5, pp. 515–522, 2019.
[6] R. D. Borcherdt, "Effects of local geology on ground motion near San Francisco Bay*," *Bull. Seismol. Soc. Am.*, vol. 60, no. 1, pp. 29–61, 1970.
[7] J. D. Bennett and Pusat Penelitian dan Pengembangan Geologi (Indonesia), "Peta Geologi Lembar Bandaaceh, Sumatra = Geologic Map of the Bandaaceh Quadrangle, Sumatra." Peta Geologi Lembar Bandaaceh, Sumatra = Geologic Map of the Bandaaceh Quadrangle, Sumatra," 1981.
[8] M. Nogoshi and T. Igarashi, "On the Amplitude Characteristics of Microtremor (Part 2)," vol. 1971, no. Part 2, pp. 59–76, 1971.
[9] Y. Nakamura, "A method for dynamic characteristics estimation of subsurface using microtremor on the ground surface," *Railw. Tech. Res. Institute, Q. Reports*, vol. 30, no. 1, 1989.
[10] Sungkono and B.J. Santosa, "Karakterisasi Kurva Horizontal-To-Vertical Spectral Ratio: Kajian Literatur Dan Permodelan," *J. Neutrino*, 2011.
[11] M. Sambridge, "Geophysical inversion with a neighbourhood algorithm - I. Searching a parameter space," *Geophys. J. Int.*, vol. 138, no. 2, pp. 479–494, 1999.
[12] M. Wathelet, "An improved neighborhood algorithm: Parameter conditions and dynamic scaling," *Geophys. Res. Lett.*, vol. 35, no. 9, pp. 1–5, 2008.
[13] B. S. Nasional, "Tata cara perencanaan ketahanan gempa untuk struktur bangunan gedung dan non gedung," *Sni*, vol. 1726, p. 2012, 2012.

IOP Conf. Series: Materials Science and Engineering **846** (2020) 012069 doi:10.1088/1757-899X/846/1/012069

Formulation of Hand Sanitizer Gel of A-Pinene Isolated from Turpentine Oil and its Antibacterial Activity

Nanik Wijayati [1*], Anisa Widiyastuti [1], Sri Mursiti [1], Senda Kartika Rakainsa [2]

[1] Faculty of Mathematics and Natural Sciences, Universitas Negeri Semarang, Semarang - Indonesia

* nanikanang@gmail.com

Abstract. Research on the antibacterial assay of the α-pinene and turpentine oil in hand sanitizer gel formulation has been carried out. Antibacterial activity assay using the disc diffusion method showed that turpentine oil has the highest antibacterial activity with a inhibition zone of 13.8 mm against Staphylococcus aureus and 8.83 mm against Escherichia coli, whereas α-pinene has the highest antibacterial activity with a 2.2 mm inhibition zone against S. aureus and 2.34 mm against E. coli.

1. Introduction

Health is an important aspect which can affect the quality of life. One effective way to preserve health is by maintaining hand hygiene [1]. Through physical contact, various kinds of viruses and bacteria attach to the hand. The most appropriate way to prevent this is to properly washing hands with soap and clean running water [2]. This effort is carried out because the hands can be germ-carrying agent and cause disease. Over time, hand washing habits have been diverted by hand sanitizer because it is more practical and easier to use. Turpentine oil is an example of natural ingredients that have the potential as an antiseptics. Turpentine oil is a clear colored liquid and has a distinctive odor derived from the distillation of tree sap which is classified as pine resin. Market demand for this oil is increasing every year due to its useful purpose for pharmaceutical raw materials, resins, polymers, perfumes and solvents [3]. Alpha-pinene is a natural terpene and is the main components of turpentine oil [4]. The higher the α-pinene content, the higher level of purity and quality of turpentine. Beside α-pinene, turpentine oil contains monoterpene hydrocarbons such as β-pinene, camphene [5] and 3-carene [6], [7].

Several studies revealed that terpenoids from turpentine oil can inhibit bacterial growth with 13 mm inhibition zones against *S. aureus* and 8 mm against *E. coli* [8]. Masruri conducted a study of antibacterial activity of α-pinene against bacterial growth with 8.3 mm inhibition zones against *S. aureus* and 8.9 mm against *E. coli* [9]. Based on these researches, we performed formulation of turpentine oil and α-pinene in the gel hand sanitizer dosage form. The purpose of this study was to determine the antibacterial activity of turpentine oil gel and α-pinene gel against *S. aureus* and *E. coli*

2. Materials and Methods

2.1. Research instruments

The instruments used in this study include set of fractional distillation devices, Fourier Transform Infra Red (FT-IR) PerkinElmer Frontier 10.03.06, Gas Chromatography-Mass Spectrophotometer (GC-MS) Perkin Elmer, autoclaves, incubators, laminar air flow and vortex mixer.

2.2. Materials

The material used include turpentine oil, anhydrous Na_2SO_4, carboxymethylcellulose (CMC), triethanolamine (TEA), propylene glycol, glycerin, distilled water, nutrient agar, McFarland 0,5 standard, bacteria cultures of *S. aureus* and *E. coli* which obtained from the Integrated Laboratory of Universitas Diponegoro, Semarang, Indonesia.

2.3. Alpha-pinene isolation

Turpentine oil was obtained from Indonesian State Forest Company Unit I Central Java, and isolated by fractional distillation. The anhydrous Na_2SO_4 was added to purify the desired results by binding to the remnants of water that is still mixed with turpentine oil. Alpha-pinene was isolated using the fractional distillation method with reduced pressure. A total of 500 mL turpentine oil is inserted in flask and filtered using filter paper. Anhydrous Na_2SO_4 is added to bind water in turpentine oil. The non-aqueous turpentine oil is put into a flask which is connected to a fractional distillation device with a temperature of 50-60°C. The results of the distillation were investigated using FT-IR and GC-MS.

2.4. Antibacterial activity assay

The disc diffusion method was used in the antibacterial assay. The testing procedures includes sterilization of tools and materials, preparation of nutrient agar media and preparation of bacterial suspensions. One ounce of each *S. aureus* and *E. coli* bacteria was added into a test tube containing physiological saline solution then homogenized and measured its turbidity according to McFarland 0.5 standard. A total of 1 mL of microbial suspension was taken using a micropipette and put into a sterile petri dish. Liquid nutrient media that was poured into the petri dish and shaken homogeneously. Paper disc that has been soaked in turpentine oil and α-pinene was placed on the surface of the nutrient agar which has been planted with bacteria. Next, the media was incubated for 24 hours into an incubator with a temperature of 37°C and the inhibition zone could be observed and measured.

2.5. Hand sanitizer gel formulation

The hand sanitizer gel formula refers to the modified formula by Manus [10]. The production of gel was carried out by adding CMC in hot distilled water then stirred. TEA was added drop by drop while stirring, then propylene glycol, turpentine oil, glycerin and distilled water was added to the desired volume. Commercial hand sanitizer (Dettol Brand) was used as positive control. The gel hand sanitizer formulations is presented in Table 1.

Table 1. Hand sanitizer gel formulations

Component	Negative control	Turpentine oil gel	α-pinene gel
Active ingredients	-	1,5 mL	1,5 mL
CMC	0,25 g	0,25 g	0,25 g
TEA	2 drops	2 drops	2 drops
Propylene glycol	0,5 mL	0,5 mL	0,5 mL
Glycerin	1 mL	1 mL	1 mL
Distilled water ad	10 mL	10 mL	10 mL

3. Result and discussion

The analysis of chemical components of turpentine oil was performed using Gas Chromatography. The chemical components including α-pinene with a percentage of 86.02%, delta-3-carene of 12.85% and β-pinene of 0.87%. The α-pinene was analyzed using an Infrared Spectrophotometer (FT-IR). The α-pinene IR spectra is presented in Figure 1.

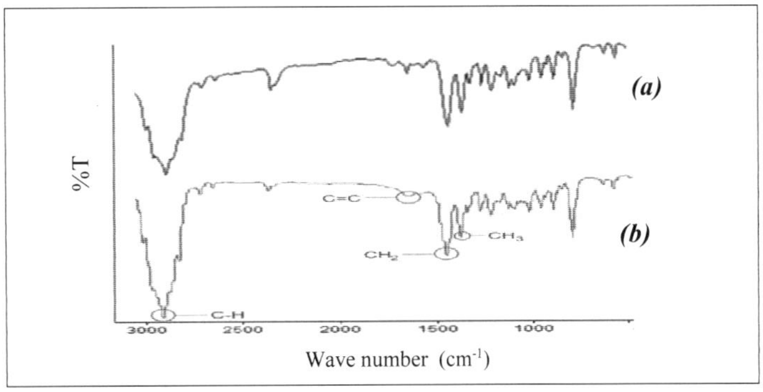

Figure 1. IR Spectra of α-pinene (a: standard α-pinene, b: isolated α-pinene)

Figure 1 showed the results of characteristic absorptions of α-pinene which was isolated previously. The area of C-H absorption, C = C, and CH_2 was occurred at wave number 2920 cm^{-1}, 1651 cm^{-1}, and 1444 cm^{-1}, respectively. Wave number 1369 cm^{-1} indicated the CH_3 absorption. Based on the results of the FT-IR above, the isolated α-pinene spectrum was the same as the standard α-pinene spectrum. Analysis of the composition of turpentine oil isolation was carried out by Gas Chromatography and Mass Spectroscopy (GC-MS) detectors. Chromatogram from the turpentine oil analysis showed four detected peaks, but only one peak with high abundance was analyzed by a mass spectrometer, that is the peak with a retention time of 3.814 min. Figure 2 shows the results of α-pinene GC-MS analysis.

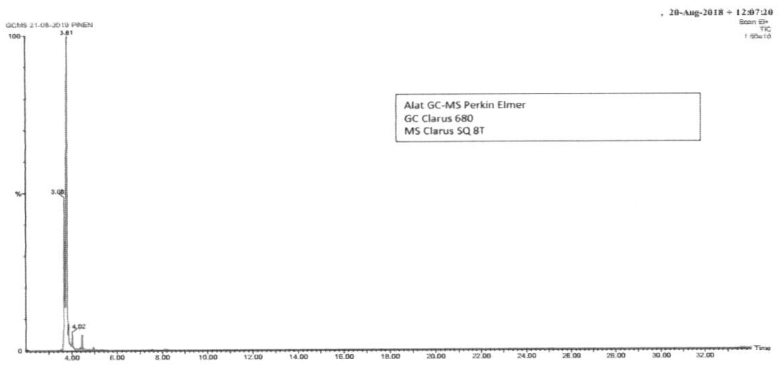

Figure 2. Chromatogram of α-pinene

The chromatogram showed the presence of several high peak. The higher the peak of a chromatogram, the greater the area (concentration) of the chemical component. The largest components contained in turpentine oil from isolation are presented in Table 2.

Table 2. Results of GC-MS analysis of α-pinene

Peak	Retention time (minutes)	Area (%)	Name of compound based on *library* MS
1	3.679	23.56	α-pinene
2	3.814	73.79	α-pinene
3	4.019	1.13	camphene
4	4.449	1.51	β-pinene

The α-pinene analysis using GC-MS showed two peaks allegedly due to equipment errors. Based on the MS library, it could not be distinguished between the first α-pinene and the second α-pinene, because it was both possessed molecular formulas of $C_{10}H_{16}$. Turpentine oil and α-pinene then tested for antibacterial activity and formulated into gel hand sanitizer. The formulation of gel hand sanitizer was carried ot by using CMC base as gelling agent, TEA as alkalizing agent, propylene glycol as humectant, preservative, and stabilizing agent. Addition of glycerin functioned as humectants and emollients [11]. Antibacterial activity test results are presented in Table 3.

Table 3. Results of antibacterial activity test

No	Sample code	Bacteria	Inhibitory zone (mm)	The average of inhibitory zone (mm)
1	A	*S. aureus*	13.5	
2			14.2	13.8
3			13.7	
4		*E. coli*	8.9	
5			8.9	8.83
6			8.7	
7	B	*S. aureus*	3.5	
8			3.2	3.47
9			3.7	
10		*E. coli*	3.3	
11			3.6	3.33
12			3.1	
13	C	*S. aureus*	8.7	
14			8.7	8.83
15			8.5	
16		*E. coli*	7.5	
17			7.8	7.57
18			7.4	
19	D	*S. aureus*	2.3	
20			2.1	2.2
21			2.2	
22		*E. coli*	2.2	
23			2.4	2.37
24			2.5	
25	E	*S. aureus*	-	
26			-	-
27			-	
28		*E. coli*	-	
29			-	-
30			-	
31	F	*S. aureus*	8.2	
32			8.3	8.3
33			8.4	
34		*E. coli*	9.4	
35			8.9	9.13
36			9.1	

A: Turpentine oil
B: Turpentine oil gel
C: α-pinene
D: α-pinene gel
E: Negative control (gel base)
F: Positive control (Dettol)

The results above indicated that turpentine oil and turpentine oil gel exhibited antibacterial activity due to the presence of secondary metabolites. The main chemical contents identified in the GC-MS analysis were terpenoid compounds, namely α-pinene, β-pinene, and camphene. Terpenoids are the largest group of secondary metabolites which has large number of compounds and variations in their basic structure and are the main constituents of essential oils. The mechanism of action of terpenoids as an antibacterial substance is by disrupting the cellular membrane [12], [13]. In the outer membrane of the cell wall, terpenoids react with porins and then damage porins and form strong polymer bonds, reducing the permeability of bacterial cell walls so that cell nutrients become deficient and will die or be inhibited Porin is the entry and exit site of nutrients, if the porin is damaged, the permeability of the bacterial cell membrane is reduced. This condition will result in the death of bacterial cells [14].

The concentration of turpentine and α-pinene oil used in this antibacterial activity test was 100% which was tested triplicates. Antibacterial activity assay showed that turpentine oil had better antibacterial activity than α-pinene, which was proven by the larger diameter of the bacterial inhibition zone in *S. aureus* and *E. coli*. This is consistent with research conducted by Mimoune and Zeynep which stated that turpentine oil had antibacterial activity against *S. aureus* and *E. coli* [8], [15].

The results of antibacterial activity assay showed that the negative control did not exhibited bacteria inhibitory zone. The inhibition zone produced by turpentine oil gel is greater than the α-pinene gel. This is due to the the presence of the other secondary metabolites in turpentine oil which has the antibacterial potency against *S. aureus* and *E. coli*, such as delta-3-carene, camphene, and β-pinene. It can be estimated that the presence of other compounds can improve the antibacterial activity [16], [17]. The inhibitory activity of gel hand sanitizer gel of turpentine oil and α-pinene was considered weak (≤ 5 mm) because the gel has thick consistency, hence it could affect the disc paper immersion.

The concentration of bacterial suspension that is opposed is one of the factors that influence whether the samples showed inhibition of bacterial growth. The high concentration of cells allows it to affect antibacterial activity [11]. In addition, bacterial properties which include age, type and state of bacteria are also influential. Bacterial dilution to reach concentrations of 10^5 and 10^6 was carried out in several studies regarding antibacterial activity test. This statement is reinforced by Pelczar & Chan that if there are more microorganisms, the more time needed to inhibit or kill [18]

4. Conclusion

Based on the results of the study, it can be concluded that turpentine oil and α-pinena in hand sanitizer gel formulations exhibit antibacterial activity against *S. aureus* and *E. coli*.

Acknowledgement

This research is supported by grant funds under 2019 Fundamental Research in Higher Education from the Directorate of Research and Public Service, Ministry of Research, Technology, and Higher Education, Republic of Indonesia

References

[1] Mathur P 2011 *Indian J. Med. Res.* 134(5) 611–620.
[2] Matar M J, Moghnieh R A, Awad L S, Kanj S S 2009 *Curr. Treat. Options Infect. Dis.* 10(2) 310–329.
[3] Mercier B, Prost J, Prost M 2009 *International Journal of Occupational Medicine and Environmental Health.* 22(4) 331–342.
[4] Wróblewska A, Miądlicki P, Tolpa J, Sreńschek-Nazzal J, Koren Z C, Michalkiewicz B 2019 *Catalysts* 9 396.
[5] Wróblewska A, Miądlicki P, Tolpa J, Sreńschek-Nazzal J, Koren Z C, Michalkiewicz 2018 *Microporous Mesoporous Mater.* 258 72–82.
[6] Haneke K E 2002 *Turpentine Oil, Wood Turpentine, Sulfate Turpentine, Sulfit Turpentine* (North Carolina: Integrated Laboratory Systems).
[7] Lindmark M H 2003 *Biotransformation of Turpentine Constituents:Oxygenation and Esterification* (Doctoral Thesis, Sweden: Sweden University)
[8] Mimoune N A, Djouher A M, Aziza Y *J. Coastal Life Med.* 1(1) 55–59.
[9] Masruri, Rahman M F, Tegas I P 2007 *Jurnal Ilmu-Ilmu Hayati,* 19(1) 32–35.
[10] Manus N, Paulina V Y, Novel S K 2016 *Jurnal Ilmiah Farmasi*, 5(3) 85–93.
[11] Ningsih A P, Nurmiati, Agustien A 2013 *Jurnal Biologi Universitas Andalas,* 2(3) 207–213.
[12] Kapros T, McDaniel S C 2009 *Allelopath. J.* 23(15) 185–192.

[13] Di Pasqua R, Betts G, Hoskins N, Edwards M, Ercolini D, Mauriello G 2007 *J. Agric. Food Chem.* 55 4863–4870.
[14] Nikaido H 2001 *Semin. Cell Dev. Biol.* 12 215–223.
[15] Zeynep U, Salih K, Fuat B, Burhan A, Selim E, Menderes C, Göksin K M 2014 *Chinese Journal of Natural Medicines,* 12(12) 901–910.
[16] Leite A M, Lima E O, Souza E L, Diniz F M, Trajano V N, Medeiros I A 2007 *Braz. J. Pharm. Sci.* 43 121–126.
[17] Ghaffari T, Kafil H S, Asnaashari S, Farajnia S, Delazar A, Baek S C, Hamishehkar H, Kim K H 2019 *Molecules*. 24 3203.
[18] Pelczar M, Chan E C S 2007 *Mikrobiologi Kedokteran* (Jakarta: University of Indonesia Press).

IOP Conf. Series: Materials Science and Engineering **846** (2020) 012070 doi:10.1088/1757-899X/846/1/012070

Liquefaction Resistance of Sand Mixed with Fines-Grained for Reclaimed Land

Yelvi[1*], Handi Sudardja[2], Shany Elizabeth[3]
[1,2,3]Civil Engineering Department, State Polytechnic of Jakarta

*yelvi@sipil.pnj.ac.id

Abstract. Liquefaction due to earthquakes can cause collapse, subsidences, landslides, *lateral spreading*, and *sand boiling*. The damage is commonly occurred at layer of water-saturayed granular sedimen with low relative density. Reclamation areas formed by loose sand are areas with high risk to liquefaction, such as damages occured in the reclamation area of Tokyo Bay Area (Fukushima Earthquake, 2011) and Wellington Port (New Zealand Earthquake, 2016). Some researches have stated that the gradation of soil grains, relative density and maximum acceleration of an earthquake affected the soil's resistance to liquefaction. This paper presents the results of the experimental studies about the influence of fines-grained in the sand that has liquefaction potential which is used as a reclamation material to review its resistance to liquefaction. The experimental studies were carried out by adding fines-grained. A mixture of sand with fines-grains was made at a relative density of 25%, 50% and 75% under earthquake load with a maximum acceleration of 0.3g. Experimental results show that the increase in the fines content up to 30% and the relative density up to 75% reduces the resistance to liquefaction of the sand–fines mixture. The liquefaction resistance increases well as the increase of fines-grained and relative density.

1. Introduction

Liquefaction is a phenomenon where the soil changes from a solid phase to a liquid phase because of effect of increase pore water pressure in the pore soil [1]. From previous liquefaction research, it is known that liquefaction generally occurs in areas formed by granular layers of water-saturated sediment with low relative density [2]–[4]. This incident caused damage to buildings, moving land, lateral spreading, and landslides. Indonesia is not spared from liquefaction events, such as those occurred in Nias earthquake (2005), Yogya (2006), Padang (2009), and Palu earthquake (2018). Reclamation area is one area that is vulnerable to liquefaction because it is usually built in a coastal or waterlogged area. Besides that, embankment material generally uses levels of sand a minimum of 80%. Such conditions have led to liquefaction in reclamation areas such as Tokyo Bay Area (Fukushima Earthquake, 2011) and Wellington Port (New Zealand Earthquake, 2016) [5]–[7]. Reclamation is not only done abroad, some regions in Indonesia have already carried out reclamation, such as Jakarta Bay, Mamuju Beach, Denpasar, Manado, Semarang, Tangerang, and also in Makassar [8]. Based on condition of Indonesia's earthquake-prone areas, it is necessary to conduct research

IOP Conf. Series: Materials Science and Engineering **846** (2020) 012070 doi:10.1088/1757-899X/846/1/012070

related to the liquefaction problem, especially in the reclamation area. Analysis of relative density is needed to prevent liquefaction by providing additional requirements that must be included in the reclamation work process to achieve the specified relative density [9]. The fine grain content affects the strength of the reclaimed soil against liquefaction, where its strength increases with the increase of fine grains content and relative density [10]. Chien, LK, et.al [11] in research for land reclamation, state that the increase of fine grain content under constant relative density will decrease of liquefaction strength. According to research conducted by Hakam (2016) [10], it stated that there is a limit value of relative density respect to its mean grain size associated with liquefaction resistance.
This study was conducted to review the effect of density relative and levels of fines content by using shaking table against the resistance of liquefaction. The sand used in this study was taken from Lampung area in which it was usually used as material for land reclamation in Indonesia. Therefore, these results might be used as a reference for reclamation material planning and ensure the safety and stability of reclaimed land from potential liquefaction.

2. Methodology

This research uses an experimental method by providing several sample specimens modeled as reclamation soils that have liquefaction potential. The test specimens were made of 12 samples, namely with a mixture of fine grain content of 0%, 10%, 20%, and 30%, and have a relative density of 25%, 50%, and 75%.

2.1 Research Materials

2.1.1 Sand sample. Selected soil sample used in this study is disturbed soil, originated from Lampung. It is one of the materials used for reclamation in Indonesia, especially in Jakarta Bay. The early stages of the test are conducted prior weeks to look at the criteria of this sand soil liquefaction of its potential as given by Tsuchida [12], shown in Figure 1.

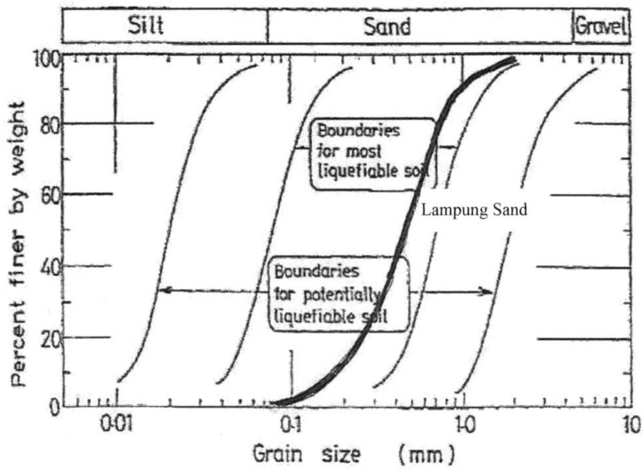

Figure 1. Gradation of Lampung sand soil and potentially liquefaction soil chart [12].

Figure 1. Shows that the distribution of Lampung sand grains is within the most potentially liquefied land boundary according to Tsuchida graph [12]. In addition to the grain size analysis test, the specific gravity and relative density tests are also carried out referring to ASTM D 1989, the results of which can be seen in Table 1.

Tabel 1. Index Properties of Lampung Sand

Parameter	Results	Unit
γd_{max}	1827	kg/m^3
γd_{min}	1456	kg/m^3
Gs	2.66	
e_{max}	0.842	
e_{min}	0.49	
D_{50}	0.4	Mm
D_{10}	0.23	Mm
D_{60}	0.45	Mm
D_{30}	0.2	Mm
C_u	1.97	
C_c	0.38	

2.1.2 Fine Grain Soil Sample. Fines-grain soil used is the soil taken from the East Canal area, East Jakarta at a depth of -7.50 m with the undisturbed sample. Then several tests are carried out to get the soil property index. Index properties of fines-grain can be seen in Table 2.

Tabel 2. Index Properties of Fines-Grain

Parameter	Notation	Result	Unit
Specific Gravity	Gs	2.56	-
Liquid Limit	LL	58.53	%
Index Plasticity	IP	23.34	%

According to the USCS classification, this type of soil is categorized as fine-grained soils with a percentage of granules passing filter No.200 with an amount of more than 50%, and a liquid limit > 50, so this land is classified as MH (inorganic silt).

2.1.3. Mixed Sand + Fine Content (FC) Sample. Research carried out as experimental research on potential liquefaction sand mixed with fine-grained soil and the fine grain content of 0%, 10%, 20%, and 30 %. Samples were prepared by mixing sand and fine grain soil that has been dried, and each was weighed its dry weight according to the relative density of 25%, 50%, and 75%. The results of the relative density testing of each mixture can be seen in Table 3.

Table 3. Results of Relative Mixed Soil Density Tests

Fine Content FC (%)	γ_{dmax} kg/m^3	γ_{dmin} kg/m^3
0%	1827	1456
10%	1762	1417
20%	1750	1408
30%	1710	1406

From the test results above, the calculation is carried out to get the weight of the mixture from mixed soil samples. After obtaining the dry weight of the mixed soil samples, the value of the dry weight of the sample to be tested into the shaking table can be obtained.

2.2. Research Procedure

2.2.1. Soil Sample Mode. Soil modeling was made like a sample box with a size of 400 x 400 x 400 mm^3. The sample box was made from acrylic material with its sides glued together well so that it allows the researcher to see firsthand the conditions that occur in the sample at the time of testing and after testing. The walls of the sample box were perforated on one side for the transmitter (measuring the pore tension). The sample box is equipped with two pore water transmitters that are installed on one side of the sample box. Installation of pore water transmitter is placed in the middle (200 mm) and bottom (320 mm) measured from the surface of the sample. The bottom side of the container was given a 5 cm thick rubber layer to allow for permanent shear deformation during vibrations, as well as improvements to avoid errors that occur in testing. The disturbed samples were tested by using shaking table testing:

a) Sand soil samples are poured carefully into a sample box that has a thickness of 12 mm and is leveled to a specified height. Then the water is put into the sample box until the surface of the water is at equal level as the sand.
b) In the pouring process, the condition is almost saturated when the water level above the sample is constant.

2.2.2. Loading. In potential liquefaction testing, the concept of loading carried out is dynamic loading. Some loading applications used in this study include the maximum earthquake speed of 0.3 g, the frequency of earthquake vibrations 1.1 Hz, and the dynamic loading time of 100 seconds.

2.2.3. Shaking Table Testing Procedure

1. Insert sand samples according to the sample preparation procedure.
2. When filling the sample take the excess water on the surface, so that the water level is the same as the sand level.
3. Shaking table is driven using an electric motor by following with the specified earthquake load ie a_{max} = 0.3 g and 1.1 Hz frequency, same with Yelvi (2018) research [13]
4. The increase in pore pressure at dynamic loading times is recorded second by second until 100 seconds.
5. The test results are in the form of a pore pressure reading which is a function of time plotted in a graphic so that it can be evaluated whether or not liquefaction occurs.

3. Experimental Results

Tests were carried out for every 12 variations in relative density and fine content, with 25%, 50%, 75% for relative density, and 0%, 10%, 20%, 30% for fine content. Loading criteria used in this study was the maximum acceleration of the earthquake of 0.3 g, and the vibration frequency of 1.1 Hz, the loading time of 100 seconds. The number of vibration cycles that occur every 100 seconds is 110 times. This research was conducted at the same water level as the soil surface. The data obtained is the value of the change in air pressure (Δuo) which is read at the layer at a depth of 20 cm (middle transmitter) and 32 cm (bottom transmitter). Evaluation of the test results is done by making a graph of the relationship between the ratio of pore pressure to the number of cycles. To measure the liquefaction resistance, a parameter ru is defined as the ratio of excess pore water pressure due to shaking to effective overburden pressure. The relationship graph is stated in Figure 2.

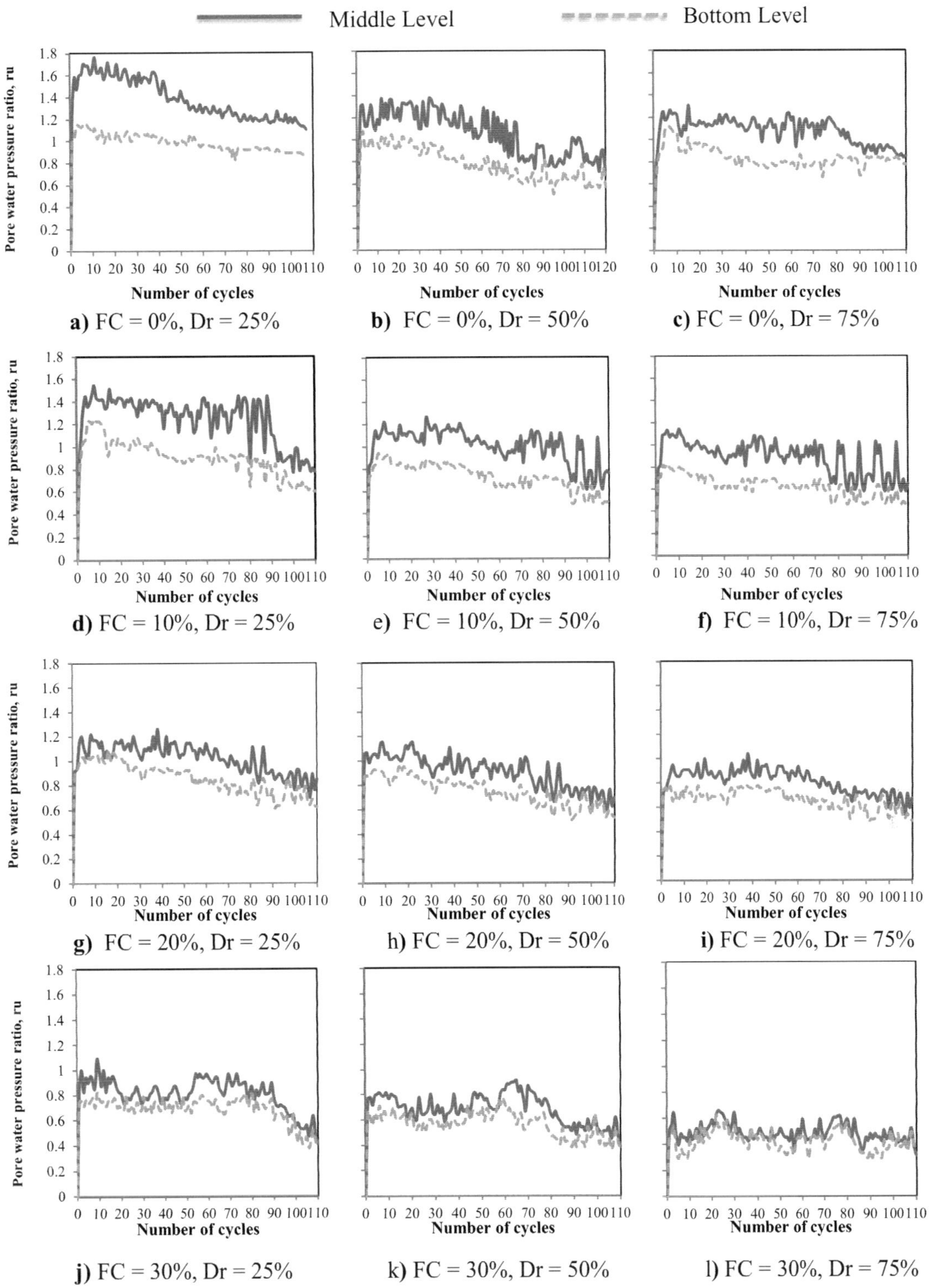

Figure 2. Correlation of Pore water pressure ration (ru) and Number of cycles for various relative density and fine content chart

Figure 2 shows that the pore water pressure ratio (ru) in middle level greater than the bottom level for all the test results. This result shows that the development of an increase in pore water pressure occurs in the middle first and followed by the bottom. This shows that in the middle the resistance to liquefaction is lower than the bottom. This behavior can be explained by the fact that the middle part of the drainage path is shorter than the bottom. Likewise, the results obtained from studies conducted in [13]–[16]. Pathak, et.al [14], explained that liquefaction initially developed from the top to bottom in a laboratory shaking table that was in accordance with previous research conducted with other laboratory tests. Figure 2 also shows that the difference between the middle and lower ru decreases due to the addition of fine grains. The addition of fine sand from 20% to 30% ru difference decreases. But the increase in relative density does not greatly affect the difference in ru values. Overall it can be concluded tahat the increasing levels of fine content and relative density increase resistance liquefaction of soil. For more details on each of these graphs taken ru maximum value at every level of fine-grain variety and relative density are presented in Table 4 and Figure 3.

Table 4. Maximum Pore Water Pressure Ratio (ru)

Fine Content	Relative Density, Dr %								
	Max Pore Water Pressure Ratio, ru								
	25%			50%			75%		
%	Middle	Bottom	$r_{u(avg)}$	Middle	Bottom	$r_{u(avg)}$	Middle	Bottom	$r_{u(avg)}$
0%	1.765	1.170	1.468	1.375	1.079	1.227	1.309	1.114	1.211
10%	1.547	1.249	1.398	1.264	0.942	1.102	1.145	0.828	0.986
20%	1.349	1.089	1.219	1.153	0.955	1.053	1.039	0.787	0.913
30%	1.091	0.821	0.956	0.911	0.770	0.841	0.659	0.628	0.641

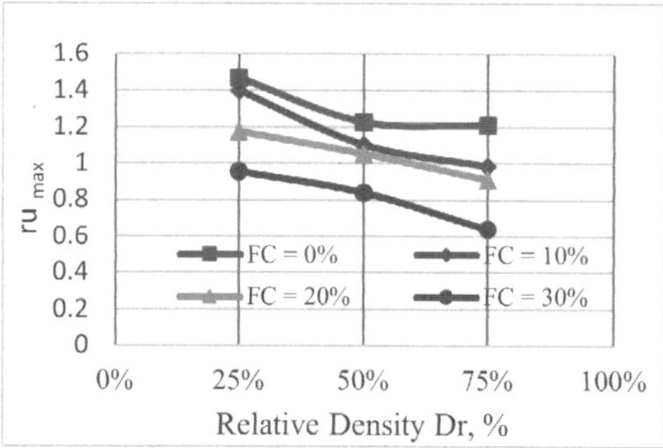

Figure 3. Correlation relative density (Dr) with maximum pore water pressure ratio (ru)

Figure 3 shows the results of a cyclic test performed on a fine-grained sand mixture with FC 0%, 10%, 20%, 30%. Tests were carried out for three values of relative density of 25%, 50%, and 75%. It can be observed that FC 0%, 10%, and 20% for Dr 25% and 50%, respectively, cause full liquefaction. It can be seen from Figure 3 that an increase in relative density can reduce the ru maximum value. This means that the denser soil will reduce the potential of the land to experience liquefaction. In this case the value of the liquefaction potential of small to relative density 75%. It is appropriate with research Chien, LK, liquefaction resistance increases well as the increase of relative density. However, at the constant relative density, liquefaction strength decreases as the fines content increase. Likewise with the results obtained by Porcino, D, et.al in [17].

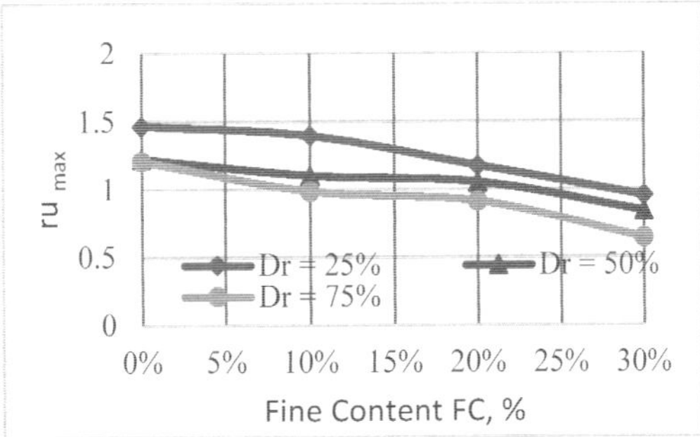

Figure 4. Correlation of Fine Content (FC) and pore water pressure ratio (ru)

Figure 4-shows that increasing levels of fine grains will reduce the max ru value. This means that the added level of fine-grains, will reduce the potential for experiencing soil liquefaction. In this case the least liquefaction potential value is at 30% fine grain content . In contrast to the results obtained from Chien, LK et al, who stated that resistance resistance increases to FC = 10% and decreases significantly if FC is greater than 10%. Karim, et.al in [18] states that the resistance of liquefaction increases to the level of fine grain boundary (LFC) and there after it is constant even though the level of fine grain increases. Different research methods and types of sand and fine grains used in the investigation can cause differences in the results obtained.

4. Conclusion

Research on liquefaction resistance in sand which is usually used as a reclamation material by adding fine grains has been carried out. The level of relative density is observed to see its effect on the resistance of liquefaction. The results obtained indicate that the liquefaction resistance will increase according to the increase in fine grain content and relative density. The difference in results obtained from other researchers is likely due to the method and type of soil used. For more accurate results, further research needs to use various methods and other types of soil so that specific conclusions can be drawn. It is expected that for areas which are prone to earthquakes, the development of the reclamation area will be well planned in order to avoid liquefaction.

5. Acknowledgments

The author would like to thank the Head of P3M Jakarta State Polytechnic through the PNJ DIPA funding for providing support for this research. The author also thanks the Head of Laboratory of the Civil Engineering Department of PNJ for granting permission to use laboratory facilities. We highly appreciate this support.

[1] H. Seed, “Design Problems in Soil Liquefaction,” *Journal of Geotechnical Engineering*, vol. 113, no. 8, pp. 827–845, 1987.
[2] T. Iwasaki, “Soil liquefaction studies in Japan: state-of-the-art,” *Soil Dynamics and Earthquake Engineering*, vol. 5, no. 1, pp. 2–68, 1986.
[3] Y. Huang and M. Yu, “Review of soil liquefaction characteristics during major earthquakes of the twenty-first century,” *Natural Hazards*, vol. 65, no. 3, pp. 2375–2384, 2013.
[4] N. Nsf and W. O. On, “L Iquefaction R Esistance of S Oils : S Ummary R Eport From the,” *Manager*, vol. 126, no. April, pp. 297–313, 2001.
[5] S. Yasuda, K. Harada, K. Ishikawa, and Y. Kanemaru, “Characteristics of liquefaction in Tokyo Bay area by the 2011 Great East Japan Earthquake,” *Soils and Foundations*, vol. 52, no.

5, pp. 793–810, 2012.

[6] I. Towhata *et al.*, "Liquefaction in the Kanto region during the 2011 off the pacific coast of Tohoku earthquake," *Soils and Foundations*, vol. 54, no. 4, pp. 859–873, 2014.

[7] Misko C, Jonathan D., P.E.,Christopher T, "Liquefaction of Reclaimed Land at Wellington Port in the 2016 Kaikoura Earthquake", Geotechnical Earthquake Engineering and Soil Dynamics V GSP 290, ASCE, pp. 367-373, 2018.

[8] http://reklamasi-pantura.com/mengintipreklamasi-di-indonesia/,6 Mei 2019

[9] B. Istijono and A. Hakam, "NEW METHOD FOR LIQUEFACTION ASSESSMENT BASED ON SOIL GRADATION AND RELATIVE DENSITY."International Journl of Advance Research in Science and Engineering, Vol No. 4, Special Issue (01), August 2015.

[10] A. Hakam, "Laboratory Liquefaction Test of Sand Based on Grain Size and Relative Density," *Journal of Engineering and Technological Sciences*, vol. 48, no. 3, pp. 334–344, 2016.

[11] L. K. Chien, Y. N. Oh, and C. H. Chang, "Effects of fines content on liquefaction strength and dynamic settlement of reclaimed soil," *Canadian Geotechnical Journal*, vol. 39, no. 1, pp. 254–265, 2002.

[12] Tsuchida, H., & Hayashi, S. (1972). Estimation of liquefaction potential of sandy soils. Publication of: Mcgraw Hill Book Company, 14.

[13] Yelvi, Sudardja H, Wiyono, E, "Effect of Relative Density to Liquefiable Sand Reinforced by Palm Fiber", International Conference of *Applied Science on Engineering, Business, Linguistics and Information Technology*, no. October, pp. 315–322, 2018.

[14] P. S. R. Pathak and M. A. Patki, "Strain Criterion For Initiation Of Liquefaction Using Shake Table Test," vol. 2, no. 5, pp. 1092–1099, 2013.

[15] B. K. Maheshwari, H. P. Singh, and S. Saran, "Effects of reinforcement on liquefaction resistance of Solani sand," *Journal of Geotechnical and Geoenvironmental Engineering*, vol. 138, no. 7, pp. 831–840, 2012.

[16] H. S. Yelvi Istiatun, "The Influence of Palm Fiber on Shear Strength of Liquefiable Sand," *International Conference of Applied Science on Engineering, Business, Linguistics and Information Technology*, no. October, pp. 63–68, 2017.

[17] D. D. Porcino and V. Diano, "The influence of non-plastic fines on pore water pressure generation and undrained shear strength of sand-silt mixtures," *Soil Dynamics and Earthquake Engineering*, vol. 101, no. March, pp. 311–321, 2017.

[18] M. E. Karim and M. J. Alam, "Effect of non-plastic silt content on the liquefaction behavior of sand-silt mixture," *Soil Dynamics and Earthquake Engineering*, vol. 65, pp. 142–150, 2014.

IOP Conf. Series: Materials Science and Engineering **846** (2020) 012071 doi:10.1088/1757-899X/846/1/012071

Study of Land Erosion and Bed Change Simulation of Danau Diatas and Danau Dibawah

Sadtim[1], Dalrino[1,*], Hartati[1], Suhendrik Hanwar[1], Aguskamar[1]

[1]Civil Engineering Department, Politeknik Negeri Padang

*dalrino350@gmail.com

Abstract. Numerical bed change study of Danau Diatas and Danau Dibawah was conducted. Hydrodynamic and sedimentation simulations are carried out using monthly discharge input limit conditions obtained from the calculation using the FJ Mock method. The RMA2 numerical model is used to obtain current patterns that are calibrated against the measurement results. Estimated land erosion is calculated using the USLE equation. Sediment discharge obtained from the USLE equation is used as an input of sediment sources to obtain sedimentation patterns using the SED2D module. The results of the modeling of the current pattern get the current velocity ranging from 0 - 0.05 m / s for Danau Diatas and between 0 - 0.04 m / s for Danau Dibawah. The bottom profile change due to the distribution of sedimentation was found to range between 0.12 ~ 0.21 m in one year simulation with a radius of 0,53 and 1,72 km. This study is a preliminary study to get an idea of the potential for sedimentation. The validity of the results is recommended to use bathymetry measurements to see the bed changes that occur with the time of measurement. The initial description obtained has shown the need for efforts to regulate land use and commensurate with lakes in anticipation of silting of lake waters due to erosion.

1. Introduction

Danau Diatas and Danau Dibawah are located in Solok Regency, West Sumatra, on the ridge of Bukit Barisan which extends along the island of Sumatra. The naming of the two lakes refers to the altitude elevation of the lake, namely Danau Diatas is at an elevation of 1,531 m above sea level and Danau Dibawah with an elevation of 1,462 m above sea level. Danau Dibawah and Danau Diatas Lakes are located side by side, separated by the nearest distance of about 2 km so that they are often dubbed Twin Lakes. Meteorological data from 2002 to 2005 indicate that the Bukit Barisan Mountains in this region have a very wet climate (super-humid) with rainfall between 2,500 - 3,000 mm / year, with 140 - 170 rainy days / year, and hyperhumid climate with rainfall > 3,000 mm / year with 180 - 220 rainy days / year. The peak rainy season occurs in April and October while the dry season is from May to June.[1] Based on PERMEN PUPR No. 04 / PRT / M / 2015 concerning the criteria and determination of river areas has defined Lake Above and Below in 2 different river areas namely the Batang Hari river area for Lake Above and the Indragiri Akuaman river area for Danau Dibawah. [2]

Figure 1. Batang Hari and Indragiri Akuaman river areas

Danau Diatas was located, at a geographical position between 1 ° 01'51 "- 1 ° 07'39" South Latitude, and between 100 ° 43 '01 "- 100 ° 50 '26" East Longitude. Based on its formation, Danau Diatas is classified as a tectonic lake with a concave shape extending from the Northwest to the Southeast as a result of the tectonic process of the large Sumatra fault. Danau Diatas waters are around 12.45 km2 with a maximum length and width of 6.4 km and 2.9 km respectively and have a coastline length of 19.9 km. Danau Diatas is a relatively shallow lake with a maximum depth of 47 m. Lake water volume is around 302 million m3 and retention time reaches 7.7 years [2]. The main outlet is Gumanti River which then integrates with Batanghari River and empties into the Berhala Strait, near the Malacca Strait. Danau Diatas has a catchment area of 40.81 km2. The rivers that enter the lake are generally in the form of intermittent rivers, which are dry in the dry season. Danau Dibawah is a tectonic lake located in the geographical position of 1 ° 0 '35 "North Latitude, and 100 ° 43 '51" East Longitude. Has a height (altitude) 1,462 meters above sea level, with an area of water around 11.2 km2 and a volume of water around 2.54 km3 and a maximum depth and averages of 309 m and 227 m. The main outflow from this lake is the Lembang River. The area of Lake Dibawah watershed is 2881.7 ha and is a steep terrain, where the edge of the lake has been utilized for agriculture, cultivation and settlement [1]. Agricultural and cultivation areas appear to be quite intensively planted with horticultural crops such as turnips, mustard greens, tomatoes, red peppers and so on. The lake waters themselves are used as a source of drinking water, and fisheries. Based on the analysis of DEM Aster, it is found that the Danau Diatas and Danau Dibawah watersheds can be divided into 18 and 27 sub-watersheds as shown in Figure 2.

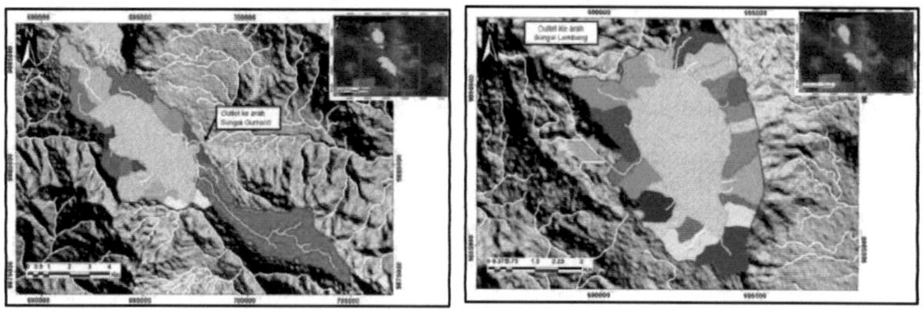

Figure 2. Catchment Area and sub-watersheds Danau Diatas (a) and Danau Dibawah (b)

With an average land slope above 25% and dominated by horticultural crop areas and planting patterns in the same direction as slope, it is feared that it will accelerate the rate of land erosion, which results in an increased risk of landslides and sedimentation in lake bodies. Land erosion in Danau Diatas and Danau Dibawah areas can be anticipated by farming activities around the lake which change the land cover condition and have an impact on surface erosion which then becomes sediment deposits in the lake body. Siltation in lake waters can be caused by increased land erosion with a number of eroded material entering the lake waters. This will reduce the capacity of the lake due to the silting process that occurs. Numerical simulations of currents and sedimentation have been carried out to see condition of current patterns and distribution of sedimentation that can have potential for the bed change of the lake that caused by siltation.

2. Methodology

Erosion is a process in which the land is detached and then moved to another place by the forces of water, wind, and gravity [3]. Erosion is the result of interactions of climate, soil, topography, vegetation, and human activities on natural resources. The eroded material then settles to a certain location and becomes the ultimate material destination of the material deposited into sedimentation [4]. Sedimentation studies to predict lake bottom siltation have been carried out at various location. Several measurement techniques have also been developed to detect the presence and age of sedimentary geology [5], [6], [7], [8]. Suggestions for handling sedimentation problems to the lake using the reservoir Saboworks to overcome the problem of sediment and debris flow entering the lake body have been given [9]. The use of GIS in estimating the potential of land sedimentation and river flow has been carried out [10], [11], [12].The use of RMA2 and SED2D in analyzing sediment flow and transport patterns has been carried out [13].Hydrodynamic modeling of Danau Diatas and Danau Dibawah is carried out to obtain the pattern of currents and sedimentation distribution that affect changes in the lake bed. Modeling results are calibrated to the flow of field measurements. The input boundary condition used in the model is the monthly discharge obtained from the calculation using the FJ Mock method (Figure 3).

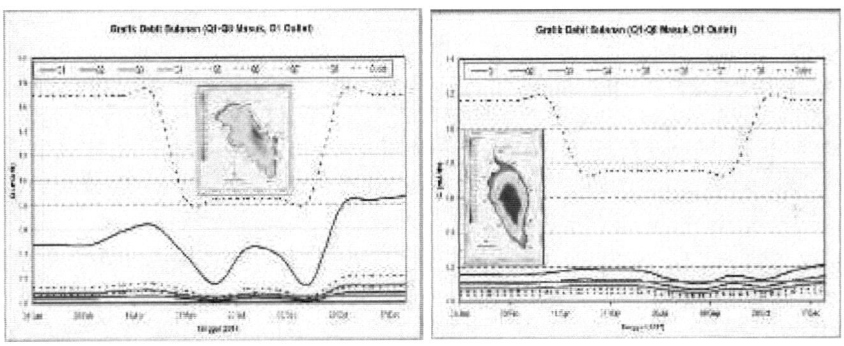

(a). Danau Diatas (b). Danau Diatas

Figure 3. Monthly Discharge as Input of model boundary conditions

The RMA2 numerical model is used to solve mass and momentum conservation equations that are integrated with depth (depth-averaged) in two horizontal dimensions. Average depth flow velocity, u used by RMA2 is expressed in the following equation :

$$U = \frac{1}{h}\int_0^h u(z)dz \qquad (1)$$

Where U is the average depth of current velocity (in the x direction), h is the depth of the elemental waters, u (z) is the flow velocity as a function of the vertical direction and z is the vertical coordinate. The form of the RMA2 governing equation that has been solved in the x and y directions is as follows:

$$h\frac{du}{dt}+hu\frac{du}{dx}+hv\frac{du}{dy}-\frac{h}{\rho}\left(E_{xx}\frac{d^2u}{dx^2}+E_{xy}\frac{d^2u}{dy^2}\right)+gH\left(\frac{da}{dx}+\frac{dh}{dx}\right)-\frac{gun^2}{\left(1{,}486h^{\frac{1}{6}}\right)^2}+(u^2+v^2)^{\frac{1}{2}}-\xi V_a^2\cos\psi-2h\omega v\sin\phi=0 \quad (2)$$

$$h\frac{dv}{dt}+hu\frac{dv}{dx}+hv\frac{dv}{dy}-\frac{h}{\rho}\left(E_{yx}\frac{d^2v}{dx^2}+E_{yy}\frac{d^2v}{dy^2}\right)+gH\left(\frac{da}{dy}+\frac{dh}{dx}\right)-\frac{gvn^2}{\left(1{,}486h^{\frac{1}{6}}\right)^2}+(u^2+v^2)^{\frac{1}{2}}-\xi V_a^2\cos\psi-2h\omega v\sin\phi=0 \quad (3)$$

where h is the vertical depth u, v: Flow velocity in the x, ρ: Coefficient of eddy viscosity, g: Acceleration of gravity a: Elevation of water bottom n: Coefficient of Manning roughness (U.S. Army Corps of Engineer, 1984). Inflows that move into the lake waters will provide sedimentation input in the lake. Sediment distribution patterns are modeled using the SED2D module. The governing equation used in the SED2D module is as follows.

$$\frac{\partial C}{\partial t}+u\frac{\partial C}{\partial x}+v\frac{\partial C}{\partial y}=\frac{\partial}{\partial x}\left(D_x\frac{\partial C}{\partial x}\right)+\frac{\partial}{\partial y}\left(D_y\frac{\partial C}{\partial y}\right)+\alpha_1 C+\alpha_2 \quad (4)$$

where C = concentration (kg / m3); t = Time (seconds); u = Direct current velocity x (m / s); x = main axis direction of flow (m); v = current direction velocity y (m / s); y = Axis perpendicular to the x-axis (m); Dx = Diffusion coefficient in the direction of x (m2 / sec) and Dy = coefficient of diffusion in the direction of y (m2 / sec). The condition of the downstream boundary is the lake water level based on data obtained when measuring the water level. Figure 11 shows the boundary conditions carried out in the modeling.

Bed changes model due to sedimentation is done using input current values obtained in RMA2 modeling using the SED2D module. The source of the sediment simulation comes from rivers that flow into the lake area and also from land erosion which was predicted previously using the USLE equation. Of the several methods for estimating the amount of surface erosion, the Universal Soil Loss Equation (USLE) method is the most commonly used method [14], [15]. Universal Soil Loss Equation (USLE) equation can be write as :

$$A = R \,.\, K \,.\, L \,.\, S \,.\, C \,.\, P \quad (5)$$

Where ; A is the amount of land eroded in t/ha/year ; R is Rainfall factor, i.e. the number of rain erosion index units, which is the multiplication of total rain energy (E) with a maximum rainfall intensity of 30 minutes (I30) ; K is soil erodibilities factor, i.e. erosion rate per unit of erosion index for a soil obtained from a standard homogeneous experimental plot, with a length of 72.6 feet (22 m) located on a slope of 9% without plants ; L is slope length factor is 9%, namely the erosion ratio of the soil with a certain slope length and the erosion of the soil with a slope length of 72.6 feet (22 m) under identical conditions ; S is slope steepness factor, which is the ratio between the amount of erosion from a slope to the steepness of a particular slope, to the amount of erosion from the soil with a slope of 9% under identical conditions ; C is factors of vegetation cover and plant management, namely the ratio between the amount of erosion from an area with vegetation cover and management of certain plants to the amount of erosion from identical soils without plants and P is factor of soil conservation measures, ie the ratio between the amount of erosion from soils treated by soil conservation measures such as contour management, planting in strips or terraces to the amount of erosion from soils treated in the same direction in the same slope.

3. Result and Discussion

With an average land slope above 25% and dominated by horticultural crop areas and planting patterns in the same direction as slope, it is feared that it will accelerate the rate of land erosion, which results in an increased risk of landslides and sedimentation in lake bodies. Image analysis in the area around the lake obtained the density of settlement conditions. The upstream boundary location of the model is determined as a monthly discharge input condition obtained from the Mock model at a location with a high topographic slope which is considered to represent field conditions (Figure 7).

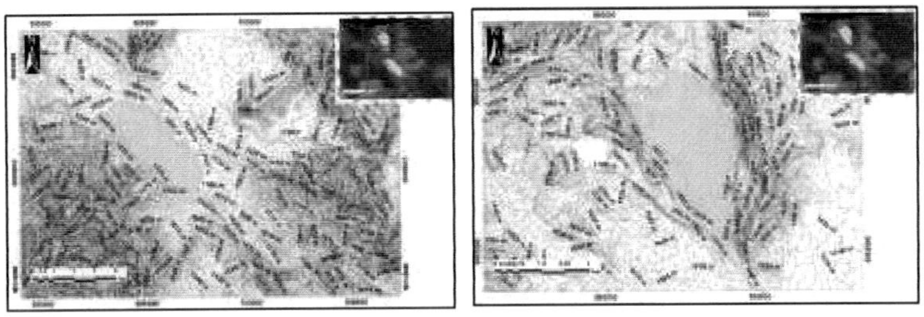

Figure 4. Topographical Conditions of Danau Diatas and Danau Dibawah

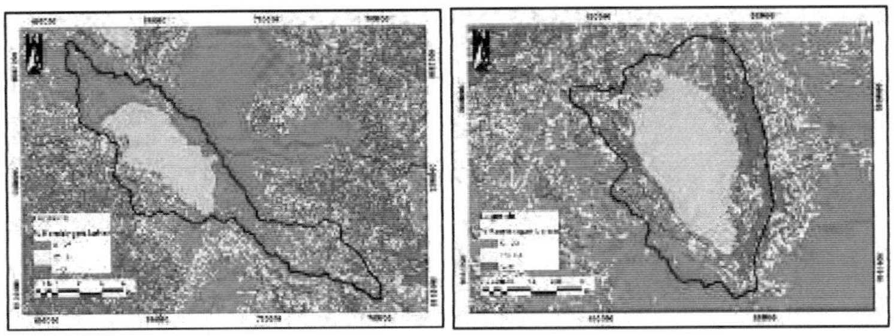

Figure 5. Slope Conditions in the Catchment Area of Danau Diatas and Danau Dibawah

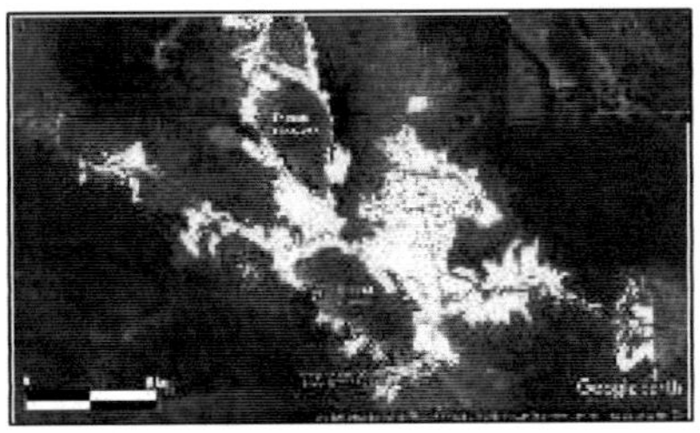

Figure 6. Existing land cover areas of Danau Diatas and Danau Dibawah

The results of current pattern modeling get the current velocity in Danau Diatas ranges from 0 - 0.05 m / s, while for the current velocity in Danau Dibawah ranges between 0 - 0.04 m / s. (Figure 8). The current survey was conducted to determine the speed and direction of the current at Danau Diatas and Danau Dibawah locations and used as calibration data on the results of hydrodynamic modeling. The calibration results show the calculated flow velocity value of the simulation results is quite close to the measurement result value at the specified location. Figure 9 and Figure 10 shows the plotting results of the current simulation against the measurement results at the location of point on Danau Diatas and Danau Dibawah.

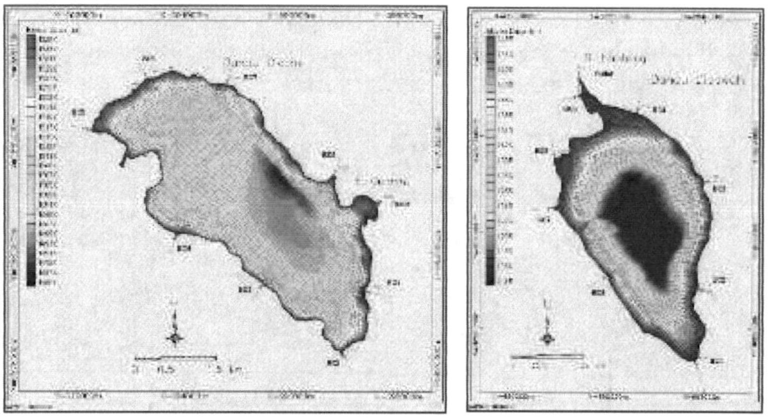

(a). Danau Diatas (b). Danau Dibawah

Figure 7. Boundary conditions of the model

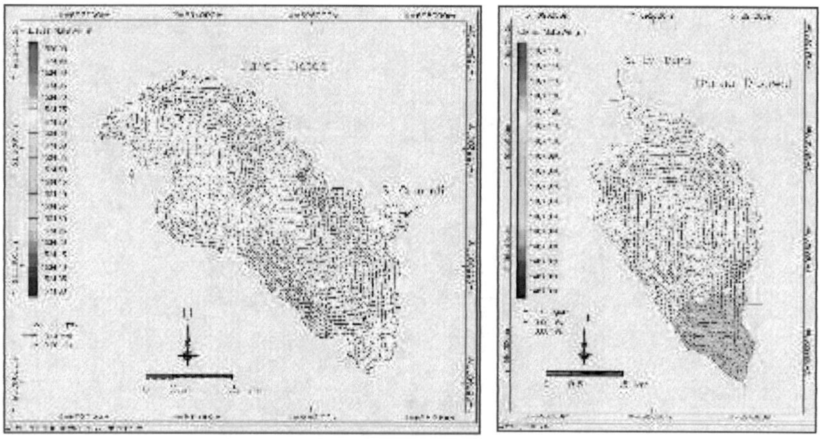

Figure 8. Current pattern of simulation results of Lake Above (a) and Lake Below (b)

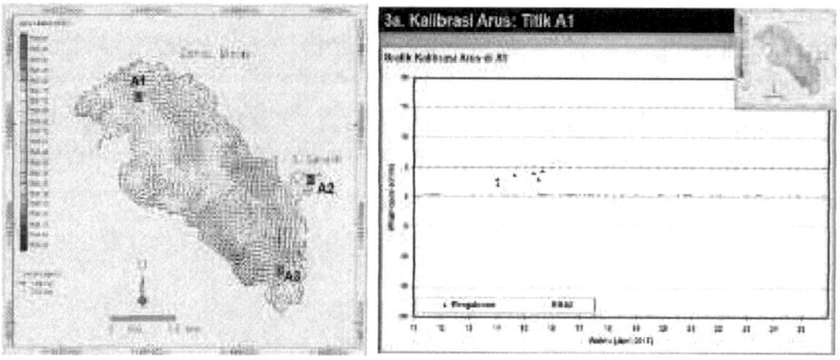

Figure 9. Location of calibration points for current speed and calibration results Danau Diatas

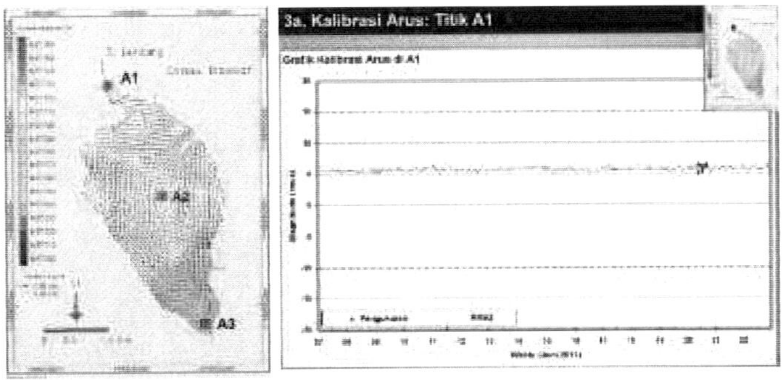

Figure 10. Location of calibration points for current speed and calibration results Danau Dibawah

Modeling of bed changes due to sedimentation is done using input current values obtained in the previous RMA2 modeling using the SED2D module. The source of the sediment simulation comes from rivers that flow into the lake area and also from land erosion which was predicted previously using the USLE equation. Of the several methods for estimating the amount of surface erosion, the Universal Soil Loss Equation (USLE) method is the most commonly used method. The simulation results obtained the distribution pattern of sedimentation originating from river mouths and land resulting in an increase in the bottom elevation at the edge of the lake shown in Figure 11 and Table 1.

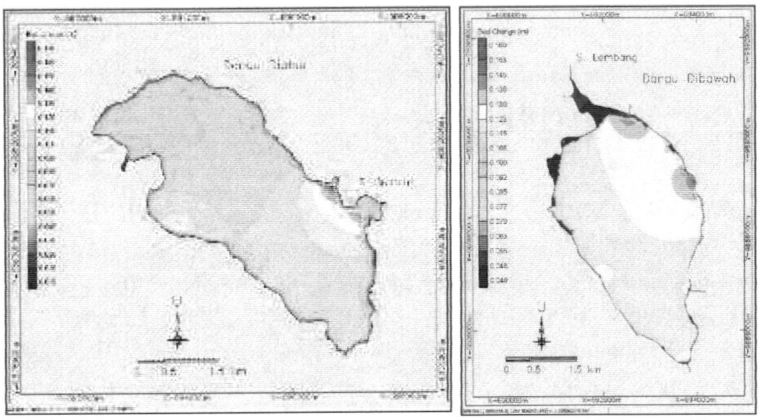

Figure 11. Bed Changes for 1 year simulation for Danau Diatas (a) and Danau Dibawah (b)

Tabel 1. Predicted sediment thickness and Radius due to simulation results

Danau Diatas		Danau Dibawah	
Predicted sediment thickness (m)	Radius (m)	Predicted sediment thickness (m)	Radius (m)
0.21	112	0.21	93
0.18	200	0.18	230
0.15	322	0.15	530
0.12	530	0.12	1720

IOP Conf. Series: Materials Science and Engineering **846** (2020) 012071 doi:10.1088/1757-899X/846/1/012071

Bed changes due to the distribution of sedimentation were found to range between 0.12 ~ 0.21 m in one simulation year and spread over an area of 0.53 km for Danau Diatas and 1.72 km for Danau Dibawah.

4. Conclusion

The cropping conditions around Danau Diatas and Danau Dibawah areas put pressure on the capacity of the existing lakes. This is caused by the occurrence of land erosion which is also influenced by the topography of the lake area. The results of current modeling using the RMA2 module have obtained good validation at the measurement location. The results of the modeling of current patterns in Danau Diatas and Danau Dibawah get the magnitude of the current speed in ranges between 0 - 0.05 m/s, and 0 - 0.04 m/s. Simulation results that show the distribution of sediments into the lake body indicate the need for efforts to regulate land use and commensurate lakes in anticipation of silting of the lake waters due to erosion. This study is a preliminary study to get an idea of the potential for sedimentation. Erosion values are obtained from the use of empirical equations and have not been validated in real terms with field conditions. It is recommended to take measurements of the sediment rating curve and changes in bathymetry in subsequent studies to obtain validated erosion and sedimentation values.

References

[1] A. Nontji, *Danau Danau Alami Nusantara*. 2016.

[2] Kementrian PUPR, *PERATURAN MENTERI PEKERJAAN UMUM DAN PERUMAHAN RAKYAT REPUBLIK INDONESIA NOMOR 04/PRT/M/2015 TENTANG KRITERIA DAN PENETAPAN WILAYAH SUNGAI DENGAN*, no. 1. 2015, pp. 1–8.

[3] S. Hardjowigeno, *Ilmu Tanah*. Jakarta: Akademika Presindo, 1995.

[4] S. Arsyad, *Konservasi Tanah Dan Air*. Bogor: Serial Pustaka IPB Press, 2000.

[5] H. Simon, S. Kelemen, and R. C. Begy, "Anthropic influences on the sedimentation rates of lakes situated in different geographic areas," *J. Environ. Radioact.*, vol. 173, pp. 11–17, 2017.

[6] A. I. Poraj-Górska, M. J. Żarczyński, A. Ahrens, D. Enters, D. Weisbrodt, and W. Tylmann, "Impact of historical land use changes on lacustrine sedimentation recorded in varved sediments of Lake Jaczno, northeastern Poland," *Catena*, vol. 153, pp. 182–193, 2017.

[7] D. G. Wren, J. M. Taylor, J. R. Rigby, M. A. Locke, and L. M. W. Yasarer, "Short term sediment accumulation rates reveal seasonal time lags between sediment delivery and deposition in an oxbow lake," *Agric. Ecosyst. Environ.*, vol. 281, no. May, pp. 92–99, 2019.

[8] I. Quintana-Cobo *et al.*, "Dynamics of floodplain lakes in the Upper Amazon Basin during the late Holocene," *Comptes Rendus - Geosci.*, vol. 350, no. 1–2, pp. 55–64, 2018.

[9] S. S. Putra, C. Hassan, Djudi, and H. Suryatmojo, "Reservoir Saboworks Solutions in Limboto Lake Sedimentations, Northern Sulawesi, Indonesia," *Procedia Environ. Sci.*, vol. 17, pp. 230–239, 2013.

[10] M. Elsahabi, A. Negm, and K. Ali, "Assessment of sedimentation capacity in Lake Nasser, Egypt, utilizing RS and GIS," *Procedia Manuf.*, vol. 22, pp. 558–566, 2018.

[11] R. Herdianto, B. Istijono, E. R. Syofyan, and Dalrino, "Investigation of Pangkalan floods: Possible reasons and future directions," *Int. J. Adv. Sci. Eng. Inf. Technol.*, vol. 8, no. 6, pp. 2510–2515, 2018.

[12] R. Herdianto, E. R. Syofyan, S. Hanwar, B. Istijono, and Dalrino, "The investigation of 1997 and 2015 El Nino events in West Sumatera, Indonesia," *Int. J. Adv. Sci. Eng. Inf.*

Technol., vol. 7, no. 2, pp. 418–423, 2017.

[13] Dalrino, Aguskamar, A. Indra, and E. R. Syofyan, "Wave and current hydrodinamics study at Batang Air Dingin river Mouthpadang, Indonesia," *Int. J. Civ. Eng. Technol.*, vol. 9, no. 11, pp. 2054–2060, 2018.

[14] Asdak Chay, *Hidrologi dan Pengelolaan Daerah Aliran Sungai*. Jogjakarta: UGM Press, 2018.

[15] Wischmeier and Smith, *Predicting Rainfall Erosion Losses— A Guide To Conservation Planning*. United States Department of Agriculture, 1978.

Numerical Modeling With HEC RAS and Scour Empiric Equation To Estimating Scour Depth on Abutment (Batang Kalu River Bridge Case)

Indra Agus[1], Dalrino, [1*], Sadtim[1], Munafri Alwys[1], Noftar Yolanda[1]

[1]Civil Engineering Department, Politeknik Negeri Padang

*dalrino350@gmail.com

Abstract. The riverbed around the abutment can experience scouring caused by flow, because its is always directly related to river flow. Abutment can obstruct the flow of rivers causing changes in flow characteristics which in turn will cause local scouring around these buildings. Toe Scour around abutment will be very dangerous and cause loss of stability on the bridge. The case of structure failure on Batang Kalu bridge in Korong Pasa Usang Nagari Kayu Tanam, Padang Pariaman Regency on Monday (10 December 2018) is a clear example of toe scouring. A case study of toe scour was on the Batang Kalu River bridge structure has been carried out. Rainfall data closest to the study location, Kandang Empat Station was used to minimize the errors in the calculation of flood discharge design. Numerical modeling with HEC RAS and scouring estimation with some empiric equation was conducted to predicting scour depth on abutment. Simulation results show scour depth results to an average of 2.731 m and close to the scour depth that occurred in the field (2.83 m). Results showed that the local scouring in the Batang Kalu River bridge structure can be suspected as the influence of the increasing river steepness. Discharge that is triggered by heavy rain with a long duration and increased river flow velocity due to steepness has the potential to create a very intense scouring. Increasing of steepness may have been caused by sand mining activities in the upper reaches. Further studies are needed to see the potential slope changes due to exploitation in the Batang Kalu River upper reaches.

1. Introduction

Changes in river morphology followed by changes in river characteristics can cause changes in flow patterns. If the river body has a building in the form of a bridge abutment, it will cause local scouring and a decrease in base elevation (degradation) around the bridge pillar. Scour is a natural phenomenon caused by the flow of water that usually occurs at the bottom of a river consisting of alluvial material but sometimes it can also occur in a hard river. Scouring can cause erosion of the soil around the foundation of a building located in a stream of water. Scour usually occurs as part of the morphological changes of the river and changes due to man-made buildings [1].

The process of erosion and deposition in rivers generally occurs due to changes in flow patterns, especially in alluvial rivers. Changes in flow patterns can occur due to obstacles or obstacles in the river flow which can be in the form of river buildings such as bridge abutments, river cribs, pillars, revetments, and so on. This kind of building is seen to be able to change the flow geometry and flow pattern, which is then followed by the emergence of local scouring around the building. Local scouring events will always be closely related to the phenomenon of river flow behavior, namely river flow hydraulics in their interactions with river geometry, geometry and layout of bridge pillars, as well as the characteristics of the subgrade where the pillars were built [2].

Abutment is a part of bridge construction which is located in a river body that functions as a bridge load. The riverbed around the abutment can experience scouring caused by flow, because its structure is always directly related to river flow. Abutment of bridges or other water structures can obstruct the flow of rivers causing changes in flow characteristics which in turn will cause local scouring around these buildings. Scouring around the bridge abutment will be very dangerous and cause loss of stability so that the collapse occurs on the bridge. Shear velocity as a component of the mechanism of sediment transport and scouring at the foot of the abutment is strongly influenced by the flow velocity that occurs in the river body, especially at locations near the abutment structure itself. The occurrence of changes in the speed around this location is actually a common thing to happen as a result of contraction of the flow due to narrowing of the channel by the reduced appearance of the river due to the existence of the structure. In general, this condition has been predicted beforehand by the bridge designer and is anticipated by efforts to protect the feet around the bridge abutment. However, changes in the character and pattern of flow due to changes in flow regime by the influence of changes in the river environment are different outside the planning which threatens the stability of the structure under the bridge. The case of the Padang - Bukittinggi transport line cut off due to failure of the structure under the Batang Kalu bridge in Korong Pasa Usang Nagari Kayu Tanam, Padang Pariaman Regency on Monday, December 10th 2018 is a clear example of structural failure due to scouring of the feet. (Figure 1).

Figure 1. Condition of the disconnection of the Batang Kalu Bridge, Kayu Tanam, Kab. Padang Pariaman due to massive toe scouring occurs by increased flow at the bottom.

One of the causes of this structural failure can be hypothesized as the impact of changes in flow character due to disturbance in the environment of the river body, as well as by an increase in flow velocity due to the amount of discharge that occurred at that time. This research will answer the problem formulation to find out how much influence the changes in flow patterns and characteristics of the river environment on the increase in transport and scour mechanism that occurs at the foot of the bridge abutment structure.

The scour is defined as an enlargement of a flow that is accompanied by the transfer of material through fluid motion [3], [4]. Local scouring occurs when the flow capacity to erode and transport sediments is greater than the sediment supply capacity [5]. Pier is part of the bridge construction which is located in the middle of the river and functions as a bridge load. Although the pier is located in the middle of the river, the river bed around the pier can also experience scouring caused by the flow, because these piers are always directly related to the river flow. Piers and abutments of bridges and other water structures can obstruct the flow of rivers, causing changes in flow characteristics which in turn will cause local scouring around these buildings. Experimental and numerical studies have been performed to investigate the flow mechanism around a bridge pier. Turbulent flow around a complex bridge pier on the scoured bed was conducted by numerical simulation approach [6]. Characterization of horseshoe vortex in a developing scour holes study was performed at a cylindrical bridge pier. Also the effect of partial blockage on multi-vents bridge pier scour was studied with experimentally and numerically methods [7]. Study that describing the effect of shape geometry of bridge pier on several conditions related on scours was conducted. Local scour around two side-by-side cylindrical bridge piers under ice-covered conditions was performed. The multi-span masonry arch bridges subjected to scour was modelled with three-dimensional mesoscale modelling [8]. Mathematical modeling approach also was performed to evaluating local scour around bridge piers for various geometrical shapes [9]. Study to investigate the responses of caisson-piles foundations to long-term cyclic lateral load and scouring was conducted [10]. The problem of flood-induced scour was evaluated with a numerical model. Model was performed to calculating the level of scour which led to the bridge failure and the corresponding collapse mechanism, and to assess the sensitivity of the bridge's modal properties (vibration frequencies and mode shapes) to different levels of scour [11]. Local scouring at bridge pier will affecting the behavior these structure subjected to flood-induced loads [12]. Also, case study of bridge scour in the UK was performed. Study was related into the effects of uncertain asset stock data on the assessment of climate change risks [13]. Some methods to reducing scour hole around bridge pier was proposed. Review on the methods used to reduce the scouring effect and also the effects of height and vertical position of slot on the reduction of scour hole depth around bridge abutments were conducted [14], [15]. Investigation related the effect of submergence ratio of parallel wall on bridge abutment scour also conducted [16]. The use of RMA2 for simulating flooding at river flow has been carried out [17].

Scouring around the piers of the bridge will be very dangerous and cause loss of stability of the piers so resulting in collapse of the bridge. The scour hole that occurred starting from the upstream of the pier, when the flow component that leads to the downstream. Near the bottom of the flow component, the direction will be revers]ed upstream, followed by the carrying of basic material so that a spiral flow is formed in the area of the scour hole that results in the scouring of the river bed around the pier as shown in Figure 2 (a). The scouring process around piers or abutments is very complex because it involves three-dimensional flow. When the flow passes through the pier, there will be a flow separation and this separation will extend to the downstream of the pier. The vortex system that occurs has a shape or characteristic such as a vortex of horse shoes so it is called the vortex horse vortex as shown in Figure 2 (b).

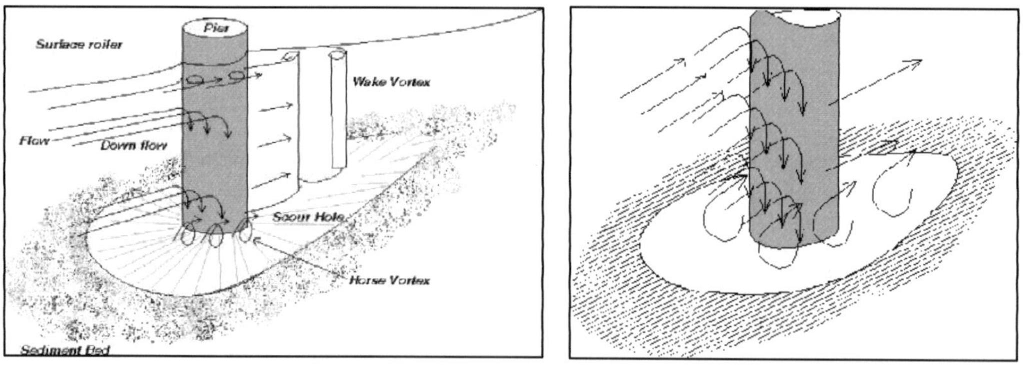

(a) (b)

Figure 2. Current Patterns Around Round Bridge Pillars and Horseshoe Vortex

The maximum scour depth occurs when the shear velocity (u*) is equal to the shear velocity of the critics in the transition area between clear-water scour and live-bed scour as shown in the following figure.

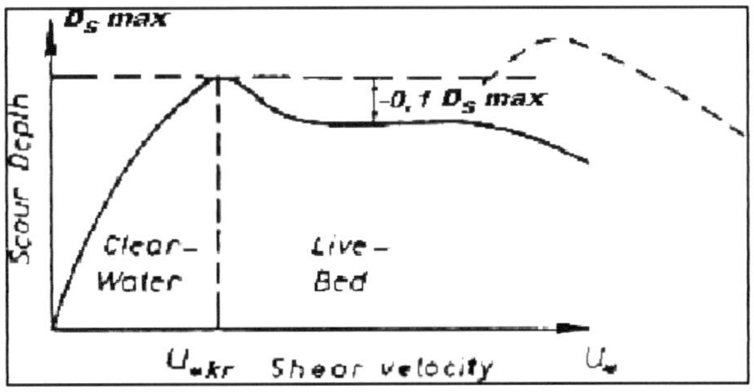

Figure 3. Depth of Scour (ds) as a Function of Shear Speed (u *) (Breusers and Raudkivi, 1991)

2. Methodology

The research location is in the Batang Kalu river, Padang Pariaman Regency. At this location there is a bridge connecting Padang - Bukittinggi, as shown in Figure 4. Toe scour is caused by the acceleration of the flow that occurs due to contraction of the flow when passing through the bridge structure. The flow pattern that occurs is influenced by the characteristics of the river profile and discharge that occurs. Determination of discharge is calculated with using the Rational, Hasper and Weduwen methods. The results of the theoretical discharge are validated against field measurement data so that the discharge values will be used in the simulation.

Figure 4. Ressearch Location

Calculation of profile and flow character using HECRAS by applying the standard step method for steady flow [18] as shown in Figure 5 with the energy equations as follow ;

$$y_1 + \frac{V^2}{2g} + z_1 = y_2 + \frac{V^2}{2g} + z_2 + h_L$$
$$y_1 + \frac{V^2}{2g} + z_1 = y_2 + \frac{V^2}{2g} + z_2 + S_e \Delta L$$

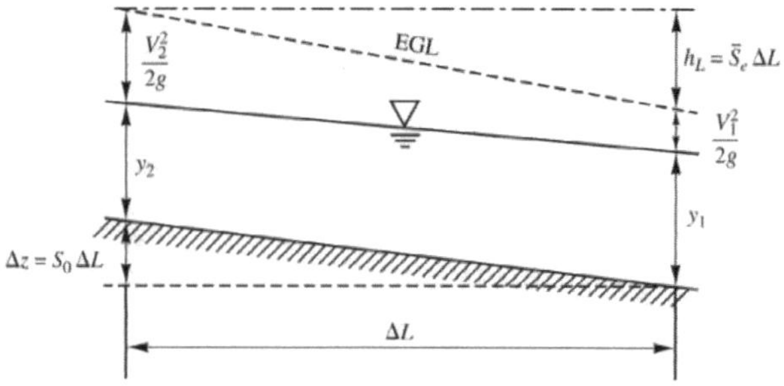

Figure 5. Standard Step Method For Steady Flow

The values of the flow variables obtained are used to determine scour predicted using Froehlich, Lacey , Blench equations and Bridge Scour Manual Method. Some of these equation as follow [4], [5], [19], [20].

- Froehlich (1987) :

$$\frac{y_s}{y_1} = 0{,}78 \times K_1 \times K_2 \times \left(\frac{a'}{y1}\right)^{0{,}63} \times F_r^{\ 1{,}16} \times \left(\frac{y1}{d_{50}}\right)^{0{,}43} G^{-1{,}87} + 1$$

Where $K_1 = 1{,}0$ (vertical wall abutment) ; $K_2 = \left(\frac{\theta}{90}\right)^{0.13}$; $Fr_e = V_e / (g\ y_1)^{0.5}$; $G = (D_{84}/D_{16})^{0.5}$

- Lacey (1930) :

$$y_{ms} = 0{,}47\left(\frac{Q}{f}\right)^{1/3}$$

Where $f = 1{,}76\ d_m^{0{,}5}$ (mm) ; y_{ms} = depth of scour from flood waters

- Blench (1969)

$$y_{ms} = 1{,}20\left[\frac{q^{2/3}}{d_{50}^{\ 1/6}}\right] \text{ for } 0{,}06 \text{ mm} < d50 > 2 \text{ mm}$$

$$y_{ms} = 1{,}23\left[\frac{q^{2/3}}{d_{50}^{\ 1/12}}\right] \text{ for } d50 > 2 \text{ mm dan } G = 2{,}65$$

- Bridge Scour Manual Method :

$$\frac{y_s}{y_1} = 4\ Fr^{0.33}\frac{K_1}{0.55}K_2$$

3. Result and Discussion

Batang Kalu River is a tributary of the Batang Anai River which belongs to the Batang Anai watershed. There are three closest rainfall stations within the Batang Anai watershed, namely Paraman Talang Station, Kandang Station IV and Lubuk Napar Station. Observation of the position of the Batang Kalu river on the location of the rainfall station found that conditions were too small to influence of the

Paraman Talang and Lubuk Napar Stations on the determination of theoretical discharge at the study location, as shown in the following figure.

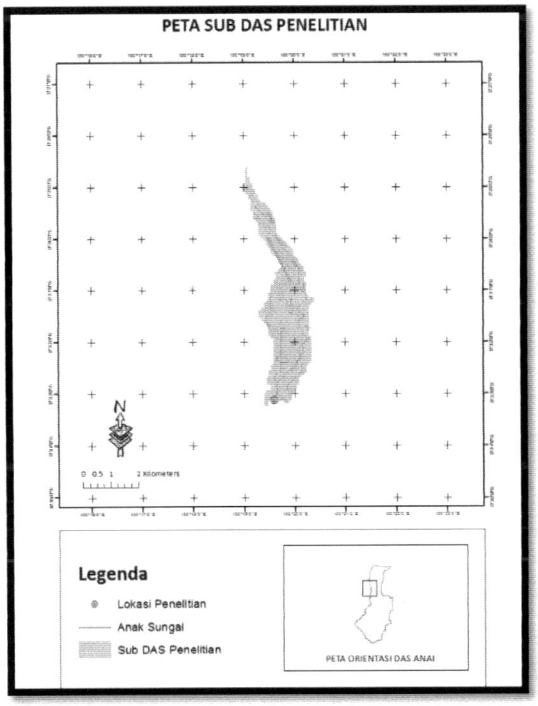

Figure 6. Location of the Batang Kalu River Against the Batang Anai watershed

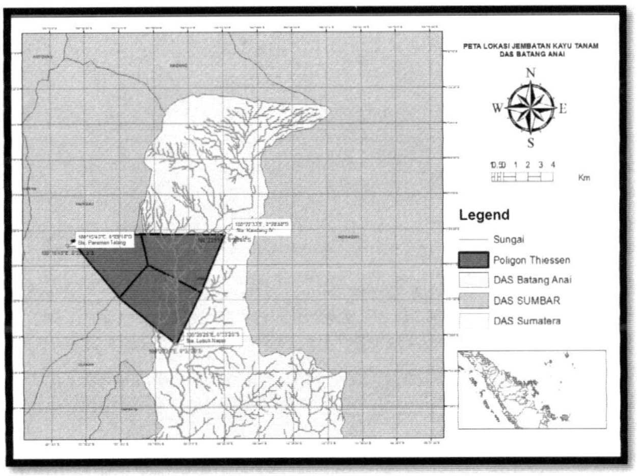

Figure 7. Position of Rainfall Station and Polygon Thiessen in Batang Anai Watershed

Considering the influence of the Paraman Talang and Lubuk Napar Stations on the study location, the hydrological analysis to obtain the flood discharge design was determined using the Kandang Empat Station (0° 30’ '48 "South and 100° 20' 6" East) with maximum rainfall data as shown in the following figure.

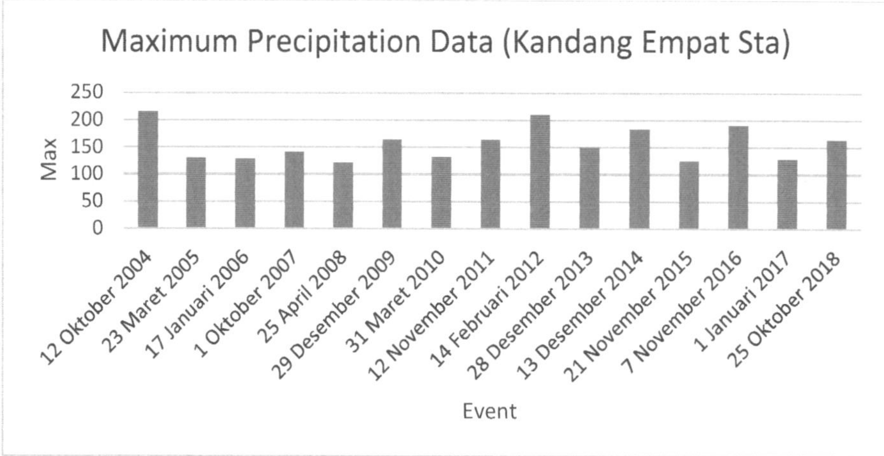

Figure 8. Maximum Precipitation Data at Kandang Empat Station

Theoretical discharge was calculated using Rational, Hasper and Weduwen methods on various Return Period with the results as shown in the following table. With calibration test by calculating the maximum capacity of river cross wet area as assume 2 years return period resulting Weduwen as the good fitted.

Flood Discharge at Batang Kalu River for variation of Return Period

T	Discharge (m^3/dt)		
	Rasional	Hasper	Weduwen
2	72.644	99.206	79.618
5	84.914	112.944	95.542
10	91.340	119.859	111.465
25	97.621	126.439	133.758
50	102.587	131.522	151.274
100	106.677	135.631	167.198

HEC RAS simulations were carried out to get the values of variable velocity and water level at several cross sections of the river being modeled. By using an actual river bed slope of 0.04 and a cross-section of the measurement river (Figure 9), velocity values at various discharge were obtained. The velocity and discharge variables obtained are input to calculate the theoretical scour depth using the empirical equation as shown in the Table 1.

Validation is done by comparing the theoretical scour depth and the actual scour depth. The results of observations show that the scour depth is 2.83 m. From the results of the validation of the model, it is obtained the condition that is close to simulation results using flood discharge with 5 years return period. This shows that flood discharge is not the main factor causing scour. Increased velocity due to increased river bed steepness can be expected as a variable that results in velocity increase resulting increasing of scour depth. The graph of the relationship of scour value to flood discharge can be seen in the graph at Figure 10.

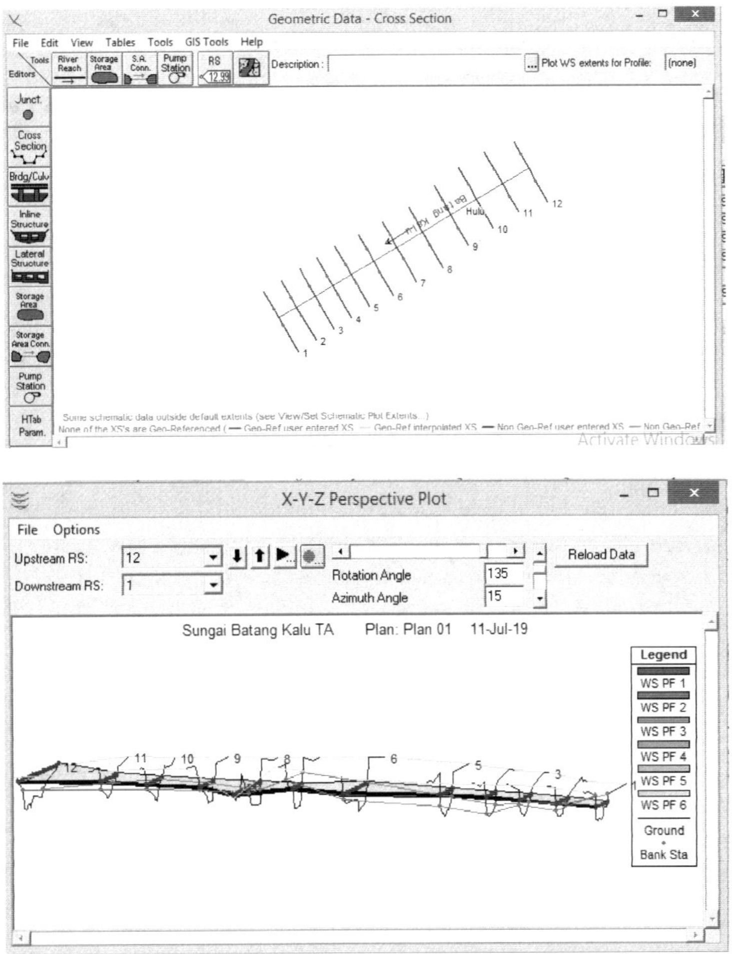

Figure 9. Layout and cross section of the river around the bridge location with HEC RAS

Table 1. Empirical Scour Depth with Several Scour Equation

Return Period (year)	Discharge (m3/s)	Velocity (m/s)	Scour Depth (m)				
			Froehlich	Bridge Scour	Blench	Lacey	Average
2	79.618	3.317	1.561	2.532	3.082	2.532	2.427
5	95.542	3.981	1.673	2.691	3.868	2.691	2.731
10	111.465	4.644	1.774	2.833	4.612	2.833	3.013
25	133.758	5.573	1.903	3.010	5.595	3.010	3.380
50	151.274	6.303	1.996	3.136	6.330	3.136	3.650
100	167.198	6.967	2.076	3.242	6.974	3.242	3.884

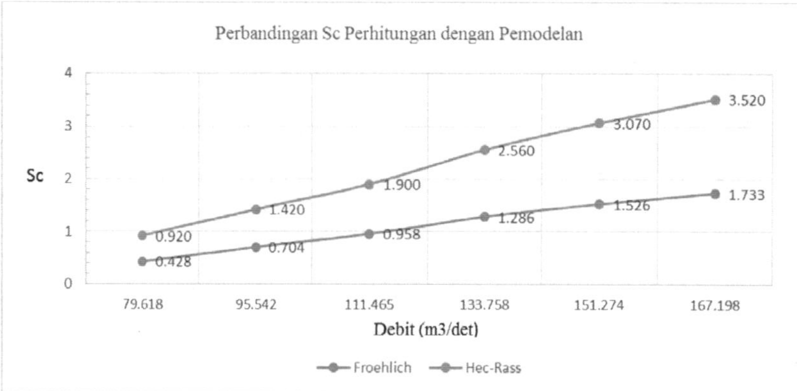

Figure 10. Comparison of Scour Depth (Sc) Against Flood Discharge (Q)

4. Conclusion

This study was a preliminary illustration to see the effect of variable flow that occurs on toe scour at the bridge abutment structure. A case study of toe scour was on the Batang Kalu River bridge structure has been carried out. Rainfall data closest to the study location, Kandang Empat Station was used to minimize the errors in the calculation of flood discharge design. Simulation results show the closeness of the scour depth results to an average of 2.731 m and close to the scour depth results that occur in the field (2.83 m). From the results of this study shows that the local scouring in the Batang Kalu River bridge structure can be suspected as the influence of the increasing river steepness. Increased discharge that is triggered by heavy rain with a long duration and increased river flow velocity due to bed steepness has the potential to create a very intense scouring around the abutment. Increasing of river steepness may have been caused by sand mining activities in the upper reaches of the river. Further studies are needed to see the potential for river slope changes due to exploitation in the Batang Kalu River upper reaches.

5. Acknowledgment

The author would like to thank Politeknik Negeri Padang for funding this research through the Hibah Unggulan Perguruan Tinggi.

References

[1] A. J. Breusers, H.N.C. and Raudkivi, *Scouring, IAHR, Hydraulic Structure Design Mannual*. Rotterdam: A.A. Balkema, 1991.

[2] D. Legono, *Hidraulika Bangunan Sungai*. Yogyakarta: Diktat Kuliah Teknik Sipil UGM Yogyakarta, 2001.

[3] H. J. Hoffmans, G.J.C.M., Verheij., *Scour Manual*. Rotterdam: A.A. Balkema.

[4] K. G. R. Garde, R.J. and Raju, *Mechanics of Sediment Transportation and Alluvial Stream Problem*. New Delhi: Willy Eastern Limited, 1977.

[5] W. H. Graf, *Fluvial Hydraulics*. England: John Wiley & Sons Ltd, 1998.

[6] M. Alemi, J. P. Pêgo, and R. Maia, "Numerical simulation of the turbulent flow around a complex bridge pier on the scoured bed," *Eur. J. Mech. B/Fluids*, vol. 76, pp. 316–331, 2019.

[7] Y. A. M. Moussa, T. H. Nasr-Allah, and A. Abd-Elhasseb, "Studying the effect of partial

blockage on multi-vents bridge pier scour experimentally and numerically," *Ain Shams Eng. J.*, vol. 9, no. 4, pp. 1439–1450, 2018.

[8] E. Tubaldi, L. Macorini, and B. A. Izzuddin, "Three-dimensional mesoscale modelling of multi-span masonry arch bridges subjected to scour," *Eng. Struct.*, 2018.

[9] A. M. A. Moussa, "Evaluation of local scour around bridge piers for various geometrical shapes using mathematical models," *Ain Shams Eng. J.*, vol. 9, no. 4, pp. 2571–2580, 2018.

[10] C. Zhang, X. Zhang, M. Huang, and H. Tang, "Responses of caisson-piles foundations to long-term cyclic lateral load and scouring," *Soil Dyn. Earthq. Eng.*, 2019.

[11] F. Scozzese, L. Ragni, E. Tubaldi, and F. Gara, "Modal properties variation and collapse assessment of masonry arch bridges under scour action," *Eng. Struct.*, 2019.

[12] C. C. Hung and W. G. Yau, "Behavior of scoured bridge piers subjected to flood-induced loads," *Eng. Struct.*, vol. 80, pp. 241–250, 2014.

[13] H. Dikanski, B. Imam, and A. Hagen-Zanker, "Effects of uncertain asset stock data on the assessment of climate change risks: A case study of bridge scour in the UK," *Struct. Saf.*, 2018.

[14] M. Zaid, Z. Yazdanfar, and H. Chowdhury, "ScienceDirect ScienceDirect ScienceDirect District the A review on the methods used to on reduce scouring effect of bridge pier bridge pier Assessing the feasibility of using the heat demand-outdoor Alam forecast temperature function for a district heat d," vol. 00, no. 2018, 2019.

[15] M. Osroush, S. A. Hosseini, A. A. Kamanbedast, and A. Khosrojerdi, "The effects of height and vertical position of slot on the reduction of scour hole depth around bridge abutments," *Ain Shams Engineering Journal*, vol. 10, no. 3. pp. 651–659, 2019.

[16] F. Sadeghi, Y. Ramezani, and H. Khozeymehnezhad, "Effect of submergence ratio of parallel wall on bridge abutment scour," *Alexandria Engineering Journal*, vol. 57, no. 4. pp. 2659–2665, 2018.

[17] Dalrino, Sadtim, Hartati, and I. Agus, "Analisis Kapasitas Penampang Sungai Batang Mahat Terhadap Besaran Debit Banjir Menggunakan Pendekatan Model Matematik," *JIRS*, vol. XV, no. 2, pp. 53–63, 2018.

[18] U.S. Army Corps of Engineer, *HEC-RAS River Analysis System*. 1984.

[19] W. H. Graf, *Hidraulics of Sedimen Transport*. Colorado: Water Resources Publications, 1984.

[20] A. J. and E. Raudkivi, *Scour at Bridge Piers*. Third Congres of the Asian and Pasific Division of the International Association for Hydraulics Research IAHR, 1982.

Ethnoecology and ethnomedicine study to ensure maritime conservation in Bangsring Underwater (Bunder) Banyuwangi, Indonesia

N Athiroh[1], I Pujiwati[2], A Hayati[3]
University of Islam Malang, East Java, Indonesia

Email: *nur_athiroh_mlg@yahoo.co.id*

Abstract. The purpose of this study was to assess ethnoecology and ethnomedicine as support for marine conservation in Bangsring Under Water (Bunder) Banyuwangi. This research used descriptive exploratory techniques. The samples were done using random sampling techniques. The results of ethnoecology studies are as follows: 63% of respondents knew the Bunder area as a conservation area of coral reefs; 61% of respondents stated that they have access to the socio-ecological Bunder, 79% of respondents knew how to use the fish in Bangsring village, 71% utilize fish in Bunder, 72% of respondents see the type of fish that is used by the community, 45% of respondents knew how to preserve local fish. The results of the ethnomedicine study showed that 83% of respondents knew about medicinal plants, 83% of respondents have used medicinal plants, 88% of respondents knew how to use medicinal plants, 72% of respondents never planted medicinal plants, 85% of respondents identified the plant parts used drugs, 9% respondents are interested in using medicinal plants, 81% of respondents need to cultivate medicinal plants. Bunder ecologically waters have a positive impact on people's lives. People use these waters for tourism and livelihood as a fisherman, as a conservation area. In the Etnomedisin aspect, the community had an enthusiastic response to the medicinal plants. There are 25 species of medicinal plants used by the community as a medicine. The survival of medicinal plants in Bangsring Village is weel guaranteed due to the high interest of the community to keep the sustainability of medicinal plants.

1. Introduction

The Bangsring Under Water (Bunder) Banyuwangi Regency, East Java are located between the Bali Sea and the Bali Strait. This area is a commercial and crossing shipping lane that has high fisheries productivity. Bangsring Under Water (Bunder) has well-known ornamental fish and coral resources. Potential of Bunder and Tabuhan Island as tourism places are very popular with both local and foreign tourists [1] [2].

Things that must be done to support Bunder as a tourism city include increasing public understanding of the threat of damage to the marine ecosystem. Lack of knowledge on ethnoecology, resulting in damage to marine ecosystems. One of the damages to the Bangsring Sea ecosystem is that it is caused by fishing exploitation activities [3].

A large number of sharks that were injured also showed the ecosystem damage in the area. The Bangsring coastal area ranged from 2-49 meters. The deepest region is in the southeast of the research area. Based on the depth of the waters, the locations that can be an alternative placement of fish apartment are located by a distance of 200- 250 meters in front of Bangsring coastal area with a total area approximately 30 Ha [4].

Fishermen treat injured sharks using chemical drugs that harm the fish and the environment. Initial observations around Bunder show that coastal communities have not utilized many ethnomedicines. Ethnomedicine herbs are sourced from the ocean and in the coastal environment around the coast. One of the ethnomedicine herbs that are often found in coastal areas is neem. Mimba has potential as an herb that can maintain human health and marine life. This phenomenon is an innovation that needs to be studied further to support maritime conservation in the area of the Bunder.

2. Method

This study uses descriptive exploratory techniques, including literature study, field observations, interviews with questionnaires, data analysis, and documentation of research objects. Problem-solving is done by describing the state of research under the actual situation in the field. Data collected in this study were obtained from the results of interviews with the public about various potential medicinal plants. The determination of the selected sample is done using random sampling techniques.

The data collection was done through the determination of respondents, interview techniques, and documentation techniques. The number of respondents was 100 people who were determined by the random sampling method. Researchers conducted direct observations in Bangsring Village and then noted the Bangsring Village community's perception of bioprospection of medicinal plants. Researchers conducted interviews directly with community leaders and the community of Bangsring Village in general predetermined to find out things that are closely related to the activities to be carried out. Researchers take pictures directly and record the important things at the research site as documentation material and as evidence that researchers investigated that place.

Social science research is central in a "reality-based community." It relies on people carefully studying experiences, events, and facts in social reality. While social research helps us answer questions about the social world, it also raises new questions and may change how we look at the world as well [5].

3. Result

3.1. Ethnoecology Study

The interview results with 100 respondents on public knowledge of Bangsring village are shown in Figure 1.

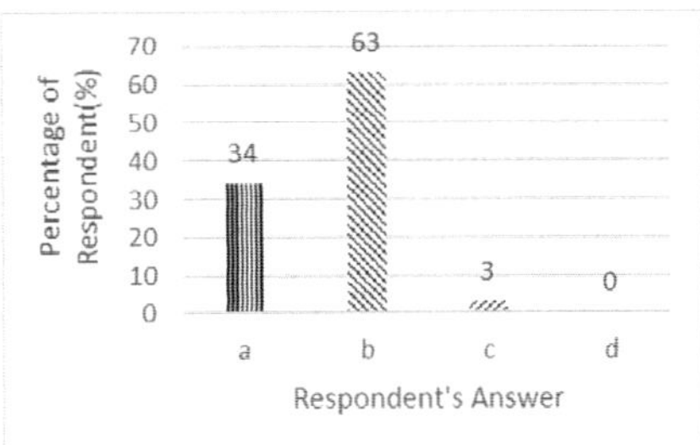

Figure 1. Percentage of knowledge about the area of Bunder

Noted: a. strongly agree; b. agree;c. disagree; d. strongly disagree

Figure 1 shows that 63% of respondents know well the Bangsring Underwater area. Bangsring Underwater is a coral reef conservation area as well as ecotourism in the Bangsring village. The existence of this conservation area causes the preservation of marine ecosystems in the town of Bangsring. The **scientific community** is a social institution of people, organizations, and roles, as well as a set of norms, behaviors, and attitudes that all operate together [5].

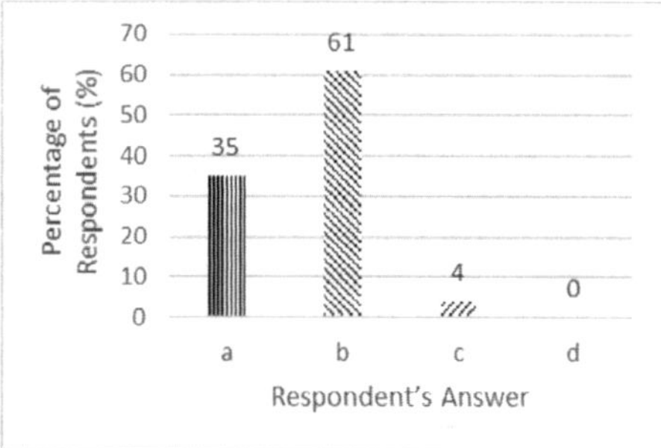

Figure 2. Percentage of benefit about the area of Bunder
Noted: a. strongly agree; b. agree;c. disagree; d. strongly disagree

Figure 2 shows that 61% of respondents answered "feeling" the benefits of the Underwater Bangsring area. According to the results of the interview, the perceived benefits include the availability of jobs, the ease of catching fish, and the increasing number of fish in the waters of Bangsring village.

Bunder creates new jobs in transportation, culinary, hospitality, and other sectors. Besides, in Bunder marine tourism have various creates new jobs that improve the economy of the local community. This study showed that the presence of Bunder had created new jobs, and as a contributor, to regional income whose income potential continues to grow [1].

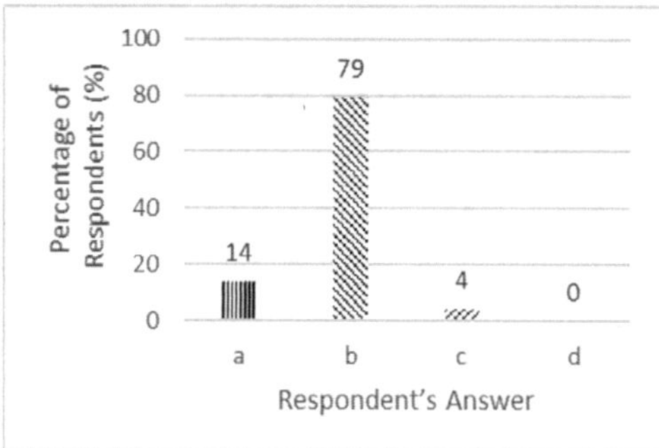

Figure 3. The use of fish in Bunder
Noted: a. strongly agree; b. agree;c. disagree; d. strongly disagree

Figure 3 shows that 79% of respondents know how to use fish in Bangsring village. From the interviews, there were two types of fish that were utilized in the town of Bangsring, namely ornamental fish and "eating" fish. Ornamental fish is used by trading. While fish "eat" are used for personal consumption or also traded to residents around Bangsring.

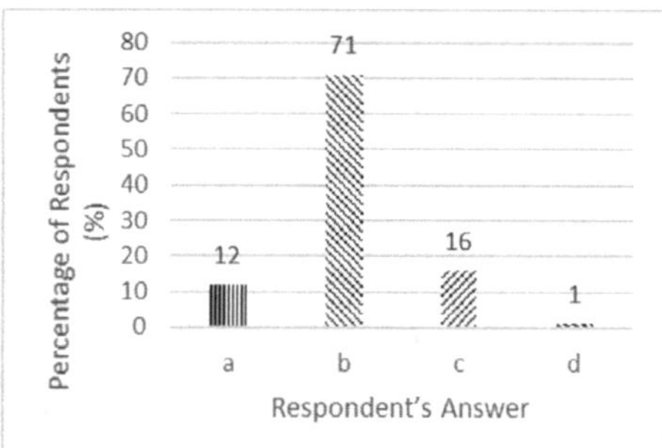

Figure 4. Percentage of fish used in Bunder

Noted: a. strongly agree; b. agree;c. disagree; d. strongly disagree

Figure 4 shows that 71% of respondents consume fish in the waters of Bangsring village. The fish is used as food and sold as ornamental fish. The majority of the Bangsring residents are fishermen. Thus, They always use fish in the waters of the Bangsring village.

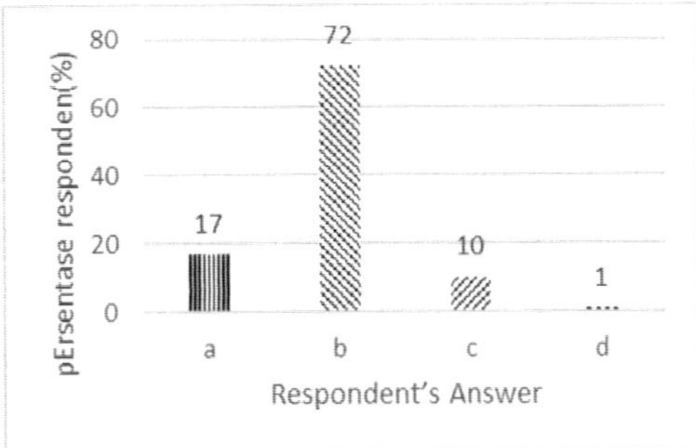

Figure 5. Knowledge of the types of fish used in Bunder
Noted: a. strongly agree; b. agree;c. disagree; d. strongly disagree

Figure 5 shows that 72% of respondents know the types of fish used in the village of Bangsring. Fish that are used are diverse, both ornamental fish and eating fish. According to the results of the interview, eating fish is the most commonly used by the Bangsring village community. Fishers catch fish only in the dry season. A kind of type of fish caught by fishermen bangsring consisting of 34 species that belong to 14 families and one new species. Fishers catch ornamental fish according to size and the amount had been determined due to maintaining ornamental diversity fish [6].

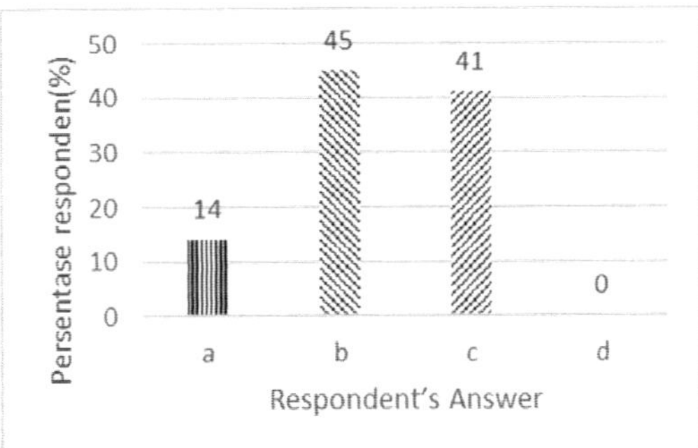

Figure 6. Knowledge of how to preserve fish in Bunder
Noted: a. strongly agree; b. agree;c. disagree; d. strongly disagree

Figure 6 shows that 45% of respondents know how to conserve fish. This knowledge was obtained through the Bangsring Underwater area, which carries out community-based marine waters conservation. It means that the Underwater Bangsring includes the community in every conservation step they take.

According to the results of interviews conducted by researchers, it is known that conservation efforts in the Bangsring waters area are related to the Samudera Bakti ornamental fish fishermen group. This group of fishermen was initially a group of ornamental fish fishermen in the village of Bangsring. After learning about the damage to the Bangsring village marine ecosystem due to excessive use of potassium, this group became the group that initiated the conservation of coral reefs in Bangsring waters. The aim is to increase fish populations in the region. After the area's ecosystem has recovered now, the group has formed an ecotourism area called the Bunder. The research was a participatory case study where primary and secondary data were collected using the purposive sampling technique [3].

The study showed that the estimated total economic value (TEV) of the coral reefs of Bangsring was IDR 38.2 billion per year or IDR 2.9 billion per hectare per year, with tourism contributing 66 %or IDR 25 billion per year of the TEV. The healthy coral reefs of Bunder support local communities and generate billions of rupiahs annually; therefore, it is critical to managing the coral reefs of Bangsring sustainably for current and future generations [7].

Social capital in Bunder has a positive influence on the success of empowerment programs conducted to the fishermen in changing fishing patterns and preserving the marine environment. Lastly, substantial social capital in the community can reduce transaction costs [8].

That besides Bangsring village, located in the district of Wongsorejo, it had excellent tourism potential. The potential of tourism in Bangsring Village needs for increased promotion and marketing management. The mobile application Bunder based on android was one of the marketing strategies and promotion of village tourism potential in Bangsring [9].

There are many steps of strategies to develop the ecotourism for Bunder those are: enhance the promotion and publication intensively, adding facilities and infrastructure, development and increasing the ecotourism potential management, division of tasks to employees in every ecotourism product, conducting the training programs to the increasing the human resources of fisherman and the effort to the processing waste to be more productive [6].

3.2. Ethnomedicine Study

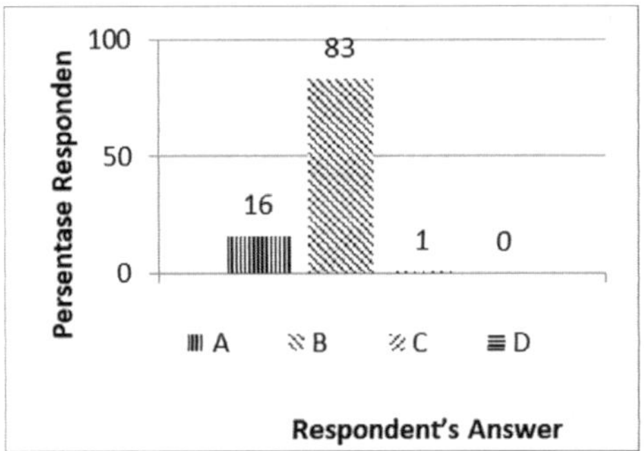

Figure 7. Knowledge about medicinal plants
Noted: a. strongly agree; b. agree;c. disagree; d. strongly disagree

Figure 7 shows that 83% of the respondents know about medicinal plants. Community knowledge about medicinal plants was passed down from parents in the past. In ancient times parents treated their sick children using medicinal herbs.

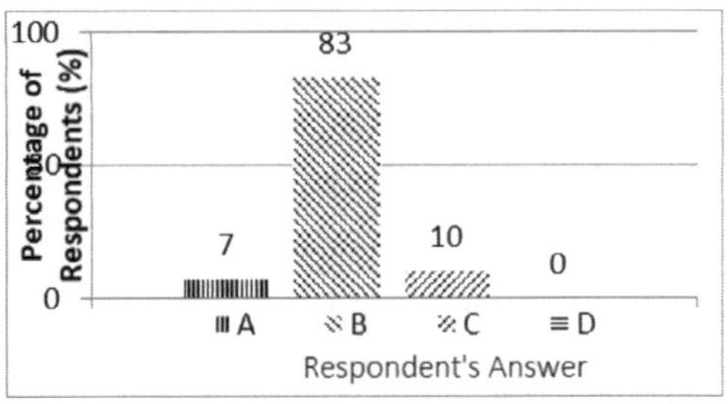

Figure 8. Percentage knowledge of medicinal plants
Noted: a. strongly agree; b. agree;c. disagree; d. strongly disagree

Figure 8 shows that 83% of Bangasring Village people have used medicinal plants. These medicinal plants are obtained from residents' yards, around agricultural gardens, and along village roads. One of the informants said that although most people have used medicinal plants, people more often use chemical drugs because they are more practical and easy to use.

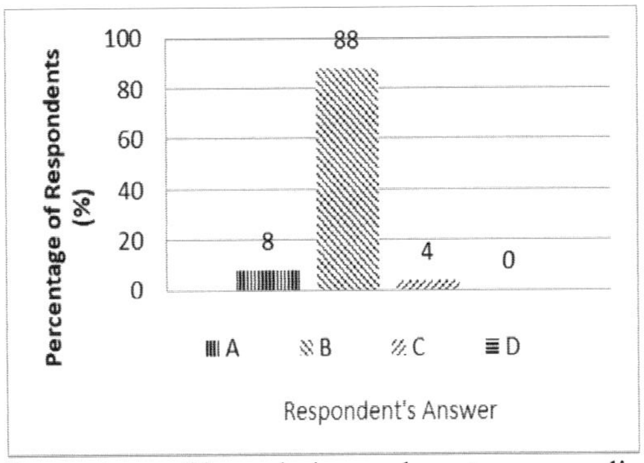

Figure 9. Percentage of knowledge on how to use medicinal plants
Noted: a. strongly agree; b. agree;c. disagree; d. strongly disagree

Figure 9 shows that 88% of respondents know how to use medicinal plants, that is, drinking boiled water from one part of the intended medicinal plant (leaves, roots, etc.).

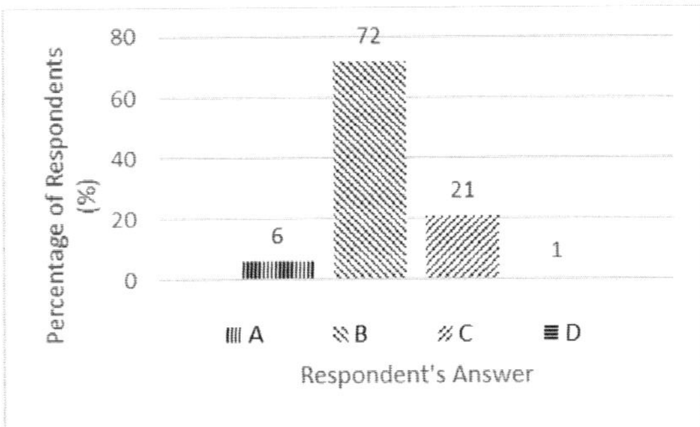

Figure 10. Percentage of planting medicinal plants by the community
Noted: a. strongly agree; b. agree;c. disagree; d. strongly disagree

Figure 10 shows that 72% of the Bangsring villagers had planted medicinal plants. These plants are mostly planted in the housing yards, but some grow wild around the residents' gardens. Plants in the village of Bangsring can grow well because the climate is suitable for the types of plants that exist so that some wild medicinal plants are found easily in this village.

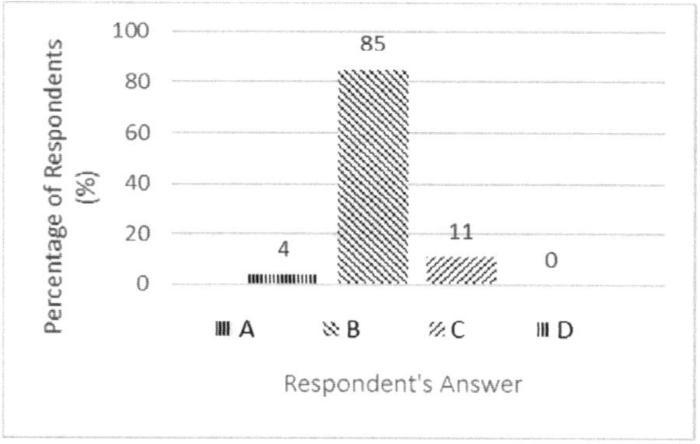

Figure 11. Percentage of knowledge of plants parts utilized
Noted: a. strongly agree; b. agree;c. disagree; d. strongly disagree

Figure 11 shows that 85% of respondents know the parts of medicinal plants that are utilized. Each plant has benefits in different parts of the plant. The part that is often used is the roots and leaves.

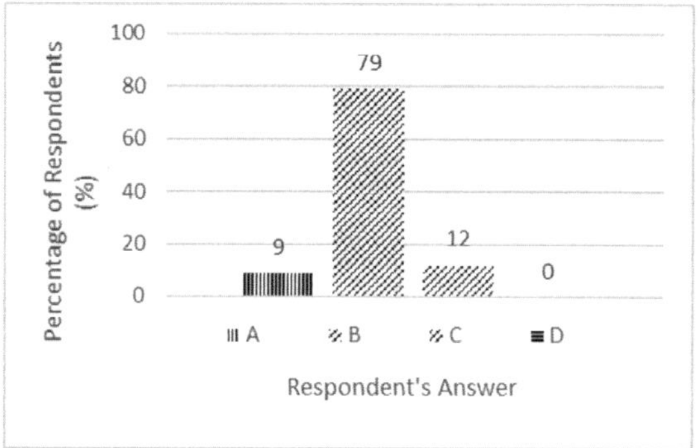

Figure 12. Percentage of interest in plant use attractiveness
Noted: a. strongly agree; b. Agree;c. disagree; d. strongly disagree

Figure 12 shows that 79% of respondents were interested in using medicinal plants. Although the use of medicinal plants is not practical, people have an interest in medicinal plants.

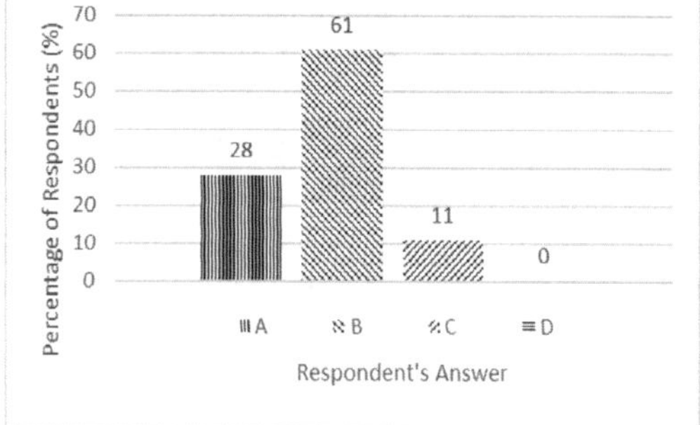

Figure 13. Percentage of the need to cultivate medicinal plants
Noted: a. strongly agree; b. agree;c. disagree; d. strongly disagree

Figure 13 shows that 81% of the respondents felt the need to cultivate medicinal plants. It shows that the community is interested in preserving medicinal plants in Bangsring village.

Tabel 1. Types of Medicinal Plants Used by the Bangsring Village Community

No.	Local name	Latin namen	Benefit
1	Kunyit	*Curcuma longa*	reduce bloating, relieve pain
2	Jahe	*Zingiber officinale*	warm body
3	Mimba	*Azadirachta indica*	increase appetite, anti worms
4	Asam	*Tamarindus indica*	launched digestion, lose weight
5	Jarak pager	*Jatropha curcas*	launched digestion treat arthritis
6	Simbukan	*Paederia feotida*	launched flue gas, treat itchy skin reduce bloating
7	Serai	*Cymbopogon citarus*	relieve pain, lower cholesterol
8	Daun salam	*Syzygium polyanthum*	lower diabetes mellitus, relieve pain
9	Lidah buaya	*Aloe vera*	lower diabetes mellitus, hair health
10	Katus gunung	*Sauropus sp*	increase endurance, treating injury
11	Binahong	*Anredera cordifolia*	treat an ulcer, cough, uric acid
12	Kemangi	*Ocimum tenuiflorum*	Increase appetite, prevent smell agencies and mouth
13	Sirsak	*Annona muricata*	strengthen the immunity of their bodies salubrious the skin
14	Pepaya	*Carica papaya*	launched digestion
15	Bawang putih	*Allium sativum*	overcome rheum, lowering blood pressure
16	Kumis kucing	*Orthosiphon aristatus*	lower diabetes mellitus, salubrious the kidneys
17	Suruh	*Peperomia pellucida*	relieve pain, headache
18	Belimbing wuluh	*Averrhoa bilimbi*	overcome rheum, cure *gondongan*
19	Pinang	*Areca catechu*	launched digestion, anti worms, increase appetite
20	Kelor	*Moringa oleifera*	healthy eyes, antioxidant, launched digestion
21	Jarak	*Ricinus communis*	launched digestion, overcome rheum sprue
22	Temulawak	*Curcuma zantorrhiza*	launched digestion
23	Kencur	*Kaempferia galanga*	Headache, cough, influenza
24	Sambiloto	*Andrographis paniculata*	cure rheum and flu, cure an infection
25	Bawang merah	*Allium cepa*	overcome constipation, increase immunity the body

Types of medicinal plants contain antioxidants, extract in our study may include quercetin, quercetin-3-Oglucoside, quercitrin (glycoside rhamnose of quercetin), and kaempferol. Quercetin exerts its antioxidant activity through scavenging reactive oxygen species [10][11][12][13].

Flavonoids such as polyphenols can increase the activity of Nitric Oxide Synthase (NOS) in vascular endothelial cells. Active substances can be diffused directly and synthesize Nitric Oxide (NO) in smooth muscle and iroendothelial further stimulate guanylate cyclase to form cGMP cause vasodilation [14].

Flavonoid compounds can function as natural antioxidants, which protect the biological system and inhibit cellular oxidation through reduction-catching active oxygen and free radical especially superoxide.

It is also supported by the previous study that the community's response to the Neem tree is quite high. The use of the Neem by the community of Bunder as a medicinal plant is used as a medication for hives, appetite enhancer, and cure for diabetes. The Bunder community manages Neem as an adjuvant medicine by drinking from the leaves. The bioprospection aspects observed in this study were: use as a drug, availability, collaborative management as a drug, conservation efforts, and its benefits as a medicinal plant for the community [15].

4. Conclusion

Bunder ecologically has a positive impact on people's lives. The conservation process involved people who participate in it. In the Ethnomedicine aspect, the community had an enthusiastic response to the medicinal plants. There are 25 species of medicinal plants used by the community as a medicine. The survival of medicinal plants in Bangsring Village is weel guaranteed due to the high interest of the community to keep the sustainability of medicinal plants.

5. Acknowledgments

Thanks to the Directorate *for* Research and devotion Society, Ministry of Research, Technology, and the Higher Education Republic of Indonesia, Nomor: 124/G164/U.LPPM/K/B.07/IV/2019, which has supported grant funding for research Penelitian Dasar Unggulan Perguruan Tinggi (PDUPT) with the title "Kajian Etnoekologi dan Etnomedisin untuk Menunjang Konservasi Maritim Bangsring Underwater (BUNDER) Banyuwangi." Thanks also to the team of research.

6. References

[1] AS Tindi, M Darwin, HA Kusworo, and PM Kutanegara 2009 Membangun Wisata Bahari: Studi Kasus Bangsring Underwater di Kabupaten Banyuwangi **2** pp. 159–176

[2] Khotimah K and Hasanudin H 2017 Desain Kapal untuk Wisata Rute Bangsring- Pulau Menjangan – Pulau Tabuhan *J. Tek. ITS* vol. **5**

[3] MA Asadi and A Andrimida 2017 Meconomic Valuation oF Coral Reefs Ecosystem of Bangsring, Banyuwangi, Indonesia *ECSOFiM Econ. Soc. Fish. Mar. J.* **04** pp. 144–152

[4] MAZ Fuad, AB Sambah, A Isdianto, and A Andira 2016 Pemetaan batimetri sebagai informasi dasar untuk penempatan fish apartment di perairan Bangsring, Kabupaten Banyuwangi, Jawa Timur (Bathymetry mapping as basic information for fish apartment placement in Bangsring waters, Banyuwangi, East Java) *Depik* **5** pp. 143–150

[5] YK Djamba and WL Neuman 2002 *Social Research Methods: Qualitative and Quantitative Approaches*, **30** (Harlow: Pearson)

[6] N Aini, H Zayadi, and S Laili 2018 Studi dan Strategi Pengembangan Produk Ekowisata Bunder (Bangsring Underwater) di Desa Bangsring Kecamatan Wongsorejo Kabupaten Banyuwangi *Biosaintropis (Bioscience-Tropic)* **3** pp. 46–52

[7] R Damayanti 2012 Pemetaan Terumbu Karang di Perairan Pulau Tabuhan Kabupaten Banyuwangi Menggunakan Citra Satelit Quickbird *J. Kelaut* **05** pp. 62–71

[8] U A Kusuma, D Satria, and A Manzilati Modal Sosial Dan Ekowisata : Studi Kasus Di Bangsring Underwater , Kabupaten Banyuwangi *Jiep* **2** pp. 1–30

[9] A A Hidayati and DK Saputra 2016 Application of Information Technology for Promotion and Sustainable Management 9n Marine Ecotourism: A Case Study in Bangsring, Banyuwangi *JIAT- J. Innov. Appl. Technol.*, **02** pp. 354–357

[10] N Athiroh, N Permatasari, D Sargowo, and M A Widodo 2014 Antioxidative and blood pressure-lowering effects of Scurrula atropurpurea on deoxycorticosterone acetate-salt hypertensive rats *Biomarkers Genomic Med.* **6** pp. 32–36

[11] N Athiroh, N Permatasari, D Sargowo, and A Widodo 2014 Effect of Scurrula atropurpurea on nitric oxide, endothelial damage, and endothelial progenitor cells of DOCA-salt hypertensive rats *Iranian journal of basic medical sciences* **17** 622

[12] N Athiroh and E Sulistyowati 2013 Scurrula atropurpurea Increases Nitric Oxide and Decreases Malondialdehyde in Hipertensive Rats *J. Nas. Terakreditasi J. Universa Med* **32** pp. 44–50

[13] N Athiroh and E Sulistyowati 2015 Evaluation Of Methanolic Extract Of Scurrula Atropurpurea (Bl.) Dans Sub-Chronic Exposure On Wistar Rat Liver *American-Eurosian Network for Scientific Information Journal* **9** pp. 245-250

[14] N Athiroh and D Wahyuningsih 2017 Study of Superoxide Dismutase and Malondialdehyde Concentrations in Mice After Administration of Methanolic Extract of Scurrula atropurpurea (BL.) *J. Kedokt. Hewan - Indones. J. Vet. Sci.* **11** pp. 19–22

[15] A Baidarus, A Hayati, N Athiroh 2019 Bioprospection of Neem (Azadirachta indica Juss) as Medicinal Plant in Bangsring Village, Wongsorejo, Banyuwangi District *Jurnal Ilm. SAINS ALAMI (Known Nature)* **2** pp. 50–56

The Integration of Retention Pond and Composite Sluice in Flood Prevention in Makassar City

Hanna Singgih
Department of Civil Engineering, Universitas Kristen Indonesia Paulus Makassar

* Corresponding author: hanna@ukipaulus.ac.id

Abstract. in 2018, Badan Pengkajian dan Penerapan Teknologi (BPPT) has developed composite sluice for controlling flood goals. The integration of retention pond and composite sluice deemed necessary to be analyzed as the integrated system as a flood disaster mitigation effort. This study aims to give a brief overview of the suitable retention pond capacity, which is integrating with composite sluice as a flood prevention effort. The retention pond designed in this study requires rainfall data, land use data and topography map. This study method uses empirical formulation based on statistical parametric on hydrology analyze and hydraulic analyze as a basis of retention pond capacity calculation and composite sluice as flood controller. The result of this study are; the maximum flood discharge in 10 years return period is 103,28 m3/s, while the riverbed discharge capacity with 8 meters width is 68,29 m3/s. To prevent flooding the retention pond volume with capacity volume in 1.117.214,24 m3 has required. In sluice practical width calculation for the retention pond, the resulting amount is 28,70 meters. It takes 32 units of composite sluice if the current needs to be maintained. The side composite sluice width has obtained 1 meter while the total composite sluice building width with four units for the retention pond is 6,4 meters.

1. Introduction

The tidal flood that hit the city of Makassar in the earlier 2019 in Bung Permai Housing Tamalanrea districts has caused around 400 families to flee. The wave was the worst since the last 20 years and inundated hundreds of home in the housing. As the flood control efforts in Makassar City, canals have been built that divide the city from Paotere in North to Mariso in west and branch in the middle to Tallo River through Pampang River. The most critical flood mitigation facility in Makassar City is the existence of two large rivers, namely Tallo River which empties in the north of the city, and Jeneberang River which leaves in the west of the town. The problem that seen nowadays is when the dry season comes, Tallo River is classified shallow and contains brackish water that feels around the PLTU.

Conversely, when the rainy season begins, the water tends to be turbid containing erosion deposits (Arifuddin et al., 2013). Tamalanrea District includes in the Medium Inundation Zone (Rudyanto et al., 2015). Based on the leading causes of flooding, retention ponds are deeming necessary to be built to temporarily accommodate flood water then streamed back to the sea after flood peak has passed. At the same time, the composite sluice is considered useful in dealing with the tidal flood that enters the river in the study area. Composite sluice system that can self-open when receding and self-close when the tide has flooding overcome occurs. In 2018, Badan Pengkajian dan Penerapan Teknologi (BPPT) had developed composite sluice for tidal flood controlling in Pekalongan City. The formulation of the problem in this study is the suitable retention pond capacity to accommodate potential flooding in the ponded area by integrating the composite sluice into the pond to give a brief description of the requirement for the function and width of the composite sluice. This study aims to provide a brief overview of the suitable retention pond capacity, which is integrating with composite sluice as a flood prevention effort. The limitation of the problem in this study is: (a). The rainfall data used is from the office of Balai Besar Wilayah Sungai Jeneberang, in the form of rainfall data from 1997 to 2016 with a data length of 20 years, (b). The area of the research site is limited to the catchment area with affected flooding Bung Permai housing and its surrounding in the research. It is namely Sub-DAS Bung Permai. Then, the pattern of analysis is reviewing into terms of hydrology and hydraulics (the geometry of the building) without a review of the embankment strength of the soil structure, economic aspect. It also only a trapezoid-shaped cross-section of simplifying the calculation.

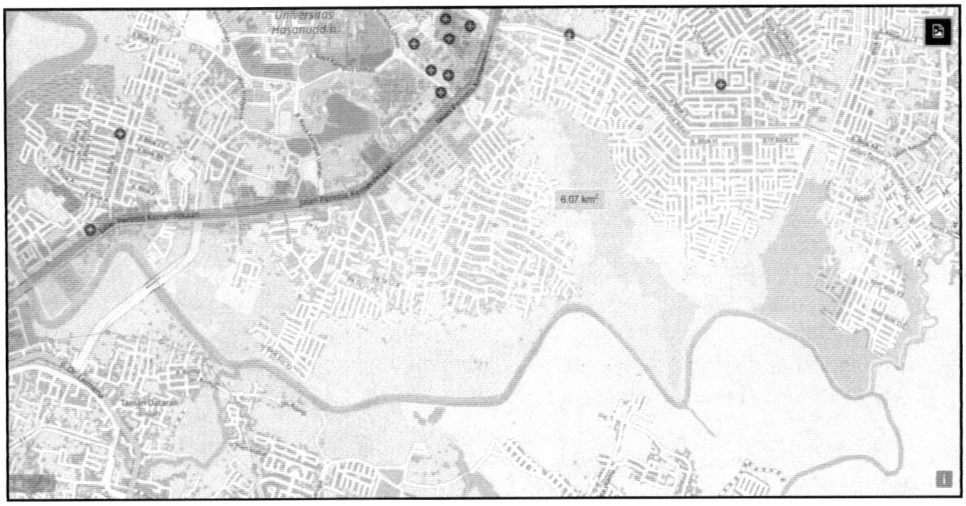

Figure 1. Research Site (Sub-WS Bung Permai)

2. Research Method

Based on the Makassar City watershed (WS) maps (source: website Geoportal Bappeda Kota Makassar). WS Tallo located between WS Jongaya and WS Bone Tanjore. Bung Permai housing located in di Tamalanrea District, included in WS Tallo. Sub-WS is determined based on UU No. 7 of 2004, which states that a watershed is a land area. It is an integral part of the river and its tributaries. And also, serves to store and drain water that comes from rainfall into lakes or the sea naturally, the boundary on land is the topographic separator and the limit in the sea to the waters that are still affected by land activities. Sub-WS which includes Bung Permai Housing can see on the following map with a scale of 1:500 shown by the yellow line with an area of 6,07 km2, and the length of the main river through which is 6,9 km. Rainfall data used in this study is the data of the maximum hourly rainfall obtained from the BBWS Pompengan Jeneberang Panakkukang Station from the year 1997 to 2016 (20 years). Makassar's topography around the Sub-WS generally shows a dark green colour which is located at an altitude of 0 to 10 meters above the sea (source: topography map of Makassar City, RBI Bakosurtanal, 1999).

3. Results and Discussion

Calculates the maximum average rainfall

The Maximum average rainfall are calculating by using Maximum Annual Series Method. Maximum annual series method calculation processed by taking one maximum data each year from the average of rain gauge.

Table I. The Maximum Average Rainfall

Year	Daily Rainfall (mm)	Year	Daily Rainfall (mm)
1997	91	2007	97
1998	103	2008	181
1999	235	2009	113
2000	376	2010	91
2001	200	2011	217
2002	161	2012	115
2003	210	2013	193
2004	128	2014	135
2005	141	2015	139
2006	110	2016	142
MAX AVERAGE RAINFALL			175,5

Frequency analysis and the suitable distribution method selection.

After frequency analysis, a suitable distribution is chosen and the following frequency analysis is the result of determination of distribution type. From table below, then the Gumbel type I and Pearson Log type III Distribution Method are chosen before it tested on the Chi Square Method

Table II. Distribution Selection Reauirements

NO	DISTRIBUTION	REQUIREMENTS	CALCULATION RESULTS	INFORMATION
1	NORMAL LOG	CS ≈ 0,00	CS = 0,735	UNFULFILL
		CK ≈ 3,00	CK = 0,416	UNFULFILL
2	NORMAL	CS = CV3 + 3CV	CS = 1,823	UNFULFILL
		CK = CV8 + 6CV6 + 15CV4 + 16CV2 + 3	CK = 4,521	UNFULFILL
3	GUMBEL tipe I	CS ≤ 1,396	CS = 1,823	FULFILL
		CK ≤ 5,4002	CK = 4,521	FULFILL
4	Pearson Log tipe III	CS ≠ 0,000	CS = 0,735	FULFILL
		SELAIN DATA DISTRIBUSI DI ATAS		

Source: Triatmodjo, 2008

Chi Square for selecting the suitable distribution

In Chi-Square Test previously determined the number of classes, which is C = 5,32 ≈ 5. Degree of Freedom (DF) is 4. The Expected Value (Ei) for 20 years is 4, and the interval class is 0,12. The requirement for acceptable distribution is Chi Square Calculation (Xh2) < Critical Chi Square (Xh2cr). From the distribution match test, Gumbel Method did not meet the requirement so that the Pearson Log type III was chosen.

IOP Conf. Series: Materials Science and Engineering **846** (2020) 012074 doi:10.1088/1757-899X/846/1/012074

From the table III the Chi Square Value is obtained in 6,50. On Degree of Freedom (DF) 4, Critical Chi Square value at the 5% significance level is 9,49. Thus the Pearson Log type III distribution can be accepted

Table III. Chi Square Method

NO	Px	Ei	Oi	Oi - Ei	$(Oi - Ei)^2$	$(Oi - Ei)^2$ / Ei
1	1,96 - 2,08	4,00	7,00	3,00	9,00	2,25
2	2,08 - 2,20	4,00	6,00	2,00	4,00	1,00
3	2,20 - 2,32	4,00	4,00	0,00	0,00	0,00
4	2,32 - 2,44	4,00	2,00	-2,00	4,00	1,00
5	2,44 - 2,56	4,00	1,00	-3,00	9,00	2,25
		20,00	20,00	0,00	26,00	6,50

Table IV. Rain Plan based on the return period

T	GT	YT	R plan
2	(0,195)	2,14	137,78
5	0,732	2,29	194,21
10	1,340	2,39	243,26
25	2,087	2,51	320,78
50	2,626	2,59	391,65
100	0,149	2,19	156,50

Rainfall Intensity Calculation

Rain intensity analysis calculation uses the Mononobe Method with a 5 hour rain duration. The equation used is:

$$I = \frac{R_{24}}{24} \times \left(\frac{R_{24}}{t}\right)^{\frac{2}{3}}$$

which; I = rainfall intensity (mm/hr, t = rainfall time (hours) and R24 = daily maximum rainfall for 24 hours (mm).

Table V. Rainfall Intensity Intensitas 5 hr duration

DURATION (hr)	MAXIMUM RAINFALL 24 HR (R24)					
	2	5	10	25	50	100
	137,78	194,21	243,26	320,78	391,65	156,50
	RAINFALL INTENSITY BY MONONOBE METHOD					
1	80,57	113,58	142,26	187,59	229,04	91,52
2	50,76	71,55	89,62	118,18	118,18	57,65
3	38,74	54,60	68,39	90,18	110,11	44,00
4	31,98	45,07	56,45	74,45	90,89	36,32
5	27,56	38,84	48,65	64,16	78,33	31,30

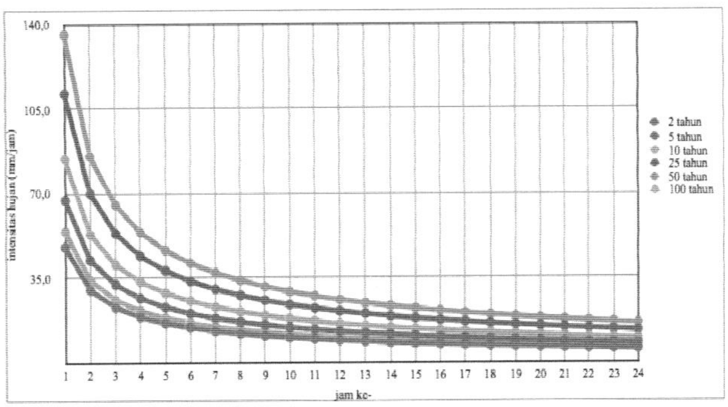

Figure 2. Rainfall Intensity (return period)

Calculation of flood discharge plan using the Nakayasu Method

Concentration Time (tc) Calculation using Kirpich Method:

$$t_c = \frac{0,06628L^{0,77}}{S^{0,385}}$$

which; L = is the length of watercourse which is 6,9 Km, S = the mean slope of the riverbed which is 0,322%. The result of Concentration Time calculation is 92,628 minutes.

Hydrograph Theory

As known before, Sub-WS area of Bung Permai Housing are 6,07 km2. Because river cross length (L)< 15 km, it is known that Tg = 0,21 x L0,7 shows 0,81 hr. Tr = (0,5 - 1)Tg = 0,41 hr. Tp = Tg + 0.8 Tr = 1,14 hr. T0,3 = α.Tg = 2,44 hr. Ro = 1 hr. Qp = A.Ro/3.6/(0,3 Tp + T0,3) = 0,61 m3/s. Because the hydrograph section rises fast but decreases slowly, then α = 3. Thus hydrograph description is as follows:

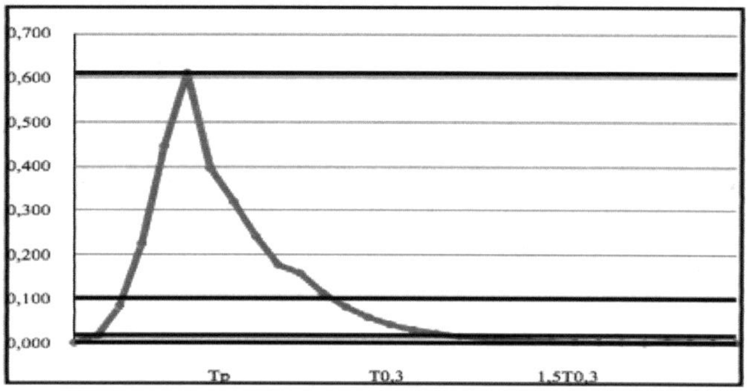

Figure 3. Synthetic Hydrograph Units(Nakayasu)

After obtaining rainfall intensity data from the area rainfall, the flood discharge plan of WS can be known using gamma 1 method. Here are the result of processing the rainfall intensity data into the flood discharge plan using Nakayasu Method.

Table 6. River Capacity Calculation

Hours	U (t,i)	Due to excess rainfall						Discharge
		142,26	89,62	68,39	56,45	48,65	0,00	m3/s
0	0,000	0,000						0,000
1	0,326	46,440	0,000					46,440
1,14	0,444	63,124	29,255	0,000				92,379
2	0,290	41,188	39,766	22,326	0,000			103,280
3	0,177	25,122	25,947	30,347	18,430	0,000		99,845
4	0,116	16,444	15,826	19,801	25,051	15,882	0,000	93,004
5	0,083	11,826	10,359	12,077	16,345	21,588	0,000	72,196
6	0,060	8,506	7,450	7,905	9,970	14,086	0,000	47,917
7	0,043	6,117	5,358	5,686	6,526	8,592	0,000	32,278
8	0,033	4,690	3,854	4,089	4,693	5,624	0,000	22,950
9	0,026	3,663	2,955	2,941	3,375	4,045	0,000	16,978
10	0,020	2,861	2,307	2,255	2,428	2,909	0,000	12,759
11	0,016	2,234	1,802	1,761	1,861	2,092	0,000	9,750
12	0,012	1,745	1,407	1,375	1,454	1,604	0,000	7,585
13	0,010	1,363	1,099	1,074	1,135	1,253	0,000	5,924
14	0,007	1,064	0,858	0,839	0,887	0,978	0,000	4,626
15	0,006	0,831	0,670	0,655	0,692	0,764	0,000	3,613
16	0,005	0,649	0,524	0,512	0,541	0,597	0,000	2,822
17	0,004	0,507	0,409	0,400	0,422	0,466	0,000	2,204
18	0,003	0,396	0,319	0,312	0,330	0,364	0,000	1,721
19	0,002	0,309	0,249	0,244	0,258	0,284	0,000	1,344
20	0,002	0,241	0,195	0,190	0,201	0,222	0,000	1,050
21	0,001	0,189	0,152	0,149	0,157	0,173	0,000	0,820
22	0,001	0,147	0,119	0,116	0,123	0,135	0,000	0,640
23	0,001	0,115	0,093	0,091	0,096	0,106	0,000	0,500
24	0,001	0,090	0,072	0,071	0,075	0,083	0,000	0,391
							max	683,016

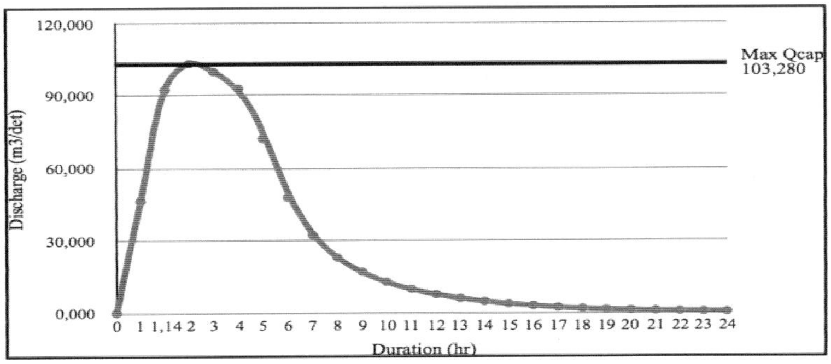

Figure 4. Total Hidrograph Discharge

Retention Pond Valome Calculation

River capacity of the downstream river in the planned reservoir has been calculated, in this case is the river cross sectional capacity of the study area.

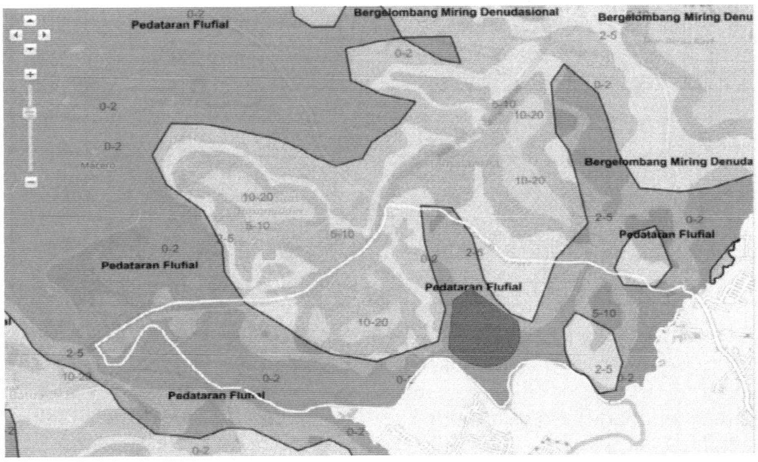

Figure 5. The Planned retention pond behind the housing

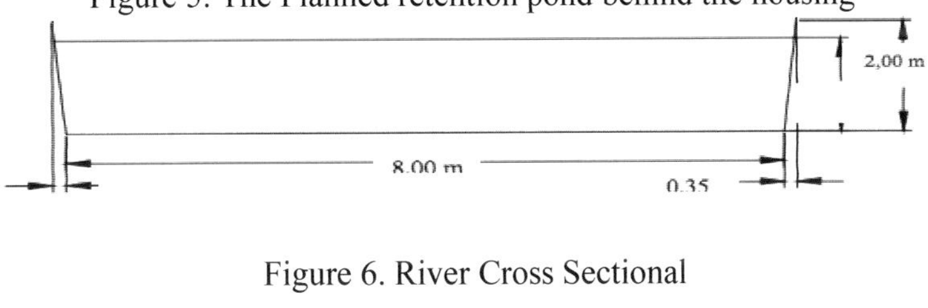

Figure 6. River Cross Sectional

Capacity calculation of the location being reviewed uses the following Manning Method:

Q = 1/n x S1/2 x R2/3 x A

Which: Q = Dischare capacity (m3/s); n = Manning Roughness Coefficient; nbase channel = 0,030 (ground), ; nchannel wall = 0,025 (installed bricks)

IOP Conf. Series: Materials Science and Engineering **846** (2020) 012074 doi:10.1088/1757-899X/846/1/012074

$$n_{ekivalen} = \left[\frac{\sum_{i=1}^{N} P_i n_i^{3/2}}{P} \right]^{2/3}$$

S	n	B	H	m	A	P	R	V	Q
2,9E-05	0,003	8	2	0,1	24,0	12,02	2,00	2,85	68,29
2,9E-05	0,003	8	1,75	0,1	20,1	11,52	1,75	2,60	52,40
2,9E-05	0,003	8	1,5	0,1	16,5	11,01	1,50	2,35	38,77
2,9E-05	0,003	8	1,25	0,1	13,1	10,51	1,25	2,08	27,31
2,9E-05	0,003	8	1	0,1	10,0	10,01	1,00	1,79	17,93
2,9E-05	0,003	8	0,8	0,1	7,1	9,51	0,75	1,48	10,55
2,9E-05	0,003	8	0,5	0,1	4,5	9,00	0,50	1,13	5,09
2,9E-05	0,003	8	0,3	0,1	2,1	8,50	0,25	0,71	1,51
2,9E-05	0,003	8	0,0	0,1	0,0	8,00	0,00	0,00	0,00

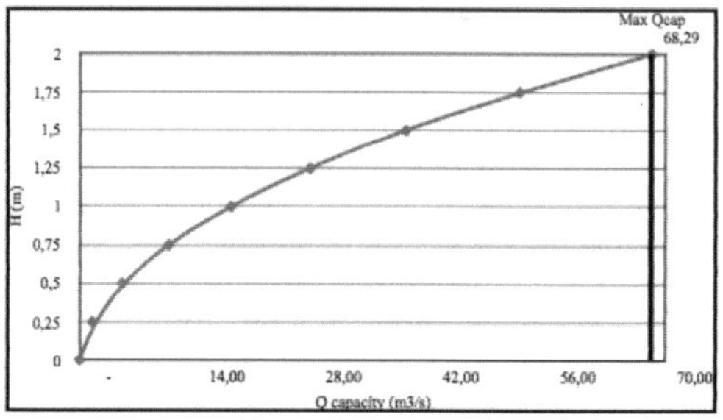

Figure 7. River capacity calculation in ponded area and air level

To know the amount of water volume, which required to store in the retention pond, the comparison between flow discharge and river capacity is needed.

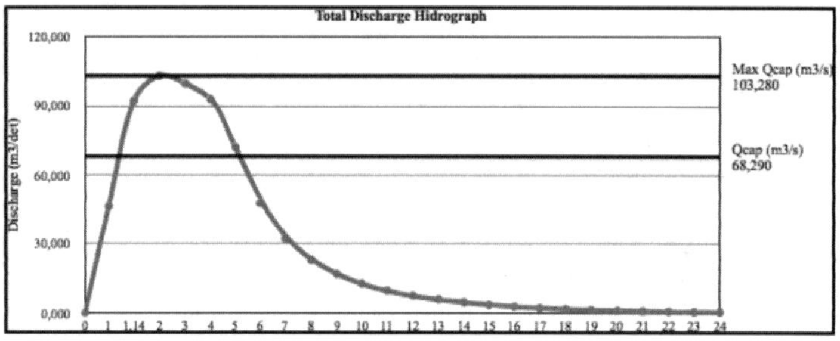

Figure 8. Flow discharge vs river capacity

Planned by the retention pond without a pump then the maximum water level of the pool = maximum water level in the channel so that the depth (H) of the retention pond is taken in 2 meters. The retention pond area is adjusted to the needs of the storage volume. Storage pool dimension has sloping wall in 1:2. then the planned pool area has predicted by land plotted on the location map.

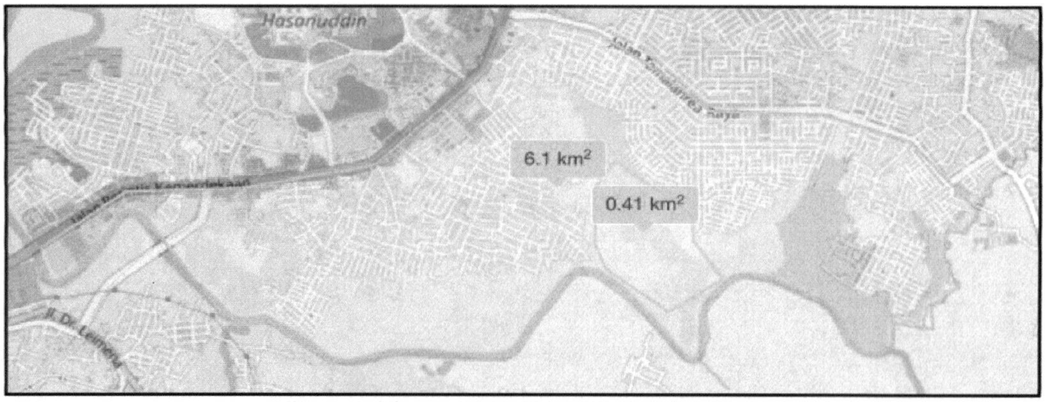

Figure 9. Planned Location of Retention Pond

IOP Conf. Series: Materials Science and Engineering **846** (2020) 012074 doi:10.1088/1757-899X/846/1/012074

The planned pool capacity can be determined by comparing the relationship between water level (H) and cumulative storage (Scum).

H(m)	A(m^2)	S(m^3)	Skum(m^3)
0,00	405.925	0	0
0,25	405.925	101.481	101.481
0,50	406.940	101.735	203.216
0,75	407.448	101.862	305.078
1,00	407.957	101.989	407.068
1,25	408.637	102.159	509.227
1,50	408.977	102.244	611.471
1,75	409.744	102.436	713.907
2,00	410.000	102.500	816.407

A

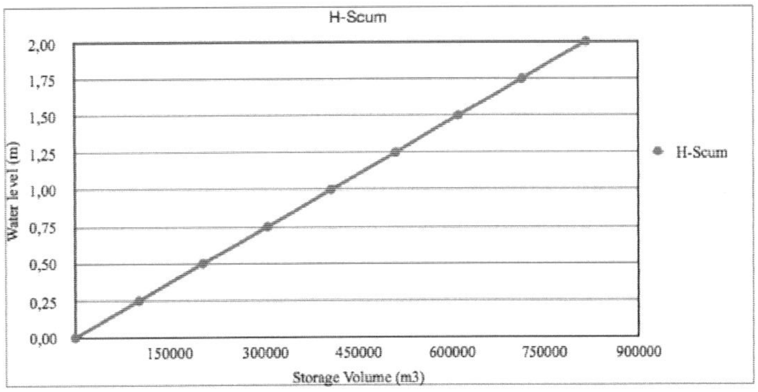

B

Figure 10. (a) H vs Scum, (b) Relationship of H-Scum

Composite Sluice Width Requirement Analysis

One of the advantages of composite sluice is being customized, which means that the sluice's width can be adjusted into the retention pond requirements. This research use Romijn Sluice Formulation as the basic to make the calculation and adjusted by the maximum discharge which has flown into the retention pond. The maximum discharge is 68,29 m3/s. To determine the dimensions of the sluice previously estimate the width of the right flood divide building. The Effective Width of Romijn Sluice (Kriteria Perencanaan (KP- 08), 2013) is:

$$Q = Cd \times Cv \times 2/3 \times \sqrt{(2/3xg)} \times B \times h1^{1,5}$$

Q = Flood discharge = 68,29 m3/s ; Cd = 0,958 (Aditya Prihantoko, et al, 2015); Cv = Inlet Velocity Coefficient = Cd * A'/A1 = Cd * ((B * h1)/(B*(h1+0,5)) = Cd * h1/(h1+0,5); g = Acceleration of Gravity = 9,81 m2/s; B = Sluice Effective Width (m); H1 = Table Energy Height (m); h1 = Upstream Table Energy Height (m) = H1 - (V12/2g). The result B = 28,70 m.

Table 7. Composite sluice effective width calculation

V	$V^2/2g$	g	H1	h1	Cd	Cv	B	Q
2,85	0,41	9,81	2,00	1,59	0,96	0,73	28,70	118,29
2,60	0,35	9,81	1,75	1,40	0,96	0,71	28,70	95,48
2,35	0,28	9,81	1,50	1,22	0,96	0,68	28,70	74,20
2,08	0,22	9,81	1,25	1,03	0,96	0,64	28,70	54,67
1,79	0,16	9,81	1,00	0,84	0,96	0,60	28,70	37,21
1,48	0,11	9,81	0,75	0,64	0,96	0,54	28,70	22,24
1,13	0,07	9,81	0,50	0,43	0,96	0,45	28,70	10,38
0,71	0,03	9,81	0,25	0,22	0,96	0,30	28,70	2,56
0,00	0,00	9,81	0,00	0,00	0,958	0,00	28,70	0,00

Total Width of composite sluice

The composite sluice has installed in Pekalongan City and has 90 cm width (0,9 m) , and assumed in this research as the effective of sluice width. By the 5 meter of river width, the total installed of composite sluices are 2 units (2 x 0,9 = 1,8 meter). For B = 28,70 meter, then total installed of composite sluices should be 32 units. This is ineffective. That is why all the composite sluices must be calculated at the same time:

1.The planned of each composite sluice width:
bp = Be + (Kp + Ka). Hmax , which: Be = 0,9 m; Kp = 0,01; Ka = Abutment coefficient = 0,1; Hmax =2 m; Then : bp = 0,9 + (0,01 + 0,1) x 2 = 1,12 m; taken bp = 1,0 m. From the width of each sluice's calculation, br = 1,0 m is taken.

2.Total Composite Sluice Building's Width: Br = N * br + Σt + Σb , which :
Br = Total Composite Sluice Building's Width; N = amount of sluice = 4 units; bp = each composite sluice width = 1,0 m; Σt = Pillar width= 0,8 m; Σb = Abutment width = 2 x 0,8 = 1,6 m; then : Bb = N x br + Σt + Σb = 4 x 1,0 + 0,8 + 1,6 = 6,4 m

4. Conclusion

Conclusion

Based on the analyses and discussion about the integration of recognition pond and composite sluice in flood prevention in Makassar City's results, the following conclusions will explain as follow with:

1. capacity with 8 meters width is 68,29 m3/s. To prevent flooding the retention pond volume with function around 1.117.214,24 m3 has required.
2. The composite conduit is customized (can be adjusted with the retention pond building condition). The existing composite sluice made in 90 centimeters width. In sluice practical width calculation for the retention pond, the resulting amount is 28,70 m, so it takes 32 units of composite sluice if the existing composite sluice needs to maintain.
3. The side composite sluice width has obtained 1 m while the total composite sluice building width with four units for the retention pond is 6,4 m.

Suggestion

Based on the conclusions of planned retention pond which the composite sluice has been integrated inside, some suggestions to be put forward are:

1. By the sluice effective 28,70 meters and the available land are quite enough, the proper retention pond to be installed is the type of riverside pond. This type advantages are there will be no disruption to the existing flow systems, easy in operational and maintenance.
2. To keep the capacity volume at 1.117.214,24 m3, redesign of width and the length of the retention pool needs to be considered to store the 10 years return period flooding.
3. The installed composite sluice into retention pond can be calculated as an inlet or outlet.
4. To keep the effective of composite sluice width in 28,70 m, the breakthrough design for the model is required by further calculating the building strength and aesthetic to be able to give the value-added for the existence of the retention pond, in addition to its effectiveness and efficiency in flood prevention. The value-added maybe from the tourism aspects which can increase the local community's economy.
5. Further calculation still requires to analyze the stability of retention pond slope, the wall reinforcement, horizontal and vertical forces and moments, gravity wall test (rolling and sliding), sluice design, sluice horizontal-vertical profile design, total sluice weight, pillar and abutment design, review to the rolling-sliding-eccentricity-soil bearing capacity, pillar dimension control and calculation of the other supporting construction such as riverside spillway, access road to the retention pond, low threshold in front of outlet sluice, trash filter and sediment catcher pool.

References

(1) Aditya Prihantoko, Marasi Deon Joubert, Dadan Rahmandani. "Penggunaan Material Komposit Sebagai Komponen Pintu Air Alternatif". Balai Irigasi, Puslitbang SDA Kementerian PU, Jakarta.2015.Sumber:https://www.researchgate.net/publication/323961956_Penggunaan_Material_Komposit_Sebagai_Komponen_Pintu_Air_Alternatif

(2) Arifuddin, M. Taufik Ishak, Roslinda Ibrahim. "Pengembangan Kawasan Sungai Tallo: Sebuah Upaya Peningkatan Kualitas Kota Makassar". Temu Ilmiah IPLBI. 2013. Sumber:https://temuilmiah.iplbi.or.id/pengembangan-kawasan-sungai-tallo-sebuah-upaya-peningkatan-kualitas-kota-makassar/

(3) A Zubair, MP Hatta, Rudyanto. "Identifikasi Zona Genangan Banjir Kota Makassar Berbasis SIG". Universitas Hasanuddin, Makassar,2015.
Sumber: http://repository.unhas.ac.id/handle/123456789/14453?show=full

(4)Dirjen Cipta Karya Kementerian Pekerjaan Umum. "Buku Sistem Drainase Perkotaan". Kementerian Pekerjaan Umum, Jakarta. 2012.

(5)Dirjen Cipta Karya Direktorat Pengembangan Penyehatan Lingkungan Permukiman. "Tata Cara Perencanaan Kolam Detensi, Kolam Retensi dan Sistem Polder (Buku Jilid I D)". Kementerian Pekerjaan Umum, Jakarta. 2012.

(6)Dirjen Sumber Daya Air Direktorat Irigasi dan Rawa. "Kriteria Perencanaan Bagian Standar Pintu Pengatur Irigasi: Perencanaan, Pemasangan, Operasi Dan Pemeliharaan (KP - 08)". Kementerian Pekerjaan Umum, Jakarta. 2013.

(7)Dirjen Sumber Daya Air Direktorat Irigasi dan Rawa. "Kriteria Perencanaan Bagian Standar Pintu Pengatur Irigasi: Spesifikasi Teknis (KP - 09)". Kementerian Pekerjaan Umum, Jakarta. 2013.

(8)Evy Harmani, M. Soemantoro. "Kolam Retensi Sebagai Alternatif Pengendali Banjir". Program Studi Teknik Sipil Universitas Dr. Soetomo Surabaya. 2017. Sumber: https://ejournal.unitomo.ac.id/index.php/sipil/article/viewFile/274/178

(9)Subarkah, Iman, "Hidrologi Untuk Perencanaan Bangunan Air". Idea Dharma, Bandung. 1978.

(10)Hartono, "Statistik Untuk Penelitian". Pustaka Pelajar - Zanafa Publishing, Riau. 2004.

(11)Chow, V.T., "Hidrolika Saluran Terbuka". Erlangga, Bandung. 1985.

(12)Soewarno, "Hidrologi Aplikasi Metode Statistik untuk Analisa Data". Nova, Bandung. 1995.

(13)Bambang Triatmodjo, "Hidrologi Terapan", Beta Offset, Bandung. 2008.

AUTHOR INDEX